UNUSUAL PATH to HIGH-ENERGY PHYSICS

From Madagascar Bush to World-Class Scientist

UNUSUAL PATH to HIGH-ENERGY PHYSICS

From Madagascar Bush to World-Class Scientist

Stephan Narison

LUPM, CNRS-IN2P3 and Faculty of Sciences, University of Montpellier - France

and

iHEPMAD, Faculty of Sciences, University of Ankatso, Madagascar

World Scientific

NEW JERSEY • LONDON • SINGAPORE • BEIJING • SHANGHAI • HONG KONG • TAIPEI • CHENNAI • TOKYO

Published by

World Scientific Publishing Co. Pte. Ltd.

5 Toh Tuck Link, Singapore 596224

USA office: 27 Warren Street, Suite 401-402, Hackensack, NJ 07601

UK office: 57 Shelton Street, Covent Garden, London WC2H 9HE

Library of Congress Cataloging-in-Publication Data
Names: Narison, S. author
Title: Unusual path to high-energy physics : from Madagascar bush to world-class scientist /
Stephan Narison, Stephan Narison, LUPM, CNRS-IN2P3 and Faculty of Sciences,
University of Montpellier, France and iHEPMAD, Faculty of Sciences,
University of Ankatso, Madagascar.
Description: New Jersey : World Scientific, [2026] | Includes bibliographical references.
Identifiers: LCCN 2025009593 | ISBN 9789819801169 hardcover |
ISBN 9789819801176 ebook for institutions | ISBN 9789819801183 ebook for individuals
Subjects: LCSH: Narison, S. | Nuclear physicists--Madagascar--Biography |
Nuclear physicists--France--Biography | Quantum chromodynamics | LCGFT: Autobiographies
Classification: LCC QC774.N37 A3 2025
LC record available at https://lccn.loc.gov/2025009593

British Library Cataloguing-in-Publication Data
A catalogue record for this book is available from the British Library.

For any available supplementary material, please visit
https://www.worldscientific.com/worldscibooks/10.1142/14051#t=suppl

Desk Editor: Nambirajan Karuppiah

Typeset by Stallion Press
Email: enquiries@stallionpress.com

UNUSUAL PATH to HIGH-ENERGY PHYSICS

From Madagascar Bush to World-Class Scientist

Stephan Narison

LUPM, CNRS-IN2P3 and Faculty of Sciences, University of Montpellier - France
and
iHEPMAD, Faculty of Sciences, University of Ankatso, Madagascar

World Scientific

NEW JERSEY • LONDON • SINGAPORE • BEIJING • SHANGHAI • HONG KONG • TAIPEI • CHENNAI • TOKYO

Published by

World Scientific Publishing Co. Pte. Ltd.

5 Toh Tuck Link, Singapore 596224

USA office: 27 Warren Street, Suite 401-402, Hackensack, NJ 07601

UK office: 57 Shelton Street, Covent Garden, London WC2H 9HE

Library of Congress Cataloging-in-Publication Data

Names: Narison, S. author

Title: Unusual path to high-energy physics : from Madagascar bush to world-class scientist / Stephan Narison, Stephan Narison, LUPM, CNRS-IN2P3 and Faculty of Sciences, University of Montpellier, France and iHEPMAD, Faculty of Sciences, University of Ankatso, Madagascar.

Description: New Jersey : World Scientific, [2026] | Includes bibliographical references.

Identifiers: LCCN 2025009593 | ISBN 9789819801169 hardcover | ISBN 9789819801176 ebook for institutions | ISBN 9789819801183 ebook for individuals

Subjects: LCSH: Narison, S. | Nuclear physicists--Madagascar--Biography | Nuclear physicists--France--Biography | Quantum chromodynamics | LCGFT: Autobiographies

Classification: LCC QC774.N37 A3 2025

LC record available at https://lccn.loc.gov/2025009593

British Library Cataloguing-in-Publication Data

A catalogue record for this book is available from the British Library.

For any available supplementary material, please visit
https://www.worldscientific.com/worldscibooks/10.1142/14051#t=suppl

Desk Editor: Nambirajan Karuppiah

Typeset by Stallion Press
Email: enquiries@stallionpress.com

Acknowledgements

First, I thank my parents who have understood that education is primordial for each human being and have continuously educated their children. Despite the only modest salary of my father, as a teacher in a primary school, they have sent us to a boarding school at Lycee Gallieni of Antananarivo (ex-academy of Aix-Marseille) which was, at that time, the best high school in Madagascar. Now, I understand their crucial initiative.

I also wish to thank my thesis supervisor Professor Eduardo de Rafael of the CPT-Marseille, my mentor, who has stimulated my interest in particle physics, for his guidance and patience. He is a researcher model for me and has influenced me a lot.

I do not also forget my high-school teacher in mathematics madam Arnaudon who helped me in my orientation and Professor Raoelina Andriambololona of the University of Antananarivo for convincing me to do theoretical physics instead of my primary wish to study engineering at the French "Grandes Ecoles" Supelec, Ecole Centrale, INSA, which accepted my application within a dedicated EU fellowship. However, we transformed, later on, this fellowship into a thesis one.

I also wish to thank all my 50 (!) international collaborators, from whom I have acquired some expertise and improved my knowledge in the field. These collaborations were also opportunities to know better their personalities, characters, cultures and ways of living which have been useful to improve my own personality and my integration into civilized society.

I also wish to thank my family and entourage for their patience during the time I spent in front of my computer doing my research works and (in some sense) sacrificing some of my family life.

Misaotra Anareo @ ny Faharetana!

Merci pour Votre Patience!

Thanks for Your Patience!

Contents

Introduction

Stephan Narison is one of the unique world-renowned theoretical high-energy physicists from Madagascar. In this book, he recounts his unusual path from bush to high-energy physics and shares his experience through his international collaborations and visits in different world institutes.

Chapter 1 describes the childhood of the author and his education in Madagascar until the master diploma at the University of Antananarivo.

Chapter 2 introduces the notion of particle physics to non-experienced audience.

Chapter 3 reports both the author's life experience as an immigrant for the preparation of his doctorate thesis within an European Union fellowship at the Center for Theoretical Physics (CPT) of Marseille-France (1974–1980).

Chapter 4 reports his works as a post-doc at the International Center for Theoretical Physics (ICTP) of Trieste, Italy (1980–1982).

Chapter 5 summarizes his works at the Laboratory of Particle Physics (LAPP) in Annecy, France (1982–1983) as a CNRS fellow.

Chapter 6 reviews his works at European Center of Nuclear Physics (CERN) in Geneva, Switzerland.

Chapter 7 describes different trips and collaborations done within some contracts or/and visiting scientists in different world institutes.

In Chapter 8, different research works produced after the permanent position at the French National Research Center (CNRS) of the Institute of Montpellier in 1984 are reported, where recent works are presented and/or reviewed.

Chapter 9 reviews the author's recent (important) papers in QCD.

At the beginning or/and at the end of each chapter, the author mentions some anecdotes or/and comments during his stay at each institute which he has visited.

The scientific part of the previous chapters is an extended abstract of each research work and is quite technical. It is addressed to PhD students and experts in the field who wish to follow the steps in deriving the results in each paper. Contrary to classical reviews, the works are classified by period of publication, where the aim is to show the evolution of the field over the year.

During the COVID-19 pandemic, the author was also interested to study the case of Madagascar and has published the first paper on this topic (Chapter 10).

Chapter 11 describes the contributions of the authors to the physics community by the organization of the QCD-Montpellier International Conference Series since 1985.

Chapter 12 reports on the development of high-energy physics (HEP) in Madagascar by the organization of the HEPMAD-Antananarivo International Conference Series since 2001, the creation of the institute of research at the University of Antananarivo in 2004 and the popularization of HEP at high schools, universities and public institutions (Academy of Sciences, Alliance Francaise, etc.) since 2010.

Chapter 13 reports on the contributions to science and society of associations initiated by the author and on the author's political commitment in Madagascar.

Chapter 14 enumerates the memberships and honours of the author.

Chapter 15 presents his publication status until 2024.

Chapter 1

From Bush to Antananarivo

1.1 The Bush Era 1951–1962

Figure 1.1. A photo with my father and mother in Antsakabary (Sofia river) where I was born.

I was born on September 17, 1951 in the small village of Antsakabary located at the north-west of Madagascar (Sofia region, ex-province of Majunga: see Figs. 1.2 and 1.3).

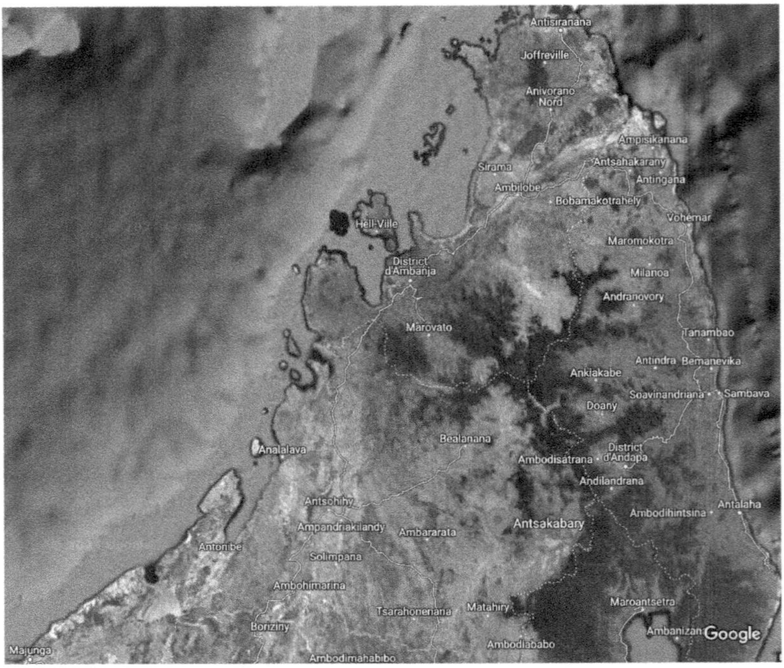

Figure 1.2. Antsakabary: a small village in the north-west of Madagascar (Sofia district, ex-province of Majunga).

Figure 1.3. Village of Antsakabary 2024.

Figure 1.4. Villages of Vohitrandriana and Amohitsoa-Antsimiasy in 2024. They belong to the Anjozorobe district, Analamanga Region, ex-province of Antananarivo.

My paternal grandfather was a high-rank pastor at a Calvin protestant church while his wife was among the first Malagasy anglican. My maternal grandfather was a teacher at a primary school and they frequented lutherian churches.

We were educated with Christianism like about 90% of Malagasy peoples with about equal number of protestants and catholics. Christianism was brought by English missionaries through good relations between the English and our queens. However, our original religion was deism, where we venerate nature (mountain, etc.). This religion is almost forgotten by modern generations.

Our parents were very involved in serving at church. During their last days, they even presided the important Lutherian Church of Antsirabe (capital of the Vakinankaratra region) and until now I have been participating in some activities for this church, so that I have good relations with some pastors in Madagascar. Churches played an important role in the politics of Madagascar such that the meaning of laicity is unclear. Politicians also exploited the churches to increase their influence towards the population. Every Sunday, when we were children, we went to the churches with our parents.

My father was a teacher at a primary school. As he was not a member of the presidential party then, but instead in the opposition, he had never got a position in a big city but always was sent to work in the bush. After transfer from one small village to another one, he got a position at the villages of Vohitrandriana and Ambohitsoa-Antsimiasy (both at the Anjozorobe district, Analamanga region: see Fig. 1.4) that I remembered.

My mother took care of the family. It was a tradition in our country that women did not work outside except to help in the cultivation of their own rice and vegetables or/and the ones of the village. In fact, the Malagasy villagers were traditionally organized in what was collectivity called "Fokonolona", which managed the social life of the village.

Living in these villages was a rich experience as we learned the ways of life (morals and customs) of the villagers who had still maintained the Malagasy ancestral morals and customs. Equally important was to learn how to live with them.

My father was not only a teacher in the village but also the must-see notable. As there was no doctor in the village, he healed the sicks. Our parents helped the villagers in different social problems. Our house was open to everybody. If someone even arrived at the lunch time, we shared food with them.

Living in these villages was also enriching as we learned to have a simple and direct contact with nature far away from the civilisation. Only a small portable radio with battery informed us of the news. We did not eat on the table but on the mats. We did not have chairs but straw poufs. We slept on the floor. We had neither electricity nor running water. We lived with simplicity and basics like the villagers themselves.

My memories of Vohitrandriana are quite a few as I was too little to remember all things. Figure 1.5 shows my two brothers, Noelin (left) and Honoré (centre), in the carriage which our father has built for us. I remember that my parents acquired in land and some few heads of zebus which they put to our names. I also remember the time we went to the school's garden to get ananas and some other fruits.

Figure 1.5. A photo with my two brothers Noelin (left) and Honoré (centre) in the carriage built by our father for us.

In Ambohitsoa-Antsimiasy, we learned how to rear volailles, zebus, porks which our parents owned, how to make wood charcoal, how to make potteries with clay, etc.

We also learned how to chase and trap birds, moles, hedgehogs, how to cultivate rice. We discovered different species of medicinal plants, etc. I loved to go to a forest of mango trees and eat the fruits on the trees.

However, despite these rustic ways of living, my father controlled our study and homework everyday rigorously after the school such that after four years of study at the primary school, he decided that I pass directly an exam at the most prestigious high school (Lycee Gallieni Andohalo) at the Capital Antananarivo without finishing the usual last two years of primary school (middle school course).

To participate in this exam, one had to walk half a day to reach the first city (Ambatomanoina on the map in Fig. 1.4), where one can take a public transport (taxi brousse) to go to Antananarivo. However, the day before I left, I had a dispute with my brother and my father punished me by hitting me with a stick such that I limped going to Ambatomanoina.

In addition to my stress for the exam, I was also worried to go to the capital and discovering the civilized world. However, contrary to what I had expected, things went smoothly. I passed the exam with good classification. However, I had applied for entering in a scientific class but after the results, the committee decided to orient me in a literary class where the main subjects are French and Latin languages but not mathematics.

(a) (b)

Figure 1.6. (a) The primary school of Antsimiasy with the students of my promotion; (b) in front of the Antsimiasy school with my mother, brothers and the two adopted girls (Soa and Sambizafy).

1.2 The High School of Lycée Galliéni 1962–1969

(a) (b)

Figure 1.7. (a) The classrooms of the Lycée Galliéni; (b) the administrative offices and the upper floors of the old dormitories of the boarding school.

1.2.1 Studies @ Lycée Galliéni

A new page of my life began. That of the bush was turned. Everything would change fast when back to school on September 1962 such that it was difficult to get adapted quickly! One often had the nostalgia of the preceding way of living. One should note the Lycée Galliéni was a school for boys until 1967. The first batch of girls who arrived were attractions as most of the students fell in love with them.

I was the most youthful in our class but also among the poorest. Most students were the sons of persons of the government (ministers, senators, deputies, senior state officials, administrators, etc.), while the others were the sons of rich business men or issued from aristocracy and bourgeoisie of the capital. However, most of these students did not care about their studies such that some of them failed during the exams.

From my side, though I did like to adapt that life style myself (skip classes, play cards in the garden or do something else), I payed attention to my studies and then succeeded brilliantly in the exams. These students from wealthy families hence started to consider me and became my friends. Each exam, they hustled to be around me to get advice. Few years ago, I met one of these minister nephews and some other friends that I have not seen since 1966–1967. They have not forgotten these boosts I gave them at school.

Everything had gone well at school. One should note that most of our teachers are French cooperants and the teaching program was the same as in France. Our Lycee belonged to the academy of Aix-Marseille in 1962–1963. The level of study was clearly higher than at present. However, we regretted that we knew more the history and geography of France than the one of Madagascar. Note the famous caricature: *Our ancestors the Gauls* which we learned at school.

One year after the literary class, one rendered me in the one of the fifth scientific schools as I was better in mathematics than in French and Latin. This ability for sciences was confirmed over the years as I was often among the first in mathematics. Since I started physical sciences at the second class, I was also among the best in the class. However, I spent most of my time with class humbugs to play cards at the garden of the Lycee by skipping French, English and some other non-scientific courses. However, I regret this at present as I had to make many personal efforts to improve my English for my research works.

The studies went smoothly and I often got good appreciation in my report card. I passed my baccalaureate exam in June 1969, but, unfortunately, I failed in the first session for some few points. I was hence obliged to revise during the school holidays to pass the second oral session in October 1969. The jury was impressed by my presentation, but, unfortunately, few points were missing to get the quite good grade (I was the first for the passable grade). At this time, getting the baccalaureate was rewarding as the level was high. In between, my brilliant class friends have been admitted at some prestigious French high schools (Grandes Ecoles) to follow the classes of superior mathematics. Later on, some have been received at the French major schools like polytechnics, but some have failed. Later, I met one of them. He is a teacher at a high school in Toulouse.

1.2.2 The life at the boarding school

I entered the Lycée Galliéni boarding school in September 1962 and stayed there until my baccalaureate was completed in June 1969. Number 43 had been attributed to me and that became my lucky number.

However, for being able to enter at the boarding school, one had to buy a bunch of clothes in a long list. My parents had to sacrifice to buy such things. In that time, I only had sandals but not shoes (see Fig. 1.8). Later, my parents bought for me Bata branded shoes, but compared to the ones of the sons of rich, these were relatively of low quality. This difference in external appearance stimulated me to find a way to be equal to the others without asking for money from my parents. Then, during the school holidays, I worked at the textile (COTONA) and beer (THB) factories of Antsirabe as a worker so that I could buy nice shoes and clothes to erase this difference in

Figure 1.8. Photo at one of the best photograph studios of Antananarivo.

external appearance: The life at the boarding school was well supervised (supervisor, study room, refectory, dormitories, ban on weekend outings for punished students, etc.). On the eve of the holidays, tatanages with cuddly toys were organized between the occupants of the dormitories on each floor as well as recreational evenings co-organized with the young girls at Lycée Jules Ferry.

The high school had its own field for sports (football, basketball, volleyball, handball, table tennis, and athletics). We note that the famous Malagasy runner Jean-Louis Ravelomanantsoa who participated in the 100 m within 10 seconds at the Tokyo Olympics in 1964 trained on the 50-m track at the Gallieni high school. Inter-high school tournaments were regularly organized on Thursday afternoons. This allowed us to meet other people and also the girls from Jules Ferry high school. There were huge competitions between high schools and often the tournaments ended in fights because of the sore losers.

I was part of our high school football team as a left winger. At this moment, we had to make a choice: continue playing football and make it a career or continue studying. The two choices could not be combined because to be better at sport, you had to train a lot and therefore skip several classes and vice versa. I made the second choice. Immediately, one could think that it was not the right choice because most of my teammates and other high-school athletes were becoming stars in our country. However, in the long term, it was not a bad idea because most of these athletes could not progress due to lack of resources from their clubs and from the Malagasy sports federation.

1.2.3 Popularisation of HEP and return to the Lycée Galliéni

Figure 1.9. Popularisation of HEP in the l'amphitheater of Lycée Galliéni in 2010 and 2014.

Since 2010, I have popularized high-energy physics (HEP) in the various Malagasy high schools, colleges, universities and other institutions (academy, alliance francaise, etc.) (see [1, 2]). The first instinct was to conduct this activity at the Galliéni high school (see

Fig. 1.9). The other activities for all Madagascar will be presented in Chapter 12.

1.3 Masters @ Antananarivo University 1969–1973

1.3.1 Studies and living

After the baccalaureate, I continued my studies at the University of Ankatso (Antananarivo) in the physics chemistry sector with a scholarship awarded by the Malagasy government. Studies were progressing normally without any repetition. I obtained my license (equivalent of Master 1) in 1971.

To support my scholarship, I did an internship in the chemistry laboratory of the Petroleum Refinery plant (currently called SOLIMA) at Tamatave in 1971 and taught in different Catholic (ESCA, Saint Michel), Protestant (Ambatonakanga), and private (Madagascar college, COPA Ankadifotsy college) college in the capital, Antananarivo. With my first salary, I bought a Renault 16 from a teacher at ESCA for my parents to thank them for all their efforts towards my success.

Figure 1.10. The university rectorate of Ankatso-Antananarivo.

However, during this period, an important event had happened in the country. The students of our university, in particular the medical students, asked the revision of the Franco-Malagasy cooperation agreement. To join the movement, we suspended our studies.

Figure 1.11. Some photos taken at the university of Ankatso-Antananarivo in 1973.

(a) (b)

Figure 1.12. (a) At the laboratory of SOLIMA (Tamatave) during an internship in 1972; (b) with my two brothers at our parents' house in Antsirabe.

After two years of struggle, this movement culminated in the famous demonstration of May 13, 1972 which led, unintentionally, to the overthrowing of President Philibert Tsiranana and his government. Taking a step back from what is happening now, we have the impression that all these nationalist movements, including that of 1947, were almost useless when we take stock of the situation and the current assessment of our country.

I had obtained my master's degree (second year) in physics in 1973 and my studies at the university had stopped there.

1.4 Transition Period and EC Scholarship 1973–1974

I continued my teachings. I lived well. I bought an apartment in the SEIMAD estate in Ambohipo. I had bought a beautiful Alfa Romeo Giulia car. However, despite this comfortable life, I was not satisfied with my knowledge and wished to continue to study.

During 1973 and 1974, I was looking for scholarships in different ministries to be able to continue my studies. By knocking on the doors of the Ministry of Mines in Tsimbazaza, I found a scholarship offer from the European Commission (EC) for engineers. I had applied to several French engineering schools with the support of my Electronics Professor Rakotomavo Bruno and my former Professor of Mathematics at the Lycée Galliéni, Ms. Arnaudon. SUPELEC (Paris), Ecole Centrale and INSA (Lyon) admitted me based on qualifications in the second year of the school without taking any competitive examination. However, Mr. Raoelina Andriambolona, my Professor of Quantum Mechanics, had persuaded me to do research in theoretical physics. It also went with my preference for freedom of work and research thinking because the engineering profession had attracted me less due to the experience I had with the SOLIMA engineers during my internship at Tamatave. He then helped me with Ms. Razaonie and Mr. Razakaboana from the Ministry of Planning to transform this engineering grant from the European Commission (EC) into a research grant in theoretical physics.

The change had been made. Professor Raoelina Andriambololona wanted me to go to Marseille where he had been trained. However, I also had an offer from the Ecole Normale Superieure de Paris to participate in the Diplome d'Etudes Approfindies (DEA) of Professor Meyer. My mind was still vacillating between the two cities.

Finally, I took the plane with the other scholarship holders from the EC and the Cooperation Agreement Fund (FAC) in October 1974. We were well taken care of. A bus picked us up at Roissy Charles de Gaule airport and brought us to the Center for Students

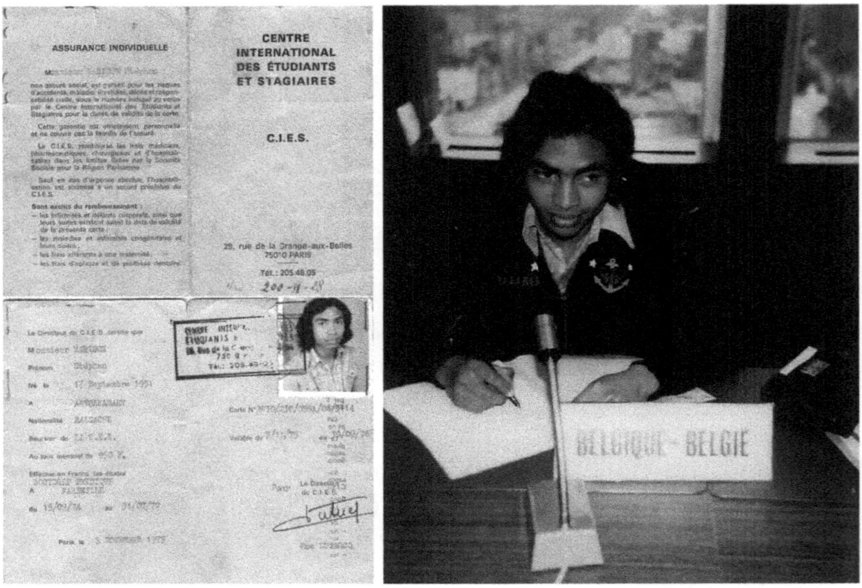

Figure 1.13. Meeting of students and trainees of the EC at Brussels in 1974.

and Trainees (CIES) in Massy (a suburb of Paris). We stayed for a month. During this period of October, the weather in the Paris Region was gloomy. This weather situation was important for me in taking the decision to go to Marseille where I hoped it will be more sunny!

Chapter 2

A Brief Notion of Particle Physics

Figure 2.1. Introduction book to particle physics.

In this chapter, I present a brief notion of particle physics addressed to non-physicists which can help to understand the different specialized research topics discussed in this book. More detailed discussions and history of particles physics from the Ionian school (~610 BC) to the present days of particle physics can be, e.g., found in the book shown in Fig. 2.1.

2.1 The Big Bang Hypothesis

There is an assumption that the present universe results from the explosion of some planets/stars or eventually a black hole (Big Bang hypothesis: Fig. 2.2) having a huge amount of energy where the matter becomes infinitely small elementary particles. With time, these elementary particles recombine to form our planet after 13.8×10^9 years. One expects that the universe is continuously expanding.

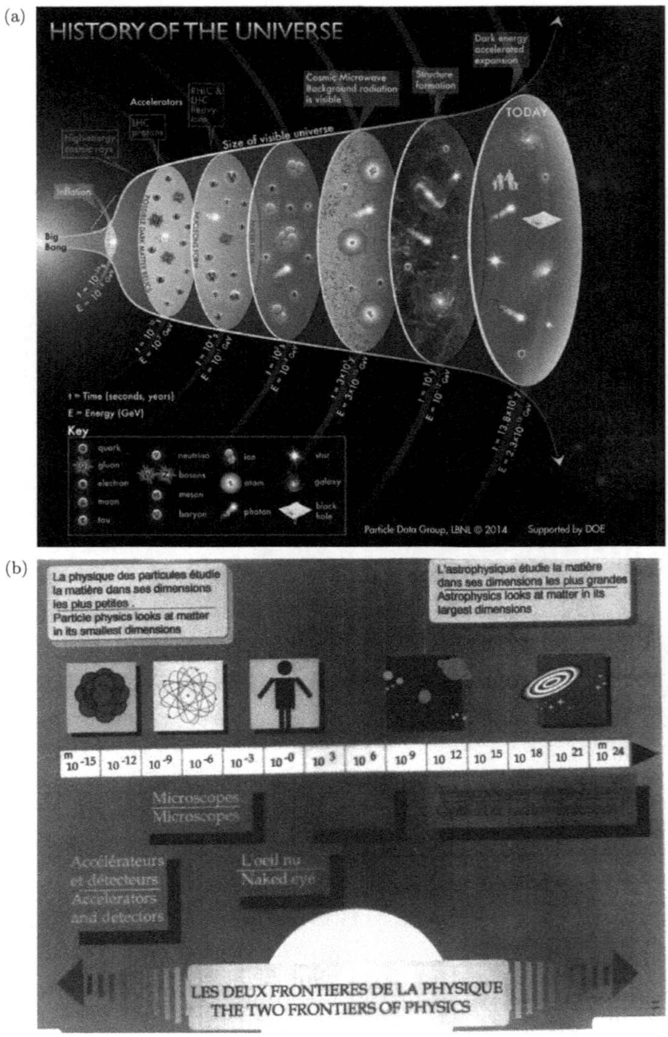

Figure 2.2. (a) The Big Bang hypothesis; (b) the two frontiers of physics.

2.2 The Two Frontiers of Physics

The aim of particle or high-energy physics is to test this hypothesis by creating the Big Bang at the laboratory scale and by studying the properties of elementary particles of a size of about 10^{-15} meters produced after the explosion.

2.3 The High-Energy Accelerators

The experience is achieved in different accelerators by colliding particle–anti-particle: electron–positron (LEP at CERN, SLAC at Stanford, BES at Beijing, BELLE at Tsukuba, etc.), proton–anti-proton (CERN, FERMILAB), proton–positron (HERA at Hamburg) or/and particle–particle: proton–proton (LHC), electron–proton (HERA) beams having (almost) the velocity of light. At the collision regions where the Big Bang energy is reached, different particles are produced. The properties of these particles are analyzed by high-energy physicists for building a theory explaining the origin of the universe.

Figure 2.3. The large hadron collider (LHC) of proton–proton at CERN, Geneva, with an energy of 13 TeV per beam.

2.4 The Electroweak Standard Model (SM)

2.4.1 Quarks and leptons

At present, the so-called Standard Model (SM) of S.L. Glashow, A. Salam and S. Weinberg (Nobel price winners in 1979) which unifies the electromagnetic (vehicle of light) and weak (responsible for radioactivity) forces (see Fig. 2.5) at high energy above the W boson mass satisfies most of the experimental precision tests.

The SM assumes that the different forms of matters come from the combination of elementary particles: three couples of quarks of fractional charges (in units of the electric charge) and three couples of leptons. Quarks have three colours (red, blue and yellow). These particles are fermions of spin 1/2 and obey Fermi statistics (Fig. 2.4).

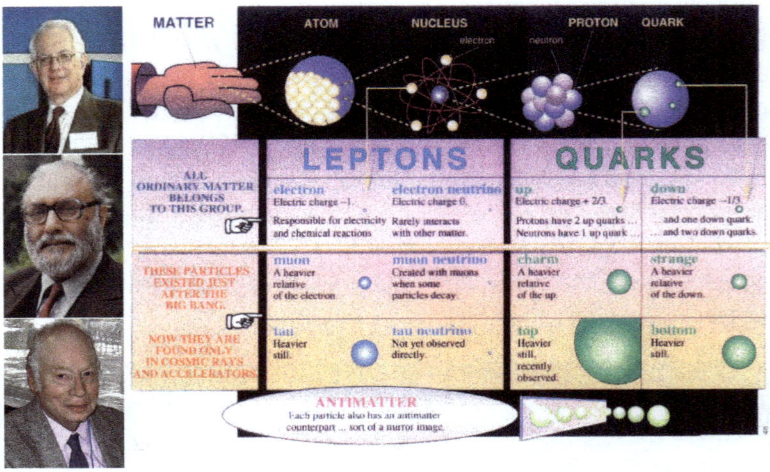

Figure 2.4. The Standard Model (SM) of leptons and quarks by S.L. Glashow, A. Salam, S. Weinberg (Nobel Laureates 1979).

2.4.2 The forces in nature

There are three microscopic forces in nature (Fig. 2.5):

- ◇ The usual electromagnetic force vehicled by an almost massless photon responsible of light.
- ◇ The weak force vehicled by the massive $W^{\pm}(80.39)$ and $Z^0(91.39)$ bosons responsible for radioactivity (β-decay, ...).
- ◇ The strong nuclear force vehicled by eight massless gluons which binds the nucleus.

Gravity, the Newton force is a macroscopic force as you see its effect by eyes like e.g. the case where you do not fall down when you are standing up.

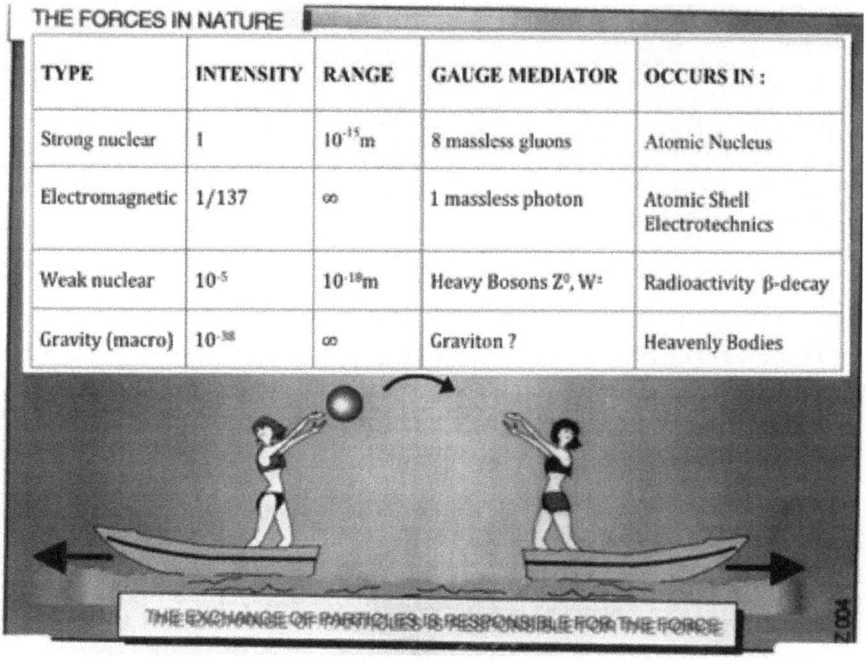

TYPE	INTENSITY	RANGE	GAUGE MEDIATOR	OCCURS IN :
Strong nuclear	1	10^{-15}m	8 massless gluons	Atomic Nucleus
Electromagnetic	1/137	∞	1 massless photon	Atomic Shell Electrotechnics
Weak nuclear	10^{-5}	10^{-18}m	Heavy Bosons Z^0, $W^{\pm}$	Radioactivity β-decay
Gravity (macro)	10^{-38}	∞	Graviton ?	Heavenly Bodies

Figure 2.5. The four forces in nature.

2.4.3 The Higgs boson of the SM

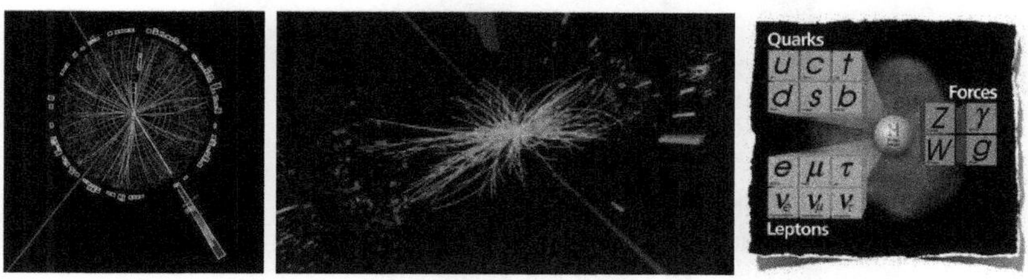

(a) (b) (c)

Figure 2.6. Discovery of the Higgs Boson at LHC from its two-photon decay (CERN 4 July 2012): (a) ATLAS, (b) CMS; (c) particles of the SM which acquire masses through the Higgs boson. European Physical Society (EPS) Prize 2013.

The SM also assumes that the Higgs boson having a mass of 125 GeV surnamed God Particle discovered at CERN in 2012 (see Fig. 2.6) gives masses to all particles. The existence of the Higgs boson has been postulated in 1964 by R. Brout, P. Higgs (Nobel

Figure 2.7. Nobel Prize 2013: Brout, Higgs; Sakurai Prize 2010: Kibble, Guralnik, Hagen, Englert, Brout.

Prize winners 2013) and some others, including Kibble, Guralnik, Hagen, Englert (Sakurai Prize 2010) (see Fig. 2.7).

However, the SM does not explain all phenomena like, e.g., the origin of the neutrino masses and some more technical problems (mass hierarchy, fine tuning, etc.) but needs to be extended where research works are underway.

2.4.4 SM as a gauge theory

't Hooft and Veltman studied the renormalization of the EW theory based on the gauge group $SU(2)_L \otimes U(1)$. They found that the theory is renormalizable and leads to finite observables. Since then, the EW phenomenological model becomes a consistent EW Theory.

Figure 2.8. G. 't Hooft, M. Veltman (Nobel Prize winners 1999).

2.5 Quantum ChromoDynamics (QCD)

QCD is the part of the SM which describes the strong nuclear force vehicled by eight massless gluons and where the quarks having spin-1/2 have three colours. It is based on the $SU(3)_c$ symmetric colour group.

2.5.1 Asymptotic freedom

Figure 2.9. D. Gross, D. Politzer, F. Wiczek (Nobel Prize winners 2004).

At high energy, one expects that the theory is asymptotically free such that quarks are liberated to form hadron jets. Due to the asymptotic freedom of the theory, the QCD coupling α_s (analogue to the electric charge squared $\alpha \equiv e^2/4\pi$ of QED) is expected to decrease like $-1/\beta_1 \log(Q^2)$ versus the energy if the β-function $\beta = \beta_1(\alpha_s/\pi) + \mathcal{O}(\alpha_s^2)$ with $\beta_1 = -(1/2)(11 - 2n_f/3)$ is negative, i.e., for the number of quark flavours $n_f \leq 16$ according to the authors shown in Fig. 2.9.

The first two quarks and one gluon jet have been first observed at PETRA-HERA, Hamburg (see Fig. 2.10(a)), supporting the presence of gluons. Different experiments performed at different energies see this $1/-\beta_1 \log(Q^2)$ behaviour from the τ-lepton to the Z-boson mass (see Fig. 2.10(b)).

2.5.2 Hadrons and the quark model

At low energy, quarks bind to form hadrons bound states. One says that quarks are confined. Within the quark model of Gell-Mann and Zweig (Fig. 2.11) introduced in 1964 before QCD, the hadrons are subdivided into two families:

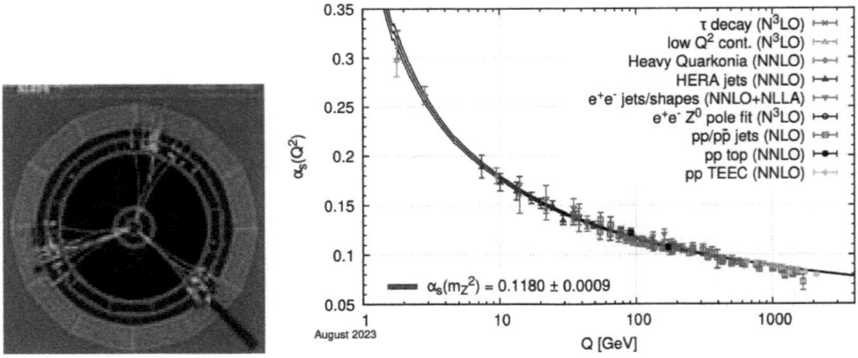

(a) (b)

Figure 2.10. (a) Two quarks ⊕ one gluon jet seen at LEP from the reaction: $e^+e^- \to \bar{q}qg \to$ three jets; (b) Q behaviour of the QCD coupling α_s.

Figure 2.11. M. Gell-Mann (Nobel Prize winner 1969), G. Zweig.

• Mesons

They are built from a quark and an anti-quark.

◇ The lightest meson is the pion built from the $\bar{u}d, \bar{u}u - \bar{d}d$ quark–anti-quarks (Fig. 2.12). It has a mass of 139.57 MeV and the quantum number $J^{PC} = 0^{-+}$ where J is the spin, P the parity and C the charge conjugation. It is the Yukawa particle responsible of the nuclear force.

◇ The lightest vector meson is the $\rho(770)$ with the quantum numbers $J^{PC} = 1^{--}$ and built from the same type of quark–anti-quark as the pion.

◇ The lightest axial-vector partner of the $\rho(770)$ meson is $A_1(1260)$ with the quantum numbers $J^{PC} = 1^{++}$ and is a $\bar{u}d$ bound state.

◇ The other mesons classified according to the quark model of Gell-Mann can be found in Fig. 2.12.

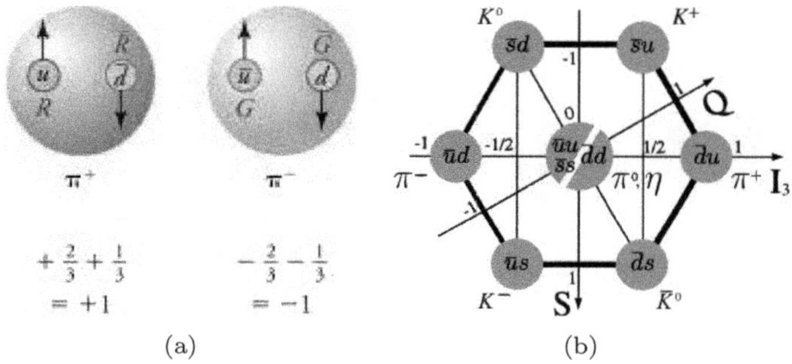

Figure 2.12. (a) Pion; (b) pseudoscalar mesons built from the u, d, s quarks.

• **Baryons**

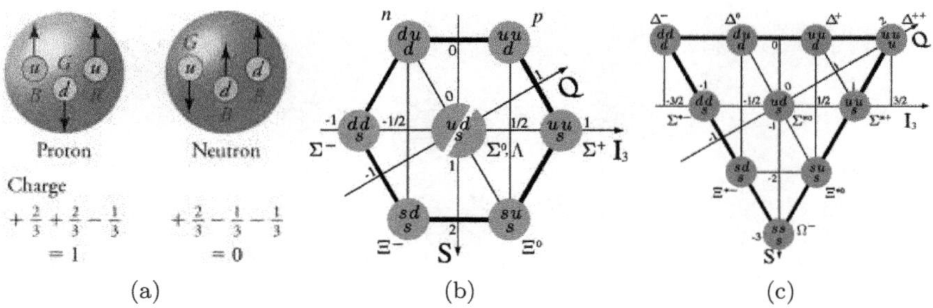

Figure 2.13. (a) Proton; (b) baryons-octet built from the u, d, s quarks; (c) baryons-decuplet.

They are built from three quarks. The lightest and well-known baryons are the proton and neutron (Fig. 2.13(a)). The proton is built from three quarks uud and the neutron from udd. They have a mass of about 1 GeV. They have spin-1/2 and belongs to the family of octet baryons (Fig. 2.13(b)) built from the quarks u, d, s. The decuplet baryons have spin-3/2 (Fig. 2.13(c)) where the lightest are $\Delta(1232)$.

• Exotic hadrons

Called exotics, these hadrons are not included in the previous quark model. Active searches for these hadrons are underway. One can have the following:

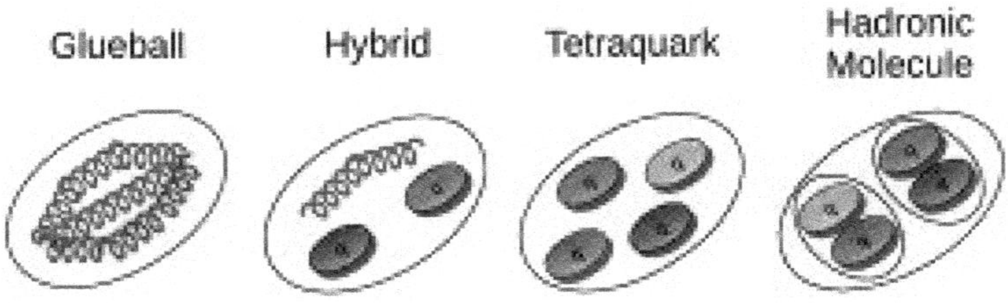

Figure 2.14. Some typical exotic mesons.

◇ Glueballs/Gluonia: bound states of two, three or more gluons,
◇ Hybrids: bound states of gluons and quark–anti-quark states,
◇ Fourquarks/Molecules,
◇ Pentaquarks/Pentamolecules.

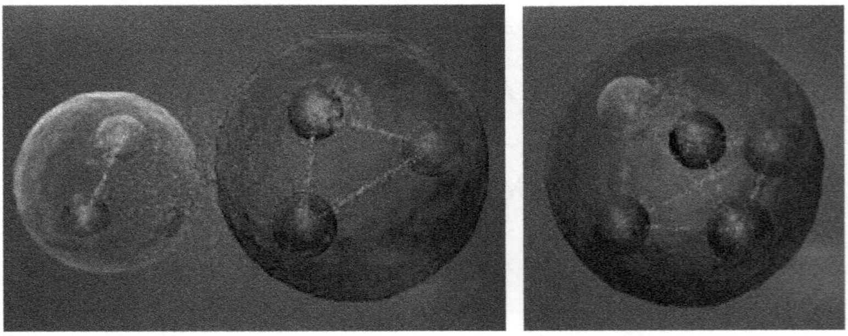

Figure 2.15. Pentamolecule and pentaquark.

Some signatures of these new states have been recently found in different experiments.

Most of the works presented in this book concern the phenomenology of hadrons (standard and exotics) using the method of QCD spectral sum rules à la Shifman–Vainshtein–Zakharov (SVZ) [3]. The results will be often compared with some other approaches, in particular, lattice calculations and potential models for heavy quarks and with experiments when available.

2.5.3 Confinement

Confinement is the most difficult aspect of QCD. There are some attempts to tackle this problem. However, they do not seem convincing. Among them, we consider the approximate SVZ approach where the non-perturbative quark and gluon condensates with higher and higher dimensions are used to parametrize the non-perturbative effects in an intermediate region where the operator product expansion (OPE) makes sense and where by duality the contribution of the lowest ground state is optimal. The applications of this approach in different channels are successful, compared to the data and to some other approaches, provided that the method is used carefully and where the optimal result from the analysis can be extracted.

2.5.4 Introductory books and reviews

An earlier useful introduction to the method of dimensional regularization and renormalization (the so-called $\overline{MS}$-scheme) by G. 't Hooft and M. Veltman [4] always used in gauge theories and calculations like the SM and QCD can, e.g., be found in Ref. [5] (see Fig. 2.16). The methods have been illustrated in some QCD and QED two-point functions.

Figure 2.16. Some introduction and reviews of the Method of Dimensional Regularization and Renormalization the $\overline{MS}$-scheme (1981).

An introduction to perturbative and non-perturbative QCD and to the method of QCD spectral (QSSR) or SVZ sum rules for Ph.D. students to QCD experts can, e.g., be found in Refs. [6, 7] (see Fig. 2.17(a) and (b)). These books also review the state of the art of the QSSR approach until 2004. Some recent results obtained after 2004 will be reviewed in Chapter 9.

(a) (b)

Figure 2.17. Some introduction and reviews on (a) QSSR (1989) and (b) QCD (2004).

2.6 Present status of Particle Physics

2.6.1 Test of the Big Bang assumption

In this section, I show the part of the Big Bang region (see Fig. 2.1) where the high-energy accelerators have reached. It is known from some model of cosmology that the universe is composed by about 5% of ordinary matter, 26.8% of dark matter and 68.2% of dark energy. The part which we shall discuss along this book corresponds to the 5% visible part of the universe.

As one can see in this figure, all particles (six leptons, six quarks, $W^{\pm}$, Z bosons, Higgs) predicted by the SM have been found and our present knowledge about the origin of the universe is about 10^{-10} s from the Big Bang explosion, after it, our universe is described successfully by the electroweak Standard Model (SM) and Quantum Chromodynamics (QCD).

2.6.2 Test of the SM and new physics

The Standard Model (SM) satisfies at present most of the high-precision experimental tests though some indirect sign of the deviation from the SM is found from the measurement of the anomalous magnetic moment (g-2) of the muon (see, e.g., the most recent

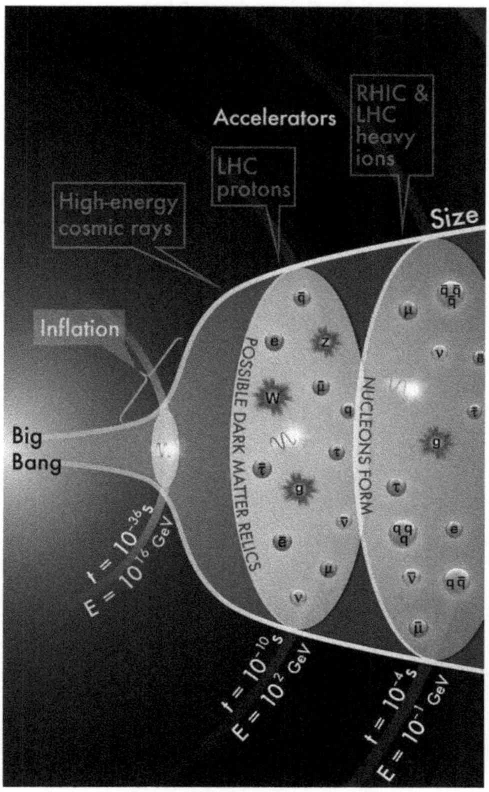

Figure 2.18. The Big Bang region reached by the high-energy accelerators.

analysis in Ref. [8] and references therein) while there may have seen some anomalies found in B, D meson semi-leptonic decays which need to be confirmed and which may indicate the presence of new particles beyond the SM. Also, there is not yet any direct production of such particles beyond the SM as probably their masses expected to be in the TeV region from different models are out of reach by the LHC except eventually a second Higgs boson which could be lighter.

2.6.3 The future of particle physics

However, the SM and QCD are not the end of the story. The three forces (electroweak and strong) are not yet unified. The Lie groups $SU(2)_W \otimes U(1)$ Model and $SU(3)_c$ describing the SM and QCD should be embedded in a larger group like, e.g., SU(5) or so many others in the current literature with the hope of reaching the origin $t = 0$ of the Big Bang explosion. Current activities in this direction have been carried out since several years (supersymmetry, superstrings, extra space–time dimensions, technicolours, composite models, etc.). However, these approaches seem (at present) to be unsuccessful and the Grand Unified Theories (GUTS) unifying the three microscopic (electroweak $\oplus$ strong) with the macroscopic (gravity) forces are still missing. Some of the approaches remain even speculative or become a new branch of mathematics far from physics.

I wish that some progress in this direction will be achieved in the near future.

2.6.4 Dark Matter and Dark Energy

However, our understanding of the origin of the universe needs more involved research beyond particle physics as it has been found by astronomers that the universe has more dark energy and dark matter than ordinary matter (see Fig. 2.19). They are invisible but Dark matter constitutes most of the mass of galaxies and galaxy clusters, and is responsible for the way galaxies are organized on large scales, while Dark energy accelerates expansion of the universe.

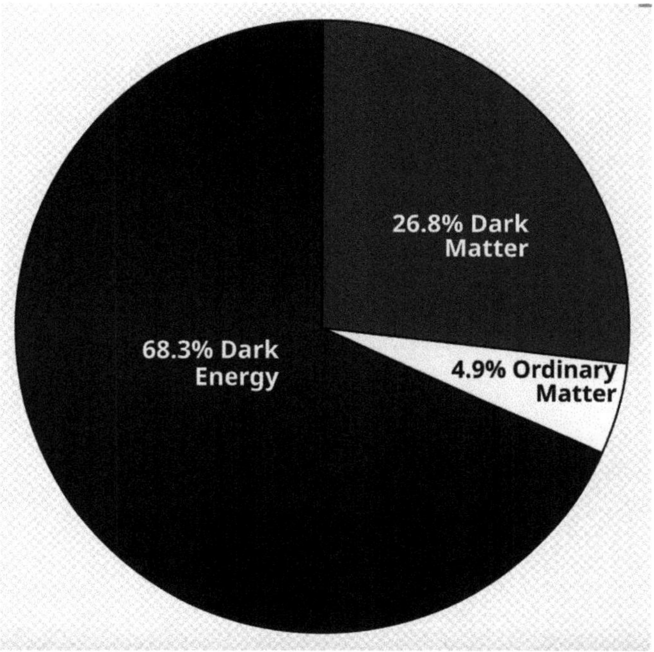

Figure 2.19. The composition of the Universe.

There is active research in particle physics on how these invisible matters couple to ordinary ones.

Chapter 3

Doctorate Thesis @ CPT Marseille 1974–1980

(a) (b)

Figure 3.1. (a) Former CPT at Joseph Aiguiller street; (b) University of Luminy, Marseille.

The bad weather in Paris pushed me to decide to go to the southern Mediterranean, to the Center for Theoretical Physics (CPT) in Marseille to find some sun. Later, the EC scholarship holders were invited to Brussels for a meeting with other students and trainees in Europe (see Fig. 1.13).

3.1 Dépaysement and Nostalgia

The scholarship holders were transferred to Marseille. We were placed in the Cité Universitaire in the centre of Marseille whose name is Madagascar. We felt less out of place, but the Usages and Customs were going to change.

Starting with the people we met in the street. We weren't used to seeing so many white people. So, we naturally felt like a stranger.

We were surprised to see white garbage collectors and cleaners even though white people were always the bosses in our country. Everyone respected them.

Then the diet. A baguette with a few pieces of ham and salad leaves (sandwich) was enough for the French while for us we needed a hill of rice on the plate with the corresponding dishes. So, we organized ourselves at the University City by buying a rice cooker and regularly eating Malagasy meals which we still continue to eat now. Next, we had to find a market with affordable prices. Fortunately, there was nearby, in the centre of Marseille, the Marché des Capucins where most foreigners did their shopping. Almost everything for food could be found there.

Few months later, we moved to the Cité Universitaire de Luminy (Aix-Marseille 2) which is outside of Marseille. Isolation encouraged nostalgia. So we got together with a few Malagasy friends to see each other regularly on weekends to organize small parties between us. It was in Luminy that I met Lydia, a student who became my current wife.

We then decided to live together and left the Cité Universitaire for larger accommodation of the Mutuelle des Etudiants (Saint Jean du Desert street, Marseille). We lived well with my EC grant, but we also wanted to give my brothers a chance to invite them to come and settle in France. So, we did any jobs (factory guard at night, etc.) Thanks to Amza, a Mauritian colleague, we were able to find work for 2–3 months during the summer holidays, in a hotel restaurant in Auron (Alpes Maritimes): me as a dishwasher and kitchen assistant and Lydia as a linen worker. It was with these salaries that we paid the plane tickets for my two brothers who joined us in December 1975.

To strengthen ties with our Malagasy compatriots, I actively participated in the Malagasy Association of Marseille which regularly organized parties and a half day of sports every Sunday morning. We also organized bus outings with the association (Andorra, Nice carnival, ski weekend, etc.). The Madagascar consulate also regularly organized the end of year's eve where all the Malagasy people from the surrounding towns participated. Thanks to these organizations, integration took place gradually.

(a) (b)

Figure 3.2. Outings of the association: (a) Carnival of Nice; (b) ski weekend.

For an anecdote: *I was invited to the European Center for Nuclear Research (CERN) in Geneva with another doctoral student from my class. We took the train. After the passport check at the Cornavin — Geneva station, the policeman banned me from staying in Switzerland and had me taken back to Bellegarde, the nearest French station. We stayed one night and CERN sent us an official car to take us back to Switzerland!*

Later, the Head of the Education Service of the University of Luminy, Ms. Gautry, who became a friend, advised me to apply for dual French nationality to facilitate my stay in France (residence card, passport) and Europe to avoid having to apply for a visa each time I leave France. The process had been made easier because her husband, a police inspector at the Marseille police station, had taken charge of the file.

So, contrary to what many Malagasy people think, having dual nationality does not mean denying your original nationality but is essential from a practical administrative point of view if you live outside of Madagascar.

3.2 Studies at the CPT of Marseille

The first exchange with my Supervisor Mr. Professor Eduardo de Rafael was not very encouraging. Before leaving Madagascar, I had

Figure 3.3. Professor E. de Rafael giving a concert with Christine Kattner (soprano singer) @ QCD 2000 — Montpellier.

mentioned to him that I had obtained a scholarship from the EC and wanted to come to the CPT to prepare my doctoral theses in theoretical physics with the recommendations of Professor Raoelina Andriambololona (hereinafter referred to as Raoelina) who had also prepared his theses of third cycle and state doctorate at the CPT of Marseille. His response was relatively negative compared to that of the Ecole Normale Superieure (ENS) in Paris. We could understand this by his assessment of the level of a Malagasy academic compared to doctoral students from ENS and the Ecole Polytechnique.

Being naturally a bit Masocious and loving the challenge, this apparent refusal motivated me to come to the CPT in addition to the Marseille sun. The first meeting was encouraging. The teacher seemed cool and normal compared to the other teachers but seemed very strict, severe and irritable. The common core courses for the Diploma of Advanced Studies (DEA) (fifth year after the baccalaureate) were beginning. It was very hard. The quantum mechanics course that we did for a year with Mr. Raoelina was done in one month at the CPT. It was also understandable because during his lesson Mr. Raoelina told us about his life and only gave his lesson a quarter of an hour before the end! The books are in English, which was a handicap because I had neglected this subject in high school. So I had to work day and late night as well as on weekends to make up for these delays. I spent these times in the library and in front of the CPT computer (see Fig. 3.4).

Figure 3.4. The computer UNIVAC and the library of CPT at Joseph Aiguiller street, Mareseille.

3.3 Doctorate of Third Cycle Thesis

3.3.1 Thesis on the $(g-2)_l$ of leptons

Finally, I was in the second year for the third cycle doctorate. I chose Mr. de Rafael as my thesis director because he also studied phenomenology apart from mathematical physics. He asked me to continue my DEA work on *hadronic contributions to the anomalous magnetic moment* $a_\mu \equiv (1/2)(g-2)_\mu$ *of the lepton* μ, the second of the family of charged leptons (e, μ, τ).

I finished my thesis work sooner than expected [9]. The results were published with Mr. de Rafael's team (Fig. 3.5) in the Elsevier Journal *Physics Letters* [10], which we later completed with a review on the subject [11].

• Hadronic vacuum polarization contribution to a_μ at LO

At the lowest order in perturbative series in $\alpha = e^2/4\pi$ where e is the electric charge, this contribution is given by the diagram in Fig. 3.6, where the loop is the vacuum polarization. It can be formulated by the dispersion relation [12–16]

$$a_\mu \equiv \frac{1}{2}(g-2)_\mu = \frac{1}{4\pi^3}\int_{4m_\pi^2}^{\infty} dt\, K_\mu(t)\sigma(e^+e^- \to \text{hadrons}), \qquad (3.1)$$

Figure 3.5. Professors E. de Rafael, J. Calmet and M. Perrottet of CPT, Marseille.

Figure 3.6. Contribution at lowest order in α of the hadronic vacuum polarisation to the anomalous magnetic moment of the muon $a_\mu \equiv (1/2)(g-2)_\mu$.

where $K_\mu(t)$ is the kernel function of QED [17]:

$$K_\mu(t) = \int_0^1 dx \frac{x^2(1-x)}{x^2 + (t/m_\mu^2)(1-x)}, \tag{3.2}$$

and $\sigma(e^+e^- \to \text{hadrons})$ is the total cross section of the reaction: $e^+e^- \to$ hadrons. We had obtained [9, 10]

$$a_\mu^{hvp}|_{lo} = (70.2 \pm 8.0) \times 10^{-9}, \tag{3.3}$$

while with the new experimental data, I recently obtained [18–21]:

$$a_\mu^{hvp}|_{lo} = (7094 \pm 37) \times 10^{-11}, \tag{3.4}$$

showing a perfect agreement of the two results after 48 years of experimental effort but with a precision improved by a factor of 20!

• Hadronic vacuum polarization contribution to a_μ at NLO

The hadronic contribution to the higher order without the contribution of the light-by-light diagram was first estimated by the authors in

Ref. [11]:

$$a_\mu^{hvp}|_\alpha = -(86 \pm 32) \times 10^{-11}, \tag{3.5}$$

in perfect agreement with recent more precise determination [22]:

$$a_\mu^{hvp}|_\alpha = -(99.34 \pm 0.91) \times 10^{-11}. \tag{3.6}$$

• Theory versus experiment for a_μ

The recent confrontation between the predictions of the standard model and the new average of experimental results [23]:

$$a_\mu^{exp} = (116592071.5 \pm 14.5) \times 10^{-11}, \tag{3.7}$$

gives [21]:

$$\Delta a_\mu^{pheno} \equiv a_\mu^{exp} - a_\mu^{th} = (86 \pm 41) \times 10^{-11}\,, \tag{3.8}$$

indicating a difference of 2.1σ between theory and experiment. If one attempts to average this result from phenomenolical approaches with the present most precise value from the lattice calculation [24]:

$$\Delta a_\mu^{lattice} = (90 \pm 56) \times 10^{-11}, \tag{3.9}$$

one obtain :

$$\langle \Delta a_\mu^{th} \rangle = (87 \pm 33) \times 10^{-11}. \tag{3.10}$$

This deviation of about 2.6σ from the SM prediction (if it persists) may signal the presence of new physics beyond the Standard Model and stimulates the needs of a much better precision for measuring the $e^+e^- \to$ Hadrons total cross-section and for the lattice calculations.

• The a_τ heavy lepton anomaly τ

After my third cycle thesis, I extended the analysis of a_μ to that of a_τ of the heavy lepton τ by varying the mass of the lepton. I had

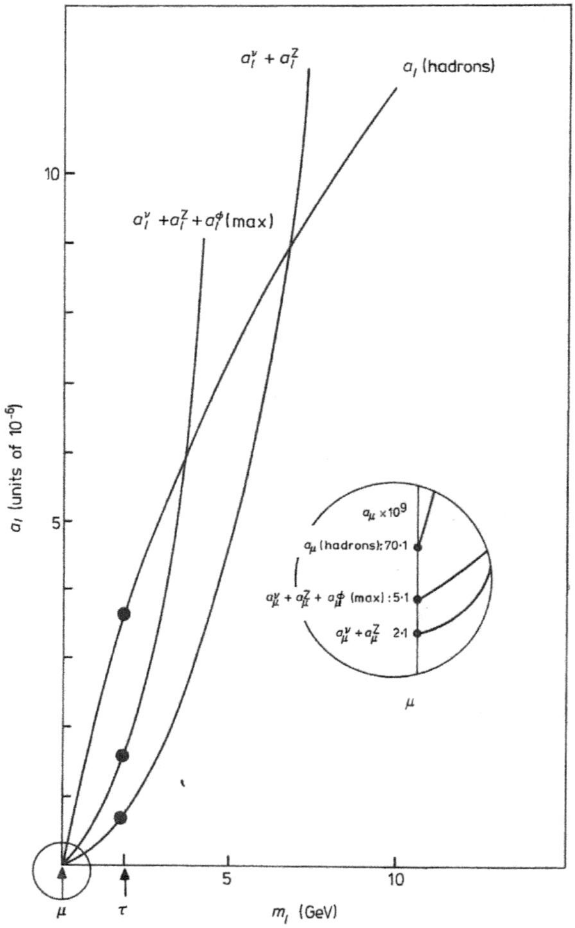

Figure 3.7. Behaviour of the different contributions to the $g-2$ of a hypothetical heavy lepton as a function of its mass.

obtained in units of 10^{-6} [23]:

$$
\begin{aligned}
a_\tau^{qed}|_{\leq\alpha^6} &= 1169.6 \\
a_\tau^{hvp}|_{lo} &= 3.7 \pm 0.4, \\
a_\tau^{weak} &= 10.8 \pm 0.2,
\end{aligned}
\tag{3.11}
$$

giving the total contribution

$$
a_\tau^{tot} = 1174.1 \pm 0.5. \tag{3.12}
$$

A new determination gives [18–20]

$$a_\tau^{hvp}|_{lo} = (3494.8 \pm 24.7) \times 10^{-9}, \tag{3.13}$$

in agreement with the previous evaluation but more precise.

It is interesting to note that the relative weights of the hadronic contributions change due to the behaviour of the kernel function in Eq. (3.2). Instead of the muon case where the contribution at low energy, that of ρ, is dominant, in the case of τ, we have a balance between the contributions for lower and higher $\sqrt{t}$ at 1.2 GeV.

The contribution of the weak interaction is no longer negligible but almost 1/4 of the hadronic one.

Thus, the measurement of a_τ will be able to tell us even more about the Standard Model although experimentally, it will be a little more difficult to make a measurement as precise as that of the muon.

• The semi-hadronic decays of the heavy lepton τ

After the Karlsruhe summer school in 1978 and after discussion with Professor M.L. Perl, the discoverer of the lepton τ (see Fig. 3.8) who had given courses in this school, I was interested in studying the influence of the mass of the neutrino τ on the branching ratios of the different hadronic decays of τ by suggesting that their more precise measurement would constrain the value of m_{ν_τ}. This effect is shown in Fig. 3.9 taken from Ref. [24], where we can note that the process $\tau \to \nu_\tau + \rho$ is very sensitive to the mass of ν_τ because of phase space. At that time, the upper limit on the mass was a few hundred MeV.

Figure 3.8. Professors S. Ting and B. Richter (Physics Nobel Price 1976), A. Lederman (Physics Nobel Price 1988), M.L. Perl (Physics Nobel Price 1995).

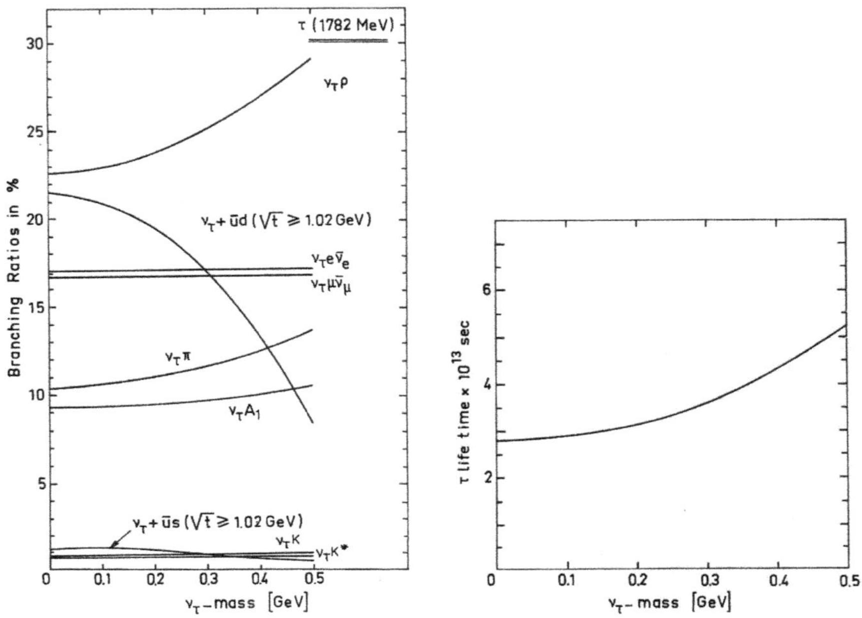

(a) (b)

Figure 3.9. (a) Behaviour of the different semi-hadronic branching ratios of the lepton τ as a function of the neutrino mass ν_τ; (b) lifespan of τ as a function of ν_τ.

3.3.2 Conferences and summer schools

• Rencontres de Moriond: Discovery of J/ψ, τ and Υ

◇ During the period 1976–1985, I regularly participated in the Moriond meetings organized by Tran Thanh Van and his team. The conferences took place in the ski resorts of Les Arcs, Tignes, etc. The idea was to spend a vacation in March and at the same time exchange each other's research results. I had discovered snow and taken my first ski lesson (see Fig. 3.10).

◇ It was impressive when S. Ting and B. Richter (Nobel Prize in Physics 1976) each presented the discovery of J (name given by Ting) in the proton–proton experiment at Brookhaven National Laboratory — Upton or ψ (name given by Richter) in the SLAC e^+e^- experiment in 1974. This particle, a bound state $\bar{c}c$ of the fourth quark c was long awaited to complete the second doublet (c, s) and validate the Glashow–Illiopoulos–Maiani mechanism (GIM) (Dirac Medal 2007) proposed in 1970 which eliminates the effects of the neutral current which changes the flavour (FCNC) in the transitions $\Delta S = 2$ weak interactions [26].

◇ A year later, Martin Perl presented the third heavy lepton τ by identifying μe events in the SLAC experiment. However, this discovery, just after that of J/ψ, took time to be accepted by the scientific community.

◇ In 1977, the discovery of the Υ bound state $\bar{b}b$ of the fifth quark was announced by Leon Lederman (Nobel Prize in Physics 1988 with M. Schwartz and J. Steinberger on the neutrino). He was so overjoyed that he started dancing on the table during the dinner party.

• The schools of physics

To complete my knowledge, I participated in different summer schools:

◇ The Karlsruhe summer school (Germany) 1978

It was the first school I attended. Teacher M.L. Perl gave us lessons on leptons including the heavy lepton τ which interested me a lot after my article on the anomalous magnetic moment of τ.

◇ The Gif school (Orsay)

I went to the Gif school around 1975–1976, where I met Professor Mr. Veltman (Nobel Prize in Physics 1999 with G. t'Hooft) and on campus my former Chemistry Professor at the University of Antananarivo, Ms. Marta Andriatsiferana. We became so friendly that Mr. Veltman thought she was my girlfriend! Later, in 1984, I was invited to give lectures on the sum rules of QCD.

◇ The Cargèse school (Corsica)

In 1979, I attended the Cargèse school where courses were given by eminent professors, including S.L. Glashow, one of the fathers of the Standard Model (Nobel Prize in Physics 1979 with A. Salam and S. Weinberg), M. Veltman (Nobel Prize 1999 with G.t'Hooft), John Ellis (Dirac Medal 2005), Mary K. Gaillard (J.J. Sakurai Prize 1993) and L. Maiani (Dirac Medal 2007). They are all experts in weak interactions.

◇ Anecdote

Before going to Cargèse, Glashow visited the CPT to see his friend E. de Rafael. We were face to face in the elevator, but I was so impressed that I couldn't say a word to him. Once in Cargèse, my wife and I met him again on a terrace where he was having lunch with Mr. Veltman and their respective wives. He then asked me where I came from. I told him that I came from Madagascar. He didn't believe it and asked me to show my passport.

Figure 3.10. My first ski lessons at "Les Arcs-Haute Savoie".

I was also able to note the somewhat particular character of Mr. Veltman: he practiced nudism on the private beach. During one evening, it seemed to me that he had drunk too much. So, he started slapping the buttocks of a physicist's wife!

3.4 The State Doctorate Thesis

3.4.1 The transition period

After the defense of my third cycle thesis, my CE scholarship ended. Mr. Raoelina asked me to return to Madagascar. An emissary of Didier Ratsiraka, President of the Republic of Madagascar, also approached me to ask me to return to Madagascar. It was a very complicated decision to take because I noticed that one of my friends had become a minister while I no longer had the money to be able to continue my studies. Fortunately, Mr. de Rafael and his colleagues at the CPT were able to find me a temporary job at the Universities of Luminy and Provence. These positions lasted 3 years while I prepared my state doctorate thesis.

3.4.2 The process $\mu \to e\gamma$

However, for the State Doctorate thesis, it was necessary to completely change the subject because our review on $g - 2$ [11] had closed the subject.

I was then interested in phenomena beyond the Standard Model. The results of experiments on the process $\mu \to e\gamma$ which violate the leptonic number were fashionable at that time. We started to calculate this process with Mr. de Rafael, but after a while, he was no longer enthusiastic.

Figure 3.11. Professors M. Veltman (Nobel Prize in Physics 1999), S.L. Glashow (Nobel Prize in Physics 1979), L. Maiani (Dirac Medal 2007), M.K. Gaillard (J.J. Sakurai Prize 1993) and J. Ellis (Dirac Medal 2005).

I continued the work with Pierre Mery where we did the calculation for a general model based on the group $SU(2)_L \otimes U(1) \otimes SU(2)_R$ [25].

3.4.3 Weinberg sum rules and chiral symmetry

- **Weinberg sum rules at the chiral limit**

Later, we studied the chiral symmetry breaking $SU(n)_L \otimes SU(n)_R$ by analyzing Weinberg's spectral sum rules. We started from the two-point hadronic spectral function:

$$W^{\mu\nu}_{V/A}(q)^i_j = \frac{1}{2} \int d^4x \, e^{iqx} \langle 0 | [J^\mu_{V/A}(x)^i_j, J^\mu_{V/A}(0)^i_j] | 0 \rangle, \tag{3.14}$$

where $J^\mu_{V,A}(x)^i_j$ designates the different vector and axial-vector hadronic currents of the group $SU(n)_L \otimes SU(n)_R$ of the algebra of currents:

$$J^\mu_V(x)^i_j = \bar{\psi}_i(x)\gamma^\mu \psi_j(x), \quad J^\mu_A(x)^i_j = \bar{\psi}_i(x)\gamma^\mu\gamma_5\psi_j(x), \tag{3.15}$$

with ψ_i the quark fields; $i, j = 1, 2, \ldots, n$, where n designates the flavour number of the quarks.

In general, the currents V and A are not conserved and the spectral functions $W^{\mu\nu}_V(q)^i_j$ and $W^{\mu\nu}_A(q)^i_j$ depend on two invariants ($q^2 \equiv t \geq 0$):

$$W^{\mu\nu}_{V/A}(q)^i_j = -(g^{\mu\nu}q^2 - q^\mu q^\nu)\mathrm{Im}F^{(1)}_{V/A}(t)^i_j + q^\mu q^\nu \mathrm{Im}F^{(0)}_{V/A}(t)^i_j, \tag{3.16}$$

where the indices (1) and (0) correspond to the spin of the associated hadrons. Introducing the notation:

$$\Delta^{(1,0)}(t)_{i,j} \equiv \mathrm{Im}F^{(1,0)}_V(t)^i_j - \mathrm{Im}F^{(1.0)}_A(t)^i_j, \tag{3.17}$$

Weinberg's sum rules stated that in the chiral limit of massless quarks ($m_i i, j = 0$) or asymptotically ($t \to \infty$), the chiral symmetry must be a good *symmetry*. Thus, we have the first sum rule:

$$\int_0^\infty dt\left(Q_{ij}(t)\equiv\Delta^{(1)}(t)_{i,j}+\Delta^{(0)}(t)_{i,j}\right)=0, \tag{3.18}$$

and the second and third sum rules:

$$\int_0^\infty dt\,t\,\Delta^{(0)}(t)_{i,j}=0,\quad \int_0^\infty dt\,t\,\Delta^{(1)}(t)_{i,j}=0, \tag{3.19}$$

which are super-convergent. By saturating the first two sum rules with hadrons of spin 1 (ρ, A_1) and 0 (π), we deduce the famous relation for $m_\pi^2=0$:

$$M_{A_1}\simeq\sqrt{2}\,M_\rho. \tag{3.20}$$

• The QCD contributions: α_s and mass corrections

During the initiation of the work, I attended a conference in the Spanish Pyrenees with QCD experts as participants to the case (E. de Rafael, F.J. Yndurain, H. Leutwyler, C. Sachrajda, etc.). It was very friendly and one afternoon we played football (the sport that I enjoy the most and that I practiced in high school). The players were impressed by my way of playing and started to consider me. After the conference, Professor Yndurain had invited me to stay for a while at the Autonomous University of Madrid. On the return road, he bought some bottles of wine in some village and continued to draw on the windshield of the car the diagrams to calculate to encourage me to take an interest in the problem. After a few days in Madrid, I was returning to Marseille. A week after, I finished my calculations and presented them to Professors de Rafael and Yndurain.

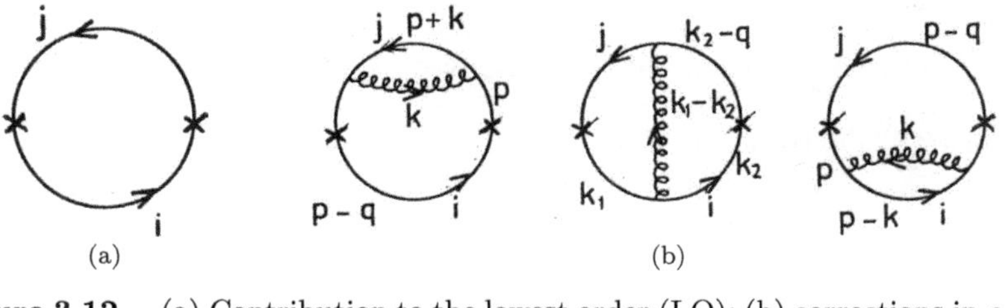

Figure 3.12. (a) Contribution to the lowest order (LO); (b) corrections in α_s/π (NLO). The cross (**x**) represents the pseudoscalar current. The wavy line denotes the gluon.

The calculation of diagram (a) did not pose any particular difficulties. For massive quarks, we obtain, after renormalization in the $\bar{M}S$-scheme:

$$\psi_5(q^2) = -\frac{3q^2}{4\pi^2}\int_0^1 dx\,(m_i^2 x + m_j^2(1-x) + m_i m_j) \times \log\frac{-q^2 x(1-x) + m_i^2 x + m_j^2(1-x)}{\nu^2}. \quad (3.21)$$

The calculation of the radiative correction due to diagrams (b) is non-trivial. At the time, we didn't have any computer programs to do the calculation. Everything was done by hand. I found:

$$\psi(q^2)^i_j = \frac{3}{8\pi^2}(m_i + m_j)^2\left[1 + \frac{17}{3}\frac{\alpha_s}{\pi} + \cdots\right](-)\log\frac{-q^2}{\nu^2}. \quad (3.22)$$

People's first reaction was to say that the result was false. Indeed, in the case of vector and axial-vector currents, the coefficient of α_s/π is 1 but not 17/3. So, they did again the calculations, and finally, they found my result.

From that moment on, everyone started to trust me!

• Breaking of the second and third Weinberg sum rules

We had evaluated the quark mass corrections to these Weinberg sum rules with Floratos [29] (Fig. 3.13) to the lowest order of the perturbation theory. Including the quark mass corrections, we found that:

$$\mathrm{Im}\,\Delta^{(1,0)}(t)_{ij} = \pm\frac{3}{2\pi}\frac{m_i m_j}{t}\left\{1 - \frac{m_i^2 + m_j^2}{t} - \frac{2m_j^2 m_i^2}{t^2} + \mathcal{O}\left(\frac{1}{t^3}\right)\right\}, \quad (3.23)$$

indicating that the first sum rule is superconvergent but not the other two. Then, we constructed more convergent combinations like[1]:

$$\mathrm{Im}\,R^{(1,0)}_{ijk} \equiv \mathrm{Im}\,\Delta^{(1,0)}(t)_{ij} - \frac{m_j}{m_k}\mathrm{Im}\,\Delta^{(1,0)}(t)_{ik} = \pm\frac{3}{2\pi}\frac{m_i m_j}{t}(m_j^2 - m_k^2)\left(1 + 2\frac{m_i^2}{t}\right) + \mathcal{O}\left(\frac{1}{t^4}\right). \quad (3.24)$$

[1]Other combinations are given in Ref. [29].

To give more meaning to the quark mass entering into the sum rules, we have introduced (for the first time) the definition of the running quark mass as a solution of renormalization group equation (RGE) [29]:

$$\bar{m}_i(q^2) = \frac{\hat{m}_i}{\log(\sqrt{-q^2}/\Lambda)^{\gamma_1/-\beta_1}}, \tag{3.25}$$

where $\gamma(\alpha_s) = 2(\alpha_s/\pi) + \mathcal{O}(\alpha_s^2)$ is the quark mass anomalous dimension, $\beta_1 = -(1/2)(11 - 2n_f/3)$ is the first coefficient of the β-function for n_f number of flavours, and $\hat{m}_i$ is the renormalization group invariant quark mass analogue to Λ associated to the α_s coupling.

Figure 3.13. Professor E.G. Floratos, University of Athens, Professor F.J. Yndurain, University of Autonoma Madrid and Professor C. Becchi, University of Genova.

• Breaking of the first Weinberg sum rule at order α_s

We also evaluated the order correction α_s to the first: sum rule and found:

$$\Delta^{(1)}(t)_{i,j} + \Delta^{(0)}(t)_{i,j} = \left(\frac{\alpha_s}{\pi}\right)\left(\frac{m_i m_j}{q^2}\right)\left(\frac{2}{\pi^2}\right)(0.500 \pm 0.001), \tag{3.26}$$

showing that it is broken by the quark masses at order α_s.

3.4.4 Das–Mathur–Okubo sum rules and flavour symmetry

An analogous analysis was made by Das–Mathur–Okubo (DMO) [28] in the case of the electromagnetic current with three different flavours of quarks:

$$J^{\mu}_{em}(x) = V^{(3)}_{\mu} + \sqrt{\frac{1}{3}} V^{(8)}_{\mu} \; : \quad V^{(a)}_{\mu} = \sqrt{\frac{1}{2}} \bar{\psi} \gamma_{\mu} \lambda_a \psi(x), \tag{3.27}$$

where λ_a are diagonal matrices of $SU(3)_F$:

$$\lambda_3 = \sqrt{\frac{1}{2}} \begin{pmatrix} 1 & & \\ & -1 & \\ & & 0 \end{pmatrix}, \quad \lambda_8 = \sqrt{\frac{1}{6}} \begin{pmatrix} 1 & & \\ & 1 & \\ & & -2 \end{pmatrix}, \tag{3.28}$$

which act on the basis of the up (u), down (d) and strange quark (s) quark fields:

$$\psi(x) = \begin{pmatrix} \psi_u(x) \\ \psi_d(x) \\ \psi_s(x) \end{pmatrix}. \tag{3.29}$$

It stipulates that asymptomatically the flavour symmetry is realized:

$$\int_{4m_\pi^2}^{\infty} dt\, t \left\{ \frac{1}{3} \sigma(e^+e^- \to \text{Hadrons})_{I=1} - \sigma(e^+e^- \to \text{Hadrons})_{I=0} \right\} = 0, \tag{3.30}$$

where $\sigma(e^+e^- \to \text{Hadrons})_{I=1,0}$ designates the total cross-section of production by e^+e^- of isovector hadrons ($I = 1$) and isoscalars ($I = 0$). The cross-section is linked to the electromagnetic spectral function by the optical theorem:

$$\sigma(e^+e^- \to \text{Hadrons}) = \frac{4\pi^2\alpha}{q^2} e^2 \frac{1}{\pi} \text{Im}\Pi_{em}(q^2), \tag{3.31}$$

where

$$-(g_{\mu\nu}q^2 - q^{\mu}q^{\nu})\, \text{Im}\Pi_j(q^2) = \frac{1}{2} \int d^4x\, e^{iqx} \left\langle 0 \middle| \left[V^j_{\mu}(x), \left(V^j_{\nu}(0)\right)^{\dagger} \right] \middle| 0 \right\rangle. \tag{3.32}$$

For two different flavours of quarks, we obtain, àt the lowest order [30]:

$$\begin{aligned} \Delta_{ij}(q^2) &\equiv \Pi_i(q^2, m_i^2) - \Pi_i(q^2, m_i^2) \\ &= \frac{3}{2\pi^2} \left[\frac{m_i^2 - m_j^2}{q^2} + m_i^2 \log \frac{-q^2}{m_i^2} - m_j^2 \log \frac{-q^2}{m_j^2} \right]. \end{aligned} \tag{3.33}$$

As in the case of the Weinberg sum rules, that of DMO is also broken by the masses of the quarks at the lowest order and are no longer convergent.

3.4.5 Finite energy sum rules à la Shankar

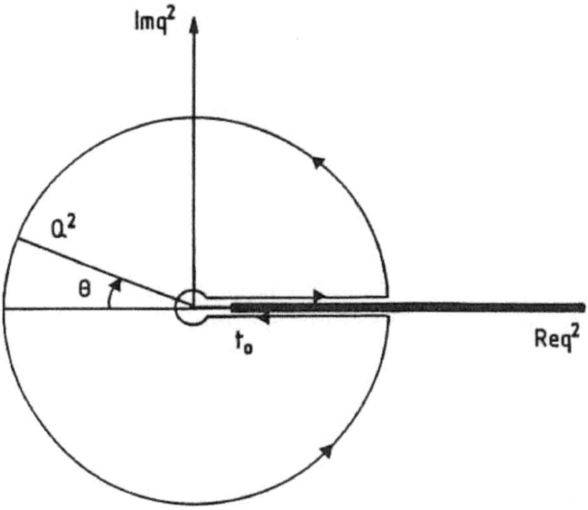

Figure 3.14. Integration contour in the complex q^2-plane.

To solve these problems of non-convergence of the Weinberg and DMO sum rules, we used the finite energy sum rules (FESR) à la Shankar [31] in the complex plane of q^2 on the circle of radius Q^2 (see Fig. 3.14). We thus obtain:

$$\int_{t_0}^{Q^2} dt\, t^p \frac{1}{\pi} \mathrm{Im}\Delta(t) = \frac{(-1)^p (Q^2)^{p+1}}{2\pi} \int_{-\pi}^{+\pi} dt\, e^{i(p+1)\theta} \Delta(-Q^2\, e^{i\theta}). \tag{3.34}$$

Thus, the QCD expressions of the FESR versions of the first, second and third Weinberg sum rules can be found in [29]. They are proportional to $\bar{m}_u \bar{m}_d$, while the combination in Eq. (3.24) is proportional to $\bar{m}_u \bar{m}_d (m_k^2 - m_j^2)$. The QCD expressions of the DMO which are proportional to $(m_j^2 - m_i^2)$ are in Ref. [30].

3.4.6 Phenomenology of FESR à la Shankar: Charged current

Some applications have been presented in Refs. [29] and [30]. In the case of $SU(2)_L \otimes SU(2)_R$, the masses of the u and d quarks being small, these corrections did not numerically modify the predictions of Weinberg and DMO but made them more rigorous.

• Sum rules of the first type

◇ *ud*, *us*

Predictions based on mass of quarks are given for the decays $\tau \to \nu_\tau A_1$ and $\tau \to \nu_\tau K^*$. Choosing $\sqrt{Q^2} = 1.2$ GeV, we obtain:

$$R_{\tau A_1} \equiv \frac{\tau \to \nu_\tau A_1}{\tau \to \nu_\tau e \bar{\nu}_e} = 0.48 \pm 0.05 + 3.5 \times 10^{-3} \frac{\hat{m}_u \hat{m}_d}{m_\pi^2},$$
$$R_{\tau K^*} \equiv \frac{\tau \to \nu_\tau K^*}{\tau \to \nu_\tau e \bar{\nu}_e} = 0.036 - 1.95 \times 10^{-4} \frac{\hat{m}_u \hat{m}_s}{m_\pi^2}. \tag{3.35}$$

◇ *cd*, *cs*

Choosing $\sqrt{Q^2} = 2.1$ GeV, we obtain:

$$\pi \frac{M_{D^*}^2}{2\gamma_{D^*}^2} - 2\pi f_D^2 = -\frac{1}{57.0} \hat{m}_d \hat{m}_c,$$
$$\pi \frac{M_{F^*}^2}{2\gamma_{F^*}^2} - 2\pi f_F^2 = -\frac{1}{60.3} \hat{m}_s \hat{m}_c, \tag{3.36}$$

with the normalizations:

$$\langle 0 | A_\mu(0) | D^+, F^+(q) \rangle = \sqrt{2} f_{D,F} q^\mu,$$
$$\langle 0 | V_\mu(0) | D^*, F^*(q) \rangle = \frac{\pi M_{D^*,F^*}^2}{\sqrt{2}\gamma_{D^*,F^*}} \epsilon^\mu. \tag{3.37}$$

• Sum rules of the second and third types

◇ By saturating the spectral functions with the pion and the kaon:

$$R_{uds}^{(0)} \quad \Longrightarrow \quad \frac{\hat{m}_s}{\hat{m}_d} \geq \frac{M_K^2 f_K^2}{m_\pi^2 f_\pi^2} = 16.8, \tag{3.38}$$

which becomes an equality if $\hat{m}_u = 0$. Combining this result with those of $\Delta^{(1)}_{ud}$ and $\Delta^{(1)}_{us}$, we obtain:

$$M_{A_1} \geq (1045 \pm 16)\,\text{MeV}. \tag{3.39}$$

◇ The other sum rules give:

$$R^{(0)}_{udc} \quad \Longrightarrow \quad \frac{\hat{m}_c}{\hat{m}_u} = \frac{M_D^2}{m_\pi^2}\frac{f_D^2}{f_\pi^2} + 0.2\frac{\hat{m}_d\hat{m}_c(\hat{m}_c^2 - \hat{m}_u^2)}{m_\pi^4} \geq 179.2\frac{f_D^2}{f_\pi^2},$$

$$R^{(0)}_{cds} \quad \Longrightarrow \quad f_F^2 \geq \frac{\hat{m}_s}{\hat{m}_d}1.18 f_D^2,$$

$$R^{(0)}_{suc} \quad \Longrightarrow \quad f_F^2 \leq \frac{\hat{m}_c}{\hat{m}_u}\frac{f_K^2}{12.56}, \tag{3.40}$$

which can be translated into lepton widths:

$$\frac{\Gamma(D \to \mu\nu)}{\Gamma(F \to \mu\nu)} = \text{tg}^2\theta_c \frac{M_D f_D^2}{M_F f_F^2} \leq 4 \times 10^{-2}\frac{\hat{m}_d}{\hat{m}_s},$$

$$\frac{\Gamma(D \to \mu\nu)}{\Gamma(F \to \mu\nu)} = \text{tg}^2\theta_c \frac{M_D f_D^2}{m_\pi f_\pi^2} \leq 4 \times 10^{-2}\frac{\hat{m}_d}{\hat{m}_s},$$

$$\frac{\Gamma(F \to \mu\nu)}{\Gamma(\pi \to \mu\nu)} = \frac{M_F f_F^2}{m_\pi f_p i^2} \leq 1.56\frac{\hat{m}_c}{\hat{m}_u}, \tag{3.41}$$

where θ_c is the Cabibbo angle: cos $\theta_c = 0.225$ [33].

3.4.7 Phenomenology of FESR à la Shankar: Neutral current

These applications are discussed in Ref. [30]. The functions $E_0(Q^2/\Lambda^2)$ and Φ_{0c} used in the following are defined in Ref. [30].

• Below the threshold of charm

The FESR version of the DMO sum rule is:

$$\mathcal{M}^{(0)}_{uds}(Q^2) \equiv \int_{4m_p i^2}^{Q^2} dt\, t\Big\{\sigma_{I=1}(t) - 3\,\sigma_{I=0}(t)\Big\}$$

$$= \frac{8}{3}\pi^2\alpha^2(2\hat{m}_s^2 - \hat{m}_u^2 - \hat{m}_d^2)E_0(Q^2/\Lambda^2), \tag{3.42}$$

where $\sigma_I(t)$ is the total cross-section for the process $e^+e^- \to$ Hadrons of isospin I. For $\hat{m}_s = 0.12$ GeV, the consistency of the theory and the data are obtained for $\sqrt{Q^2} \simeq 1.5$ GeV.

• Between the thresholds of charm and beauty

In this region, we used the moment:

$$\begin{aligned} \mathcal{M}^{(0)}_{h-c}(Q^2) &\equiv \int_{4m_pi^2}^{Q^2} dt\, t \Big\{ \sigma_{u,d,s}(t) - \frac{3}{5}\sigma_{u,d,s,c}(t) \Big\} \\ &= \left[\frac{32}{45}\pi^2\alpha^2(6\hat{m}_c^2 - \hat{m}_s^2 - \frac{5}{2}(\hat{m}_u^2 + \hat{m}_d^2) \right] \\ &\quad \times \Big[E_0(Q^2/\Lambda^2) + \Phi_{0c}(Q^2/\Lambda^2) \Big], \end{aligned} \tag{3.43}$$

The analysis gave us a value for the mass constituting the charm:

$$M_c \equiv \overline{m}_c(2m_c) = \simeq (1.55 \pm 0.26) \text{ GeV}. \tag{3.44}$$

• Above the threshold of beauty

An analogous sum rule was used for b:

$$\begin{aligned} \mathcal{M}^{(0)}_{h-b}(Q^2) &\equiv \int_{4m_pi^2}^{Q^2} dt\, t \Big\{ \sigma_{u,d,s}(t) - \frac{2}{3}\frac{1}{Q_b^2}\Big[\sigma_{u,d,s,c,b}(t) - \sigma_{u,d,s,c}(t) \Big] \Big\} \\ &= \left[\frac{16}{9}\pi^2\alpha^2(6\hat{m}_b^2 - \hat{m}_s^2 - \frac{5}{2}(\hat{m}_u^2 + \hat{m}_d^2) \right] \\ &\quad \times \Big[E_0(Q^2/\Lambda^2) + \Phi_{0b}(Q^2/\Lambda^2) \Big], \end{aligned} \tag{3.45}$$

where $Q_b = -1/3$ is the b-quark charge in units of e. Using the mass of the charm obtained previously, we obtained:

$$M_b/M_c \geq 3.87. \tag{3.46}$$

3.4.8 Bounds on $m_{u,d,s}$ and on the D, F lepton couplings

To complete the thesis work, we studied with F. Yndurain (see Fig. 3.13), the pseudoscalar two-point function:

$$\psi(q^2)^i_j = i \int d^4x \, \langle 0|\mathcal{T} J_5(x)^i_j \, J_5^\dagger(0)^i_j|0\rangle \; : \qquad u \equiv u, d, s, c, \quad (3.47)$$

where $J_5(x)^i_j \equiv (m_i + m_j)\bar{\psi}_i(i\gamma_5)\psi_j(x)$ is the pseudoscalar current and ψ_i is the quark field i of mass m_i.

• Lower bounds on quark masses

Using the positivity of the spectral function, we obtained bounds on the quark masses and on the leptonic couplings of the D and F mesons:

$$(\bar{m}_u + \bar{m}_d)(Q^2) \geq \frac{4\pi}{\sqrt{3}} \frac{m_\pi f_\pi}{\sqrt{Q^2}} \left(1 - \frac{17}{6}\frac{\alpha_s}{\pi} + \cdots\right). \qquad (3.48)$$

However, the problem with this lower bound is the arbitrary value of Q^2. Attempts for an optimal choice of Q^2 are made in Ref. [32]. In the ud and us channels, we found $\sqrt{Q^2} = 2$ GeV, leading to:

$$(\bar{m}_u + \bar{m}_d)(2\,\text{GeV}) \geq 40 \text{ MeV}, \quad (\bar{m}_u + \bar{m}_s)(3\,\text{GeV}) \geq 130 \text{ MeV}. \qquad (3.49)$$

$(\bar{m}_u + \bar{m}_d)$ is too large compared to the 12 MeV value of the current algebra, while $\bar{m}_s$ is consistent with the current algebra prediction of 150 MeV. However, one should note that current algebra predictions do not specify the scale at which the masses are evaluated. We note later on that $m_{u,d}$ does not behave here like m_π^2 but linearly.

• Lower bounds on the lepton couplings of D and F

Extending the analysis to the case of D and F and using the value of the charm quark invariant mass $\hat{m}_c = (2.1 \pm 0.3)$ GeV obtained in

Ref. [30], we obtain:

$$|f_D| \leq 815 \text{ MeV}, \qquad |f_F| \leq 900 \text{ MeV}, \tag{3.50}$$

which are not very restrictive. So, we exploited the fact that $m_c \gg \Lambda$ and took the limit $Q^2 = 0$. So:

$$\begin{aligned}\Phi(Q^2)^i_j \equiv \frac{d\,\psi(Q^2)^i_j}{d\,Q^2}|_{Q^2=0} &= \frac{1}{4\pi^2} : m_i^2 = m_j^2 \equiv m_c^2 \\ &= \frac{1}{4\pi^2}\frac{1}{2} : m_i^2 \equiv m_c^2, \quad m_j^2 = 0. \end{aligned} \tag{3.51}$$

This leads to leading order in α_s to:

$$|f_D| \leq \frac{M_D}{2\sqrt{2}\pi} \simeq 211 \text{ MeV}|f_F| \leq \frac{M_D}{2\sqrt{2}\pi} \simeq 228 \text{ MeV}, \tag{3.52}$$

where f_D and f_F are normalized as $f_\pi = 93.2$ MeV. These results completed my doctoral thesis work (see Fig. 3.15).

Sadly, the Marseille adventure is over although the work started in Marseille will continue with the participation of Professor C. Becchi (see Fig. 3.13).

I returned to ICTP-Trieste where I had already been taken on as a post-doc a year before because I did not obtain a position at the CNRS in France. In some sense, it looks normal in the French system

RÉPUBLIQUE FRANÇAISE

MINISTÈRE DE L'ÉDUCATION NATIONALE

(1) UNIVERSITE D'AIX-MARSEILLE II

DIPLÔME DE DOCTEUR D'ÉTAT

(2) SCIENCES

(arrêté du 16 avril 1974)

VU les titres initiaux produits par Monsieur NARISON Stéphan

né le 17 Septembre 1951 à ANTSAKABARY (Madagascar)

VU les pièces constatant que l'intéressé a présenté en soutenance, conformément aux règlements, à la date du 27 JUIN 1980

une thèse ou un ensemble de travaux (3) portant sur le sujet suivant :

" Quelques applications des théories de JAUGE à des processus fondamentaux."

devant un jury constitué au sein de (4) l'Université d'AIX-MARSEILLE II

et composé de MM. Henri BACRY : Président du jury

MM. FURLAN - GOURDIN - MEBKHOUT - YNDURAIN : Membres du jury

VU la décision du dit jury prononçant l'admission de l'intéressé avec la mention TRES HONORABLE

LE DIPLÔME DE DOCTEUR D'ÉTAT (2) " SCIENCES "

est conféré à Monsieur NARISON Stéphan

pour en jouir avec les droits et prérogatives qui y sont attachés.

Fait à Marseille le 18 Avril 1995

Le Président de l'Université,

Didier RAOULT

Signature du titulaire :

Vu, pour le Ministre et par délégation :

LE RECTEUR DE L'ACADÉMIE, CHANCELIER DES UNIVERSITÉS,

Le Recteur

N° 13 BIS/158

Figure 3.15. State Doctorate degree awarded in Marseille on June 27, 1980, with the distinction *Very Honorable.*
Juries: Professors M. Bacry, M. Mebkhout and E. de Rafael (Marseille), G. Furlan (Trieste), M. Gourdin (Paris) and F.J. Yndurain (Madrid).

that the priority for the application to the CNRS position comes from the famous *Grandes Ecoles.* Then, coming from a developing country, you cannot compete with the French elites.

A new battle begins for demonstrating my value at the international level!

Chapter 4

ICTP Trieste Post-Doc 1980–1982

Figure 4.1. Return to ICTP Trieste and to Venice in September 2019.

I took the train in Marseille on October 1, 1979 for Trieste with a change in Ventimiglia. At the station, I understood that I was leaving for a new page in my life because I did not understand the Italian language. After a long journey of 10 hours, I arrived in Trieste where I had taken a small hotel for a few days before finding an apartment. Then, I took a course for learning the Italian language which was necessary to integrate to the Italian community and way

Figure 4.2. Professor A. Salam, Founding Director of the ICTP (Nobel Prize in Physics in 1999 with S. Weinberg and S.L. Glashow (Fig. 3.11)), Professor G. Furlan (Co-founder with Professors A. Salam and P. Budinich of ICTP-Trieste in 1964); Professor A. Salam invited me to ICTP.

of life. Professor A. Salam, Director of ICTP (Nobel Prize in Physics in 1999 with S. Weinberg (Fig. 4.2 and S. Glashow (Fig. 3.11)) and Professor G. Furlan welcomed me. Salam divided his time between ICTP and Imperial College London. He had his private residence at ICTP. It was said that he was a digamus (one wife in London and one in Trieste). He had a compact group on the study of weak interaction models of Pati–Salam and supergravity. Moreover, Pati regularly came to ICTP to supervise the group. Salam had a table reserved for lunch (at the back to the right when you enter the refectory). The members of the group joined him. As I was not doing these research subjects but quantum chromodynamics (QCD), I rarely joined them.

In the meantime, I continued to work with the Marseille group to finish my thesis work. We then wrote the article on the mass of quarks [32]. Later, after the defense of my thesis, Professor C. Becchi (see Fig. 3.13) joined us. We improved our previous work by adding non-perturbative effects in the pseudoscalar two-point function [34].

4.1 Research Works at ICTP

4.1.1 The proton and neutron mass difference

To be able to join the ICTP group, I had to change the subject. With N. Craigie, Head of the QCD group at ICTP and N. Riazzudin (author of the famous book with his brother Fayyazzudin on weak interactions) from the Islam Abad University of Pakistan, we studied the proton–neutron electromagnetic mass difference in the framework of QCD [35] by reconciling the operator product expansion (OPE) approach in QCD with that of the Dyson-Schwinger equations. The expression for the proton–neutron electromagnetic mass difference is given by the OPE version of the Cottingham formula [37]:

$$\delta M_{\infty} = \frac{3\alpha}{4\pi}\int_{Q_0^2}^{\infty}\frac{dQ^2}{Q^2}\sum_f e_f^2\, m_f(Q^2)\langle N|\bar{\psi}_f\psi_f|N\rangle$$
$$+ \text{ higher twists} : f \equiv u, d, \qquad (4.1)$$

where $m_f(Q^2)$ is the light quark running masses. The matrix element $\langle N|\bar{u}u - \bar{d}d|N\rangle$ can be determined from the baryon mass differences $(\Sigma^+ - \Sigma^0, \Xi^+ - \Xi^0, \ldots)$.

We find, within the minimal subtraction scheme ($\bar{MS}$) and renormalization group improved Feynman graphs, that the electromagnetic mass shift is finite with the same sign as the quark mass difference $m_u - m_d$. This result reconciles the conflicting conclusions reached by respectively using the OPE and the Dyson equations.

4.1.2 Dimensional regularization and renormalization

To have a compact literature on dimensional regularization and renormalization techniques introduced by 't Hooft and Veltman [4] usually named $\bar{MS}$-scheme, I wrote a review of this technique for QCD and QED practitioners taking the examples of renormalization of the coupling constant α_s, the quark mass, the renormalization group and the explicit calculation of some two-point functions built by bilinear currents of quarks [5]. This review has been completed and improved thanks to the various advices of Professor A. Peterman (referee) during my stay as a fellow at CERN.

This approach consists of working in an imaginary space of dimension d and then taking the limit $d = 4 - \epsilon$ by making ϵ tend to zero. In this space, the divergences encountered during the evaluations of the Feynman diagrams which illustrate different physical processes become poles in $1/\epsilon^n$ $(n = 1, 2, \ldots)$. Renormalization consists of absorbing the ultra-violet (UV) poles into renormalization constants in order to have a finite observable.

4.1.3 The positivity of the QCD moments

I continued the studies on the QCD sum rules by extending the analysis to vector and axial-vector mesons made with light quarks u, d, s [38]. This work was partly done at the ICTP in Trieste and at the University of Madagascar where I was invited by Professor Raoelina Andriambolona to deliver courses in the nuclear physics sector.

For this research, I worked with the moments of the two-point function constructed with the electromagnetic vector currents $V^\mu(x)^i_i$ defined in Eq. (3.13). and the vector charged currents $V^\mu(x)^i_j$ and axial $A^\mu(x)^i_j$ defined in Eq. (3.13):

$$\begin{aligned}\mathcal{M}^{(n)}_{ij}(x) \equiv (-1)^n \frac{1}{n!}(t_c+Q^2)^n \frac{\partial^n \Pi_{ij}(Q^2)}{(\partial Q^2)^n} &= \text{Pole} \\ + \left(P^{(n)}(x) \equiv \int_0^1 dx\, x^{n-1}\left(\frac{t_c+Q^2}{t+Q^2}\right)\frac{1}{\pi}\text{Im}\Pi_{ij}\right. \\ \times \left.\left[\frac{t+Q^2(1-x)}{x}\right]\right),\end{aligned} \tag{4.2}$$

where

$$x = \frac{t_c+Q^2}{t+Q^2}. \tag{4.3}$$

Using the positivity of $\text{Im}\Pi_{ij}$ and therefore of $P^{(n)}_{ij}(x)$, the analysis becomes the known problems of the moments. We then obtain the constraints

$$P^{(n)}_{ij}(x) \geq 0, \qquad \det\begin{vmatrix} P^{(n)}_{ij} & P^{(n+1)}_{ij} \\ P^{(n+1)}_{ij} & P^{(n+2)}_{ij}\end{vmatrix} \geq 0, \ldots \tag{4.4}$$

However, as mentioned in Refs. [32, 34], the error on $P^{(n)}_{ij}$ increases with n such that for a matrix $n \geq 3$, the analysis is insignificant.

• The expression of electromagnetic moments

In the case of the electromagnetic current associated with the ρ meson, the expression of the moment after the application of the renormalization group is

$$\mathcal{M}_{ii}^{(n)}(x) = \frac{1}{n}\left(\frac{t_c + Q^2}{Q^2}\right)^n \sum_{D=0,2,4,\cdots} \frac{C_D\langle \mathcal{O}_D\rangle}{(Q^2)^{D/2}}, \tag{4.5}$$

$C_0\langle \mathcal{O}_0\rangle = 1/4\Pi^2[1 + \bar{a}_s + \bar{a}_s^2(-\beta_1/2(1 + 1/2 + \ldots 1/n - 1 + F_3)]$ is the perturbative contribution of the unit operator; $\overline{a}_s \equiv \overline{\alpha}_s/\pi$; $F_3 = 1.986 - 0.115 n_f$ for n_f quark flavours; $-\beta_1 = (1/2)(11 - 2n_f/3)$ is the first function β of the renormalization group.

The non-perturbative corrections are

$$C_4\langle \mathcal{O}_4\rangle = -\frac{n(n+1)}{Q^4}\left[2m_i\langle 0|\bar{\psi}_i\psi_i|0\rangle + \frac{1}{12\pi}\langle 0|\alpha_s G_a^{\mu\nu} G_{\mu\nu}^a|0\rangle\right], \tag{4.6}$$

where ψ and $G_{\mu\nu}^a$ are the quark and gluon fields.

• The theoretical expression of charged moments

In the case of the charged current, the mass corrections to the moments become

$$\mathrm{LO} \times \left\{1 - 3\left(\frac{n}{Q^2}\right)\left[(\overline{m}_i^2 + \overline{m}_j^2) - \frac{1}{2}(\overline{m}_i + \overline{m}_j)^2\left(1 + \frac{1}{2} + \cdots + \frac{1}{n-1}\right)\right]\right\}, \tag{4.7}$$

where LO designates the perturbative contribution to the lowest order. $\bar{m}_i$ is the running quark mass.

• Phenomenological applications

In Ref. [38], we used the constraints

$$-\frac{\beta_1}{2}\left(1 + \frac{1}{2} + \cdots \frac{1}{n-1}\right) \equiv a \ll 1, \qquad 6\overline{m}_i^2 \frac{n}{Q^2} \equiv b \ll 1, \tag{4.8}$$

to fix n and Q^2 by taking $a \simeq 2.5\%$ and $b \simeq 10\%$. By taking these constraints, upper bounds on the width $\Gamma(V \to e + e^-)$ and $\Gamma(A \to e\nu_e)$ were obtained in Table 1 of Ref. [38]. We obtain for the neutral vector mesons in units of keV,

$$\Gamma(\rho \to e + e^-) \le (5.3 \pm 1.5) \qquad [\exp 6.5 \pm 0.8],$$
$$\Gamma(\omega \to e + e^-) \le (0.78 \pm 0.22) \qquad [\exp 0.68 \pm 0.07]. \tag{4.9}$$

For the charged vector mesons, we obtain

$$\frac{\Gamma(\tau \to \nu_\tau A_1)}{\Gamma(\tau \to \nu_\tau e \bar{\nu}_e)} \le (0.7 \pm 0.2) \quad [\exp 0.56 \pm 0.15],$$
$$\frac{\Gamma(\tau \to \nu_\tau K^*)}{\Gamma(\tau \to \nu_\tau e \bar{\nu}_e)} \le (3.3 \pm 0.9) \times 10^{-2}. \tag{4.10}$$

These results can also be compared with those of FESR à la Shankar in Ref. [29].

4.2 SVZ Sum Rules

In parallel with our work [29, 30, 32, 34] appeared the sum rules of SVZ (Shifman–Vainshtein–Zakharov) [3, 40] (see Fig. 4.3) which are currently very well known. An introduction and review of work on this topic can be found in Refs. [5–7]. Other journals and conference papers can see found in Refs. [41–68].

4.2.1 The expansion of SVZ

One of the important contributions of SVZ is the addition to the perturbative contributions of non-perturbative effects by means of quark and gluon condensates within the Wilson OPE. In the example of the two-point function made by the pseudoscalar bilinear currents of the quarks (Eq. (3.45)), we can write

$$\psi_5(Q^2)^i_j = \sum_{D=0,2,4,\ldots} \frac{C_D \langle \mathcal{O}_D \rangle}{(Q^2)^{D/2}}, \tag{4.11}$$

where $Q^2 \equiv -q^2 \ge 0$, i, j represent the flavours of the quarks, D is the dimension of the (non) perturbative operator $\langle \mathcal{O}_D \rangle$ entering the operator product expansion (OPE) of Shifman–Vaibshtein–Zakharov

Figure 4.3. From left to right: Arkady Vainshtein, SN, Valya Zakharov, Mikhael Shifman at the Ringberg Castle Workshop Understanding Confinement in the honour of Professor V.I. Zakharov (2005).

(SVZ) (see Eq. (4.5) in the case of electromagnetic current). C_D is the perturbatively calculable Wilson coefficient. SVZ assumed that there is no ambiguity in separating these two quantities. This assumption was later demonstrated where the IR renormalon effects cancel out each other in the product.

- ◇ $D = 0$: Usual perturbative series in $a_s \equiv \alpha_s/\pi$,
- ◇ $D = 2$: $\bar{m}_q^2$: quark mass corrections included in C_0,
- ◇ $D = 4$: $\langle \alpha_s\, G^2 \rangle$, $m_q \langle \bar{\psi}_q \psi_q \rangle$: condensates of gluons and quarks,
- ◇ $D = 6$: $m_{Q,\ g} \langle \bar{\psi}_q G_a^{\mu\nu} \lambda_a \psi_q \rangle$ mixed condensate of quark gluons,
  $\alpha_s \langle \bar{\psi}_q \psi_q \rangle^2$: four-quark condensate,
  $g^3 f_{abc} \langle G^a G^b G^c \rangle$: three-gluon condensate,
- ◇ $D = 8$: $g^4 \langle GGGG \rangle$: four-gluon condensate $\oplus \cdots$.

4.2.2 The Borel/Laplace exponential sum rules

They can be obtained from the moments in Eq. (4.2) by taking the limit Q^2, n large but leaving the ratio $n/Q^2 \equiv \tau$ fixed [3][1]:

$$\mathcal{L}_0^c(\tau,\mu) \equiv \lim_{\substack{Q^2, n \to \infty \\ n/Q^2 \equiv \tau}} \frac{(-Q^2)^n}{(n-1)!} \frac{\partial^n \Pi}{(\partial Q^2)^n} = \int_{t>}^{t_c} dt\ e^{-t\tau} \frac{1}{\pi} \mathrm{Im}\,\psi_5(t,\mu),$$

$$\mathcal{R}_{10}^c(\tau) \equiv \frac{\mathcal{L}_1^c}{\mathcal{L}_0^c} = \frac{\int_{t>}^{t_c} dt\ e^{-t\tau} t\, \mathrm{Im}\,\psi_5(t,\mu)}{\int_{t>}^{t_c} dt\ e^{-t\tau} \mathrm{Im}\,\psi_5(t,\mu)}, \tag{4.12}$$

where τ is the LSR variable, $t>$ is the hadronic threshold and μ the arbitrary subtraction point of QCD. Here, t_c is the threshold of the "QCD continuum" which parameterizes, from the discontinuity of the Feynman diagrams, the spectral function $\mathrm{Im}\,\psi_5(t,\mu^2)$.

• Why do we have different names for these sum rules?

SVZ called them Borel transforms because each term in the OPE has suppression factors in 1/(D-2)! which would correspond to a Borel series. These factors accelerate the convergence of the OPE.

The non-relativistic version of these sum rules applied in heavy quark systems (charmonium and bottomium) was discussed in Refs. [64, 70] who called them Magic Moments.

However, the inclusion of the PT correction α_s in the QCD expression showed that the sum rules have instead the property of an inverse Laplace transform, [69]. From now on, we will call it LSR in this book.

---

[1]SVZ take originally as a variable $M^2 = 1/\tau$. However, we prefer to work with τ as this variable can be related to the imaginary time one in the non-relativistic version of the sum rule [64, 70].

• Role of exponential weight

Thanks to the exponential weight and for moderate values of τ around $(0.5 - 1)$ GeV^{-2}, the exponential factor entering in the integral enhances the contribution of the lowest mass resonance and suppresses one of the QCD continuum. We will see in some examples of phenomenological applications the role of this factor.

• The minimum duality ansatz (MDA)

In *the minimum duality ansatz (MDA)*, the spectral function is written as

$$\frac{1}{\pi}\text{Im}\,\psi_5(t,\mu) = 2f_P^2 M_P^4 + \theta(t-t_c)\frac{1}{\pi}\text{Im}\,\psi_5(t,\mu)|_{QCD}, \tag{4.13}$$

where $\text{Im}\,\psi_5(t,\mu)|_{QCD}$ comes from the discontinuity of the Feynman diagrams, while t_c is the QCD continuum threshold.

In this case, we can clearly see that the moment $\mathcal{L}$ is often used to determine the leptonic coupling f_P of the hadron while the ratio

$$\mathcal{R}_{10}^c(\tau) \simeq M_P^2 \tag{4.14}$$

is used to estimate its mass. We shall see in several hadronic channels that with this minimum duality ansatz, we are able to predict the mass of the lowest resonance with very good precision (5–10% and sometimes even better).

Later, this parameterization was generalized with two and three resonances.

• Meeting with A.I. Vainshtein

During my stay at ICTP, Professor A.I. Vainshtein had been invited for a few weeks. He delivered an interesting seminar on the application of LSR using experimental data $e^+e^- \to I = 1$ hadrons, where he showed the determination of α_s and gluon condensates $\langle\alpha_s G^2\rangle$ and quarks $\langle\bar\psi\psi\bar\psi\psi\rangle$ [71].

We became such good friends that every evening we went to dinner together in different restaurants in the city. He then explained to me that his name was derived from the name "wine" and that he had vineyards in Novosibirsk (Russia). Through this friendship, my interest in the sum rules had increased.

4.2.3 The sum of the quark masses u, d with LSR

• The original results

We continued our work on light quark masses [32,34] with Professor E. Rafael using the Laplace sum rule (LSR) (Eq. (4.12)). In the case of the two-point function of the pseudoscalar current, the LSR is written as [69]

$$\begin{aligned}&\int_{t_0}^{\infty} dt\, e^{-t\tau}\frac{1}{\pi}\mathrm{Im}\,\psi_5(t)^u_s\\&= \frac{3}{8\pi^2}(\bar{m}_u+\bar{m}_s)^2\tau^{-2}\Bigg\{1+\bar{a}_s\left(\frac{11}{3}+2\gamma_E\right)\\&-\left[\bar{m}_u^2+\bar{m}_s^2+(\bar{m}_u-\bar{m}_s)^2\right]\tau+\frac{\pi}{3}\tau^2\langle\alpha_s G^2\rangle\\&-\frac{8\pi^2}{3}\tau^2\left[\left(m_s-\frac{m_u}{2}\right)\langle\bar{u}u\rangle+\left(m_u-\frac{m_s}{2}\right)\langle\bar{s}s\rangle\right]\Bigg\}, \qquad (4.15)\end{aligned}$$

where the non-perturbative contributions have been added. I invited him to deliver a seminar at the ICTP and finish our work. After dinner, he joked that the quark masses were too heavy in his pocket. We found for the sum of the invariant masses

$$(\hat{m}_u+\hat{m}_d) \geq (13\pm 3)\ \mathrm{MeV}\ :\quad \Lambda = 210\ \mathrm{MeV}, \qquad (4.16)$$

which we deduced from Fig. 4.4, and compared with that of the algebra of currents based on the symmetry $SU(6)$ that connects the matrix elements $\langle 0|\bar{u}i\gamma_5 d|\pi\rangle$ and $\langle 0|\bar{u}i\gamma_\mu d|\rho\rangle$ [72]:

$$(m_u+m_d) \simeq \frac{4}{3}f_\pi\left(\frac{m_\pi^2}{M_\rho^2}\right)\gamma_\rho = 11\,\mathrm{MeV}. \qquad (4.17)$$

However, it should be noted that the bound obtained in Eq. (4.16) depends a lot on the value of Λ (it decreases when Λ increases. The current value of Λ for three quarks is of the order of 340 MeV [33]),

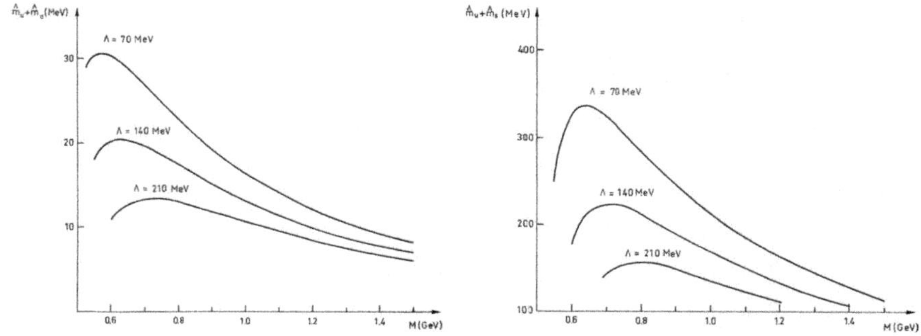

(a) (b)

Figure 4.4. Lower sounds of the sum of invariant quark masses as a function of $M^2 \equiv 1/\tau$ for different choices of Λ: (a) $(\hat{m}_u + \hat{m}_d)$; (b) $(\hat{m}_u + \hat{m}_s)$.

while the value in Eq. (4.17) is not defined in any scale. We have given a QCD version of this mass from the current algebra by evaluating the current quark mass at M_ρ [69]:

$$(\overline{m}_u+\overline{m}_d)(M_\rho^2) = \left(\frac{4e}{3}\right)^{1/2} f_\pi \left(\frac{m_\pi^2}{M_\rho^2}\right) \gamma_\rho \left[1-8\pi^2 \frac{f_\pi^2 m_\pi^2}{M_\rho^4}\right] = 21.4\,\mathrm{MeV}. \tag{4.18}$$

However, we must be careful when using these values because we must add the effect of the correction in α_s and which are large in the LHS.

• Recent works

Recent works have been published to improve this bound [73] as well as the estimation of masses [74] with QCD sum rules by including perturbative and non-perturbative QCD corrections and the contribution of the radial excitation $\pi'(1300)$ of the pion. In Ref. [74], we obtain the sum

$$\begin{aligned}(\overline{m}_u + \overline{m}_d)(2\ \mathrm{GeV}) &\geq (6.56 \pm 0.70)\ \mathrm{MeV} \\ &= (7.90 \pm 0.56)\ \mathrm{MeV}.\end{aligned} \tag{4.19}$$

Extending the analysis into the $\bar{u}s$ channel gives

$$\begin{aligned}(\overline{m}_u + \overline{m}_s)(2\ \mathrm{GeV}) &\geq (84.2 \pm 4.5)\ \mathrm{MeV} \\ &= (101.1 \pm 5.5)\ \mathrm{MeV}.\end{aligned} \tag{4.20}$$

These new values are in agreement with those of the recent lattice calculations [75].

4.2.4 SU(3) breaking for the kaon PCAC and $\langle\bar{s}s\rangle$

The analysis of Ref. [69] has been extended to estimate the value of the two-point pseudoscalar function at zero momentum: $\psi_5(0)$ using the subtracted dispersion relation [76]. Its Laplace transform leads to

$$\psi_5(0)^u_s = \int_{t_0}^{\infty}\frac{dt}{t}\,e^{-t\tau}\frac{1}{\pi}\mathrm{Im}\,\psi_5(t)^u_s - \frac{3}{8\pi^2}(\bar{m}_u+\bar{m}_s)^2\tau^{-1}$$
$$\times\left\{1+\bar{a}_s\left(\frac{11}{3}+2\gamma_E\right)+2\bar{m}_s^2\tau\log(m_s^2\tau)-\frac{8\pi^2}{3}\tau^2\right.$$
$$\left.\left[\left(m_s-\frac{m_u}{2}\right)\langle\bar{u}u\rangle+\left(m_u-\frac{m_s}{2}\right)\langle\bar{s}s\rangle\right]+\frac{\pi}{3}\tau^2\langle\alpha_s G^2\rangle\right\}. \tag{4.21}$$

We show in Fig. 4.5 the variation of $\psi_5(0)^u_s$ as a function of $M \equiv 1/\sqrt{\tau}$. Requiring that the correction in α_s is less than 40% and that of the QCD continuum is less than 10% (we fixed $\sqrt{t_c} \simeq M_K + 2m_\pi$), the optimal value is obtained for $\sqrt{\tau} \simeq 2M_K$, and we obtain, for $\Lambda \simeq$ 300 MeV [76],

$$\psi_5(0) \simeq 4.1\times 10^{-3}\ \mathrm{GeV}^4, \tag{4.22}$$

where $f_K = 1.16 f_\pi \simeq 107$ MeV is the coupling constant of K. This result indicates a violation of the SU(3) symmetry of PCAC for the kaon of 27%. This result was confirmed by the more recent analyses [6, 41] reviewed in Ref. [7].

4.2.5 Review of my stay in Trieste and at the ICTP

My post-doc was about to end. I had to go to LAPP-Annecy-FR for a scholarship awarded by CNRS-IN2P3.

◇ I learned a lot during this stay, including Italian, which I didn't know a word about two years ago. I had also discovered the Italian

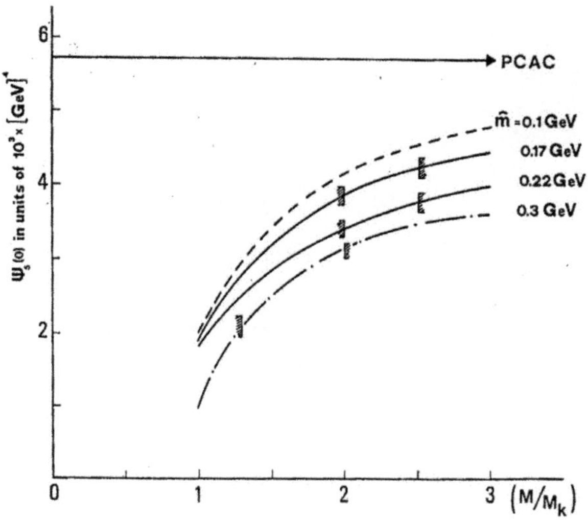

Figure 4.5. Behaviour of $\psi_5(o)^u_s$ for different values of Λ as a function of $M \equiv 1/\sqrt{\tau}$.

way of life: the food, the ritual of going for a coffee at the Piazza Unità d'Italia, known in the centre of the city, of going to the wine cellars in the evening, places that did not yet exist in France at the time and where I had started to discover wine.

◇ I had also met different people which was also a great experience.
◇ I had discovered Venice which had always made me dream.
◇ For work, I got closer to the Italian researchers at the university, including Professor G. Furlan who was part of my thesis committee and expert in current algebra.
◇ I had started work with people from his group (Professor N. Paver and Dr. D. Treleani) on the difference in quark masses and the breaking of chiral symmetry.
◇ After a brief discussion with E. Witten in my office on my question about glueballs and $\eta'(958)$ in relation to his work on the $U(1)$ channel, I was starting to become interested in this branch of physics.
◇ In the meantime, I married Lydia, the Malagasy friend whom I met in Marseille. A new life as a couple and globe trotters was about to begin.

Figure 4.6. Marriage in Marseille (September 2019).

Chapter 5

CNRS Fellow @ LAPP, Annecy 1982–1983

Figure 5.1. LAPP laboratory @ Annecy le Vieux, France.

I arrived in the group in October 1981. In particular, I thank Professors P. Aurenche, G. Girardi, P. Sorba and R. Stora for their welcome.

Figure 5.2. Some photos of my hosts @ Annecy le Vieux, France (Professors G. Girardi, P. Sorba and R. Stora).

The group had a special status with CERN because it had a dedicated working room thanks to the help (I think) of Mary K. Gaillard and John Ellis of CERN (see Fig. 3.11). The members of the group (theorists and experimentalists) went to CERN every day, which was beneficial to us.

As everyone in the group was interested in some electroweak models beyond the Standard Model, I started to learn supersymmetry, technicolour and composite models. I even gave seminars on the superfields which I learned in Trieste from the Salam group!

However, I still continued, in parallel, with my previous work in QCD which, in that time, was the determination of the light quark masses from QCD spectral sum rules — A work which I started in Marseille for my doctorate thesis.

One morning, I was alone in the Annecy working room at CERN. Someone knocked on the door and entered. He said: Are you the gangster of Marseille? Then, he presented himself as Murray Gell-Mann! I was very surprised and greatly impressed just by hearing his name. He was very polite and then he asked me to explain how we (the Maresille group) got the running quark masses. He was interested in understanding the relation of these masses with his current algebra ones. I wrote on the blackboard the relation from the pion sum rule:

$$(\bar{m}_u + \bar{m}_d)(M_0^2) \geq \frac{4\pi}{\sqrt{3}} \left(\frac{m_\pi^2}{M_0^2}\right) f_\pi, \tag{5.1}$$

where M_0^2 is the sum rule scale which (*a priori*) is arbitrary, but we fix its value at the optimal point (see Fig. 4.4). The other notations are standard. He listened, took some notes and said that he liked the behaviour $m_q \sim m_\pi^2$ as expected from the pion PCAC relation, but I did not know his final opinion about the result. Since then, my interest in the subject increased.

5.1 Weinberg's Sum Rules with Laplace

I reanalyzed [77] our previous work in Ref. [29] using Laplace transform Weinberg sum rules.

5.1.1 The first sum rule

It is written as:

$$\int_{t_0}^{\infty} dt\, e^{-t\tau} \frac{1}{\pi} \operatorname{Im} Q_{ij}(t)$$
$$= \frac{\tau^{-1}}{\pi^2} \left(\frac{\alpha_s}{\pi}\right) \left\{ -\bar{m}_u \bar{m}_d \tau + \frac{8}{3}\pi^2 \left(m_j \langle \bar{\psi}_i \psi_j \rangle + m_i \langle \bar{\psi}_j \psi_j \rangle\right) \tau^2 \right.$$
$$\left. - \frac{32}{9} \pi^3 \langle \bar{\psi}_i \psi_j \rangle^2 \tau^3 \right\}. \tag{5.2}$$

5.1.2 The second and third sum rules

Their expressions in QCD are:

$$\int_{t_0}^{\infty} dt\, e^{-t\tau}\, t\, \frac{1}{\pi} \operatorname{Im} \Delta_{ij}^{(1,0)}(t) = \pm \frac{6}{\pi^2} \bar{m}_i \bar{m}_j \left(-\frac{5}{2} + \frac{\pi}{\alpha_s}\right)$$
$$\pm 2 \left(m_j \langle \bar{\psi}_i \psi_j \rangle + m_i \langle \bar{\psi}_j \psi_j \rangle\right) \tau + \mathcal{O}(\alpha_s), \tag{5.3}$$

where we corrected the expressions in Ref. [77] using the two-point function expressions in [79].

5.1.3 The sum rule of FNR

A more convergent combination of the spectral function was proposed in Ref. [29] (see Eq. (3.22)). Its Laplace transform is:

$$\begin{aligned}\int_{t_0}^{\infty} dt\, e^{-t\tau}\, t\, \frac{1}{\pi}\,\mathrm{Im}\, R^{(1,0)}_{ijk}(t) \\ = \pm \frac{3}{2\pi^2}\bar{m}_i\bar{m}_j \left[\bar{m}_k^2 \log\left(m_k^2\tau\right) - \bar{m}_j^2 \log\left(m_j^2\tau\right)\right] \\ \pm 2\left[m_j\langle\bar{\psi}_i\psi_i\rangle + m_i\langle\bar{\psi}_j\psi_j\rangle - (j \longleftrightarrow k)\right]\tau. \qquad (5.4)\end{aligned}$$

- **Broadhurst's sum rule**

Reference [78] observed that the combination:

$$\Delta\psi_{uds} = \psi(Q^2)^u_s - \left(\frac{m_s - m_u}{m_s + m_u}\right)\psi_5(Q^2)^u_s + \left(\frac{m_s - m_u}{2m_u}\right)\psi_5(Q^2)^u_d \qquad (5.5)$$

does not depend on the value of $\psi(0)^i_j$ and $\psi_5(0)^i_j$, where $\psi^i_j(Q^2)$ is the two-point function constructed with the scalar current:

$$J_S(x)^i_j = (m_i - m_j)\bar{\psi}_i\psi_j(x). \qquad (5.6)$$

The Laplace transform of this combination is:

$$\int_{t_0}^{\infty} dt\, e^{-t\tau}\, t\, \frac{1}{\pi}\,\mathrm{Im}\,\Delta\psi_{ijk}(t) = \frac{3}{2\pi^2}\bar{m}_i\bar{m}_j^3\left\{1 + \frac{\alpha_s}{\pi}\left(\frac{19}{3} + \gamma_E\right)\right\}. \qquad (5.7)$$

5.1.4 Phenomenology

In the original article [77], the results are given as bounds. Here, we will consider them more as estimates.

• The first sum rule

Saturating the spectral function by π, ρ, A_1, we obtain, by choosing $\tau = 1/M_\rho^2$:

$$f_\pi \simeq \frac{M_\rho}{2\gamma_\rho}\frac{1}{\sqrt{e}}\left\{1 - \frac{M_{A_1}^2}{M_\rho^2}\frac{\gamma_\rho^2}{\gamma_{A_1}^2}e^{1-M_{A_1}^2/M_\rho^2}\right\}^{1/2} \simeq (92 \pm 7)\ \text{MeV}, \quad (5.8)$$

while by choosing $\tau = 1/M_{A_1}^2$, we can predict:

$$\frac{\pi M_{A_1}^2}{2\gamma_{A_1}^2} = e\left\{\frac{\pi M_\rho^2}{2\gamma_\rho^2}e^{-M_\rho^2/M_{A_1}^2} - 2\pi f_\pi^2\right\} \simeq 6.4 \times 10^{-2}\ \text{GeV}^2 \quad (5.9)$$

in agreement with the experimental values: 92 MeV and $(7.3 \pm 2.2) \times 10^{-2}$ GeV^2.

The analysis was extended in the case of K^* to predict:

$$R_{\tau K^*} \equiv \frac{\Gamma(\tau \to \nu_\tau K^*)}{\Gamma(\tau \to \nu_\tau e\bar{\nu}_e)} \simeq 0.06. \quad (5.10)$$

• The second and third sum rules

As in Weinberg's [27] original article, the second sum rule combined with the first gives the mass of A_1. Choosing $\tau \simeq 1/M_{A_1}^2$, we obtain:

$$M_{A_1} \simeq M_\rho\left\{1 - 2f_\pi^2\frac{2\gamma_\rho^2}{M_\rho^2}\, e^{M_\rho^2/M_{A_1}^2}\right\}^{-1/2} \simeq 1.62 M_\rho \simeq 1.25\ \text{GeV}. \quad (5.11)$$

This mass is larger than that of Weinberg [27] but in agreement with recent experimental results of 1.26 GeV [33].

The third sum rule was used to estimate the product $\hat{m}_i\hat{m}_j$. By saturating the spectral functions with low mass resonances, we obtain:

$$(\hat{m}_u\hat{m}_d)^{1/2} \simeq (9 \pm 2)\,\text{MeV}. \quad (5.12)$$

The mass product is too large compared to the recent value: $\hat{m}_u \simeq$ 3 MeV, $\hat{m}_d \simeq 6$ MeV $\hat{m}_s \simeq 114$ MeV [74].

• The FNR and Broadhurst sum rules

We estimated the mass ratio m_s/m_d with the FNR sum rule and we obtain:

$$\frac{\hat{m}_s}{\hat{m}_d} = \frac{m_s}{m_d} \simeq 2\frac{f_K^2 M_K^2}{f_\pi^2 m_\pi^2} e^{-M_K^2 \tau} \simeq 23 \sim 27 \tag{5.13}$$

by taking $\sqrt{1/\tau} \simeq M_\rho \sim M_\phi$. That of Broadhurst gives:

$$\frac{m_s}{m_d} \simeq 2\frac{f_K^2 M_K^2}{f_\pi^2 m_\pi^2} e^{-M_K^2 \tau} - 1 \simeq 22 - 26 \tag{5.14}$$

to compare with the recently obtained report: (19 ± 2) in [74].

• Comments on results

Note that the results on quark masses are not precise compared to recent approaches (see, e.g., [74]). One of the greatest sources of uncertainty is the neglect of the contribution of scalar mesons in the extraction of masses, which can possibly explain these differences.

5.2 The f_P Couplings of Pseudoscalar Meson

5.2.1 f_π and f_K from the Weinberg's first sum rule

The couplings are normalized as:

$$\langle 0|J_5(x)|P\rangle = \sqrt{2}\, M_P^2 f_P : \qquad P \equiv \pi, K, D, F, \tag{5.15}$$

where $f_\pi = 92.2$ MeV controls the decay $\pi^+ \to \mu^+\nu_\mu$ and:

$$J_5(x) = (m_i + m_j)\bar{\psi}_i(i\gamma_5)\psi_j, \tag{5.16}$$

An estimate of f_π can be obtained from Weinberg's first sum rule (Eq. (5.8)):

$$f_\pi \simeq (92 \pm 7)\ \text{MeV}. \tag{5.17}$$

Using this sum rule in the channel $\bar{u}s$, we can also obtain a value of f_K:

$$f_K \simeq 116\ \text{MeV} = 1.26\, f_\pi, \tag{5.18}$$

Comparing with the lattice results: $f_K \simeq= 1.19\, f_\pi$.

5.2.2 f_D and f_F from the pseudoscalar LSR

Using the Laplace transform of the pseudoscalar current sum rules in the channels $\bar{d}c$ and $\bar{s}c$ and the positivity of the spectral function, we obtain the expression [74]:

$$f_D \leq \frac{\sqrt{3}}{4\pi}(\bar{m}_c + \bar{m}_d) e^{M_D^2\tau/2} \frac{1}{\tau M_D^2}\left\{1 + \left(\frac{\bar{\alpha}_s}{\pi}\right)\left(\frac{11}{6} + \gamma_E\right) - \bar{m}_c^2\tau \right.$$
$$\left. - \left(\frac{4}{3}\pi^2 m_c \langle\bar{\psi}_d\psi_d\rangle - \frac{\pi}{6}\langle\alpha_s G^2\rangle\right)\tau^2 + \mathcal{O}(\bar{m}_c^4\tau^2)\right\}, \tag{5.19}$$

where $\langle\alpha_s G^2\rangle = 0.04\ \text{GeV}^4$ is the gluon condensate [3]. Choosing $1/\sqrt{\tau} \simeq 2 \sim 3$ GeV corresponding to a correction of the quark mass lower than 50%, we obtain:

$$f_D \leq (240 \sim 450)\ \text{MeV}, \tag{5.20}$$

where $\hat{m}_c = (2.08 \pm 0.36)$ GeV [30] was used. In the channel $\bar{s}c$, we can use the relation:

$$f_F \leq f_D\left(1 + \frac{m_s}{m_c}\right)\left(\frac{M_D}{M_F}\right)^2 \simeq 1.1 f_D. \tag{5.21}$$

5.2.3 f_D and f_F from the derivative of the pseudoscalar function

Considering the charm as a heavy quark ($m_c \gg \Lambda$), we can work with the derivative:

$$\Theta_{ij}^{(2)} \equiv \frac{1}{2!}\frac{\partial^2 \psi_5(Q^2)_j^i}{(\partial Q^2)^2}\bigg|_{Q^2=0} = \int_{t_0}^{\infty}\frac{dt}{t^3}\frac{1}{\pi}\,\mathrm{Im}\,\psi_5(t)_j^i. \tag{5.22}$$

In the case, $m_i \gg m_j$, one has:

$$\Theta_{ij}^{(2)} = \frac{1}{8\pi^2}\left\{1 + 3\frac{\hat{m}_i}{\hat{m}_j} + \left(\frac{\bar{\alpha}_s}{\pi}\right)\frac{2}{3}\left(\frac{1}{4}+\frac{\pi^2}{3}\right) - \frac{8\pi^2}{\bar{m}_i^4}m_i\langle\bar{\psi}_j\psi_j\rangle \right.$$
$$\left. + \frac{2\pi}{3}\frac{1}{\bar{m}_i^4}\langle\alpha_s G^2\rangle\right\}, \tag{5.23}$$

which leads to:

$$f_D \leq \frac{M_D}{4\pi}\times 1.13 \simeq 200 \text{ MeV}, \quad f_F \leq \frac{M_F}{4\pi}\left(1+3\frac{\hat{m}_s}{\hat{m}_c}\right)^{1/2}$$
$$\times 1.35 \simeq 257 \text{ MeV}, \tag{5.24}$$

where we took the recent values of $\hat{m}_s$ and $\hat{m}_c$ (see Ref. [47]) for $\Lambda =$ 0.34 GeV. The bounds of these coupling constants can be compared with more recent predictions [80]:

$$f_D = (144 \pm 4) \text{ MeV}, \quad f_F = (174 \pm 4) \text{ MeV}, \tag{5.25}$$

and some other predictions from different methods reviewed in Ref. [33].

5.2.4 Gluon component of the mass of the $U(1)$ meson

• Analysis in Ref. [77]

We continued the analysis by considering the two-point function $\psi_-(q^2)$ of the gluonic current [77]:

$$J_P(x) = \frac{3\alpha_s}{4\pi}G^{\mu\nu}\tilde{G}_{\mu\nu}(x), \tag{5.26}$$

where $\tilde{G}_{\mu\nu} = (1/2)\epsilon_{\mu\nu\rho\sigma}G^{\rho\sigma}$ avec $G_{\mu\nu} = \partial_\mu A_\nu - \partial_\nu A_\mu$ is the gluon field tensor. Using the QCD expressions [81], we obtain the sum rule independent of its value at $q^2 = 0$:

$$\int_{t_0}^{\infty} dt\, e^{-t\tau} \frac{1}{\pi} \mathrm{Im}\, \psi_P(t) = \frac{9}{4\pi^2} \left(\frac{\alpha_s}{\pi}\right)^2 \tau^{-3} \left\{ 1 + \mathcal{O}\left(\alpha_s \tau^2, \tau^4\right) \right\}. \quad (5.27)$$

The coupling constant f_P is defined as:

$$\langle 0|J_P|P\rangle = \sqrt{2} f_P M_P^2, \quad (5.28)$$

which we can deduce from the decay $\psi \to \eta' \gamma$:

$$f_P \simeq (0.5 \sim 0.7)\sqrt{3}\, f_\pi. \quad (5.29)$$

Using the positivity of the spectral function and taking $\tau \simeq 1$ GeV^{-2}, we get:

$$M_P \leq (0.72 \sim 0.85) \text{ GeV} \quad \text{for } \Lambda = 210 \text{ MeV}. \quad (5.30)$$

We can consider this result as the gluon component of the mass of the $\eta'(958)$ meson.

- **Revision of the analysis in Ref. [82]**

This analysis was revised in Ref. [82] where other sum rules were used:

$$\mathcal{F}^{(n)} \equiv \tau^n \int_{t_0}^{\infty} \frac{dt}{t^{3-n}}\, e^{-t\tau} \frac{1}{\pi} \mathrm{Im}\, \psi_P(t): \quad n = 1, 2, 3, \quad (5.31)$$

as well as their ratio:

$$\mathcal{R}_1 = \frac{\mathcal{F}^{(3)}}{\mathcal{F}^{(2)}}, \quad \mathcal{R}_2 = \frac{\mathcal{F}^{(2)}}{\mathcal{F}^{(1)}}, \quad (5.32)$$

such that in the presence of zero-mass quarks, the subtraction points [83, 84]:

$$\psi_P(0) = \psi'_P(0) = 0. \quad (5.33)$$

◇ Upper bound on the mass M_P

For $n = 3$, we can extract an upper bound on M_P in the chiral limit. Using the experimental value of f_P in Eq. (5.29) and the stability point $1/\sqrt{\tau} \simeq 1$ GeV, we deduce:

$$M_P \leq (0.75 \sim 0.9)\ \text{GeV}, \tag{5.34}$$

as obtained previously in Eq. (5.30).

◇ Upper bound on the coupling constant f_P

For $n = 1$, we can extract an upper bound on f_P in the chiral limit and at the point of stability in $1/\sqrt{\tau} \simeq 2$ GeV:

$$f_P \leq (150 \sim 200)\ \text{MeV}. \tag{5.35}$$

◇ Radial excitation mass

The moment ratios $\mathcal{R}_{1,2}$ were used to give a bound on the mass of the radial excitation. At the stability point, $1/\sqrt{\tau} \simeq 1.8$ GeV, where the two ratios of sum rules also meet, we deduce the common solution:

$$M_{P'} \simeq 2.2\ \text{GeV}, \tag{5.36}$$

which could be the mass of the *true gluonium*.

◇ $\psi_P(0)$ in the world without quarks

For this, we used the moment $n = 2$ which gives the relation:

$$\psi_P(0) = \int_{t_0}^{\infty} \frac{dt}{t} e^{-t\tau} \frac{1}{\pi} \operatorname{Im} \psi_P(t) - \mathcal{F}^{(2)}|_{\text{QCD}}, \tag{5.37}$$

leading to:

$$\psi_P(0) \simeq 0.5(2 f_P^2 M_P^2), \tag{5.38}$$

indicating that the prediction with large N_c may overestimate this constant by a factor of 2.

5.3 Scalar Current Sum Rules

5.3.1 The mass difference of light quarks

We have extended our work in the case of the scalar current:

$$\partial_\mu V^\mu(x)^q_u = i(m_q - m_u) : \bar{\psi}_q(x)\psi_u(x) : \quad \text{with :} \quad q \equiv d, s. \quad (5.39)$$

This work was carried out with the collaboration of E. de Rafael (Marseille), N. Paver and D. Treleani from the University of Trieste [85].

• **The sum rule ($\bar{m}_s - \bar{m}_u$)**

In this channel, we used the experimental information on the phase shift $I = 1/2$ of $K\pi$ of the wave S to parameterize the scalar spectral function:

$$\frac{1}{\pi}\text{Im}\psi(t)^u_s \geq \frac{3}{32\pi^2}\frac{1}{t}\sqrt{(t-t_0)(t-t_1)}|d(t)|^2\theta(t-t_0), \quad (5.40)$$

where:

$$t_0 = (M_K + m_\pi)^2, \quad t_1 = (M_K - m_\pi)^2, \quad (5.41)$$

and write his Omnes representation:

$$D(t) = \exp\left(\frac{t}{\pi}\int_{t_0}^{\infty}\frac{dt'}{t'(t'-t-i\epsilon)}\phi(t')\right), \quad (5.42)$$

with:

$$\phi(t) = \text{tg}^{-1}\frac{\eta(t)\sin 2\delta(t)}{1+\eta(t)\cos 2\delta(t)}, \quad (5.43)$$

where $\delta(t)$ is the S-wave of the phase shift $I = 1/2$ of $K\pi$ and $\eta(t)$ the corresponding elasticity which have been precisely measured [87]. We assumed that $\delta(t)$ has no pole in the complex plane and that if asympotically $\delta(t) \to \beta\pi$ when $t \to \infty$, then $D(t) \to t^{-\beta}$. We will choose $\beta = 1$ in our analysis which is compatible with the counting rules of the form factors of scalars [88]. The data show a smooth behaviour below 1.3 GeV. A resonance appears at 1500 (K_0^*) with a width of 250 MeV and another at 1850 MeV with the same width. Exponential kills the high-energy effect of the form factor.

In the QCD part, we worked with the Laplace transform:

$$\int_0^\infty dt\, e^{-t\tau}\,\frac{1}{\pi}\,\mathrm{Im}\,\psi(t)^u_s$$
$$= \frac{3}{8\pi^2}(\bar{m}_u+\bar{m}_s)^2\tau^{-2}\Bigg\{1+\bar{a}_s\left(\frac{11}{3}+2\gamma_E\right)$$
$$-\left[\bar{m}_u^2+\bar{m}_s^2+(\bar{m}_u+\bar{m}_s)^2\right]\tau+\frac{\pi}{3}\tau^2\langle\alpha_s G^2\rangle$$
$$+\frac{8\pi^2}{3}\tau^2\left[\left(m_s+\frac{m_u}{2}\right)\langle\bar{u}u\rangle+\left(m_u+\frac{m_s}{2}\right)\langle\bar{s}s\rangle\right]$$
$$-\frac{1408}{81}\pi^3\alpha_s\langle\bar{\psi}\psi\rangle^2\tau^3\Bigg\}, \tag{5.44}$$

where the contribution of the dimension 6 four-quark condensate $\alpha_s\langle\bar{\psi}\psi\rangle^2$ has been added.

- **The sum rule $(\bar{m}_d - \bar{m}_u)$**

Introducing the scalar meson $a_0(980) \equiv \delta$ into the spectral function and using its positivity, we deduce:

$$\frac{1}{\pi}\mathrm{Im}\psi(t)^u_s \geq \left(\sqrt{2}f_\delta M_\delta^2\right)^2 \delta(t-M_\delta^2). \tag{5.45}$$

In order to determine f_δ, we assume a dominance of meson δ on the form factor $\langle K^0|\partial_\mu V^\mu(0)^d_u|K^+\rangle$ which is normalized at $t=0$ to the "tadpole" component of the mass difference $K^0 - K^+$:

$$f(t) = \frac{\sqrt{2}f_\delta M_\delta^2\, g_{\delta\bar{K}K}}{-t+(M_\delta+i\Gamma_\delta/2))^2} \quad : \qquad |f(0)|^2 = \left(M_{K^0}^2 - M_{K^+}^2\right)^2_{\mathrm{TAD}}. \tag{5.46}$$

We use:

$$\left(M_{K^0}^2 - M_{K^+}^2\right)^2_{\mathrm{TAD}} \simeq 6.0\times10^{-3}\ \mathrm{GeV}^2, \tag{5.47}$$

which is an average of the different predictions [89]. We can estimate $g_{\delta\bar{K}K}$ using the result of Irving [90] compatible with that of SU(3) symmetry:

$$g_{\delta\bar{K}K}/g_{\eta\pi} = \sqrt{3/2}, \tag{5.48}$$

and the experimental width:

$$\Gamma(\delta \to \eta\pi) = \frac{|g_{\eta\pi}|^2}{8\pi M_\delta^2} \frac{\lambda^{1/2}(M_\delta^2, M_\eta^2, m_\pi^2)}{2M_\delta} \simeq (51 \pm 4) \text{ MeV}. \tag{5.49}$$

- **Phenomenology**

With the previous data, we obtain, for $\Lambda = 200$ MeV:

$$\hat{m}_s - \hat{m}_u \geq 170 \text{ MeV}, \quad \hat{m}_d - \hat{m}_u \geq 6 \text{ MeV}, \tag{5.50}$$

These results are big compared to the recent ones [74]. However, it should be noted that these values depend a lot on the optimal choice of τ, where these masses were evaluated as well as on the value of Λ relatively small compared to the current value $\Lambda = 340\,\text{MeV}$ [33].

5.3.2 Improved extraction of m_s/m_d and $\langle\bar{s}s\rangle/\langle\bar{d}d\rangle$

Different combinations of pseudoscalar and scalar sum rules were studied in Ref. [91] to re-extract m_s/m_d and $\langle\bar{s}s\rangle/\langle\bar{d}d\rangle$. In particular, we work with a combination of Eq. (4.21):

$$\psi_5(0)^u_s = \int_{t_0}^{\infty} \frac{dt}{t} e^{-t\tau} \frac{1}{\pi} \text{Im}\, \psi_5(t)^u_s \left[1 - t\tau \left(1 - 2\bar{m}_i^2 \tau \log \bar{m}_i^2 \tau + 2\frac{\bar{\alpha}_s}{\pi}\right)\right], \tag{5.51}$$

which is less sensitive to the higher state contributions than Eq. (4.21). The contributions of the radial excitations π' and K' have been added. We obtain:

$$f_{\pi'} \simeq (0.02 \sim 0.04) f_\pi, \quad f_{\pi'} \simeq (0.34 \sim 0.44) f_K. \tag{5.52}$$

With these values and adding the contribution of the scalar meson, we deduce:

$$\frac{m_s}{m_d} \simeq 27, \quad \frac{\langle\bar{s}s\rangle}{\langle\bar{d}d\rangle} \simeq (0.44 \sim 0.61). \tag{5.53}$$

These results confirm those obtained previously.

5.4 The End of the Contract with LAPP, Annecy

My fellowship at LAPP ended on September 1, 1982. Fortunately, I had obtained another two-year fellowship to work in the Theoretical Division of CERN in Geneva from that date. I am grateful to all the anonymous people who supported this application because it was an important step in my career and in my life.

The working atmosphere at LAPP was stimulating (seminars and conviviality) and the contact with CERN was very important for initiating some international collaborations.

Chapter 6

CERN Geneva Fellow and Associate 1983–1994

Figure 6.1. CERN, Geneva, Switzerland.

6.1 Living in Annecy and Working in Geneva

We chose to stay and live in the village of Menthon St. Bernard on the boarder of the lake of Annecy because we were starting to integrate socially and we liked the city and the people.

I drive between Annecy and Geneva every day and pass through the Swiss countryside to exit at the small custom office Soral near Saint Julien en Genevois. During these trips, I shared a sympathy with a wild rabbit who crossed the road each time I passed. However, one day, I didn't know what happened to him. He rushed at my car, but although I avoided him, he hit the door of the car and was killed instantly. I was very sad and picked him up. I gave it to our neighbours who said the next day that they had enjoyed eating it, which made me even sadder.

(a) (b)

Figure 6.2. Larry in **(a)** Annecy; **(b)** Genève in 1983.

Integration into the CERN theoretical group was a little more complicated because it is a large international laboratory. There is an incessant passage of researchers from all countries. Each country had its own group to take lunch. I always went to eat or have a coffee with my colleagues Patrick Aurenche, Gonzales-Mestre from Annecy, the group of Mary K. Gaillard and John Ellis and sometimes with some permanent members of CERN (R. Stora, J. Prentki, A. Martin, J.S. Bell, G. Veneziano, etc.), with some of the secretaries (Nanie, Marie-Noelle, Michelle, etc.) or with some passing physicists and visiting collaborators. This moment in the CERN cafeteria/restaurant was unavoidable for discussions on hot topics in research.

The secretaries of the TH division were also welcoming and helped us a lot (Tatiana, Marie-Noelle, Nanie and later Michelle and Jeanne). At that time, it was they who typed out researchers' articles.

For my first stay at CERN as a fellow, I completed my review on Techniques of Dimensional Renormalization and Applications to the Two Point Functions of QCD and QED initiated in Trieste (*Phys. Rep.* 84 (1982) 263–399) where the Editor-in-Chief was Maurice Jacob. I have benefited much from the advice of André Peterman who phoned me almost each morning to give some comments about the manuscript, especially on the reference to his paper with E.C.G. Stueckelberg, *Helv. Phys. Acta.*, 26 (1953) 499 where they introduced the concept of Renormalization Group Invariance in order to understand the short-distance behaviour of QED before the famous paper of Gell-Mann and Low published in *Phys. Rev.* 95 (1954) 1300. The anecdote is that A. Peterman is physically invisible at the theory group and sometimes he passed by during the weekend to pick up his mails while E.C.G. Stueckelberg goes to the TH seminar with his dog.

After completing this review paper and in parallel to my works on the search for some New Physics, I continued my works on QCD spectral sum rules (QSSR) started from my PhD thesis in Marseille. During that period, two other groups (L.J. Reinders, H. Rubinstein, J. Weyers and J. Bell, R.A. Bertlmann) working on QSSR were also at the TH Division while L.B. Okun chief of the ITEP-Moscow, the institute of the sum rules inventors M.A. Shifman, A.I. Vainshtein and V.I. Zakharov visits CERN regularly. In principle, it is a favourable environment for me for doing the sum rules research but, unfortunately, not, as there were some disagreements among the groups, especially between the Rubinstein and Marseille groups. Then, to keep activities in this field, I have started new collaborations with N. Paver from Trieste, the Barcelona group (P. Pascual, R. Tarrach, J. Latorre, A. Bramon and E. Bagan) where some of them visited regularly CERN.

However, to integrate into the TH group, I tried to interact with Prof. J. Bell (permanent member) by showing him regularly my QSSR analysis in the glueball sector where he provided some useful advice for improving the analysis. I have also attempted to interact with Prof. G. Veneziano (permanent member) by presenting to him my QSSR estimate of the η' mass and QCD topological charge related to his famous paper on the $U(1)_A$ channel within a large N_c expansion. Since that, I had regular discussions with him on related hadron physics which have led later on to some collaborations including G. Shore. I sincerely appreciated his patience and kindness.

On June 27, 1983, our first child Larry was born in Annecy (see Fig. 6.2). The birth of the child changed the organization of our lives.

The good news continued because in 1984 before the end of my contract at CERN, I finally got a permanent position as a first class researcher at the Mathematical Physics Laboratory of the CNRS in Montpellier. These news were of course encouraging to continue my research activities with serenity. I sincerely thank all persons who have supported my applications, and among them (I guess) E. de Rafael and G. Menessier.

However, we had difficulty to leave Annecy as we became attached to the region and the people. We have especially appreciated the relation among neighbours, reminding us our tradition in Madagascar and the sincerity of the Savoyard citizens.

Being in Montpellier, I continued to keep in touch with CERN, and visited it regularly as a paid associate researcher from 1986 to 1994.

6.2 Composite Models

The W and Z bosons have just been discovered at CERN by the UA1 and UA2 [92] groups. Physicists asked themselves the question of their nature, whether they are elementary or bound states of more elementary particles. Among these models, we were attracted by the haplon model of Fritzsch and Mandelbaum [93], where the W and Z bosons are bound states of spin 1/2 haplons, and the Abbott–Farhi model [94], where they are built with scalar elementary particles. This model looked interesting as it mimics the low-energy hadron spectrum at higher energies where the W, Z are the replica of the ρ-meson of QCD. However, the model is incomplete as it does not address the question of the chirality of the Standard Model. I had several discussions with Prof. H. Fritzsch on his model each time he visited CERN and he has invited me to give a seminar at the Munich University. He told me how he escaped from East Germany from the beach by boat via Turkey to West Germany. His exceptional adventure had impressed Prof. M. Gell-Mann with whom he collaborated at CALTECH-California. We had a good relationship and he belonged later on to the QCD-Montpellier conference committee. He was also fascinated by Madagascar and came two times there with P. Minkowski for the HEPMAD conference.

6.2.1 Models with composite *W* and *Z*

The difference in mass between the W and Z bosons is due to the $\gamma - Z$ mixing, which breaks the symmetry of the weak SU(2) isospin:

$$M_z = M_W/\sqrt{1-\lambda^2}: \quad \lambda = \frac{g}{e}\sin^2\theta_W, \tag{6.1}$$

where g is the coupling of W to the fermions: $G_F/\sqrt{2} = g^2/(8M_W^2)$ is the Fermi constant and $\sin^2\theta_W = 0.231$ [33] is the electroweak mixing angle.

We then extended the applications of the QCD sum rules to give constraints on the scales of these theories [95–97].

• *W* as a bound state of fermions

Assuming that these bound states are described by a hyperstrong theory analogous to QCD (called Quantum Haplo Dynamics (QHD)) at the TeV scale, we studied the two-point function associated with the vector current:

$$V^\mu(x) = \bar{\psi}_i\gamma^\mu\psi_j, \tag{6.2}$$

where ψ_i is the field of haplons which are the constituents of W. We obtained constraints by the Laplace sum rule in Ref. [95]. Constraints obtained by the finite energy sum rule (FESR) method for $SU(2)_h$ are [96]:

$$\begin{aligned} F_W^2 &\simeq \frac{t_c}{8\pi^2}, \\ F_W^2 M_W^2 &\simeq \frac{t_c^2}{16\pi^2} - \frac{1}{24\pi}\langle\alpha_h G_h^2\rangle, \\ F_W^2 M_W^4 &\simeq \frac{t_c^2}{16\pi^2} - \frac{112}{81}\pi\alpha_h\langle\bar{\psi}\psi\rangle^2, \end{aligned} \tag{6.3}$$

where α_h is the QHD coupling and $\langle\alpha_h G_h^2\rangle$ the hypergluon condensate. $\sqrt{t_c}$ matches approximately to the mass of the radial excitation of W. The F_W coupling is normalized as:

$$\langle 0|\bar{\psi}_i\gamma^\mu\psi_j|W\rangle = \epsilon^\mu\sqrt{2}F_W M_W. \tag{6.4}$$

F_W can be connected to the coupling g of W to the fermions at $t = 0$:

$$F_W \simeq M_W g \simeq 1.6 M_W. \tag{6.5}$$

The first equation gives the compositeness scale:

$$\sqrt{t_c} \simeq 14\, M_W \simeq 1.15 \text{ TeV}, \tag{6.6}$$

while the other two equations give the values of the condensates:

$$\begin{aligned} \langle \alpha_h G_h^2 \rangle^{1/4} &\simeq 12\, M_W \simeq 965\,\text{GeV}, \\ \left(-\sqrt{\alpha_h}\langle \bar{\psi}\psi\rangle\right)^{1/3} &\simeq 4.5\, M_W \simeq 362\,\text{GeV}, \end{aligned} \tag{6.7}$$

to compare with the case of QCD:

$$\begin{aligned} \sqrt{t_c} \simeq 1.3\,\text{GeV} \simeq 2M_\rho, \quad \langle \alpha_s G^2 \rangle^{1/4} \simeq 503\,\text{MeV}, \\ \left(-\sqrt{\alpha_s}\langle \bar{\psi}\psi\rangle\right)^{1/3} \simeq 286\,\text{MeV}. \end{aligned} \tag{6.8}$$

We note that the dynamics of QHD and QCD are very different, because the mass difference between the ground state and the radial excitation ($\sqrt{t_c}$) are very different, while the contribution of the hypergluon condensate is relatively more important compared to the haplon/preon condensate. By taking the ratio of the first two sum rules of Eq. (6.3), we deduce the mass formula:

$$M_W^2 \simeq \frac{t_c}{2}\left\{1 - \frac{2\pi}{3}\frac{\langle \alpha_h G_h^2 \rangle}{t_c^2}\right\}, \tag{6.9}$$

indicating the large value of the hypergluon condensate to explain the large mass difference $\sqrt{t_c} - M_W$.

• *W* as bound state of scalar particles

In this case, we obtained the constraints for $SU(2)_h$:

$$\begin{aligned} 2F_W^2 &\simeq \frac{t_c}{12\pi^2} + 4\langle \Phi^\dagger \Phi \rangle, \\ 2F_W^2 M_W^2 &\simeq \frac{t_c^2}{24\pi^2} - \frac{1}{6\pi}\langle \alpha_h G_h^2 \rangle + 6\pi\alpha_h \langle \Phi^\dagger \Phi \rangle^2. \end{aligned} \tag{6.10}$$

Using $F_W \simeq 1.6$ GeV in Eq. (6.5) from the Standard Model relation and the Higgs mechanism:

$$M_W = \frac{g}{\sqrt{2}}\langle\Phi^\dagger\Phi\rangle^{1/2}, \tag{6.11}$$

we deduce:

$$\langle\Phi^\dagger\Phi\rangle^{1/2} \simeq 85.7 \text{ GeV}, \tag{6.12}$$

which implies:

$$\sqrt{t_c} \simeq 24.5 M_W \simeq 2\,\text{TeV}. \tag{6.13}$$

Taking the ratio of the two equations in Eq. (6.10), we deduce:

$$M_W^2 \simeq \frac{t_c}{2}\left\{1 - 48\pi^2\frac{\langle\Phi^\dagger\Phi\rangle}{t_c} - 4\pi\frac{\langle\alpha_h G_h^2\rangle}{t_c^2} + \mathcal{O}\left(\frac{1}{t_c^3}\right)\right\}, \tag{6.14}$$

similar to that of Eq. (6.9). In this case, the large difference in mass of the W with its radial excitation requires a relative high value of the hypergluon condensate:

$$\langle\alpha_h G_h^2\rangle \simeq (1.04\,\text{TeV})^4\,. \tag{6.15}$$

- **Case of a supersymmetric model**

The associated Noether current is:

$$J_\mu(x) = \bar\psi_i\gamma_\mu\psi_i + \Phi_i^\dagger \overleftrightarrow{D}_\mu \Phi_i. \tag{6.16}$$

With the expression of the corresponding two-point function [66], and taking $\langle\alpha_h G_h^2\rangle = 0$ for a supersymmetry not broken spontaneously, we deduce, for $SU(2)_h$:

$$F_W^2 \simeq \frac{t_c}{4\pi^2} + 2\langle\Phi^\dagger\Phi\rangle,$$

$$M_W^2 \simeq \frac{t_c}{2}\left\{1 - \frac{8\pi^2}{t_c}\langle\Phi^\dagger\Phi\rangle + \mathcal{O}\left(\frac{1}{t_c^2}\right)\right\}, \tag{6.17}$$

Using the value of $\langle\Phi^\dagger\Phi\rangle$ and F_W from the Standard Model relation, we deduce:

$$\sqrt{t_c} \simeq 10 \text{ M}_\text{W}. \tag{6.18}$$

• The hyperpion $\tilde{\pi}$

In these models, we expect to naturally have Goldstone bosons ($\tilde{\pi}$) associated with the divergence of the axial current:

$$\langle 0|\partial^{\mu}\bar{\psi}\gamma_{\mu}\gamma_5\psi|\tilde{\pi}\rangle = \sqrt{2}F_{\tilde{\pi}}M_{\tilde{\pi}}^2, \tag{6.19}$$

where we considered the non-supersymmetric case of fermions [95, 96]. We have fixed the scale Λ_h of QHD analogous to Λ of QCD by the $\tilde{\pi}$ PCAC relation:

$$F_{\tilde{\pi}}^2 M_{\tilde{\pi}}^2 \simeq -2m_{\psi}\langle\bar{\psi}\psi\rangle \approx \frac{1}{2}\Lambda_h^2. \tag{6.20}$$

Using the constraint that comes from LSR:

$$m_{\psi}^2 \approx \frac{16\pi^2}{9}\left(F_{\tilde{\pi}}^2 M_{\tilde{\pi}}^4\right)\tau_0^2, \tag{6.21}$$

where τ_0 is the value of the optimization scale that we set around $1/t_c$, we deduce:

$$F_{\tilde{\pi}} \approx \frac{8}{3}\sqrt{2}\pi\langle\bar{\psi}\psi\rangle\tau_0 \approx (400 \sim 640)\ \text{GeV}, \tag{6.22}$$

if we take $\langle\bar{\psi}\psi\rangle$ from Eq. (6.7) or the value around $-(3-4)$ M_W from Ref. [95]. We deduce [93]:

$$15\ \text{GeV} \leq M_{\tilde{\pi}} \leq (80-85)\ \text{GeV}, \quad 90\,\text{GeV} \leq \Lambda_h \simeq\leq\ 250\,\text{GeV}, \tag{6.23}$$

where the lower bound on the mass comes from the PETRA [98] experiment. We studied the decay modes of $\tilde{\pi}$. Its decays in $\gamma\gamma$ and in $\mu^+\mu^-$ are given by:

$$\Gamma(\tilde{\pi} \to \mu^+\mu^-) \simeq \frac{g_{\tilde{\pi}\mu\mu}^2}{8\pi}M_{\tilde{\pi}}\left(1-\frac{4m_{\mu}^2}{M_{\tilde{\pi}}^2}\right)^{3/2},$$

$$\Gamma(\tilde{\pi} \to \gamma\gamma) \simeq \left(\frac{\alpha}{\pi}\right)^2 (Q_i^2 - Q_j^2)(N_c N_h)^2 \frac{1}{64\pi}\frac{M_{\tilde{\pi}}^2}{F_{\tilde{\pi}}^2}, \tag{6.24}$$

where N_c and N_h are the colour and hypercolour numbers of haplons/preons. We obtain:

$$\Gamma(\tilde{\pi}^0 \to \gamma\gamma) \simeq (0.1 - 4) \text{ keV}, \ \ \Gamma(\tilde{\pi}^0 \to \mu^+\mu^-) \simeq (2-4)\times 10^{-2} \text{ keV}, \tag{6.25}$$

with the previous mass region. Other decay modes are given in Ref. [96] where we used the Golberger–Treiman type relation with:

$$g_{\tilde{\pi}ll'} \simeq \frac{(m_f^2 + m_{f'}^2)^{1/2}}{\sqrt{2}F_{\tilde{\pi}}} U_{ff'}, \tag{6.26}$$

with $U_{ff'}$ the mixing angle of the Cabibbo–Kobayashi–Maskawa type with the two fermions f and f'. We also assumed that the $\tilde{\pi}$ has Higg-type couplings with pairs of fermions.

We continued these analyzes in Ref. [99], but looking at the case of a very light $\tilde{\pi}$ hyperpion. The branching ratio:

$$\frac{\Gamma(\psi \to \tilde{\pi}\gamma)}{\Gamma(\psi \to e^+e^-)} \simeq \frac{g^2_{\tilde{\pi}e^+e^-}}{2\pi\alpha} \leq 10^{-8} \tag{6.27}$$

gives [96]:

$$F_{\tilde{\pi}} \geq 2.2 \text{ TeV}. \tag{6.28}$$

Using Eqs. (6.20) and (6.23), we deduce:

$$M_{\tilde{\pi}} \leq 80 \text{ MeV}. \tag{6.29}$$

The decay rate $\tilde{\pi}^0 \to \gamma\gamma$ behaves like $M^3_{\tilde{\pi}}$, and this process is almost unobservable for a light hyperpion. The one in $\mu^+\mu^-$ can be written as:

$$\Gamma(\tilde{\pi} \to \mu^+\mu^-) \simeq 90\,\text{eV} \left(\frac{M_{\tilde{\pi}}}{15\,\text{GeV}}\right)\left(1 - \frac{4m_\mu^2}{M_{\tilde{\pi}}^2}\right)^{3/2}. \tag{6.30}$$

Other phenomenological aspects of $\tilde{\pi}$ are discussed in Refs. [96, 99]:

$$\Gamma(W \to \tilde{\pi}^+\tilde{\pi}^-) \leq 200 \text{ MeV}, \quad \Gamma(Z \to \tilde{\pi}^+\tilde{\pi}^-) \leq 310 \text{ MeV}. \tag{6.31}$$

The most promising test of the substructure of W, Z is the detection of hyperpions because their decays differ from those of supersymmetric particles.

6.2.2 Hypergluonium $\tilde{\eta}'$

Stimulated by the excess event in the process $Z \to \gamma e^+e^-$ with a very energetic photon seen by the groups UA1, UA2, we considered the case of the particle $X \equiv \tilde{\eta}'$ which couples universally to lepton pairs. Experiment suggests that the invariant mass of e^+e^- is of the order of 50 GeV.

For this, we consider the two-point function in the channel $U(1)_A$ for two flavours of preons:

$$\psi_G(q^2) = i \int d^4x e^{iqx} \langle 0|J_G(x)J_G(0)^\dagger|0\rangle :$$
$$J_G = \left(\frac{\alpha_h}{2\pi}\right) G^{\mu\nu}\tilde{G}_{\mu\nu}(x), \tag{6.32}$$

with $\tilde{G}_{\mu\nu} \equiv (1/2)\epsilon_{\mu\nu\rho\sigma}G^{\rho\sigma}$, where $G_{\mu\nu}$ is the hypergluon tensor of $SU(N)_h$. The coupling of the $\tilde{\eta}'$ particle associated with the current is normalized as:

$$\langle 0|J_G|\tilde{\eta}'\rangle = \sqrt{2}F_{\eta'}M^2_{\tilde{\eta}'}. \tag{6.33}$$

Assuming, as in the case of QCD, that $\psi_G(0) = 0$ in the limit of massless haplon/preon, we obtain the constraint of FESR [99]:

$$2F^2_{\eta'}M^2_{\eta'} \simeq \left(\frac{\alpha_h}{2\pi}\right)^2 \left\{\frac{t_c^2}{\pi^2 \log(t_c/\Lambda_h^2)} + 4\langle G_h^2\rangle\right\}. \tag{6.34}$$

Using $\Lambda_h \approx 200\,\text{GeV}$, $\sqrt{t_c} \simeq 1.15$ TeV, $\langle\alpha_h G_h^2\rangle \simeq 965\,\text{GeV}$ and $F_{\tilde{\pi}} \approx 520\,\text{GeV}$ with ~ 2.2 TeV, we obtain:

$$M_{\tilde{\eta}'} \simeq 176 \text{ GeV} \quad \text{or} \quad 42 \text{ GeV}. \tag{6.35}$$

The case of a scalar hypergluon current:

$$J_G = \alpha_h G_h^2 \tag{6.36}$$

leads to the same result as that of the pseudoscalar current. Our result is not in favour of the observed mass of 50 GeV.

6.2.3 The isoscalar scalar $\tilde{\sigma}$

We finally considered the two-point function corresponding to the scalar isoscalar current [99]:

$$J_S = \frac{m}{\sqrt{2}}\left(\bar{\psi}_i\psi_i + \bar{\psi}_j\psi_j\right). \tag{6.37}$$

The associated particle is:

$$\langle 0|J_S|\tilde{\sigma}\rangle = \sqrt{2}F_{\tilde{\sigma}}M_{\tilde{\sigma}}^2. \tag{6.38}$$

The corresponding FESR constraint is:

$$2F_{\tilde{\sigma}}^2 M_{\tilde{\sigma}}^4 \simeq \frac{3}{16\pi^2}m_i^2 t_c^2 + \frac{m_i^2}{8\pi}\langle\alpha_h G_h^2\rangle. \tag{6.39}$$

In the case of scalar preon, we obtain:

$$2F_{\tilde{\sigma}}^2 M_{\tilde{\sigma}}^2 \simeq \frac{1}{8\pi^2}m_i^2 t_c + 2m_i^2\langle\Phi_i^\dagger\Phi_i\rangle. \tag{6.40}$$

Using the experimental mass $M_{\tilde{\sigma}} = 50$ GeV, $F_{\tilde{\sigma}} \approx F_{\tilde{\pi}} \approx 500$ MeV, $\sqrt{t_c} \simeq 1$ TeV, $\langle\alpha_h G_h^2\rangle^{1/4} = 965$ GeV and $\langle\Phi^\dagger\Phi\rangle^{1/2} = 86$ GeV, we obtain:

$$m_i \simeq 7\,\text{GeV} \quad (\text{respectively}\,207\,\text{GeV}) \tag{6.41}$$

for preons of spin 1/2 (respectively spin 0) whose hypothesis of good SU(2) symmetry favours preons of spin 1/2. The phenomenology of $\tilde{\sigma}$ (decay and production) was studied in detail in Ref. [99], assuming that it couples to the Higgs with the pair of fermions.

6.2.4 Models with composite excited leptons

• Case of a non-supersymmetric model

We analyze the possibility that quarks and leptons are bound states of more elementary preons [100] to also try to explain the process $Z \to e^+e^{-\gamma}$ from an electron excited $e^* \to e\gamma$. We consider the two-point function:

$$S_f(q) = i\int d^4x\, e^{iqx}\langle 0|\mathcal{T}J_f(x)J_f^\dagger(0)|0\rangle, \tag{6.42}$$

associated with the current:

$$J_f(x) \equiv \Phi^\dagger \frac{1}{2}(1-\gamma_5)\psi, \tag{6.43}$$

where Φ (respectively ψ) is the scalar (respectively fermionic) field. Parameterizing the spectral function as:

$$S_f(q) = \frac{1}{2}(1-\gamma_5)\hat{q}\left\{\sum_{f,f^*}\lambda_i^2 \frac{1}{Q^2+m_i^2} + \text{“QHD continuum”}\right\}, \tag{6.44}$$

we deduce the FESR constraints for $SU(N)_h$:

$$\sum_{e,fe*}\lambda_i^2 = \frac{N}{32\pi^2}t_c + \langle\Phi^\dagger\Phi\rangle,$$
$$\sum_{e,e^*}\lambda_i^2 m_i^2 = \frac{N}{64\pi^2}t_c^2 + \frac{1}{64\pi}\langle\alpha_h G_h^2\rangle, \tag{6.45}$$

Using the positivity of the electron contribution in the first equation and neglecting m_e^2 in the second, we obtain, for $N_h = 2$:

$$m_{e^*}^2 \geq \frac{t_c}{2}\left(\frac{1+\pi\langle\alpha_h G_h^2\rangle/(2t_c^2)}{1+16\pi^2\langle\Phi^\dagger\Phi\rangle/t_c}\right). \tag{6.46}$$

Using the values of the parameters obtained previously, we obtain

$$m_{e^*} \geq 790 \text{ GeV}. \tag{6.47}$$

The experimental candidate of 50 GeV suggests a value of $\langle\Phi^\dagger\Phi\rangle$ of the order of 2 TeV while keeping the values of t_c and $\langle\alpha_h G_h^2\rangle$.

- **Case of a supersymmetric model**

In this case, we obtain [100]:

$$m_{e^*}^2 \geq \frac{t_c}{2}\left(\frac{1+84\pi^3\langle\alpha_h\Phi^\dagger\Phi\rangle^2/(t_c^2)}{1+16\pi^2\langle\Phi^\dagger\Phi\rangle/t_c}\right). \tag{6.48}$$

This constraint shows that with the experimental value $M_{\tilde{\sigma}} = 50$ GeV and the values $\sqrt{t_c} \simeq 1.15$ TeV and $\alpha_h \simeq 1$, we have no real solution for $\langle\Phi^\dagger\Phi\rangle$. Note that the relationships between the different condensates, the fermion masses and the parameters of the Godstone boson in supersymmetric QCD (SQCD) have been studied with the sum rules in Ref. [101].

• Conclusion

Our analysis, which is almost independent of the model, does not favour the existence of an excited electron around 50 GeV but rather around 790 GeV. However, after a few months, these abnormal events disappeared from the experiment.

6.3 Testing the Compositeness Hypothesis

Constraints on the composite scale have been studied in Ref. [102] and the references in this article, among which we have retained the strongest constraints. The effects of the composite particle loop on the propagators of the W and Z bosons have been systematically analyzed in Ref. [103].

6.3.1 Anomalous magnetic moment a_l and electric dipole d_l

We looked at the effect of the four fermion terms and the magnetic coupling induced by the substructure of the excited lepton l^* which we add to the standard Lagrangian of QED:

$$\mathcal{L}_{f'fff'} = \sum_{f,f'} \frac{g^2}{2\Lambda^2_{ff'}} \bar{\psi}_{f'}\Gamma(c_j - \gamma_5 d_j)\psi_f \times \bar{\psi}_f\Gamma(e_j - \gamma_5 f_j)\psi_{f'},$$

$$\mathcal{L}_{l^*l\gamma} = -e\frac{(M_l^* + m_l)}{2\Lambda^2_{ll^*}} \times \bar{\psi}_{l^*}\sigma^{\mu\nu}(a - \gamma_5 b)\psi_l F_{\mu\nu}, \tag{6.49}$$

where $\Lambda_{ll'}$ is the compositeness scale. Γ is a generic notation for Dirac γ matrices. M_l^* and m_l are the masses of l^* and l and $F_{\mu\nu}$ is the tensor of the electromagnetic field. Considering the lepton–photon–lepton vertex:

$$V_l^\mu = -e\Big\{\gamma^\mu F_1(q^2) + \frac{i}{2m_l}\sigma^{\mu\nu}q_\nu\Big[F_2(q^2) - \gamma_5 F_3(q^2)\Big]\Big\}, \tag{6.50}$$

where $F_i(q^2)$ are the Dirac form factors, we have evaluated the effect of these terms on the anomalous magnetic moment and the electric dipole moment of the lepton:

$$a_l \equiv (g-2)/2 = F_2(0), \quad d_l = \frac{F_3(0)}{2m_l}. \tag{6.51}$$

In the $\Gamma \equiv \gamma^\mu$ case, the contributions are zero in the case where the four-fermion couplings are of the $V-A$ type and the magnetic coupling is of the $V+A$ type. However, in the case where $c_f = -d_f = 1$ and the other couplings $a = b = e_f = f_f = 1$, the contributions are non-zero. We deduce the constraints:

$$\Lambda_{\mu l^*} \geq 2 \times 10^4\,\text{TeV from} \quad (\text{g}-2)_\mu, \quad \Lambda_{\mu l^*} \geq 51\,\text{TeV from} \quad \text{d}_\mu \tag{6.52}$$

in the case $\Lambda_{\mu l^*} \gg M_{l^*}$ and:

$$\Lambda_{\mu l^*} \geq 55\,\text{TeV from} \quad (\text{g}-2)_\mu, \quad \Lambda_{\mu l^*} \geq 3\,\text{TeV from} \quad \text{d}_\mu \tag{6.53}$$

in the case $\Lambda_{\mu l^*} \gg M_{l^*}$.

6.3.2 $K^- \to \pi^-\pi^0$

To the contribution of the Standard Model, we can add the four-quark operator $\Delta S = 1$, $\Delta I = 1/2$:

$$\mathcal{O}_{3/2} = \frac{g^2}{2\Lambda_{3/2}^2}\left[(\bar{d}_L\gamma_\mu u_L)(\bar{u}_L\gamma^\mu s_L) + (\bar{u}_L\gamma_\mu u_L - \bar{d}_L\gamma_\mu d_L)(\bar{d}_L\gamma^\mu s_L)\right]. \tag{6.54}$$

Tolerating a deviation from the SM of 10, one obtains:

$$\Lambda_{3/2} \geq 10^3\,\text{TeV}. \tag{6.55}$$

6.3.3 Effect of loops of composite particles at M_Z

A systematic study of the effects on the propagators of the W and Z bosons of the loops of composite particles having different spin (0, 1/2, 3/2, 2) was made in Ref. [103]. We considered the quantity:

$$\frac{\delta M_Z}{M_Z} = \frac{1}{2}\left[\left(\Pi_{WW}(q^2) + \Pi_{\gamma\gamma}(q^2) + \frac{\cos\theta_W}{\sin\theta_W}(1-\text{tg}^2\theta_W)\Pi_{\gamma Z}(q^2)\right)\frac{1}{q^2}\bigg|_{q^2=0} - \frac{1}{M_Z^2}\text{Re}\Pi_{ZZ}(-M_Z^2)\right], \tag{6.56}$$

where $\Pi_{ij}(q^2)$ are the two-point functions associated with the vector currents $V^\mu = \bar{\psi}\gamma^\mu\psi$ and axial $A^\mu = \bar{\psi}\gamma^\mu\gamma_5\psi$:

$$\Pi^{\mu\nu}_{ij}(q^2) \equiv i\int d^4x\, e^{iqx}\langle 0|T J^\mu_{ij}(x)\left(J^\mu_{ij}\right)^\dagger(0)|0\rangle$$
$$= -\left(g^{\mu\nu} - q^\mu q^\nu/q^2\right)\Pi_{ij}(q^2) + ..., \tag{6.57}$$

such as:

$$J^\mu_i = \lambda_i(a_i V^\mu - b_i A^\mu), \tag{6.58}$$

with

$$\lambda_\gamma = g\,\sin\theta_W, \quad \lambda_W = g/\sqrt{8}, \quad \lambda_Z = \frac{1}{4}\cos\theta_W$$

and, for example, with the quark u:

$$a_\gamma = \frac{2}{3}, \quad a_W = b_W = 1, \quad a_z = 1 - \frac{8}{3}\sin^2\theta_W, \quad b_Z = 1. \tag{6.59}$$

- **Loop of $\tilde{\pi}\tilde{\pi}$**

We obtain this contribution starting from the effective Lagrangian:

$$\mathcal{L}_\pi = g_\pi \tilde{\pi} \overrightarrow{\partial}_\mu \tilde{\pi} A^\mu + g_\pi^2 \tilde{\pi}^\dagger \tilde{\pi} A_\mu A^\mu, \tag{6.60}$$

Assuming that $g_\pi = g_W = e/\sin\theta_W$, we deduce:

$$\frac{\delta M_Z}{M_Z} \simeq 10^{-3} \quad \text{for } M_{\tilde{\pi}} = 0.5\,\text{TeV}. \tag{6.61}$$

Conversely, by tolerating a deviation from the prediction of the SM and the experiment of ± 1‰, we obtain:

$$g_\pi \leq 0.3. \tag{6.62}$$

- **Effective coupling**

We consider the coupling with three bosons $(Z, \gamma, \tilde{\pi})$:

$$\mathcal{L} = g_{PV_jV_k}\epsilon^{\mu\nu\rho\sigma} p_{1\rho} p_{2\sigma} V^\mu_j V^\nu_k P. \tag{6.63}$$

Estimating the coupling $\tilde{\pi}\gamma Z$ using SU(2) symmetry and arguments based on vector meson dominance (VDM), we deduce:

$$\frac{\delta M_Z}{M_Z} \simeq 10^{-5} \quad \text{for } M_{\tilde{\pi}} = 0.5\,\text{TeV}. \tag{6.64}$$

For an experimental precision of 1/mile, we deduce:

$$g_{P\gamma\gamma} \leq \alpha \times 0.07\,\text{GeV}^{-1} \quad \text{for } M_{\tilde{\pi}} = 0.5\ ,\text{TeV}. \tag{6.65}$$

Constraints on the anomalous couplings to three and four vecterus bosons were also obtained in Ref. [103].

- **Excited fermion of spin 1/2**

We consider the effective Lagrangian:

$$\mathcal{L}_* = g\left(\frac{M_*}{2\nu^2}\right)\bar{\psi}^*\sigma^{\mu\nu}(a + b\gamma_5)\psi F_{\mu\nu}, \tag{6.66}$$

where ν is the ultra-violet (UV) subtraction point. For $\nu = 1$ TeV and $M_* = 0.5$ TeV, we obtain:

$$\frac{\delta M_Z}{M_Z} \simeq N_c\left(\frac{g^2}{4\pi}\right) 6 \times 10^{-3}. \tag{6.67}$$

- **Excited fermion of spin 3/2**

It can be described by the field:

$$\psi^*_\mu = \left(g_{\mu\nu} - \frac{1}{4}\gamma_\mu\gamma_\nu\right) D^\mu\psi, \tag{6.68}$$

where D^μ is the covariant derivative and ψ the field of the spin 1/2 particle. The fermion–fermion–gauge boson interaction Lagrangian is:

$$\mathcal{L} = \frac{g_{3/2}}{\nu}\left(\bar{\psi}^*_\mu\psi A^\mu + \epsilon^{\mu\nu\rho\sigma}\bar{\psi}^*_\mu\gamma_5\gamma_\nu\psi^*_\sigma A_\rho\right). \tag{6.69}$$

For $\nu \simeq 1$ TeV, $M^* \simeq \nu/2$ and for an experimental precision of 1‰, we obtain the constraint:

$$g_{3/2} \leq 0.3. \tag{6.70}$$

- **Spin 2 boson**

The vertex: vector boson-tensor boson pair can be described by:

$$V^{\rho\sigma} = -(i/\nu)\Big[g_2 F_1^{\rho\mu} F_{2\mu}^{\sigma} + (g_1/\nu^2) F_1^{\mu\nu} F_{2\nu\mu} p_1^{\rho} p_2^{\sigma}\Big]. \tag{6.71}$$

Its propagator is:

$$D^{\rho\sigma\alpha\beta} = \frac{1}{2}\Big[g^{\rho\alpha}g^{\sigma\beta} + g^{\rho\beta}g^{\sigma\alpha} - \frac{2}{3}g^{\rho\sigma}g^{\alpha\beta}\Big]\frac{1}{q^2 - M_2^2}, \tag{6.72}$$

with: $q \equiv p_1 + p_2$. For $\nu \simeq 1$ TeV, $M^* \simeq \nu/2$ and for an experimental precision of 1‰, we obtain the constraint:

$$g_2 \leq 0.3, \tag{6.73}$$

similar to the case of spin 3/2.

6.4 Tests of BSM in e^+e^-

6.4.1 Production of (pseudo)scalars

In Ref. [104], we proposed to test the nature of a (pseudo)scalar S boson of the type $\tilde{\pi}, \tilde{\eta}'$ and $\tilde{\sigma}$ if they are produced in polarized e^+e^- by measuring the longitudinal and transverse asymmetries by the production of a single or a pair of (pseudo)scalars from the reactions:

$$e^+e^- \to Z\,S, \quad e^+e^- \to \bar{S}\,S. \tag{6.74}$$

These polarizations are defined as follows:

$$\tilde{A}_{\parallel} = \frac{\sigma(P_{\parallel}^- = +1,\, P_{\parallel}^+ = 0) - (P_{\parallel)}^- = -1,\, P_{\parallel}^+ = 0)}{\sigma(P_{\parallel}^- = +1,\, P_{\parallel}^+ = 0) + (P_{\parallel}^- = -1,\, P_{\parallel}^+ = 0)},$$

$$\tilde{A}_{\perp} = \frac{\sigma(P_{\perp}^+ = P_{\perp}^- = 1) - (P_{\perp}^+ = -1,\, P_{\perp}^- = -1)}{\sigma(P_{\perp}^+ = P_{\perp}^- = 1) + (P_{\perp}^+ = -1,\, P_{\perp}^- = -1)},$$

$$A_{FB} = \frac{\int_0^1 d\cos\theta\,\frac{d\sigma}{d\cos\theta} - \int_{-1}^0 d\cos\theta,\,\frac{d\sigma}{d\cos\theta}}{\int_0^1 d\cos\theta\,\frac{d\sigma}{d\cos\theta} + \int_{-1}^0 d\cos\theta\,\frac{d\sigma}{d\cos\theta}}, \tag{6.75}$$

where $P_{\parallel}$ and $P_{\perp}$ are the degrees of longitudinal and transverse polarization of the beams e^+e^-. $A_{\parallel}$ can be observed if one or both beams $e^{\pm}$ are polarized and it measures the difference in collision probabilities e^+e^- of opposite helicities. $A_{\perp}$ arises from amplitude interference with opposite helicities and can be observed if the two beams $e^{\pm}$ are transversely polarized. A_{FB} measures odd powers of $\cos\theta$ in the cross-section resulting from the interference effects of the two neutral gauge bosons. Thus, A_{FB} counts the number of lepton pairs emitted in the front and rear hemispheres.

6.4.2 Models with extra Z'

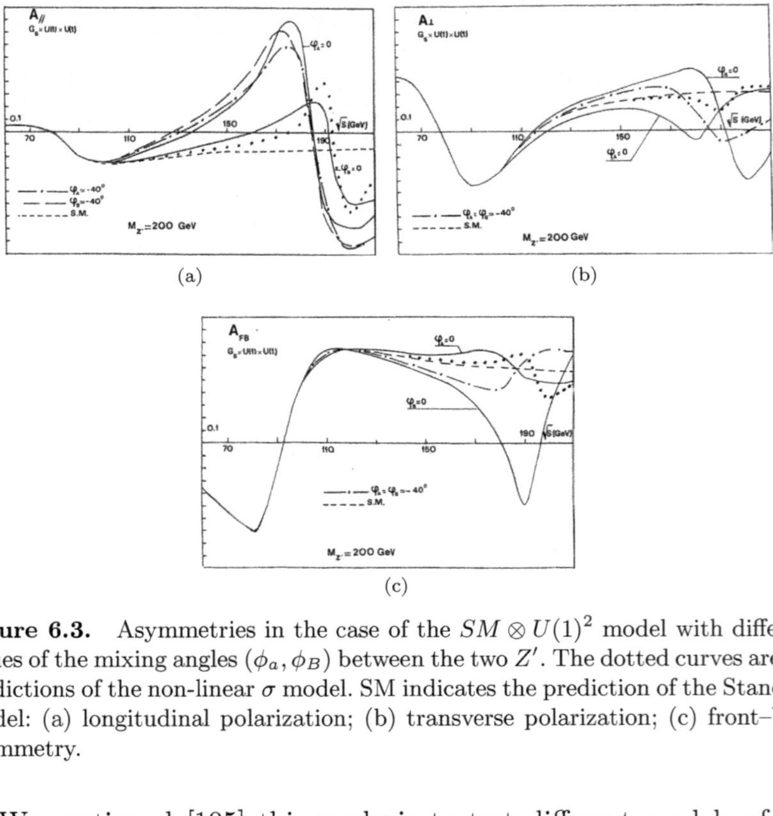

Figure 6.3. Asymmetries in the case of the $SM \otimes U(1)^2$ model with different values of the mixing angles (ϕ_a, ϕ_B) between the two Z'. The dotted curves are the predictions of the non-linear σ model. SM indicates the prediction of the Standard Model: (a) longitudinal polarization; (b) transverse polarization; (c) front–back asymmetry.

We continued [105] this analysis to test different models of the type $SM \otimes U(1)$ [106], $SM \otimes U(1)^2$ [107] with an extra-Z' which are based on superstrings and models of the type $SU(2)_L \otimes SU(2)_R \otimes SU(2)_V$ [108]. In the latter case, the scalar sector of SM becomes a non-linear σ model gauged in the formal limit where the Higgs mass tends towards infinity.

We show in Fig. 6.3, the behaviour of these asymmetries in the case $SM \otimes U(1)^2$ as a function of the mixing angles (ϕ_a, ϕ_B) between the two Z'. The case $SM \otimes U(1)$ is very similar and given in Ref. [105]. We also show that of the gauged non-linear σ model.

6.5 Tests of BSM in pp and $\bar{p}p$

Once established in Montpellier in 1984, I decided to supervise two doctoral students, including S. Larbi and J.L. Kneur. I gave them the topic of Testing Composite Models in proton (anti-proton)–proton hadronic collisions $p(\bar{p})\,p$ for the Tevatron and SSC projects. P. Chiappetta from Marseille later joined our group.

6.5.1 Production of $Z\gamma$

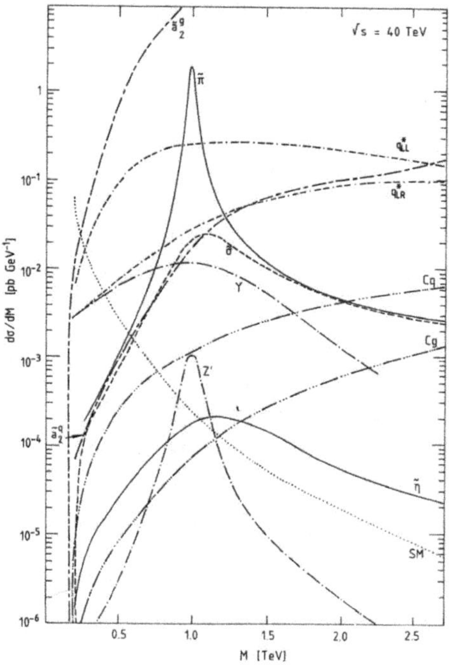

Figure 6.4. Invariant mass distribution at SSC energy for a resonance mass of 1 TeV: $\bar{q}q$ process, gluon–gluon fusion, for a rapidity cut $Y = 2.5$ and P_{TY} cut of 50 GeV.

We studied the effects of intermediate composite bosons of spin 0 $(\tilde{\pi}, \tilde{\sigma}, \tilde{\eta}')$, spin one (Z', Y), spin 2 $(\tilde{a}_2)$, excited spin 1/2 quarks (q^*) and contact terms [109] in the process:

$$pp,\, \bar{p}p \to X \to Z\gamma. \tag{6.76}$$

This reaction occurs via a loop in the Standard Model. On can see in Fig. 6.4 that the behaviour of the exotic contributions differ completely from the SM one, such that an excess of events of this type must reveal physics beyond SM. To illustrate the spin 0 bosons, we have considered technicolour-type couplings at the $Z\gamma$ and gg vertices. In this way, $\tilde{\pi}, \tilde{\sigma}$ and $\tilde{\eta}$ have total width of 52, 100 and 500 GeV. Note that this reaction is experimentally among the best and most feasible.

6.5.2 Drell–Yan process

We studied the process:

$$pp,\, \bar{p}p \to X \to \mu^+\mu^-. \tag{6.77}$$

The mass distribution of the pair $\mu^+\mu^-$ as a function of Q^2 is shown in Fig. 6.5(a) at the energy of the SSC $\sqrt{s} = 40\,\text{TeV}$ for a resonance of mass 1 TeV. Note that at this energy, the composite bosons with spin 0 $\tilde{\eta}$ and spin two $\tilde{a}_2^g$ and $\tilde{a}_2^q$ present characteristic signatures, while the bosons $Y_{A,B}$, Z' and V are diluted in the continuum. Only the bosons of the $S_{1,2,3}$ superstrings manifest as narrow resonances.

The front–back asymmetry is shown in Fig. 6.5(b) at the Tevatron energy $\sqrt{s} = 2$ TeV for a resonance of mass $3M_Z$. In this case, we see that the effect of the $Y_{A,B}$ bosons differs from the others below the threshold. Spin zero and two bosons do not have A_{FB} asymmetry.

6.5.3 Production of pairs W^+W^- and $\bar{Z}Z$

Motivated by two anomalous events $W \to e\nu$ in the UA1 experiments which are more numerous than those predicted by SM, we conjectured that these events can come from a non-standard V boson in the s pathway which decays to W^+W^-:

$$\bar{p}p \to \bar{q}q \to (V^0, V^\pm) \to (W^+W^-.\, W^-Z^0). \tag{6.78}$$

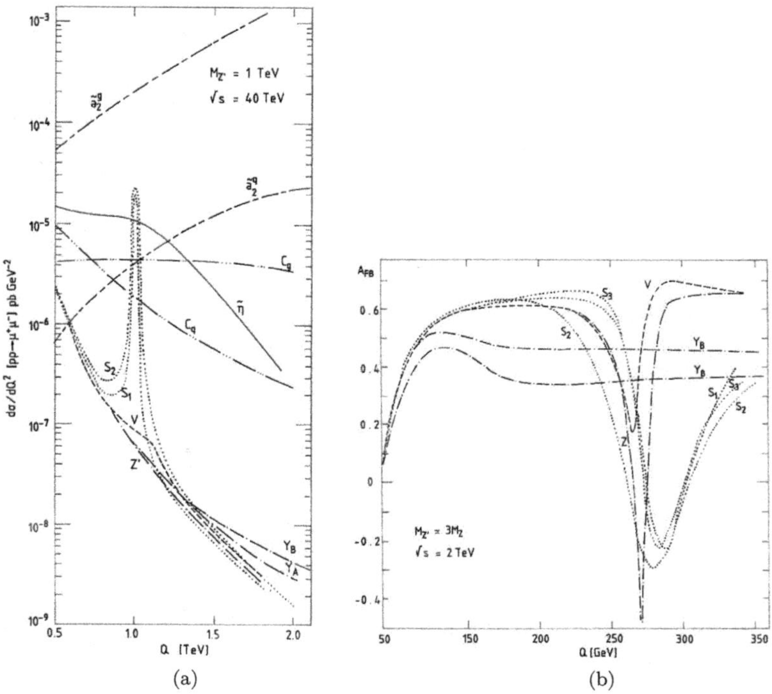

Figure 6.5. (a) Lepton mass distribution as a function of the lepton mass pair Q at SSC energy ($\sqrt{s} = 40$ TeV). The mass of the resonances is equal to 1 TeV; (b) Forward–backward asymmetry A_{FB} as a function of the lepton mass pair Q at Tevatron energy ($\sqrt{s} = 2$ TeV).

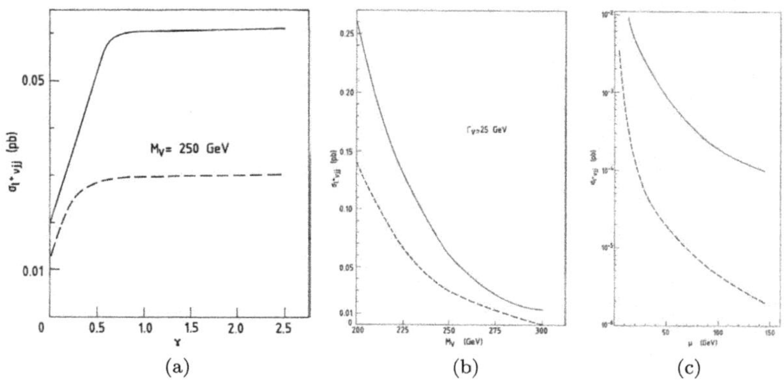

Figure 6.6. Behaviour of the total cross-section: $\bar{p}p \to V^0 \to W^+W^- \to l^+\nu jj$ (full curve) and $\bar{p}p \to V^+ \to W^+Z \to l^+\nu jj$ (dashed curve) for $b = 0$ and $g'/g'' = 0.22$: (a) versus the trilinear coupling γ for $M_V = 250$ GeV and (b) versus M_V for $\Gamma_V = 25\,\text{GeV}$; (c) behaviour of the total cross-section: $\bar{p}p \to \tilde{\sigma} \to W^+W^- \to l^+\nu jj$ for $M_{\tilde{\sigma}} = 200\,\text{GeV}$ (full curve), 300 GeV (dashed curve).

if $M_V \geq 2M_W$. We also looked at the contact interaction $(\bar{q}q)(W^+W^-)$. The analysis for $M_V = 250$ GeV and $\Gamma_V = 25$ GeV is shown in Fig. 6.6 with a choice for couplings constrained by the V boson width of 25 GeV allowed. The events of UA1, if confirmed, would have indicated the presence of $\tilde{\sigma}$ having a Higgs coupling with a pair of fermions and at the vertex of W. Unfortunately, these events were not confirmed.

6.5.4 Production at HERA

These analyses were extended in the case of the electron–proton collision at HERA [112] by the reaction:

$$e\,\mathrm{p} \to \mathrm{e} + \mathrm{jet\,of\,hadrons}, \tag{6.79}$$

where the X bosons are produced in the t pathway. We studied the differential production cross-section and longitudinal asymmetries. We noted that the largest deviations of SM come from the $\tilde{\sigma}$, the isoscalar vector boson Y and the tensor boson T. Lower bounds on the masses are given in Table 3 of Ref. [112].

6.6 The QCD Condensates

6.6.1 Renormalization of high-dimension QCD condensates

With Rolf Tarrach (Fig. 7.5) [113], we have studied the renormalization of high-dimension ($D = 5, 6$) QCD vacuum condensates and tried to find combinations which are renormalization group invariant (RGI) and independent of the subtraction constant. In so doing, we use the background field techniques [114] and classify systematically all possible operators having dimension $D = 5, 6$, including those induced by the equations of motion.

• Dimension 5 condensates

For massive quarks, we find that one can form a combination:

$$\phi_5 = \alpha_s^{\gamma_5/-\beta_1} \langle O_4 + x\, O_1 + y\, O_2 \rangle, \tag{6.80}$$

with:

$$O_4 = ig\bar{\psi}\sigma^{\mu\nu}\frac{1}{2}\lambda_a G^a_{\mu\nu}\psi, \quad O_1 = im^2\bar{\psi}\psi, \quad O_2 = -\frac{1}{4}im\, G^a_{\mu\nu}G^{\mu\nu}_a, \tag{6.81}$$

which is renormalization group invariant. For $SU(3)_c \otimes SU(3)_f$, one has [6] [1]:

$$x = -\frac{1944}{315}, \quad y = -\frac{72}{63}, \quad \gamma_5 = -1/3 \; : -\beta_1 = -\frac{11}{2} + \frac{n_f}{3}. \tag{6.82}$$

• Dimension 6 gluon condensate

We have classified different dimension 6 condensates where we found 10 independent scalar gauge invariant operators in the chiral limit $m_j = 0$ and for one quark flavour. We have not considered operators which do not contribute at zero momentum. We show that one can build a RGI three-gluon condensate:

$$\phi_G = \alpha_s^{\gamma_G/-\beta_1}\langle g\, f_{abc}G^a_{\mu\nu}G^{b,\nu\rho}G^{c,\mu}_{\rho}\rangle, \tag{6.83}$$

with:

$$\gamma_G = -\frac{23}{6} \tag{6.84}$$

for $SU(3)_c$.

• Dimension 6 four-quark condensates

For the four-quark condensates, we note that the operator $O_2 = g^2\bar{\psi}\psi\bar{\psi}\psi$ mixes with some other four-quark operators under renormalization and vice versa such that the factorization hypothesis [3]:

$$\langle\bar{\psi}\Gamma_1\psi\bar{\psi}\Gamma_2\psi\rangle = \frac{1}{N^2}\left[Tr\Gamma_1 Tr\Gamma_2 - Tr(\Gamma_1\Gamma_2)\right]\langle\bar{\psi}\psi\rangle^2 \tag{6.85}$$

is incompatible with the renormalization group equation. Here, the normalization factor N is defined as $\langle\bar{\psi}_\alpha\psi_\beta\rangle = (\delta_{\alpha\beta}/N)\langle\bar{\psi}\psi\rangle$. The subscripts α, β include spin, colour and flavour.

We shall test phenomenologically the validity of the vacuum saturation in the following section.

---

[1] Values for $SU(N)_c$ and for one flavour is given in Ref. [113].

6.6.2 QCD condensates from the ratio of moments

In Ref. [115], we use the ratio of moments $\mathcal{R}_\tau$ for extracting the gluon $\langle\alpha_s G^2\rangle$ and four-quark condensates $\alpha_s\langle\bar\psi\psi\rangle^2$ using the $e^+e^- \to$ Hadrons total cross-section. The advantage of the ratio of moments compared to the moment $\mathcal{L}$ used in Ref. [71] is that the leading PT α_s correction disappears while the strength of the dimension 6 four–quark condensates increase. Then, we expect that $\mathcal{R}_\tau$ is a good place to extract these two condensates. We write:

$$R_\tau \equiv -\frac{d}{d\tau}\log\mathcal{L}(\tau) = \tau^-1\{1+C_4\langle O_4\rangle\tau^2+C_6\langle O_6\rangle\tau^3+\cdots\}, \quad (6.86)$$

where $\mathcal{L}(\tau)$ is the lowest moment associated to the two-point vector correlator $\Pi(q^2)$:

$$\mathcal{L}(\tau) = \int_0^\infty dt\, e^{-t\tau}\frac{1}{\pi}\mathrm{Im}\Pi(t), \quad (6.87)$$

$$C_4\langle O_4\rangle = \frac{2}{3}\langle\alpha_s G^2\rangle, \quad C_6\langle O_6\rangle = \rho\frac{448}{27}\pi^3\alpha_s\langle\bar{u}u\rangle^2, \quad (6.88)$$

where ρ indicates the eventual deviation from the factorization hypothesis.

We obtain using a two-parameter fit [115]:

$$C_4\langle O_4\rangle = -(6.9\pm 0.9)\times 10^{-2}\,\mathrm{GeV}^4, \;\; C_6\langle O_6\rangle = (0.18\pm 0.02)\,\mathrm{GeV}^6. \quad (6.89)$$

We analyze the sensitivity of the previous estimate by using as input the value of $C_4\langle O_4\rangle$ obtained previously and by adding the contribution of the dimension 8 condensates in the fit. We obtain:

$$C_6\langle O_6\rangle = (0.18\pm 0.02)\,\mathrm{GeV}^6, \quad C_8\langle O_8\rangle = -(1.9\pm 1.6)\times\,\mathrm{GeV}^8, \quad (6.90)$$

which shows that, in the region below 1 GeV^{-2}, the OPE converges.

We study the effect of the data cut on the analysis. Our final result is:

$$C_4\langle O_4\rangle = -(6.9\pm 1.7)\times 10^{-2}\,\mathrm{GeV}^4, \quad C_6\langle O_6\rangle = (0.18\pm 0.06)\,\mathrm{GeV}^6. \quad (6.91)$$

Then, we deduce from the analysis:

$$\langle \alpha_s G^2 \rangle = (3.9 \pm 1.0) \times 10^{-2} \text{GeV}^4, \tag{6.92}$$

which corresponds to the standard SVZ-value, while the four-quark condensate is about a factor 2 larger than the four-quark factorization estimate. This result goes in the same direction as in Ref. [113] (previous section) where the factorization assumption is inconsistent with the renormalization group.

6.7 Light Mesons from QCD Spectral Sum Rules

6.7.1 SU(2) breaking of $\langle \bar{d}d \rangle / \langle \bar{u}u \rangle$ and $\omega\rho\pi$ coupling

This work is the beginning of a large international collaboration between CERN, Trieste and Barcelona [116].

We have extended the analysis in Ref. [76] of $\psi_5(0)^u_s$ in Eq. (4.21) to the $\bar{u}d$ channel. We have worked with the combination in Eq. (5.51) in the scalar channel. We have used the value of the δ coupling from [85]. We have obtained:

$$1 - \langle \bar{d}d \rangle / \langle \bar{u}u \rangle = 0.022 - .027, \tag{6.93}$$

indicating an isospin violation of about 2% compatible with the one of SVZ from $\omega - \rho$ mixing [117]. We have applied this result to calculate the $\omega\rho\pi$ coupling using vertex sum rule and obtained:

$$\frac{|g_{\omega\rho^0\pi^0}|}{|g_{\omega\rho^+\pi^-}|} \simeq \frac{m^2_{\pi^+}}{m^2_{\pi^0}} \left\{ 1 - \frac{\langle \bar{u}u - \bar{d}d \rangle}{3\langle \bar{u}u + \bar{d}d \rangle} - \frac{I_{xy}}{24\pi^2} \frac{\hat{m}_u Q^2}{\langle \bar{u}u \rangle} [\log(Q/\Lambda)]^{8/9} \right\}$$
$$\simeq 1.07, \tag{6.94}$$

where $I_{xy} \simeq 2.34$ is the Feynman integral from the perturbative triangle contribution. This value is in reasonable agreement with the one from the observed ratio $\Gamma(\rho^- \to \pi^- \gamma)$ and $\Gamma(\omega^- \to \pi^0 \gamma)$:

$$\frac{|g_{\omega\rho^0\pi^0}|}{|g_{\omega\rho^+\pi^-}|} \simeq 0.97 \frac{\gamma_\rho}{\gamma_\omega} \left[\frac{\Gamma(\omega^- \to \pi^0 \gamma)}{\Gamma(\rho^- \to \pi^- \gamma)} \right]^{1/2} \simeq (1.02 \pm 0.08), \tag{6.95}$$

where the experimental errors come mainly from the ρ width.

6.7.2 Gell-Mann Okubo like-mass formulae

We have analyzed the perturbative and non-perturbative corrections and the QCD continuum effects to the meson mass squared differences à la Gell-Mann Okubo (GMO) [118] using the ratio of LSR:

$$\mathcal{R}_H(\tau) \simeq -\frac{d}{d\tau}\log \int_0^\infty dt\, e^{-t\tau}\frac{1}{\pi}\mathrm{Im}\Pi_H(t), \tag{6.96}$$

where $\Pi_H(q^2)$ is a generic notation for a two-point function built from the T-product of two hadronic currents. One can remind that at the sum rule optimization scale τ_0, one expects:

$$\mathcal{R}^H(\tau) \simeq M_H^2. \tag{6.97}$$

• Vector mesons

Parametrizing the spectral function by the minimal duality ansatz, one has, for example, for the ρ meson:

$$\frac{1}{\pi}\mathrm{Im}\Pi_H(t) \simeq \frac{M_\rho^2}{2\gamma_\rho^2} + \theta(t-t_c)\text{"QCD continuum"}. \tag{6.98}$$

We fix t_c from the asymptotic coincidence of the two sides of $\mathcal{R}_H$ or from FESR:

$$t_c \simeq 2\pi^2\frac{M_\rho^2}{\gamma_\rho^2} \simeq (1.4\,\mathrm{GeV})^2. \tag{6.99}$$

Fixing $\tau_0 \simeq 1/M_\rho^2$, we obtain the two sides of the GMO mass formula:

$$\mathcal{R}_\phi(\tau_0) - \mathcal{R}_\rho(\tau_0) \simeq 6\bar{m}_s^2\left\{1 + (2\bar{m}_s^2)\,\tau\,\log\,(m_s^2\tau_0)\right\} - \frac{16\pi^2}{3}m_s\langle\bar{s}s\rangle\tau_0 + B_\phi - B_\rho, \tag{6.100}$$

with:

$$B_V = \frac{881}{27}\pi^3\alpha_s\langle\bar{q}q\rangle^2. \tag{6.101}$$

For $\tau \simeq 1$ GeV^{-2}, we have:

$$\begin{aligned} M_\phi^2 - M_\rho^2 &\simeq (64 + 78 - 0.3)\times 10^{-3}\ \mathrm{GeV}^2 \\ &\Longrightarrow M_\phi \simeq 0.86\ \mathrm{GeV}\ \ [\exp\ 1.02\ \mathrm{GeV}], \end{aligned} \tag{6.102}$$

where the QCD corrections come, respectively, from PT, $\langle \bar{s}s \rangle$ and four-quark condensate. We have used $\bar{m}_s(1\text{ GeV}) \simeq 120$ MeV. We have also used a violation of kaon PCAC by 40% and $\langle \bar{s}s \rangle / \langle \bar{d}d \rangle \simeq 0.6$ which gives $m_s \langle \bar{s}s \rangle \simeq 0.5 M_K^2 f_K^2$. For the four-quark condensate, we take a value about three times the vacuum saturation estimate [18–20]. The QCD continuum correction contributed by a factor 1.1 in the LHS.

A similar relation is obtained for the K^* meson where the PT and NP QCD corrections are half and leads to:

$$M_K^* \simeq 850 \text{ MeV} \quad [\text{exp } 892 \text{ MeV}]. \tag{6.103}$$

The underestimate of the ϕ meson mass can be due to the assumption that the ϕ and ρ sum rules optimize at the same value of τ which can be a bad approximation due to their relative large mass splittings.

- **Scalar mesons**

We have derived similar GMO mass formulae for scalar mesons $\delta, \kappa, S_8, S_1, S_X, S_Y$ associated, respectively, to the currents:

$$J_\delta = (m_u - m_d)\, i : \bar{u}d :, \quad J_\kappa = (m_u - m_s)\, i : \bar{u}s :,$$
$$J_8 = \frac{i}{\sqrt{6}} : \bar{u}u + \bar{d}d - 2\bar{s}s :, \quad J_1 = \frac{i}{\sqrt{3}} : \bar{u}u + \bar{d}d + \bar{s}s :,$$
$$J_X = i : \bar{s}s :, \quad J_Y = \frac{i}{\sqrt{2}} : \bar{u}u + \bar{d}d. \tag{6.104}$$

We obtain the following sum rules:

$$M_\kappa^2 \simeq M_\delta^2 + 2\bar{m}_s^2 - \frac{8\pi^2}{3} m_s \langle \bar{s}s \rangle \tau_0 \left(1 + 2\frac{\langle \bar{u}u \rangle}{\langle \bar{s}s \rangle}\right) \Longrightarrow M_\kappa \simeq 1.1\,\text{GeV},$$
$$M_1^2 \simeq M_\kappa^2 - \frac{8\pi^2}{3} m_s \langle \bar{s}s \rangle \tau_0 \left(1 - 2\frac{\langle \bar{u}u \rangle}{\langle \bar{s}s \rangle}\right) \Longrightarrow M_1 \simeq 1.06\,\text{GeV},$$
$$M_8^2 \simeq 2M_\kappa^2 - M_\delta^2 - \frac{16\pi^2}{3} m_s \langle \bar{s}s \rangle \tau_0 \left(1 - 2\frac{\langle \bar{u}u \rangle}{\langle \bar{s}s \rangle}\right) \Longrightarrow M_8 \simeq 1.13\,\text{GeV}. \tag{6.105}$$

For $M_{1,8}$, we have eliminated the quark mass correction by appropriately taking the combination with the M_κ sum rule. We note that the κ mass is lower than the one 1.43 GeV given by PDG [33]. However, one may better compare our prediction with the effective mass $M_\kappa^{eff} \simeq 1.2$ GeV obtained from the analysis of the $K\pi$ phase shift data [87].

6.8 Vertex Sum Rules

Here, we continue our former works using QCD spectral sum rules à la SVZ discussed in the previous chapters. In Refs. [119, 121], we considered the three-point function:

$$T(p,q) = \int d^4x\, d^4y\, e^{i(px+qy)} \langle 0|\mathcal{T} J_1(x) J_2(0) J_3(y)|0\rangle, \tag{6.106}$$

built from three generic hadronic currents J_i. From Ref. [122], it can have the spectral representation:

$$T(p^2, q^2, (p+q)^2) = \int dt_1\, dt_2\, dt_3 \frac{\rho(t_1, t_2, t_3)}{t_1 - p^2)(t_2 - q^2)(t_3 - (p+q)^2}. \tag{6.107}$$

Contrary to what is often used in the current literature, we evaluate the vertex at the symmetric point:

$$p^2 = q^2 = (p+q)^2 \equiv -Q^2 \gg \Lambda^2, \tag{6.108}$$

where the vertex has a good analytic property. Using a Feynman parametrization, it can be written into a single denominator:

$$T(Q^2) = 2\int_0^1 x\, dx \int_0^1 dy \int_0^\infty dt_1\, dt_2\, dt_3 \times \frac{\rho(t_1, t_2, t_3)}{[Q^2 + (t_1 - t_2)xy + (t_2 - t_3)x + t_3]^3}, \tag{6.109}$$

with the Laplace transform:

$$\mathcal{L}(\tau) = \tau^3 \int_0^1 x\, dx \int_0^1 dy \int_0^\infty dt_1\, dt_2\, dt_3\, e^{-(t_1 - t_2)xy + (t_2 - t_3)x + t_3)\tau} \times \rho(t_1, t_2, t_3). \tag{6.110}$$

With this expression, one can express a duality between the QCD and hadronic sides of the vertex, as illustrated in Fig. 6.7.

Figure 6.7. Duality between the QCD and hadronic sides of the vertex.

6.8.1 Three-meson couplings

In Ref. [119], we considered the $\omega\rho\pi$, $\phi K^* K$, $\delta K K$ and $\rho\pi\pi$ couplings. In the example of the $\omega\rho\pi$ coupling, the currents are:

$$J_\mu^{(\rho)} = :\bar{\psi}_u \gamma_\mu \psi_d:, \quad J_\nu^{(\omega)} = \frac{1}{6}\left(:\bar{\psi}_u \gamma_\nu \psi_u + \bar{\psi}_d \gamma_\nu \psi_d:\right),$$
$$J^{(\pi^+)} = (m_u + m_d):\bar{\psi}_d i\gamma_5 \psi_u. \tag{6.111}$$

The Lorentz structure of the vertex gives:

$$T_{\mu\nu}(p.q) = i\epsilon_{\mu\nu\rho\sigma} p^\rho q^\sigma A(p,q). \tag{6.112}$$

The QCD evaluation of the vertex shows that to leading order in the light quark mass, its contribution is given by the quark condensate:

$$A(p,q)|_{QCD} = -\frac{1}{Q^4}\frac{1}{6}(m_u + m_d)\langle 0|2\bar{\psi}_u\psi_u + \bar{\psi}_d\psi_d|0\rangle$$
$$+ (u \longrightarrow d) + \mathcal{O}\left(\frac{m^2}{Q^2}\right), \tag{6.113}$$

while saturating the vertex spectral function by the ρ, ω and π mesons, one obtains:

$$A(p,q)|_{exp} = \frac{\sqrt{2} m_\pi^2 f_\pi}{Q^2} \frac{\sqrt{2} M_\rho^2}{2\gamma_\rho} \frac{M_\omega^2}{2\gamma_\omega} \frac{g_{\omega\rho\pi}}{(Q^2 + M_\rho^2)(Q^2 + M_\omega^2)}. \tag{6.114}$$

After taking the Laplace transform of the two sides of $A(Q^2)$, the relation between f_π and γ_ρ, pion PCAC:

$$(m_u + m_d)\langle 0|\bar{\psi}_u\psi_u + \bar{\psi}_d\psi_d|0\rangle = -2m_\pi^2 f_\pi^2, \tag{6.115}$$

and the first Weinberg sum rule,

$$f_\pi^2 = \left(\frac{M_\rho}{2\gamma_\rho}\right)^2 e^{-\tau M_\rho^2}, \tag{6.116}$$

one obtains:

$$g_{\omega\rho\pi} \approx \frac{3}{2f\pi} = 16\,\text{GeV}^{-1}. \tag{6.117}$$

Similar dominance of the quark condensate is obtained in the other channels. One obtains:

$$g_{\phi K^* K} \approx 0.7\sqrt{2} g_{\omega\rho\pi}, \quad g_{\delta K^+ K^0} \approx \frac{8\pi^2}{3\sqrt{2}} \frac{m_s\langle 0|\bar{\psi}_s\psi_s - 2\bar{\psi}_d\psi_d|0\rangle}{m_K^2 f_K}$$
$$\approx \frac{8\pi^2}{3\sqrt{2}}\left(\frac{m_s}{m_d}\right)\frac{m_\pi^2 f_\pi^2}{m_K i^2 f_K^2}\left(1-\frac{r}{2}\right) f_K \tag{6.118}$$

if one uses the pion PCAC relation in Eq. (6.115) and $r = \langle 0|\bar{\psi}_s\psi_s|0\rangle/\langle 0|\bar{\psi}_d\psi_d|0\rangle$. Using $m_s/m_d \simeq 25$, $f_K/f_\pi = 1.2$ and $r \simeq 0.6$, one obtains:

$$g_{\delta\bar{K}K} \simeq 2\,\text{GeV} \;:\quad 2.4\,\text{GeV} \;\; [\text{exp}]. \tag{6.119}$$

We also obtain:

$$g_{\kappa K\pi} \approx \frac{8\pi^2}{3\sqrt{2}} \frac{m_s\langle 0|\bar{\psi}_s\psi_s|0\rangle}{m_K^2 f_K}, \quad \frac{g_{\rho\pi\pi}}{2\gamma_\rho} \approx \frac{3}{8\pi^2}\frac{(\bar{m}_u+\bar{m}_d)^2}{f_\pi^2}$$
$$\times \frac{e^{M_\rho^2\tau}}{\tau^3 m_\pi^4 M_\rho^2} \approx 1.5, \tag{6.120}$$

to be compared with the universality relation $g_{\rho\pi\pi} = 2\gamma_\rho$.

6.8.2 πNN coupling

Here, we consider the vertex:

$$T(p,q) = \int d^4x\, d^4y\, e^{i(px+qy)} \langle 0|\mathcal{T} O(x) J_P(0) \bar{O}(y)|0\rangle, \tag{6.121}$$

where $O(x) = O_1(x) + t\, O_2(x)$ is the operator describing the nucleon and t is an arbitrary mixing parameter. The hadronic currents are:

$$\begin{aligned} O_1(x) &= \epsilon^{abc}[u_a^T(x) C d_b(x)]\gamma_5 u_c(x), \\ O_2(x) &= \epsilon^{abc}[u_a^T(x) C\gamma_5 d_b(x)] u_c(x), \end{aligned} \tag{6.122}$$

where C is the charge conjugation matrix. $J_P(x)$ is the current describing the π and η mesons:

$$\begin{aligned} J_\pi &= m : \bar{u}(i\gamma_5)u - \bar{d}(i\gamma_5)d :, \\ J_\eta &= m : \bar{u}(i\gamma_5)u + \bar{d}(i\gamma_5)d : -2m_s : \bar{s}(i\gamma_5)s :, \end{aligned} \tag{6.123}$$

The couplings of the resonances to the currents are normalized as:

$$\langle 0|O(x)|N\rangle = \lambda_N u_N(x), \quad \langle 0|J_P(x)|N\rangle = m_P^2 f_P, \tag{6.124}$$

with $f_\eta = \sqrt{3} f_\pi$. We evaluate the vertex at the symmetric point. Like in the case of three-meson coupling, the QCD contribution is dominated by the $\langle\bar{\psi}\psi\rangle$ condensate. We use the constraint on λ^2 from the nucleon two-point function [6, 120], one obtains:

$$g_{\pi\bar{N}N} \approx \frac{f_\pi}{M_N}(16\pi^2)\frac{(8+8t+2t^2)}{(5+2t+5t^2)}. \tag{6.125}$$

Stability in t is obtained for $t \geq 10$, leading to[2]:

$$g_{\pi\bar{N}N} \approx 10 \qquad [\text{exp } 13]. \tag{6.126}$$

However, predictions of some other couplings ($\eta\bar{N}N, \rho\bar{N}N, \omega\bar{N}N$) given in Ref. [121] are not accurate.

[2]Note that the data is reproduced for $t = -0.51$ or $t = 3.38$ [121]. These values differ from the one $t = -0.2$ of the nucleon two-point function [120].

6.9 Scalar Quarkonia

To complete the study of the scalar channel, we have also analyzed the properties of the $\bar{q}q'$ light scalar mesons.

6.9.1 $I = 1, 1/2$ $a_0(980)$ and K^{*0} states

• Light quark mass differences

They have been studied previously in Refs. [6, 123] and in previous sections in connection with the light quark mass difference, where they have been assumed to be related to the divergence of the vector currents:

$$\begin{aligned} a_0 &\rightarrow \partial_\mu V^\mu_{\bar{u}d} = (m_u - m_d)\ :\bar{u}(i)d:, \\ K^{*0} &\rightarrow \partial_\mu V^\mu_{\bar{u}s} = (m_u - m_s)\ :\bar{u}(i)d:. \end{aligned} \tag{6.127}$$

• Masses and widths

The a_0 and K^*0 hadronic couplings have been obtained using either a vertex sum rule [121, 124] or/and $SU(3)$ symmetry rotation [125]. The leading order vertex sum rule results are

$$g_{a_0K^+K^-} \simeq 3 \text{ GeV}, \quad \frac{g_{a_0K^+K^-}}{g_{K^{*0}K^+\pi^-}} \simeq 1.17, \tag{6.128}$$

while the $SU(3)$ relation gives:

$$g_{a_0\eta\pi} \simeq \sqrt{\frac{2}{3}} g_{a_0K^+K^-}, \tag{6.129}$$

which leads to

$$\Gamma\,(a_0 \rightarrow \eta\pi) \simeq 84 \text{ MeV}, \tag{6.130}$$

in agreement with the range of data from 50 to 100 MeV given by PDG [33]. Using the previous value of the K^*0 coupling, one can deduce:

$$\Gamma(K^{*0} \rightarrow K\pi) \simeq \frac{3}{2}\Gamma(K^{*0} \rightarrow K^+\pi^-) \simeq 104 \text{ MeV}, \tag{6.131}$$

which is about a factor 4 smaller than the present data [33], but is a typical value for the width of a $\bar{q}q$ state.

6.9.2 Isoscalar $f_0(980)$ state

• Ground states

The ones of the $I = 0$ isoscalar channel $S_2 \equiv (\bar{u}u + \bar{d}d)$ and $S_3 \equiv \bar{s}s$ have been studied in Ref. [126] where one found the masses and dominant hadronic widths which come from the analysis of the vertex sum rules shown in Fig. 6.8:

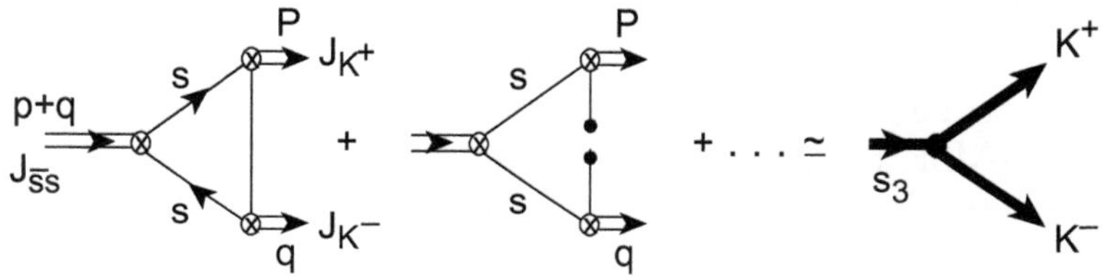

Figure 6.8. Duality between the QCD vertex (LHS) and $S_3K^+K^-$ coupling (RHS).

$$M_{S_2} \simeq 1\,\text{GeV}, \quad \Gamma[S_2 \to \pi^+\pi^-] \simeq 110\,\text{MeV},$$
$$M_{S_3} \simeq (1.47 \pm 0.04)\,\text{GeV}, \quad \Gamma[S_3 \to K^+K^-] \simeq (73 \pm 27)\,\text{MeV}. \tag{6.132}$$

• Radial excitations

◇ Their corresponding radial excitations are

$$M_{S'_2} \simeq 1.3\,\text{GeV}, \quad \Gamma[S'_2 \to \pi^+\pi^-] \simeq (300 \pm 150)\,\text{MeV},$$
$$M_{S'_3} \approx 1.7\,\text{GeV}, \quad \Gamma[S'_3 \to K^+K^-] \simeq (112 \pm 50)\,\text{MeV}, \tag{6.133}$$

where one can remark that S'_2 is degenerated with $\pi'(1.3)$ as expected from SU(2) symmetry.

◇ The $\gamma\gamma$ widths of these quarkonia are in units of keV [126]:

$$S_2: \ 0.67, \quad S_3: \ (0.4 \pm 0.04), \quad S'_2: \ (4 \pm 2), \quad S'_3: \ (1.1 \pm 0.5). \tag{6.134}$$

We shall analyze later on the mixing of these isoscalar scalar states with gluonia/glueballs.

6.10 Reviews on light quark systems until 2004

Previous works on light meson systems have been reviewed, improved and updated in Refs. [6, 7, 41]. These references complement this short overview. These books also aim to introduce the readers to the method of QCD spectral sum rules (QSSR).

6.11 Gluonia/Glueballs Prior to 1989

During my one year stay in Annecy, I met Mike Teper who worked on the Yang–Mills gauge theory using lattice QCD. He was interested in the study of scalar glueball. In parallel, I tried to tackle the same problem with the QSSR approach by analyzing Refs. [127, 128] and by completing my previous work on the $U(1)_A$ meson [82].

6.11.1 Digluonia currents

The basic inputs of the analysis are the digluonia currents:

$$
\begin{aligned}
J_s &= \beta(\alpha_s) G^{(a)}_{\alpha\beta} G^{\alpha\beta}_{(a)}, \\
Q(x) &= \left(\frac{\alpha_s}{8\pi}\right) \operatorname{tr} G^{(a)}_{\alpha\beta} \tilde{G}^{\alpha\beta}_{(a)}, \\
\theta^{\mu\nu}_g &= -G^{\mu}_{(a),\alpha} G^{(a),\alpha\nu} + \frac{1}{4} g^{\mu\nu} G^{(a)}_{\alpha\beta} G^{\alpha\beta}_{(a)}, \qquad (6.135)
\end{aligned}
$$

with:

$$
\tilde{G}^{(a)}_{\mu\nu} \equiv \frac{1}{2}\epsilon_{\mu\nu\alpha\beta} G^{(a),\alpha\beta}, \quad G^{(a)}_{\mu\nu} \equiv \partial_\mu A^{(a)}_\nu - \partial_\nu A^{(a)}_\mu . \qquad (6.136)
$$

These currents have, respectively, the quantum numbers of a scalar 0^{++}, pseudoscalar 0^{-+} and tensor 2^{++} gluonia. (a) is the colour index and the greek letters are tensor indices; $A^{(a)}_\mu$ is the gluon field and $G^{(a)}_{\mu\nu}$ the gluon field strengths; $Q(x)$ is the famous anomaly of the $U(1)_A$ channel [83]. $J_s(x)$ is the gluon component of the trace of the energy–momentum tensor:

$$
\theta^\mu_\mu = \beta(\alpha_s) G^{(a)}_{\alpha\beta} G^{\alpha\beta}_{(a)} + (1 + \gamma_m(\alpha_s)) \sum_{u,d,s} m_i \bar{\psi}_i \psi_i, \qquad (6.137)
$$

where:

$$\beta(\alpha_s) - \beta_1(\alpha_s/\pi) + \cdots, \qquad \gamma_m(\alpha_s) = (\alpha_s/\pi) + \cdots, \qquad (6.138)$$

are, respectively, the β-function $(-\beta_1 = -(11/2 - n_f/3))$ and the quark anomalous dimension $(\gamma_1 = 2)$. The corresponding QCD two-point correlator is represented in Fig. 6.9 within the SVZ–expansion up to dimension 8.

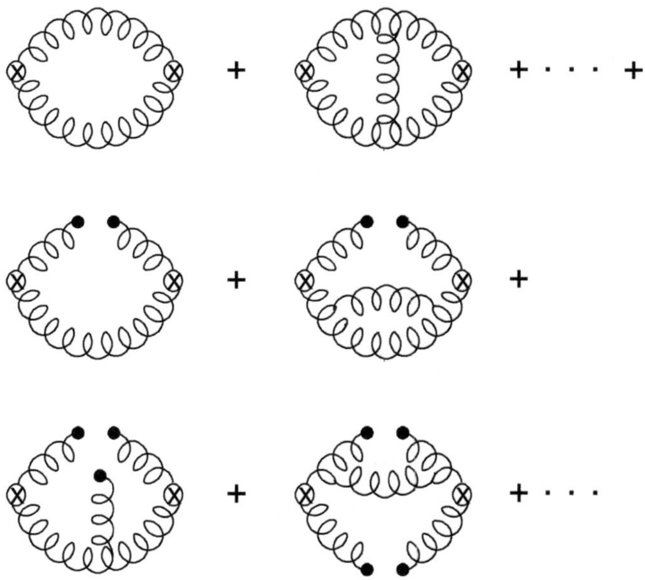

Figure 6.9. Feynman diagrams representing the gluonic two-point correlator within the SVZ-expansion where the contributions of the gluon condensates up to dimension 8 are shown.

6.11.2 Gluonia spectra from unsubtracted sum rule (USR)

- **The unsubtracted sum rule (USR)**

We used [82, 129] the unsubtracted Laplace sum rules (USR) and their ratios to their corresponding two-point correlators $\psi_G(q^2)$:

$$\mathcal{L}_n(\tau) = \int_0^\infty dt\, e^{-t\tau}\, t^n\, \frac{1}{\pi}\text{Im}\psi_G(t), \quad \mathcal{R}_{n,n-1}(\tau) = -\frac{d}{d\tau}\mathcal{L}_n \simeq M_G^2,$$
$$n = 0, 1, 2, \ldots, \tag{6.139}$$

to extract the couplings and masses of the gluonia ground states, where the couplings are normalized as $f_\pi = 92.3$ MeV:

$$\langle 0|J_G|G\rangle = \sqrt{2}M_G^2 f_G. \tag{6.140}$$

We consider the OPE within the SVZ expansion and without instanton effects for the QCD expressions up to dimension-six gluon condensate contributions. We use the values:

$$\langle \alpha_s G^2\rangle = (0.066 \pm 0.033)\ \text{GeV}^4,$$
$$g^3 f_a bc \langle G^a G^b G^c\rangle = (1.1 \pm 0.2)\,\text{GeV}^2 \langle \alpha_s G^2\rangle, \tag{6.141}$$

where we fix $\langle \alpha_s G^2\rangle$ in a generous range, including the SVZ estimate [3] and some others [70]. The $\langle G^3\rangle$ condensate comes from the dilute gas instanton model [3, 127, 128, 130].

• Gluonia masses and couplings

This work is published in Ref. [129], where I have especially benefitted from the assistance of John Bell and Gabriele Veneziano, while Patrick Aurenche provided me with the fitting program FUMILI for the numerical analysis. The resulting gluonia masses from $\mathcal{R}_{10}$ are:

$$M_{2^{++}} \simeq M_{0^{-+}} = (1.64 - 1.82)\,\text{GeV}, \quad M_{0^{++}} \simeq (0.95 - 1.1)\,\text{GeV}, \tag{6.142}$$

and couplings:

$$41 \le f_{2^{++}} \le 74\,\text{MeV}, \quad 23 \le f_{0^{-+}} \le 34\,\text{MeV},$$
$$60 \le f_{2^{++}} \le 100\,\text{MeV}. \tag{6.143}$$

6.11.3 Topological susceptibility and its slope from LSR

We have also used the subtracted sum rule in the $U(1)_A$ channel for extracting the topological susceptibility $\psi_P(0)$ [82] (see Section 5.2.4) and in Ref. [129] in a pure Yang–Mills, where the subtraction constant and its slope are not zero. The LSR result is about a factor 2 lower than the one from the $U(1)_A$ current algebra [84]. Here, we extract its slope from:

$$\psi'_P(0)|_{\text{No quarks}} = \int_0^\infty \frac{dt}{t^2}\, e^{-t\tau}\, \frac{1}{\pi}\text{Im}\psi_P(t) - \mathcal{L}_{-2}|_{\text{QCD}} + \tau\psi_P(0). \tag{6.144}$$

In Ref. [129], it was convenient to express the subtraction constant with the combination of $\mathcal{L}_0$ and $\mathcal{L}_{-1}$ for $\psi_P(0)$ and $\mathcal{L}_0$, $\mathcal{L}_{-1}$ and $\mathcal{L}_{-2}$ for $\psi'_P(0)$ in order to reduce the effects of the QCD continuum and PT corrections. One obtains:

$$\psi'_P(0)|_{\text{No quarks}} \simeq -[0.05\,\text{GeV}]^2. \tag{6.145}$$

6.11.4 Conformal charge and its slope from LSR

The conformal charge is a scholarly name for the subtraction constant of the scalar gluonium two-point function. We have estimated this quantity $\psi_S(0)$ and its slope $\psi'_S(0)$ in the same way as done for the pseudoscalar channel in Ref. [129] from the subtracted sum rules. We found for pure Yang–Mills:

$$\psi_S(0)|_{\text{No quarks}} = (2.05 \pm 0.44)\,\text{GeV}^4, \quad \psi'_S(0)|_{\text{No quarks}} \simeq -0.3\,\text{GeV}^2. \tag{6.146}$$

The value of the conformal charge is slightly higher than the one from some LET [127, 128]:

$$\psi_S(0) \simeq -\frac{16}{\pi}\beta_1\langle\alpha_s G^2\rangle = 1.46\,\text{GeV}^4, \tag{6.147}$$

where $\beta_1 = -11/2$ (pure Yang–Mills without quarks) and $\langle\alpha_s G^2\rangle = 0.07\,\text{GeV}^4$.

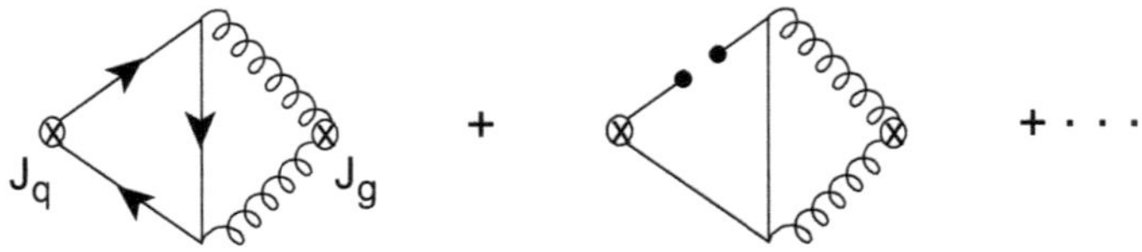

Figure 6.10. Off-diagonal two-point correlator controlling the meson–gluonium mass mixing.

6.11.5 Meson–digluonium mixing

• Off-diagonal (pseudo)scalar two-point correlator

In Ref. [131], we have estimated the meson–gluonium mixing using LSR from the off-diagonal two-point correlator shown in Fig. 6.10.

We found that, in both scalar and pseudoscalar channels, the PT contribution to the correlator behaves like $m_s^2\,(\alpha_s/\pi)^2\,q^2\log^{1,2}(-q^2/\nu^2)$ while the next dimension 4 condensates contribute as

$$(\alpha_s/\pi)^2\,m_s\langle\bar\psi\psi\rangle\log(-q^2/\nu^2)\quad\text{and}$$
$$(\alpha_s/\pi)^2\,(m_s^2/q^2)\log(-q^2/\nu^2)\langle\alpha_s G^2\rangle,$$

which vanishes in the chiral limit $m_s^2=0$ and starts to order α_s^2. An estimate of the mixing angle in the pseudoscalar channel gives [131]:

$$\theta=(1.6-16)^0, \tag{6.148}$$

indicating that the mixing angle is small such that the observed gluonium states after mixing are expected to have masses around the pure Yang–Mills ones obtained in the previous section.

• Off-diagonal 2^{++} tensor two-point correlator

In Ref. [134], we study the off-diagonal meson–gluonium two-point correlator for the 2^{++} tensor mesons by including the contribution of the $\langle\alpha_s G^2\rangle$ condensate. This contribution is α_s suppressed compared to the diagonal correlator and corresponds to a mixing angle:

$$\theta=-10^0. \tag{6.149}$$

Within a two-component mixing formalism and assuming a SU(3) symmetry, one can express the f_2 coupling to Goldstone pairs as:

$$
\begin{aligned}
g^2_{f_2\pi\pi} &= \frac{9}{10}\left[1 + 2\sin\theta\, r + \mathcal{O}(\sin^2\theta)\right],\\
g^2_{f_2\bar{K}K} &= \frac{3}{10}[1 + 4\sin\theta\, r + \mathcal{O}(\sin^2\theta)],\\
g^2_{f_2\bar{K}K} &= \frac{3}{5}[1 + 2\sqrt{2}\sin\theta\, r + \mathcal{O}(\sin^2\theta)],
\end{aligned}
\tag{6.150}
$$

where $r \equiv g_{G\pi\pi}/g_{f_2\pi\pi}$ is the ratio of the gluonium over the f_2 couplings, while g is the unmixed coupling which one can deduce from the a_2 and $K_2^*(1430)$ one. The couplings are normalized as:

$$
\Gamma(T \to PP) = \frac{g^2_{TPP}}{30\pi\, M_T^2}|p_P|^5, \tag{6.151}
$$

where p_P is the momentum. Using the data for $a_2 \to \bar{K}K$ and $K_2^* \to K\pi$, one deduces [125]:

$$
g^2 = (205 \pm 10)\ \text{GeV}^{-2}, \tag{6.152}
$$

while the data on $f_2 \to \pi\pi, \bar{K}K$ give:

$$
r \simeq (0.68 \pm 0.22). \tag{6.153}
$$

Considering this value of r as the maximal effects of the gluonia on the $f_2 - G$ systems, we deduce, for $M_G = 1.7$ GeV:

$$
\Gamma(G \to \pi\pi + \bar{K}K + \eta\eta) \leq (200 \pm 60)\,\text{MeV}. \tag{6.154}
$$

We also deduce the $\gamma\gamma$ width from an effective Lagrangian [127, 135]:

$$
\Gamma(G \to \gamma\gamma) \simeq \left(\frac{M_G}{M_{0^{++}}}\right)^3 \frac{4}{15}\Gamma(0^{++} \to \gamma\gamma) \simeq 0.26\,\text{keV}, \tag{6.155}
$$

taking $M_{0^{++}} \simeq 1$ GeV.

6.12 Scalar Gluonia Spectra Until 1989

6.12.1 Scalar digluonia spectra

• Subtracted sum rules (SSR)

Instead of the USR used in Section 6.11.2, one can also study the spectra using a subtracted sum rule (SSR) similar to the case of pseudoscalar channel in Eq. (5.2.4), where the subtraction point can be estimated using an LET (see (6.147)). In this case, using the standard minimal duality ansatz, "one resonance ⊕ QCD continuum", one obtains a mass.

We clarify the origin of the discrepancy of the results from the subtracted and unsubtracted sum rules in Ref. [132], where a parametrization of the spectral function with two resonances is necessary for resolving this puzzle.

The presence of $\psi_S(0)$ pushes the value of the sum rule scale $\tau = 1/M^2$ to lower value which enhances the contribution of the high mass resonance, while the unsubtracted sum rule (USR) optimizes at larger value of τ where the contribution of the low mass resonance dominates the sum rule. Therefore, using as inputs the mass of σ and $G(1.6)$ [132]:

$$M_\sigma \approx 1 \text{ GeV}, \quad M_G \simeq (1.5 \pm 0.2)\,\text{GeV}, \tag{6.156}$$

we need the couplings:

$$f_\sigma \approx (760 \sim 931)\,\text{MeV}, \quad f_G \simeq (544 \sim 656)\,\text{MeV} \tag{6.157}$$

to get the consistency of the two sum rules. An upper bound on the decay constant of $f_0(1.3) \equiv \sigma'$ has also been derived [132]:

$$f_{f_0} \leq (139 \sim 224)\,\text{MeV}. \tag{6.158}$$

• Decay to pairs of pseudoscalar bosons

We analyze their hadronic widths by using a low-energy theorem for the vertex function:

$$V(q^2) = \langle \pi_1 | \theta_\mu^\mu | \pi_2 \rangle : \quad q = p_1 - p_2, \tag{6.159}$$

with the low-energy theorem (LET):

$$V(0) = m_\pi^2 \quad \text{and} \quad V'(0) = 1. \tag{6.160}$$

From its dispersion relation:

$$V(q^2) = \int_0^\infty \frac{dt}{t - q^2 - i\epsilon} \frac{1}{\pi} \mathrm{Im} V(t), \tag{6.161}$$

then, we derive two sum rules from $V(0) = 0 = V'(0)$ in the chiral limit to get the coupling of σ_B and σ'_B to $\pi\pi$. Similar sum rules from current algebra constraints can be deduced for $G \to \eta\eta, \eta'\eta$. We obtain in units of MeV [132]:

$$\Gamma[\sigma_B \to \pi^+\pi^- + 2\pi^0] \simeq (410 \pm 220), \quad \Gamma[G_1(1.55) \to \eta\eta'] \leq (44 \pm 8), \tag{6.162}$$

and:

$$\frac{\Gamma[G_1(1.55) \to \eta\eta]}{\Gamma[G_1(1.55) \to \eta\eta']} \simeq 0.26, \tag{6.163}$$

while the coupling of σ, σ' to $\eta\eta'$and $\eta\eta$ are negligible.

- **Decay to 4π**

We study the decays of $\sigma'(1.37)$ and $G(1.504)$ by assuming that they come for a pair of $\sigma(0.5)$ where each decays into 2π. Then, we extend the previous LET for the pseudoscalar to the case of scalar mesons:

$$\langle \sigma_B | \theta^\mu_\mu | \sigma_B \rangle = 2M^2_{\sigma_B}. \tag{6.164}$$

Using the observed width of σ', we deduce:

$$\Gamma(G \to 4\pi) \simeq (60 \sim 138)\,\mathrm{MeV}. \tag{6.165}$$

- **$\gamma\gamma$ and $J/\psi \to \gamma\sigma, G$ widths**

Their $\gamma\gamma$ couplings can be obtained from the Euler–Heisenberg effective Lagrangian box diagram (Fig. 6.11(a)) or/and triangle anomaly (Fig. 6.11(b)). One obtains in units of keV [126]:

$$\sigma_B: \quad (0.03 \sim 0.08), \quad \sigma'_B: \quad (0.01 \sim 0.03), \quad G: \quad (0.3 \sim 0.6). \tag{6.166}$$

The previous parameters should help to understand the complex spectra of the scalar mesons.

- One striking feature of the glueball nature of σ_B and σ'_B is the expected large decay of σ'_B to $(4\pi)_S$ through a pair of σ_B.

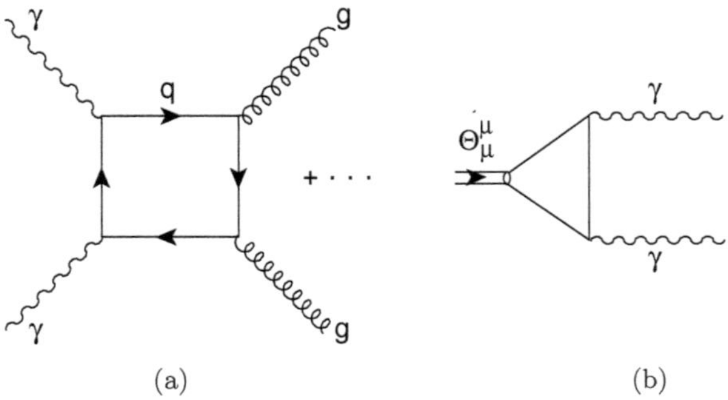

Figure 6.11. Vertex controlling the gluonium couplings to $\gamma\gamma(J/\psi)$: (a) box diagram, (b) triangle anomaly.

• Nature of the $a_0(980)$ and $f_0(975)$

In Ref. [125], we assume that the $a_0(980)$ is a $\bar{q}q$ state associated to the divergence of the vector current, while $f_0(975)$ result from a quarkonium–gluonium mixing, as discussed in Ref. [191]. In so doing, we estimate the $a_0 \to \gamma\gamma$ width using vertex sum rule and find:

$$\Gamma[a_0 \to \gamma\gamma] = (0.3 \sim 2.1)|\,\text{keV} \quad \text{Exp} : (0.09 - 0.44)\,\text{keV}. \tag{6.167}$$

A more refined analysis and data of the $\pi\pi$ and $\gamma\gamma$ widths of σ and $f_0(975)$ leads to a maximal meson–gluonium mixing, which improves the result in Ref. [191] (see Section 7.1.2).

6.12.2 Scalar trigluonium spectrum

To complete our analysis, we have also analyzed [133] the mass and coupling of the 0^{++} scalar trigluonium associated to the current:

$$J_{3g} = g^3\, f_{abc} G^{(a)} G^{(b)} G^{(c)}. \tag{6.168}$$

We work to leading order in α_s and include the contributions of the gluon condensates up to dimension 6. We extract the optimal result using stability with respect to the sum rule variable and continuum threshold t_c. We obtain:

$$M_{3g} = 3.1\,\text{GeV}, \quad f_{3g} = 62\,\text{MeV} \quad \text{for } \sqrt{t_c} \simeq 3.4\,\text{GeV}, \tag{6.169}$$

where the coupling is normalized as:

$$\langle 0|J_{3g}|G\rangle = \sqrt{2} M_{3g}^4 f_{3g}. \tag{6.170}$$

We have also studied the digluonium–trigluonium mixing from the off-diagonal two-point gluonium correlator:

$$\psi_{23} = i \int d^4x \, \langle 0|\mathcal{T} J_{2g}(x) J_{3g}(0)^\dagger |0\rangle, \tag{6.171}$$

where $J_{2g}(x) = \alpha_s G^2(x)$ is the digluonium current. We find a tiny mixing angle:

$$\theta = 0.2^0, \tag{6.172}$$

where we have used the digluonium mass $M_\sigma = 0.5$ GeV and coupling $f_\sigma = 100$ MeV where $M_G = 1.6$ GeV has been considered as the radial excitation of σ.

6.13 Gluonia/Glueballs from 1996 to 2021

6.13.1 Masses and couplings

In Ref. [126], we updated and completed our previous works on gluonia/glueball spectra discussed in Section 6.12 using new values of the QCD parameters:

$$\Lambda = (375 \pm 125)\,\text{MeV}, \qquad \langle \alpha_s G^2 \rangle = (0.07 \pm 0.01)\,\text{GeV}^4, \tag{6.173}$$

while we continued to use the estimate of the $g^3 f_{abc}\langle G^a G^b G^c\rangle = 1.5\,\text{GeV}^2\langle \alpha_s G^2\rangle$ from the instanton liquid model [3].

In addition to the ratio of USR moments $\mathcal{R}_{10}$ used in Ref. [129], we have added the analysis of the ratio of higher moments $\mathcal{R}_{21}$ which is more sensitive to higher mass gluonium.

The results of the analysis are summarized in Table 6.1, which we consider as a complement of the ones in Ref. [129]. $\mathcal{R}_{10}$ used in Ref. [129] gives the σ_B mass, while $\mathcal{R}_{21}$ used in Ref. [126] gives the $G(1.5)$ one.

6.13.2 Scalar gluonia

Then, we have also revised our estimate of the hadronic and $\gamma\gamma$ widths of the puzzling scalar gluonium channel (see Table 6.2).

Some attempts to interpret the observed scalar mesons as a result of a maximal meson–gluonium mixing like in Ref. [125] are discussed in Ref. [126].

Table 6.1. Unmixed digluonia masses and couplings from QSSR.

J^{PC}	Name	Mass [GeV] Estimate	Upper Bound	f_G [MeV]	$\sqrt{t_c}$ [GeV]
0^{++}	G	1.5 ± 0.2	2.16 ± 0.22	390 ± 145	$2.0 \sim 2.1$
	σ_B	1.00 [129]		1000	
	σ'_B	1.37 (input)		600	
	$3G$	3.1		62	
2^{++}	T	2.0 ± 0.1	2.7 ± 0.4	80 ± 14	2.2
0^{-+}	P	2.05 ± 0.19	2.34 ± 0.42	$8 \sim 17$	2.2
	E/ι	1.44 (input)		$7\ :\ J/\psi \to \gamma\iota$	

Table 6.2. Unmixed scalar gluonia decays. In our approach, the σ_B, σ'_B couplings to Goldstone pairs are universal modulo SU(3) breakings.

Name	Mass [GeV]	$\pi^+\pi^-$ [GeV]	$\eta\eta$ [MeV]	$\eta\eta'$ [MeV]	$(4\pi)_S$ [MeV]	$\gamma\gamma$ [keV]
σ_B	$0.75 - 1.0$	$0.2 - 0.5$	$SU(3)$			$0.2 - 0.3$
σ'_B	1.37 (input)	$0.5 - 1.3$	$SU(3)$		$43 - 316$	$0.7 - 1.0$ (exp)
G	1.5	≈ 0	$1.1 - 2.2$	$5 - 10$	$60 - 138$	1.0 ± 0.8

6.14 Light Hybrid Mesons

The two-point function associated to the hybrid currents:

$$\mathcal{O}_V(x) \equiv: g\bar{\psi}(x)\lambda_a\gamma_\nu\psi(x)G_a^{\mu\nu}(x),$$
$$\mathcal{O}_A(x) \equiv: g\bar{\psi}(x)\lambda_a\gamma_\nu\gamma_5\psi(x)G_a^{\mu\nu}(x), \tag{6.174}$$

is:

$$\Pi(q) \equiv i\int d^4x\, e^{iqx}\langle 0|\mathcal{T}\mathcal{O}^\mu_{V/A}(x)\left(\mathcal{O}^\nu_{V/A}\right)^\dagger(0)|0\rangle$$
$$= -(q^2g^{\mu\nu} - q^\mu q^\nu)\Pi^{(1)}(q^2) + q^\mu q^\nu\Pi^{(0)}(q^2), \tag{6.175}$$

where $\Pi^{(1)}(q^2)$ and $\Pi^{(0)}(q^2)$ have the quantum numbers of a spin 1^{-+} and spin 0^{--} hybrid mesons. ψ is the light quark field. Reference [196] has first calculated this quantity, but Ref. [197] found a mistake on the QCD expression. We have rechecked independently this expression [198] and found errors in both papers (missing diagrams, sign). These results affect the phenomenological analysis. Using the Laplace transform:

$$\mathcal{R}^{(1,0)} = -\frac{d}{d\tau}\log\int_0^\infty dt\, e^{-t\tau}\frac{1}{\pi}\mathrm{Im}\Pi^{(1,0)}(t) \tag{6.176}$$

and introducing the coupling to the current:

$$\frac{1}{\pi}\mathrm{Im}\Pi(t) \simeq 2f_H^2 M_H^6 \delta(t - M_H^2) + \theta(t - t_c)\,\mathrm{QCD\ continuum}, \tag{6.177}$$

a least square fit of the two sides of the sum rules with the two parameters (M_H^2, t_c) gives the masses:

$$M_{\tilde{\rho}} \simeq \sqrt{t_c} \simeq 1.7\ \mathrm{GeV}, \quad M_{\tilde{\eta}} \simeq \sqrt{t_c} \simeq 3.1\ \mathrm{GeV}, \tag{6.178}$$

and the couplings:

$$f_{\tilde{\rho}} \simeq 46.4\ \mathrm{MeV}, \quad M_{\tilde{\eta}} \simeq 23.4\ \mathrm{MeV}. \tag{6.179}$$

The values of the masses are much higher than the previous estimates. We shall come back to this point in Section 6.1.

6.15 Collaboration with G. Veneziano (1988–1998) and G. Shore (1993–1998)

(a) (b)

Figure 6.12. Professor G. Veneziano visiting Montpellier in 2012.

During my stay at CERN, I have benefitted from several discussions and collaborations with Professor G. Veneziano on different topics related to hadron physics, especially on his famous $U(1)$ paper and on scalar glueballs. Then, we initiated a work on the nature of scalar mesons, especially to understand the nature of $\sigma/f_0(500)$ and $G(1.6)$.

6.15.1 Scalar glueballs

Our collaboration started when I mentioned to him the contradictory results obtained by different groups on the masses of scalar glueballs. Some authors (like me) found a mass below 1 GeV when working with the $\mathcal{L}_0$ unsubtracted moment and while working with the lowest subtracted moment $\mathcal{L}_{-1}$, one found a higher mass which is unexpected as naïvely one expects that $\mathcal{L}_{-1}$ is more sensitive to a low meson mass compared to $\mathcal{L}_0$ due to the less power of τ. In the same period, the experimental group GAMS found a narrow width scalar $G(1.6)$ which looks to be a glueball as it decays into a pair of singlet pseudoscalars $\eta\eta'$, $\eta\eta$, while the nature of the low mass scalar $\sigma/f_0(500)$ remains unexplained.

Then, we tried to understand the origin of the discrepancy between different sum rules estimates which motivates our work described in Section 6.12, where the unsubtracted sum rule (USR) is more sensitive to a low mass, while the subtracted one (SSR) is instead sensitive to a higher mass. In fact, the presence of the subtraction constant $\psi(0)_S$ pushes the value of the sum rule optimization scale τ and lower value (or higher energy) which renders this SSR sensitive to higher mass glueball like for the case of higher sum rule moments not sensitive to $\psi(0)_S$.

6.15.2 Photon spin

Deep inelastic, polarised μp scattering data from the EMC collaboration [139] (see also Ref. [140]) have found that the flavour singlet part of the first moment of the polarised proton structure function $g_1(x, Q^2)$ was found to be anomalously small. $g_1(x, Q^2)$ is related via the OPE to the forward proton matrix element of the axial U(1) current $J^0_{\mu 5}$, the so-called "spin" of the proton. Its small value reflects a strong violation of the OZI approximation in the $U(1)_A$ singlet axial-vector channel and the proton "spin" problem has been extensively discussed from this point of view using the U(1) generalisation of the Goldberger–Treiman (GT) relation [141, 142].

Unfortunately, in the case of a proton target, it is not so easy to disentangle target-independent dynamical features of QCD, such as the realisation of the anomaly, with specific properties of the proton itself.

• The g_1^γ $n = 1$ moment sum rule

In Ref. [143], we replace the proton target by a photon where we expect that the essential features of the QCD dynamics could be studied independently of the non-perturbative bound-state physics of the target. Deep inelastic scattering from a photon target is realised experimentally by the two-photon process which at high energies dominates the reaction $e^+e^- \to e^+e^- X$ in electron–proton colliders.

For $n_f = 3$, we find, using the OPE, the $n = 1$ moment sum rule:

$$\int_0^1 dx\, g_1^\gamma(y, Q^2; \kappa^2) = \frac{1}{18}\frac{\alpha}{\pi}\left(1 - \frac{\alpha_s(Q^2)}{\pi}\right)$$
$$\left[f^{(3)}(\kappa^2) + f^{(8)}(\kappa^2) + 8f^{(0)}(\kappa^2; Q^2)\right], \tag{6.180}$$

where $f^{(3,8)}(\kappa^2)$ are the non-singlet triplet, octet non-perturbative form factors. The singlet form factor $f^{(0)}(\kappa^2; Q^2)$ has a non-trivial Q^2 dependence due to its anomalous dimension. Their κ^2-dependence can be studied from the AVV vertex function which is related to the $P \to \gamma^*\gamma^*$ decay amplitude for off-shell photons. The form factors read:

$$f^{(0)}(\kappa^2; Q^2) = 1 - \frac{\pi}{2N_c a^{(0)}}\frac{1}{\alpha} f_{\eta^0}(0; Q^2)$$
$$\times \left[g_{\pi^{(3,8)}\gamma\gamma}(0, ; Q^2) - g_{\pi^{(3,8)}\gamma^*\gamma^*}(\kappa^2; Q^2)\right],$$
$$f^{(3,8)}(\kappa^2) = 1 - \frac{\pi}{2N_c a^{(3,8)}}\frac{1}{\alpha} f_\pi \left[g_{\pi^{(3,8)}\gamma\gamma}(0) - g_{\pi^{(3,8)}\gamma^*\gamma^*}(\kappa^2)\right], \tag{6.181}$$

with y the Bjorken variable, $N_c = 3$ the number of colours, $a^{(0)} = 2/3$, $a^{(3)} = 1/6, a^{(8)} = 1/6\sqrt{3}$ and $\pi^{(0)} \equiv \eta^{(0)} \equiv \pi^0$ and $\pi^{(8)} \equiv \eta$, where $\eta^{(0)}$ is the singlet component of the observed η'-meson, while $f_{\eta^0}(0,0)$ is related to the topological charge $\psi_P(0) \equiv \chi(0)$ (see Section 5.2.4 for its definition).

In the limit $\kappa^2 = 0$, the previous relation gives:

$$\int_0^1 g_1^\gamma(y, Q^2; \kappa^2 = 0) = 0, \tag{6.182}$$

while in the asymptotic limit, $m_\rho^2 \ll \kappa^2 \ll Q^2$, we have, for n_f flavours and $N_c = 3$ colours:

$$\int_0^1 g_1^\gamma(y, Q^2; \kappa^2) = N_c \frac{\alpha}{\pi} \sum_f \hat{e}_f^4 \left[1 - \frac{2\gamma_1}{\beta_1} \frac{1}{n_f} \frac{1}{\log(Q^2/\Lambda^2)} \right.$$
$$\left. - c \frac{2\gamma_1}{\beta_1^2} \left(\frac{1}{\log(Q^2/\Lambda^2)} - \frac{1}{\log(\kappa^2/\Lambda^2)} \right) \right], \tag{6.183}$$

with:

$$\beta_1 = -\frac{1}{6}\left(11N_c - 2n_f\right), \quad \gamma_1 = -\frac{3}{N_c} \frac{N_c^2 - 1}{8} n_f,$$

$$c = n_f^{-1} \left(\sum_f \hat{e}_f^2 \right)^2 \left(\sum_f \hat{e}_f^4 \right)^{-1}, \tag{6.184}$$

$\hat{e}$ being the quark electric charge normalized to e; $\beta(\alpha_s) = \beta_1 (\alpha_s/\pi) + \cdots$, $\gamma(\alpha_s) = \gamma_0(\alpha_s/\pi) + \gamma_1(\alpha_s/\pi)^2 + \cdots$, where $\gamma_0 = 0$. In between these two extremes, the first moment of $g_1^\gamma(y, Q^2, \kappa^2)$ is governed by the expression in Eq. (6.181), with the form factors interpolating smoothly between the limits.

The important prediction here is that because the anomaly is realised non-perturbatively in the spontaneously chiral breaking phase of QCD, the turnover mass scale for the form factors is a hadronic mass of $\mathcal{O}(M_\rho^2)$ but not the perturbative light quark mass scale $\mathcal{O}(m_q^2)$.

- **Measurement of g_1^γ first moment ($n = 1$) sum rule**

We have also shown that the nth moments of the structure functions F_2^γ and g_1^γ are proportional to the $(n + 1)$st moments of the differential cross-sections unpolarized $d\sigma/dQ^2dxd\kappa^2$ and polarized $d\Delta\sigma/dQ^2dxd\kappa^2$, respectively, where $\sigma = \frac{1}{2}(\sigma_{++} + \sigma_{+-})$ and $\Delta\sigma = \frac{1}{2}(\sigma_{++} - \sigma_{+-})$.

We have explored the possibility to measure the differential cross-sections at LEP 200 within some different cuts. We conclude that a measurement of g_1^γ at polarised LEP200 may be feasible provided an appropriate cut is made on $Q^2_{\rm min}$ to maximise the statistical significance of the polarisation asymmetry. In practice, however, the situation is likely to be much more difficult, since the value $L = 500\,\text{pb}^{-1}$ may be optimistic and we have not included the effect of polarisation efficiency.

It is clear from this analysis that the optimum conditions for measuring g_1^γ are a combination of high luminosity and moderate s. In this respect, the best existing e^+e^- machine would be PEP, which has (without polarisation) already achieved a luminosity equivalent to $600\,\text{pb}^{-1}$per year.

The same conclusion as for g_1^γ happens for the case of the moment sum rules in Eq. (6.180).

6.15.3 Proton spin in the chiral limit

In Ref. [144], we study the hadronic part of the amplitude in a deep inelastic scattering process in the large Q^2 limit which comes from the imaginary part of the two-current matrix element $\langle N|J_\mu(q)J_\nu(-q)|N\rangle$, where J_μ is the current coupling to the exchanged hard photon or W/Z bosons and $|N\rangle$ denotes the target proton as illustrated in Fig. 6.13(a).

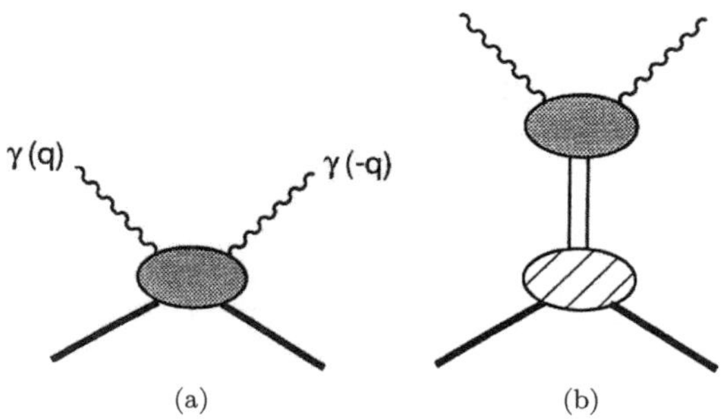

Figure 6.13. (a) The two-current matrix element $\langle N|J_\mu(q)J_\nu(-q)|N\rangle$; (b) decomposition of the matrix element into a composite operator propagator (denoted by double line) and a proper vertex (hatched).

Using OPE, the dominant contributions to the amplitude come from the lowest twist operators for large Q^2. Within this set of lowest twist operators, those of spin n contributes to the nth moments of the structure functions:

$$\int_0^1 dx x^{n-1} F(x, Q^2) = \sum_i C_i^n(Q^2)\langle N|\mathcal{O}_i^n(0)|N\rangle, \tag{6.185}$$

where $C_i^n(Q^2)$ are Wilson coefficients calculable within perturbative QCD at large Q^2, while $\mathcal{O}_i^n$ are non-perturbative composite operators. We decompose (see Fig. 6.13(b)) the matrix elements into proper vertices $\Gamma_{\tilde{\mathcal{O}}_k NN}$ which are 1PI with respect to a physically motivated basis set $\tilde{\mathcal{O}}_k$ of renormalized composite operators:

$$\langle N|\mathcal{O}_i^n(0)|N\rangle = \sum_k \langle 0|\mathcal{O}_i \tilde{\mathcal{O}}_k|0\rangle \Gamma_{\tilde{\mathcal{O}}_k NN}. \tag{6.186}$$

• The first moment sum rule for g_1^p

Here, we start from the Ellis–Jaffe [145] sum rule which is known in PT QCD to order α_s^3 in the $\bar{MS}$ scheme where the QCD corrections are less than 10% of the LO contribution (see Ref. [144] for the original references). It reads to order α_s:

$$\begin{aligned}\Gamma_1^p(Q^2) &\equiv \int_0^1 dx g_1^p(x, Q^2) \\ &= \frac{1}{6}\left\{\left[G_A^{(3)(0)} + \frac{1}{\sqrt{3}} G_A^{(8)}(0)\right]\left(1 - \frac{\alpha_s}{\pi} + \cdots\right)\right. \\ &\qquad \left. + \frac{2}{3} G_A^{(0)}(0, Q^2)\left(1 - \frac{1}{3}\frac{\alpha_s}{\pi} + \cdots\right)\right\}.\end{aligned} \tag{6.187}$$

$G^{(a)}$ are form factors in the proton matrix elements of the renormalized (index R) axial current:

$$\langle P|J_{\mu 5R}^{(a)}(k)|P\rangle = G_A^{(a)}(k^2)\bar{u}\gamma_\mu\gamma_5 u + G_A^{(a)} P(k^2) k_\mu \bar{u}\gamma_5 u, \tag{6.188}$$

where, in our normalisations:

$$G_A^{(3)} = \frac{1}{2}(\Delta u - \Delta d),$$

$$G_A^{(8)} = \frac{1}{2\sqrt{3}}(\Delta u + \Delta d - 2\Delta s),$$

$$G_A^{(0)} = \frac{1}{2\sqrt{3}}(\Delta u + \Delta d + \Delta s) \equiv \Delta\Sigma, \tag{6.189}$$

where Δq is the light quark polarized part on distribution. $G_A^{(3)}$ is known from neutron η decay, while $G_A^{(8)}$ is measured in hyperon decay. Here, we shall concentrate on the singlet form factor $G_A^{(0)}$. It has been measured by EMC [139] to be:

$$G_A^{(0)}(Q^2 = 11\,\text{GeV}^2) = 0.19 \pm 0.17, \tag{6.190}$$

which is much smaller than the quark model OZI prediction ($\Delta s = 0$):

$$G_A^{(0)}(0)_{\text{OZI}} = G_A^{(8)}(0)_{\text{OZI}} = 0.58 \pm 0.02. \tag{6.191}$$

Assuming the absence of massless boson in the $U(1)_A$, it follows from Eq. (6.188):

$$G_A^{(0)}(0, Q^2)\bar{u}\gamma_5 u = \frac{1}{2M}\langle P|\partial^\mu J^{(a)}_{\mu 5R}(k)|P\rangle = \frac{1}{2M}2n_f\langle P|Q_R(0)|P\rangle, \tag{6.192}$$

where $Q_R(0)$ is the renormalized topological charge density. The second equality comes from chiral Ward identities. Using Eq. (6.186), one can rewrite:

$$\langle P|Q_R(0)|P\rangle = \langle 0|Q_R(0)Q_R(0)|0\rangle\Gamma_{Q_R\bar{P}P} + \langle 0|Q_R(0)\Phi_{5R}(0)|0\rangle\Gamma_{\Phi_{5R\bar{P}P}}, \tag{6.193}$$

where the propagators are at zero momentum. The composite operator in the first term of Eq. (6.193) is the topological susceptibility or charge $\chi(0)$ which is the value $\psi(0)_P$ of the pseudoscalar current $Q(x)$ defined in Eq. (6.135). In the chiral limit, $\chi(0) = 0$.

It was shown in Ref. [142] that the second term can be related to the slope of the topological charge:

$$G_A^{(0)}(0, Q^2)\bar{u}\gamma_5 u = \frac{1}{2M}2n_f\sqrt{\chi'(0)}\Gamma_{\Phi_{5R\bar{P}P}}. \tag{6.194}$$

In this chapter, we re-estimate the slope of the topological charge previously done in Ref. [129] using here both Laplace and finite energy sum rules. We find the average at the sum rule scale $\tau = 0.5$ GeV^{-2}:

$$\sqrt{\chi'(0)} = (25.3 \pm 2.6)\,\text{MeV}. \tag{6.195}$$

We run this result up to the EMC energy $Q^2 = 10$ GeV^2. At this scale, the result becomes:

$$\sqrt{\chi'(0)}|_{\text{MC}} = (23.2 \pm 2.4)\,\text{MeV}, \tag{6.196}$$

which is smaller by a factor (1.64 ± 0.17) than the OZI estimate $(1/\sqrt{6})f_\pi$. This result leads to:

$$G_A^{(0)}(0, Q^2 = 10\,\text{GeV}^2) = G_A^{(0)}|_{\text{OZI}} \frac{\sqrt{\chi'(0)}}{(1/\sqrt{6})f_\pi} = (0.35 \pm 0.05), \tag{6.197}$$

compared to the OZI prediction in Eq. (6.191). It corresponds to the value:

$$\Gamma_1^p(10\,\text{GeV}^2) = (0.143 \pm 0.005), \tag{6.198}$$

where we have used $\alpha_s(M_\tau) = 0.347 \pm 0.030$ from τ-decay. After the completion of our work, new data from SMC [146] appeared leading to the world average [147]:

$$\Gamma_1^p(10\,\text{GeV}^2) = (0.145 \pm 0.014), \tag{6.199}$$

which implies:

$$G_A^{(0)}(0, Q^2 = 10\,\text{GeV}^2) = (0.37 \pm 0.12) \tag{6.200}$$

in excellent agreement with our predictions.

Then, we conclude, as has been emphasised in previous works [141, 142, 144], that the so-called *proton spin problem* is not a problem of spin — rather, it is a question of understanding the dynamical origin of the OZI violation, making the suppression of $G_A^{(0)}$ relative to $G_A^{(3)}$ and $G_A^{(8)}$.

6.15.4 Proton spin beyond the chiral limit

We extend the previous analysis to the case of massive light quarks [149]. The formal basis of our analysis is the derivation of the new, unified Goldberger–Treiman relations:

$$G_A^a = \frac{1}{2M_N} F_{ab} \hat{\Gamma}_{\eta^b NN}, \tag{6.201}$$

where F is determined from

$$F_{ac} F_{cb}^T = \lim_{\mathrm{k}=0} \frac{d}{dk^2} i \int d^4x e^{ikx} \langle 0|\mathcal{T}\partial^\mu J_{\mu 5}^a(x) \partial^\nu J_{\mu 5}^b(0)|0\rangle. \tag{6.202}$$

Apart from the mixing between the flavour singlet and octet channels induced by the non-vanishing strange quark mass, the most significant change is the generalisation of the slope of the topological susceptibility $\chi'(0)$ to the equivalent correlation function involving the total divergence of the singlet axial current, viz. $\frac{d}{dk^2}\langle 0|\mathcal{T}\partial^\mu J_{\mu 5}^a(x)\partial^\nu J_{\mu 5}^b(0)|0\rangle$. This is the quantity which displays the smoothest approach to the chiral limit, with the explicit quark mass dependence present in the individual correlators cancelling in the sum.

We have also checked that the SU(3) breakings increase by about 20% (respectively 10%) the values of $\chi'(0)$ (respectively $f_{\eta'}$) obtained in the chiral limit. We obtain:

$$\sqrt{\chi'(0)|\mathrm{SU(3)}} = (33.5 \pm 3.9)\,\mathrm{MeV}, \quad f_{\eta'} = (27.4 \pm 3.7)\,\mathrm{MeV}. \tag{6.203}$$

The value of $\chi'(0)$ is still lower than the OZI value and does not support the conclusion of Ref. [150].

6.15.5 $\chi'(0)$ at N2LO and the proton spin in the chiral limit

More recently, the slope of the topological susceptibility $\chi'(0)$ has been estimated [151] with a value 24.3(3.4) MeV at 2 GeV^2 in the chiral limit (see Section 9.14), including N2LO perturbative corrections and using the recent value of the three-gluon condensate:

$$g^3 f_{abc}\langle G^a G^b G^c\rangle = (8.2 \pm 1.0)\,\mathrm{GeV}^2 \langle \alpha_s G^2\rangle \tag{6.204}$$

from a fit of the charmonium data [152], which notably differs from the instanton liquid model estimate $g^3 f_{abc}\langle G^a G^b G^c\rangle \approx (1.5 \pm 0.5)\,\mathrm{GeV}^2\langle\alpha_s G^2\rangle$ [3, 127, 128, 130].

6.15.6 Bjorken sum rule

In Ref. [144], we study the effect of our result to the Bjorken sum rule $\Gamma_1^p - \Gamma_1^n$ [153] and attempt to extract the coefficient of the higher twist coefficients $a_{n,p}/Q^2$:

$$a_p - a_n = 0.135 \pm 0.145. \tag{6.205}$$

Pursuing the analysis with the moment of the neutron, one obtains

$$\Gamma_1^p(2\,\text{GeV}^2) = -(0.031 \pm 0.006) + \frac{a_n}{Q^2}, \tag{6.206}$$

which we compare with the SMC [146, 147] and SLAC data [148]:

$$\Gamma_1^p(2\,\text{GeV}^2) = -(0.022 \pm 0.011). \tag{6.207}$$

The previous analysis gives:

$$a_p = -(0.117 \pm 0.145), \quad a_n = -(0.018 \pm 0.025), \tag{6.208}$$

which is not conclusive.

6.16 Heavy Quark Masses

6.16.1 Running versus the heavy quark pole mass

My interest in this topic comes from the recent discoveries of the J/ψ and Υ families. One unclear point in that time was the definition of the heavy quark mass and the relation between the on-shell quark mass and the running one. In fact, unlike QED where a free electron with a physical mass can be observed, the situation is not the same for quarks due to QCD confinement. Then, one has to build a framework (renormalization and regularization scheme: see, e.g., Ref. [5] for a review), where the quark mass is consistently defined.

The running quark mass has been introduced in Ref. [29] within dimensional renormalization ($\bar{MS}$ scheme) of 't Hooft and Veltman [4] where it has been related to the renormalization group invariant (RGI) mass $\hat{m}_q$ as in Eq. (3.23).

In Refs. [154, 155], its relation with the on-shell mass has been given:

$$M_{o.s}(\nu) = \bar{m}_{MS}(\nu)\left[1 + \left(\frac{4}{3} + 2\mathrm{Log}\frac{\nu}{M}\right)\left(\frac{\alpha_s}{\pi}\right) + \mathcal{O}(\alpha_s^2)\right], \quad (6.209)$$

where the pole mass is defined at the pole of the quark propagator ($\nu^2 = M^2$). The coefficient of α_s comes from the finite part of the quark self-energy of the propagator (see the details in Ref. [6]). Reference [154] has shown that this definition is infrared-safe.

To get a relation between the running and pole masses, Ref. [157] has also evaluated the spectral function associated to the electromagnetic current which is a physical observable associated to the $e^+e^- to$ hadron total cross-section in two ways: the on-shell scheme of QED and the $\bar{M}S$ off-shell scheme. Requiring that a physical observable should not depend on the renormalization scheme used to get it, i.e., the spectral function has the same radiative correction, they obtain the relation between the two mass definitions.

In Ref. [158], we have studied the regularization and renormalization independence of the relation between the running and pile mass. In this way, we first use dimensional regularization for evaluating the quark propagator and renormalize the result on shell. In the second way, we use dimensional regularization and renormalize the mass in the $\bar{M}S$ scheme. We deduce to order α_s that the relation between the running and pole mass is *regularization* and *renormalization scheme-independent.*

The expression including higher order PT corrections is reviewed in Ref. [156], where one can note that the higher order corrections to the pole mass is large. In some other papers quoted in Ref. [7], the pole mass may also be affected by renormalons, while some alternative definitions have been proposed to escape this problem. These features disfavour its use compared to the running mass.

6.16.2 Determination of the charm and bottom masses

During my stays at CERN, I have written different papers [50, 158–160] for determinating these masses using the charmonium and bottomium data within the QCD spectral sum rule approach ($q^2 = 0$

moments or/and Laplace sum rules):

$$\mathcal{M}^{(n)} \equiv \frac{1}{n!}\frac{\partial^n \Pi_b(q^2)}{(\partial q^2)^n}|_{q^2=0} = \int_{4M_b^2} \frac{dt}{t^{n+1}}\frac{1}{\pi}\mathrm{Im}\Pi_b(t),$$

$$\mathcal{L}_{c,b}(\tau, M_{c,b}^2) = \int_{4M_b^2}^{\infty} dt\, e^{-t\tau}\frac{1}{\pi}\mathrm{Im}\Pi_{c,b}(t). \tag{6.210}$$

The main feature of the OPE is the only presence of gluon condensate contributions known up to dimension 8 where the heavy quark condensate contributions have also been expressed in terms of the gluon condensates:

$$M_Q\langle\bar{Q}Q\rangle = \frac{1}{12\pi^2}\langle\alpha_s G^2\rangle + \mathcal{O}(1/M_Q^2) + \cdots. \tag{6.211}$$

These works at CERN are summarized in the most recent review [42] where to order α_s^2 of PT series, the running masses are:

$$\bar{m}_c(M_c^{(2)}) = (1.23^{+0.02}_{-0.04} \pm 0.03)\,\mathrm{GeV},$$

$$\bar{m}_b(M_b^{(2)}) = (4.23^{+0.03}_{-0.04} \pm 0.02)\,\mathrm{GeV}, \tag{6.212}$$

where the errors are, respectively, due to α_s and the gluon condensate. One can transform these results to the pole mass to order α_s^2:

$$M_c^{(2)} = (1.42 \pm 0.03)\,\mathrm{GeV}, \quad M_b^{(2)} = (4.62 \pm 0.02)\,\mathrm{GeV}. \tag{6.213}$$

6.17 Heavy-Light Meson Couplings and Masses

6.17.1 The decay constants f_B and f_D from LSR

My interest on this topic has been stimulated by the request of G. Altarelli during my visit at CERN to clarify the different QCD sum rule predictions of the decay constant f_B of the B-meson available in the literature. Indeed, f_B is a fundamental input for understanding the $\bar{B}^0 - B^0$ mass difference for CP-violation in the $\bar{B}B$ system.

In so doing, one starts from the two-point correlator $\psi_5(q^2)$ associated to the pseudoscalar heavy–light current:

$$J_5(x) = (m_c + m_u)\ :\bar{u}(i\gamma_5)c: \tag{6.214}$$

and work with its Laplace transform. Parametrizing the spectral function using the minimal duality ansatz: "B meson" $\oplus\ \theta(t - t_c) \times$

QCD continuum, one obtains the sum rule [161]:

$$2f_B^2 M_B^4\, e^{-M_B^2\tau} \simeq \int_{M_b^2}^{t_c} dt\, e^{-t\tau}\frac{1}{\pi}\text{Im}\psi_5|_{\text{Pert}}(t) + \big[\langle C_4\langle O_4\rangle + \langle C_6\langle O_6\rangle\tau\big], \tag{6.215}$$

where f_B is normalized as $f_\pi = 92.3$ MeV. The PT expression is known to order α_s and to all order in the quark mass expansion at that α_s order. We have included in the RHS the contribution due to the dimension 4 and 6 condensates. Unlike the case of heavy quark systems, the light quark $\langle\bar\psi\psi\rangle$ and quark-gluon mixed $g\langle\bar\psi(\lambda^a/2)G_a^{\mu\nu}\sigma_{\mu\nu}\psi\rangle$ condensates play an important role in the analysis.

We study the τ and t_c-behaviour of f_B and consider as conservative optimal value the one from the τ-stability (minimum) to the t_c-stability (plateau). The same strategy is used for the D-meson. In this way, we obtain:

$$f_B = (132 \pm 17)\,\text{MeV} = (1.43 \pm 0.18)\,f_\pi,$$
$$f_D = (122 \pm 11)\,\text{MeV} = (1.32 \pm 0.12)\,f_\pi. \tag{6.216}$$

We have used the pole mass:

$$M_c(M_c^2) = (1.46 \pm 0.06)\,\text{GeV}, \quad M_b(M_b^2) = (4.71 \pm 0.13)\,\text{GeV}, \tag{6.217}$$

deduced from the Euclidean mass $M_Q(p^2 = -M_Q^2) = 1.26$ GeV obtained in Refs. [3, 62] in order to be consistent with the on-shell expression of the spectral function. For the B-meson which is relatively heavy ($m_b^2 \gg \Lambda^2$), we can also work with the moment $\mathcal{M}^{(n)}(Q_0^2)$ where we take $Q_0^2 = 0, 0.5M_b^2, M_b^2$. We note that the choice of Q_0^2 does not almost affect the value of f_B but instead shifts the position of the τ-minimum. We obtain from the moment:

$$f_B = (130 \pm 22)\,\text{MeV} \tag{6.218}$$

in complete agreement with the one from LSR. We pursue the analysis of f_B by working with the non-relativistic sum rule and obtain:

$$f_B = (125 \pm 28)\,\text{MeV}. \tag{6.219}$$

Taking the average of different determinations, we obtain the final sum rule prediction:

$$f_B = (129 \pm 13)\,\text{MeV}. \tag{6.220}$$

6.17.2 The decay constant f_B from HQET sum rule

The previous results indicate that $f_B \approx f_D$ which contradicts the one from heavy quark effective theory (HQET) which predicts that the coupling behaves as $1/\sqrt{M_Q}$. It suggests a large mass correction at the scale of M_b.

In Refs. [164,165], we extract f_B using new HQET moments. However, the results are inaccurate due to the large α_s corrections. Taking the best result from LSR, one obtains from Table 1 of Ref. [167]:

$$f_B = (1.49 \pm 0.06 \pm 0.05)\, f_\pi, \quad \frac{f_{B_s}}{f_B} = 1.16 \pm 0.04, \tag{6.221}$$

which is an update of the results in the previous section.

6.17.3 Static limit and $1/M_Q$ correction

In Refs. [162–165], we tried to estimate this large $1/M_Q$ correction using non-relativistic HQET sum rules:

$$\mathcal{L}(\omega) = \int_0^{\omega_c} d\omega\, e^{-\omega\tau}\, \frac{1}{\pi}\mathrm{Im}\psi(\omega), \tag{6.222}$$

where ω is defined as $t = M_Q^2 + \omega M_Q$, while in Ref. [166], we use a semi-local duality sum rule:

$$\int_0^{\omega_c} d\omega\, \mathrm{Im}\psi_5^{res}(t) = \int_0^{\omega_c} d\omega\, \mathrm{Im}\psi_5^{QCD}(t). \tag{6.223}$$

Using the value of f_B in the static limit [167], the $1/M_Q$ mass correction is [164]:

$$f_B^\infty = (1.98 \pm 0.31)\, f_\pi. \tag{6.224}$$

Then, using the previous estimate of f_D and f_B in the full theory, we can parametrize the quark mass dependence of the decay constant as:

$$f_B\sqrt{M_b} \simeq (0.33 \pm 0.06)\,\mathrm{GeV}^{3/2}\alpha_s^{1/\beta_1}\Big\{1 - \frac{2}{3}\frac{\alpha_s}{\pi} - \frac{A}{M_Q} + \frac{B}{M_Q^2}\Big\}. \tag{6.225}$$

Within a linear parametrization:

$$A = (0.6 \pm 0.1)\,\mathrm{GeV}, \tag{6.226}$$

while including the quadratic mass correction, one obtains:

$$A \simeq 1.1\,\text{GeV}, \quad B \simeq 0.7\,\text{GeV}^2, \tag{6.227}$$

in accordance with previous findings [163, 164].

6.17.4 SU(3) breaking for f_{D_s} and f_{B_s}

• Relativistic $Q^2 = 0$ moment

In Ref. [161], we estimated the SU(3) breaking from the $n = 2$ and $Q^2 = 0$ relativistic moment used in Ref. [77]. We obtain the relation:

$$\begin{aligned} f_{P_s} \simeq f_P(M_{P_s}/M_P)\big[1 + (3m_s/M_Q)(1 - M_Q^2/t_c) \\ - m(8\pi^2/M_Q^3)\langle \bar{s}s - \bar{u}u\big]^{1/2}. \end{aligned} \tag{6.228}$$

We obtain:

$$f_{D_s} = 1.26\, f_D, \qquad f_{B_s} = 1.09\, f_B. \tag{6.229}$$

where we have used:

$$\langle \bar{s}s\rangle/\langle \bar{u}u\rangle = 0.6, \quad m_s/M_c = 0.12, \quad m_s/M_b = 0.04, \tag{6.230}$$

and $t_c = 9$ (respectively 50) GeV^2 for D (respectively B) meson as the QCD continuum threshold.

• Non-relativistic $Q^2 = 0$ moment within HQET

In Ref. [167], we estimate the SU(3) breaking from the ratio $R_P \equiv f_{P_s}/f_P$ using a compact analytic two-loop expression within a non-relativistic HQET sum rule:

$$\begin{aligned} R_P^2 = \rho_P^2\Big\{1 + 2(2.2 \pm 0.2)\left(\frac{m_s}{\omega_c}\right)\left(1 - \frac{m_s}{M_Q}\right) \\ - 2(8.2 \pm 1.6)\left(\frac{m_s}{\omega_c}\right)^2 + \cdots\Big\}, \end{aligned} \tag{6.231}$$

where:

$$\rho_P = \left(\frac{M_P}{M_{P_s}}\right)^2\left(1 + \frac{m_s}{M_Q}\right) \tag{6.232}$$

is an overall factor due to the current and τ is the non-relativistic sum rule variable. ω_c is related to the QCD continuum threshold.

Including the PT α_s corrections, one obtains:

$$R_D = 1.15 \pm .04, \quad R_B = 1.16 \pm 0.04, \tag{6.233}$$

where the error comes mainly from the estimate of the α_s^2 correction.

6.17.5 Mixed condensate from the $B^* - B$ splitting

Due to the important role of the light quark $\langle \bar{u}u \rangle$ and mixed quark-gluon $\langle \bar{u}Gu \rangle$ condensates in this channel, we expect that it is a good place to extract $\langle \bar{u}Gu \rangle$. We have done the analysis in Refs. [6,168]. We fix its value from a simultaneous fit of the B and B^* masses. Using a factor violation of the four-quark estimate of the vacuum saturation, the value $\Lambda_5 = (100 \pm 50)$ MeV, one obtains the correlated values:

$$\begin{aligned} M_b(p^2 = M_b^2) &= (4.56 \pm 0.05)\,\text{GeV}, \\ M_0^2 = (0.80 \pm 0.01\,\text{GeV}^2) :& \quad g\langle \bar{u}(\lambda_a/2)G^a_{\mu\nu}\sigma_{\mu\nu}u \rangle \equiv M_0^2 \langle \bar{u}u \rangle. \end{aligned} \tag{6.234}$$

The value of M_0^2 confirms and complements the estimate from light baryon sum rules [67, 120, 194, 222].

6.17.6 Spectra and couplings of other B-like mesons

In Refs. [6, 168], we have extended the analysis to study the masses and couplings of some other heavy–light mesons:

$$B_\delta(0^{++}),\ B_A^*(1^{++}),\ B_s^*(1^{--}),\ B_s(0^{-+}),\ B_c(0^{-+},\ B_c^*(1^{--}). \tag{6.235}$$

The mass splittings and their corresponding couplings are given in Table 10.2 of Ref. [6], where we note that:

$$\begin{aligned} B_\delta - B &\simeq B_A^* - B^* \simeq (415 \pm 211)\,\text{MeV}, \\ B_c - B &\simeq B_c^* - B^* \simeq (1.6 \pm 0.2)\,\text{GeV} \approx M_c. \end{aligned} \tag{6.236}$$

We improve the results of B_c and B_c^* masses and couplings, and extend the analysis to the heavy baryons in Ref. [217] by combining QSSR results with the ones of potential models. Potential model predictions are more accurate than that of QSSR for the masses. We obtain:

$$\begin{aligned} M_{B_c} &= (6255 \pm 20)\,\text{MeV}, \quad M_{B_c^*} = (6330 \pm 20)\,\text{MeV}, \\ M_{\Lambda_{bcu}} &= (6.93 \pm 0.05)\,\text{MeV}, \quad M_{\Omega_{bcs}} = (7.00 \pm 0.05)\,\text{GeV}, \\ M_{\Xi^*_{ccu}} &= (3.63 \pm 0.05)\,\text{MeV}, \quad M_{\Xi^*_{bbu}} = (10.21 \pm 0.05)\,\text{GeV}. \end{aligned} \tag{6.237}$$

Our prediction for M_{B_c} is in perfect agreement with the recent experimental mass [33]:

$$M_{B_c} = (6276 \pm 1)\,\text{MeV}. \tag{6.238}$$

More discussions on B_c-like mesons can be found in Section 7.3.11.

6.17.7 HQET parameters

The meson-quark mass gap $\bar{\Lambda}$, the heavy quark kinetic K and chromomagnetic Σ energies are defined as:

$$M_B = M_b + \bar{\Lambda} - \frac{1}{2M_b}(K + 3\Sigma), \tag{6.239}$$

where:

$$K = \frac{1}{2M_B}\langle B(v)|\mathcal{K}|B(v)\rangle, \quad \Sigma = \frac{1}{6M_B}\langle B(v)|\mathcal{S}|B(v)\rangle, \tag{6.240}$$

with:

$$\mathcal{K} \equiv \bar{h}(iD)^2 h, \quad \mathcal{S} \equiv \frac{1}{2}\bar{h}\sigma_{\mu\nu}F^{\mu\nu}h, \tag{6.241}$$

where h is the heavy field and $F^{\mu\nu}$ is the electric field tensor. The sum rule estimate of the mass gap and the kinetic energy gives [42, 164, 165]:

$$\bar{\Lambda} = (0.65 \pm 0.05)\,\text{GeV}, \quad K = (0.5 \pm 0.2)\,\text{GeV}^2. \tag{6.242}$$

Combining these results with the estimate:

$$\Sigma = \frac{1}{4}(M^2_{B^*} - M^2_B), \tag{6.243}$$

one can deduce the value of the pole mass to order α_s:

$$M_b = (4.61 \pm 0.05)\ \text{GeV}, \tag{6.244}$$

in good agreement with the one from the relativistic sum rule.

6.18 Heavy to Light Exclusive Decays

6.18.1 Hybrid sum rule (HSR) for vertex functions

This analysis has been detailed in Refs. [169, 173] and reviewed in Refs. [42, 49]. For this purpose, we shall analyse the form factors of the $\bar{B} \to \pi(\rho)$ semi-leptonic processes defined in a standard way as:

$$\begin{aligned}
&\langle \rho(p')|\bar{u}\gamma_\mu(1-\gamma_5)b|B(p)\rangle \\
&\quad = (M_B + M_\rho)A_1\epsilon^*_\mu - \frac{A_2}{M_B+M_\rho}\epsilon^* p'(p+p')_\mu \\
&\quad + \frac{2V}{M_B+M_\rho}\epsilon_{\mu\nu\rho\sigma}p^\rho p'^\sigma, \\
&\langle \pi(p')|\bar{u}\gamma_\mu b|B(p)\rangle = f_+(p+p')_\mu + f_-(p-p')_\mu,
\end{aligned} \tag{6.245}$$

and the $B \to \rho\gamma$ rare processes:

$$\begin{aligned}
&< \rho(p')|\bar{s}\sigma_{\mu\nu}\left(\frac{1+\gamma_5}{2}\right)q^\nu b|B(p) >= i\epsilon_{\mu\nu\rho\sigma}\epsilon^{*\nu}p^\rho p'^\sigma F_1^{B\to\rho} \\
&\quad + \left\{\epsilon^*_\mu(M_B^2 - M_\rho^2) - \epsilon^* q(p+p')_\mu\right\}\frac{F_1^{B\to\rho}}{2}.
\end{aligned} \tag{6.246}$$

In the QCD spectral sum rules (QSSR) evaluation of the form factors, we shall consider the generic three-point function:

$$V(p,p',q) = -\int d^4x \int d^4y \exp(ip'x - ipy)\langle 0|TJ_L(x)O(0)J_b(y)|0\rangle, \tag{6.247}$$

whose Lorentz decompositions are analogous to the previous hadronic amplitudes. Here, $J_L \equiv \bar{u}\gamma_\mu d$ $(J_\pi \equiv (m_u + m_d)\bar{u}i\gamma_5 d)$ is the bilinear quark current having the quantum numbers of the ρ (π) mesons; $J_b \equiv (M_b + m_d)\bar{d}i\gamma_5 b$ is the quark current associated to the B-meson; $O \equiv \bar{b}\gamma_\mu u$ is the charged weak current for the semi-leptonic transition, while $O \equiv \bar{b}\frac{1}{2}\sigma^{\mu\nu}q_\nu$ is the penguin operator for the rare decay. The vertex function obeys the double dispersion relation:

$$V(p^2, p'^2, q^2) = \frac{1}{4\pi^2}\int_{M_b^2}^\infty \frac{ds}{s-p^2}\int_0^\infty \frac{ds'}{s'-p'^2}\,\mathrm{Im}\,V(s,s',q^2) + \cdots. \tag{6.248}$$

We shall work with the hybrid sum rule (HSR):

$$\mathcal{H}(n,\tau) \equiv \frac{1}{n!}\left(\frac{\partial}{\partial p^2}\right)^n_{p^2=0} \mathcal{L}\left(V(p^2,p'^2,q^2)\right)$$
$$= \frac{1}{\pi^2}\int_{M_b^2}^{\infty}\frac{ds}{s^{n+1}}\int_0^{\infty} ds' \exp(-\tau' s')\mathrm{Im}\,V(s,s',q^2), \tag{6.249}$$

rather than with the DLSR ($\mathcal{L}$ is the Laplace transform operator). This sum rule guarantees that terms of the type:

$$\frac{M_b^{2l}}{\left(M_b^2-p^2\right)^k p'^{2k'}}, \tag{6.250}$$

which appear in the successive evaluation of the Wilson coefficients of high-dimension operators, will not spoil the OPE for $M_b \to \infty$ unlike the case of the double Laplace transform sum rule, which blows up in this limit for some of its applications in the heavy-to-light transitions.

It is important to note that *to leading order in* α_s, *the* $B \to K^*\gamma$ *rare and* B/D *semi-leptonic processes are dominated by the contribution of the light* $\langle\bar{u}u\rangle$ *condensate.*

The details of the analysis can be found in the original papers quoted previously. We just quote and comment the results.

6.18.2 $\bar{B} \to \rho$ semi-leptonic decay

The QCD-analytic q^2 behaviour of the form factors is a polynomial in q^2/M_b^2, which mimics quite well the usual pole parametrization, except in the case of the A_1^B form factor, where there is a significant deviation from this polar form:

$$A_1(q^2) = A_1(0)\left(1-\frac{q^2}{M_b^2}\right). \tag{6.251}$$

The q^2-dependence of the normalized V^B and A_2^B form factors is very weak since the dominant light-quark condensate contribution has no q^2-dependence. The small increase with q^2 is due to the q^2-dependence of the small and non-leading contribution from the perturbative graph. Lattice points in the Euclidean q^2-region agree with our results.

The M_b-dependence of the form factors expected from HQET and lattice results is recovered:

$$A_1^B(q^2_{\rm max}) \sim \frac{1}{\sqrt{M_b}}, \quad A_2^B(q^2_{\rm max}) \simeq V^B(q^2_{\rm max}) \sim \sqrt{M_b}. \tag{6.252}$$

We extract with a good accuracy the ratios of form factors at zero momentum:

$$\frac{V^B(0)}{A_1^B(0)} \simeq \frac{A_2^B(0)}{A_1^B(0)} \simeq 1.11 \pm 0.01, \quad \frac{A_1^B(0)}{F_1^B(0)} \simeq 1.18 \pm 0.06. \tag{6.253}$$

6.18.3 $\bar{B} \to \pi$ semi-leptonic decay

The corresponding form factors have been estimated in Refs. [163, 172] using HSR and double Laplace sum rule (DLSR) with the value:

$$f_+(0) = 0.20 \pm 0.05, \tag{6.254}$$

while talking of different values from some other approaches lead to the average [169]:

$$f_+(0) = 0.25 \pm 0.02. \tag{6.255}$$

The q^2-dependence of the form factor is given in Ref. [169], from which one can deduce:

$$f_+^B(q^2_{\rm max}) \simeq f_+^B(0) \sim \sqrt{M_b}. \tag{6.256}$$

6.18.4 $\bar{B}-$ semi-leptonic decay rates

Now, we are in a position to predict the decay rates:

$$\Gamma_{\bar{B}\to\pi e\bar{\nu}} \simeq (3.6 \pm 0.6) \times |V_{ub}|^2 \times 10^{12}\ {\rm s}^{-1}. \tag{6.257}$$

We also obtain the following ratios:

$$\frac{\Gamma_{\bar{B}\to\rho e\bar{\nu}}}{\Gamma_{\bar{B}\to\pi e\bar{\nu}}} \simeq 0.9 \pm 0.2, \quad \frac{\Gamma_+}{\Gamma_-} \simeq 0.20 \pm 0.01,$$

$$\alpha \equiv 2\frac{\Gamma_L}{\Gamma_T} - 1 \simeq -(0.60 \pm 0.01). \tag{6.258}$$

Thanks to a better control of the ratios of form factors, the ratio of the $\bar{B}$ decays into π over ρ can be predicted to a good accuracy. It becomes compatible with the prediction obtained by only retaining the contribution of the vector component of the form factors.

6.18.5 SU(3) breakings in $\bar{B}/D \to Kl\bar{\nu}$ and V_{cd}/V_{cs}

The previous analysis is extended to the estimate of the SU(3) breakings in the ratio:

$$R_P \equiv \frac{f_+^{P\to K}(0)}{f_+^{P\to\pi}(0)} \quad : \ P \equiv B, D. \tag{6.259}$$

Using again the HSR, we obtain:

$$R_B = 1.007 \pm 0.020, \qquad R_D = 1.102 \pm 0.007. \tag{6.260}$$

These values have been used in Ref. [171] for an estimate of the ratio:

$$\frac{V_{cd}}{V_{cs}} = 0.322 \pm 0.056. \tag{6.261}$$

Using the recent value of $|V_{cd}| = 0.225 \pm 0.008$, one would obtain:

$$|V_{cs}| = 0.70, \tag{6.262}$$

which is lower than the recent value of 0.986 ± 0.016 quoted by PDG [33].

6.18.6 $\bar{B} \to \rho\gamma$ rare decay

The $B/D \to \pi\gamma$ form factors are found numerically to be [170]:

$$F_1^B(0) \simeq (27.0\pm3.4)\times10^{-2}, \quad F_1^B(0) \simeq (62.0\pm10.0)\times10^{-2}, \tag{6.263}$$

and the ratio:

$$r_\gamma \equiv \frac{A_1^B(0)}{F_1^B(0)} \simeq 1.18 \pm 0.06. \tag{6.264}$$

A semi-analytical M_b behaviour including a $1/M_b^2$ correction reads [169]:

$$F_1^B(0) = \frac{1.6 \times 10^{-2}}{f_B}\left(1 + \frac{(20 \pm 4)}{M_b^2}\right), \tag{6.265}$$

where (20 ± 4) comes from a numerical integration of the PT contribution. Its q^2-dependence is:

$$F_1^B(q^2) \simeq F_1^B(0) \sim \sqrt{M_b}. \tag{6.266}$$

6.18.7 $\bar{B} \to K^*\gamma$ rare decay

Adding the SU3 breaking corrections dominantly due to m_s, one obtains [170]:

$$\frac{F_1^{B\to K^*}}{F_1^{B\to\rho}} = 1.14 \pm 0.02, \quad \frac{F_1^{D\to K^*}}{F_1^{D\to\rho}} = 1.22 \pm 0.04, \tag{6.267}$$

where the systematic errors cancel in the ratios. This implies the branching ratio:

$$Br(B \to K^*\gamma) \simeq (4.45 \pm 1.13) \times 10^{-5}, \tag{6.268}$$

in agreement with CLEO data of $(4.5 \pm 1.5 \pm 0.9) \times 10^{-5}$ [174].

6.18.8 $\Upsilon\bar{B}B$ coupling, slope of the Isgur–Wise function, V_{cb}

Within HQET, it is expected that the $\bar{B}$ meson decay is independent of the process.

With the advent of the heavy quark symmetry, there has been considerable interest and progress in the understanding of the semi-leptonic form factors of the transition of a heavy quark into another heavy quark, as in this infinite mass limit all semi-leptonic form factors reduce to the single Isgur–Wise (IW) function [175] $\xi(y \equiv vv')$, where v, v' is the b, c quark velocity. The question is whether this result in the heavy quark infinite mass limit can be applied to the physical B and D mesons.

As discussed in Ref. [176], the bound on the slope of the form factor depends on the residues and on the positions of Υ poles. In principle, the couplings of the three lightest states are unconstrained. However, using the location of the three Υ poles and the fact that the b-number form factor of the B-meson is 1 at $q^2 = 0$, Ref. [176] obtains the following bounds:

$$|\eta_{\Upsilon_1}| \leq 9.9 \times 10^3, \quad |\eta_{\Upsilon_2}| \leq 17.1 \times 10^3, \quad |\eta_{\Upsilon_3}| \leq 8.1 \times 10^3, \tag{6.269}$$

which are quite weak compared to the one for η_4 obtained from the data:

$$|\eta_{\Upsilon_4}| \leq 0.75 \pm 0.15, \tag{6.270}$$

with the definition:

$$\eta_{\Upsilon_i} \equiv \frac{g_{\Upsilon_i \bar{B}B}}{2\gamma_{\Upsilon_i}} : \quad < 0| - \bar{b}\gamma^\mu b|\Upsilon_i >= \epsilon^\mu \frac{M^2_{\Upsilon_i}}{2\gamma_{\Upsilon_i}}, \tag{6.271}$$

where we have included in the definition of the current the negative sign related to the b-quark charge.

In Ref. [177], we estimate the $\Upsilon\bar{B}B$ couplings using QSSR vertex sum rules within a $\bar{b}q$ nature of the $\bar{B}B$ meson pairs. These couplings play an important role on the upper bound of the slope of the form factor which is related to the one of the IW function as [178]:

$$\xi'(1) \simeq F'(1) - \frac{16}{75}\log \alpha_s(M_b), \tag{6.272}$$

We obtain the sum of couplings:

$$\sum_{i=1}^{4} \eta_{\Upsilon_i} \simeq -(0.224 \pm 0.064) \longrightarrow \xi'(1) \geq -1.04. \tag{6.273}$$

Using the QCD sum rule expression of the slope of the IW function [179,180], we obtain the optimal result at the stability point $\tau^{-1} \simeq 1.7$ GeV:

$$\xi'(1) \simeq -(1.00 \pm 0.02). \tag{6.274}$$

We have used the data from $\bar{B}^0 \to D^{*-}l\nu$ [181, 182] (respectively $\bar{B}^- \to D^{*0}l\nu$ [181]) and obtained in units of 10^{-3}:

$$|V_{cb}| = \left(\frac{1.48\text{ps}}{\tau_B}\right)^{1/2} \times \left\{36.5 \pm 1.5 \quad [\text{resp. } 40.6 \pm 2.3]\right\}. \tag{6.275}$$

This old value can be compared with the present value:

$$|V_{cb}| = 40.8 \pm 1.4. \tag{6.276}$$

6.19 CP Violation in the Standard Model (SM)

6.19.1 $\bar{B} - B$ mixing at NL2O from QSSR

In Ref. [183], we compute the leading α_s corrections to the two-point correlator of the $O_{\Delta B=2}$ operator:

$$O_{\Delta B=2} = \bar{b}\gamma^L_\mu d\bar{b}\gamma^L_\mu d, \tag{6.277}$$

which controls the $B^0\bar{B}^0$ mixing [183]:

$$\langle \bar{B}^0 | O_{\Delta B=2} | B^0 \rangle = \frac{4}{3} f_B^2 M_B^2 B_B(\nu). \tag{6.278}$$

We have tested the validity of the vacuum saturation $B_B = 1$ of the bag constant using a sum rule analysis of the four-quark two-point correlator to two loops. We found that the radiative corrections due to the non-factorizable contributions are quite small. Under some physically reasonable assumptions for the spectral function, we found that the vacuum saturation estimate is only violated by about 15%, giving:

$$B_B \simeq 1 \pm 0.15. \tag{6.279}$$

By combining this result with the one for f_B, we deduce:

$$f_B \sqrt{B_B} \simeq (197 \pm 18)\ MeV, \tag{6.280}$$

if we use the normalization $f_\pi = 132$ MeV.

6.19.2 $\bar{K} - K$ mixing at NL2O from QSSR

We have also estimated the B_K-parameter associated to the $K^0 - \bar{K}^0$ mixing, using the four-quark two-point correlator [184]. Using the Laplace sum rule (LSR) and adopting the parametrization of the spectral function in Ref. [185], we have obtained the *conservative* estimate:

$$B_K \simeq (0.58 \pm 0.22), \tag{6.281}$$

in agreement (within the errors) with the FESR result (0.39 ± 0.10) [186].

6.19.3 Estimate of the CP-violation parameters (ρ, η)

We are now ready to discuss the implications of the previous results for the estimate of the CP-violation parameters (ρ, η) defined in the standard way within the Wolfenstein parametrization [33]. Using the previous values of f_B, B_B and B_K, which are all of them obtained from a Laplace sum rule analysis, and using the other input used in Ref. [187], one obtains the *best* fit [184]:

$$(\rho, \eta) \approx (0.09,\ 0.41) \tag{6.282}$$

in very good agreement with the expectation in Ref. [188] derived from an alternative method (see also Refs. [186,187]). Here, the value of ρ is very sensitive to the change of B_K and f_B.

The present determination is [33]:

$$(\rho, \eta) \approx (0.159 \pm 0.010),\ (0.348 \pm 0.010). \tag{6.283}$$

6.20 The $\tau \to \nu_\tau \eta\pi$ Process in and Beyond QCD

The collaboration with Albert Bramon and Toni Pich started in Ref. [189] when I visited CERN. We estimate various sources of the $\tau \to \nu_\tau \eta\pi$ process using QCD-duality information on the decay constants of standard $\bar{q}q$ and exotic four-quark and hybrid mesons.

6.20.1 Standard $\bar{q}q$ mesons

We consider that $a_0(980)$ and $K_0^*(1350)$ are associated to the divergence of the vector current:

$$\langle 0|\bar{u}\gamma_\mu d|a_0\rangle = \sqrt{2} f_{a_0} q_\mu, \quad \langle 0|\bar{u}\gamma_\mu s|K_0^*\rangle = \sqrt{2} f_{K_0^*} q_\mu, \tag{6.284}$$

where the decay constants come from QCD spectral sum rules [6]:

$$f_{a_0} \simeq (1.3 \pm 0.2)\,\text{MeV}, \quad f_{K_0^*} \simeq (31 \pm 3)\,\text{MeV}. \tag{6.285}$$

The ratio:

$$f_{K_0^*}/f_{a_0} \simeq (24 \sim 30) \tag{6.286}$$

is almost independent of the approaches such that one can predict with a good precision the ratio of branching ratios:

$$\frac{B(\tau \to \nu_\tau K_0^*)}{B(\tau \to \nu_\tau a_0)} = \tan\,\theta_c^2 \left(\frac{f_{K_0^*}}{f_{a_0}}\right)^2 \left(\frac{1 - M_{K_0^*}^2/M_\tau^2}{1 - M_{a_0}^2/M_\tau^2}\right)^2. \tag{6.287}$$

Expressing that the $\tau \to \nu_\tau K_0^*$ process is smaller than the sum of all Cabibbo suppressed decays of τ, one can deduce the upper bound:

$$B(\tau \to \nu_\tau \eta\pi) \leq 0.28\%. \tag{6.288}$$

Stating that the known $\tau \to \nu_\tau K\pi$ branching ratio in the 0^{++} channel is smaller than the sum of all known $K\pi$ branching ratio of about 1%, one can derive the improved bound:

$$B(\tau \to \nu_\tau \eta\pi) \leq 0.1\%, \tag{6.289}$$

while using the value of f_{a_0} in Eq. (6.285), one can predict a ridiculously small branching ratio:

$$B(\tau \to \nu_\tau \eta\pi) \simeq 1.5 \times 10^{-5}\%. \tag{6.290}$$

6.20.2 Four-quark/molecule state

We also analyze the possibility that a_0 is a four-quark/molecule state described by the operator:

$$O_4 = \frac{1}{M_H^2}\left(\sum_{1,\gamma_5} \bar{u}\Gamma d\bar{s}\Gamma s + t\sum_{1,\gamma_5} \bar{u}\lambda_a\Gamma d\bar{s}\lambda^a\Gamma s\right), \tag{6.291}$$

where $M_H \simeq 1$ GeV is a typical hadronic scale, while t is an operator mixing estimated to be about 0.2–0.4 [124]. Defining the coupling as:

$$\langle 0|O_4|a_0\rangle = \sqrt{2} f_{a_0} M_{a_0}^4/M_H^2, \tag{6.292}$$

one obtains [190]:

$$f_{a_0} \simeq 2.7\,\text{MeV} \Longrightarrow B(\tau \to \nu_\tau \eta\pi) \simeq 4 \times 10^{-4}\left(\frac{1\,\text{GeV}}{M_H}\right)^4, \tag{6.293}$$

while using $g_{a_0 K^- K^0} = 5 \sim 8$ GeV [124], one obtains:

$$B(\tau \to \nu_\tau \bar{K}K) \simeq (3-8) \times 10^{-5}\left(\frac{1\,\text{GeV}}{M_H}\right)^4, \tag{6.294}$$

and the ratio independent of M_H:

$$\frac{B(\tau \to \nu_\tau \eta\pi)}{B(\tau \to \nu_\tau \bar{K}K)} \simeq 0.5 \sim 1.3. \tag{6.295}$$

Using the experimental upper limit on $\tau \to \nu_\tau \bar{K}K$, one deduces:

$$B(\tau \to \nu_\tau \eta\pi) \leq 0.5\%. \tag{6.296}$$

6.20.3 Hybrid meson state

It has been shown in Ref. [199] that the $\tilde{\rho}(1.4 \sim 2.2)$ hybrid meson decays to $\eta\pi$. The ratio of branching ratios reads:

$$\frac{B(\tau \to \nu_\tau \eta\pi)_{\tilde{\rho}}}{B(\tau \to \nu_\tau \pi)} = \left(\frac{f_{\tilde{\rho}}}{f_\pi}\right)^2 \left(\frac{M_{\tilde{\rho}}}{M_H}\right)^4 \left(1 - \frac{M_{\tilde{\rho}}^2}{M_H^2}\right)^2 \times \left(1 + \frac{2M_{\tilde{\rho}}^2}{M_\tau^2}\right) B(\tilde{\rho} \to \eta\pi), \tag{6.297}$$

which for $M_{\tilde{\rho}} \leq 1.4$ GeV gives:

$$B(\tau \to \nu_\tau \eta\pi) \leq 4.3 \times 10^{-4} \left(\frac{1\,\text{GeV}}{M_H}\right)^4, \tag{6.298}$$

which is still negligible for a reasonable value of M_H.

6.20.4 Contact term beyond the standard model

We also analyze the eventual contact term obeying the chirality constraint which may contribute to the $\tau \to \nu_\tau \eta\pi$ process:

$$\mathcal{L}_S = \left(\frac{g_S}{M_S}\right)^2 \bar{\nu}_\tau (1 + \gamma_5) \tau\, \bar{u}(1 + g_A \gamma_5) d, \tag{6.299}$$

where g_S and M_S are coupling and mass of a hypothetical exotic scalar of some models beyond the SM.

◇ For a universal coupling at the fermion vertex, we can deduce from $\pi \to e\bar{\nu}_e$ data the bound:

$$\frac{\sqrt{2} g_A g_S^2}{M_S^2 G_F} \leq 3.6 \times 10^{-6} \left(\frac{m_u + m_d}{1\,\text{MeV}}\right), \tag{6.300}$$

which implies:

$$\frac{B(\tau \to \nu_\tau \eta\pi)_{\text{scalar}}}{B(\tau \to \nu_\tau \eta\pi)_{\text{standard}}} \leq \frac{4.7 \times 10^{-9}}{g_A^2}, \tag{6.301}$$

which is negligible in the chiral theory, where $g_A = 1$.

◇ For a Higgs-type coupling, where $g_l \sim g_S m_l / M_S$, the 1‰ accuracy of $\pi^- \to \mu \bar{\nu}_\mu$ leads to:

$$\frac{B(\tau \to \nu_\tau \eta \pi)_{\text{scalar}}}{B(\tau \to \nu_\tau \eta \pi)_{\text{standard}}} \leq \frac{3 \times 10^{-3}}{g_A^2}, \qquad (6.302)$$

which is still very small.

Chapter 7

Globe Trotters Era and Collaborations

7.1 Collaboration with Trieste

Figure 7.1. Nello Paver and Daniele Treleani (University Trieste).

7.1.1 Scalar current sum rules

This collaboration initiated when I left Trieste to Annecy where we have written papers with N. Paver and D. Treleani (University of Trieste) and E. de Rafael (University of Marseille) on the light quark mass differences [85] and later on the SU(3) breaking of light quark vacuum condensates [91] (see Section 4.3).

This collaboration has been continued at CERN with A. Bramon (University of Barcelona), where we studied the SU(2) breaking of the chiral condensate $\langle\bar{d}d\rangle/\langle\bar{u}u\rangle$ [116] (see Section 5.6).

7.1.2 Gluonium and the 0^{++} spectrum

In Reference [191], we study the lowest mass isoscalar 0^{++} spectrum by assuming that the broad $\sigma(0.5)$ and $S^*(0.98)$ emerge from the mixing of a scalar meson S_2 and gluonium state G:

$$|\sigma\rangle = \cos\theta|G\rangle - \sin\theta|`S_2\rangle,$$
$$|S^*\rangle = \sin\theta|G\rangle + \cos\theta|`S_2\rangle, \tag{7.1}$$

where θ is the mixing angle. We evaluate the $S_2 \to \pi\pi$ width using vertex sum rules evaluated at the symmetric Euclidean point and found:

$$\Gamma(S_2 \to \pi^+\pi^-) \simeq 43\,\text{keV}. \tag{7.2}$$

The G coupling to $\pi^+\pi^-$ has been obtained from an effective Lagrangian [135]:

$$g_{G\pi^+\pi^-} \simeq g_{GK^+K^-} \simeq M_G^2/f_G \simeq 5\,\text{GeV},$$
$$\Rightarrow \Gamma(G \to \pi^+\pi^-) \simeq 0.64 \pm 0.30\,\text{GeV}, \tag{7.3}$$

where we have used the value of $f_G \approx (0.08 \pm 0.02)$ GeV and $M_G \simeq 1$ GeV from Ref. [129] and after an appropriate change of the normalization. As $S_2 \to \pi^+\pi^-$ is much smaller than $G \to \pi^+\pi^-$, we can deduce the mixing angle:

$$\theta = \text{tg}^{-1}(g_{S^*\pi^+\pi^-}/g_{\sigma\pi^+\pi^-}) \simeq 14\text{–}19^\circ, \tag{7.4}$$

where we have used the data: $\Gamma(S^* \to \pi^+\pi^-) = (30\text{–}40)$ MeV and $\Gamma(\sigma \to \pi^+\pi^-) = (500\text{–}800)$ MeV.

This result suggests that the decays of σ and S^* to $\pi^+\pi^-$ are dominated by the gluonium contribution, while there is an indication from $\pi\pi$ and $\bar{K}K$ scattering data [87] about the universality of their couplings to pseudoscalar pairs:

$$|g_{S^*\pi^+\pi^-}| = 1.56\,\text{GeV}, \quad |g_{S^*K^+K^-}| = 3\,\text{GeV},$$
$$|g_{\sigma\pi^+\pi^-}| = 1.56\,\text{GeV}, \quad |g_{\sigma K^+K^-}| = 3\,\text{GeV}. \tag{7.5}$$

7.1.3 Nature of $a_0(980)$ from QSSR

• $\bar{q}q$ nature of $a_0(980)$

◇ Mass and coupling

Here, we study the $\bar{q}q$ nature of the $a_0(980)$ using QSSR. In this approach, a_0 is associated to the divergence of the vector current:

$$\langle 0|\partial_\mu \bar{u}\gamma^\mu d|a_0\rangle = \sqrt{2} f_{a_0} M_{a_0}^2. \tag{7.6}$$

Using the ratio of moments and LSR moments, its mass and coupling are found to be [6]:

$$M_{a_0} \simeq (1 \sim 1.05)\,\text{GeV}, \quad f_{a_0} \simeq (0.5 \sim 1.2)\,\text{MeV}. \tag{7.7}$$

◇ $a_0 \to \eta\pi$ and $\bar{K}K$

The $a_0 \to \bar{K}K$ coupling has been estimated in Ref. [119] using vertex sum rule subtracted at the symmetric point:

$$g_{a_0\bar{K}K} \simeq 2 \text{ GeV} \Longrightarrow g_{a_0\eta\pi} \simeq \sqrt{2}3 g_{a_0\bar{K}K}, \tag{7.8}$$

which leads to:

$$\Gamma(a_0 \to \eta\pi) \simeq 37\,\text{MeV}, \tag{7.9}$$

compared with the data of 55 MeV.

◇ $a_0 \to \gamma\gamma$

We estimate the $\gamma\gamma$ width through the $g_{a_0\rho\omega}$ coupling $\oplus$ VDM by using QSSR vertex shown in Fig. 7.2, where we note that in the

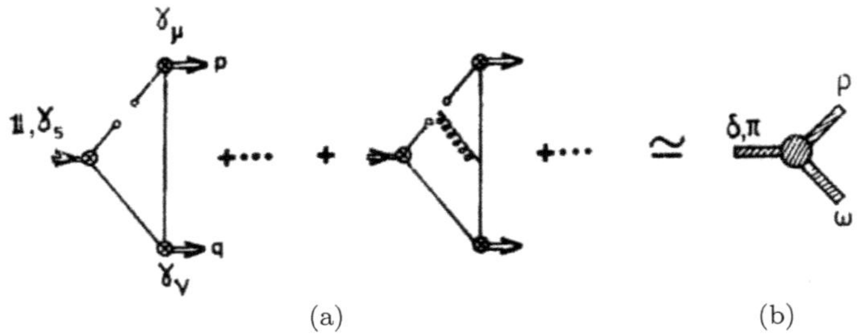

Figure 7.2. Duality between QCD and experiment for the vertex function: (a) QCD quark and mixed quark-gluon condensates; (b) $g_{a_0\rho\omega}$ hadronic coupling.

chiral limit $m_q = 0$, the dominant contribution comes from the $\bar{q}q$ condensate. Then, we obtain:

$$g_{a_0\rho\omega} \simeq (8.0 \sim 9.4)\,\text{GeV}^{-1}. \tag{7.10}$$

Using vector meson dominance, we deduce:

$$\Gamma_{a_0\to\gamma\gamma} \simeq (1.5 \sim 2.2)\,\text{keV}. \tag{7.11}$$

This prediction is higher by a factor of 4.5–5.2 than the data:

$$\Gamma_{a_0\to\gamma\gamma} B(a_0 \to \eta\pi) \simeq \left(0.19 \pm 0.07^{+0.10}_{-0.07}\right)\,\text{keV}. \tag{7.12}$$

Therefore, one may question the $\bar{q}q$ nature of a_0.

• Four-quark/molecule nature of $a_0(980)$

The four-quark/molecule scheme is more popular than the meson–gluonium one. It has motivated one to explain the unusual properties of scalar mesons below/around 1 GeV. In Ref. [124], we study these light four-quark states using QSSR. We assume that a_0 is described by the mixing of two operators:

$$\mathcal{O}_{a_0} = \mathcal{O}_1^{\pm} + t\,\mathcal{O}_2^{\pm}, \tag{7.13}$$

where t is an arbitrary mixing angle and:

$$\mathcal{O}_1^{\pm} = \frac{1}{\sqrt{2}} \sum_{\Gamma=1,\pm\gamma_5} \bar{s}\Gamma s\left(\bar{u}\Gamma u - \bar{d}\Gamma d\right),$$

$$\mathcal{O}_2^{\pm} = \frac{1}{\sqrt{2}} \sum_{\Gamma=1,\pm\gamma_5} \bar{s}\Gamma\lambda_a s\left(\bar{u}\Gamma u - \bar{d}\Gamma\lambda^a d\right), \tag{7.14}$$

with λ^a the colour matrix.

◇ Mass and coupling

To leading order in α_s, the effect of the $\mathcal{O}_2^{\pm}$ operator does not affect the result of Ref. [6, 136] for the mass:

$$M_{a_0} \simeq 1\,\text{GeV}, \tag{7.15}$$

but rescale the coupling as:

$$f_{a_0} = 2.5\,\text{MeV}\left(1 + \frac{32}{9}t^2\right)^{1/2}. \tag{7.16}$$

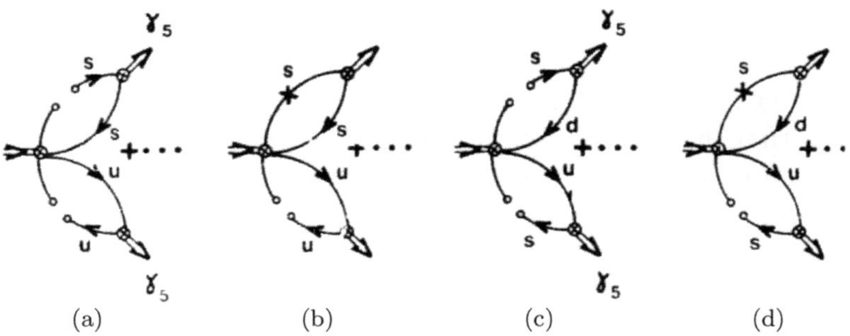

Figure 7.3. Four-quark operator contributions to the vertex function: (a, b) $a_0 \to \eta\pi$; (c, d) $a_0 \to \bar{K}K$.

◇ $a_0 \to \eta\pi$

To leading order in α_s, only the $\mathcal{O}_1^-$ operator can contribute to the vertex describing the $a_0 \to \eta\pi$ width due to chirality and to the zero trace of λ_a. An LSR analysis gives [124]:

$$\Gamma(a_0 \to \eta\pi) = (62 \sim 88)\,\text{MeV}/\left(1 + \frac{32}{9}t^2\right), \tag{7.17}$$

where the dominant numerical contribution comes from the $\langle \bar{u}u \rangle$ condensate. We fix the mixing angle:

$$t \simeq 0.2 \sim 0.4 \tag{7.18}$$

from the data $\Gamma(a_0 \to \eta\pi) \simeq 55$ MeV.

◇ $a_0 \to K^+K^-$

Here, both operators $\mathcal{O}_1^-$ and $\mathcal{O}_2^-$ contribute to the vertex (non-zero trace of two λ_a). We obtain [124]:

$$g_{a_0\bar{K}K} \simeq \sqrt{\frac{3}{2}} g_{a_0\eta\pi} \times \left(1 + \frac{16}{3}t\right) \simeq (5\text{–}8)\,\text{GeV}, \tag{7.19}$$

where the last factor indicates the departure from SU(3) symmetry due to the operator $\mathcal{O}_2^-$. So, it is interesting to also have an accurate measurement of this coupling.

• $a_0 \to \gamma\gamma$

Neglecting to a first approximation the contribution of $\mathcal{O}_2^-$, one can inspect that, in the four-quark scheme, the super-Zweig allowed disconnected diagram vanishes at lowest order.

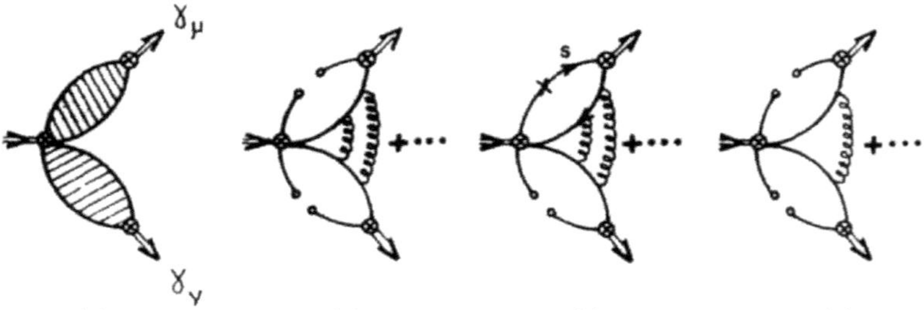

Figure 7.4. Four-quark operator contributions to $a_0 \to \gamma\gamma$: (a) full-disconnected quark blobs due to $\mathcal{O}_1^-$; (b, c) quark blobs connected by two gluon lines due to $\mathcal{O}_1^-$; (d) quark blobs connected by one gluon line due to $\mathcal{O}_2^-$.

Here, we test the scheme in the case of the scalar isovector $I = 1$ $a_0(980)$, which should not be affected by the gluonium contribution using QSSR.

The model can explain the a_0 mass [124, 136].

The a_0-decay to $\eta\pi$ agrees with the data [124], but the coupling to K^+K^- strongly depends on the operator and ranges from 1.6 [137] to (5–8) GeV [138].

However, its width to $\gamma\gamma$ is expected to be ridiculously small [124] due to the fact that it behaves as $\mathcal{O}(\alpha_s/\pi)^2$:

$$\Gamma[a_0(4q) \to \gamma\gamma] \simeq (2 \sim 5) \times 10^{-4}\,\text{keV}, \tag{7.20}$$

compared to the one for a $\bar{q}q$ state:

$$\Gamma[a_0(\bar{q}q) \to \gamma\gamma] \simeq (1.6 \sim 2.6)\,\text{keV}. \tag{7.21}$$

From the previous analysis of the $a_0 \to \gamma\gamma$ width, we conclude that the nature of a_0 is more complex than expected. It may emerge from a $\bar{q}q$-four-quark mixing.

7.2 Collaboration with the Barcelona Universities

Once established in Montpellier, I initiated an international collaboration with the universities of Montpellier, Barcelona (Autonoma and Central), ICTP and INFN of Trieste within a DRCI contract from the French CNRS. Later on, the collaboration has been extended to the University of Heidelberg. I also used a part of this grant for some of my visits at CERN and at the ICTP.

H.G. Dosch remarked that the triangle collaboration Barcelona–Heidelberg–Montpellier reminded the old-time cooperation in medicine between these three Universities where (to my knowledge) Montpellier was the oldest European one.

(a) (b)

Figure 7.5. Barcelona: (a) University Central; (b) University Autonoma.

Figure 7.6. (a) P. Pascual, R. Tarrach, J. Latorre (University Central); (b) A. Bramon, E. Bagan (University Autonoma).

7.2.1 Light hybrid mesons

The work on hermaphrodite mesons which we started in Ref. [198] (see Section 6.14) has been revisited and extended here [199]. In particular, we add the contribution of the four-quark condensate and do the analysis using the sum rule stability criteria. For a 1^{-+} and 0^{--} light hybrid mesons, we found [199]:

$$\begin{aligned} M_{\tilde{\rho}} &\simeq \sqrt{t_c} \simeq (1.6 \sim 2.1)\ \text{GeV}, \quad f_{\tilde{\rho}} \simeq (17 \sim 21)\ \text{MeV}, \\ M_{\tilde{\eta}} &\simeq 3.8\ \text{GeV} : \sqrt{t_c} \simeq 4.1\ \text{GeV}, \end{aligned} \tag{7.22}$$

where the error is mainly due to the estimate of the four-quark condensate (vacuum saturation or 2×(vacuum saturation)). The mass of $\tilde{\rho}$ confirms the result in Ref. [198], but the one of $\tilde{\eta}$ is 30% larger. One can note the large splitting between the $\tilde{\rho}$ and ρ meson. The SU(3) mass splitting between the $\tilde{\rho}$ and $\tilde{\phi}$ mesons is found to be:

$$M^2_{\tilde{\phi}} - M^2_{\tilde{\rho}} \simeq 0.3\,\text{GeV}^2. \tag{7.23}$$

These results improve and extend previous results in Ref. [198] discussed in Section 6.14. $\tilde{\rho}$ is expected to decay mainly into $\rho\pi$ [199]:

$$\Gamma(\tilde{\rho} \to \rho\pi) \simeq 0.6 \text{ GeV} \Longrightarrow \Gamma(\tilde{\rho} \to \gamma\pi) \simeq (3.8 \sim 7.5) \text{ MeV}. \quad (7.24)$$

We have used the vector meson dominance (VDM) relation, $g_{\tilde{\rho}\gamma\pi} \simeq g_{\tilde{\rho}\rho\pi}(e/\gamma_\rho)$, where the $\tilde{\rho}$ coupling is normalized as in Ref. [198] and $\gamma_\rho \simeq 2.55$ as in previous chapters.

The other hadronic modes are:

$$\Gamma(\tilde{\rho} \to K^*K) \simeq 0.3 \text{ GeV}, \quad \Gamma(\tilde{\rho} \to \eta'\pi) \simeq (3 \sim 5) \text{ MeV},$$
$$\Gamma(\tilde{\rho} \to \eta\pi) \simeq \Gamma(\tilde{\rho} \to \eta'\pi)\sin^2\theta_{\eta'\eta} \simeq (0.2 \sim 0.4) \text{ MeV}, \quad (7.25)$$

where we have used $\theta_{\eta'\eta} \simeq 17^0$ [192].

We shall improve again the determination of the masses in Section 7.6.1.

7.2.2 Scalar trigluonium

This work done at CERN [133] has been discussed in Chapter 5 (Section 11). We are presently improving this analysis by including α_s-corrections [200].

7.2.3 Tensor meson and digluonium

• Tensor meson–digluonium mixing

In Section 6.11.5, we have studied the mixing between a tensor and digluonium meson.

• Vector tensor meson

In Reference [193], we calculate the α_s and quark mass corrections to the two-point correlator of the 2^{++} meson, where we show that the quark and gluonic currents:

$$J_q^{\mu\nu} = i : \bar{q}(x)\left(\gamma^\mu \overrightarrow{D^\nu} + \gamma^\nu \overrightarrow{D^\mu}\right) q(x),$$

$$J_g^{\mu\nu} = -G^{\alpha\mu}G^\nu_\alpha + \frac{1}{4}G^{\mu\nu}G^{\alpha\beta}G_{\alpha\beta} \quad (7.26)$$

mix under renormalization. Therefore, the quark current acquires an anomalous dimension:

$$\gamma_{11}^{1} = \frac{16}{9} \tag{7.27}$$

for $SU(3)_c$ colour. Defining the RGI tensor meson coupling $\hat{g}_T$ as:

$$g_T(\nu) = \frac{\hat{g}_T}{\left(\log \nu/\Lambda)^{\gamma_{11}^{1}}/-\beta - 1\right)}, \tag{7.28}$$

where the spectral function is:

$$\frac{1}{\pi}\text{Im}\psi_2(t) = 2g_T^2 M_T^6 \delta(t - M_T^2) + \text{“QCD continuum”} \times \theta(t - t_c), \tag{7.29}$$

we obtain using Laplace sum rule moment for two flavours:

$$\hat{g}_T|_{\text{SU(2)}_\text{f}} = 0.13 - 0.18. \tag{7.30}$$

Including SU(3) breaking, we get:

$$\hat{g}_T|_{\text{SU(3)}_\text{f}} = 0.11 - 0.15, \tag{7.31}$$

suggesting that SU(3) breakings do not affect the value of the coupling much. Using the ratio of moments, we deduce, for $t_c = 2.5 \sim 3.5$ GeV^2:

$$M_{T_2} = (1.4 \sim 1.6)\,\text{GeV} \tag{7.32}$$

to be compared with the experimental mass $M_{f_2} = 1.27\,\text{GeV}$. However, one should note from Fig. 7.2(b) of Ref. [193] that the mass prediction depends strongly with the value of t_c. Using the double ratio of moments, we can study the mass ratio due to SU(3) breakings:

$$r(\tau) \equiv \frac{M_{T_3}^2}{M_{T_2}^2}. \tag{7.33}$$

Parametrizing $t_c|_{\text{SU(3)}_\text{f}} \simeq t_c|_{\text{SU(2)}_\text{f}} + \delta$ with $\delta \simeq (0.5 \sim 1.5)$ GeV^2, we can deduce from Fig. 3.3(b) of [193]:

$$r(\tau \simeq 0.6\,\text{GeV}^{-2}) \simeq 1.33 \sim 1.50, \tag{7.34}$$

giving:

$$M_{T_3} \simeq (1.33 \sim 1.56)\,\text{GeV}, \tag{7.35}$$

to be compared with the data: $M_{K_2^*} = 1430$ MeV. We have used the experimental mass $M_{f_2} = 1.27\,\text{GeV}$.

- **Axial tensor meson**

We extend the analysis to the axial tensor meson described by the current:

$$J_q^{\mu\nu} = i : \bar{q}(x) \left(\gamma^\mu \overrightarrow{D^\nu} + \gamma^\nu \overrightarrow{D^\mu} \right) \gamma_5 q(x). \qquad (7.36)$$

Here, the contribution of the four-quark condensate changes sign. Then, the coupling has no τ-stability, while one has a minimum in the analysis of the mass leading to the absolute upper bound:

$$M_{\pi_2} \leq 1.89\,\text{GeV}. \qquad (7.37)$$

From τ where $t_c = 6$ GeV^2 to the previous upper bound, we deduce the range:

$$M_{\pi_2} \simeq (1.84\text{–}1.89)\,\text{GeV}, \qquad (7.38)$$

which is slightly higher than the experimental mass: $M_{\pi_2} = 1.8\,\text{GeV}$.

7.3 Heidelberg–Montpellier Exchange Program

The collaboration started from the Deutsche Forschungsgemeinschaf (DFG) support for visiting the University of Heidelberg. Then, within the DRCI-CNRS French exchange program, Professor H.G. Dosch visited the University of Montpellier in 1988. In 1991, I went to Heidelberg within the Procope program of the French–German exchange, and in 1992, I got a Von Humboldt senior fellowship for continuing the collaboration. Our works are based on different applications of QCD spectral sum rules.

7.3.1 ρ meson at finite temperature

In this work, we reanalyzed [6,202] the work of Ref. [203] on the longitudinal part of the retarded commutator of the light vector current

(a) (b)

Figure 7.7. (a) Institute of Theoretical Physics, Philosophenweg, Heidelberg (DE); (b) Professor H.G. Dosch (host and collaborator) with his wife, my wife Lydia and our son Larry at the black forest.

where the spectral function is parametrized using the minimal duality ansatz: ρ meson $\oplus$ QCD continuum. Using the LSR, we obtain the constraint:

$$4\pi^2 \frac{M_\rho^2}{\gamma_{\rho L}^2} e^{-M_\rho^2 \tau} \simeq \int_0^{t_c} dt\, e^{t\tau}\, \mathrm{th}(\sqrt{t}/4\tau) + 2\int_0^{\infty} dt\,(\sqrt{t/2T}) \times \left(n_F - \frac{1}{3}n_B\right) + \frac{\pi}{3}\langle\alpha_s G^2\rangle\tau - 2\pi^3\alpha_s\langle\bar\psi\psi\rangle^2\tau^2, \quad (7.39)$$

and its FESR version. Using the ratio of LSR, one can derive a sum rule for the ρ meson mass. We show its behaviour and the one of the coupling in Fig. 7.8 for the set of QCD parameters:

$$\langle\alpha_s G^2\rangle = 0.04\ \mathrm{GeV}^2, \quad \alpha_s\langle\bar\psi\psi\rangle^2 \simeq 3.5\times 10^{-4}\,\mathrm{GeV}^6, \quad (7.40)$$

and versus the change of the temperature T. We see a smooth behaviour of the ρ meson mass and coupling versus T for $T \leq 200$ MeV. Similar behaviour is observed from the FESR constraints. This result indicates that the hot fermion gas for $T \leq 200$ MeV does not lead to a drastic change of the spectrum contrary to the claim of Ref. [203].

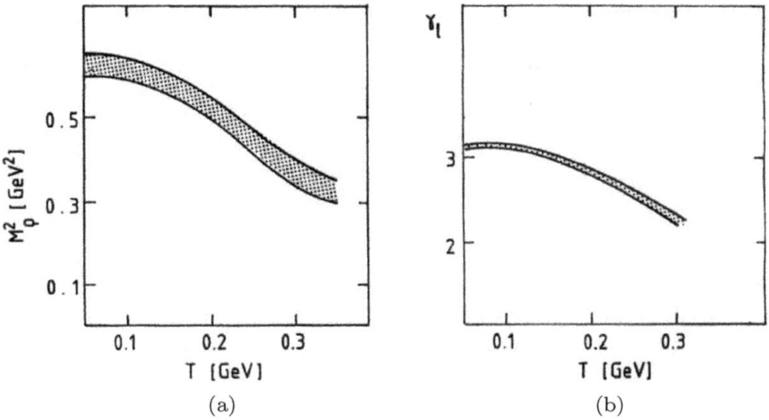

Figure 7.8. T-behaviour of the ρ meson: (a) mass; (b) coupling for given values of the QCD parameters in Eq. (7.40).

7.3.2 Light baryons

We have reanalyzed [120] the baryon sum rules [194, 195] by emphasizing the search for the optimal choice of the interpolating operators and the inclusion of NLO PT corrections for the decuplet and octet light baryons. The decuplet spin 3/2 current (Δ baryon) is chosen as:

$$\Delta^\mu = \frac{1}{\sqrt{2}} : \psi^T C \gamma_\nu \psi (g^{\mu\nu} - \gamma^\mu \gamma^\nu/4)\psi :, \tag{7.41}$$

while the one for the octet (Nucleon) is:

$$N = \frac{1}{\sqrt{2}} : (\psi C \gamma_5 \psi) + b\,(\psi C \psi)\,\gamma_5 \psi :. \tag{7.42}$$

The optimal choice of the nucleon current is obtained by taking the b derivative of the expression of the two-point function which gives:

$$b = -\frac{1}{5}. \tag{7.43}$$

This value confirms the choice of Ref. [222] but not the value -1 chosen by Ref. [194]. For the Ioffe's choice, we do not also find any τ-stability for the LSR and then no optimal result. It was emphasized in Ref. [67, 222] that around the Ioffe's choice $b = -1$, the result depends strongly on the value of b which is not the case of $b = -1/5$, where there is a smooth dependence.

We perform in Ref. [120] a systematic analysis of all light baryons spectra including the PT α_s corrections and $SU(3)$ breakings due to the strange quark mass and condensate where the value $\langle\bar{s}s\rangle/\langle\bar{d}d\rangle \simeq 0.6$ from the light meson systems has been used.

- ◇ The obtained constraint on the SU(3) breaking of the mixed quark-gluon condensate is not accurate.
- ◇ The $N - \Delta$ mass splitting is constrained by the mixed condensate which is fixed to be:

$$g\langle\bar{\psi}\,(\lambda_a/2)\sigma^{\mu\nu}\psi G^a_{\mu\nu}\rangle \equiv M_0^2\langle\bar{\psi}\psi\rangle, \quad M_0^2 \simeq (0.8 \pm 0.2)\ \text{GeV}^2. \tag{7.44}$$

  A larger value of M_0^2 makes the nucleon too heavy and Δ too light and vice versa for smaller values.
- ◇ The consistency of the overall fit needs a violation of the vacuum saturation assumption by a factor of about 3.6 [67, 120, 222] as also requested from the $e^+e^- \to$ hadron data analysis [6, 115].

7.3.3 Skyrme model parameter from QSSR

In Ref. [204], we compare the sum rule prediction of the nucleon mass at LO [67, 120]:

$$M_N = -24\pi^2\langle\bar{u}u\rangle\tau_{0N}Z_N, \tag{7.45}$$

with:

$$Z_N = 1 - \frac{2}{5}M_0^2\tau_{0N} - \frac{23}{30}\pi\langle\alpha_s G^2\rangle\tau_{0N}^2 - 64\pi^4\langle\bar{u}u\rangle^2\tau_{0N}^3, \tag{7.46}$$

with the one of the Skyrme model [205]:

$$M_N \simeq f_\pi(\mu/e + 9e^3/8\pi\, L), \tag{7.47}$$

where e is the Skyrme model parameter, $\mu = 73,\ L = 51$ come from the radial integrals over the soliton distribution. We eliminate f_π and $\langle\bar{u}u\rangle$ using the pion PCAC relation, $(m_u + m_d)\langle\bar{u}u\rangle \simeq -m_\pi^2 f_\pi^2$: $f_\pi = 92.3$ MeV, and introduce the sum rule expression of the quark mass [69]:

$$(m_u + m_d) \simeq (4\pi/\sqrt{3})m_\pi^2 f_\pi \tau_{0\pi} Z_\pi^{1/2}, \tag{7.48}$$

with:

$$Z_\pi^{1/2} = 1 - \frac{\pi}{6}\langle\alpha_s G^2\rangle\tau_{0\pi}^2 - \frac{448}{81}\pi^3\alpha_s\langle\bar{u}u\rangle^2\tau_{0\pi}^2, \tag{7.49}$$

ending with the relation:

$$e/[\mu + (9e^4/8\pi L)] = (1/6\pi\sqrt{3})(Z_\pi^{1/2}/Z_N)(\tau_{0\pi}/\tau_{0N}). \tag{7.50}$$

Asking the τ stability of Eqs. (7.46) and (7.49), one fixes:

$$\tau_{0\pi} \simeq (2 \sim 3)\tau_{0N}, \tag{7.51}$$

where the contributions of the radial excitation $\pi'(1300)$ has been added. Then, we deduce:

$$e \simeq (5 \sim 8). \tag{7.52}$$

This result is stable against the variation of the nucleon mixing operator b around $-1/5$, but the choice $b = -1$ of Ref. [194] leads to $e \simeq 0.13$. We can independently use the Δ sum rule:

$$M_\Delta = -(84\pi^2/3)\langle\bar{u}u\rangle\tau_{0\Delta}Z_\Delta, \tag{7.53}$$

with:

$$Z_\Delta = 1 + \frac{1}{2}M_0^2\tau_{0\Delta} + \frac{11}{9}\pi\langle\alpha_s G^2\rangle\tau_{0\Delta}^2 - \frac{320}{3}\pi^4\langle\bar{u}u\rangle^2\tau_\Delta^3, \tag{7.54}$$

together with the previous sum rules into the expression of the soliton mass M and moment of inertia I [205]:

$$M \equiv \frac{\mu f_\pi}{e} = \frac{1}{4}(5M_N - M_\Delta), \quad I = \frac{\pi L}{3e^3 f_\pi} = \frac{3}{2}/(M_\Delta - M_N), \tag{7.55}$$

and derive the constraint:

$$e \simeq \left(\mu L \frac{16}{90}\pi \frac{10\, Z_\Delta - 9Z_N}{9\, Z_N - 2Z_\Delta}\right)^{1/4} \simeq (5 \sim 7), \tag{7.56}$$

where we have assumed $\tau_{0N} \simeq \tau_{0\Delta}$ We note that the result is almost independent of the choice of the mixing current b. The sum rule result agrees with the one:

$$e \simeq 5.45 \tag{7.57}$$

from some phenomenological determinations [205]. One should note from our analysis that the skyrme parameter does not depend on the chiral symmetry breaking term $\langle \bar{u}u \rangle$.

7.3.4 N_c behaviour and the Skyrme model

In this section, we study the sum rule for large number N_c of the u, d quark fields. The nucleon N $(I = J = 1/2)$ and Δ $(I = J = N_c/2)$ interpolating currents are defined as [206]:

$$\begin{aligned} \mathcal{O}_\alpha^N &= \epsilon_{A_1 \cdots A_{N_c}} \times : u_\alpha^{A_1} (u^{A_2} C \gamma_5 d^{A_3}) \cdots (u^{A_{N_c-1}} C \gamma_5 d^{A_{N_c}}) :, \\ \mathcal{O}_\alpha^{\Delta \mu_1 \cdots \mu_\nu} &= \epsilon_{A_1 \cdots A_{N_c}} \times : u_\alpha^{A_1} (u^{A_2} C \gamma^{\mu_1} d^{A_3}) \cdots (u^{A_{N_c-1}} C \gamma^{\mu_\nu} d^{A_{N_c}}) :, \end{aligned} \tag{7.58}$$

where $\nu \equiv (N_c - 1)/2$. The corresponding two-point correlator for the nucleon N has the Lorentz decomposition:

$$S_{\alpha\bar{\alpha}}(p^2) = \hat{p}_{\alpha\bar{\alpha}} F_1^N(p^2) + \delta_{\alpha\bar{\alpha}} F_2^N(p^2), \tag{7.59}$$

while for Δ, one has a similar decomposition by retaining the terms proportional to the overall factor:

$$\frac{1}{\nu} \sum_\sigma g^{\mu_1 \bar{\mu}_{\sigma(1)}} \cdots g^{\mu_\nu \bar{\mu}_{\sigma(\nu)}}. \tag{7.60}$$

To leading order, one obtains, after taking the Laplace transform:

$$M_B = \frac{\mathcal{L}[F_2^B]}{\mathcal{L} F_1^B} = -(4\pi^2) \frac{1}{4} (3N_c - 1)(3N_c - 3) I \frac{\langle \bar{\psi}\psi \rangle}{N_c} \tau_0 C_B, \tag{7.61}$$

where:

$$C_N = \frac{1}{5}(N_c - 4), \quad C_\Delta = \frac{1}{3}(N_c + 2). \tag{7.62}$$

The difference in C_B comes from the combinatorics of quark lines for evaluating the perturbative graphs. Assuming that $\tau_0 \simeq M_B^{-1/2}$, one can deduce in the large N_c limit:

$$\frac{M_\Delta}{M_N} \simeq \left(\frac{5}{3}\right)^{1/3}. \tag{7.63}$$

In the large N_c limit, the Skyrme model gives:

$$M_N = \frac{\mu f_\pi \sqrt{N_c}}{e\sqrt{3}} + \frac{3 f_\pi \sqrt{N_c}}{8L e^3 \sqrt{3}}, \quad M_\Delta = \frac{\mu f_\pi \sqrt{N_c}}{e\sqrt{3}} + N_c(N_c+2)\frac{f_\pi \sqrt{N_c}}{8L e^3 \sqrt{3}}. \tag{7.64}$$

Working with the quantity $(M_N - M_\Delta)/M_N$ from the two approaches, one obtains:

$$e \simeq \frac{(9 \sim 12)}{\sqrt{N_c}}, \tag{7.65}$$

which is consistent with the expectation that the Skyrme parameter decreases as $1/\sqrt{N_c}$, while the baryon masses increases with N_c. The error takes into account the uncertainty for fixing the LSR scale τ_0. Extrapolating this result to $N_c = 3$, one obtains $e \simeq (5.2 \sim 6.9)$, which is larger than the one $e \simeq (3.5$–$4.7)$ from Eq. (7.61).

7.3.5 Chiral $\langle\bar\psi\psi\rangle$ condensate from QSSR

In Ref. [207], we select some processes from which we can directly extract the chiral condensate from QCD spectral sum rules.

• $B^* - B$ mass difference

Though the contribution of $\langle\bar\psi\psi\rangle$ is masked by the one of the quark mass in the $B^* - B$ mass difference, one can instead extract quite accurately the size of the mixed condensate [168]:

$$g\langle\bar\psi(\lambda_a/2)\sigma^{\mu\nu}G^a_{\mu\nu}\rangle \equiv M_0^2\langle\bar\psi\psi\rangle = -(9\pm 1)\times 10^{-3}\ \text{GeV}^5. \tag{7.66}$$

Combined with $M_0^2 \simeq (0.8 \pm 0.1)\,\text{GeV}^2$ from nucleon sum rule, one deduces:

$$\langle\bar\psi\psi\rangle(M_N) = -[(225\pm 9)\ \text{MeV}]^3. \tag{7.67}$$

• $D \to K^* l_\nu$

Using a vertex sum rule approach, this process is dominated to leading order by the chiral $\langle\bar\psi\psi\rangle$ condensate. Including different QCD corrections up to dimension six and using as inputs the values of the D, K^* decay constants and the experimental value of the vertex at zero momentum $V(0) = 1.1 \pm 0.2$, one deduces:

$$0.6 \leq \langle\bar\psi\psi\rangle/(-0.229\,\mathrm{GeV})^3 \leq 1.5. \tag{7.68}$$

• Bound and values on light quark masses

Using the GMOR relation $m_\pi^2 f_\pi^2 = -(m_u + m_d)\langle\bar\psi\psi\rangle + \mathcal{O}(m_q^2)$ ($f_\pi =$ 92.3) MeV, we translated the range in Eq. (7.68) into limits on the light quark masses:

$$9.4\,\mathrm{MeV} \leq (\bar m_u + \bar m_d)(1\,\mathrm{GeV}) \leq 15.7\ \mathrm{MeV}. \tag{7.69}$$

Using the ratio: $2m_s/(m_u + m_d) \simeq (25.7 \pm 2.6)$ [208], we deduce:

$$\bar m_s(1\,\mathrm{GeV}) \geq (121 \pm 12)\ \mathrm{MeV}. \tag{7.70}$$

If instead, we use the value of the chiral condensate in Eq. (7.67), we obtain:

$$(\bar m_u + \bar m_d)(1\,\mathrm{GeV}) \simeq 15\ \mathrm{MeV}, \quad \bar m_s(1\,\mathrm{GeV}) \simeq 182\ \mathrm{MeV}. \tag{7.71}$$

These values can be compared with the ones obtained in Refs. [69, 73, 74, 85] from QSSR.

7.3.6 Scalar meson and the sea quark content of the nucleon

• Scalar density

In Ref. [209], we use QSSR to estimate the scalar sea quark content of the nucleon:

$$\langle\bar s s\rangle_N \equiv \langle N|\bar s s|N\rangle. \tag{7.72}$$

Introducing as an intermediate state the scalar meson S, it can be rewritten as:

$$\langle\bar s s\rangle_N \simeq \sigma_{S\equiv\sigma,f_0,\cdots}\langle 0|\bar s s|S\rangle\langle S|\bar N N\rangle\frac{1}{M_S^2}. \tag{7.73}$$

We shall be concerned with the gluon and quark components of the energy–momentum tensor:

$$\theta_q = \sum_{u,d,s} m_i \bar{\psi}_i \psi_i, \quad \theta_g \equiv \frac{1}{4}\beta(\alpha_s) G^2, \tag{7.74}$$

where m_i is the quark current mass and $\beta(\alpha_s)$ is the Gell-Mann β-function. Using the low-energy theorem:

$$\langle N|\theta_q + \theta_g|N\rangle = M_N, \tag{7.75}$$

and neglecting the u, d quark contributions, we obtain:

$$m_s \langle \bar{s}s\rangle_N \leq \frac{M_N}{1 + f_\sigma/(4 f_{f0})}, \tag{7.76}$$

where we have assumed that the coupling of σ and $f_0(1.4)$ to $\bar{N}N$ is the same. The decay constants are normalized as:

$$\langle 0|\theta_g|\sigma\rangle = \frac{1}{4}\sqrt{2} M_\sigma^2 f_\sigma, \quad \langle 0|m_s \bar{s}s|f_0\rangle = \sqrt{2} M_{f0}^2 f_{f0}. \tag{7.77}$$

We have neglected the coupling of $f_0(1.4)$ to θ_g and that of σ to $\bar{s}s$. We estimate the f_0 coupling by relating it to the one of $f_{K_0^*(\bar{u}s)}$ which is [6]:

$$f_f \simeq f_{K_0^*(\bar{u}s)} \simeq (28 \pm 5)\ \text{MeV}. \tag{7.78}$$

Using the value of $f_\sigma \simeq (0.47 \pm 0.19)$ GeV [132], we deduce

$$\langle \bar{s}s\rangle_N (1\,\text{GeV}) \leq 1.1, \tag{7.79}$$

where the inequality is related to an overestimate of the $f_0\bar{N}N$ coupling assumed to be the same as of the lowest ground state σ. Using the value $\langle \bar{u}u + \bar{d}d\rangle_N \simeq 60$ MeV from the σ-term, one deduces the sea quark content of the nucleon:

$$\rho_s \equiv \frac{\langle \bar{s}s\rangle_N}{\langle \bar{u}u + \bar{d}d\rangle_N} \leq 0.1. \tag{7.80}$$

• Golbeger–Treiman $U(1)_S$ relation

Then by neglecting the θ_q contribution in Eq. (7.75), we deduce:

$$M_N \simeq g_{\sigma \bar{N} N}\langle 0|\theta_g|\sigma\rangle \Longrightarrow g_{\sigma \bar{N} N} \simeq \frac{4M_N}{\sqrt{2}f_\sigma} \simeq (3.6 \pm 0.9), \qquad (7.81)$$

which is a generalization of the Golbeger–Treiman relation in the $U(1)_S$ channel.

• Heavy quark scalar density

To estimate the heavy quark density, we consider the relation [3]:

$$\bar{Q}Q \simeq -\frac{1}{12\pi M_Q}\alpha_s G^2 + \mathcal{O}\left(\frac{1}{M_Q^3}\right). \qquad (7.82)$$

Then we can deduce from Eq. (7.75):

$$\langle \bar{Q}Q\rangle_N \simeq \frac{2}{27M_Q}(M_N - m_s\langle \bar{s}s\rangle_N) \simeq \frac{2}{27M_Q}, \qquad (7.83)$$

which exhibits a clear $1/M_Q$ scaling for the heavy quark density. One obtains:

$$\rho_c \simeq 5 \times 10^{-3}, \quad \rho_b \simeq 1.5 \times 10^{-3}, \qquad (7.84)$$

which are negligible. We have used the value of the constituent quark mass given in Ref. [6].

7.3.7 Gluon content of σ from $D_{(s)}$ semi-leptonic decays

In Ref. [210], we test the gluon component of the σ meson from $D_{(s)}$ semi-leptonic decays. Several groups predict:

$$\Gamma(D \to S_2 l\nu) = (8 \pm 2) \times 10^{-16}\,\text{GeV}, \qquad (7.85)$$

where $S_2 \equiv (\bar{u}u + \bar{d}d)$ and $M_{S_2} \simeq 0.6$ GeV. However, the enigmatic nature of σ has led Ref. [132] to consider the possibility that it can contain a large gluon component which is natural if this $U(1)_S$ scalar state is described by the trace of the energy–momentum tensor in

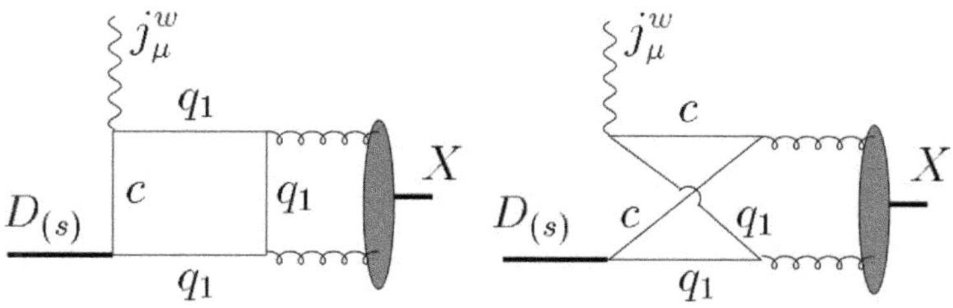

Figure 7.9. Gluonium formation in semi-leptonic $D_{(s)}$ decays.

Eq. (7.74). Here, we consider that the σ decay production is dominated by its gluon component as shown in Fig. 7.9 via the process:

$$D_s \to \sigma_B l\nu \to \pi\pi l\nu, \quad D_s \to \sigma_B l\nu \to \bar{K}Kl\nu, \tag{7.86}$$

with about the same rate in a SU(3) symmetry.

Using a Euler–Heisenberg effective interaction, we have:

$$\mathcal{L}_{eff} \sim \frac{g_W \alpha_s}{p^2 M_c^2} F^2 G^2, \tag{7.87}$$

where F is the electroweak field strength tensor and G the gluon one. Using dispersion techniques like the one used in Refs. [127, 128], we obtain:

$$\Gamma(D_s \to \sigma_B(gg)\, l\nu) \sim |V_{cs}|^2 G_F^2 M_c^3 \frac{1}{M_c^4 M_\sigma^4} |\langle 0|\alpha_s G^2|\sigma_B\rangle|^2, \tag{7.88}$$

while using the result of Ref. [211] for the production of a quarkonium:

$$\Gamma(D_s \to S_q(\bar{q}q)l\nu) \sim |V_{cq}|^2 G_F^2 M_c^5 |f_+(0)|^2, \tag{7.89}$$

we deduce:

$$\frac{\Gamma(D_s \to \sigma_B(gg)\, l\nu)}{\Gamma(D_s \to S_q(\bar{q}q)l\nu)} \sim \frac{1}{|f_+(0)|^2} \left(\frac{f_\sigma}{M_c}\right)^2 \sim \mathcal{O}(1). \tag{7.90}$$

This qualitative estimate indicates that the gluonium production can be of the same order as the $\bar{q}q$ one contrary to the naïve α_s PT counting rule suppression of order α_s^2. This result being a consequence of the OZI-rule violation of the σ_B decay.

7.3.8 Baryons with one heavy quark

In this section, we estimate the masses and couplings of baryons having one heavy quark.

- **$\Sigma_Q : J^P = (1/2)^+$**

We describe these octet states with the current [212]:

$$J^{\Sigma_Q} = (u^T C\gamma_5\, Q)u + k\,(u^T C\, Q)\gamma_5\, u, \quad Q \equiv \mathrm{c, b}, \tag{7.91}$$

where k is an arbitrary mixing parameter. We include the contribution of the condensates up to dimension six in the OPE and work to LO of PT series. We use the LSR for charm states and LSR and Moments evaluated at $Q^2 = 0$ for bottom state for estimating the mass and the residue. The k behaviour of the mass indicates a stability in the range:

$$-0.5 \leq k \leq 0.5 \tag{7.92}$$

to be compared with $\equiv k = -1/5$ of the nucleon sum rule. We deduce the masses and residues:

$$\begin{aligned} M_{\Sigma_c} &\simeq (2.54 \sim 2.94)\,\mathrm{GeV}, \quad |Z_{\Sigma_c}|^2 \simeq (4.2 \sim 7.7) \times 10^{-4}\,\mathrm{GeV}^6, \\ M_{\Sigma_b} &\simeq (5.70 \sim 6.62)\,\mathrm{GeV}, \quad |Z_{\Sigma_b}|^2 \simeq (1.0 \sim 4.5) \times 10^{-3}\,\mathrm{GeV}^6, \end{aligned} \tag{7.93}$$

and the mass difference:

$$M_{\Sigma_b} - M_{\Sigma_c} \simeq 3.4\,\mathrm{GeV}. \tag{7.94}$$

Using the experimental value of M_{Σ_c}, we then predict:

$$M_{\Sigma_b} = 5.85\ \mathrm{GeV}, \tag{7.95}$$

in remarkable agreement with the present experimental data of 5.81 GeV [33].

- **$\Sigma_Q^* : J^P = (3/2)^+$**

Here, we consider the interpolating current [213]:

$$J_\mu^{\Sigma_Q^*}(x) = \frac{1}{\sqrt{3}}\epsilon^{\alpha\beta\lambda} : 2(q_\alpha^T\, C\, \gamma_\mu\, Q_\beta)q_\lambda + (q_\alpha^T\, C\, \gamma_\mu\, q_\beta)Q_\lambda : . \tag{7.96}$$

Using τ and t_c stability criteria, we obtain:

$$M_{\Sigma_c^*} = (2.15 \sim 2.92)\ \text{GeV}, \quad M_{\Sigma_b^*} = (5.4 \sim 6.4)\ \text{GeV}, \tag{7.97}$$

from which we estimate the mass difference:

$$M_{\Sigma_b^*} - M_{\Sigma_c^*} \simeq 3.3\,\text{GeV}, \tag{7.98}$$

in remarkable agreement with the present data of 3.31 GeV [33]. We obtain the residue:

$$|Z_{\Sigma_c^*}|^2 \simeq (1.1 \sim 2.2)\times 10^{-3}\,\text{GeV}^6, \quad |Z_{\Sigma_b^*}|^2 \simeq (2.0 \sim 5.4)\times 10^{-3}\,\text{GeV}^6. \tag{7.99}$$

- **Λ_Q : $J^P = (1/2)^+$**

The corresponding interpolating current is [213]:

$$J_\mu^{\Lambda_Q}(x) = \epsilon^{\alpha\beta\lambda} : (q_\alpha^T\, C\, \gamma_5\, q'_\beta) Q_\lambda + b\,(q_\alpha^T\, C\, q'_\beta)\gamma_5\, Q_\lambda : . \tag{7.100}$$

We fix the mixing parameter in the range given by the Σ_Q sum rule. We do not find a τ-stability within the OPE used up to dimension six. Then, we cannot extract a reliable result. However, comparing the value of M_{Σ_Q} and M_{Λ_Q} at the value of τ and t_c, where the Σ_Q sum rules optimizes, we can approximately deduce:

$$M_{\Sigma_c} - M_{\Lambda_c} \approx 207\ \text{MeV}, \quad M_{\Sigma_b} - M_{\Lambda_b} \approx 163\ \text{MeV}, \tag{7.101}$$

which are not too bad compared with the present data [33]:

$$M_{\Sigma_c}^{exp} - M_{\Lambda_c}^{exp} \approx 168\ \text{MeV}, \quad M_{\Sigma_b}^{exp} - M_{\Lambda_b}^{exp} \approx 191\ \text{MeV}, \tag{7.102}$$

7.3.9 Baryons from HQET sum rules

- **The current**

In Ref. [214], we study the behaviour of the correlator in the limit $M_Q \to \infty$ within an effective heavy quark theory (HQET). In HQET, the heavy quark field is replaced by the velocity-dependent effective field h_v^+. Choosing $v^0 = 1$ and $v^k = 0$ for $k = 1, 2, 3$, the effective field satisfies $\gamma^0 h_v^+ = h_v^+$. Hence, we are left with two spin $1/2$ Λ-type (antisymmetric in the u and d quarks) and two Σ-type (symmetric in u and d) currents:

$$\tilde{J}_{\Lambda_1} = (u^t C\,\gamma^5\, d) h_v^+, \quad \tilde{J}_{\Lambda_2} = (u^t C\,\gamma^0\gamma^5\, d) h_v^+ \quad (\Lambda - \text{type}),$$
$$\tilde{J}_{\Sigma_1} = (u^t C\,\gamma^k\, d)\gamma^k\gamma^5\, h_v^+, \quad \tilde{J}_{\Sigma_2} = (u^t C\,\gamma^0\gamma^k\, d)\gamma^k\gamma^5 h_v^+ \quad (\Sigma - \text{type}), \tag{7.103}$$

where the summation over matched indices are understood. For spin 3/2 current, one can construct the two symmetric (Σ-type):

$$\tilde{J}^l_{\Sigma_1^*} = (u^t C\,\gamma^k\, d)\left(\delta^{kl} + \frac{1}{3}\gamma^l\gamma^k\right) h_v^+,$$
$$\tilde{J}_{\Sigma_2^*} = (u^t C\,\gamma^0\gamma^k\, d)\left(\delta^{kl} + \frac{1}{3}\gamma^l\gamma^k\right) h_v^+. \tag{7.104}$$

Some general combinations of these currents which transform under SU(3) flavour symmetry and reproduce the standard quark currents have been built in Ref. [214].

• Positive and negative parity states

If one considers baryonic currents $\mathcal{O}^+$ and $\mathcal{O}^-$ which couple strongly with positive and negative parity states, i.e., $\mathcal{P}\mathcal{O}^\pm = \gamma^0\mathcal{O}^\pm$, one has:

$$i\int d^4x\, e^{iqx}\langle 0|\mathcal{T}\mathcal{O}(x), \bar{\mathcal{O}}(0)|0\rangle = F_1(p^2)\gamma\cdot,p + F_2(p^2)$$
$$= Z_\pm \frac{\gamma\cdot p + m_\pm}{p^2 - m_\pm^2} + \cdots. \tag{7.105}$$

In the case of mixed current $\Psi^+ = \mathcal{O}^+\alpha\gamma^5\mathcal{O}^-$ which also has a positive parity $\mathcal{P}\Psi^+ = \gamma^0\Psi^+$, one has:

$$i\int d^4x\, e^{iqx}\langle 0|[\Psi(x)\bar{\Psi}(0)]|0\rangle = Z_+\frac{\gamma\cdot p + m_+}{p^2 - m_+^2} + \alpha^2 Z_-\frac{\gamma\cdot p - m_-}{p^2 - m_-^2}, \tag{7.106}$$

i.e., away from the pole the admixture of the wrong current shifts the weights of $F_{1,2}$. However, one knows in HQET that the baryon spectral function proportional to $\gamma\cdot v$ should be of the same order in M_Q as that proportional to 1. So, the spectral function built with any of the previous currents $\mathcal{O}$ can be separated into positive and negative parity state contributions as:

$$i\int d^4x\, e^{iqx}\langle 0|[\mathcal{O}(x),\bar{\mathcal{O}}(0)]|0\rangle = M_Q\,\gamma\cdot v\,\mathrm{Im}F_1 + \mathrm{Im}F_2$$

$$= \frac{1+\gamma\cdot v}{2}(M_Q\,\mathrm{Im}F_1 + \mathrm{Im}F_2) - \frac{1-\gamma\cdot v}{2}(M_Q\,\mathrm{Im}F_1 - \mathrm{Im}F_2), \tag{7.107}$$

where we have written $p_\mu = M_Q\,v_\mu + \bar{p}_\mu$ with $\bar{p}_\mu \ll M_Q$. The $(1+\gamma\cdot v)$ term gives the positive parity contribution, while $(1-\gamma\cdot v)$ the negative one. This result indicates that the separation of states of positive and negative parities is unambiguous in HQET but not in a standard QCD approach.

• LSR analysis

In HQET, it is convenient to introduce the LSR variables:

$$\tau\,[\mathrm{GeV}]^{-1} = 2M_Q\tau_R, \quad E_c = \sqrt{t_c} - M_Q, \tag{7.108}$$

and the mass gap between the resonance and the quark:

$$\delta M_B = M_B - M_Q. \tag{7.109}$$

In the case of the positive parity state Σ_b^+, one has τ-stability by retaining the condensate contributions up to dimension 5, but there is no E_c-stability. Taking $E_c = 1.5$ GeV from the lowest τ-stability and multiplying this value by a factor 2, one obtains

$$\delta M_{\Sigma^+} \simeq (1.1 - 2.1)\ \mathrm{GeV}. \tag{7.110}$$

Applied to Λ^+, we do not have τ-stability like in the case of standard sum rule discussed previously. The coupling Z_{Λ^+} presents τ-stability and agrees with the previous one in Eq. (7.93).

In the case of negative parity state, we have the sane feature as for the positive parity case. We obtain, for $E_c \simeq 2-3$ GeV,

$$\delta M_{\Sigma^-} \simeq (1.8 - 2.5)\ \mathrm{GeV}. \tag{7.111}$$

One should mention that the value of the baryon mass gap obtained here is much larger than the one from HQET meson sum rules of about (0.6–0.8) GeV [215].

7.3.10 Baryons with two heavy quarks

In this section, we estimate the mass and coupling of baryons having two heavy quarks [216].

• $\Xi^*_Q : J^P = (3/2)^+$

We describe this state by the current:

$$J_\mu^{\Xi^*_Q}(x) = \frac{1}{\sqrt{3}}\epsilon^{\alpha\beta\lambda} : 2(Q^T_\alpha\, C\,\gamma_\mu\, u_\beta)Q_\lambda + (Q^T_\alpha\, C\,\gamma_\mu\, Q_\beta)u_\lambda : . \quad (7.112)$$

We obtain:

$$M_{\Xi^*_c} = (3.58 \pm 0.05)\ \text{GeV}, \quad M_{\Xi^*_b} = (10.33 \pm 1.09)\ \text{GeV},$$
$$|Z_{\Xi^*_c}|^2 \simeq (2.9 \sim 7.8) \times 10^{-3}\,\text{GeV}^6, \quad |Z_{\Sigma^*_b}|^2 \simeq (5 \sim 23) \times 10^{-3}\,\text{GeV}^6. \quad (7.113)$$

• $\Xi_Q : J^P = (1/2)^+$

We describe this state by the current:

$$J_\mu^{\Xi_Q}(x) = \epsilon^{\alpha\beta\lambda} : (Q^T_\alpha\, C\,\gamma_5\, u_\beta)Q_\lambda + b\,(Q^T_\alpha\, C\, u_\beta)\gamma_5\, Q_\lambda : . \quad (7.114)$$

We obtain:

$$M_{\Xi_c} = (3.48 \pm 0.06)\ \text{GeV}, \quad M_{\Xi_b} = (9.94 \pm 0.91)\ \text{GeV},$$
$$|Z_{\Xi_c}|^2 \simeq (0.7 \sim 1.8) \times 10^{-3}\,\text{GeV}^6, \quad |Z_{\Xi_b}|^2 \simeq (4.5 \sim 13.2) \times 10^{-3}\,\text{GeV}^6. \quad (7.115)$$

• $\Lambda(bcu)\ : J^P = (1/2)^+$

We choose the corresponding interpolating current:

$$J_\mu^{\Lambda}(x) = \epsilon^{\alpha\beta\lambda} : (u^T_\alpha\, C\,\gamma_5\, c_\beta)b_\lambda + k\,(u^T_\alpha\, C\, c_\beta)\gamma_5\, b_\lambda : . \quad (7.116)$$

and obtain:

$$M_\Lambda = (6.44 \pm 0.19)\ \text{GeV}, \quad |Z_\Lambda|^2 \simeq (4.6 \sim 19) \times 10^{-3}\,\text{GeV}^6. \quad (7.117)$$

A more general choice of the current is given in Refs. [216, 217], which leads to a slightly higher value for the mass:

$$M_\Lambda = (6.86 \pm 0.28)\ \text{GeV}. \quad (7.118)$$

7.3.11 Hadrons with charm and beauty

Reference [217] ends our collaboration.

• $B_c(0^-)$ meson

We start by estimating the mass and coupling of the $B_c(0^-)$ meson associated to the quark current:

$$J_5 = (m_b + m_c) : \bar{b}(i\gamma_5)c : \text{ with } \langle 0|J_5|B_c\rangle = M_{B_c} f_{B_c}, \tag{7.119}$$

from the corresponding two-point correlator. Including the α_s PT corrections and non-perturbative terms up to dimension 6, we deduce from the LSR and moments sum rules the average:

$$M_{B_c} = (6.5 \pm 0.4)\ \text{GeV}, \quad f_{B_c} = (2.94 \pm 0.22) f_\pi, \tag{7.120}$$

and the corresponding QCD continuum threshold value:

$$t_c \equiv (m_b + m_c + E_c)^2 : \quad E_c = (1 \sim 2)\ \text{GeV}. \tag{7.121}$$

This value agrees with the potential model predictions [217]:

$$M_{B_c} = (6255 \pm 20)\ \text{MeV}, \quad f_{B_c} = (3.86 \pm 1.31) f_\pi. \tag{7.122}$$

The mass of the $B_c^*(1^-)$ is also predicted by the potential model to be:

$$M_{B_c^*} = (6330 \pm 20)\ \text{MeV}. \tag{7.123}$$

• Baryons

We extend the analysis to the baryon sector and revisit the QSSR analysis of the $\Lambda(bcu)$ given in Eq. (7.117). The ratios of sum rules for fixing the mass does not show any τ or n stability, we fix these values at the one where the sum rules for coupling stabilizes. In this way, we obtain:

$$|Z_\Lambda|^2 \simeq (4 \sim 20) \times 10^{-3}\ \text{GeV}^6, \quad M_\Lambda = (6.86 \pm 0.28)\ \text{GeV}. \tag{7.124}$$

Table 7.1 summarizes your results in the baryon sector.

Table 7.1. Spin 1/2 Baryon masses from Potential Models (PM) and QCD Spectral Sum Rules (QSSR). Averaged spin $1/2 \oplus 3/2$ masses are given in Ref. [217].

Spin 1/2 Baryons	PM [217]	QSSR
(ccq)	3.63	3.48 ± 0.06 [216]
(ccs)	3.72	–
(bbq)	10.21	9.94 ± 0.91 [216]
(bbs)	10.27	–
(bcq)	6.93	6.86 ± 0.28 [217]
(bcs)	7.00	–

• Semi-leptonic decays of the B_c meson

We estimate the semi-leptonic decay of the B_c meson using vertex sum rules.

◇ $B_c \to Pl\nu$

The semi-leptonic decay matrix element we are interested in can be decomposed as:

$$\langle P|J_\mu|B_c\rangle = F_+(t)(p+p')_\mu + F_-(t)(p-p')_\mu, \tag{7.125}$$

where:

$$J_\mu =: \bar{\psi}\gamma_\mu(1-\gamma_5)Q :, \quad J_5 = (m_+ m_c) : \bar{b}(i\gamma_5)c : . \tag{7.126}$$

P is a pseudoscalar meson. For semi-leptonic decays, only the form factor F_+ contributes as the one of F_- is proportional to the mass squared of the lepton. Then, we use vertex sum rules for estimating $F_+(0)$. We use a double dispersion relation and parametrize the higher state contributions by the QCD continuum from thresholds:

$$t_c = (m_Q + m_{Q'} + E_c)^2, \quad t'_c = (m_Q + m_{Q'} + E'_c)^2, \tag{7.127}$$

where in many cases [6, 218]:

$$E_c \simeq (1-2)\,\text{GeV} \tag{7.128}$$

yields optimal results for the QSSR analysis. Then we obtain the LSR:

$$\int_{(m_Q+m_{Q'})^2}^{t_c} ds \int_{(m_\psi+m_{Q'})^2}^{t'_c} ds' \rho_+(s,s',t) e^{-s\tau} e^{-s'\tau'}$$
$$+\text{"non-pert. terms"} = e^{-M_F^2\tau'} e^{-M_{B_c}^2\tau} \langle 0|J_F|H_F\rangle\langle B_c|J_5^\dagger|0\rangle F_+(t). \tag{7.129}$$

In the evaluation, we do not take the (small) contribution from the gluon condensate into account and we hence come to the following sum rules:

$$F_+(t) = \frac{e^{-M_F^2\tau'} e^{-M_{B_c}^2\tau'}}{\langle 0|J_F|H_F\rangle\langle B_c|J_5^\dagger|0\rangle} \int_{(m_Q+m_{Q'})^2}^{(m_Q+m_{Q'}+E_c)^2}$$
$$\times\, ds \int_{(m_\psi+m_{Q'})^2}^{(m_Q+m_{Q'}+E'_c)^2} ds' \rho_+(s,s',t) e^{-s\tau} e^{-s'\tau'}. \tag{7.130}$$

In the next step, the matrix elements $\langle 0|J_F|H_F\rangle$ and $\langle B_c|J_5^\dagger|0\rangle$ are expressed through sum rules. By choosing the parameters τ and τ' to be 1/2 of the corresponding parameter in the two-point sum rule, the exponential dependence drops out if we evaluate $F_+(t)$ from (7.129). Note that the sum rule for the two-point functions yield an expression for $|\langle 0|J_F|H_F\rangle|^2$ and $|\langle B_c|J_5^\dagger|0\rangle|^2$. We furthermore choose the continuum threshold the same for the two- and three-point functions, i.e., t_c for the B_c channel and t'_c for the H_F channel. There is a very subtle point in the t-dependence of the perturbative double spectral function. For $t < 0$, there is no problem in applying the Cutkosky rules in order to determine $\rho_+(s,s',t)$ and the limits of integration. For $t > 0$,

Table 7.2. Estimates of the form factors from vertex sum rules.

	$B_c \to \eta_c$ $B_c \to J/\psi$	$B_c \to B_s$ $B_c \to B_s^*$	$B_c \to B$ $B_c \to B^*$	$B_c \to D$ $B_c \to D^*$
$F_+(0)$	0.55 ± 0.10	0.60 ± 0.12	0.48 ± 0.14	0.18 ± 0.08
$F_V(0)$ [GeV^{-1}]	0.048 ± 0.007	0.15 ± 0.02	0.11 ± 0.02	0.02 ± 0.01
$F_+^A(0)$ [GeV^{-1}]	-0.030 ± 0.003	-0.005 ± 0.005	0.005 ± 0.005	0.010 ± 0.010
$F_0^A(0)$ [GeV]	3.0 ± 0.5	3.3 ± 0.7	1.7 ± 0.7	0.8 ± 0.4

which is the physical region for decays, non-Landau-type singularities appear [219, 220], which make the determination of the double spectral function very cumbersome. For finite values of t_c, t'_c, the non-Landau singularities do not contribute to the sum rule (7.129) if t is smaller than a certain value t_{cr}, which depends on t_c and t'_c, and hence, the Cutkosky rules may be applied in a straightforward way.

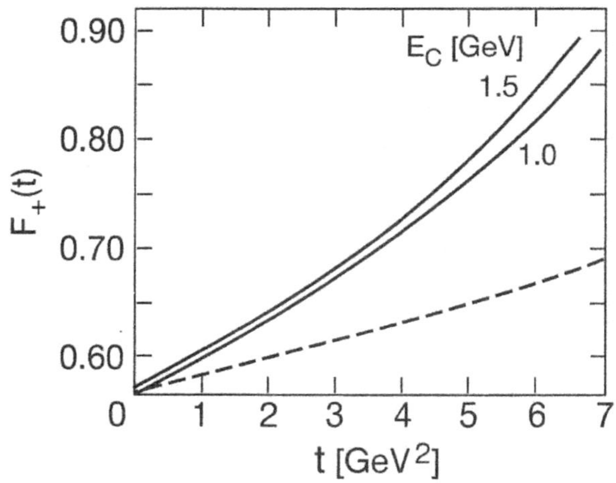

Figure 7.10. t-behaviour of the $F^+(t)$ form factor for the process $B_c \to \eta_c$: Solid lines — QCD prediction using a cubic polynomial fit for two different values of E_c; Solid lines — VDM prediction using a pole mass $M_{B_c^*} = 6.33$ GeV.

For the determination of the ratios, we extrapolate the t-dependence of that range to the full range with a cubic extrapolation. We show the t-behaviour of the $F_+(t)$ form factor in Fig. 7.10 by comparing the QSSR prediction with the VDM one:

$$F_+(t) = \frac{1}{1 - t/M^2_{\text{pole}}}. \tag{7.131}$$

For $B_c \to J/\pi$, we find in units of GeV:

$$\begin{aligned} &F_V : M_{\text{pole}} \simeq 4.08\,\text{GeV}, \quad F_+^A : M_{\text{pole}} \simeq 4.44\,\text{GeV}, \\ &F_0^A : M_{\text{pole}} \simeq 4.62\,\text{GeV}. \end{aligned} \tag{7.132}$$

The corresponding decay rate is:

$$\frac{d\Gamma_+}{dt} = \frac{G_F^2 |V_{Q\psi}|^2}{192\pi^3 M_{B_c}^3} \lambda^{3/2}(M_{B_c}^2, M_F^2, t) F_+^2(t). \qquad (7.133)$$

Table 7.3. Partial decay rates for B_c and B_c^* mesons.

Channels	Reference	Rates in $10^{10}\mathrm{s}^{-1}$
$B_s l\nu$	This paper	0.35 ± 0.10
	[221]	0.18
$B_s^* l\nu$	This paper	0.35 ± 0.10
	[221]	0.87
$\eta_c l\nu$	This paper	0.27 ± 0.07
	[221]	0.03
$J/\psi l\nu$	This paper	0.32 ± 0.08
	[221]	0.21

⋄ $B_c \to V l\nu$

For the case of a vector meson V in the final state, the relevant amplitudes are:

$$\langle H_F \epsilon^{(\lambda)} | J_\mu | B_c \rangle = -iF_0^A(t)\epsilon_\mu^* + iF_+^A(t)\epsilon^{*(\lambda)} \cdot p\,(p+p')_\mu + 2F_V(t)\epsilon_\mu^{\nu\rho\sigma}\epsilon_\nu^{*(\lambda)} p_\rho p'_\sigma + \cdots \qquad (7.134)$$

The amplitudes F_+ and F_V receive their contributions from the vector currents, while F_0^A and F_+^A do so from the axial-vector one. For the vector final state:

$$\frac{d\Gamma_+}{dt} = \frac{G_F^2 |V_{Q\psi}|^2}{192\pi^3 M_{B_c}^3} \lambda^{1/2}(M_{B_c}^2, M_F^2, t) \times \left[2(F_0^A)^2 + \lambda F_V^2 + \frac{1}{4M_F^2}\left((M_{B_c}^2 - M_F^2 - t)F_0^A + \lambda F_+^A\right)^2\right],$$

$$\lambda = \lambda(M_{B_c}^2, M_F^2, t). \qquad (7.135)$$

We proceed in a similar way as in the case of the pseudoscalar meson to estimate the form factors. We show the t behaviour of the form factors The results are given in Table 7.2. We find that the t-dependence of the form factors is stronger than expected from VDM. It is important that experiments check this feature.

7.3.12 $B^*B\pi(\gamma)$ and $D^*D\pi(\gamma)$ within $1/M$ in full QCD

In Ref. [222], we evaluate the leading and non-leading $1/M_b$ corrections to the $B^*B\pi$ and $B^*B\gamma$ couplings using QCD spectral moment sum rules in the full theory. These couplings are defined as:

$$\langle B^*(p)B(p')\pi(q)\rangle = g_{B^*B\pi}q_\mu\epsilon^\mu,$$
$$\langle B^*(p)B(p')\gamma(q)\rangle = -eg_{B^*B\gamma}p_\alpha p'_\beta\epsilon^{\mu\nu\alpha\beta}\epsilon_\mu\epsilon'_\nu, \qquad (7.136)$$

where $q \equiv p' - p$ and $-Q^2 \equiv q^2 \leq 0$, while ϵ_μ are the polarization of the vector particles.

- **Vertex sum rules**

We shall be concerned with the vertex function:

$$V^{\nu(\mu)}(p,p',q) = -\int d^4x\, d^4y\, e^{i(p'y-px)}\langle\mathcal{T}J_L^{5(\mu)}(x)J_{B^*}^\nu(0)J_B(y)\rangle, \qquad (7.137)$$

where the currents are:

$$J_L^\mu = \sum_{u,d} e_q\bar{q}\gamma^\mu q + \sum_{c,b} e_Q\bar{Q}\gamma^\mu Q, \quad J_L^5 = (m_u+m_d)\bar{u}\gamma^5 d,$$
$$J_{B^*}^\nu = \bar{u}\gamma^\nu b, \quad J_B = (M_b+m_d)\bar{d}\gamma^5 b, \qquad (7.138)$$

u, d, c, b are the quark fields and e_q, e_Q their electric charge in units of e. The vertex obeys the double dispersion relation[1]:

$$V(p,p',q) = -\frac{1}{4\pi^2}\int_{M_b^2}^\infty \frac{ds}{s-p^2}\int_{M_b^2}^\infty \frac{ds'}{s'-p'^2}\,\mathrm{Im}V(s,s') + \cdots \qquad (7.139)$$

[1]Here, the dispersion relation is done with respect to the two heavy meson momenta, which is not the case of $B \to \pi(\rho)l\bar{\nu}$. This different configuration is important for the M_b-behaviour of the different QCD contributions.

Exploiting the fact that M_b is much larger than the QCD scale Λ, where the LHS can be evaluated using the operator product expansion à la SVZ [3, 6, 7], one can work, in the chiral limit, with the double moment sum rule:

$$\mathcal{M}^{(n,n')} = -\frac{1}{4\pi^2}\int_{M_b^2}^{\infty}\frac{ds}{s^{n+1}}\int_{M_b^2}^{\infty}\frac{ds'}{s'^{n'+1}}\,\mathrm{Im}V(s,s'), \tag{7.140}$$

where n and n' are *finite* numbers of derivatives evaluated at $p^2 = p'^2 = 0$.

◇ We find that, for large M_b and contrary to the heavy-to-light $B \to \pi(\rho)l\bar{\nu}$ form factors, which are dominated by the *soft* light quark vacuum condensate, these couplings are governed by the *hard* perturbative graph, like other heavy-to-heavy transitions.

◇ We also find that for the $B^* \to B\gamma$, the $1/M_b$ correction is mainly due to the perturbative and light quark condensate contributions originating from the graphs involving the heavy quark part of the electromagnetic current, which are essential for explaining the large charge dependence in the observed $D^{*-} \to D^-\gamma$ and $D^{*0} \to D^0\gamma$ decays.

• B^*-meson

We obtain from the moments:

$$g_{B^*B\pi} \simeq 14.5 \pm 3.3, \quad g_{D^*D\pi} \simeq 7.1 \pm 1.6. \tag{7.141}$$

We cross-check the validity of the previous result by invoking semi-local duality sum rules for the two- or three-point functions which correspond, respectively, to the particular cases, where $(n_3, n_2) = (-1/2, -1)$ or $(n_3, n_2) = (-1, -2)$.[2] We obtain:

$$g_{B^*B\pi} \simeq 12.7 \pm 2.9, \quad g_{D^*D\pi} \simeq 5.0 \pm 1.1. \tag{7.142}$$

Our *best* numerical predictions *without any free parameters* for the B^*-meson are:

$$g_{B^{*-}B^0\pi^-} \simeq 14 \pm 4, \quad \Gamma_{B^{*-}\to B^-\gamma} \simeq (0.10 \pm 0.03)\,\mathrm{keV}, \tag{7.143}$$

[2]One should also note that the use of the Laplace sum rules leads to a small value of the optimization scale τ, which is *practically* similar to the semi-local duality sum rules used here.

and the large charge dependence of the ratio:

$$\Gamma_{B^{*-}\to B^-\gamma}/\Gamma_{B^{*0}\to B^0\gamma} \simeq 2.5. \tag{7.144}$$

- ***D**-meson**

For the D^*-meson, we find:

$$\Gamma_{D^{*-}\to D^0\pi^-} \simeq 1.54\Gamma_{D^{*0}\to D^0\pi^0} \simeq (8 \pm 5)\,\text{keV}, \tag{7.145}$$

$$\Gamma_{D^{*-}\to D^-\gamma} \simeq (0.09^{+0.40}_{-0.07})\,\text{keV}, \quad \Gamma_{D^{*0}\to D^0\gamma} \simeq (3.7 \pm 1.2)\,\text{keV}, \tag{7.146}$$

where the branching ratios agree within the errors with the present data, while the total widths:

$$\Gamma_{D^{*0}\to all} \simeq (11 \pm 4)\,\text{keV}, \quad \Gamma_{D^{*-}\to all} \simeq (12 \pm 7)\,\text{keV} \tag{7.147}$$

are much smaller than the present experimental upper limits.

7.4 Visiting Scientist at Vienna University 1995

Figure 7.11. My host R. Bertlmann (Institute for Theoretical Physics, Vienna).

7.4.1 Anecdotes

My first contact with Vienna was my participation in an annual workshop between Austria and the neighbouring East countries from an invitation by Reinhold Bertlmann (Institute for Theoretical Physics) in 1982–1983. The participants stood in line to get some money from the organizers for covering their expenses. Arrived at my turn, the chairman (I do not remember exactly the person, but he might be presumably Professor H. Pietschmann) asked Reinhold "who is this guy..." before giving me the money! It seems that I did not look like a physicist!

In 1995, I initiated a collaboration with the Institute Theoretical Physics, Vienna and got a joint grant from the French Ministry of Foreign Affairs, French embassy in Vienna and the Austrian Ministry of Research. I have started this work [223], where Reinhold Bertlmann (see Fig. 7.11) my host has initially participated partially. The stay in Vienna is certainly nice if one likes opera and traditional coffee places (Reinhold brought me to one of these). Another day, he brought me to the famous country of Tyrol for dinner and and the same time listening to the traditional tyrolian music with accordion. I did not expect to see a Malagasy person playing the tyrolian music, while he was surprised to see a Malagasy in this countryside.

7.4.2 $\langle\alpha_s G^2\rangle$ and α_s from heavy quarkonia mass splittings

During my stay, I studied, using the sum rules approach, the mass splittings between different charmonium and bottomium states within a $1/M$ expansion with the aim to extract the gluon condensate and the QCD coupling α_s [223].

• The double ratio of LSR

In so doing, I use the double ratio of LSR moments:

$$\begin{aligned}\mathcal{R}_{HH'}(x) &\equiv \frac{\mathcal{R}_H}{\mathcal{R}_{H'}} \simeq \frac{M_H^2}{M_{H'}^2}\\ &= \Delta_0^{HH'}\left[1+\alpha_s\Delta_{\alpha_s}^{HH'}+\frac{4\pi}{9}\langle\alpha_s G^2\rangle\sigma^2x^2\Delta_G^{HH'}\right],\end{aligned} \tag{7.148}$$

provided that each ratio of moment:

$$\begin{aligned}\mathcal{R}_H(\sigma) &\equiv -\frac{d}{d\sigma}\log\mathcal{L}_H(\sigma,m^2)\\ &= 4m^2F_H(\omega)\left[1+\alpha_s P_H(\omega)+\frac{\pi}{36}\frac{\langle\alpha_s G^2\rangle}{m^4}Q_H(\omega)\right], \qquad (7.149)\end{aligned}$$

stabilizes at about the same value of $\sigma = \omega/(4m^2)$, as in this case, there is a cancellation of the different leading terms, such as the heavy quark mass (and its ambiguous definition used in some previous literatures), the negligible continuum effect (which is already small in the ratio of moments), and each leading QCD corrections. $\mathcal{L}_H(\sigma, m^2)$ is the LSR moments where its expression comes from Refs. [64, 70, 224, 225]. We shall limit ourselves here to the α_s-correction for the perturbative contribution and to the leading order one in α_s for the gluon condensate effects.

We perform our analysis semi-analytically. Among different estimates, we mention the following:

- **Gluon condensate $\langle\alpha_s G^2\rangle$ from $M_{\psi(S_1^3)} - M_{\eta_c(S_0^1)}$**

We extract the gluon condensate from $M_{\psi(S_1^3)} - M_{\eta_c(S_0^1)}$ mass splitting and found:

$$\langle\alpha_s G^2\rangle \simeq (0.10 \pm 0.04)\ \text{GeV}^4. \qquad (7.150)$$

We have estimated the error due to higher order effects by replacing the coefficient of α_s with the one obtained from the effective Coulombic potential, which tends to reduce the estimate to 0.07 GeV^4. We have tested the convergence of the QCD series in σ by using the numerical estimate of the dimension 6 gluon condensate $g\langle f_{abc}G^aG^bG^c\rangle$ contributions given in Ref. [225]. This effect is about 0.1% of the zeroth-order term and does not influence the previous estimate in Eq. (22), which also indicates the good convergence of the ratio of exponential moments already at the charm mass scale in contrast with the $q^2 = 0$ moments studied in Ref. [3]. We also expect that in the double ratio of moments used here, the radiative corrections to the gluon condensate effects (their expression for the two-point correlator is however available in the literature [226]) are much smaller than in the individual moments, such that they will give a negligible effect in the estimate of the gluon condensate.

• α_s from the P_1^1–P_1^3 axial mass splitting

The corresponding double ratio of moments has the nice feature to be independent of the gluon condensate to leading order in α_s and reads:

$$\frac{M^2_{P_1^1}}{M^2_{P_1^3}} \simeq 1 + \alpha_s \left[\Delta_\alpha^{13}(\text{exact}) = 0.014^{-0.004}_{+0.008} \right]. \tag{7.151}$$

We use the experimental value of the $h_c(1P)$ mass of 3526.1 MeV and a naïve exponential resummation of the higher order α_s terms. Then, we deduce:

$$\alpha_s(\sigma^{-1} \simeq 1.3 \text{ GeV}) \simeq 0.64^{+0.36}_{-0.18} \pm 0.02$$
$$\Longrightarrow \alpha_s(M_Z) \simeq 0.127 \pm 0.009 \pm 0.002, \tag{7.152}$$

where the error is much bigger than the one from LEP and τ decay data, but its value is perfectly consistent with the latter. The theoretical error is mainly due to the uncertainty in Δ_α, while a naïve resummation of the higher order α_s terms leads to the second error.

• Gluon condensate from the $\Upsilon - \chi_b$ mass splittings

As the S and P wave ratios of moments are optimized at the same value of σ, we can compare directly, with a good accuracy, the different P states with the Υ (S_1^3) one. As the coefficient of the α_s^2 corrections, after inserting the expression of σ_{min}, are comparable with the one from the Coulombic potential, we expect that the prediction of this splitting is more accurate than in the case of the hyperfine. We obtain:

$$\langle \alpha_s G^2 \rangle \simeq (6.9 \pm 2.5) \times 10^{-2} \text{ GeV}^4. \tag{7.153}$$

We expect that this result is more reliable than the one obtained from the $M_\psi - M_{\eta_c}$ as the latter can be more affected by the non-calculated next-next-to-leading perturbative radiative corrections than the former. An average of the two results from the $\psi - \eta_c$ and $\Upsilon - \chi_b$ mass splittings leads to:

$$\langle \alpha_s G^2 \rangle \simeq (7.5 \pm 2.5) \times 10^{-2}\ \text{GeV}^4, \tag{7.154}$$

where we have retained the most precise error.

• $\Upsilon - \eta_b$ mass splitting

Some mass splittings are given in Table 2 of Ref. [223]. Among these, we mention the $\Upsilon - \eta_b$ mass splitting:

$$M_\Upsilon - M_{\eta_b} \approx \left(63^{-29}_{+51}\right)\ \text{MeV}, \tag{7.155}$$

implying the possible observation of η_b from the Υ radiative decay.

7.5 Associate @ University of Michigan and MPI Munich

(a) (b)

Figure 7.12. (a) Michigan University (Ann Arbor); (b) Max-Planck Institute for Physics (Munich).

Figure 7.13. My host and collaborator Professor V.I. Zakharov (ITEP Moscow, Max Planck Munich).

As a follower of the SVZ works, I was quite happy to meet Professors Valya I. Zakharov and Arkady I. Vainshtein (who I already met in Trieste in 1981) during my visit at CERN in 1985. I have invited them to visit our institute in Montpellier and they have accepted. I brought them to Montpellier with my new Mercedes car. The meeting was quite pleasant and our group was happy to have prestigious guests. Then, we try to invite Valya for a permanent position after his contract in Michigan, but (unfortunately) we did not succeed. The same problem happens for our collaborator, K.G. Chetyrkin. From such an experience, I realize that one should be involved in more politics. However, we succeed to invite Valya for few months contract position in Montpellier during July–September 1998, September–December 2001 and June–August 2002. Since 1990, Valya participates regularly in the QCD-Montpellier conference and is until now a conference committee member.

We have started some collaborative works during his stay in Montpellier which we continued in Michigan (Ann Arbor) (1997) and Munich (2003). These works are centred on the analysis of the OPE within the SVZ-expansion in terms of sum of condensates with increasing dimensions which are expected to parametrize the non-perturbative effects of QCD. Due to gauge invariance, one cannot build a dimension two condensate $\langle A^2\rangle$ (A_μ is the gluon field) (see, however, Ref. [227]), such that one does not expect to have such a term in the OPE except the quark mass squared m_q^2. However, this term is present in some holographic models [228] and in a lattice calculation of the QCD potential [229] and gluon condensates [230, 231, 530]. It has also been proposed [232, 233] for a phenomenological parametrization of UV renormalon as an alternative to the large β-approximation. The large β-approximation is not yet fully justified due to the non-observation of the factorial growth and alternate signs of the calculated coefficients of the Adler D-function and some other QCD observables known to order α_s^4. Instead, these calculated coefficients grow as a geometric sum [234].

7.5.1 $1/Q^2$ corrections due to a tachyonic gluon mass λ

This term has been also proposed [232, 233] for a phenomenological parametrization of UV renormalon. In Ref. [235], we do a systematic analysis of the effects of this term in different processes involving two-point correlators of hadronic currents.

• Estimate of λ^2

Here, we have extended the estimate of the value of λ^2 from $e^+e^- \to$ hadrons [236, 237] to the pion channel and found a consistent value leading to (see also [238]):

$$(\alpha_s/\pi)\lambda^2 \simeq -(0.06 \pm 0.07)\ \text{GeV}^2, \tag{7.156}$$

which has been improved to:

$$(\alpha_s/\pi)\lambda^2 \simeq -(6.7 \pm 1.5) \times 10^{-2}\ \text{GeV}^2 \tag{7.157}$$

from a recent analysis of the $e^+e^- \to$ hadrons data to order α_s^4 of the PT series [239].

• Effects of λ^2 in some observables

◇ We found that this term reduces the value of the light quark mass from pseudoscalar sum rules by about -5% and the value of m_s from $e^+e^- \to$ hadrons and τ decay data by about$-$(1–3)%.

◇ Its effect on α_s from the τ-decay lowest moment is about $-(5\pm 10)\%$. The recent analysis in Ref. [239] indicates that its effect does not exceed -4%.

• The sum rule scales puzzle

◇ ρ-meson channel

Considering the ρ-meson two-point function:

$$\Pi_\rho(M^2) \approx (\text{parton model})\left[1 + \left(\frac{\alpha_s}{\pi}\right)\left(1 - 1.05\,\frac{\lambda^2}{M^2}\right) + \frac{\pi}{3M^4}\langle\alpha_s G^2\rangle + \cdots\right], \tag{7.158}$$

and assuming a 10% correction from the non-perturbative contribution, one can fix $\lambda^2 = 0$ the critical scale:

$$M^2_{\text{crit}}(\rho\,\text{channel}) \approx \sqrt{10\frac{\pi}{3}\langle\alpha_s G^2\rangle} \approx (0.6 - 0.8)\,\text{GeV}^2, \tag{7.159}$$

where the factor 10 quantifies the 10% correction. This result is not much affected by the value of λ^2 and agrees with a more involved phenomenological analysis. Since the gluon condensate parametrises the contribution of the infrared region, one may say that at least in the ρ-channel, the breaking of the asymptotic freedom is due to infrared phenomena.

◇ Pseudoscalar meson channel

In the pion channel, the lower bound on $M^2_{\rm crit}$ can be determined by evaluating the pion contribution to the dispersive integral and equating it to the parton-model result. Since all other hadronic states give positive-definite contributions, at lower M^2, the parton model is certainly violated. In more detail, one gets:

$$M^2_{\rm crit}(\pi\,\text{channel}) \geq \frac{4\pi}{\sqrt{3}}\frac{f_\pi m_\pi^2}{(m_u+m_d)} \approx 2\,\text{GeV}^2, \tag{7.160}$$

where the running light quark masses are evaluated at the scale M^2 and $f_\pi = 92.3$ MeV. Adding the PT and λ^2 corrections:

$$\rho_\pi = (\text{parton level})\left[1+\frac{\alpha_s}{\pi}\left(4.821 - 4\lambda^2\tau\right)\right], \tag{7.161}$$

the value of $M^2_{\rm crit}(\pi\,\text{channel})$ becomes 1.65 GeV^2 without λ^2 and 1.6 GeV^2 for the value of λ^2 obtained in Eq. (7.157). One should note that although the difference of about a factor 3 in the values of $M^2_{\rm crit}$ in the ρ and π channels might seem not so big, it cannot be matched by the standard IR contributions.

◇ Scalar meson channel

The scalar channel differs from the pseudoscalar one by the contribution of the four-quark condensate with flips sign and with a slightly different value of the coefficient such that in this approach, one expects to have

$$M^2_{\rm crit}(a_0\,\text{channel}) \approx M^2_{\rm crit}(\pi\,\text{channel}). \tag{7.162}$$

This is not the case of direct instanton models where the leading corrections act in opposite signs for the scalar and pseudoscalar channels.

◇ Scalar glueball channel

In case of the scalar-gluonium channel, the value of $M^2_{\rm crit}$ follows from a low-energy theorem for the correlator associated with the current $\alpha_s G^2$ [128]. This low-energy theorem which estimates the subtraction constant $\Pi_G(0)$ can be translated into correction to the parton model predictions at large M^2 as:

$$\Pi_G(M^2) = (\text{parton model})\left(1 - \frac{8\pi}{\beta_1}\left(\frac{\pi}{\alpha_s}\right)^2 \frac{\langle\alpha_s G^2\rangle}{M^4}\right) \tag{7.163}$$

compared with the ρ channel. This changes dramatically the estimate of $M^2_{\rm crit}$ based on 10% correction to the parton model:

$$M^2_{\rm crit}(0+\text{ gluonium}) \approx 20 M^2_{\rm crit}(\rho\,\text{channel}) \approx 15\,\text{GeV}^2. \tag{7.164}$$

However, one should mention that the effect of $\Pi_G(0)$ can be avoided when working with unsubtracted sum rules, while the optimal value of M^2 from the subtracted sum rule is lower of about 3 GeV^2 [55, 132, 249].

In this channel, λ^2 contributes without any α_s factor as:

$$\Pi_G(M^2) = (\text{parton model})\left(1 - \frac{3\lambda^2}{M^2} + \cdots\right). \tag{7.165}$$

Asking a 10% correction to the parton model leads to a value of $M^2_{\rm crit}(0+\text{ gluonium})$ consistent with the one in Eq. (7.164).

◇ Comments

We conclude that the introduction of the tavhyonic gluon mass into the standard SVZ expansion does not disturb the known phenomenology. Instead, it brings some improvements and an explanation of the different values of scales appearing in the QCD spectral sum rule analysis.

7.5.2 Power corrections from current correlators in x space

In Ref. [240], we study the effects of power corrections within the SVZ-expansion on the sum of the pseudoscalar and scalar two-point

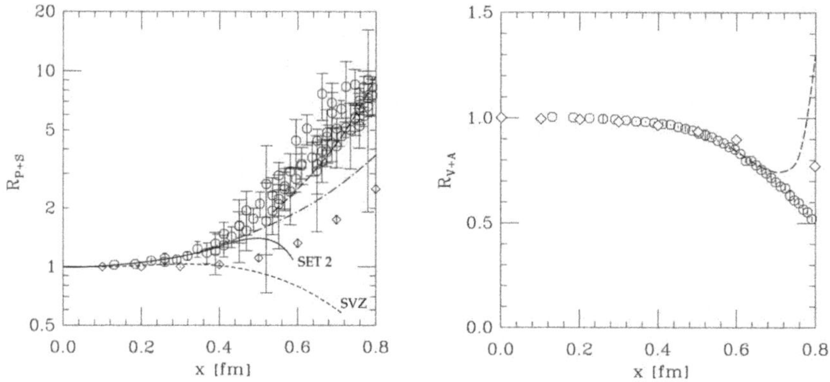

(a) (b)

Figure 7.14. (a) S+P correlator; (b) V+A correlator. Comments are in the text.

Table 7.4. Different parameters used in the analysis of the S + P data in units of GeV^d (d is the dimension of the operator).

Sources	$\langle \alpha_s G^2 \rangle$	$\alpha_s \langle \bar{\psi}\psi \rangle^2$	$(\alpha_s/\pi)\lambda^2$
SET 2 [237]	0.07	5.8×10^{-4}	0
SET 3 [235, 237]	0.07	5.8×10^{-4}	-0.12

correlators (S+P) and the vector plus axial-vector correlators (V+A) in the x space. We confront our results with the lattice data and some predictions from the instanton liquid model.

• S + P channel

In Fig. 7.14(a), we compare the lattice data from Ref. [241] with the OPE predictions for the two sets of QCD condensate values given in Table 7.4. The dot-dashed curve is the prediction for SET 3 where the contribution of the x^2-term has been added to SET 2. The bold dashed curve is SET 3 + a fitted value of the $D = 8$ condensate contributions. The diamond curve is the prediction from the instanton liquid model of Ref. [242].

• V+A channel

In Fig. 7.14, we compare the lattice data from Ref. [241] with the OPE predictions for the SET 3 QCD condensates values given in Table 7.4, including a fitted value of the $D = 8$ contributions. The diamond curve is the prediction from the instanton liquid model of Ref. [242].

• Discussions: Two-step QCD

We note that the corrections introduced in Ref. [235] drastically improve the agreement of the theoretical predictions for the current correlator in the (S + P) channel with the lattice data (without affecting the other channels measured) in the moderate x-region, which can be extended to higher x-values by including higher dimension condensates. The main uncertainty of the analysis is due to the neglect of the pure perturbative corrections, not detected so far on the lattice. Further checks could be provided by measuring the current correlators in the other channels discussed in Ref. [235].

More detailed discussions on the comparison with some other approaches can be found in Ref. [240].

7.5.3 Duality between PT series and power corrections

In Ref. [234], we discuss the duality relation between the λ^2 contribution and the large order terms of PT series. We observe that the existing coefficients of the calculated PT series grow geometrically but neither with alternate signs nor a factorial growth which do not favour the large β approximation. We also demonstrate that the geometric long PT series is dual to a tachyonic gluon mass. Then, we conclude that to avoid a double counting, one should use either a tachyonic gluon mass or a geometric estimate of the coefficients but not both to avoid a double counting. One expects that the size of the tachyonic gluon contribution decreases when more and more terms are added in the PT series. This feature has been checked in a recent paper [239], as illustrated in Fig. 7.15.

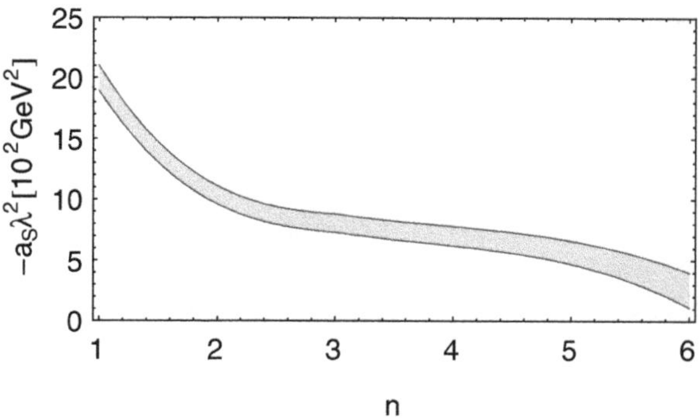

Figure 7.15. Tachyonic gluon contribution versus the degree of the terms in the PT series.

7.6 Visiting Scientist at KEK, Tsukuba, 2000

(a) (b)

Figure 7.16. KEK Institute, Tsukuba.

Figure 7.17. Professor K. Hagiwara, KEK Institute, Tsukuba.

I have been invited by Professor Kaoru Hagiwara from 20/12/99 to 19/06/2000 as a Visiting Scientist of the Japanese Ministry of Education, Science, Sports and Culture. It was exciting as it is my first visit to an oriental country. Kaoru organized quite well my stay, such that I had the possibility to deliver seminars in different universities (Tsukuba, Ochanomizu University and Riken Institute (Tokyo), Hiroshima, Sendai, Kanazawa, Kyoto) during the *Japan Tour* and to spend a week at the KIAS Institute (Seoul, South Korea). At the same time, I discovered the Japanese culture and their ways of living. I was impressed by the cohabitation between traditional life and modernity and the advanced technology of Japan when visiting the Sony building in Tokyo such that after living about one year there, I had difficulty to readapt to Europe. I also realized that the Japanese researchers and people in general are hard workers.

7.6.1 Light hybrid mesons

We have again continued the hybrid work [201] by adding PT α_s corrections and the contribution due to a tachyonic gluon mass. Using LSR stability criteria, we found:

$$\begin{aligned} &M_{\tilde{\rho}} \simeq (1.6 \sim 1.7)\ \text{GeV}, \quad f_{\tilde{\rho}} \simeq (25 \sim 50)\ \text{MeV}, \\ &M_{\tilde{\eta}} \simeq 2.8\ \text{GeV} \simeq \sqrt{t_c}, \end{aligned} \tag{7.166}$$

and:

$$M^2_{\tilde{\phi}} - M^2_{\tilde{\rho}} \simeq 0.3\,\text{GeV}^2, \tag{7.167}$$

which confirms and improves the results in Eqs. (7.22) and (7.23). Here, we have used the value of the four-quark condensate $\rho\langle\bar{\psi}\psi\rangle^2 \simeq 5.8 \times 10^{-4}$ from $e^+e^- \to$ hadrons data [236].

7.6.2 Weak CP-violation and ϵ'/ϵ

• Generalities and data

CP–violation is one of the most important weak interactions phenomena in particle physics, where in their presence, stable (under strong interactions) K^0 $(\bar{s}d)$ and $\bar{K}^0(\bar{d}s)$ particles with definite strangeness eigenvalues ± 1 become unstable. Therefore, the decay of a long-lived kaon into a two-pion final state is an evidence for CP-violation.

We have studied the weak CP-violation in Ref. [243] through the decay of $K_{L,S} \rightarrow \pi^+\pi^-$, $\pi^0\pi^0$. The (indirect) CP-violation can be measured from the $K^0 - \bar{K}^0$ mixing through the parameter:

$$\epsilon \equiv \frac{A[K_L \rightarrow (\pi\pi)_{I=0}]}{A[K_S \rightarrow (\pi\pi)_{I=0}]}, \tag{7.168}$$

where $K_{L(S)}$ are long (short)-lived kaon. Experimentally, one finds [33]:

$$\epsilon \simeq (2.228 \pm 0.011) \times 10^{-3} e^{i(43.5\pm0.1)^0}. \tag{7.169}$$

Direct CP-violation can be measured from the ratio [33]:

$$\text{Re}\left(\epsilon'/\epsilon\right) \simeq (16.5 \pm 2.6) \times 10^{-4}, \tag{7.170}$$

where:

$$\epsilon' \equiv \frac{1}{\sqrt{2}}\left\{\frac{A[K_L \rightarrow (\pi\pi)_{I=2}]}{A[K_S \rightarrow (\pi\pi)_{I=0}]} - \epsilon \times \left[\omega \equiv \frac{A[K_S \rightarrow (\pi\pi)_{I=2}]}{A[K_S \rightarrow (\pi\pi)_{I=0}]}\right]\right\}. \tag{7.171}$$

One has:

$$\omega_{exp} \simeq \frac{1}{22} e^{-i(42\pm4)^0}, \tag{7.172}$$

where the phase comes from $\pi\pi$-scattering and the amplitude from the ratio $K_S \rightarrow (\pi^+\pi^-)_{I=2}$ over $K_S \rightarrow (\pi^0\pi^0)_{I=0}$. It is usual to introduce the amplitudes:

$$A\left(K^0 \rightarrow (\pi\pi)_I\right) = iA_I\, e^{i\delta_I}, \quad A\left(\bar{K}^0 \rightarrow (\pi\pi)_I\right) = i - A_I^*\, e^{i\delta_I}, \tag{7.173}$$

the $K_L - K_S$ mass difference $\Delta m \equiv m_L - m_S$ and the off-diagonal part M_{12} of the complex mass matrix: $\text{Re}\, M_{12} \simeq \Delta m/2$. Therefore, one can write within a good approximation:

$$\epsilon \simeq \frac{1}{\sqrt{2}} e^{i\frac{\pi}{4}}\left[\frac{\text{Im}\, M_{12}}{\Delta m} + \frac{\text{Im}\, A_0}{\text{Re}\, A_0}\right], \quad \omega \simeq e^{i(\delta_2 - \delta_0)}\frac{\text{Re}\, A_2}{\text{Re}\, A_0}, \tag{7.174}$$

and:

$$\epsilon' \simeq e^{i(\delta_2 - \delta_0 + \frac{\pi}{2})}\frac{\text{Re}\, A_2}{\text{Re}\, A_0}\left[\frac{\text{Im}\, A_2}{\text{Re}\, A_2} - \frac{\text{Im}\, A_0}{\text{Re}\, A_0}\right]. \tag{7.175}$$

• Theory of ϵ'/ϵ

From the theory, it is customary to study the $\Delta S = 1$ process from the weak Hamiltonian:

$$\mathcal{H}_{eff} = \frac{G_F}{\sqrt{2}} V_{ud} V_{us}^* \sum_{i=1}^{10} C_i(\mu) \mathcal{Q}_i(\mu), \tag{7.176}$$

where V_{ij} are Cabibbo–Kobayashi–Maskawa mixing matrices, $C_i(\mu)$ are Wilson coefficients calculable perturbatively and known to NLO. $\mathcal{Q}_i(\mu)$ are four-quark operators which will be estimated using non-perturbative approaches.

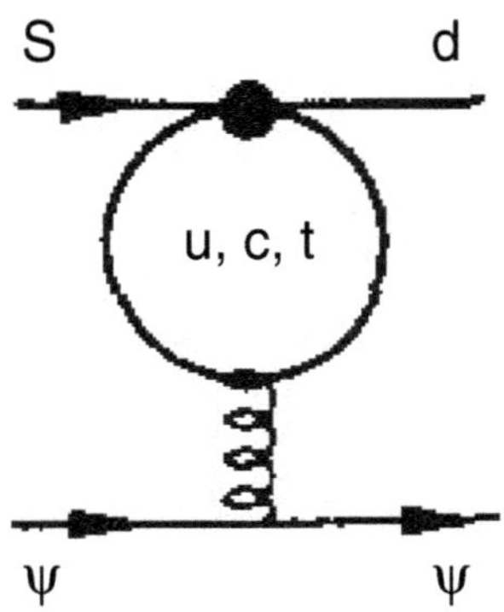

Figure 7.18. Penguin operator entering in the process $K \to \pi\pi$, where the wavy line denotes gluon for QCD and photon for electroweak theory.

It has been claimed in Ref. [244] that the so-called penguin operators (see Fig. 7.18) give the dominant contributions to the ratio ϵ'/ϵ (see, however, Ref. [245] using an effective Lagrangian approach):

$$\mathcal{Q}_6^{1/2} \equiv \frac{3}{2}(\bar{s}_\alpha d_\beta)_{V-A} \sum_{u,d,s} (\bar{\psi}_\beta \psi_\alpha)_{V+A},$$

$$\mathcal{Q}_8^{3/2} \equiv \frac{3}{2}(\bar{s}_\alpha d_\beta)_{V-A} \sum_{u,d,s} (e_\psi \bar{\psi}_\beta \psi_\alpha)_{V+A}. \tag{7.177}$$

To leading order in $1/N_c$, one has:

$$\mathcal{Q}_6^{1/2} \simeq B_6^{1/2}/m_s^2, \quad \mathcal{Q}_8^{3/2} \simeq B_6^{3/2}/m_s^2. \tag{7.178}$$

In this way, one can write the approximate expression [246]:

$$\frac{\epsilon'}{\epsilon} \approx 13\,\mathrm{Im}\lambda_t \left(\frac{110}{\bar{m}_s(2)\,[\mathrm{MeV}]}\right)^2 \times \left[B_6^{1/2}(1-\Omega_2) - 0.4\,B_8^{3/2}\left(\frac{m_t}{165\,\mathrm{GeV}}\right)\right]\left(\frac{\Lambda^{(4)}_{\overline{MS}}}{340\,\mathrm{MeV}}\right), \tag{7.179}$$

with $\Omega_2 \simeq (0.16 \pm 0.03)$ as the SU(2) isospin breaking, including the $\pi^0 - \eta$ mixing and:

$$\mathrm{Im}\lambda_t = |V_{ub}||V_{cb}|\sin\delta \simeq (1.33 \pm 0.14) \times 10^{-4}, \tag{7.180}$$

where δ is the CKM phase. The average $\hat{B}_K = (0.80 \pm 0.15)$ of the $\Delta S = 2$ process has been used.

◇ **$I = 3/2$ transition**

Using soft pion techniques and chiral symmetry, one can reduce the $\langle K|\mathcal{Q}_i^{3/2}|\pi\pi\rangle$ matrix element to the vacuum condensates $\langle 0|\mathcal{Q}_i^{3/2}|0\rangle$:

$$\langle K|\mathcal{Q}_7^{3/2}|\pi\pi\rangle = -\frac{4}{f_\pi^3}\langle \mathcal{O}_7^{3/2}\rangle,$$

$$\langle K|\mathcal{Q}_8^{3/2}|\pi\pi\rangle = -\frac{4}{f_\pi^3}\left\{\frac{1}{3}\langle \mathcal{O}_7^{3/2}\rangle + \frac{1}{2}\langle \mathcal{O}_8^{3/2}\rangle\right\}, \tag{7.181}$$

with:

$$\mathcal{Q}_7 \equiv \frac{3}{2}(\bar{s}d)_{V-A}\sum_{u,d,s} e_\psi(\bar{\psi}\psi)_{V+A}. \tag{7.182}$$

We estimate $\langle \mathcal{Q}_{7,8}^{3/2}\rangle$ using Weinberg and DMO sum rules:

$$2\pi\langle\alpha_s\mathcal{O}_8^{3/2}\rangle(\mu^2) = \int_0^\infty dt\, t^2\frac{\mu^2}{t+\mu^2}\,\frac{1}{\pi}\Big[\mathrm{Im}\Pi_V(t) - \mathrm{Im}\Pi_A(t)\Big],$$

$$\frac{16\pi^2}{3}\langle\alpha_s\mathcal{O}_7^{3/2}\rangle(\mu^2) = \int_0^\infty dt\, t^2\log\left(\frac{t+\mu^2}{s}\right)\frac{1}{\pi}[\mathrm{Im}\Pi_V(t) - \mathrm{Im}\Pi_A(t)]. \tag{7.183}$$

Using the ALEPH/OPAL data, we obtain to NLO in the $\overline{MS}$ scheme:

$$\langle \mathcal{O}_7^{3/2} \rangle_{\overline{MS}} \simeq -(0.7 \pm 0.2), \quad \langle \mathcal{O}_8^{3/2} \rangle_{\overline{MS}} \simeq -(9.1 \pm 6.4). \tag{7.184}$$

Using further soft pion techniques, we translate these results into the B-factors:

$$B_7^{3/2} \simeq (0.7 \pm 0.2) \left(\frac{m_s(2)\ [\mathrm{MeV}]}{119} \right)^2 \left(\frac{92.4}{f_\pi\ [\mathrm{MeV}]} \right)^4,$$
$$B_8^{3/2} \simeq (1.7 \pm 0.4) \left(\frac{m_s(2)\ [\mathrm{MeV}]}{119} \right)^2 \left(\frac{92.4}{f_\pi\ [\mathrm{MeV}]} \right)^4. \tag{7.185}$$

◇ $\boldsymbol{I = 1/2}$ **transition**

From Ref. [244], this process is expected to be dominated by the operator $\mathcal{O}_6^{1/2}$. We estimate this contribution as:

$$\langle (\pi\pi)_{I=0} | \mathcal{O}_6^{1/2} | K^0 \rangle \simeq -\Big[2\langle \pi | \bar{u}\gamma_5 d | 0 \rangle \langle \pi | \bar{s}u | K^0 \rangle + \langle \pi\pi | \bar{d}d + \bar{u}u | 0 \rangle \langle 0 | \bar{s}\gamma_5 d | K^0 \rangle \Big]. \tag{7.186}$$

○ The first term is well known and can be estimated from the $K \to \pi\mu\nu$ form factors [244]:

$$\langle \pi | \bar{s}u | K^0 \rangle = \Big[f_+(M_K^2 - m_\pi^2) + f_- m_\pi^2 \Big] / (ms - m_u), \tag{7.187}$$

with $f_+(0) = 1$ and $f_-(0) = 0$. Taking $\mu = m_c = 1.2$ GeV, Ref. [243] obtains:

$$\langle \pi\pi | \bar{d}d + \bar{u}u | 0 \rangle \langle 0 | \bar{s}\gamma_5 d | K^0 \rangle = \frac{\sqrt{2} f_\pi m_\pi^2}{(m_u + m_d)} \frac{M_K^2 - m_\pi^2}{(m_s - m_u)}$$
$$\simeq (0.32 \pm 0.03) \left(\frac{142.6}{(m_s - m_u)\ [\mathrm{MeV}]} \right)^2 \mathrm{GeV}^3, \tag{7.188}$$

where $m_u(m_c) = (4.4 \pm 0.5)$, $m_d(m_c) = (8 \pm 1), m_s(m_c) = (147 \pm 15)$ in units of MeV.

∘ For the second term, we assume that it is dominated by the direct production of the S_2 scalar meson in the s-channel for an on-shell kaon $p^2 = M_K^2$. Therefore, it can be decomposed as:

$$\begin{aligned}\langle\pi\pi|\bar{d}d + \bar{u}u|0\rangle &= \langle\pi\pi|S_2\rangle\langle S_2|\bar{d}d + \bar{u}u|0\rangle \\ &\equiv \frac{g_{S\pi^+\pi^-}}{(p^2 - M_S^2)} \frac{2f_S}{(m_u + m_d)} M_S^2. \end{aligned} \tag{7.189}$$

With the values of the parameters [6, 249]:

$$f_S = (0.32 \pm 0.08)(m_u + m_d), \quad g_{\pi\pi} \simeq 2.5\,\text{GeV}, \tag{7.190}$$

we conclude that the scalar meson contribution is:

$$\langle\pi\pi|\bar{d}d + \bar{u}u|0\rangle\langle 0|\bar{s}\gamma_5 d|K^0\rangle, \tag{7.191}$$

where we take into account the fact that QSSR cannot fix the signs of the S_2 coupling and decay constant, which will be fixed later on from chiral constraints on the weak amplitude.

Using soft pion techniques, one obtains [246]:

$$\langle K|Q_6^{1/2}|\pi\pi\rangle = 4\sqrt{\frac{3}{2}}\left(\frac{M_K^2}{m_s + m_d}\right)^2 \sqrt{2}(f_K - f_\pi)B_6^{1/2}, \tag{7.192}$$

leading to:

$$\begin{aligned}B_6^{1/2} &\simeq 3.7\left(\frac{m_s + m_d}{m_s - m_u}\right)^2 \\ &\quad \times\left[(0.65 \pm 0.09) - (0.53 \pm 0.13)\left(\frac{(m_s - m_u)\ [\text{MeV}]}{142.6}\right)\right].\end{aligned} \tag{7.193}$$

Using the previous values of $B_8^{3/2}$ and $B_6^{1/2}$ into Eq. (7.179) and the updated value of the quark masses from [74], one deduces:

$$B_6^{1/2} \simeq (1.24 \pm 0.35), \quad B_8^{3/2} \simeq (1.17 \pm 0.13), \tag{7.194}$$

which implies:

$$\frac{\epsilon'}{\epsilon} \simeq (10.5 \pm 5.9), \tag{7.195}$$

to be compared with the data in Eq. (7.170). Our estimate at $\mu = 2$ GeV updates our previous results and are comparable with the ones from the lattice calculations quoted in Ref. [250].

Since then, some progresses have been made for improving the theoretical expression of ϵ'/ϵ, where N2LO QCD and some chiral corrections have been added. Its recent parametrization at $\mu = 1$ GeV is [250]:

$$\frac{\epsilon'}{\epsilon} \simeq \mathrm{Im}\lambda_t \left\{ (1-\Omega_2)\left[15.4 B_6^{1/2} - 2.9\right] - \left[8 B_8^{3/2} - 2\right] \right\}, \tag{7.196}$$

where Ω_2 becomes $(29 \pm 7) \times 10^{-2}$ and $\lambda_t \simeq 1.4 \times 10^{-4}$. Using our values of $B_6^{1/2}$ and $B_8^{3/2}$ obtained at 2 GeV[3] into this expression, one obtains:

$$\frac{\epsilon'}{\epsilon} \simeq \left(5.8^{+5.6}_{-2.1}\right). \tag{7.197}$$

I left Japan and came back home, but, finally, we initiated a new collaboration with the Kaoru group on the $\bar{B}^0 - B^0$ mass difference.

7.6.3 The $\bar{B}^0 - B^0$ mass difference

Here, we analyze the $\bar{B}^0 - B^0$ mass differences and the SU(3) breakings using a two-point function:

$$\psi_H(q^2) = i \int d^4x e^{iqx} \langle 0|\mathcal{T}\mathcal{O}_q(x) \left(\mathcal{O}_q(0)\right)^\dagger |0\rangle, \tag{7.198}$$

built from the four-quark $\Delta B = 2$ operator ($q \equiv d, s$):

$$\mathcal{O}_q(x) \equiv \bar{b}\gamma_\mu \frac{1}{2}(1-\gamma_5) q \bar{q} \gamma_\mu \frac{1}{2}(1-\gamma_5) b. \tag{7.199}$$

Radiative factorizable and non-factorizable corrections to the perturbative graphs in the massless light quark case have been obtained in Ref. [183]. The B-meson contribution is introduced via:

$$\langle \bar{B}_q^0 | \mathcal{O}_q | B_q^0 \rangle = \frac{4}{3} \frac{f_B^2}{2} M_{B^0}^2 B_{B_q}, \tag{7.200}$$

---

[3]We plan to update our estimate in a future work.

where f_B is normalized as $f_\pi = 130.56$ MeV and B_{B_q} is the bag parameter which is equal to one if one uses a factorization of the four-quark operator. Introducing the renormalization group invariant (RGI) quantity:

$$\hat{B}_{B_q} = B_{B_q}(\nu)\alpha_s^{\gamma_0/\beta_1}\left\{1+\left(\frac{5165}{12696}\right)\left(\frac{\alpha_s}{\pi}\right)\right\}, \tag{7.201}$$

with $\gamma_0 = 1$ being the anomalous dimension of the operator $\mathcal{O}_q$ and $\beta_1 = -23/6$ for five flavours. ν is the subtraction point. Using Laplace and moments sum rules, we obtain [251]:

$$f_B\sqrt{\hat{B}_B} \simeq (247 \pm 59)\,\text{MeV}, \quad \xi \equiv \frac{f_{B_s}\sqrt{\hat{B}_{B_s}}}{f_B\sqrt{\hat{B}_B}} \simeq (1.18 \pm 0.03), \tag{7.202}$$

which is comparable with the global fit of the CKM mixing angle:

$$f_B\sqrt{\hat{B}_B} \simeq (231 \pm 15)\,\text{MeV}, \tag{7.203}$$

while ξ is comparable with the lattice value (1.14 ± 0.06). Using $f_B \simeq (192 \pm 19)$ MeV to order α_s for consistencies, we deduce

$$\hat{B}_{B_d} \simeq \hat{B}_{B_s} \simeq (1.65 \pm 0.38) \Longrightarrow B_{B_d}(4M_b^2) \simeq (1.1 \pm 0.25), \tag{7.204}$$

indicating a small SU(3) breaking for the bag parameter and a good realization of the vacuum saturation. Using the experimental values of the CKM parameters expressed in terms of the Wolfenstein λ, ρ and η ones, we deduce the $\bar{B}^0 - B^0$ mass difference:

$$\Delta M_s \simeq 18.6(2.2)\,ps^{-1}, \tag{7.205}$$

consistent with the experimental lower bound of 13.1 ps^{-1}.

7.6.4 Visit at KIAS Seoul 2000

Kaoru has also organized my visit to KIAS, Seoul and I was impressed by the efficiency of the triangle collaboration: Japan–South Korea–Taiwan. I stayed one week for giving a series of seminars.

Figure 7.19. Korea Institute for Advanced Study (KIAS), Seoul.

7.7 Associate Scientist at Hsinschu and Taipei 2001

(a) (b)

Figure 7.20. (a) Replica of the old gate in Beijing: symbol of the Tsingua University (Hsinschu); (b) National Taiwan University (Taipei).

(a) (b)

Figure 7.21. (a) Professor Li Hsiang-nan (Tsinghua University); (b) Professor W.Y. Pauchy Hwang (Taipei University).

Professor Li Hsiang-nan participated in the QCD 2000 conference in Montpellier and invited me to visit his institute from 01/12/2000 to 31/02/2001 and Professor W.Y. Pauchy Hwang also invited me to visit his university from 01/09/2001 to 11/12/2001. With Pauchy, we had a common interest in QCD spectral sum rules, but, unfortunately, I was more interested in the test of the Standard Model during this stay, while he started to build a group of cosmology. This visit in Taiwan was again an enriching experience to discover another oriental country and its corresponding culture. Here, I realized that the environment of the institute is organized to stimulate you to work hard (lunch brought in the office, tea room, etc.). Indeed, like them, I had to work hard like I did when I was a Ph.D. student.

7.7.1 Hadronic contribution to the muon and tau anomalies

Here, I revise my Thèse de 3ème cycle work on the g-2. To lowest order in the QED coupling α, the hadronic contributions to the vacuum polarization can be obtained from the well-known dispersion relation given in Eq. (3.2). Using the recent $e^+e^- \to$ hadrons and τ-decay data, we obtained for the muon anomaly [252]:

$$\begin{aligned} a_\mu|^{\rm lo}_{\rm hvp} \times 10^{11} &= 6949.1(118.0): \quad e^+e^-, \\ &= 6969.5(76.4): \quad e^+e^- \oplus \tau - \text{decays}, \end{aligned} \tag{7.206}$$

while for τ:

$$a_\tau|^{\rm lo}_{\rm hvp} \times 10^8 = 348.0(3.8): \quad e^+e^- \oplus \tau - \text{decays}, \tag{7.207}$$

which are improvements of our results quoted in Eqs. (3.3) and (3.10). The errors in the determinations have reduced by a factor 8.

7.7.2 The QED coupling

The hadronic contributions to the QED running can be quantified as:

$$\Delta\alpha_{\rm had}(M_Z^2) = -\left(\frac{\alpha}{3\pi}\right) M_Z^2 \int_{4m_\pi^2}^{\infty} \frac{R_{ee}(t)}{t(t-M_Z^2)}, \tag{7.208}$$

where M_Z is the Z-boson mass and R_{ee} is the ratio of the $e^+e^- \to$ hadrons over the $e^+e^- \to \mu^+\mu^-$ total cross-sections. Using combined $e^+e^- \to$ hadrons and τ-decays data below 3 GeV^2, and e^+e^- data $\oplus$ QCD continuum above, we obtained [253]:

$$\Delta\alpha^{\rm lo}_{\rm hvp} = (2763.4 \pm 16.5) \times 10^{-5}. \tag{7.209}$$

We have estimated the radiative corrections due to $\pi^0\gamma, \eta\gamma, \pi^+\pi^-\gamma, \ldots$ as:

$$\Delta\alpha^{\rm nlo}_{\rm had} = (6.4 \pm 2.7) \times 10^{-5}. \tag{7.210}$$

Adding the QED contribution to three loops [254]:

$$\Delta\alpha_{\rm QED} = 3149.7685 \times 10^{-5}, \tag{7.211}$$

we obtained:

$$\alpha^{-1}(M_Z) = \alpha^{-1}(0)\big[1 - \Delta\alpha_{\rm QED} - \Delta\alpha_{\rm had}\big] = 128.926(25). \tag{7.212}$$

7.7.3 The muon hyperfine splittings

During my stay, I started to study the hadronic contributions to the QED running coupling and muon hyperfine splittings. These processes are related to the $e^+e^- \to$ hadron total cross-section times a kernel $K(t)$ specific to each process via a dispersion relation similar to the one of $g-2$ given in Eq. (3.2):

$$\Delta\nu_{\rm hvp}^{\rm lo} = \frac{-\nu_F}{2\pi^3}\frac{m}{M}\int_{4m_\pi^2}^{\infty} dt\, H(t)\sigma(e^+e^- \to \text{hadrons}), \qquad (7.213)$$

where m and M are the electron and muon masses and $\nu_F = 445$ 903 192 0.(511)(34) Hz the Fermi energy splitting. The QED kernel function reads:

$$H(t) = \left(\frac{t}{4M^2}+2\right) v \log\frac{1+v}{1-v} - \left(\frac{t}{M^2}+\frac{3}{2}\right),$$

$$\log\frac{t}{M^2}+\frac{1}{2}\ :\quad v = \sqrt{1-\frac{4M^2}{t}}. \qquad (7.214)$$

Using combined $e^+e^- \to$ hadrons and τ-decays data below 3 GeV^2 and e^+e^- data $\oplus$ QCD continuum above, we obtained [253]:

$$\Delta\nu_{\rm hvp}^{\rm lo} = (232.5 \pm 3.2)\ \text{Hz}. \qquad (7.215)$$

Adding the QED to fourth order and the lowest electroweak contributions:

$$\Delta\nu_{\rm QED}^{(4)} = 4\,270\,819(220)\text{Hz},$$

$$\Delta\nu_{\rm weak}^{\rm lo} = -\frac{G_F}{\sqrt{2}} m_e m_\mu \left(\frac{3}{4\pi\alpha}\right)\nu_F \simeq -65\,\text{Hz}, \qquad (7.216)$$

and an estimate of the higher order ones:

$$|\Delta\nu_{\rm weak}^{\rm ho}| \approx 0.7\,\text{Hz}, \quad |\Delta\nu_{\rm had}^{\rm ho}| \approx 7(2)\,\text{Hz}, \qquad (7.217)$$

one obtains the Standard Model predictions:

$$\nu_{\rm SM} \equiv \nu_{\rm F} + \Delta\nu_{\rm weak} + \Delta\nu_{\rm had} + \Delta\nu_{\rm QED}. \qquad (7.218)$$

ν_F can be estimated from the relation:

$$\nu_F = \rho_F\left(\frac{\mu_\mu}{\mu_e}\right)\frac{1}{(1+m_e/m_\mu)^3}\ :\rho_F = \frac{16}{3}(Z\alpha)^2 Z^2 c\, R_\infty. \qquad (7.219)$$

Then, we deduce:

$$\nu_{SM} = 4\,463\,302\,913(511)(34)(220)\ \text{Hz}, \qquad (7.220)$$

where the largest error comes from the ratio of magnetic moments μ_l. The third error comes from QED which contrary to the case of the muon anomaly gives the largest theoretical error. The data is:

$$\nu_{\rm exp} = 4\,463\,302\,776(51)\ \text{Hz}. \qquad (7.221)$$

7.7.4 c, b quark masses and $f_{D_{(s)}}$, $f_{B_{(s)}}$ decay constants

I continued my work on the c and b quark masses and the leptonic decay constants of the $D_{(s)}$ and $B_{(s)}$ heavy–light mesons [255]. Values of the running and pole quark masses have been obtained to order α_s^2 from the pseudoscalar heavy–light channels as well as the leptonic decay constants $f_{D_{(s)}}$, $f_{B_{(s)}}$. Using some Laplace sum rules, one obtains:

$$\bar{m}_c(m_c) = (1.10 \pm 0.04)\,\text{GeV}, \quad M_c \simeq (1.46 \pm 0.04)\,\text{GeV},$$
$$\bar{m}_b(m_b) = (4.05 \pm 0.06)\,\text{GeV}, \quad M_c \simeq (4.69 \pm 0.06)\,\text{GeV}, \tag{7.222}$$

where the values are slightly lower than from the J/ψ and Υ systems [256]:

$$\bar{m}_c(m_c) = \left(1.23^{+0.04}_{-0.05}\right)\,\text{GeV}, \quad \bar{m}_b(m_b) = (4.23 \pm 0.05)\,\text{GeV}, \tag{7.223}$$

while the values of the running and pole masses satisfy quite well the 3-loop QCD perturbative relation.

The values of the decay constants are:

$$f_D = (205 \pm 20)\ \text{MeV}, \quad f_B = (203 \pm 23)\ \text{MeV},$$
$$f_{D_s} = (235 \pm 24)\ \text{MeV}, \quad f_{B_s} = (236 \pm 30)\ \text{MeV}, \tag{7.224}$$

where they are normalized as $f_\pi = 130.56$ MeV. A comparison of f_D and f_B confirms a large $1/\sqrt{M_Q}$ correction to the heavy quark symmetry, while the one of $f_{D,B}$ with f_{D_s,B_s} indicates SU(3) breakings of about 15% on the leptonic decay constant.

7.8 Montpellier–Sao Paulo Collaboration 2004–2011

F. Navarra and M. Nielsen (University of Sao Paulo) participated regularly in the QCD-Montpellier conferences. In 2004, R. Matheus, one of their students come to Montpellier within a Sao Paulo State Foundation for Research Support (FAPESP) fellowship. Since then, we have done a series of works on different heavy quarks (baryons, four-quarks and molecules states, pentaquark states) using QCD spectral sum rules.

(a) (b)

Figure 7.22. (a) Institute of Physics, University of Sao Paulo; (b) with Professors M. Nielsen and F. Navarra.

(a) (b)

Figure 7.23. (a) With Marina Nielsen and Jean-Marc Richard at Sao Paulo; (b) at the Freedom Square of Sao Paulo.

7.8.1 Pentaquark sum rules

We write a paper on the pentaquark sum rules [257] using different configurations of currents: diquark–diquark–quark, diquark–three quarks. We shall list the following some possible operators describing the isoscalar $I = 0$ and $J = 1/2$ channel which would correspond to the experimental candidate $\Theta(1540)$ [265].

• Interpolating currents

Defining the pseudoscalar (ps) and scalar (s) diquark interpolating fields as:

$$Q^{ps}_{ab}(x) = [u^T_a(x)Cd_b(x)], \quad Q^{s}_{ab}(x) = [u^T_a(x)C\gamma_5 d_b(x)], \tag{7.225}$$

where a, b, c are colour indices and C denotes the charge conjugation matrix, the lowest dimension current built by two diquarks and one antiquark describing Θ as a $I = 0$, $J^P = 1/2^+$ S-wave resonance is [266] (see also Ref. [267])[4]:

$$\eta^{\Theta}_{[266]} = \epsilon^{abc}\epsilon^{def}\epsilon^{cfg}Q^{ps}_{ab}Q^{s}_{de}C\bar{s}^T_g, \tag{7.226}$$

and the one with one diquark and three quarks is [268]:

$$\eta^{\Theta}_{[268]} = \frac{1}{\sqrt{2}}\epsilon^{abc}Q^{s}_{ab}\left\{u_e\bar{s}_e i\gamma_5 d_c - (u \leftrightarrow d)\right\}. \tag{7.227}$$

This latter choice can be interesting if the instanton repulsive force arguments [258] against the existence of a pseudoscalar diquark bound state apply. Alternatively, a description of $\Theta(1.54)$ as a $I = 0$, $J^P = 1/2^+$ P-wave resonance has been proposed in Ref. [270] and used in Ref. [271] in the sum rule analysis:

$$\eta^{\Theta}_{[271]} = \left(\epsilon^{abd}\delta^{ce} + \epsilon^{abc}\delta^{de}\right)[Q^{s}_{ab}(D^{\mu}Q^{s}_{cd}) - (D^{\mu}Q^{s}_{ab})Q^{s}_{cd}]\gamma_5\gamma_{\mu}C\bar{s}^T_e. \tag{7.228}$$

We have generalized this current by considering its mixing with the following one having the same dimension and quantum numbers:

$$\eta^{\Theta}_{new} = \epsilon^{abc}\epsilon^{def}\epsilon^{cfg}Q^{ps}_{ab}Q^{s}_{de}\gamma_{\mu}(D^{\mu}C\bar{s}^T_g). \tag{7.229}$$

---

[4] A negative parity state can be obtained by multiplying by γ_5, the diquark operator.

• Results

Using both LSR and FESR sum rules within stability criteria, we found for an S-wave $I = 0$, $j = 1/2$ and negative parity state:

$$M_\Theta \simeq (1513 \pm 114)\ \text{MeV}, \quad \lambda_\Theta^2 \simeq (0.14 \sim 0.49) \times 10^{-9}\ \text{GeV}^{12}, \tag{7.230}$$

where the residue is normalized in the standard way (see Ref. [257]).

For the $I = 0$, P-wave channel, we found:

$$M_P \simeq (1.99 \pm 0.19)\ \text{GeV}, \quad \lambda_\Theta^2 \simeq (0.7 \sim 7.1) \times 10^{-9}\ \text{GeV}^{14}. \tag{7.231}$$

7.8.2 Four-quark nature of $X(3872)$ state

• $X(3872)$ as a $\bar{3} - 3$ colour four-quark state

Later on, we write another paper [259] on the test of the nature of the $X(3872)$ found by BELLE [260] from $B^+ \to X(3872)\,K^+ \to J/\psi\pi^+\pi^-K^+$. The data have been confirmed by some other experiments. Assuming that $X(3872)$ is a $\bar{c}c\bar{d}d$ four quark bound state with $J^{PC} = 1^{++}$, we describe it with the interpolating current in the $3 - 3$ colour triplet configuration:

$$j_3^\mu = \frac{i\epsilon_{abc}\epsilon_{dec}}{\sqrt{2}}[(q_a^T C\gamma_5 c_b)(\bar{q}_d\gamma^\mu C\bar{c}_e^T) + (q_a^T C\gamma^\mu c_b)(\bar{q}_d\gamma_5 C\bar{c}_e^T)]. \tag{7.232}$$

We consider the corresponding two-point correlator up to some part of dimension seven condensates and to lowest order of PT series.

Using LSR within stability criteria, we find $M_X = (3925 \pm 127)$ MeV which is compatible, within the errors, with the experimental candidate X(3872), while the SU(3) breaking terms lead to an unusual mass splitting $M_{X^s} - M_X = -(61 \pm 30)$ MeV. The mass difference between the neutral states due to isospin violation of about $(2.6 - 3.9)$ MeV is smaller than the value (8 ± 3) MeV proposed in Ref. [261]. For the b-quark, we predict $M_{X_b} = (10144 \pm 106)$ MeV for $X_b(b\bar{b}q\bar{q})$, which is much below the $\bar{B}B^*$ threshold in contrast to the $\bar{B}B^*$ molecule prediction [262], and for $X_b^s(b\bar{b}s\bar{s})$, a mass splitting $M_{X_b^s} - M_{X_b} = -(121 \pm 182)$ MeV. Our analysis also indicates that the mass splitting between the ground state and the radial excitation of about $(225{-}250)$ MeV is much smaller than in the case of ordinary mesons and is (within the errors) flavour-independent. We also extract the decay constants, analogous to f_π, of such mesons, which are useful for further studies of their leptonic and hadronic decay widths. Using LSR and FESR, we found the masses, in units of MeV:

$$M_X(\bar{c}c\bar{d}d) = (3925 \pm 127), \quad M_{X_s}(\bar{c}c\bar{s}s) - M_X(\bar{c}c\bar{d}d) = -(61 \pm 30), \tag{7.233}$$

where M_X is compatible with the data. For the b-quark analogue, we found:

$$\begin{aligned} M_X(\bar{b}b\bar{d}d) = (10144 \pm 106), \quad & M_{X_s}(\bar{b}b\bar{s}s) - M_X(\bar{b}b\bar{d}d) \\ & = -(121 \pm 30), \end{aligned} \tag{7.234}$$

much below the BB^* threshold of 10604 MeV. We also found that the mass splitting between the radial excitation and the lowest ground state is approximately (225–250) MeV and is almost flavour-independent. It is much smaller than the one of ordinary mesons like $\rho' - \rho$.

• $X(3872)$ as $\bar{6} - 6$ colour four-quark and molecule states

In Ref. [263], we have extended our analysis by studying some other colour sextet ($\bar{6} - 6$) configurations of interpolating currents:

$$j_6^\mu = \frac{i}{\sqrt{2}}[(q_a^T C\gamma_5\lambda_{ab}^S c_b)(\bar{q}_d\gamma^\mu C\lambda_{de}^S \bar{c}_e^T) + (q_a^T C\gamma^\mu\lambda_{ab}^S c_b)(\bar{q}_d\gamma_5 C\lambda_{de}^S \bar{c}_e^T)], \tag{7.235}$$

where a, b, c, etc. are colour indices, C is the charge conjugation matrix, q denotes a u or d quark and λ^S stands for the six symmetric Gell-Mann matrices: $\lambda^S = (\lambda_0,\ \lambda_1,\ \lambda_3,\ \lambda_4,\ \lambda_6,\ \lambda_8)$.

These tetraquark currents can be compared with the one describing the X state as a $D^* - D$ molecule:

$$j_{\text{mol}}^\mu(x) = \left(\frac{g}{\Lambda}\right)^2_{\text{eff}} \frac{1}{\sqrt{2}}[(\bar{q}_a\gamma_5 c_a)(\bar{c}_b\gamma^\mu q_b) - (\bar{q}_a\gamma^\mu c_a)(\bar{c}_b\gamma_5 q_b)]. \tag{7.236}$$

Using double ratios of sum rules, we found that all different configurations lead to the value:

$$M_3(\bar{c}c\bar{d}d) \simeq M_6 \simeq M_{\rm mol} = (3925{\pm}127), \quad \sqrt{t_c|_3} \simeq (4.15{\pm}0.03)\ \text{GeV}, \tag{7.237}$$

while the relative low value of t_c indicates that the next radial excitation of the X-meson can be in the range:

$$M_{X'} \approx M_X + (225 \pm 127)\ \text{MeV}. \tag{7.238}$$

This low value of t_c suggests that the $\bar{3}-3$ resonance may be difficult to separate from the QCD continuum and also suggests that it can be a wide resonance.

Therefore, we use vertex sum rules to calculate the hadronic couplings. We found:

$$\Gamma^{3,6,\rm mol}_{X\to J/\psi+n\pi} \approx 50\,\text{MeV}, \tag{7.239}$$

which is too big compared with the data:

$$\Gamma_{X\to \rm all} \leq 2.3\,\text{MeV}. \tag{7.240}$$

• $X(3872)$ as $\lambda - J/\psi$ molecule

We consider the so-called $\lambda - J/\psi$-like molecule current [263]:

$$j^\mu_\lambda = \left(\frac{g'}{\Lambda}\right)^2_{eff'} (\bar{c}\lambda^a\gamma^\mu c)(\bar{q}\lambda_a\gamma_5 q), \tag{7.241}$$

where λ_a is a colour matrix. Using a double ratio of sum rules, we obtain:

$$r_{\lambda/3} \equiv \frac{M_\lambda}{M_3} = 0.96 \pm 0.03, \tag{7.242}$$

where the errors come from the stability regions and m_c. From vertex sum rules, we obtain:

$$\Gamma_{X\to J/\psi\pi\pi} \approx 0.3\,\text{MeV}, \quad \frac{\Gamma_{X\to J/\psi\gamma}}{\Gamma_{X\to J/\psi\pi\pi}} \approx 0.07, \tag{7.243}$$

which better reproduces the previous data of the total width and the radiative branching fraction:

$$\frac{\Gamma_{X\to J/\psi\gamma}}{\Gamma_{X\to J/\psi\pi\pi}} = 0.14 \pm 005, \tag{7.244}$$

than some other configurations discussed previously.

7.8.3 $T_{cc,bb}$ versus $X_{c,b}$

In Ref. [264], we analyse the four-quark/molecule states built from a $QQ\bar{q}\bar{q}$ operators, where $Q \equiv c, b$ and $q \equiv u, d$ quarks. We assume that the $X(3872)$ and T_{cc} states are described by the molecular currents:

$$j^{\mu}_{X}(x) = \left(\frac{g}{\Lambda}\right)^2_{\text{eff}} \frac{1}{\sqrt{2}}\Big[(\bar{q}_a(x)\gamma_5 c_a(x)\bar{c}_b(x)\gamma^{\mu} q_b(x)) - (\bar{q}_a(x)\gamma^{\mu} c_a(x)\bar{c}_b(x)\gamma_5 q_b(x)) \Big], \tag{7.245}$$

and:

$$j^{\mu}_{T_{cc}}(x) = \left(\frac{g'}{\Lambda}\right)^2_{\text{eff}} (\bar{q}_a(x)\gamma_5 c_a(x)\bar{q}_b(x)\gamma^{\mu} c_b(x)), \tag{7.246}$$

where a and b are colour indices.

We use a double ratio of sum rules involving the ratio of the T_{QQ} and X masses. We found that X and T_{cc} are almost degenerated:

$$r_{T_{cc}/X} = 1.00 \pm 0.01. \tag{7.247}$$

The same result is obtained for X_b and T_{bb}. Then, we conclude that $D\bar{D}^*$ and $\bar{D}\bar{D}^*$ should be degenerated in mass if they exist.

For the $SU(3)$ breaking, we found:

$$\frac{M_{T_{cc\bar{s}\bar{s}}}}{M_{T_{cc\bar{d}\bar{d}}}} \simeq (0.95\text{--}0.98), \tag{7.248}$$

which acts in the opposite sign like in the ratio. An analogous feature has been observed for the X state.

7.8.4 Heavy baryons

We have renewed our collaboration contract within the CNRS, France and FAPESP, Brazil in 2010. Then, R. Albuquerque came to Montpellier for a Cotutelle Ph.D. thesis. We started to work on baryon sum rules.

- **SU(3) breakings of baryons with one heavy quark**

We wrote a paper [272].

◇ Octet (spin 1/2) baryons

Following Ref. [212], we work with the lowest dimension currents:

$$\begin{aligned}
\eta_{\Xi_Q} &= \epsilon_{\rm abc}\left[(q_a^T C\gamma_5 s_b) + b(q_a^T C s_b)\gamma_5\right] Q_c, \\
\eta_{\Lambda_Q} &= \eta_{\Xi_Q} \qquad (s \to q), \\
\eta_{\Omega_Q} &= \epsilon_{\rm abc}\left[(s_a^T C\gamma_5 Q_b) + b(s_a^T C Q_b)\gamma_5\right] s_c, \\
\eta_{\Sigma_Q} &= \eta_{\Omega_Q} \qquad (s \to q),
\end{aligned} \tag{7.249}$$

where b is *a priori* an arbitrary mixing parameter. Its value has been found to be:

$$b = -1/5, \tag{7.250}$$

in the case of light baryons [67, 120, 222] and in the range [212–214]:

$$-0.5 \le b \le 0.5, \tag{7.251}$$

for non-strange heavy baryons. The corresponding two-point correlator reads:

$$\begin{aligned}
S(q) &= i\int d^4x\ e^{\rm iqx}\ \langle 0|\mathcal{T}\overline{\eta}_Q(x)\eta_Q(0)|0\rangle \\
&\equiv \hat{q}F_1 + F_2,
\end{aligned} \tag{7.252}$$

where F_1 and F_2 are two invariant functions.

One can estimate the baryon masses from the following ratios:

$$\mathcal{R}_i^q = \frac{\int_{t_q}^{t_c} ds\ s\ e^{-s\tau}\ {\rm Im}F_i(s)}{\int_{t_q}^{t_c} ds\ e^{-s\tau}\ {\rm Im}F_i(s)}, \quad i = 1, 2, \quad \mathcal{R}_{21}^q = \frac{\int_{t_q}^{t_c} ds\ e^{-s\tau}\ {\rm Im}F_2(s)}{\int_{t_q}^{t_c} ds\ e^{-s\tau}\ {\rm Im}F_1(s)}, \tag{7.253}$$

where at the τ-stability point:

$$M_{B_q^{(*)}} \simeq \sqrt{\mathcal{R}_i^q} \simeq \mathcal{R}_{21}^q. \tag{7.254}$$

◇ Decuplet (spin 3/2) baryons

We follow Ref. [213] and work with the interpolating currents:

$$\eta^\mu_{\Xi^*_Q} = \sqrt{\frac{2}{3}}\left[(q^T C\gamma_\mu Q)s + (s^T C\gamma_\mu Q)q + (q^T C\gamma_\mu s)Q\right],$$

$$\eta^\mu_{\Omega^*_Q} = \frac{1}{\sqrt{2}}\eta^\mu_{\Xi^*_Q} \quad (q \to s),$$
$$\eta^\mu_{\Sigma^*_Q} = \frac{1}{\sqrt{2}}\eta^\mu_{\Xi^*_Q} \quad (s \to q), \tag{7.255}$$

Table 7.5. QSSR predictions of the strange heavy baryon masses in units of MeV from the double ratio (DR) of sum rules with the QCD input parameters in Table 7.10 and using as input the observed masses of the associated non-strange heavy baryons. We have used $\kappa \equiv \langle\bar{s}s\rangle/\langle\bar{d}d\rangle = 0.74 \pm 0.03$ fixed from the experimental $\Xi_{c,b}$ and Ω_c masses.

Baryons	I	$r^{sd}_{B^*_Q}$	Mass	Data
$\boldsymbol{J^P = \frac{1}{2}^+}$				
$\Xi_c(cqs)$	$\frac{1}{2}$		input	2467.9 ± 0.4 [PDG]
$\Omega_c(css)$	0		input	2697.5 ± 2.6 [PDG]
$\Xi_b(bqs)$	$\frac{1}{2}$		input	5792.4 ± 3.0 [PDG]
$\Xi'_c(cqs)$	$\frac{1}{2}$	1.043(10)	2559(25)	2575.7 ± 3.1 [PDG]
$\Xi'_b(bqs)$	$\frac{1}{2}$	1.014(7)	5893(42)	–
$\Omega_b(bss)$	0	1.0455(64)	6076(37)	6165.0 ± 13 [D0]
				6054.4 ± 6.9 [CDF]
$J^P = \frac{3}{2}^+$				
$\Xi^*_c(cqs)$	$\frac{1}{2}$	1.049(8)	2641(21)	2646.1 ± 1.3 [PDG]
$\Omega^*_c(css)$	0	1.109(17)	2792(38)	2768.3 ± 3.0 [PDG]
$\Xi^*_b(bqs)$	$\frac{1}{2}$	1.024(8)	5961(21)	–
$\Omega^*_b(bss)$	0	1.040(9)	6066(49)	–

where an anti-symmetrization over colour indices is understood. The normalization in Eq. (7.255) is chosen in such a way that in all cases, one gets the same perturbative contribution. The corresponding two-point correlator reads:

$$S^{\mu\nu}(q) = i\int d^4x\; e^{iqx}\; \langle 0|\mathcal{T}\overline{\eta}^\mu_Q(x)\eta^\nu_Q(0)|0\rangle$$
$$\equiv g^{\mu\nu}\left(\hat{q}F_1 + F_2\right) + \cdots \tag{7.256}$$

Table 7.6. QSSR predictions of the strange heavy baryon hyperfine splittings in units of MeV from the double ratio (DR) of sum rules with the QCD input parameters in Table 7.10 and using as input the predicted values in Table 7.5. We have added the errors quadratically.

Hyperfine Splittings	Data
$M_{\Xi_c^*} - M_{\Xi_c} = 173(21)$	179(1)
$M_{\Xi_c^*} - M_{\Xi'_c} = 82(33)$	70(3)
$M_{\Omega_c^*} - M_{\Omega_c} = 95(38)$	70(3)
$M_{\Xi_b^*} - M_{\Xi_b} = 169(21)$	–
$M_{\Xi_b^*} - M_{\Xi'_b} = 68(47)$	–
$M_{\Omega_b^*} - M_{\Omega_b} = -10(61)$	$M_{\Sigma_b^*} - M_{\Sigma_b} = 22$

• Mass splittings of baryons with two heavy quarks

In Ref. [273], we study the mass splittings of baryons with two heavy quarks using a double ratio of LSR. In so doing, we study the two-point correlator associated to the currents:

◇ Spin 1/2 QQq baryons

$$J_{\Xi_Q} = \epsilon_{\alpha\beta\lambda}\left[(Q^T_\alpha C\gamma_5 q_\beta) + b(Q^T_\alpha C q_\beta)\gamma_5\right] Q_\lambda, \tag{7.257}$$

where $q \equiv d, s$ are light quark fields, $Q \equiv c, b$ are heavy quark fields, b is *a priori* an arbitrary mixing parameter fixed to the optimal value [212–214, 272]:

$$-0.5 \le b \le 0.5, \tag{7.258}$$

for non-strange heavy baryons.

◇ Spin 3/2 QQq baryons

$$J^\mu_{\Xi^*_Q} = \sqrt{\frac{1}{3}}\epsilon_{\alpha\beta\lambda}\left[2(Q^T_\alpha C\gamma^\mu d_\beta)Q_\lambda + (Q^T_\alpha C\gamma^\mu Q_\beta)q_\lambda\right]. \tag{7.259}$$

Table 7.7. QSSR predictions for the doubly heavy baryon mass ratios and splittings, which we compare with the Potential Model (PM) range of results in Refs. [217, 274]. The PM prediction for the spin 3/2 is an average with the one for spin 1/2. The mass inputs are in GeV and the mass splittings are in MeV.

Mass Ratios	Mass Inputs	Mass Splittings	PM
$\Xi^*_{cc}/\Xi_{cc} = 1.0167(19)$	$\Xi_{cc} = 3.52[33]$	$\Xi^*_{cc} - \Xi_{cc} = 59(7)$	70–93
$\Xi^*_{bb}/\Xi_{bb} = 1.0019(3)$	$\Xi_{bb} = 9.94[217]$	$\Xi^*_{bb} - \Xi_{bb} = 19(3)$	30–38
$\Omega_{cc}/\Xi_{cc} = 1.0260(70)$	$\Xi_{cc} = 3.52[33]$	$\Omega_{cc} - \Xi_{cc} = 92(24)$	90–102
$\Omega_{bb}/\Xi_{bb} = 1.0049(13)$	$\Xi_{bb} = 9.94[217]$	$\Omega_{bb} - \Xi_{bb} = 49(13)$	60–73
$\Omega^*_{cc}/\Xi^*_{cc} = 1.0260(75)$	$\Xi^*_{cc} = 3.58^a$	$\Omega^*_{cc} - \Xi^*_{cc} = 94(27)$	91–100
$\Omega^*_{bb}/\Xi^*_{bb} = 1.0050(15)$	$\Xi^*_{bb} = 9.96^a$	$\Omega^*_{bb} - \Xi^*_{bb} = 50(15)$	60–72
$\Omega_{bc}/\Xi_{bc} = 1.0060(17)$	$\Xi_{bc} = 6.86[217]$	$\Omega_{bc} - \Xi_{bc} = 41(7)$	70–89

Note: [a]We have combined your results for the mass splittings with the experimental value of $M_{\Xi_{cc}}$ and the central value of $M_{\Xi_{bb}}$ in Eq. (7.115).

We work to lowest order of PT series and include the contribution of the $\langle\bar{s}s\rangle$ condensate. The results for the ratio of masses are given in Table 7.7 and compared with the potential model estimates.

7.9 France–China Collaboration 2008–2011

Our collaboration belongs to a large-scale collaboration between France and China. Our group is hadron physics and is composed by Montpelier, Grenoble (Professor J.M. Richard) and ICHEP of Beijing (Professor Q. Zhao) (see third photo in Fig. 7.24). The exchange concerns a participation to workshops (Beijing and Wuhan universities), seminars (Sangahai, Beijing, and Nankin universities). This collaboration is also interesting to me for discovering some aspects of China.

7.9.1 Isospin violating decay of $\psi(3770) \to J/\psi + \pi^0$

We study this process through the DD^* loop shown in Fig. 7.25 [275].

Among the three couplings at the vertices, the ψDD^* ones cannot be directly measured as the J/ψ mass is below the $\bar{D}D$ threshold. Thus, we use QSSR for estimating these couplings. Introducing the QCD continuum threshold via the mass gap Δ which (*a priori*) is a

(a)

(b)

(c)

(d)

Figure 7.24. (a) The IHEP institute — Academy of Sciences (Beijing); (b) dinner with some members of the IHEP group; (c) with Professors Q. Zhao (IHEP Beijing) and J.M. Richard (IPN Lyon) coordinators of the collaboration; (d) the Great Wall of China.

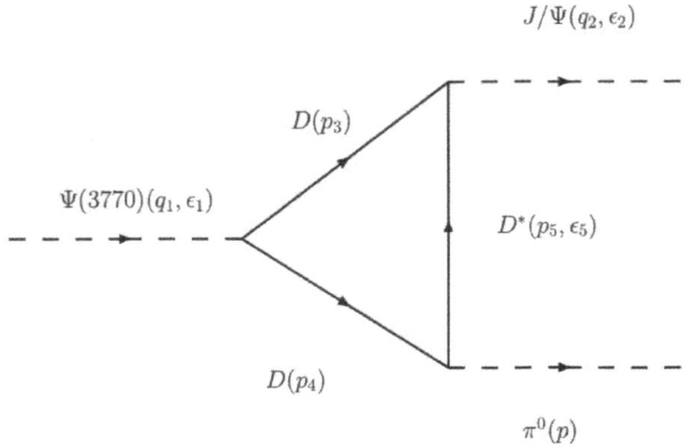

Figure 7.25. $\psi(3770) \to J/\psi + \pi^0$ through DD^* loop within an effective Lagrangian approach.

(a) (b) (c)

Figure 7.26. (a) Tiananmen square (Beijing); (b) Forbidden City (Beijing).

free parameter to be fixed from the sum rule:

$$\sqrt{u_0}(\sqrt{s_0}) \equiv M_D(M_\psi) + \Delta_D(\Delta_\psi). \tag{7.260}$$

We first extract the leptonic decay constants f_D and f_ψ using two-point function sum rule and extract the optimal values using τ and Δ stability criteria. The optimal results are obtained for:

$$\tau_\psi/\tau_D \simeq 0.28 \sim 0.30 \text{ GeV}^{-2}, \quad \Delta_\psi \simeq \Delta_D \simeq (0.8 \sim 1.2) \text{ GeV}, \tag{7.261}$$

compared to the (*ad hoc*) choice: $M_D^2/M_\psi^2 \simeq 0.36$. In order to estimate the 3-hadron coupling, we use vertex sum rule and assume

(a) (b)

Figure 7.27. (a) Old Shanghai city; (b) Modern Shanghai Tower.

like in the case of $B \to \psi l \nu_l$ semi-leptonic decay discussed in Section 7.3.11 that the optimal value of the three-point function τ_3 is half of the one from the two-point function τ_2.

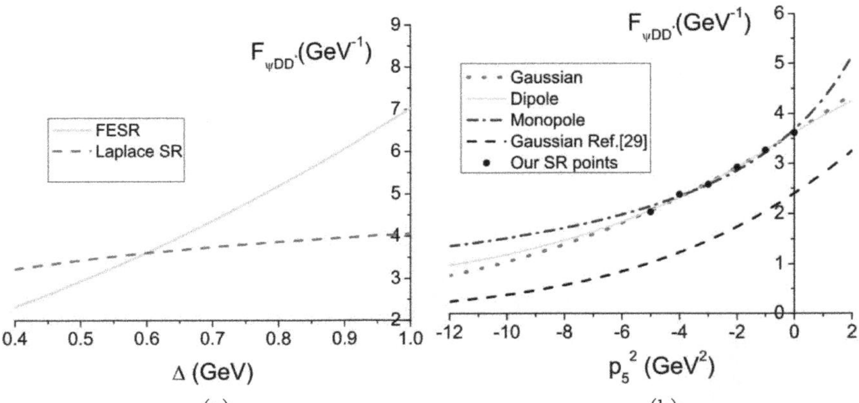

(a) (b)

Figure 7.28. ψDD^* form factor : (a) At zero momentum versus the mass gap Δ from LSR and FESR; (b) p_5^2 behaviour of the form factor for $\Delta = 0.6$ GeV, and for different pole parametrizations.

However, we observe that, at these τ-stability points, the value of the form factor increases slightly with $\Delta_{D,\psi}$. Adding constraints from FESR which are reached by letting $\tau_3 \to 0$, we obtain a common solution for $\Delta_{D,\psi} \simeq 0.6$ GeV, as shown in Fig. 7.28. We also show the p_5^2 behaviour of the form factor for different parametrizations.

The $D^*D\pi$ coupling can be related to $f_+(0)$ which is known to be about 0.6 from different approaches:

$$F_{D^*D\pi^0}(p_5^2) = \frac{\sqrt{2}f_+(0)}{1-\alpha_0\, p_5^2/M_{D^*}^2}\frac{M_{D^*}}{f_{D^*}}, \tag{7.262}$$

where $\alpha_0 \simeq 0.21(0.24)$ for $D^0(D^+)$ from the data [276]. Then, using $f_{D^*} \simeq 0.24$ GeV, one deduces:

$$F_{D^*D\pi^0}(M_{D^*}^2) \simeq 10. \tag{7.263}$$

Using these information on the couplings, we predict within the uncertainties on the extracted coupling $\psi''D^+D^-$:

$$\begin{aligned} Br[\psi(3770) \to J/\psi\pi^0]_{\text{total}} &\simeq (0.33 \sim 1.36) \times 10^{-4}, \\ Br[\psi(3770) \to J/\psi\pi^0]_{\text{absorptive}} &\simeq (0.15 \sim 0.56) \times 10^{-4}. \end{aligned} \tag{7.264}$$

Measurements of this process constitute an important test of strength of the isospin violation.

7.10 Montpellier–Rio–Tanà Collaboration 2012–2023

Figure 7.29. University of Antananarivo (MG).

Figure 7.30. From left to right: R. Albuquerque (University of Rio de Janeiro), ex-students of iHEPMAD (University of Antananarivo): A. Rabemananjara, D. Rabetiarivony, G. Randriamanatrika.

Figure 7.31. University of Rio de Janeiro (BR).

This collaboration involves R. Albuquerque (University of Rio de Janeiro) who had been a cotutelle Ph.D. student in Montpellier and the Ph.D. students of the iHEPMAD Research Institute, which I created in 2004 at the University of Antananatrivo (Madagascar).

This work is a continuation of the program started with the University of Sao Paulo on the four-quark and molecule states using

QCD spectral sum rules. It has been motivated by the discovery of many exotic states in e^+e^- collisions at BELLE, BABAR and BES and at hadron colliders, LHC and Tevatron.

In all these papers, we emphasize the fact that working with LO PT expression, the mass of the heavy quark is ill defined such that working with the $\overline{MS}$ mass is not justified. One should also look at the contribution of a pole quark mass such that the quoted prediction should take into account the error due to this quark mass definition ambiguity. This ambiguity does not affect much the double ratio of moments like the one used in Refs. [272, 273] as the quark mass does not give the leading order contribution.

In our paper, we use factorization of the spectral function associated to the quark antiquark as suggested in Ref. [286] to estimate the NLO corrections to the four-quarks and molecule correlators. This factorization is valid to leading order in $1/N_c$.

7.10.1 XYZ states

In Refs. [257, 259, 263], we have studied the four-quarks/molecules nature of the X(3872) and in Ref. [277], the ones of a 1^{--} and 0^{++} to lowest order of QCD. Here, we extend the previous analysis to a systematic study of the XYZ states [278, 279] summarized recently in Ref. [287].

• Interpolating currents

The interpolating currents $\mathcal{O}$ for the molecules (respectively four-quark states) are given in Table 7.8 (respectively Table 7.9).

• Spectral Function within MDA

We shall use the Minimal Duality Ansatz (MDA) given in Eq. (7.265) for parametrizing the spectral function:

$$\frac{1}{\pi}\,\mathrm{Im}\Pi(t) \simeq f_H^2 M_H^8 \delta(t - M_H^2) + \text{“QCD continuum”}\,\theta(t - t_c), \tag{7.265}$$

Table 7.8. Interpolating currents with a definite C-parity describing the molecule-like states. $Q \equiv c$ (respectively b) for the $\bar{D}D$ (respectively $\bar{B}B$)-like molecules. $q \equiv u, d$.

States	J^{PC}	Molecule Currents $\equiv \mathcal{O}_{mol}(x)$
	$\mathbf{0^{++}}$	
$\bar{D}D,\ \bar{B}B$		$(\bar{q}\gamma_5 Q)(\bar{Q}\gamma_5 q)$
$\bar{D}^*D^*,\bar{B}^*B^*$		$(\bar{q}\gamma_\mu Q)(\bar{Q}\gamma^\mu q)$
$\bar{D}_0^*D_0^*,\ \bar{B}_0^*B_0^*$		$(\bar{q}Q)(\bar{Q}q)$
	$\mathbf{1^{++}}$	
$\bar{D}^*D,\ \bar{B}^*B$		$\frac{i}{\sqrt{2}}\Big[(\bar{Q}\gamma_\mu q)(\bar{q}\gamma_5 Q) - (\bar{q}\gamma_\mu Q)(\bar{Q}\gamma_5 q)\Big]$
$\bar{D}_0^*D_1,\ \bar{B}_0^*B_1$		$\frac{1}{\sqrt{2}}\Big[(\bar{q}Q)(\bar{Q}\gamma_\mu\gamma_5 q) + (\bar{Q}q)(\bar{q}\gamma_\mu\gamma_5 Q)\Big]$
	$\mathbf{0^{-\pm}}$	
$\bar{D}_0^*D,\ \bar{B}_0^*B$		$\frac{1}{\sqrt{2}}\Big[(\bar{q}Q)(\bar{Q}\gamma_5 q) \pm (\bar{Q}q)(\bar{q}\gamma_5 Q)\Big]$
$\bar{D}^*D_1,\ \bar{B}^*B_1$		$\frac{1}{\sqrt{2}}\Big[(\bar{Q}\gamma_\mu q)(\bar{q}\gamma^\mu\gamma_5 Q) \mp (\bar{Q}\gamma_\mu\gamma_5 q)(\bar{q}\gamma^\mu Q)\Big]$
	$\mathbf{1^{-\pm}}$	
$\bar{D}_0^*D^*,\ \bar{B}_0^*B^*$		$\frac{1}{\sqrt{2}}\Big[(\bar{q}Q)(\bar{Q}\gamma_\mu q) \mp (\bar{Q}q)(\bar{q}\gamma_\mu Q)\Big]$
$\bar{D}D_1,\ \bar{B}B_1$		$\frac{i}{\sqrt{2}}\Big[(\bar{Q}\gamma_\mu\gamma_5 q)(\bar{q}\gamma_5 Q) \pm (\bar{q}\gamma_\mu\gamma_5 Q)(\bar{Q}\gamma_5 q)\Big]$

Table 7.9. Interpolating currents with a definite P-parity describing the four-quark states. $Q \equiv c$ (respectively b) in the charm and bottom channels. $q \equiv u, d$.

J^P	Four-Quark Currents $\equiv \mathcal{O}_{4q}(x)$
$\mathbf{0^+}$	$\epsilon_{abc}\epsilon_{dec}[(q_a^T\, C\gamma_5\, Q_b)(\bar{q}_d\, \gamma_5 C\, \bar{Q}_e^T) + k(q_a^T\, C\, Q_b)(\bar{q}_d\, C\, \bar{Q}_e^T)]$
$\mathbf{1^+}$	$\epsilon_{abc}\epsilon_{dec}[(q_a^T\, C\gamma_5\, Q_b)(\bar{q}_d\, \gamma_\mu C\, \bar{Q}_e^T) + k(q_a^T\, C\, Q_b)(\bar{q}_d\, \gamma_\mu\gamma_5 C\, \bar{Q}_e^T)]$
$\mathbf{0^-}$	$\epsilon_{abc}\epsilon_{dec}[(q_a^T\, C\gamma_5\, Q_b)(\bar{q}_d\, C\, \bar{Q}_e^T) + k(q_a^T\, C\, Q_b)(\bar{q}_d\, \gamma_5 C\, \bar{Q}_e^T)]$
$\mathbf{1^-}$	$\epsilon_{abc}\epsilon_{dec}[(q_a^T\, C\gamma_5\, Q_b)(\bar{q}_d\, \gamma_\mu\gamma_5 C\, \bar{Q}_e^T) + k(q_a^T\, C\, Q_b)(\bar{q}_d\, \gamma_\mu C\, \bar{Q}_e^T)]$

where $f_H^{(i)}$ is the decay constant defined as:

$$\langle 0|\mathcal{O}^{(s,p)}|H\rangle = f_H^{(s,p)} M_H^4, \quad \langle 0|\mathcal{O}^{\mu}|H\rangle = f_H^{(1)} M_H^5 \epsilon_\mu, \tag{7.266}$$

respectively for spin 0 and 1 hadronic states H with ϵ_μ the vector polarization. The higher state contributions are smeared by the

"QCD continuum" coming from the discontinuity of the QCD diagrams and starting from a constant threshold t_c.

• NLO and N2LO PT corrections using factorization

Assuming a factorization of the four-quark interpolating current as a natural consequence of the molecule definition of the state, we can write the corresponding spectral function as a convolution of the spectral functions associated to quark bilinear current for the $\bar{D}D^*$ and $\bar{D}_0^*D^*$ spin 1 states:

$$\frac{1}{\pi}\mathrm{Im}\Pi_H^{(1)}(t) = \theta(t-4M_Q^2)\left(\frac{1}{4\pi}\right)^2 t^2 \int_{M_Q^2}^{(\sqrt{t}-M_Q)^2} dt_1 \int_{M_Q^2}^{(\sqrt{t}-\sqrt{t_1})^2} dt_2 \times \lambda^{3/2}\frac{1}{\pi}\mathrm{Im}\Pi^{(1)}(t_1)\frac{1}{\pi}\mathrm{Im}\psi^{(s,p)}(t_2). \tag{7.267}$$

For the $\bar{D}D$ spin 0 state, one has:

$$\frac{1}{\pi}\mathrm{Im}\psi_H^{(s)}(t) = \theta(t-4M_Q^2)\left(\frac{1}{4\pi}\right)^2 t^2 \int_{m_Q^2}^{(\sqrt{t}-M_Q)^2} dt_1 \int_{m_Q^2}^{(\sqrt{t}-\sqrt{t_1})^2} dt_2 \times \lambda^{1/2}\left(\frac{t_1}{t}+\frac{t_2}{t}-1\right)^2 \times \frac{1}{\pi}\mathrm{Im}\psi^{(p)}(t_1)\frac{1}{\pi}\mathrm{Im}\psi^{(p)}(t_2), \tag{7.268}$$

and for the $\bar{D}^*D^*$ spin 0 state:

$$\frac{1}{\pi}\mathrm{Im}\psi_H(t) = \theta(t-4M_Q^2)\left(\frac{1}{4\pi}\right)^2 t^2 \int_{m_Q^2}^{(\sqrt{t}-M_Q)^2} dt_1 \int_{m_Q^2}^{(\sqrt{t}-\sqrt{t_1})^2} dt_2 \times \lambda^{1/2}\left[\left(\frac{t_1}{t}+\frac{t_2}{t}-1\right)^2+\frac{8t_1t_2}{t^2}\right] \times \frac{1}{\pi}\mathrm{Im}\Pi^{(1)}(t_1)\frac{1}{\pi}\mathrm{Im}\Pi^{(1)}(t_2), \tag{7.269}$$

where:

$$\lambda = \left(1 - \frac{(\sqrt{t_1} - \sqrt{t_2})^2}{t}\right)\left(1 - \frac{(\sqrt{t_1} + \sqrt{t_2})^2}{t}\right) \tag{7.270}$$

is the phase space factor and M_Q is the on-shell heavy quark mass. Im $\Pi^{(1)}(t)$ is the spectral function associated to the bilinear $\bar{c}\gamma_\mu(\gamma_5)q$ vector or axial-vector current, while Im $\psi^{(5)}(t)$ is associated to the $\bar{c}(\gamma_5)q$ scalar or pseudoscalar current.[5] An analogous convolution is assumed for the four-quark states.

- **The Laplace sum rule (LSR)**

The exponential or Laplace sum rule (LSR) and its ratio read[6]:

$$\mathcal{L}_H(\tau, t_c, \mu) = \int_{4M_Q^2}^{t_c} dt\; e^{-t\tau} \frac{1}{\pi} \mathrm{Im}\Pi_H^{(1,0)}(t, \mu), \tag{7.271}$$

$$\mathcal{R}_H(\tau, t_c, \mu) = \frac{\int_{4M_Q^2}^{t_c} dt\; t\; e^{-t\tau} \frac{1}{\pi} \mathrm{Im}\Pi_H^{(1,0)}(t, \mu)}{\int_{4M_Q^2}^{t_c} dt\; e^{-t\tau} \frac{1}{\pi} \mathrm{Im}\Pi_H^{(1,0)}(t, \mu)} \simeq M_R^2, \tag{7.272}$$

where μ is the subtraction point which appears in the approximate QCD series when radiative corrections are included and τ is the sum rule variable replacing q^2. Similar sum rules are obtained for the (pseudo)scalar two-point function $\psi^{(s,p)}(q^2)$.

- **QCD input parameters**

The QCD parameters which shall appear in the following analysis will be the charm and bottom quark masses $m_{c,b}$ (we shall neglect the light quark masses $q \equiv u, d$), the light quark condensate $\langle \bar{q}q \rangle$, the gluon condensates $\langle \alpha_s G^2 \rangle \equiv \langle \alpha_s G^a_{\mu\nu} G^{\mu\nu}_a \rangle$ and $\langle g^3 G^3 \rangle \equiv \langle g^3 f_{abc} G^a_{\mu\nu} G^{b,\nu}_\rho G^{c,\rho\mu} \rangle$, the mixed condensate $\langle \bar{q}Gq \rangle \equiv$

[5] In the chiral limit $m_q = 0$, the PT expressions of the vector (respectively scalar) and axial-vector (respectively pseudoscalar) spectral functions are the same.

[6] The last equality in Eq. (7.273) is obtained when one uses MDA in Eq. (7.266) for parametrizing the spectral function.

$\langle\bar{q}g\sigma^{\mu\nu}(\lambda_a/2)G^a_{\mu\nu}q\rangle = M_0^2\langle\bar{q}q\rangle$ and the four-quark condensate $\rho\alpha_s\langle\bar{q}q\rangle^2$, where $\rho \simeq 3$–4 indicates the deviation from the four-quark vacuum saturation. Their values are given in Table 7.10. We shall work with the running light quark condensates and masses, which read to leading order in α_s:

$$\langle\bar{q}q\rangle(\tau) = -\hat{\mu}_q^3\,(-\beta_1 a_s)^{2/\beta_1}, \langle\bar{q}Gq\rangle(\tau) = -M_0^2\hat{\mu}_q^3\,(-\beta_1 a_s)^{1/3\beta_1}, \tag{7.273}$$

where $\beta_1 = -(1/2)(11-2n_f/3)$ is the first coefficient of the β function for n_f flavours; $a_s \equiv \alpha_s(\tau)/\pi$; $\hat{\mu}_q$ is the spontaneous RGI light quark condensate [29].

Table 7.10. QCD input parameters: The original errors for $\langle\alpha_s G^2\rangle$, $\langle g^3G^3\rangle$ and $\rho\langle\bar{q}q\rangle^2$ have been multiplied by about a factor 3 for a conservative estimate of the errors (see also the text).

Parameters	Values	References
$\alpha_s(M_\tau)$	0.325(8)	[79, 288, 289]
$\overline{m}_c(m_c)$	1261(12) MeV	Average [33, 290, 291, 293, 453, 454]
$\overline{m}_b(m_b)$	4177(11) MeV	Average [33, 290, 291, 453, 454]
$\hat{\mu}_q$	(253 ± 6) MeV	[6, 7, 123, 294–296]
M_0^2	(0.8 ± 0.2) GeV^2	[297–301]
$\langle\alpha_s G^2\rangle$	$(7 \pm 3) \times 10^{-2}$ GeV^4	[64, 70, 115, 236, 289–291, 303, 304]
$\langle g^3G^3\rangle$	(8.2 ± 2.0) GeV$^2 \times \langle\alpha_s G^2\rangle$	[290, 291]
$\rho\alpha_s\langle\bar{q}q\rangle^2$	$(5.8 \pm 1.8) \times 10^{-4}$ GeV^6	[115, 289, 297, 298]

• Molecule states masses and couplings

The results are given in Table 7.11 (respectively Table 7.12) for the charm (respectively bottom) channel where the corresponding hadronic threshold and experimental candidates are shown in the last two columns. The errors come from the QCD parameters and from the range of τ, t_c and μ where the optimal results are extracted.

• Four-quark states masses and couplings

The results are given in Table 7.13 (respectively Table 7.14) for the charm (respectively bottom) channel where the experimental candidates are shown in the last column. The sources of errors are the same as in the molecules case.

Table 7.11. $\bar{D}D$-like molecules masses, invariant and running couplings from LSR within stability criteria at LO to N2LO of PT.

	$\hat{f}_X$ [keV]			$f_X(4.5)$ [keV]			Mass [MeV]				
Nature	LO	NLO	N2LO	LO	NLO	N2LO	LO	NLO	N2LO	Threshold	Exp.
$\mathbf{J^{PC} = 0^{++}}$											–
$\bar{D}D$	56	60	**62(6)**	155	164	**170(15)**	3901	3901	**3898(36)**	3739	
$\bar{D}^*D^*$	–	–	–	269	288	**302(47)**	3901	3903	**3903(179)**	4020	
$D_0^*D_0^*$	27	42	**50(8)**	74	116	**136(22)**	4405	4402	**4398(54)**	4636	
$\mathbf{J^{PC} = 1^{+\pm}}$											X_c, Z_c
$\bar{D}^*D$	87	93	**97(10)**	146	154	**161(17)**	3901	3901	**3903(62)**	3880	
$\bar{D}_0^*D_1$	48	71	**83(10)**	81	118	**137(16)**	4394	4395	**4401(164)**	4739	
$\mathbf{J^{PC} = 0^{-\pm}}$											–
$\bar{D}_0^*D$	68	88	**94(7)**	190	240	**257(19)**	5956	5800	**5690(140)**	4188	
$\bar{D}^*D_1$	–	–	–	382	490	**564(38)**	6039	5898	**5797(141)**	4432	
$\boldsymbol{J^{PC} = 1^{-}}$											Y_c
$\bar{D}_0^*D^*$	112	143	**157(10)**	186	238	**261(17)**	6020	5861	**5748(101)**	4328	
$\bar{D}D_1$	98	126	**139(13)**	164	209	**231(21)**	5769	5639	**5544(162)**	4291	
$\mathbf{J^{PC} = 1^{-+}}$											Y_c
$\bar{D}_0^*D^*$	105	135	**150(13)**	174	224	**249(22)**	6047	5920	**5828(132)**	4328	
$\bar{D}D_1$	97	128	**145(15)**	162	213	**241(25)**	5973	5840	**5748 (179)**		

• Comments

We have shown in the following our analysis:

◇ The introduction of PT radiative corrections is mandatory to solve the ambiguity of the heavy quark mass definitions (pole and running quark masses) in the analysis. Here, we have estimated these corrections using a factorization of two correlators built from bilinear quark currents. We have tested the factorization by an estimate of the first $1/N_c$ correction and found that this correction is small.

◇ We found that the PT corrections affect more coupling than the mass determination which is mainly due to the fact that these corrections tend to cancel out in the ratio of moments used to extract the hadron masses.

◇ More comments on the comparison of our predictions with some experimental candidates can be found in Refs. [278,279] and the summary review in Ref. [287].

7.10.2 Nature of the X(5568)

• Interpolating currents

In the following, we shall analyze different assumptions on the nature of $X(5568)$ which can be specified by the form of the minimal QCD interpolating currents given in Table 7.15. C is the charge conjugation matrix; u, d, s, b are the quark fields and the summation over colour indices is understood; k is a free mixing parameter.

• Higher Order (HO) PT corrections to the spectral function

We extract the higher order (HO) PT corrections by considering that the molecule two-point spectral function is the convolution of the two ones built from two quark bilinear currents (factorization) as a consequence of the molecule definition of the state[7]:

$$J_K(x) \equiv \bar{d}i\gamma_5 s; \quad \frac{1}{\pi}\mathrm{Im}\psi_K(t),$$

[7]A similar approach has been used in Ref. [278] for estimating *for the first time* the higher order perturbative corrections to the spectral functions of molecule states.

Table 7.12. $\bar{B}B$-like molecules masses, invariant and running couplings from LSR within stability criteria from LO to N2LO of PT.

	$\hat{f}_X$ [keV]			$f_X(5.5)$ [keV]			Mass [MeV]				
Nature	LO	NLO	N2LO	LO	NLO	N2LO	LO	NLO	N2LO	Threshold	Exp.
$\mathbf{J^{PC} = 0^{++}}$											–
$\bar{B}B$	4.0	4.4	**5(1)**	14.4	15.6	**17(4)**	10605	10598	**10595(58)**	10559	
$\bar{B}^*B^*$	–	–	–	27	30	**32(5)**	10626	10646	**10647(184)**	10650	
$B_0^*B_0^*$	2.1	3.2	**4(1)**	7.7	11.3	**14(4)**	10653	10649	**10648(113)**	–	
$\mathbf{J^{PC} = 1^{+\pm}}$											X_b, Z_b
$\bar{B}^*B$	7	8	**9(3)**	14	16	**17(5)**	10680	10673	**10646(150)**	10605	
$\bar{B}_0^*B_1$	4	6	**7(1)**	8	11	**14(2)**	10670	10679	**10692(132)**	–	
$\mathbf{J^{PC} = 0^{-\pm}}$											–
$\bar{B}_0^*B$	11	16	**20(3)**	39	55	**67(10)**	12930	12737	**12562(260)**	–	
$\bar{B}^*B_1$	–	–	–	71	105	**136(19)**	12967	12794	**12627(225)**	11046	
$\mathbf{J^{PC} = 1^{--}}$											Y_b
$\bar{B}_0^*B^*$	21	29	**35(6)**	39	54	**66(11)**	12936	12756	**12592(266)**	–	
$\bar{B}B_1$	21	29	**35(7)**	39	54	**65(12)**	12913	12734	**12573(257)**	11000	
$\mathbf{J^{PC} = 1^{-+}}$											Y_b
$\bar{B}_0^*B^*$	20	29	**34(4)**	38	54	**64(8)**	12942	12774	**12617(220)**	–	
$\bar{B}B_1$	20	29	**35(5)**	37	53	**65(9)**	12974	12790	**12630(236)**	11000	

$$J_B(x) \equiv \bar{b} i\gamma_5 u; \quad \frac{1}{\pi}\text{Im}\psi_B(t),$$
$$J^{\mu}_{B^*}(x) \equiv \bar{b}\gamma^{\mu} u; \quad \frac{1}{\pi}\text{Im}\Pi_{B^*}(t). \qquad (7.274)$$

Then we use the convolution integral in Eq. (7.267) for estimating the NLO PT corrections:

Table 7.13. Four-quark masses, invariant and running couplings from LSR within stability criteria from LO to N2LO of PT.

	$\hat{f}_X$ [keV]			$f_X(4.5)$ [keV]			Mass [MeV]			
Nature	LO	NLO	N2LO	LO	NLO	N2LO	LO	NLO	N2LO	Exp.
c-**quark**										
$S_c(0^+)$	62	67	**70(7)**	173	184	**191(20)**	3902	3901	**3898(54)**	–
$A_c(1^+)$	100	106	**112(18)**	166	176	**184(30)**	3903	3890	**3888(130)**	X_c, Z_c
$\pi_c(0^-)$	84	106	**113(5)**	233	292	**310(13)**	6048	5872	**5750(127)**	–
$V_c(1^-)$	123	162	**178(11)**	205	268	**296(19)**	6062	5904	**5793(122)**	Y_c

- **Results and nature of X(5568)**

The results of the analysis for the masses and couplings based on stability criteria are summarized in Table 7.14.

As one can see from Table 7.16, our predictions do not support the assignments for the $X(5568)$ being a pure BK molecule or four-quark state. Assuming, for instance, that it can result from the mixing of the BK molecule and four-quark state, we consider the two-component mixing:

$$|X\rangle = |B\bar{K}\rangle \cos\theta + |(bu)(\bar{d}\bar{s})\rangle \sin\theta,$$
$$|X_{\perp}\rangle = -|B\bar{K}\rangle \sin\theta + |(bu)(\bar{d}\bar{s})\rangle \cos\theta. \qquad (7.275)$$

Using our result, $M_{BK} \simeq M_{(bu)(\bar{d}\bar{s})} \simeq 5196$ MeV, one can deduce, for reproducing $M_X = 5568$ MeV:

$$\sin 2\theta \simeq 0.15, \qquad (7.276)$$

and its orthogonal component, $M_{X_{\perp}} \simeq 4791$ MeV, which is much below the $\bar{B}K$ threshold.

Table 7.14. Four-quark masses, invariant and running couplings from LSR within stability criteria from LO to N2LO of PT.

	$\hat{f}_X$ [keV]			$f_X(5.5)$ [keV]			Mass [MeV]			
Nature	LO	NLO	N2LO	LO	NLO	N2LO	LO	NLO	N2LO	Exp.
***b*-quark**										
$S_b(0^+)$	4.6	5.0	**5.3(1.1)**	16	17	**19(4)**	10652	10653	**10654(109)**	–
$A_b(1^+)$	8.7	9.5	**10(2)**	16	18	**19(3)**	10730	10701	**10680(172)**	Z_b
$\pi_b(0^-)$	18	23	**27(3)**	62	83	**94(11)**	13186	12920	**12695(254)**	–
$V_b(1^-)$	24	33	**40(5)**	45	62	**75(9)**	12951	12770	**12610(242)**	Y_b

Table 7.15. Minimal interpolating currents J_X describing $X(5568)$.

Nature	J^P	Current
Molecule		
BK	0^+	$(\bar{b}\, i\gamma_5\, u)(\bar{d}\, i\gamma_5\, s)$
$B_s\pi$	0^+	$(\bar{b}\, i\gamma_5\, s)(\bar{d}\, i\gamma_5\, u)$
B^*K	1^+	$(\bar{b}\, i\gamma_\mu\, u)(\bar{d}\, i\gamma_5\, s)$
$B_s^*\pi$	1^+	$(\bar{b}\, i\gamma_\mu s)(\bar{d} i\gamma_5\, u)$
Four-quark		
	1^-	$(s^T C\gamma_5\, u)(\bar{b}\,\gamma_\mu\gamma_5 C\,\bar{d}^T) + k(s^T C\, u)(\bar{b}\,\gamma_\mu\, C\,\bar{d}^T)$
	1^+	$(s^T C\gamma_5\, u)(\bar{b}\,\gamma_\mu C\,\bar{d}^T) + k(s^T C\, u)(\bar{b}\,\gamma_\mu\gamma_5 C\,\bar{d}^T)$

7.10.3 Doubly hidden scalar states

In Ref. [281], we consider doubly hidden molecule $(\bar{Q}Q)(Q\bar{Q})$ and four-quark $(\bar{Q}\bar{Q})(QQ)$ states. A recent review and summary can be found in Ref. [305].

• The QCD molecule local interpolating currents

We shall be concerned with the following QCD local interpolating currents of dimension six:

$$\langle 0|\mathcal{O}^H_{\mathcal{M}}(x)|\mathcal{M}\rangle = f^H_{\mathcal{M}} M^4_{\mathcal{M}} : \; \mathcal{O}^H_{\mathcal{M}}(x) \equiv (J^H_{\mathcal{M}} \bar{J}^H_{\mathcal{M}})(x), \tag{7.277}$$

where $f^H_{\mathcal{M}}$ is the meson decay constant, $J^H_{\mathcal{M}}(x)$ is the lowest dimension, bilinear quark currents and $H \equiv S, P, V, A$.

For the scalar (0^{++}) molecule states, these currents are:

$$J^{[S,P,V,A]}_{\mathcal{M}} = \bar{Q}[1, \gamma_5, \gamma_\mu, \gamma_5\gamma_\mu]Q. \tag{7.278}$$

Table 7.16. Exotic hadron masses and couplings from LSR within stability at N2LO.

Nature	J^P	Mass [MeV]	$\hat{f}_X$ [keV]	$f_X(4.5)$ [keV]
b-quark channel				
Molecule				
B^*K	1^+	5186 ± 13	4.48 ± 1.45	8.02 ± 2.60
BK	0^+	5195 ± 15	2.57 ± 0.75	8.26 ± 2.40
$B_s^*\pi$	1^+	5200 ± 18	5.61 ± 0.87	10.23 ± 1.59
$B_s\pi$	0^+	5199 ± 24	3.15 ± 0.70	10.5 ± 2.30
Four-quark $(su)(\overline{bd})$				
A_b	1^+	5186 ± 16	5.05 ± 1.32	9.04 ± 2.37
S_b	0^+	5196 ± 17	2.98 ± 0.70	9.99 ± 2.36
c-quark channel				
Molecule				
D^*K	1^+	2395 ± 48	155 ± 36	226 ± 52
DK	0^+	2402 ± 42	139 ± 26	254 ± 48
$D_s^*\pi$	1^+	2395 ± 48	215 ± 35	308 ± 49
$D_s\pi$	0^+	2404 ± 37	160 ± 22	331 ± 46
Four-quark $(su)(\overline{cd})$				
A_c	1^+	2400 ± 47	192 ± 41	260 ± 55
S_c	0^+	2395 ± 68	122 ± 26	221 ± 47

• The scalar tetraquark states

In the case of tetraquark states, we use with same choice of diquark currents as in [306]:

$$J_{\mathcal{T}}^{[P,S,A,V]} = Q_a^T C[1, \gamma_5, \gamma_\mu, \gamma_5\gamma_\mu]Q_b, \tag{7.279}$$

in order to make a direct comparison with their LO results. We do not consider the current associated to $\sigma_{\mu\nu}$ which corresponds to a two-point correlator of higher dimension. We shall also consider the four-quark operator:

$$\mathcal{O}_{\mathcal{T}} = \epsilon_{abc}\epsilon_{cde}(Q_a^T C\gamma_\mu Q_b)(Q_d^T C\gamma^\mu Q_e), \tag{7.280}$$

in order to make a direct comparison with Ref. [307].

• Results

The results are summarized in Table 7.17.

• Some phenomenological implications

Our results do not favour the ones from some potential models where the exotic states are below the $\overline{\eta}_c\eta_c$ meson thresholds. Instead, our results may explain the existence of a 0^{++} broad structure around (6.2–6.7) GeV which can be due to $\overline{\eta}_c\eta_c$, $\overline{\chi}_{c1}\chi_{c1}$ and $\overline{J/\psi J/\psi}$ molecules or /and to scalar-scalar, vector–vector and axial–axial scalar tetraquark states.

If the new LHCb peak candidate [308, 309] around (6.8–6.9) GeV is a 0^{++} state, the value of its mass suggests that it is likely a $\overline{\chi}_{c0}\chi_{c0}$ molecule or a pseudoscalar–pseudoscalar tetraquark states. Its signature from a $\overline{J/\psi} J/\psi$ invariant mass may come from the di-χ_{c0} decaying to di-$\gamma J/\psi$.

In the case of a $\overline{\chi}_{c1}\chi_{c1}$ molecule, the predicted mass is below the $\chi_{c1}\chi_{c1}$ threshold while our NLO predictions for the beauty states indicate that all of them are above the $\eta_b\eta_b$ and $\Upsilon(1S)\Upsilon(1S)$ thresholds.

7.10.4 $X_{0,1}(2900)$ states and DK invariant mass

In Ref. [282], we revisit, improve and complete some recent estimates of the 0^+ and 1^- open charm $(\bar{c}\bar{d})(us)$ tetraquarks and the corresponding molecules masses and decay constants from QCD spectral sum rules (QSSR) by using QCD Laplace sum rule (LSR) within stability criteria where the factorised perturbative NLO corrections and the contributions of quark and gluon condensates up to dimension 6 in the OPE are included.

• Interpolating currents

We shall be concerned with the interpolating operators given in Table 7.18 and its corresponding two-point correlator:

$$\Pi_{\mathcal{M},\mathcal{T}}(q^2) = i\int d^4x\; e^{-iqx}\langle 0|\mathcal{T}\mathcal{O}^J_{\mathcal{M},\mathcal{T}}(x)\left(\mathcal{O}^J_{\mathcal{M},\mathcal{T}}(0)\right)^\dagger|0\rangle, \quad (7.281)$$

where the superscript J refers to the spin of the particles. We work with the LSR moment for getting the couplings and with their ratio for extracting the mass of the lowest ground states.

Table 7.17. Comparison of the values of the 0^{++} scalar tetraquark masses and couplings from different QSSR approaches. Our predictions are at LO ($\bar{m}_Q$ mass and only the central value is quoted) and up to NLO of PT series $\oplus\langle G^3\rangle$ condensates. $\langle G^3$ only affects the last digit of the central values but increases the error by a factor (1.5–2) (see Ref. [281]). The predictions of Ref. [306] are from Moments at LO and of Ref. [307] from LSR at LO. As already mentioned earlier, the choice of the numerical values of the $\overline{MS}$ running quark masses used at LO is not justified due to the ambiguous quark mass definition to that order.

Scalar	$M_{\bar{c}\bar{c}cc}$ [GeV]				$M_{\bar{b}\bar{b}bb}$ [GeV]			
$\bar{q}\bar{q}qq$	LO	NLO $\oplus$ G3	LO [306]	LO [307]	LO	NLO $\oplus$ G3	LO [306]	LO [307]
Eq. (7.279)								
$\overline{S}_q S_q$	6.59	6.41 ± 0.08	6.44 ± 0.15		19.51	19.22 ± 0.12	18.45 ± 0.15	
$\bar{A}_q A_q$	6.52	6.45 ± 0.08	6.46 ± 0.16	–	19.51	19.87 ± 0.16	18.46 ± 0.14	–
$\overline{V}_q V_q$	6.55	6.46 ± 0.18	6.59 ± 0.17		19.49	19.49 ± 0.08	18.59 ± 0.17	
$\overline{P}_q P_q$	7.37	6.80 ± 0.27	6.82 ± 0.18		19.96	19.75 ± 0.08	19.64 ± 0.14	
Eq. 7.280								
$\bar{A}_q A_q$	6.50	6.47 ± 0.07		5.99 ± 0.08	19.49	19.72 ± 0.12		18.84 ± 0.09

Table 7.18. Interpolating operators describing the scalar (0^+) and vector (1^-) molecules and tetraquark states.

Scalar states (0^+)	Vector states (1^-)
Tetraquarks	
$\mathcal{O}^0_{SS} = \epsilon_{ijk}\,\epsilon_{mnk}\,(u_i^T\,C\gamma_5\,d_j)\,(\bar{c}_m\,\gamma_5 C\,\bar{s}_n^T)$	$\mathcal{O}^1_{AP} = \epsilon_{mnk}\,\epsilon_{ijk}(\bar{c}_m\,\gamma_\mu C\,\bar{s}_n^T)(u_i^T\,C\,d_j)$
$\mathcal{O}^0_{PP} = \epsilon_{ijk}\,\epsilon_{mnk}\,(u_i^T\,C\,d_j)\,(\bar{c}_m\,C\,\bar{s}_n^T)$	$\mathcal{O}^1_{PA} = \epsilon_{mnk}\,\epsilon_{ijk}\,(\bar{c}_m\,C\,\bar{s}_n^T)\,(u_i^T\,C\gamma_\mu\,d_j)$
$\mathcal{O}^0_{VV} = \epsilon_{ijk}\,\epsilon_{mnk}\,(u_i^T\,C\gamma_5\gamma_\mu\,d_j)\,(\bar{c}_m\,\gamma^\mu\gamma_5 C\,\bar{s}_n^T)$	$\mathcal{O}^1_{SV} = \epsilon_{ijk}\,\epsilon_{mnk}\,(u_i^T\,C\gamma_5\,d_j)\,(\bar{c}_m\,\gamma_\mu\gamma_5 C\,\bar{s}_n^T)$
$\mathcal{O}^0_{AA} = \epsilon_{ijk}\,\epsilon_{mnk}\,(u_i^T\,C\gamma_\mu\,d_j)\,(\bar{c}_m\,\gamma^\mu C\,\bar{s}_n^T)$	$\mathcal{O}^1_{VS} = \epsilon_{ijk}\,\epsilon_{mnk}\,(u_i^T\,C\gamma_5\gamma_\mu\,d_j)\,(\bar{c}_m\,\gamma_5 C\,\bar{s}_n^T)$
Molecules	
$\mathcal{O}^0_{DK} = (\bar{c}\gamma_5 d)(\bar{s}\gamma_5 u)$	$\mathcal{O}^1_{D_1K} = (\bar{c}\gamma_\mu\gamma_5 d)\,(\bar{s}\gamma_5\,u)$
$\mathcal{O}^0_{D^*K^*} = (\bar{c}\gamma^\mu d)(\bar{s}\gamma_\mu u)$	$\mathcal{O}^1_{DK_1} = (\bar{c}\gamma_5 d)\,(\bar{s}\,\gamma_\mu\gamma_5\,u)$
$\mathcal{O}^0_{D_1K_1} = (\bar{c}\gamma^\mu\gamma_5 d)(\bar{s}\gamma_\mu\gamma_5 u)$	$\mathcal{O}^1_{D^*K_0^*} = (\bar{c}\gamma_\mu d)\,(\bar{s}\,u)$
$\mathcal{O}^0_{D_0^*K_0^*} = (\bar{c}d)(\bar{s}u)$	$\mathcal{O}^1_{D_0^*K^*} = (\bar{c}\,d)\,(\bar{s}\,\gamma_\mu\,u)$

• Results from the LSR analysis

Our results within the sum rule variable τ, the continuum threshold t_c and the subtraction constant μ stability criteria are shown in Tables 7.19 and 7.20.

One can note from the analysis that the NLO PT, the $\langle\alpha_s G^2\rangle$ and $\langle g^3 G^3\rangle$ corrections are small, while the ones of the chiral condensates $\langle\bar{\psi}\psi\rangle$, $\langle\bar{\psi}G\psi\rangle$, $\langle\bar{\psi}\psi\rangle^2$ are important in this open-charm channel. This feature is typical for the case of the open-charm and beauty states.

• Confrontation with the LHCb data

We confront our results with the D^-K^+ invariant mass recently reported by LHCb from $B^+ \to D^+(D^-K^+)$ decays shown in Fig. 7.32.

◇ We expect that the bump near the D^-K^+ threshold can be originated from the $0^{++}(D^-K^+)$ molecule and/or D^-K^+ scattering. The prominent $X_0(2900)$ scalar peak and the bump $X_J(3150)$ (if $J = 0$) can emerge from a *minimal mixing model*, with a tiny mixing angle $\theta_0 \simeq (5.2 \pm 1.9)^0$, between a scalar *tetramole* ($\mathcal{T}_{\mathcal{M}0}$) (superposition of nearly degenerated hypothetical molecules and compact tetraquarks states with the same quantum numbers) having a mass $M_{\mathcal{T}_{\mathcal{M}0}}$=2743(18) MeV and the first radial excitation of the D^-K^+ molecule with mass $M_{(DK)_1} = 3678(310)$ MeV.

Table 7.19. Predictions from LSR at NLO for the decay constants and masses of the (0^+) scalar and (1^-) vector molecules and tetraquark states.

States	$f_{(G)}$ [keV]	$M_{(G)}$ [MeV]
0^+ States		
Tetraquark		
SS	345(28)	2736(21)
PP	538(41)	5917(98)
VV	713(66)	5704(149)
AA	498(43)	2675(65)
Molecule		
DK	254(48)	2402(42) (Ref. [280])
D^*K^*	405(33)	2808(41)
D_1K_1	664(57)	5258(113)
$D_0^*K_0^*$	249(18)	6270(160)
1^- States		
Tetraquark		
AP	416(38)	5542(139)
PA	285(29)	2666(32)
SV	259(25)	2593(31)
VS	412(43)	5698(175)
Molecule		
D_1K	191(21)	2676(47)
DK_1	351(31)	5377(166)
$D_0^*K^*$	216(22)	2744(41)
$D^*K_0^*$	255(23)	5358(153)

Table 7.20. Predictions from LSR at NLO for the decay constants and masses of the first radial excitations of the *low mass* scalar (0^+) and (1^-) vector molecules and tetraquarks states.

States	$f_{(G)_1}$ [keV]	$M_{(G)_1}$ [MeV]
0^+ States		
Molecule		
$(DK)_1$	211(51)	3678(310)
$(D^*K^*)_1$	568(167)	4626(252)
Tetraquark		
$(SS)_1$	359(81)	4586(268)
$(AA)_1$	547(95)	4593(289)
1^- States		
$(D_1K)_1$	157(71)	4582(414)
$(D_0^*K^*)_1$	237(63)	4662(269)
Tetraquark		
$(PA)_1$	258(82)	4571(213)
$(SV)_1$	243(68)	4541(345)

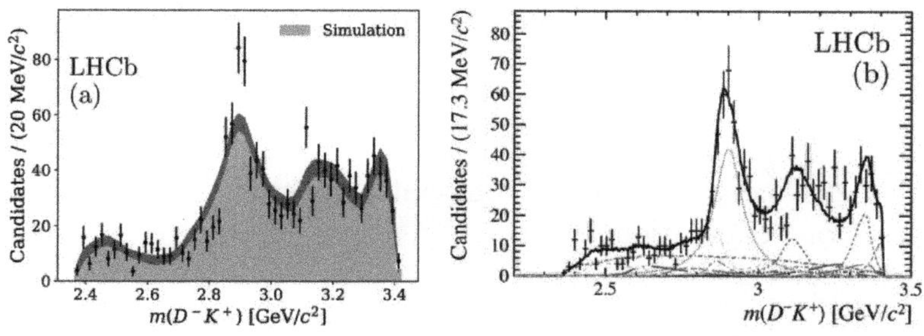

Figure 7.32. LHCb preliminary results for the D^-K^+-invariant mass from $B \to D^+D^-K^+$-decays.

◇ In an analogous way, the $X_1(2900)$ and the $X_J(3350)$ (if $J = 1$) could be a mixture between the vector *tetramole* ($\mathcal{T}_{\mathcal{M}1}$) with a mass $M_{\mathcal{T}_{\mathcal{M}1}} = 2656(20)$ MeV and its first radial excitation having a mass $M_{(\mathcal{T}_{\mathcal{M}1})_1} = 4592(141)$ MeV with an angle $\theta_1 \simeq (9.1 \pm 0.6)°$.

◇ A (non)-confirmation of the previous *minimal mixing models* requires an experimental identification of the quantum numbers of the bumps at 3150 and 3350 MeV.

7.10.5 Z_c state

• The BES III data

Motivated by the BES III data where the K^+ recoil or invariant $D_s^*D \oplus D^*D_s$ mass (see Fig. 7.33) [310] is a good $Z_{cs}(1^+)$-like state candidate with a narrow peak is experimentally found at (see Fig. 7.33):

$$M = (3982.5^{+1.8}_{-2.6} \pm 2.1) \text{ MeV}, \quad \Gamma = (12.8^{+5.3}_{-4.4} \pm 3.0) \text{ MeV}, \quad (7.282)$$

we re-estimate the Z_c-like states using LSR within stability criteria.

• The QCD molecule and tetraquarks currents

We shall be concerned with the QCD local currents $\mathcal{O}_H^\mu(x)$ of dimension 6 given in Table 7.21 for 1^+ axial-vector molecules and tetraquarks where $q \equiv d, s$. b is a free mixing parameter where its optimal value was found to be zero [278,279]. The appropriate hadron

$\mathcal{H}$ (here, it will be an axial vector) couples to the current as:

$$\langle 0|\mathcal{O}^{\mu}_{\mathcal{H}}(x)|\mathcal{H}\rangle = f_{\mathcal{H}} M^5_{\mathcal{H}} \epsilon^{\mu}, \tag{7.283}$$

where $f_{\mathcal{H}}$ is the hadron decay constant analogue to f_π and ϵ^μ is the axial-vector polarisation.

- **Results of the analysis**

The results for the couplings and masses for different interpolating currents are summarized in Table 7.22.

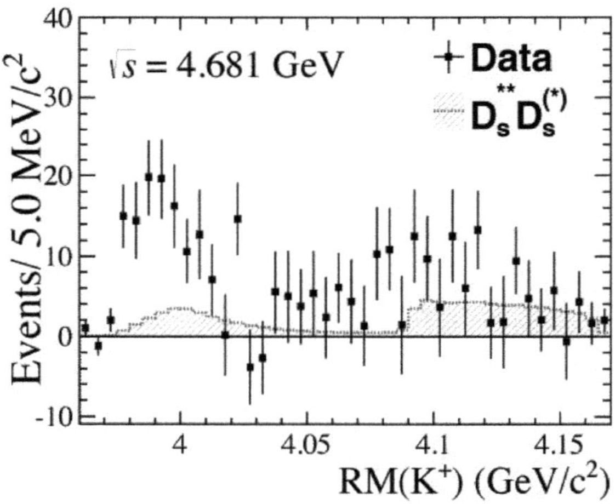

Figure 7.33. K^+ recoil-mass at $\sqrt{s} = 4.681$ GeV after background subtractions from BESIII [310].

- **Tetramole States and Phenomenology**

◇ We note from the tables that the approach cannot discriminate the molecule and four-quark states having the same quantum numbers. Then we introduce the *tetramole* state which have the mean of their masses and couplings. Then, we identify $Z_c(3900)$ with tetramole state:

$$M_{\mathcal{T}_c} = 3900(42)\ \text{MeV}, \quad f_{\mathcal{T}_c} = 155(11)\ \text{keV}. \tag{7.284}$$

Table 7.21. (1^+) molecules and tetraquarks currents ($q \equiv d, s$).

Molecules	Currents
$D_q^* D$	$(\bar{c}\gamma_\mu q)(\bar{d}\, i\gamma_5 c)$
$D^* D_q$	$(\bar{q}\gamma_\mu c)(\bar{c}\, i\gamma_5 q)$
$D_{q0}^* D_1$	$(\bar{c}q)(\bar{d}\gamma_\mu\gamma_5 c)$
$D_0^* D_{q1}$	$(\bar{c}q)(\bar{q}\gamma_\mu\gamma_5 c)$
$D_s^* D_s$	$(\bar{c}\gamma_\mu s)(\bar{s}\, i\gamma_5 c)$
$D_{s0}^* D_{1s}$	$(\bar{c}s)(\bar{s}\gamma_\mu\gamma_5 c)$
Tetraquarks	**Currents**
A_{cq}	$\epsilon_{abc}\epsilon_{dec}\big[(q_a^T\, C\gamma_5\, c_b)(\bar{d}\gamma_\mu C\, \bar{c}_e^T)$ $+b\,(q_a^T\, C\, c_b)(\bar{d}\gamma_\mu\gamma_5 C\, \bar{c}_e^T)\big]$
A_{css}	$\epsilon_{abc}\epsilon_{dec}\big[(s_a^T\, C\gamma_5\, c_b)(\bar{s}\gamma_\mu C\, \bar{c}_e^T)$ $+b\,(s_a^T\, C\, c_b)(\bar{s}\gamma_\mu\gamma_5 C\, \bar{c}_e^T)\big]$

- $Z_c(4025)$ found by BES [311] or $Z_c(4050)$ found by BELLE [312] can be identified with the $D_0^* D_1$ molecule with the parameters:

$$M_{D_0^* D_1} = 4023(130)\ \text{MeV}, \quad f_{D_0^* D_1} = 96(23)\ \text{keV}. \tag{7.285}$$

- The $Z_c(4430)$ might be interpreted as a weakly coupled radial excitation with a coupling $(f_{(D^*D)_0} = 46(56)$ MeV.
- The $Z_{cs}(3983)$ can be a tetramole combination of the $D_s^* D$, $D^* D_s$ molecules and A_{cs} tetraquark states with:

$$M_{\mathcal{T}_{cs}} = 3973(31)\ \text{MeV}, \quad f_{\mathcal{T}_{cs}} = 136(9)\ \text{keV}. \tag{7.286}$$

- The $Z_{cs}(4100)$ bump can be a combination of a $D_{s0}^* D_1 \oplus D_0^* D_{s1}$ molecules with the mean mass:

$$M_{D_{s0}^* D_1 \oplus D_0^* D_{s1}} = 4067(94)\ \text{MeV}, \quad f_{D_{s0}^* D_1 \oplus D_0^* D_{s1}} = 88(16)\ \text{keV}. \tag{7.287}$$

- One can also define a tetramole $\mathcal{T}_{css}$ which is a superposition of the $D_s^* D_s$ molecule and A_{css} tetraquark states with:

$$M_{\mathcal{T}_{css}} = 4068(48)\ \text{MeV}, \quad f_{\mathcal{T}_{css}} = 114(10)\ \text{keV}, \tag{7.288}$$

- and a higher mass $D_{s0}^* D_{s1}$ molecule:

$$M_{D_{s0}^* D_{s1}} = 4198(129)\ \text{MeV}, \quad f_{D_{s0}^* D_{s1}} = 79(14)\ \text{keV}, \tag{7.289}$$

where the coupling is relatively weak.

Table 7.22. Summary of the coupling and mass predictions of the lowest ground states and comparison with our previous results in Refs. [278,279,287].

	Couplings			Masses		
States	Ref. [278]	Ref. [279]	New	Ref. [278]	Ref. [279]	New
Molecule $(\bar{c}d)(c\bar{u})$						
D^*D	154(7)	–	140(15)	3901(62)	–	3912(61)
$D_0^*D_1$	96(15)	–	96(23)	4394(164)	3854(182)	4023(130)
Tetraquark						
$(\bar{c}\bar{d})(cu)$						
A_{cd}	176(30)	–	173(17)	3890(130)	–	3889(42)
Molecule $(c\bar{s})(\bar{c}u)$						
D_s^*D	–	–	130(15)	–	–	3986(51)
D^*D_s	–	–	133(16)	–	–	3979(56)
$D_{s0}^*D_1$	–	–	86(23)	–	–	4064(133)
$D_0^*D_{s1}$	–	–	89(22)	–	–	4070(133)
Tetraquark						
$(\bar{c}\bar{s})(cu)$						
A_{cs}	–	–	148(17)	–	–	3950(56)
Molecule $(\bar{c}s)(c\bar{s})$						
$D_s^*D_s$	–	114(13)	–	–	3901(62)	4091(57)
$D_{s0}^*D_{s1}$	–	79(14)	70(16)	–	4269(205)	4198(129)
A_{css}	–	114(16)	131(14)	–	4209(112)	4014(86)

7.10.6 XTZ states

We wrote the paper "Improved XTZ masses and mass ratios from Laplace sum rules at NLO" [284], where we use ratio and double ratios of Laplace sum rules (LSR) to improve our previous determinations of the masses and mass spitting of Z_Q, X_Q and $T_{QQ\bar{q}\bar{q}'}$ 1^{++} and 0^{++} states ($Q \equiv c, b$; $q, q' \equiv u, d, s$).

• Interpolating currents

We shall work with the two-point correlators associated to the interpolating currents in Table 7.23, which are the same as in previous sections, but we remind here. We shall include the factorized NLO PT corrections and truncate the OPE up to dimension 6 condensate contributions. The sum rule analysis is similar to the previous ones where the optimal results are obtained within stability criteria.

Table 7.23. Interpolating operators describing the $Z_c, X_c, T_{cc\bar{q}'\bar{q}}$ states discussed in this paper, where $b = 0$ is the optimized mixing parameter [283].

States	$I(J^P)$	$\bar{3}_c 3_c$ Four-quark Currents	References
Z_c	(1^+)	$\mathcal{O}_{A_{cq}} = \epsilon_{ijk}\epsilon_{mnk}[(q_i^T\, C\gamma_5\, c_j)(\bar{q}'_m\gamma_\mu C\,\bar{c}_n^T)$ $+b, (q_i^T\, C\, c_j)(\bar{q}'_m\gamma_\mu\gamma_5 C\,\bar{c}_n^T)]$ $\mathcal{O}_{D_q^* D_q} = (\bar{c}\gamma_\mu q)(\bar{q}'\, i\gamma_5 c)$	[283]
X_c	(1^+)	$\mathcal{O}_X^3 = \epsilon_{ijk}\,\epsilon_{mnk}[(q_i^T\, C\gamma_5\, c_j)(\bar{c}_m\,\gamma^\mu C\,\bar{q}_n^T) +$ $(q_i^T\, C\gamma^\mu\, c_j)(\bar{c}_m\,\gamma_5 C\,\bar{q}_n^T)]$ [259, 263], $\mathcal{O}_X^6 = \epsilon_{ijk}\,\epsilon_{mnk}[(q_i^T\, C\gamma_5\lambda_{ij}^a\, c_j)(\bar{c}_m\,\gamma^\mu C\lambda_{mn}^a\,\bar{q}_n^T)$ $+(q_i^T\, C\gamma^\mu\lambda_{ij}^a\, c_j)(\bar{c}_m\,\gamma_5 C\lambda_{mn}^a\,\bar{q}_n^T)]$ $\mathcal{O}_{D_q^* D_q} = \frac{1}{\sqrt{2}}[(\bar{q}\gamma_5 c)(\bar{c}\gamma_\mu q) - (\bar{q}\gamma_\mu c)(\bar{c}\gamma_5 q)]$ $\mathcal{O}_{\psi\pi} = (\bar{c}\gamma_\mu\lambda^a c)(\bar{q}\gamma_5\lambda^a q)$	[264, 278]
$T_{cc\bar{u}\bar{d}}$	$0(1^+)$	$\mathcal{O}_T^{1^+} = \frac{1}{\sqrt{2}}\epsilon_{ijk}\,\epsilon_{mnk}\,(c_i^T\, C\gamma^\mu\, c_j)\,[(\bar{u}_m\,\gamma_5 C\,\bar{d}_n^T)$ $-(\bar{d}_m\,\gamma_5 C\,\bar{u}_n^T)]$	[263]
$T_{cc\bar{u}\bar{s}}$	$\frac{1}{2}(1^+)$	$\mathcal{O}_{T_{us}^{1+}} = \epsilon_{ijk}\,\epsilon_{mnk}\,(c_i\, C\gamma^\mu c_j^T)\,(\bar{u}_m\,\gamma_5 C\bar{s}_n^T)$	
$T_{cc\bar{u}\bar{d}}$	$1(0^+)$	$\mathcal{O}_T^{0^+} = \frac{1}{\sqrt{2}}\epsilon_{ijk}\,\epsilon_{mnk}\,(c_i^T\, C\gamma^\mu\, c_j)\,[(\bar{u}_m\,\gamma_\mu C\,\bar{d}_n^T)$ $+(\bar{d}_m\,\gamma_\mu C\,\bar{u}_n^T)]$	[263]
$T_{cc\bar{u}\bar{s}}$	$\frac{1}{2}(0^+)$	$\mathcal{O}_{T_{us}^{0+}} = \epsilon_{ijk}\,\epsilon_{mnk}\,(c_i\, C\gamma_\mu c_j^T)\,(\bar{u}_m\,\gamma^\mu C\bar{s}_n^T)$	
$T_{cc\bar{s}\bar{s}}$	$0(0^+)$	$\mathcal{O}_T^{0^+} = \epsilon_{ijk}\,\epsilon_{mnk}\,(c_i\, C\gamma_\mu c_j^T)\,(\bar{s}_m\,\gamma^\mu C\bar{s}_n^T)$	

• Summary of the results

Our results are summarized in Table 7.24, where one can note along the text that the combined uses of the ratio and double ratio of sum rules have improved the accuracy of the masses.

◇ $X_c(3872)$ and $Z_c(3900)$ as tetramoles

The analysis confirms the interpretation of $Z_{c,b}$ and $X_{c,b}$ as tetramole states (superposition of four-quark and molecule states having the same masses and equally coupled to the currents) presented in Section 7.10.5.

◇ $T_{QQ\bar{q}\bar{q}'}$ states

In the second part of the paper, we study the molecule nature of the $T_{QQ\bar{q}\bar{q}'}$ states and the mass splittings due to SU(3) breakings. One can notice the following from Table 7.24 that:

- The results from the DRSR (double ratio of sum rules) are more accurate than the one from R (ratio of sum rules).
- In the case of the charm channel, the SU(3) breaking increases the mass of a single strange state by about 50 MeV, while the double strange state is about 100 MeV higher than the non-strange one. The effect is about half in the bottom channel.
- The bottom states are all below the hadronic thresholds though the errors in the difference is large, while the charm ones are almost all above them.

7.10.7 $T_{QQ\bar{q}\bar{q}'}(1^-, 0^-)$ states

More recently, we have completed the analysis of the $T_{QQ\bar{q}\bar{q}'}$ states by analyzing the axial-vector (1^-) and pseudoscalar (0^-)states [313]. In addition to the usual stability criteria versus the sum rule variable τ, the QCD continuum threshold t_c and the perturbative subtraction constant μ, we add the (rigorous) condition: $R_{P/C} \equiv$ Pole (Resonance) over the QCD continuum contributions ≥ 1. We include factorized PT NLO correction and the controlled condensates up to dimension 6.

• The interpolating operators

In this paper, we choose to work with the $\bar{3}_c 3_c$ lowest dimension interpolating currents of the four-quark states given in Table 7.25. We shall not consider the molecule currents as our approach cannot distinguish a four-quark from a molecule state.

There are other possibilities for constructing such states by using the currents proposed in Ref. [314].

• Ground state masses and couplings

Our results are compiled in Tables 7.27–7.30 and compared with some LO existing ones.

- ◇ One can note that the mass splittings due to SU3 breakings $(m_s, \langle\bar{s}s\rangle)$ are tiny $(\leq 50$ MeV).
- ◇ We found that the pseudoscalar and vector states can be divided into two classes:

Table 7.24. Summary of the results of the XZT states masses in units of MeV obtained in this paper from LSR and DRSR using the currents in Table 7.23. Our final values are in the column "LSR ⊕ DRSR" which are the mean from LSR with the ones deduced from DRSR. $\Delta E_B \equiv M_H - M_{\text{th}}$ where M_{th} is the hadronic threshold.

				Mass				ΔE_B
States	J^P	Decay	Thresholds	Data	Config.	LSR	LSR ⊕ DRSR	
Z_c	1^+	$\bar{D}^0 D^{*+}$	3876	3900	D^*D	3912(61) [278, 283]		
					$\bar{3}_c 3_c$	3889(58) [278, 283]		
					$\mathcal{T}_{Z_c}$		*3900(42)* [278, 283]	+24(42)
Z_b	1^+	$\bar{B}^0 B^{*+}$	10605		$\mathcal{T}_{Z_b}$		*10579(99)* [278, 283]	−26(99)
X_c	1^+	$\bar{D}^0 D^{*+}$	3876	3872	$\bar{3}_c 3_c$	3876(76)		
					$\bar{6}_c 6_c$	3864(76)		
					$\psi\pi$		3889(76)	
					D^*D		3912(61)	
					$\mathcal{T}_{X_c}$		*3876(44)*	+0(44)
X_b	1^+	$\bar{B}^0 B^{*+}$	10605		$\bar{3}_c 3_c$	10545(131)		−60(131)
$T_{cc\bar{u}\bar{d}}$	1^+	$\bar{D}^0 D^{*+}$	3876	3875	$\bar{3}_c 3_c$	3885(74)	3886(4)	+14(4)
$T_{cc\bar{u}\bar{s}}$	1^+	$\bar{D}^0_s D^*$	3975		–	3940(89)	3931(7)	−44(7)
$T_{cc\bar{u}\bar{d}}$	0^+	$\bar{D}^0 D^0$	3730		–	3882(81)	3883(3)	+153(3)
$T_{cc\bar{u}\bar{s}}$	0^+	$\bar{D}^0_s D$	3833		–	3936(90)	3927(6)	+94(6)
$T_{cc\bar{s}\bar{s}}$	0^+	$D^+_s D^-_s$	3937		–	4063(72)	3993(11)	+56(11)
$T_{bb\bar{u}\bar{d}}$	1^+	$\bar{B}^0 B^{*+}$	10605		–	10441(147)	10501(98)	−104(98)
$T_{bb\bar{u}\bar{s}}$	1^+	$\bar{B}^0_s B^{*+}$	10692		–	10476(154)	10521(83)	−171(83)
$T_{bb\bar{u}\bar{d}}$	0^+	$\bar{B}^0 B^0$	10559		–	10419(146)	10484(63)	−75(63)
$T_{bb\bar{u}\bar{s}}$	0^+	$\bar{B}^0_s B^0$	10646		–	10454(153)	10511(58)	−135(56)
$T_{bb\bar{s}\bar{s}}$	0^+	$\bar{B}^0_s B^0_s$	10734		–	10538(129)	10567(57)	−167(57)

Table 7.25. Interpolating operators describing the $T_{cc\bar{q}\bar{q}'}$-like states. We note that the current $\mathcal{O}_{T^{0-}_{ss}}$ gives a null contribution.

States	$\bar{3}_c 3_c$ Four-quark Currents
Pseudoscalar $\mathbf{0^-}$	
$T_{cc\bar{u}\bar{d}}$	$\mathcal{O}_{T^{0-}_{ud}} = \frac{1}{\sqrt{2}}\epsilon_{ijk}\,\epsilon_{mnk}\left(c_i^T\,C\gamma^\mu\,c_j\right)\left[\left(\bar{u}_m\,\gamma_5\,\gamma_\mu C\,\bar{d}_n^T\right) - \left(\bar{d}_m\,\gamma_5\,\gamma_\mu C\,\bar{u}_n^T\right)\right]$
$T_{cc\bar{u}\bar{s}}$	$\mathcal{O}_{T^{0-}_{us}} = \epsilon_{ijk}\,\epsilon_{mnk}\left(c_i\,C\gamma_\mu c_j^T\right)\left(\bar{u}_m\,\gamma_5\,\gamma^\mu C\bar{s}_n^T\right)$
$T_{cc\bar{s}\bar{s}}$	$\mathcal{O}_{T^{0-}_{ss}} = \epsilon_{ijk}\,\epsilon_{mnk}\left(c_i\,C\gamma_\mu c_j^T\right)\left(\bar{s}_m\,\gamma_5\,\gamma^\mu C\bar{s}_n^T\right)$
Vector 1^-	
$T_{cc\bar{u}\bar{d}}$	$\mathcal{O}_{T^{1-}_{ud}} = \frac{1}{\sqrt{2}}\epsilon_{ijk}\,\epsilon_{mnk}\left(c_i^T\,C\gamma^\mu\,c_j\right)\left[\left(\bar{u}_m\,C\,\bar{d}_n^T\right) - \left(\bar{d}_m\,C\,\bar{u}_n^T\right)\right]$
$T_{cc\bar{u}\bar{s}}$	$\mathcal{O}_{T^{1-}_{us}} = \epsilon_{ijk}\,\epsilon_{mnk}\left(c_i\,C\gamma^\mu c_j^T\right)\left(\bar{u}_m\,C\bar{s}_n^T\right)$

Table 7.26. Some other interpolating operators describing the $T_{cc\bar{q}\bar{q}'}$-like states used in Ref. [314].

States	Four-quark Currents	Light quarks q
Pseudoscalar 0^-		
$T_{cc\bar{q}\bar{q}}$	$\eta_1 = (c_a^T C c_b)\,[\,(\bar{q}_a\gamma_5 C \bar{q}_b^T) + (\bar{q}_b\gamma_5 C \bar{q}_a^T)\,]$	u, d, s
$T_{cc\bar{q}\bar{q}}$	$\eta_2 = (c_a^T C\gamma_5 c_b)\,[\,(\bar{q}_a C \bar{q}_b^T) + (\bar{q}_b C \bar{q}_a^T)\,]$	u, d, s
$T_{cc\bar{q}\bar{u}}$	$\eta_4 =$ $(c_a^T C\gamma_\mu c_b)\,[\,(\bar{u}_a \gamma^\mu\gamma_5 C \bar{q}_b^T) - (\bar{u}_b \gamma^\mu\gamma_5 C \bar{q}_a^T)\,]$	d, s
$T_{cc\bar{q}\bar{u}}$	$\eta_5 = (c_a^T C\gamma_\mu\gamma_5 c_b)\,[\,(\bar{u}_a \gamma^\mu C \bar{q}_b^T)$ $+ (\bar{u}_b \gamma^\mu C \bar{q}_a^T)\,]$	d, s
Vector 1^-		
$T_{cc\bar{q}\bar{q}}$	$\eta_1 =$ $(c_a^T C\gamma^\mu\gamma_5 c_b)\,[\,(\bar{q}_a\gamma_5 C \bar{q}_b^T) + (\bar{q}_b\gamma_5 C \bar{q}_a^T)\,]$	u, d, s
$T_{cc\bar{q}\bar{q}}$	$\eta_2 = (c_a^T C\gamma_5 c_b)\,[\,(\bar{q}_a \gamma_\mu\gamma_5 C \bar{q}_b^T) +$ $(\bar{q}_b \gamma_\mu\gamma_5 C \bar{q}_a^T)\,]$	u, d, s
$T_{cc\bar{u}\bar{q}}$	$\eta_5 = (c_a^T C\gamma^\mu c_b)\,[\,(\bar{u}_a C \bar{q}_b^T) - (\bar{u}_b C \bar{q}_a^T)\,]$	d, s
$T_{cc\bar{u}\bar{q}}$	$\eta_6 = (c_a^T C c_b)\,[\,(\bar{u}_a \gamma^\mu C \bar{q}_b^T) + (\bar{u}_b \gamma^\mu C \bar{q}_a^T)\,]$	d, s

- ∘ Class H (Heavy) states where the masses of the $T_{cc\bar{q}\bar{q}'}$ (respectively $T_{bb\bar{q}\bar{q}'}$) states are around 6 (respectively 13) GeV which are 2 GeV higher than the ones of their axial partners $(1^+, 0^+)$ [284]. This feature is in line with our previous findings for XYZ states [278–280].
- ∘ L (Light) states around (3.8–4.4) GeV, where the pseudoscalar $T_{cc\bar{q}\bar{q}}$ (respectively all vector states $T_{cc\bar{q}\bar{q}'}$) are below the $\bar{D}D_0, \bar{D}_sD_{s0}$ (respectively $\bar{D}D_1, \bar{D}_sD_{s1}$) open charm thresholds.
- ∘ Class L $T_{bb\bar{q}\bar{q}'}$ pseudoscalar and vector states have masses around 10.4 GeV where all of them are below the open beauty thresholds. Though more accessible experimentally, Class L states have weaker couplings with the currents than the Class H ones and may be difficult to observe.

• First radial excitation masses and couplings

The results for the first radial excitations of $T_{QQ\bar{u}\bar{d}}$ and $T_{QQ\bar{u}\bar{d}}$ are summarized in Tables 7.29 and 7.30.

◇ The radial excitations of the Class L states are expected to be in the region $(5.8 - 6.4)$ GeV (respectively $(12.7 - 13.1)$ GeV) for the charm (respectively bottom) channels. As these radial excitations couple strongly with the corresponding interpolating currents, their

Table 7.27. Our predictions of the couplings [keV] and masses [MeV] are compared with the ones in Ref. [314]. For the *Class H* currents, we found that the results quoted in Ref. [314] using factorization of the four-quark condensates do not satisfy stability criteria, while the ones from non-factorization have imaginary couplings. More details on the comparison with Ref. [314] are given in Ref. [313].

	Our Work						Ref. [314]	
States	Current	t_c [GeV^2]	τ [GeV^{-2}]	$f^{NLO}_{T_{ccqq'}}$	τ [GeV^{-2}]	$M^{NLO}_{T_{ccqq'}}$	t_c	$M^{LO}_{T_{ccqq'}}$
Class H								
$\mathbf{0^-}$								
$T_{cc\bar{u}\bar{d}}$	$\mathcal{O}_T^{0^-}$, η_4	$50 \to 60$	$0.12 \to 0.18$	1559(133)	$0.16 \to 0.18$	6303(123)	22	4430(130)
$T_{cc\bar{d}\bar{d}}$	η_2	$48 \to 60$	$0.14 \to 0.18$	1348(165)	$0.16 \to 0.18$	6267(148)		—
$T_{cc\bar{u}\bar{s}}$	$\mathcal{O}_{T_{us}^{0^-}}$, η_4	$44 \to 58$	$0.16 \to 0.20$	983(147)	$0.15 \to 0.20$	6102(172)		—
$T_{cc\bar{s}\bar{s}}$	η_2	$42 \to 58$	$0.16 \to 0.22$	1164(168)	$0.18 \to 0.22$	5946(180)		—
1^-								
$T_{cc\bar{u}\bar{d}}$	$\mathcal{O}_T^{1^-}$, η_5	$46 \to 60$	$0.14 \to 0.18$	841(117)	$0.15 \to 0.18$	6212(172)		—
$T_{cc\bar{d}\bar{d}}$	η_2	$48 \to 60$	$0.13 \to 0.16$	1165(107)	$0.14 \to 0.16$	6362(140)		—
$T_{cc\bar{u}\bar{s}}$	$\mathcal{O}_{T_{us}^{1^-}}$, η_5	$44 \to 58$	$0.16 \to 0.20$	558(76)	$0.17 \to 0.20$	6064(167)		—
$T_{cc\bar{s}\bar{s}}$	η_2	$46 \to 58$	$0.15 \to 0.20$	1085(109)	$0.17 \to 0.20$	6110(154)		—
Class L								
$\mathbf{0^-}$								
$T_{cc\bar{u}\bar{d}}$	η_5	$24 \to 36$	$0.32 \to 0.45$	289(42)	$0.41 \to 0.47$	4380(128)	23	4410(140)
$T_{cc\bar{d}\bar{d}}$	η_1	$24 \to 36$	$0.25 \to 0.35$	402(45)	$0.25 \to 0.35$	3965(76)	24	4430(120)
$T_{cc\bar{u}\bar{s}}$	η_5	$24 \to 36$	$0.50 \to 0.56$	236(51)	$0.50 \to 0.56$	4251(127)	24	4500(160)
$T_{cc\bar{s}\bar{s}}$	η_1	$24 \to 36$	$0.25 \to 0.36$	322(42)	$0.25 \to 0.36$	4096(96)	25	4460(130)
1^-								
$T_{cc\bar{u}\bar{d}}$	η_6	$26 \to 40$	$0.22 \to 0.31$	293(51)	$0.22 \to 0.31$	4113(108)	23	4340(160)
$T_{cc\bar{d}\bar{d}}$	η_1	$26 \to 40$	$0.23 \to 0.32$	380(39)	$0.23 \to 0.32$	4170(110)	23	4350(140)
$T_{cc\bar{u}\bar{s}}$	η_6	$26 \to 40$	$0.23 \to 0.32$	262(29)	$0.23 \to 0.32$	4157(100)	23	4350(160)
$T_{cc\bar{s}\bar{s}}$	η_1	$26 \to 40$	$0.26 \to 0.35$	289(37)	$0.26 \to 0.35$	4210(103)	24	4470(130)

Table 7.28. Our predictions of the couplings [keV] and masses [MeV] compared with the ones in Ref. [314] (more details are given in the text). Note that in this b-quark channel, the η_5 current becomes a Class H_0.

		Our Work				Ref. [314]	
States	Current	t_c [GeV^2]	τ [GeV^{-2}]	$f^{NLO}_{T_{bbqq'}}$	$M^{NLO}_{T_{bbqq'}}$	t_c	$M^{LO}_{T_{ccqq'}}$
Class H							
0^-							
$T_{bb\bar{u}\bar{d}}$	$\mathcal{O}_{T^{0-}_{ud}}, \eta_4$	$185 \to 230$	$0.060 \to 0.075$	424(108)	13100(278)	–	?
$T_{bb\bar{u}\bar{d}}$	η_5	$180 \to 220$	$0.08 \to 0.09$	165(31)	12730(207)	115	10300(300)
$T_{bb\bar{d}\bar{d}}$	η_2	$185 \to 230$	$0.060 \to 0.080$	373(90)	13039(258)	—	
$T_{bb\bar{u}\bar{s}}$	$\mathcal{O}_{T^{0-}_{us}}, \eta_4$	$185 \to 230$	$0.070 \to 0.080$	283(65)	13005(247)		–
$T_{bb\bar{u}\bar{s}}$	η_5	$175 \to 210$	$0.085 \to 0.090$	144(40)	12553(252)	115	10400(200)
$T_{bb\bar{s}\bar{s}}$	η_2	$185 \to 230$	$0.06 \to 0.085$	302(82)	12745(274)		—
1^-							
$T_{bb\bar{u}\bar{d}}$	$\mathcal{O}_{T^{1-}_{ud}}, \eta_5$	$190 \to 230$	$0.065 \to 0.080$	246(59)	13053(262)		?
$T_{bb\bar{d}\bar{d}}$	η_2	$190 \to 230$	$0.065 \to 0.075$	361(75)	13131(240)		–
$T_{bb\bar{u}\bar{s}}$	$\mathcal{O}_{T^{1-}_{us}}, \eta_5$	$180 \to 230$	$0.070 \to 0.085$	158(41)	12910(268)		–
$T_{bb\bar{s}\bar{s}}$	η_2	$180 \to 230$	$0.075 \to 0.090$	294(69)	12834(246)		–
Class L							
0^-							
$T_{bb\bar{d}\bar{d}}$	η_1	$130 \to 160$	$0.120 \to 0.160$	14(2)	10407(124)	125	10600(300)
$T_{bb\bar{s}\bar{s}}$	η_1	$130 \to 160$	$0.130 \to 0.170$	12(2)	10437(109)	125	10600(300)
1^-							
$T_{bb\bar{u}\bar{d}}$	η_6	$130 \to 160$	$0.110 \to 0.150$	11(2)	10469(110)	120	10400(200)
$T_{bb\bar{d}\bar{d}}$	η_1	$130 \to 160$	$0.120 \to 0.160$	12(2)	10412(99)	125	10600(300)
$T_{bb\bar{u}\bar{s}}$	η_6	$130 \to 160$	$0.120 \to 0.160$	10(2)	10438(97)	120	10400(200)
$T_{bb\bar{s}\bar{s}}$	η_1	$130 \to 160$	$0.130 \to 0.170$	10(2)	10445(116)	125	10600(300)

Table 7.29. Masses and couplings of the first radial excitations of $T_{cc\bar{u}\bar{d}}$ and $T_{cc\bar{d}\bar{d}}$.

States	Current	t_c [GeV^2]	τ [GeV^{-2}]	$f^{NLO}_{T_{ccqq'}}$ [keV]	τ [GeV^{-2}]	$M^{NLO}_{T_{ccqq'}}$ [MeV]
Class H						
0^-						
$T_{cc\bar{u}\bar{d}}$	$\mathcal{O}^{0^-}_T$, η_4	$76 \to 94$	$0.07 \to 0.12$	2101(215)	$0.08 \to 0.12$	8214(214)
$T_{cc\bar{d}\bar{d}}$	η_2	$74 \to 130$	$0.07 \to 0.12$	2026(514)	$0.08 \to 0.12$	8212(353)
1^-						
$T_{cc\bar{u}\bar{d}}$	$\mathcal{O}^{1^-}_T$, η_5	$74 \to 94$	$0.10 \to 0.10$	1262(224)	$0.10 \to 0.10$	8110(325)
$T_{cc\bar{d}\bar{d}}$	η_2	$74 \to 94$	$0.12 \to 0.12$	1571(423)	$0.12 \to 0.12$	8111(316)
Class L						
0^-						
$T_{cc\bar{u}\bar{d}}$	η_5	$38 \to 50$	$0.20 \to 0.31$	596(135)	$0.20 \to 0.31$	5787(213)
$T_{cc\bar{d}\bar{d}}$	η_1	$40 \to 60$	$0.17 \to 0.17$	580(169)	$0.16 \to 0.23$	5957(251)
1^-						
$T_{cc\bar{u}\bar{d}}$	η_6	$44 \to 62$	$0.26 \to 0.21$	451(100)	$0.14 \to 0.21$	6217(349)
$T_{cc\bar{d}\bar{d}}$	η_1	$46 \to 62$	$0.14 \to 0.16$	731(135)	$0.13 \to 0.16$	6415(264)

presence in these regions may mask the searches for some Class H ground states having similar masses.

◇ The Class H $T_{cc\bar{u}\bar{d}}$ 0^- pseudoscalar ground state is exceptionally (300–500) MeV above the radial excitation of the corresponding Class L state. Its eventual detection can be an alternative test of the violation of the factorization of the four-quark condensates.

◇ One may finally assume that some physical states emerge from some non-trivial mixing of the different interpolating operators discussed here. However, looking at these eventual mixings is beyond the aim of the present paper.

7.11 Short visits and International Conferences

It was enriching to go to different international conferences for physics and also discover the cultures of the countries. Presenting a talk is also an important experience. Some key conferences are covered in the following sections.

7.11.1 ICHEP Conferences

Initiated in 1950 at Rochester as a Rochester Conference, this Series of Conference is organized regularly every two years. This is the most

Table 7.30. Masses and couplings of the first radial excitations of $T_{bb\bar{u}\bar{d}}$ and $T_{bb\bar{d}\bar{d}}$.

States	Current	t_c [GeV^2]	τ [GeV^{-2}]	$f^{NLO}_{T_{ccqq'}}$ [keV]	τ [GeV^{-2}]	$M^{NLO}_{T_{ccqq'}}$ [MeV]
Class H						
0^-						
$T_{bb\bar{u}\bar{d}}$	$\mathcal{O}_T^{0^-}$, η_4	$250 \to 310$	$0.0425 \to 0.0675$	1009(314)	$0.0425 \to 0.0675$	15319(645)
$T_{bb\bar{u}\bar{d}}$	η_5	$230 \to 310$	$0.0450 \to 0.0600$	457(151)	$0.0475 \to 0.0600$	14829(350)
$T_{bb\bar{d}\bar{d}}$	η_2	$250 \to 310$	$0.0400 \to 0.0700$	905(308)	$0.0425 \to 0.0700$	15266(509)
1^-						
$T_{bb\bar{u}\bar{d}}$	$\mathcal{O}_T^{1^-}$, η_5	$250 \to 310$	$0.0425 \to 0.0700$	587(208)	$0.0425 \to 0.0700$	15275(628)
$T_{bb\bar{d}\bar{d}}$	η_2	$250 \to 310$	$0.0400 \to 0.0700$	816(303)	$0.0400 \to 0.0700$	15373(576)
Class L						
0^-						
$T_{bb\bar{d}\bar{d}}$	η_1	$170 \to 250$	$0.08 \to 0.10$	100(37)	$0.08 \to 0.10$	12596(405)
1^-						
$T_{bb\bar{u}\bar{d}}$	η_6	$170 \to 250$	$0.07 \to 0.11$	69(29)	$0.075 \to 0.10$	12679(437)
$T_{bb\bar{d}\bar{d}}$	η_1	$170 \to 250$	$0.08 \to 0.10$	88(35)	$0.075 \to 0.11$	12667(408)

important and largest scale conference in HEP where most relevant results are presented.

• 21st ICHEP Conference (Paris, 1982)

This was my first participation as an auditor in a large-scale conference. I was very happy to hear that Professor Maini mentioned my work on the D and B meson leptonic decay constants at the plenary talk.

• 22nd ICHEP Conference (Leipzig, 1984)

I presented for the first time at the international level our work on light quark masses: QCD parameters for light meson systems. However, Professor B.L. Ioffe reacted vigorously claiming that working with the pseudoscalar pion channel does not make sense as small size instantons break the OPE at the scale around 1 GeV^2 we are working.

We see with time that this argument is not confirmed. I also remember driving my car to Leipzig from CERN. This was the opportunity for me to visit a socialist country.

• 23rd ICHEP Conference (Berkeley, 1986)

Professor N. Isgur invited me to give a talk on my work on glueballs. I also took the opportunity to visit SLAC within a Montpelier–SLAC exchange program. This is my discovery of the US with his foods, his large scale space, Nappa valley, San Francisco, etc. It was unfortunately the period when the VIH appeared in San Francisco such that everybody was careful with the contact with another person. The atmosphere was boring. Even going to public toilet becomes delicate.

• 24th ICHEP Conference (Munich, 1988)

This was the discovery of Munich, the Marienplatz, the German beer and foods. I presented our works on tensor mesons and gluonium and their mixings [315].

• 28th ICHEP Conference (Warsaw, 1996)

I presented my work on heavy quarkonia mass splittings in QCD [316].

• 32nd ICHEP Conference (Beijing, 2004)

Discovery of Beijing – Eat Snake.

7.11.2 EPS conferences

• EPS Conference on HEP (Bari, 1985)

I gave the talk: Compositeness Versus Local Duality Sum Rules [317].

• EPS Conference on HEP (Jerusalem, 1997)

Convenor: Hadron spectroscopy including glueballs, exotics, states with c and b quarks.

Talk: Gluonia in QCD with Massless Quarks [318].

7.11.3 Workshops

Besides the *Rencontre de Moriond on QCD* at the ski stations (Flaine, les Arcs, Tigne) in Haute Savoie (France) where I regularly participated, e.g., in Ref. [319], where I presented: QCD sum rules for light quarks, I remember these following workshops:

• 3rd LEAR Workshop (Tignes, France, 1985)

I presented a review: Low Mass Hadrons of QCD [320].

• 4th LEAR Workshop (Villars-sur-Ollon, Switzerland, 1987)

I presented a review: Exotic Hadrons [321].

• Workshop on Gueball, Hybrids and Exotic Hadrons (Upton, NY, 1988)

I presented a review: QCD Tests of G(1.6) = Glueball [322].

• QCD 20 years later (Aachen, Germany, 1992)

No talk.

• Tau charm factory (Marbella, Spain, 1993)

It was a pleasant workshop in the village of Marbella. I gave two talks:

1. The Decay Constants of the Heavy – Light Mesons from QCD Spectral Sum Rules [323],
2. How the Tau Reveals the Secret of QCD [324].

• Tau 94 (Montreux, Switzerland, 1994)

Another pleasant workshop was the one at Montreux among different experts in τ-decays. I gave a review talk on α_s from τ-decays [325].

• Fubini Fest (Torino, Italy, 1994)

The title of the workshop in honour of Professor Fubini is *Symposium on Symmetry and Simplicity in Physics.* I presented the talk on the extraction of α_s from τ-decay: simple but efficient tests of QCD from τ-decays [326].

• Ringberg Workshop on Understanding Confinement 2005

This workshop was organized in the honour of Professor Zakharov where the three inventors of SVZ sum rules were there (see Fig. 4.3). I gave the talk, *The SVZ-expansion and Beyond* [327], where I discuss the effects of the tachyonic gluon mass on the QCD spectral sum rule phenomenology.

• Oberwoelz, Austrian Alps, 2018

This workshop is organized regularly by late Professor H. Fritzsch *et al.* who invited me to come to his meeting. This was again an opportunity to discover the Austrian Alps region. I presented a talk on QCD spectral sum rules.

Chapter 8

Permanent Position in Montpellier

Figure 8.1. QCD14 group at the CNRS Montpellier; QCD22 group at the Montpellier University.

After a tour around the world, I finally got a permanent position at the Centre National de la Recherche Scientifique (CNRS) in Montpellier-FR as a "Chargé de Recherche" first class in 1984. I was then promoted as a Directeur de Recherche second class in 1995 and first class in 2010. Since my retirement in 2018, I have been Directeur de Recherche Emerite.

As mentioned in Chapter 6, once in Montpellier, I initiated a collaboration with the University of Barcelona (Autonoma and

Centrale) and with Trieste and Heidelberg within a DRCI contract from the French CNRS. I also continued to be in contact with CERN where I went regularly for 1–3 months until 1991. Some of the works, which I have done travelling, have been already presented in Chapter 6. In this chapter, I present different works completed when I stayed home.

To put Montpellier in the QCD map, I also started to organize the QCD-Montpellier Conference Series in 1985, which became an unavoidable conference for the QCD community as it saw the participation of some Nobel Prizes and famous QCD experts.

8.1 $\alpha_s^{2,3}$ and $m_f^{2,4}$ Corrections to the Z-widths

In Ref. [329], we estimate, in the $\bar{M}S$-scheme, the strengths of the higher-order QCD (α_s^2, α_s^3) and quark mass ($m_f^2, m_f^4, \alpha_s(m_f^2, m_f^4)$) corrections to the Z-widths in the standard model (SM). Then, we give improved SM predictions of these widths. The observables Γ_H/Γ_e and σ_H, which can be predicted and measured accurately are almost insensitive to the changes of the Higgs (40 GeV $\leq M_H \leq$ 1 TeV) and top quark ($M_Z \leq M_t \leq$ 230 GeV) masses. Then, the results should be useful in the high precision tests of the SM. Improved estimates of some quantities related to the Z-width which are sensitive to the top quark mass are also given. The results are summarized in Table 8.1.

Table 8.1. SM predictions of the different Z-branching ratios and widths for two values of M_t and for $M_Z \leq M_H \leq 1$ TeV.

M_t [GeV]	90	220
$R_H \equiv \Gamma_H/\Gamma_e$	20.70 ± 0.10	20.53 ± 0.10
$R_b \equiv \Gamma_b/\Gamma_e$	4.49 ± 0.02	4.27 ± 0.02
$R_{tot} \equiv \Gamma_{tot}/\Gamma_e$	29.68 ± 0.10	29.48 ± 0.10
$B_b \equiv \Gamma_b/\Gamma_H$	0.217 ± 0.006	$0.208 \pm 0.0.06$
Γ_e [MeV]	84.01 ± 1.10	84.20 ± 1.10
Γ_Z [MeV]	2496.1 ± 33.7	2482.2 ± 33.5
σ_H [nb]	41.48 ± 0.37	41.71 ± 0.37

Note: We use $M_Z = (91.14 - 91.20)$ GeV, $\alpha_s(M_Z^2) = (0.116 \pm 0.016)$ and $\bar{m}_b(M_Z^2) = (3.0 \pm 0.1)$ GeV.

8.2 Evanston–Geneva–Montpellier Collaboration

Figure 8.2. E. Braaten (Evanston) and A. Pich (CERN).

My interest in τ physics began during my doctorate thesis preparation, where I have estimated (for the first time) the anomalous magnetic moment of the τ-lepton and some other hypothetical heavier leptons [23]. I also studied the dependence of the τ semi-hadronic widths on the ν_τ mass in Ref. [24] discussed in Chapter 2.

8.2.1 The initial paper on α_s from τ-decays with A. Pich

The experimental data [330]:

$$R_\tau \equiv \frac{\Gamma(\tau \to \nu_\tau + \text{Hadrons})}{\Gamma(\tau \to \nu_\tau + \nu_\tau e^- \bar{\nu}_e)} = 3.68 \pm 0.13 \tag{8.1}$$

clearly deviates from the parton model prediction:

$$R_\tau|_{\text{parton}} = 3. \tag{8.2}$$

Inspired by the paper of Eric Braaten [331], Toni Pich and Î studied the inclusive $\tau \to$ Hadrons $+\nu_\tau$ process which can be expressed as an integral over the invariant mass of the final hadrons [332]:

$$R_\tau \equiv \frac{\Gamma(\tau \to \nu_\tau + \text{hadrons})}{\Gamma(\tau \to \nu_\tau + \nu_\tau e^- \bar{\nu}_e)} = 12\pi \int_0^{M_\tau^2} dx(1-x)^2 \times \{(1+2x)\text{Im}\Pi^{(1)}(x) + \text{Im}\Pi^{(0)}(x)\}, \tag{8.3}$$

where $x \equiv s/M_\tau^2$, $\text{Im}\Pi^{(1,0)}(x)$ are spectral functions associated to the spin one vector and axial-vector and spin 0 pseudo(scalar) hadrons.

In the complex s plane, we have the equivalent Cauchy integral:

$$R_\tau = 6i\,\pi \int_{|s|=M_\tau^2} dx(1-x)^2\{(1+2x)\Pi^{(1+0)}(x) - 2x\Pi^{(0)}(x)\}. \tag{8.4}$$

We parametrize the QCD expression of R_τ using the SVZ expansion by including the contributions of the QCD condensates up to dimension 8 and the PT corrections up to order α_s^3. Using the values of the QCD condensates from some other sources, we deduce by confronting the theory with the data:

$$\Lambda_3 = (100 \sim 200)\,\text{MeV}. \tag{8.5}$$

8.2.2 The completed paper with E. Braaten and A. Pich

After our initial paper [332], Eric Braaten participated in the QCD90 Montpellier conference. The discussions in this conference led to our common paper [79] named BNP in this book. In this paper, we present in detail and improve the different QCD contributions to $R_{\tau,V}$, $R_{\tau,A}$ and $R_{\tau,S}$ given in our previous work [332].

• The Cauchy integral

In the complex s plane, we have the equivalent Cauchy integral:

$$R_\tau = 6i\,\pi \int_{|s|=M_\tau^2} dx(1-x)^2\Big\{(1+2x)\text{Im}\Pi^{(1+0)}(x) - 2x\text{Im}\Pi^{(0)}(x)\Big\}, \tag{8.6}$$

where we have rewritten the integrand in terms of the sum $\Pi^{(1)} + \Pi^{(0)}$ which has a smooth behaviour for $s \to 0$ if $\Pi^{(1)}$ and $\Pi^{(0)}$ can have a pole. If $\Pi^{(0)}$ has a pole at $s = 0$, it cancels in $\Pi^{(1)} + \Pi^{(0)}$ and contributes as a constant to $s\Pi^{(0)}(s)$ which integrates to 0 by Cauchy's residue theorem. It is known that Π_V^J does not have a pole while Π_A^J has a Goldstone pole in the chiral limit $m_q = 0$. However, this pole does not affect the calculation of R_τ as it is the combination $\Pi^{(1)}+\Pi^{(0)}$ and $s\Pi^{(0)}(s)$ which enter in the calculation. The branching ratio can be parametrized as:

$$R_{\tau,V/A} = \frac{3}{2}|V_{ud}|^2 \left(1 + \delta^{(0)} + \sum_{D=2,4,\cdots} \delta^{(D)}_{ud,V/A}\right),$$

$$R_{\tau,S} = 3|V_{us}|^2 \left(1 + \delta^{(0)} + \sum_{D=2,4,\ldots} \delta^{(D)}_{us}\right), \tag{8.7}$$

where $\delta_{ij}^{(D)}$ is the average of the vector (V) and axial-vector (A) corrections: $\delta_{ij}^{(D)} = (\delta_{ij,V}^{(D)} + \delta_{ij,A}^{(D)})/2$. V_{ud}, V_{us} are the CKM mixing angles:

$$V_{ud} = 0.9753 \pm 0.0006, \quad V_{us} = 0.221 \pm 0.003. \tag{8.8}$$

- **Perturbative corrections**

$\delta^{(0)}$ is the PT corrections in the chiral limit $m_q = 0$. It reads for three flavours and to order α_s^3:

$$\delta^{(0)} = a_s(M_\tau) + 5.2023a_s^2(M_\tau) + 26.366a_s^3(M_\tau) \pm 130a_s^4(M_\tau), \tag{8.9}$$

where $a_s(M_\tau) \equiv \alpha_s(M_\tau)/\pi$. We estimate the non-calculated α_s^4 term using an approximate geometrical growth of the PT series. One should note that a recent calculation for three flavours leads to [334]:

$$\delta^{(0)}_{\alpha_s^4} = 128.5a_s^4(M_\tau), \tag{8.10}$$

in perfect agreement with the previous estimate.

Quadratic quark mass corrections

The $m_i m_j$ light quark mass corrections are given to order α_s in [79]. It reads:

$$M_\tau^2 \delta_{ij,V/A}^{(D=2)} = -8\left[1 + \frac{16}{3} a_s(M_\tau)\right][m_i^2 + m_j^2] \pm 4\left[1 + \frac{25}{3} a_s(M_\tau)\right] m_i m_j. \quad (8.11)$$

• Dimension 4 corrections

Corrections to order α_s^2 to the dimension four quark $m\langle\bar\psi\psi$ and gluon $\langle\alpha_s G^2\rangle$ condensates are taken into account while the quartic $m_{i,j}^4, m_i^2 m_j^2$ corrections are given to leading order in α_s [79]:

$$\begin{aligned} M_\tau^4 \delta_{ij,V/A}^{(D=4)} &= \frac{11}{4}\pi a_s^2(M_\tau)\langle\alpha_s G^2\rangle + 16\pi^2\left[1 + \frac{9}{2} a_s^2(M_\tau)\right](m_i \mp m_j) \\ &\quad \times \langle\bar\psi_i \mp \bar\psi_j\psi_j\rangle - 18\pi^2 a_s^2(M_\tau)\langle m_i\bar\psi_i\psi_i + m_j\bar\psi_j\psi_j\rangle \\ &\quad - 8\pi^2 a_s^2(M_\tau)\sum_k\langle m_k\bar\psi_k\psi_k\rangle + \left[-\frac{48}{7}a_s^{-1}(M_\tau) + \frac{22}{7}\right] \\ &\quad \times [m_i(M_\tau) \mp m_j(M_\tau)][m_i^3(M_\tau) \mp m_j^3(M_\tau)] \\ &\quad \pm 6 m_i(M_\tau) m_j(M_\tau)[m_i(M_\tau) \mp m_j(M_\tau)]^2 \\ &\quad + 36 m_i^2(M_\tau) m_j^2(M_\tau), \\ M_\tau^4 \delta_{ij}^{(D=4)} &= \frac{11}{4}\pi a_s^2(M_\tau)\langle\alpha_s G^2\rangle + 16\pi^2\left[1 + \frac{27}{8} a_s^2(M_\tau)\right] \\ &\quad \times \langle m_i\bar\psi_i + m_j\bar\psi_j\psi_j\rangle - 8\pi^2 a_s^2(M_\tau)\sum_k\langle m_k\bar\psi_k\psi_k\rangle \\ &\quad + 24 m_i^2(M_\tau) m_j^2(M_\tau) + \left[-\frac{48}{7}a_s^{-1}(M_\tau) + \frac{22}{7}\right] \\ &\quad \times [m_i^4(M_\tau) + m_j^4(M_\tau)]. \end{aligned} \quad (8.12)$$

We shall use [6]:

$$\langle\alpha_s G^2\rangle = (6.28 \pm 3.14) \times 10^{-2}\ \text{GeV}^4. \quad (8.13)$$

We introduce the renormalization group invariant quark mass and condensate:

$$m_i\langle\bar{\psi}_i\psi_i\rangle = -\hat{m}_i\hat{\mu}_i^3 : \quad \bar{m}_i(M_\tau) = \frac{\hat{m}_i}{[\log(M_\tau/\Lambda)]^{2/-\beta_1}}, \tag{8.14}$$

with $-\beta_1 = 11/2 - n_f/3$ for n_f flavours. Using the values [6]:

$$\hat{m}_u = (8.7{\pm}1.5)\,\text{MeV}, \quad \hat{m}_d = (15.4{\pm}1.5)\,\text{MeV}, \quad \hat{m}_s = (270{\pm}30)\,\text{MeV}, \tag{8.15}$$

we deduce:

$$\hat{\mu}_u = \hat{\mu}_d = (189 \pm 7)\,\text{MeV}, \quad \hat{\mu}_s = (160 \pm 10)\,\text{MeV}. \tag{8.16}$$

• Dimension 6 corrections

These are the m_q^6 quark masses, the $m_q\langle\bar{\psi}G_a^{\mu\nu}\lambda^a\sigma_{\mu\nu}\psi\rangle$ quark gluon and the $\langle g^3 f_{abc}G^aG^bG^c\rangle$ three gluon and the $\langle\bar{\psi}\psi\bar{\psi}\psi\rangle$ four-quark condensates. The leading contribution comes from the four-quark condensate. We shall neglect the quark mass corrections. The mixed quark-gluon condensate multiplied by m_q can also be neglected. The $\langle G^3\rangle$ condensate contribution vanishes to leading order in α_s [335] and will be neglected. The leading order contributions of the dimension 6 condensates $\alpha_s\langle\bar{\psi}\psi\rangle^2$ is only kept where we use a numerical value which deviates from the vacuum saturation estimate. Using vacuum saturation of the four-quark condensates, one can parametrize their contribution as [3]:

$$M_\tau^6\delta_{ij,V/A}^{(D=6)} = \begin{pmatrix} 7 \\ -11 \end{pmatrix} \frac{256\pi^3}{27}\rho\alpha_s\langle\bar{\psi}\psi\rangle^2, \tag{8.17}$$

where ρ indicates the deviation from the vacuum saturation estimate. However, one should note that the vacuum saturation estimate is inconsistent with the scaling properties of the four-quark operators [113, 336]. We shall use [6]:

$$\rho\alpha_s\langle\bar{\psi}\psi\rangle^2 = (3.8 \pm 2.0) \times 10^{-4}\,\text{GeV}^6. \tag{8.18}$$

• Dimension 8 corrections

The dimension 8 gluon condensates contribution have been calculated in the chiral limit by [337]. Using the estimate of their size within a $1/M_Q$ expansion [338], one obtains [79]:

$$M_\tau^8 \delta_{ij,V/A}^{(D=8)} = -\frac{39}{162}\langle\alpha_s G^2\rangle^2 \sim 10^{-5}, \tag{8.19}$$

which is negligible.

• Phenomenology

Using the experimental average:

$$R_\tau^{\rm exp} = 3.61 \pm 0.05, \tag{8.20}$$

we obtain:

$$\alpha_s(M_\tau) = 0.34 \pm 0.04 \Longrightarrow \alpha_s(M_Z) = 0.120^{+0.004}_{-0.005}. \tag{8.21}$$

The comparison of these two determinations of $\alpha_s(\mu)$ in two extreme energy regimes, M_τ and M_Z, provides a beautiful test of the predicted running 1/log behaviour of the QCD coupling constant.

8.2.3 α_s from τ-decays and $e^+e^- \to I = 1$ hadrons

In Ref. [340], we study the sum of the τ-exclusive widths of τ for V and A from which we extract the value of α_s and compare with the total inclusive modes.

• Sum of semi-inclusive modes

The sum of semi-inclusive modes can be directly measured from the branching ratio:

$$R_\tau^B \equiv \frac{1 - B_e - B_\mu}{B_e} : \quad B_l \equiv \frac{\Gamma(\tau \to \nu_\tau l \bar\nu_l)}{\Gamma_\tau}, \tag{8.22}$$

where Γ_τ is the total decay rate, or independently from the lifetime $1/\Gamma_\tau$:

$$R_\tau^\Gamma \equiv \frac{\Gamma_\tau - \Gamma_{\tau\to e} - \Gamma_{\tau\to\mu}}{\Gamma}_{\tau\to e}, \tag{8.23}$$

Their mean leads to:

$$R_\tau^{\rm inclusive} = 3.62 \pm 0.03, \tag{8.24}$$

giving:

$$\alpha_s(M_\tau^2) = 0.36 \pm 0.03 \implies \alpha_s(M_Z^2) = 0.122 \pm 0.004. \qquad (8.25)$$

• Sum of sem-exclusive modes

For the axial-vector and $\Delta S = 0$ sum of semi-exclusive modes, we obtain:

$$R_{\tau,A} = 1.64 \pm 0.03, \quad R_{\tau,S} = 0.014 \pm 0.03, \qquad (8.26)$$

where we have used the PDG92 [341] and OHIO92 [342] data. In addition to these data, we have used the $e^+e^- \to I = 1$ hadrons one to estimate the vector semi-exclusive mode. We obtain the average:

$$R_{\tau,V} = 1.78 \pm 0.03. \qquad (8.27)$$

They lead to the total sum of semi-exclusive widths:

$$R_\tau^{\text{exclusive}} = 3.56 \pm 0.04, \qquad (8.28)$$

giving:

$$\alpha_s(M_\tau^2) = 0.33 \pm 0.04. \qquad (8.29)$$

This result agrees with the average of the direct measurement of the semi-inclusive modes from the τ-lifetime and the leptonic branching ratios given previously. A separate analysis of the sum of the semi-exclusive V modes gives:

$$\alpha_s(M_\tau^2) = 0.33 \pm 0.05, \qquad (8.30)$$

while the A ones give a smaller value:

$$\alpha_s(M_\tau^2) = 0.31 \pm 0.05. \qquad (8.31)$$

Both results agree within the errors with the one from the semi-inclusive mode in Eq. (8.25).

• M_τ-dependence of $R_{\tau,V}$

Using the $e^+e^- \to I = 1$ hadrons data, we also study the M_τ dependence of $R_{\tau,V}$ in Fig. 8.3. We note a good agreement with the data for $M_\tau \geq 1.2$ GeV. This feature confirms the suppression of errors near the real axis due to the threshold factor $(1 - s/M_\tau^2)^2$.

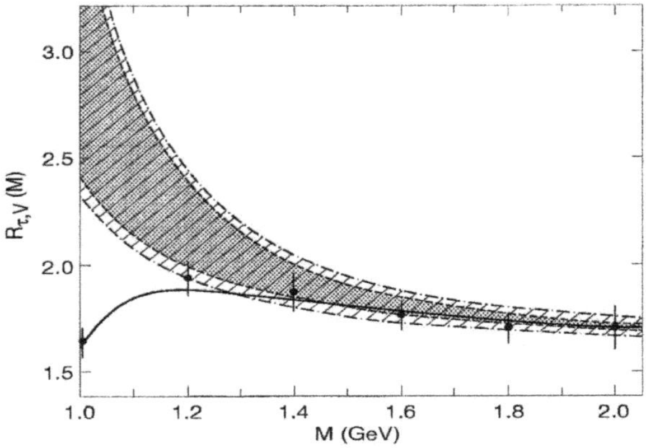

Figure 8.3. M_τ dependence of $R_{\tau,V}$ using $e^+e^- \to I = 1$ hadrons data.

8.3 V–A Condensates from τ-Decays

In Ref. [343], we have used the ALEPH/OPAL data on the VV–AA spectral functions from hadronic τ-decays for fixing the size of the V–A QCD vacuum condensates up to dimension 18 associated to the Π^{LR} QCD correlator:

$$\Pi^{LR}_{\mu\nu}(q) \equiv i \int d^4x \; e^{iqx} \langle 0|\mathcal{T} J^L_\mu(x) \left(J^R_\nu(0)\right)^\dagger |0\rangle \overset{m_q \to 0}{=} -(g_{\mu\nu}q^2 - q_\mu q_\nu)\Pi^{LR}(q^2), \tag{8.32}$$

built from the left- and right-handed components of the local weak current:

$$J^L_\mu = \bar{u}\gamma_\mu(1-\gamma_5)d, \quad J^R_\mu = \bar{u}\gamma^\mu(1+\gamma_5)d. \tag{8.33}$$

8.3.1 Set of Weinberg-like Laplace transform sum rules (LSR)

We shall work with:

$$\mathcal{L}_n \equiv (-1)^n \frac{d^n \mathcal{L}_0}{d\tau^n} \simeq \int_0^\infty dt \; t^n \; \mathrm{e}^{-t\tau} \; \frac{1}{\pi} \mathrm{Im}\Pi_{LR} \simeq (-1)^n \times \sum_{d \geq (n+1)} \frac{\tau^{(d-n-1)}}{(d-n-1)!} \mathcal{O}_{2d}, \tag{8.34}$$

where $\mathcal{L}_0$ is the first Weinberg sum rule:

$$\mathcal{L}_0(\tau) = \int_0^\infty dt\ \mathrm{e}^{-t\tau}\ \frac{1}{\pi}\mathrm{Im}\Pi_{LR} - 2f_\pi^2 \simeq \sum_{d\geq 2} \frac{\tau^{(d-1)}}{(d-1)!}\ \mathcal{O}_{2d}, \qquad (8.35)$$

Unlike some fitting procedures often used in the current literature, our results are based on optimization criteria within τ and t_c stability.

8.3.2 Finite energy sum rules (FESR)

We have also used finite energy sum rules (FESR):

$$\mathcal{M}_n \equiv \int_0^{t_c} dt\ t^n \frac{1}{\pi}\mathrm{Im}\Pi_{LR} = (-1)^n \mathcal{O}_{2n+2}, \qquad (8.36)$$

and some weighted FESR:

$$J_{\omega_n} \equiv \int_0^{t_c} dt\ \omega_n \left(\frac{t}{t_c}\right) \frac{1}{\pi}\mathrm{Im}\Pi_{LR}, \qquad (8.37)$$

where the weight factor ω_n is:

$$\omega_n(x) = x\left[1 - \left(\frac{n}{n-1}\right)x + \left(\frac{1}{n-1}\right)x^n\right] \qquad (8.38)$$

for $n = 2, 3, 4, 5, 6$ to check the validity of the Laplace sum rule (LSR) results. However, unlike the case of LSR analyzed in previous sections, the FESR results depend crucially on the choice of t_c at which one extracts the optimal results. Requiring the zero of the first and second Weinberg sum rules, one gets:

$$t_c \simeq (1.4 \sim 1.5)\ \mathrm{GeV}^2 \quad \text{and} \quad (2.5 \sim 2.6)\ \mathrm{GeV}^2, \qquad (8.39)$$

where the predictions for $t_c \simeq (1.4 \sim 1.5)$ GeV^2 are more consistent with the one of LSR. The weighted FESR gives more stable results which are optimal for $t_c \simeq 1.7{-}2.2$ GeV^2.

Our results are compiled in Table 8.2. A more complete discussion on the comparison with related estimate is given in Table 8.1 of Ref. [343].

Table 8.2. Estimated values of the $D \equiv 2d \leq 18$-dimension $\langle \mathcal{O}_D \rangle$ condensates in units of 10^{-3} GeV^D for the V–A channel. We have ordered the condensates from $\mathcal{O}_{18}$ to $\mathcal{O}_6$ according to their chronological estimate and obtained $\mathcal{O}_4 = -0.5 \pm 0.1 \times 10^{-3}$ GeV^4.

Authors	$\mathcal{O}_{18}$	$\mathcal{O}_{16}$	$\mathcal{O}_{14}$	$\mathcal{O}_{12}$	$\mathcal{O}_{10}$	$\mathcal{O}_8$	$\mathcal{O}_6$
This work							
LSR [343]							
Eq. (8)	-1 ± 0.6	$+4.3 \pm 1.9$	-9.6 ± 3.1	$+14.7 \pm 3.7$	-17.1 ± 4.4	$+15.4 \pm 4.8$	-9.7 ± 4.1
Eqs. (14,15)						$+15.8 \pm 3.2$	-8.4 ± 1.6
Eq. (16)							-8.0 ± 1.1
Average						$\mathbf{+15.6 \pm 4.0}$	$\mathbf{-8.7 \pm 2.3}$
wFESR							
Rev.[1]		$+30 \pm 10$	-28 ± 8	$+25 \pm 5$	-22 ± 3	$+16.8 \pm 2.0$	$-10.2 \pm .4$
FESR[2]							
BG Rev.[3]						$+12 \pm 1.5$	-6.6 ± 0.2
DS Rev.[4]						$+10 \pm 2$	-8.0 ± 2.0
LR Rev.[5]				$+53 \pm 16$	-39 ± 12	$+26.0 \pm 8.4$	-14.7 ± 4.8
Others							
MHA+ρ'[348]		$+11.5 \pm 3.5$	-12.5 ± 3.4	$+13.2 \pm 3.3$	-13.1 ± 3.0	$+11.7 \pm 2.6$	-7.9 ± 1.6
MHA[348]		$+11.9 \pm 3.9$	-12.8 ± 3.9	$+13.3 \pm 3.9$	-13.2 ± 3.6	$+11.7 \pm 3.1$	-7.9 ± 2.0
ZYA[349]					-4.5 ± 3.4	$+7.8 \pm 3.0$	-7.1 ± 1.5
IZ [350]						$+7.0 \pm 4.0$	-6.4 ± 1.6
ALEPH [247]						$+11.0 \pm 1.0$	-7.7 ± 0.8
OPAL [248]						$+7.5 \pm 1.3$	-6.0 ± 0.6
DGHS [351]						$+8.7 \pm 2.4$	-6.0 ± 0.6
CS3 [352]							-4.0 ± 2.8
Average	-1 ± 0.6	$+14.4 \pm 4.6$	-15.7 ± 3.7	$+23.8 \pm 6.4$	-18.2 ± 5.9	12.2 ± 2.9	-7.8 ± 1.6

Notes: [1]We have redone the analysis in Ref. [344] using t_c-stability criterion.
[2]The revised (Rev.) FESR results have been obtained at $t_c \approx 1.5$ GeV^2, and the original (Orig.) ones at $t_c \approx 2.5$ GeV^2.
[3]These results have been obtained in Ref. [345] at 1.5 GeV^2.
[4]We have corrected the value of $\mathcal{O}_8$ and rescaled the results in Ref. [346].
[5]The central values come from Ref. [347]. Inspired from the results in Ref. [347] at 2.5 GeV^2, we have roughly estimated the systematic errors to be about 30%.

8.4 QCD Parameters from $e^+e^- \to I = 1$ Hadrons

8.4.1 Dimension two ($D = 2$) operator

Within the standard SVZ expansion and due to gauge invariance, the only possible $D = 2$ contribution comes from the quark mass squared which is already taken into account in the PT expression of the correlator. However, there are different arguments (see Section 7.5.1) suggesting that this term can be present (questionable gauge invariance [227], holographic models [228], phenomenological parametrization of UV renormalon [232, 233]).

In Ref. [236], we assume the existence of this term and estimate its strength using $e^+e^- \to I = 1$ hadrons data where the contributions of light quark masses m_u, m_d are neglected. In so doing, we work with the previous ratio of moments $\mathcal{R}(\tau)$ and use as inputs the correlated values of $D = 4, 6$ condensates obtained above and two times their values. We extract C_2 at the stability region obtained at the continuum threshold conservative range from 1.8 to 3.1 GeV^2 and for $\tau \simeq 0.6$ GeV^{-2} (minimum):

$$C_2|_{LSR} \simeq -(0.03 \sim 0.08)\ \text{GeV}^2. \tag{8.40}$$

We have also used FESR to extract C_2 and obtains the minimum in t_c:

$$C_2|_{FESR} \simeq -(0.02 \pm 0.12)\ \text{GeV}^2, \tag{8.41}$$

which is less accurate. Then, we deduced the value of the chiral invariant quark mass parametrized as:

$$C_2 \equiv 6m_{\text{eff}}^2 \implies m_{\text{eff}} \simeq i\,(71 \sim 116)\,\text{MeV}. \tag{8.42}$$

This imaginary mass will be interpreted later on as a tachyonic gluon mass [235] (see Section 7.5.1).

We have checked that the effect of this extra contribution only affects slightly the ρ-meson sum rule ($\leq$10%) and the estimate of α_s from τ-decays ($\leq$3.6%).

8.4.2 QCD condensates and α_s from τ-like decays

In Ref. [237], we shall work with the vector component of the τ-decay-like quantity deduced from CVC [353]:

$$R_{\tau,1} \equiv \frac{3\cos^2\theta_c}{2\pi\alpha^2} S_{\rm EW} \int_0^{M_\tau^2} ds \left(1-\frac{s}{M_\tau^2}\right)^2 \left(1+\frac{2s}{M_\tau^2}\right) \times \frac{s}{M_\tau^2}\,\sigma_{e^+e^-\to\ I=1}, \tag{8.43}$$

where $S_{\rm EW} = 1.0194$ is the electroweak correction from the summation of the leading-log contributions and $\delta_{\rm EW} = 0.0010$ is the electroweak correction coming from the constant term [354]. The QCD expression of $R_{\tau,1}$ reads:

$$R_{\tau,1} = \frac{3}{2}\cos^2\theta_c \times S_{\rm EW}\left(1+\delta_{\rm EW}+\delta^{(0)}+\sum_{D=2,4...}\delta_1^{(D)}\right), \tag{8.44}$$

where $\delta_{\rm EW} = 0.0010$ is the electroweak correction coming from the constant term [355]. The expression of the QCD corrections can be obtained from Ref. [79] (see Section 8.2.1). Neglecting the small quark mass contribution and the one of $\langle\alpha_s G^2\rangle$ which is to order α_s, we do a two-parameter fit of the $D=6$ and $D=8$ contributions and find, for $\Lambda_3 \simeq 290-480$ MeV:

$$d_8 \equiv M_\tau^8\delta_1^{(8)} = -(0.85\pm0.18){\rm GeV}^8 \quad d_6 \equiv M_\tau^6\delta_1^{(6)} = (0.34\pm0.20){\rm GeV}^6. \tag{8.45}$$

To improve the value of the four-quark condensate, we shall work with the ratio of LSR. We also remark that the ratio:

$$r_{46}[{\rm GeV}^{-2}] \equiv \frac{\langle\alpha_s G^2\rangle}{\rho\alpha_s\langle\bar{u}u\rangle^2} \simeq (105.9\pm11.9)\,{\rm GeV}^{-2} \tag{8.46}$$

is quite stable from different approaches where the number comes from the average [237]. Then, we redo the analysis of $D=4$ and $D=6$, and obtain after some iterations:

$$\langle\alpha_s G^2\rangle = (7.1\pm0.7)10^{-2}\ {\rm GeV}^4 \quad d_8 = -(1.5\pm0.6)\ {\rm GeV}^8. \tag{8.47}$$

Using the mean value of r_{46}, we also obtain:

$$\rho\alpha_s\langle\bar{u}u\rangle^2 = (5.8 \pm 0.9)10^{-4}\ \text{GeV}^6, \tag{8.48}$$

which we consider as an improvement of the analysis in 1993.

8.4.3 Instanton contribution

Let us now extract the size of the instanton-like contribution by assuming that it acts like a $D \geq 9$ operator. A good place for doing it is $R_{\tau,1}$ as, in the Laplace sum rules, this contribution is suppressed by an 8! factor. Using the previous values of the $D = 6$ and $D = 8$ condensates, we deduce:

$$\delta_1^{(9)} = -(7.0 \pm 26.5)10^{-4}(1.78/M_\tau)^9, \tag{8.49}$$

which, though inaccurate, indicates that the instanton contribution is negligible for the vector current.

Further cancellations in the sum of the vector and axial-vector components of the τ widths are however expected [356, 357] ($\delta^{\text{inst}} \approx \frac{1}{20}\delta_V^{\text{inst}}$).

8.4.4 Sum of the non-perturbative corrections to R_τ

Using our previous estimates, it is also informative to deduce the sum of the non-perturbative contributions to the decay widths of the observed heavy lepton of mass 1.78 GeV. In so doing, we add the contributions of operators of dimensions $D = 4$ to $D = 9$ and we neglect the expected small $\delta^{(2)}$-contribution. For the vector component of the τ hadronic width, we obtain[6]:

$$\delta_V^{NP} \equiv \sum_{D=4}^{9} \delta_1^{(D)} = (2.38 \pm 0.89)10^{-2}, \tag{8.50}$$

while using the expression of the corrections for the axial-vector component given in [79], we deduce:

$$\delta_A^{NP} = -(7.95 \pm 1.12)10^{-2}, \tag{8.51}$$

[6]We have used, for $M_\tau = 1.78$ GeV, the conservative values: $\delta_V^{(9)} \approx -\delta_A^{(9)} \simeq -(0.7 \pm 2.7)10^{-3}$ and $\delta^{(9)} \approx 1/20\delta_V^{(9)}$ [356].

and then:

$$\delta^{NP} \equiv \frac{1}{2}(\delta_V^{NP} + \delta_A^{NP}) = -(2.79 \pm 0.62)10^{-2}. \tag{8.52}$$

8.4.5 Implication on the value of α_s from R_τ

Using the experimental average:

$$R_\tau = 3.56 \pm 0.03, \tag{8.53}$$

we deduce, by including an estimate of the terms up to order α_s^5:

$$\alpha_s(M_\tau) = 0.33 \pm 0.03. \tag{8.54}$$

8.5 Strange Quark Mass from e^+e^- and τ-Decay

The running strange quark mass $\bar{m}_s$ plays an important role for controlling SU(3) breaking. Besides its extraction from (pseudo)scalar sum rules (see previous chapters), one also attempts to extract $\bar{m}_s$ from $e^+e^- \to$ hadrons and τ-decay data using different forms of the sum rules (see Ref. [358] for a review).

8.5.1 Difference between the $I = 1$ and $I = 0$ of e^+e^- data

In Ref. [359], the difference of $I = 1\,\tau$ like sum rule in Eq. (8.43) with the isoscalar $I = 0$ component of the $e^+e^- \to$ hadron data have been considered for extracting $\bar{m}_s$:

$$\Delta_{10} \equiv R_{\tau,1} - 3R_{\tau,0}, \tag{8.55}$$

where 3 is an isospin factor. Including the α_s^2 corrections and using $\Lambda_3 = 375^{+105}_{-85}$ MeV, one obtains:

$$\bar{m}_s(1\,\mathrm{GeV}) = (197 \pm 29)\,\mathrm{MeV}. \tag{8.56}$$

This result has been used to re-estimate the effect of SU(3) breaking to kaon PCAC:

$$\psi_5(0) = 2M_K^2 f_K^2(1 - \delta_K): \qquad \delta_K = 0.34^{+0.24}_{--0.17}, \tag{8.57}$$

confirming a large violation of kaon PCAC [6,41]. The resulting value of the *non-normal ordered quark condensates* is:

$$\langle \bar{s}s\rangle/\langle \bar{d}d\rangle = 0.87^{+0.03}_{-0.28}, \tag{8.58}$$

confirming previous results in Refs. [6, 41].

8.5.2 Difference between the $I = 1$ and $e^+e^- \to \phi, \phi', \ldots$ data

The analysis has been pursued in Ref. [295], but taking the difference of the $I = 1$ and ϕ-families data:

$$\Delta_{1\phi} \equiv R_{\tau,1} - R_{\tau,\phi}. \tag{8.59}$$

We obtain to order α_s^3:

$$\bar{m}_s(1\,\text{GeV}) = (186 \pm 25_{\text{exp}} \pm 19_{\text{th}})\,\text{MeV}. \tag{8.60}$$

We have also updated the previous value in Eq. (8.56) by adding the α_s^3 corrections and found:

$$\bar{m}_s(1\,\text{GeV}) = (176 \pm 24_{\text{exp}} \pm 19_{\text{th}})\,\text{MeV}. \tag{8.61}$$

8.5.3 Final result from e^+e^- and isospin violation

The agreement of the two results rule out some results, indicating a huge SU(2) violation of the isospin symmetry due to the $\omega - \rho$ mixing.

As a final result, we have taken the mean value from the two determinations:

$$\bar{m}_s(1\,\text{GeV}) = (178\pm33)\,\text{MeV} \quad \Longrightarrow \quad \bar{m}_s(2\,\text{GeV}) = (129\pm24)\,\text{MeV}, \tag{8.62}$$

which agrees within the error with the one from pseudoscalar sum rules.

8.5.4 Effects of tachyonic gluon mass λ^2 to $\bar{m}_s$ from e^+e^-

As the contribution of the tachyonic gluon mass is flavour-independent, it does not contribute to the previous difference of sum rules from e^+e^- data. In Ref. [360], we consider a given flavoured $\bar{u}s$ spectral function of τ-decay data involving the sum of spin zero and one mesons, but at the same time, our analysis will be affected by λ^2. To LO, the QCD expression of the $VV + AA$ two-point function is:

$$\Pi^{(0+1)}_{V+A}(Q^2) = \frac{1}{2\pi^2}\left\{-\log\frac{Q^2}{\nu^2} - \frac{\mathcal{O}^0_2}{Q^2} + \frac{\mathcal{O}^0_4}{Q^4} + \frac{\mathcal{O}^0_6}{Q^6}\right\}, \quad (8.63)$$

with obvious notations, where:

$$\begin{aligned}
\mathcal{O}^0_2 &= 3m_s^2 + \frac{\alpha_s}{\pi}\lambda^2,\\
\mathcal{O}^0_4 &= 4\pi^2\left(m_s\langle\bar{s}s\rangle + m_u\langle\bar{u}u\rangle\right) + \frac{\pi}{3}\langle\alpha_s(G^a_{\mu\nu})^2\rangle,\\
\mathcal{O}^0_6 &= \frac{256\pi^3}{81}\alpha_s\langle\bar{u}u\rangle^2.
\end{aligned} \quad (8.64)$$

From this expression, one can derive the combination of Laplace and finite energy sum rules:

$$\begin{aligned}
\mathcal{N}_{10} &\equiv \mathcal{L}_0 - \tau\mathcal{L}_1 \equiv \int_0^{t_c} dt\,(1 - t\tau)\,e^{-t\tau}\frac{1}{\pi}\mathrm{Im}\Pi^{(0+1)}_{V+A},\\
\mathcal{S}_{10} &\equiv \mathcal{M}_0 - \frac{2}{t_c}\mathcal{M}_1 \equiv \int_0^{t_c} dt\left(1 - 2\frac{t}{t_c}\right)\frac{1}{\pi}\mathrm{Im}\Pi^{(0+1)}_{V+A},
\end{aligned} \quad (8.65)$$

which are sensitive, to leading order, to m_s^2 and λ^2. Unlike the individual sum rules, these combinations of sum rules are less sensitive to the high-energy tail of the spectral functions (effect of the t_c continuum threshold). Then, one may expect that they are more accurate than the former. We shall also work with the ratio of moments:

$$\mathcal{R}_{10} \equiv 2\frac{\mathcal{M}_1}{\mathcal{M}_0}, \quad (8.66)$$

which will also be useful for testing the duality between the LHS (experiment) and the RHS (QCD theory). We shall use the estimate:

$$d_2 \equiv (\alpha_s/\pi)\lambda^2 \simeq -(0.07 \pm 0.03)\ \mathrm{GeV}^2. \quad (8.67)$$

The best optimal result is obtained from FESR at $t_c = 3.15$ GeV^2:

$$\bar{m}_s(2\ \mathrm{GeV}) \simeq (93^{+29}_{-32})\ \mathrm{MeV}. \quad (8.68)$$

This result has been compared in Ref. [360] with some other determinations.

8.5.5 Estimate of $|V_{us}|$

We can estimate the CKM angle by working with the FESR $\mathcal{M}_0$ or/and the LSR $\mathcal{L}_0$. FESR gives the result:

$$|V_{us}| \simeq 0.215 \pm 0.017, \tag{8.69}$$

while the LSR gives:

$$|V_{us}| \simeq 0.219 \pm 0.021, \tag{8.70}$$

which one can compare with the more precise PDG average: $|V_{us}| = 0.2253 \pm 0.0008$. This result indicates the consistency of the approach used to get m_s.

8.6 Neutron EDM and Magnetic Susceptibility

8.6.1 Improved chiral estimate of the NEDM

In Ref. [361], we re-examine the estimate of the neutron electric dipole moment (NEDM) D_N from chiral and QCD spectral sum rules (QSSR) approaches. It is defined as:

$$\begin{aligned} V_\mu \equiv \mathcal{T}\langle n(p_f)|J_\mu(0)\; i\int d^4x\; \delta\mathcal{L}_{CP}(x)|n(p_i)\rangle \\ = -\,(iD_N)\,\bar{u}(p_f)\sigma_{\mu\nu}k^\nu\gamma_5 u(p_i) + \mathcal{O}(k^2), \end{aligned} \tag{8.71}$$

where $k \equiv p_f - p_i$ is the photon momentum and $\delta\mathcal{L}_{CP}(x) = \bar{q}(i\gamma_5)\mathcal{A}q$, with $\mathcal{A}$ three-dimension hermitian matrix acting on the flavour space (u, d, s) and $\sigma_{\mu\nu} \equiv (1/2)\,(\gamma_\mu\gamma_\nu - \gamma_\nu\gamma_\mu)$.

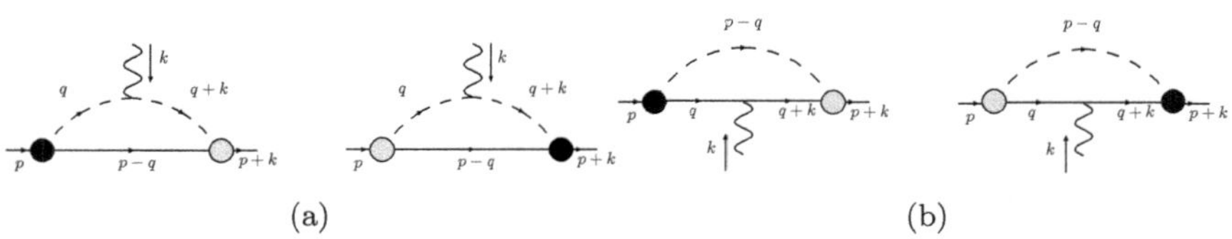

Figure 8.4. Diagrams contributing to D_N: (a) first class; (b) second class.

The lowest order contribution to D_N comes from the diagrams in Fig. 8.4. Their evaluation leads to [361]:

$$D_N|_{\text{chiral}} = \frac{\bar{g}_{\pi NN} g_{\pi NN}}{M_N} \left(\frac{1}{4\pi^2}\right) \times \left\{ -\frac{\log a}{2} + 0 + \frac{1}{4}\left(\pi\sqrt{a} + 3a\right) + \cdots \right\}, \tag{8.72}$$

where $a \equiv (m_\pi/M_N)^2$. One can note that the $\log a$ term appears naturally in the unexpanded full expression without an arbitrary choice of the cut-off M_N and the constant and $\log a$ terms cancel out from the two classes of diagrams. The leading log contribution is similar to the one of chiral perturbation theory (ChPT) where the addition of counterterms in the effective Lagrangian induces a $\log(M_N/\nu)^2$-term and leads to the (renormalized) NEDM expression [363]:

$$D_N(\nu)|_{\text{ChPT}} = \frac{\bar{g}_{\pi NN} g_{\pi NN}}{M_N} \left(\frac{1}{4\pi^2}\right) \left\{ \log \frac{\nu}{m_\pi} + k \right\}, \tag{8.73}$$

where ν is an arbitrary hadronic scale and k is an unknown model-dependent constant. The πNN couplings are defined as:

$$\mathcal{L}_{\pi NN} = \sqrt{2}\pi \bar{N} \tau \left(i\gamma_5 g_{\pi NN} + \bar{g}_{\pi NN}\right) N, \tag{8.74}$$

where τ are isospin Pauli matrices. Keeping only the model-independent longterm and ν in the range $M_N/3$ (constituent quark mass) to M_N (typical hadron scale), we obtain in units of the electric charge e:

$$D_N(\nu) = (2.7 \sim 23.8) \times 10^{-3}\theta \ \text{GeV}^{-1} = (5.4 \sim 47.6) \times 10^{-17}\theta \ \text{cm}. \tag{8.75}$$

We have used the well-measured CP-conserving coupling $g_{\pi NN} = 13.4$ and the estimate of the CP-violating coupling $\bar{g}_{\pi NN} = (0.0282 \pm 0.0071)\theta$ from an effective Lagrangian approach [362], where θ is the strong CP-violating phase coefficient of the topological charge $Q(x) = (g^2/32\pi^2) G^a_{\mu\nu} G^{\mu\nu}_a$ in the QCD Lagrangian.

8.6.2 Magnetic Susceptibility from QSSR

The analysis is based on the baryon two-point correlator put in a background with non-zero θ and electromagnetic field $F_{\mu\nu}$:

$$S(q^2)|_{\theta,\ F} = i\int d^4x\ e^{iqx}\langle 0|\mathcal{T}N(x)\bar{N}(0)|0\rangle|_{\theta,F}. \tag{8.76}$$

N is the nucleon operators, which can be written in general as:

$$N(x) \equiv 2\left\{\left(\psi C\gamma^5\psi\right)\psi + b\left(\psi C\psi\right)\gamma^5\right\}, \tag{8.77}$$

where C is the charge conjugate, ψ is the quark field, and b is (*a priori*) an arbitrary mixing between the two operators. Its optimal value from the analysis of the nucleon mass is $b = -1/5$ [67,120,195], which differs from the Ioffe's choice $b = -1$ [194] and the non-relativistic value $b = 0$ used in lattice calculations. The QCD expression reads [364]:

$$S(Q^2)|^{\text{th}}_{\theta,\ F} = -\theta m^*\langle\psi\psi\rangle\left(\tfrac{1}{32\pi^2}\right)\Big\{-\chi C_0\log\tfrac{Q^2}{\nu^2}+$$
$$\Big[C_{2a}\left(\log\tfrac{Q^2}{\nu_\chi^2}-1\right)+C_{2b}\Big]\tfrac{1}{Q^2}+\mathcal{O}\left(\tfrac{1}{Q^4}\right)\Big\}. \tag{8.78}$$

ν and ν_χ correspond to an UV and a small mass arbitrary subtractions; $\langle\psi\psi\rangle$ is the quark condensate. The coefficient functions are:

$$C_0 = (b+1)^2(4e_d - e_u),$$
$$C_{2a} = -4(b-1)^2 e_d\left(1+\frac{1}{4}(2\kappa+\xi)\right),$$
$$C_{2b} = -\frac{\xi}{2}\Big[(4b^2-4b+2)e_d + (3b^2+2b+1)e_u\Big], \tag{8.79}$$

where e_u and e_d are the electric charge of the u and d quarks in units of e. χ, ζ, ξ are the magnetic susceptibilities of the QCD condensates which encode the electromagnetic field dependence of the two-point correlator. In units of e, they are defined as [365]:

$$\langle 0|\bar{\psi}\sigma_{\mu\nu}\psi|0\rangle|_F = e_q\chi F_{\mu\nu}\langle\bar{\psi}\psi\rangle,$$
$$g\langle 0|\bar{\psi}\frac{\lambda_a}{2}G^a_{\mu\nu}\psi|0\rangle|_F = e_q\kappa F_{\mu\nu}\langle\bar{\psi}\psi\rangle,$$
$$g\langle 0|\bar{\psi}\gamma_5\frac{\lambda_a}{2}G^a_{\mu\nu}\psi|0\rangle|_F = ie_q\xi F_{\mu\nu}\langle\bar{\psi}\psi\rangle. \tag{8.80}$$

The size of these magnetic susceptibilities have been estimated in the literature using different methods. The values [366]:

$$\kappa \simeq -(0.34 \pm 0.10), \quad \xi \simeq -(0.74 \pm 0.2) \tag{8.81}$$

induce (*a posteriori*) small numerical corrections in the present analysis and will not be reconsidered. On the contrary, the dominant contribution comes from χ, which is not known with a good accuracy.

The phenomenological part of the correlator can be parametrized in the zero width approximation by:

$$S(q^2)|^{\text{exp}}_{\theta,\ F} = \frac{\lambda_N^2 M_N D_N}{(q^2 - M_N^2)^2} + \frac{A}{q^2 - M_N^2} + \text{“QCD continuum”}, \tag{8.82}$$

where λ_N is the nucleon coupling to its corresponding current, A is an arbitrary coupling which parametrizes the single pole contributions, and “QCD continuum” stands for the QCD smearing of excited state contributions and comes for the discontinuity of the QCD diagrams.

Using two different QSSR approaches, we estimate the magnetic susceptibility to be:

$$\chi \simeq -(8.5 \pm 1.0)\ \text{GeV}^{-2}, \tag{8.83}$$

in good agreement with the LSR previous value [365] and the one in Ref. [367] using the quark triangle anomaly.

8.6.3 NEDM from QSSR

Using the previous QCD expression, we have also estimated the NEDM through the value of χ and directly from QSSR. We have obtained the conservative estimate:

$$\begin{aligned} D_N|_{\rm QSSR} &\simeq (2.15 \pm 0.71) \times 10^{-3}\theta \text{ GeV}^{-1} \\ &\simeq (4.24 \pm 1.40) \times 10^{-17}\theta \text{ cm}. \end{aligned} \quad (8.84)$$

8.6.4 Final range of the NEDM values

We have also re-estimated the value of NEDM from a constituent quark model. Combining these previous results with the present experimental upper limit (in units of e) [368]:

$$D_N|_{\rm exp} \leq 6.3 \times 10^{-26} \text{ cm}, \quad (8.85)$$

one can deduce in units of 10^{-10}:

$$\begin{aligned} \theta &\leq (1.6 \pm 0.4) \text{ [Chiral]} : \ \nu = M_N, \\ &\leq (1.3 \sim 11.7) \text{ [ChPT]} : \ M_N/3 \leq \nu \leq M_N, \\ &\leq (6.9 \pm 3.5) \ \text{ [Constituent quark]}, \\ &\leq (14.9 \pm 4.9) \ \text{ [QSSR]}. \end{aligned} \quad (8.86)$$

These results indicate that the weakest upper bound comes from QSSR, while the strongest upper bound comes from the chiral estimate evaluated at the scale $\nu = M_N$.

8.7 Open Charm and Beauty Chiral Multiplets

The recent BABAR, BELLE and CLEO observations of two new states $D_{sJ}(2317)$ and $D_{sJ}(2457)$ [369] in the $D_s\pi$, $D_s\gamma$ and $D_s\pi\gamma$ final states have stimulated a renewed interest in the spectroscopy of open charm states which one can notice from different recent theoretical attempts to identify their nature [369].

In fact, we have already addressed a similar question in the past [168], where we have predicted using QSSR the mass splitting of the $0^+ - 0^-$ and $1^- - 1^+$ $\bar{b}u$ mesons using double ratio of moment sum rules based on an expansion in the inverse of the b quark mass. We found that the value of the mass splittings between the chiral multiplets were about the same and approximately independent of the spin of these mesons signalling a heavy quark-type approximate symmetry:

$$M_{B(0^+)} - M_{B(0^-)} \approx M_{B^*(1^+)} - M_{B^*(1^-)} \approx (417 \pm 212)\ \text{MeV}. \tag{8.87}$$

The effect and errors on the mass splittings are mainly due to the chiral condensate $\langle\bar{\psi}\psi\rangle$ and the value of the b quark mass.

In Ref. [370], we use an analogous approach to the open charm states. However, the same method in terms of the $1/m_c$ expansion and some other non-relativistic sum rules will be dangerous here due to the relatively light value of the charm quark mass. Instead, we shall work with relativistic exponential sum rules used successfully in the light quark channels [3, 6, 7].

We shall work here with the (pseudo)scalar two-point correlators:

$$\psi_{P/S}(q^2) \equiv i \int d^4x\ e^{iqx}\ \langle 0|\mathcal{T} J_{P/S}(x) J^\dagger_{P/S}(0)|0\rangle \tag{8.88}$$

built from the (pseudo)scalar and (axial)-vector heavy–light quark currents:

$$J_{P/S}(x) = (m_Q \pm m_q)\bar{Q}(i\gamma_5)q, \qquad J^\mu_{V/A} = \bar{Q}\gamma^\mu(\gamma_5)q. \tag{8.89}$$

If we fix $Q \equiv c$ and $q \equiv s$, the corresponding mesons have the quantum numbers of the $D_s(0^-)$, $D_s(0^+)$ mesons. m_Q and m_s are the running quark masses. We work with the Laplace transform sum rules:

$$\mathcal{L}^H_{P/S}(\tau) = \int_{t_\leq}^{\infty} dt\ \mathrm{e}^{-t\tau}\ \frac{1}{\pi}\ \mathrm{Im}\psi^H_{P/S}(t), \quad \text{and}$$

$$\mathcal{R}^H_{P/S}(\tau) \equiv -\frac{d}{d\tau}\log \mathcal{L}^H_{P/S}(\tau), \tag{8.90}$$

where $t_\leq$ is the hadronic threshold and H denotes the corresponding meson. The QCD expression of the correlator is well known to two-loop accuracy (see, e.g., [6, 7] and the explicit expressions given in Refs. [161, 255]), in terms of the perturbative pole mass M_Q, and including the non-perturbative condensates of dimensions less than or equal to six. For a pedagogical presentation, we write the sum rule in the chiral limit ($m_s = 0$), where the expression is more compact:

$$\mathcal{L}^H_{P/S}(\tau) = M_Q^2 \left\{ \int_{M_Q^2}^{\infty} dt\ \mathrm{e}^{-t\tau}\ \frac{1}{8\pi^2} \left[3t(1-x)^2 \left[1 + \frac{4}{3}\left(\frac{\alpha_s}{\pi}\right) f(x) \right] \right. \right.$$

$$\left. \left. + \mathcal{O}(\alpha_s^2) \right] C_4\langle O_4\rangle_{P/S} + \tau C_6 \langle O_6\rangle_{P/S}\ \mathrm{e}^{-M_Q^2\tau} \right\}, \qquad (8.91)$$

The different terms are:

$$x \equiv M_Q^2/t,$$

$$f(x) = \frac{9}{4} + 2\mathrm{Li}_2(x) + \log x \log(1-x) - \frac{3}{2}\log(1/x - 1)$$
$$- \log(1-x) + x\log(1/x-1) - (x/(1-x))\log x,$$

$$C_4\langle O_4\rangle_{P/S} = \mp M_Q \langle \bar{d}d\rangle\ \mathrm{e}^{-M_Q^2\tau} + \langle \alpha_s G^2\rangle \left(\frac{3}{2} - M_Q^2\tau\right)/12\pi, \qquad (8.92)$$

$$C_6\langle O_6\rangle_{P/S} = \mp\frac{M_Q}{2}\left(1 - \frac{M_Q^2\tau}{2}\right) g\left\langle \bar{d}\sigma_{\mu\nu}\frac{\lambda_a}{2}G_a^{\mu\nu}d\right\rangle$$
$$- \left(\frac{8\pi}{27}\right)\left(2 - \frac{M_Q^2\tau}{2} - \frac{M_Q^4\tau^2}{6}\right)\rho\alpha_s\langle\bar{\psi}\psi\rangle^2, \qquad (8.93)$$

where we have used the contribution of the gluon condensate given in Ref. [371], which is IR finite when letting $m_q \to 0$.[7] The previous sum rules can be expressed in terms of the running mass $\bar{m}_Q(\nu)$ through the perturbative two-loop relation [154]:

$$M_Q = \bar{m}_Q(p^2)\left[1 + \left(\frac{4}{3} + \log\frac{p^2}{M_Q^2}\right)\left(\frac{\alpha_s}{\pi}\right) + \mathcal{O}(\alpha_s^2)\right], \qquad (8.94)$$

where M_Q is the pole mass. We shall use the QCD input parameters given in Table 7.10 of Chapter 7.

[7]The numerical change is negligible compared with the original expression obtained in Ref. [372].

8.7.1 Calibration of the sum rule and re-estimate of $\bar{m}_c(m_c)$

First, we calibrate the sum rule by predicting the observed value of $M_{D(0^-)}$ and $M_{D_s(0^-)}$. The optimal values are obtained at the minimum of the cures in Fig. 8.5 to which correspond:

$$t_c = (7.5 \pm 1.5)\ \text{GeV}^2,$$
$$\bar{m}_c(m_c^2) = (1.13^{+0.07}_{-0.02} \pm 0.02 \pm 0.02 \pm 0.02)\ \text{GeV}. \qquad (8.95)$$

8.7.2 The $0^+/0^-$ meson mass splittings

• We study in Fig. 8.6(a), the τ-dependence of the $D_s(0^+)$ mass at the values of t_c and m_c obtained previously. In this way, we obtain

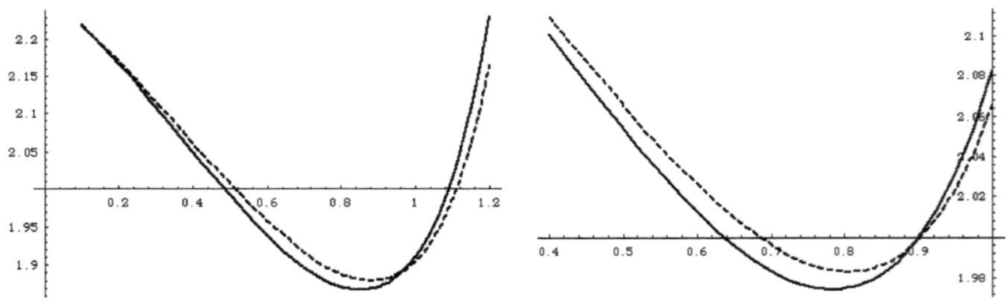

(a) (b)

Figure 8.5. τ in GeV^{-2}-dependence of the (a) $M_D(0^-)$ in GeV for $\bar{m}_c(m_c) =$ 1.11 GeV and (b) $M_{D_s(0^-)}$ in GeV for $\bar{m}_c(m_c) = 1.15$ GeV at a given value of $t_c = 7.5$ GeV^2. The dashed line denotes the result including the leading $\langle\bar{\psi}\psi\rangle$ contribution. The full line denotes the result including non-perturbative effects up to dimension 6.

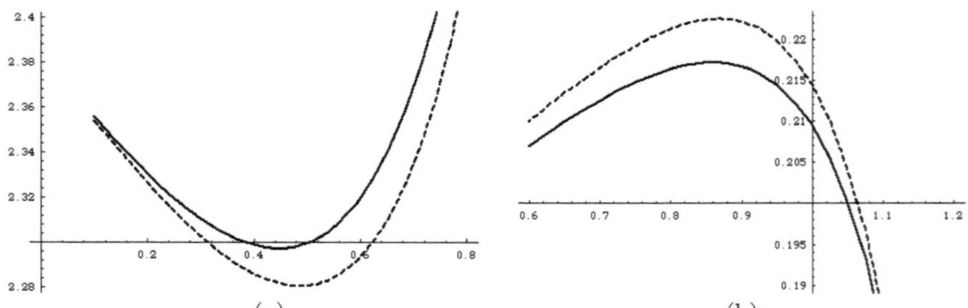

(a) (b)

Figure 8.6. Similar to Fig. 8.5: (a) τ-behaviour of $M_{D_s(0^+)}$; (b) τ-behaviour of $f_D(0^+)$ for given values of $t_c = 7.5$ GeV^2 and $m_c(m_c) = 1.15$ GeV.

in units of MeV:

$$M_{D_s(0^+)} \simeq (2297^{+81+63}_{-98-70} \pm 11 \pm 2 \pm 11)$$
$$\Longrightarrow M_{D_s(0^+)} - M_{D_s(0^-)} = (328 \pm 113), \qquad (8.96)$$

where the errors come respectively from t_c, m_c, $\langle\bar{\psi}\psi\rangle$, m_s, and Λ. We have used the experimental value of $M_{D_s(0^-)}$. The reduction of the theoretical error needs precise values of the continuum threshold[8] and of the charm quark mass which are not within the present reach of the estimate of these quantities. Further discoveries of the continuum states will reduce the present error in the splitting. One should also note that in the ratio of sum rules with which we are working, we expect that perturbative radiative corrections are minimized though individually large in the expression of the correlator and the quark mass.

- The value of the mass splittings obtained previously is comparable with the one of $B(0^+)$-$B(0^-)$ given in Eq. (8.87) and suggests an approximate heavy-flavour symmetry of this observable.
- We also derive the result in the limit of $SU(3)_F$ symmetry where the strange quark mass is put to zero and where the $\langle\bar{\psi}\psi\rangle$ condensate is chirally symmetric ($\langle\bar{s}s\rangle = \langle\bar{d}d\rangle$). In this case, one can predict an approximate degenerate mass within the errors:

$$M_{D_s(0^+)} - M_{D(0^+)} \simeq 25 \text{ MeV}, \qquad (8.97)$$

  which indicates that the mass splitting between the strange and non-strange 0^+ open charm mesons is almost not affected by $SU(3)$ breakings contrary to the case of the 0^- mesons with a splitting of about 100 MeV.
- We extend the analysis to the case of the $B(0^+)$ meson. Here, it is more informative to predict the ratio of the 0^+ over the 0^- masses as the predictions on the absolute values though presenting

[8]The range of t_c values 6–9 GeV^2 obtained previously for the $D(0^-)$ mesons coincides *a posteriori* with the corresponding range for the $D(0^+)$ meson if one assumes that the splitting between the radial excitations is the same as the one between the ground states, i.e., about 300 MeV. We have checked during the analysis that this effect is unimportant and is inside the large error induced by the range of t_c used.

stability in τ tend to overestimate the value of M_B. We obtain:

$$\frac{M_{B(0^+)}}{M_{B(0^-)}} \simeq (1.08 \pm 0.03 \pm 0.03 \pm 0.02 \pm 0.02)$$
$$\Longrightarrow M_{B(0^+)} - M_{B(0^-)} \simeq (422 \pm 196)\ \text{MeV}, \quad (8.98)$$

where the errors come respectively from t_c taken in the range 43–60 GeV^2, m_b, $\langle\bar{\psi}\psi\rangle$, and τ. It agrees with the result in Eq. (8.87) obtained from moment sum rules [168]. We have used the value of $\bar{m}_b(m_b)$ given in Table 7.10 of Chapter 7.

8.7.3 The $D(0^+)$ and $D_s(0^+)$ decay constants

For completing our analysis, we estimate the decay constant $f_{D(0^+)}$ analogue to $f_\pi = 130.6$ MeV. We show the behaviour of $f_{D(0^+)}$ versus τ in Fig. 8.6(b), where a good stability is obtained. Adopting the range of t_c-values obtained previously and using $\bar{m}_c(m_c) = 1.11^{+0.08}_{-0.04}$ GeV required for the best fit of the non-strange $D(0^+)$ meson mass, we deduce to two-loop accuracy:

$$f_{D(0^+)} = (217^{+5+15}_{-15-19} \pm 10 \pm 10)\ \text{MeV}, \quad (8.99)$$

where the errors come respectively from the values of t_c, m_c, $\langle\bar{\psi}\psi\rangle$ and Λ. We have fixed $M_{D(0+)}$ to be about 2272 MeV from our previous fit. It is informative to compare this result with the one of $f_D = (205{\pm}20)$ MeV, where the main difference can be attributed by the sign flip of the quark condensate contribution in the QCD expression of the corresponding correlators. A numerical study of the $SU(3)$ breaking effect leads to:

$$r_s \equiv \frac{f_{D_s(0^+)}}{f_{D(0^+)}} \simeq 0.93 \pm 0.02, \quad (8.100)$$

which is reverse to the analogous ratio in the pseudoscalar channel $f_{D_s}/f_D \simeq 1.14 \pm 0.04$ given semi-analytically in Ref. [167]. In order to understand this result, we give a semi-analytic parametrization of this $SU(3)$ breaking ratio. Keeping the leading term in m_s and $\langle\bar{\psi}\psi\rangle$, one obtains:

$$r_s \simeq \left(1 - \frac{m_s}{m_c}\right)\left[1 - 7.5\langle\bar{s}s - \bar{d}d\rangle\right]^{1/2}\left(\frac{M_{D_s(0^+)}}{M_{D(0^+)}}\right)^2 \simeq 0.9, \quad (8.101)$$

where the main effect comes from the negative sign of the m_s contribution in the overall normalization of the scalar current, while the meson mass ratio does not compensate this effect because of the almost equal mass of $D_s(0^+)$ and $D(0^+)$ obtained in previous analysis. This feature is opposite to the case of $f_{D(0^-)}$.

8.7.4 The $1^+/1^-$ meson mass splitting

Our previous results in Eqs. (8.87), (8.96)–(8.98) suggest that the mass splittings are approximately (heavy and light) flavour and spin-independent. Therefore, one can write to a good approximation the empirical relation:

$$\begin{aligned} M_{D_s(0^+)} - M_{D_s(0^-)} &\approx M_{D(0^+)} - M_{D(0^-)} \approx M_{B(0^+)} - M_{B(0^-)}, \\ &\approx M_{B(1^+)} - M_{B(1^-)} \approx M_{D_s^*(1^+)} - M_{D_s^*(1^-)}. \end{aligned} \quad (8.102)$$

Using the most precise number given in Eq. (8.96), one can deduce:

$$\begin{aligned} &M_{D_s^*(1^+)} - M_{D_s^*(1^-)} \simeq (328 \pm 113)\ \text{MeV} \\ &\Longrightarrow M_{D_s^*(1^+)} \simeq (2440 \pm 113)\ \text{MeV}. \end{aligned} \quad (8.103)$$

This result is consistent with the 1^+ assignment of the $\bar{c}s$ meson $D_{sJ}(2457)$ discovered recently [369].

8.8 Divergences in Non-commutative Theories

8.8.1 Introduction

Using dimensional regularization (DR) for some two-point functions of a prototype non-commutative (NC) ϕ^4 scalar theory in four dimensions [374], we explicitly analyze, to one-loop, the IR and UV divergences of non-planar diagrams having quadratic divergences and compare to the case of the Pauli–Villars cut-off regularization (PVR).

We also note that the IR structure $1/p \circ p$ obtained from DR is reproduced by PVR in the limit where the UV cut-off Λ is set to infinity.

We study the phenomenological implications of this result by rederiving bounds from low-energy data on the violation of Lorentz invariance based on the existence of the quadratic divergence. The most stringent (and regularization-independent) bound on Lorentz violation from low-energy data is $1/\sqrt{\theta} \approx \nu \geq 10^{15}$ GeV for NCQCD and 10^{10} GeV for NCQED, which comes from the absence of sidereal variations between the Cs and Hg atomic clocks.

One of the peculiar features of the NC approach is the correlation between the UV and IR divergences of the theories, as explicitly shown in some scalar models [373], using the Pauli–Villars cut-off regularization (PVR). In Ref. [374], we test if the correlation of the UV and IR divergences is an articraft of PVR or a more general phenomena independent of the regularization procedure. In so doing, we use 't Hooft–Veltman dimensional regularization (DR) [6, 7, 41].

In so doing, we start from a Lagrangian density of the NC ϕ^4 theory:

$$\mathcal{L}(x) = \frac{1}{2}\left(\partial_\mu \phi\right)^2 + \frac{1}{2} m^2 \phi^2 + \frac{g^2}{4!} \phi \star \phi \star \phi \star \phi, \tag{8.104}$$

where the $\star$ product is defined as:

$$\left(\phi_1 \star \phi_2\right)(x) = e^{\frac{i}{2}\theta_{\mu\nu}\partial_\mu^y \partial_\nu^z} \phi_1(y)\phi_2(z)|_{y=z=x}. \tag{8.105}$$

Let's consider the 1PI two-point function, which, to lowest order, corresponds to the inverse propagator:

$$S^{(0)}(p^2) = p^2 + m^2. \tag{8.106}$$

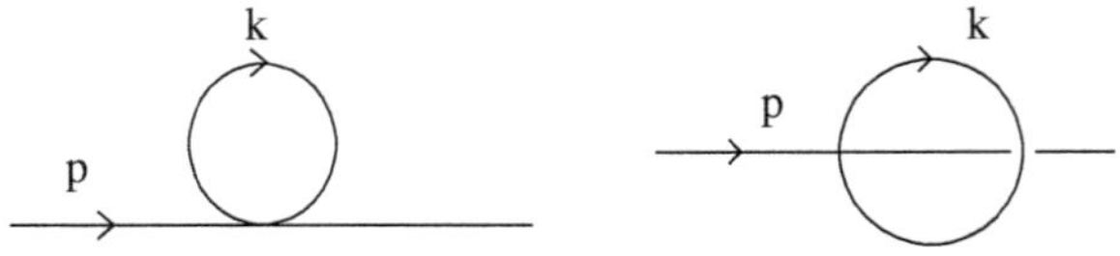

Figure 8.7. Planar and non-planar one-loop corrections to the inverse propagator in ϕ^4 theory.

In the NC theory, the one-loop corrections come from the planar and non-planar diagrams shown in Fig. 8.7, leading to the corrections in n-space–time dimensions:

$$S^{(1)}_{\text{planar}} = \frac{g^2}{3}\int \frac{d^n k}{(2\pi)^n}\frac{1}{k^2+m^2},$$
$$S^{(1)}_{\text{non-planar}} = \frac{g^2}{3!}\int \frac{d^n k}{(2\pi)^n}\frac{1}{k^2+m^2}e^{ik\times p}, \tag{8.107}$$

where the NC-phase argument is $k\times p \equiv k_\mu\theta^{\mu\nu}p_\nu$. Parametrizing the propagator à la Schwinger:

$$\frac{1}{p^2+m^2} = \int_0^\infty dz\ e^{-z(k^2+m^2)}, \tag{8.108}$$

and using in n-dimensions the property:

$$\int \frac{d^n k}{(2\pi)^n}e^{-zk^2} = \frac{1}{(4\pi z)^{n/2}}, \tag{8.109}$$

one obtains for $n = 4-\epsilon$ dimension.

8.8.2 Planar diagram

$$S^{(1)}_{\text{planar}} = \frac{g^2}{48\pi^2}m^2\Gamma(\epsilon/2)\left(\frac{m^2}{\nu^2}\right)^{\epsilon/2} \simeq \frac{g^2}{48\pi^2}m^2\left[\frac{2}{\epsilon} - \log\frac{m^2}{\nu^2} + \cdots\right], \tag{8.110}$$

where ν is the typical UV scale of dimensional regularization and $\cdots$ indicate regular terms. In effective theories, like, e.g., the well-known chiral perturbation theory of QCD, the $m^2\log(m^2/\nu^2)$ is a typical one-loop correction induced by a pion loop with a mass m, while ν is fixed to be about the hadronic scale of about 1 GeV, where beyond this value the effective approach is expected to be not valid.

8.8.3 Non-planar diagram: $p\circ p \to 0$ after integration

One obtains in n-dimensions:

$$S^{(1)}_{\text{non-planar}} = \frac{g^2}{96\pi^2}\int_0^\infty \frac{dz}{z^{n/2}}\ e^{-zm^2-\frac{p\circ p}{4z}}, \tag{8.111}$$

where $p \circ q \equiv -p^\mu \theta^2_{\mu\nu} q^\nu = |p_\mu \theta^2_{\mu\nu} q_\nu|$. We evaluate the integral in $n = 4-\epsilon$ space–time dimension. Taking the limit $\epsilon \to 0$, one obtains after expanding the Bessel function for $p \circ p \to 0$:

$$S^{(1)}_{\text{non-planar}} \simeq \frac{g^2}{24\pi^2}\left[\frac{1}{p \circ p} + \frac{m^2}{2}\log m^2 p \circ p\right]. \tag{8.112}$$

This expression demonstrates that the non-planar diagram is UV finite in the $\epsilon \to 0$ limit and there is no need for introducing an UV cut-off. One may also deduce from this expression that the UV ($\epsilon \to 0$) limit and the IR ($p \circ p \to 0$) one are independent at this order of perturbation theory. This result has to be contrasted with the one:

$$S^{(1)}_{\text{non-planar}} \simeq \frac{g^2}{96\pi^2}\left[\Lambda^2_{\text{eff}} + m^2 \log\left(\frac{m^2}{\Lambda^2_{\text{eff}}}\right)\right] \tag{8.113}$$

obtained using the PV cut-off scheme [373] in four dimensions, where:

$$\Lambda^{-2}_{\text{eff}} \equiv \Lambda^{-2} + p \circ p. \tag{8.114}$$

8.8.4 Non-planar diagram: $p \circ p \to 0$ before integration

However, one may also ask the question whether the operation taking the limit $\epsilon = 0$ and $p \circ p = 0$ is commutative. Therefore, we put $p \circ p = 0$ in Eq. (8.111) before integration. We obtain:

$$S^{(1)}_{\text{non-planar}} \simeq \frac{g^2}{96\pi^2}(m^2)^{1-\epsilon/2}\Gamma(-1+\epsilon/2), \tag{8.115}$$

which has now a $1/\epsilon$ pole.[9] This result indicates that the $p \circ p$ term also acts (in an indirect way) as a regulator of the UV integral, which may be another aspect of the UV/IR mixing within dimensional regularization. However, the question is to know if this second case is

[9]In the case of massless integral, we do not have such a pole if we use the standard prescription that a tadpole-like integral $\int d^n k/k^2 = 0$ within dimensional regularization.

physically interesting? Indeed, taking $p \circ p = 0$ inside the integral is equivalent to reduce the non-planar into the planar one, and the result in Eq. (8.115) is nothing else (modulo normalization) than the planar one in Eq. (8.110).

8.8.5 Non-planar diagram: $m = 0$ limit

In this case, the planar diagram vanishes identically in DR due to the standard property of the tadpole integral.

For the non-planar diagram, one can put immediately $m = 0$ in the integral in Eq. (8.111) or in the result in Eq. (8.112) and obtain:

$$S^{(1)}_{\text{non-planar}} \simeq \frac{g^2}{24\pi^2}\left(\frac{1}{p \circ p}\right)^{1-\epsilon/2} \Gamma(1-\epsilon/2). \tag{8.116}$$

One can see that the presence of the IR-phase leads to a non-zero UV finite result in the $\epsilon \to 0$ limit, indicating the direct role of the IR phase on the UV behaviour of momentum integral.

8.8.6 Phenomenology

The absence of quadratic divergences within dimensional regularization can (*a priori*) affect different phenomenological constraints on NC models based on the existence of this term. In general, in a theory with a massive particle m, one should do the replacement:

$$\Lambda^2 \to m^2 \log \frac{\nu^2}{m^2}, \tag{8.117}$$

where ν is the UV scale of DR and corresponds to the scale until which the effective theory approach is expected to be valid. This log term is also present in PVR but non-leading compared with the Λ^2 one and then is regularization-independent. Using this result, one expects that most of the stringent bounds on the scale of space–time non-commutativity derived from the low-energy tests of Lorentz violation, which are based on the quadratic term, are regularization-dependent. Regularization-independent constraints should then come from the log term.

• NC-QCD

One expects the presence of the Lorentz violating term [375] induced by the magnetic operator $\bar{q}\theta_{\mu\nu}\sigma^{\mu\nu}q$ acts like a $\vec{\sigma}\cdot\vec{B}$ interaction with a fixed $\vec{B}$ and leads to sidereal variations in, e.g., hyperfine splittings. Such variations in the differences between Cs and Hg atomic clocks sensitive to external $\vec{\sigma}\cdot\vec{B}$ interactions are bounded at the 10^{-31} GeV level. The authors in Ref. [375] estimated this operator in a nuclear environment and obtained:

$$\left(\frac{\alpha_s}{12\pi}\right) p_0\Lambda^2\theta \le \Delta E \Longrightarrow \theta\Lambda^2 \le 10^{-29}, \tag{8.118}$$

where $p_0 \approx m_{\rm con} \approx 300$ MeV is the off-shell light quark momentum, $\alpha_s \approx 1$, and ΔE is the bound on the sidereal variation.

Using our previous result in Eq. (8.117), the constraint in Eq. (8.118) becomes:

$$\theta\ m^2_{\rm con} \log \frac{\nu^2}{m^2_{\rm con}} \le 10^{-29}, \tag{8.119}$$

which leads to the tight lower bound on the scale of NCQCD:

$$1/\sqrt{\theta} \approx \nu \ge 10^{15}\ \text{GeV}. \tag{8.120}$$

• NC-QED

The effect of a similar operator has been analyzed in Ref. [376]. In the case of the electron and light quark constituents, the bound has typically the size:

$$\theta\Lambda^2 \le 10^{-19}. \tag{8.121}$$

The one from the electron is expected to be much weaker due to the electron mass suppression. Using again our result in Eq. (8.117), one obtains, for constituent light quarks:

$$\theta\ m^2_{con} \log \frac{\nu^2}{m^2_{con}} \le 10^{-19}, \tag{8.122}$$

which leads to the lower bound on the scale of NCQED:

$$1/\sqrt{\theta} \approx \nu \ge 10^{10}\ \text{GeV}. \tag{8.123}$$

This bound is again much stronger than the one from collider data.

8.9 Beijing–Cracow–Montpellier–Munich Collaboration

The following works are done in collaboration with R. Kaminski (IPN, Cracow), G. Mennessier (University of Montpellier), W. Ochs (MPI, Munich) and X.-G. Wang (University of Montpellier, University of Peking)[10] where R. Kaminski and W. Ochs have been invited in Montpellier within the CNRS visiting scientist program, while X.-G. Wang was preparing a cotutelle thesis in Montpellier.

Figure 8.8. G. Mennessier, W. Ochs, and X.-G. Wang.

8.9.1 Glueball nature of the $\sigma/f_0(600)$ from $\pi\pi$ and $\gamma\gamma$

In Refs. [56, 377],[11] we estimate the $I = 0$ scalar meson $\sigma/f_0(600)$ parameters from $\pi\pi$ and $\gamma\gamma$ scattering data below 700 MeV using an improved analytic K-matrix model.

• The analytic K-matrix model

The general formalism has been applied by Mennessier [378] (see also Refs. [379, 380]) to the calculation of the electromagnetic processes $\gamma\gamma \to \pi\pi, K\bar{K}$, given the strong processes $\pi\pi \to \pi\pi, K\bar{K}$. The strong processes are represented by a K-matrix model representing the amplitudes by a set of resonance poles. In this case, the dispersion

[10] This section is dedicated to my friend and colleague Gerard Mennessier, who left us on 8 April 2016.

[11] For some reviews on glueballs/gluonia, see, e.g., Refs. [54]–[61].

relations in the multi-channel case can be solved explicitly, which is not possible otherwise. This model can be reproduced by a set of Feynman diagrams, including resonance (bare) couplings to $\pi\pi$ and $K\bar{K}$ and four point $\pi\pi$ and $K\bar{K}$ interaction vertices. A subclass of bubble pion loop diagrams including resonance poles in the s-channel are resumed (unitarized Born). The model also includes contributions from the exchange of vector mesons in the t-channel which become important at the higher energies above 700 MeV. The $\gamma\gamma$ scattering amplitudes also fulfill the constraints at $s = 0$ required by the Born approximation. The radiative width of the resonances cannot be predicted due to the polynomial ambiguity, which can be taken into account by introducing as free parameters the "direct couplings" of the resonances to $\gamma\gamma$ in an effective interaction vertex.

- **Amplitude for elastic $\pi\pi$ scattering**

In the K-matrix fits to the $\pi\pi$ elastic scattering data for the range of energy [(0.5–1.8) GeV], a pole is found in the isoscalar S-wave amplitude $T_0^{(0)}$ with a large imaginary part which corresponds to a state [381–384]:

$$M \approx \Gamma \approx 1 \text{ GeV}, \tag{8.124}$$

though this value may depend on the treatment of the other resonances [385–387]. The mass value in Eq. (8.124) is close to the one where the observed S-wave phase shift goes through 90° as in a simple Breit Wigner form without background. This broad object has been called $f_0(1000)$ [384], for a while $f_0(400\text{–}1200)$ by the PDG and now $f_0(600)/\sigma$. In general, the complex resonance self-energy $\Sigma = M - (i/2)\Gamma$ is energy-dependent and determines the poles of the S-matrix corresponding to the mass and width of the physical particles; they do not usually coincide with the Breit Wigner masses. In recent determinations, where analyticity and unitarity properties are used to continue the amplitude into the deep imaginary region, one obtains, for the complex σ pole[12]:

$$441 - \text{i } 272 \text{ MeV} \quad [388] \quad \text{or} \quad 489 - \text{i } 264 \text{ MeV} \quad [389]. \tag{8.125}$$

[12] A similar result (see Eq. (8.140)) is found in the analytic model [378].

We shall see that the σ pole can be referred to the same broad object defined above.

In the present analysis, we have extended the model by the introduction of a *shape function* $f_0(s)$ which multiplies the $\sigma\pi\pi$ coupling and takes explicitly into account left-handed cut singularities for the strong interaction amplitude. This allows a more flexible parametrisation of the $\pi\pi$ data at low energies and improves the high energy behaviour. For simplicity, we don't include the four-point coupling term. The real analytic function $f_0(s)$ is regular for $s > 0$ and has a left cut for $s \leq 0$. For our low-energy approach, a convenient approximation, which allows for a zero at $s = s_A$ and a pole at $\sigma_D > 0$ simulating the left-hand cut, is:

$$f_0(s) = \frac{s - s_{A0}}{s + \sigma_{D0}}. \tag{8.126}$$

The unitary $\pi\pi$ amplitude is then written as:

$$T_0^{(0)}(s) = \frac{G f_0(s)}{s_R - s - G\tilde{f}(s)} = \frac{G f_0(s)}{\mathcal{D}(s)}. \tag{8.127}$$

$T_\ell^{(I)} = e^{i\delta_\ell^{(I)}} \sin\delta_\ell^{(I)}/\rho$ with $\rho(s) = (1 - 4m_\pi^2/s)^{1/2}$; $G = g_{\pi,B}^2$ is the bare coupling squared and:

$$\text{Im } \mathcal{D} = \text{Im}(-G\tilde{f}_0) = -(\theta\rho)G\ f_0, \tag{8.128}$$

with $(\theta\rho)(s) = 0$ below and $(\theta\rho)(s) = \rho(s)$ above threshold $s = 4m_\pi^2$. The amplitude near the pole s_0 where $\mathcal{D}(s_0) = 0$ and $\mathcal{D}(s) \approx \mathcal{D}'(s_0)(s - s_0)$ is:

$$T_0^{(0)}(s) \sim \frac{g_\pi^2}{s_0 - s}, \qquad g_\pi^2 = \frac{G f_0(s_0)}{-\mathcal{D}'(s_0)}. \tag{8.129}$$

The real part of $\mathcal{D}$ is obtained from a dispersion relation with subtraction at $s = 0$ and one obtains:

$$\tilde{f}_0(s) = \frac{2}{\pi}\Big[h_0(s)\ - h_0(0)\Big], \tag{8.130}$$

where $h_0(s) = f_0(s)\tilde{L}_{s1}(s)$–$(\sigma_{N0}/(s + \sigma_{D0}))\tilde{L}_{s1}(-\sigma_{D0})$, σ_{N0} is the residue of $f_0(s)$ at $-\sigma_{D0}$ and $\tilde{L}_{s1}(s) = \left[\left(s - 4m_\pi^2\right)/m_\pi^2\right]\tilde{L}_1(s, m_\pi^2)$ with $\tilde{L}_1$ from Ref. [378].

- **Amplitudes for $\gamma\gamma \to \pi\pi$ scattering**

Starting from the S-wave amplitude in Eq. (8.127), we derive the amplitude $T_\gamma^{(I)}$ for the electromagnetic process for isospin $I = 0$ as:

$$T_\gamma^{(0)} = \sqrt{\frac{2}{3}}\alpha\left(f_0^B + G\,\frac{\tilde{f}_0^B}{\mathcal{D}}\right) + \alpha\frac{\mathcal{P}}{\mathcal{D}}. \tag{8.131}$$

Here, the contribution from the Born term of $\gamma\gamma \to \pi^+\pi^-$ is given by $f_0^B = 2L_1$ as defined in Ref. [378], a real analytic function in the s-plane with left cut $s \leq 0$. The function $\tilde{f}_0^B$ represents $\pi\pi$ rescattering; it is regular for $s < 4m_\pi^2$, but has a right cut for $s \geq 4m_\pi^2$ with:

$$\text{Im}\ \tilde{f}_0^B(s + i\epsilon) = (\theta\rho\ f_0\ f_0^B)(s), \tag{8.132}$$

which vanishes at $s = 0$. With this definition, the Watson theorem is fulfilled, i.e., the phase of $T_\gamma^{(0)}$ is the same as the one of the elastic amplitude $\mathcal{D}^{-1}$ in Eq. (8.127). The real part is derived from a dispersion relation with subtraction at $s = 0$ for $\tilde{f}_0^B$ to satisfy the Thomson limit and has a representation similar to the one in Eq. (8.130), but by replacing $(\tilde{f}_0, h_0)$ by $(\tilde{f}_0^B, h_0^B)$. The function h_0^B is defined as h_0 below Eq. (8.130) but with $\tilde{L}_1$ replaced by $-\tilde{L}_1^2$ everywhere. It vanishes at $s = -\sigma_{D0}$ and $\tilde{f}_0^B(s)$ is regular at this point. Finally, the polynomial $\mathcal{P}$ reflects the ambiguity from the dispersion relations and is set here to $\mathcal{P} = sF_\gamma\sqrt{2}$. It represents the direct coupling of the resonance to $\gamma\gamma$. The residues at the pole s_0 of the rescattering and direct contributions to $T_\gamma^{(0)}$ in Eq. (8.131), respectively, are obtained as:

$$g_\gamma^{\text{resc}}\,g_\pi = \sqrt{\frac{2}{3}}\alpha\frac{G\tilde{f}_0^B(s_0)}{-\mathcal{D}'(s_0)},\quad g_\gamma^{\text{dir}}\,g_\pi = \alpha\frac{s_0F_\gamma\sqrt{2}}{-\mathcal{D}'(s_0)}, \tag{8.133}$$

from which one can deduce the branching ratio:

$$\frac{\Gamma_{\sigma\to\gamma\gamma}}{\Gamma_{\sigma\to\pi\pi}} \simeq \frac{1}{|\rho(s_0)|}\left|\frac{g_\gamma}{g_\pi}\right|^2 \simeq \frac{2\alpha^2}{|\rho(s_0)|}\left|\frac{s_0}{Gf_0(s_0)}\right|^2 F_\gamma^2. \tag{8.134}$$

Similarly, we parametrize the $I = 2$ S-wave amplitude $T_0^{(2)}$ by introducing the shape function f_2:

$$T_0^{(2)} = \frac{\Lambda f_2(s)}{1 - \Lambda\tilde{f}_2(s)},\quad f_2(s) = \frac{s - s_{A2}}{(s + \sigma_{D1})(s + \sigma_{D2})}, \tag{8.135}$$

and obtain:

$$T_\gamma^{(2)} = \frac{\alpha}{\sqrt{3}} \left(f_2^B + \frac{\Lambda \tilde{f}_2^B(s)}{1 - \Lambda \tilde{f}_2(s)} \right), \tag{8.136}$$

where $f_2^B = f_0^B$ and $\text{Im}\tilde{f}_2(s) = (\theta\rho) f_2(s)$, $\text{Im}\tilde{f}_2^B(s) = (\theta\rho\ f_2\ f_2^B)(s)$. These amplitudes are again both subtracted at $s = 0$ as in the case of $I = 0$ and one finds in analogy:

$$\tilde{f}_2(s) = \frac{2}{\pi} \Big[h_2(s) - h_2(0) \Big], \tag{8.137}$$

where $h_2(s) = f_2(s)\tilde{L}_{s1}(s) - (\sigma_{N1}/(s + \sigma_{D1}))\ \tilde{L}_{s1}(-\sigma_{D1}) - (\sigma_{N2}/(s + \sigma_{D2}))\tilde{L}_{s1}(-\sigma_{D2})$; σ_{N1}, σ_{N2} are the residues of $f_2(s)$ at $-\sigma_{D1}, -\sigma_{D2}$ and $\tilde{f}_2^B(s)$ is defined as $\tilde{f}_2(s)$ in Eq. (8.137) but with $\tilde{L}_{s1}$ replaced by $-\tilde{L}_{s1}^2$. The cross-sections for the $\pi\pi$ and $\gamma\gamma$ scattering processes are obtained from the previous expressions of the amplitudes as described in the appendices of Ref. [378], where the $\cos\theta$ integration should be done from 0 to 1 for the neutral cases and from -1 to 1 for the charged cases.

• Results of the analysis

In the present analysis, we repeat the calculation but take the new version of the model [378] by introducing the *shape function* $f_0(s)$ defined in Eq. (8.126). We first determine the parameters of the model for $\pi\pi$ scattering below 700 MeV from the best approximation of our formula to the phase shifts $\delta_0^{(0)}$ and $\delta_0^{(2)}$ obtained in the Roy equation analysis in Ref. [388], which takes into account the high-energy behaviour suggested in Ref. [389]. For $I = 0$, we obtain in units of GeV^2:

$$s_{A0} = 0.016, \quad G \simeq 1.184, \quad s_R = 0.823, \quad \sigma_{D0} = 0.501, \tag{8.138}$$

and for $I = 2$:

$$s_{A2} = 0.039, \quad \Lambda \simeq -4.136, \quad \sigma_{D1} = 0.693, \quad \sigma_{D2} = 4.384, \tag{8.139}$$

from which we find the pole mass and residue:

$$M_\sigma \simeq 422 - \text{i}\ 290\ \text{MeV}; \quad g_\pi \simeq 0.06 - \text{i}\ 0.50\ \text{GeV}, \tag{8.140}$$

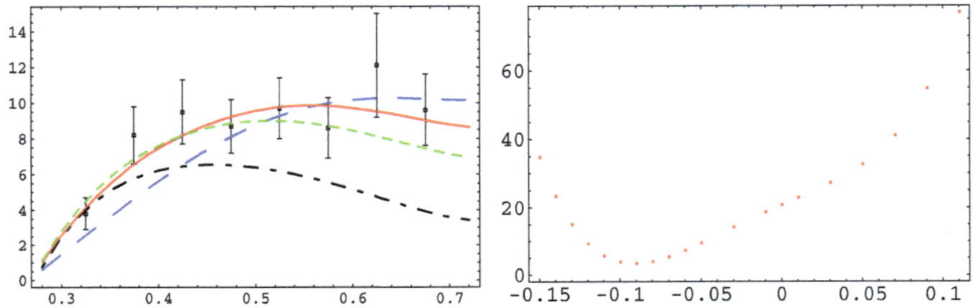

(a) (b)

Figure 8.9. (a) Fit of the $\pi^0\pi^0$ cross-section in nanobarn (nb) versus $\sqrt{s}$ using unitarized Born amplitude: $F_\gamma = 0$ (dot-dashed); $F_\gamma = -0.09$: I $= 0$ (large dashed), I $= 0 + 2$ (continuous); $F_\gamma = -0.07$: I $= 0 + 2$ (small dashed). The data are from Crystal Ball [390] for $|\cos\theta| \leq 0.8$. (b) Corresponding total χ^2 for eight number of data points versus F_γ.

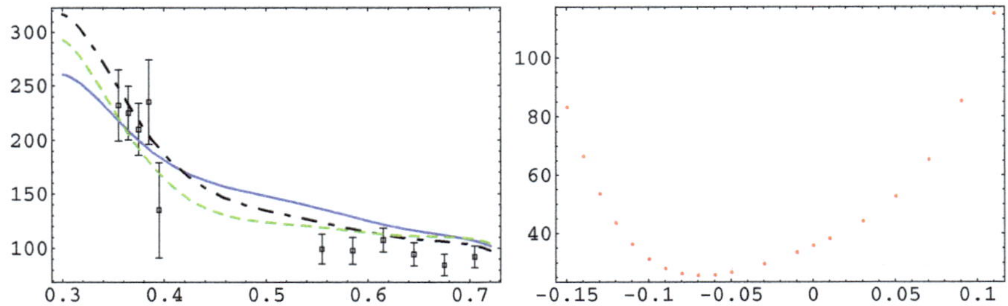

(a) (b)

Figure 8.10. (a) Same as in Fig. 8.9 but for $\pi^+\pi^-$ using unitarized Born amplitude: $F_\gamma = 0$ (dot-dashed); $F_\gamma = -0.07$ and I $= 0 + 2$ (small dashed). The continuous line corresponds to the non-unitarized Born amplitude with $F_\gamma = 0$. The data are from MARKII [391] for $|\cos\theta| \leq 0.6$. (b) Total χ^2 from a simultaneous fit of the $\pi^+\pi^-$ and $\pi^0\pi^0$ total cross-sections shown in the previous figures for 19 number of data points versus F_γ.

which is close to the mass M_σ in Eq. (8.125) and in Ref. [378]. Given the $\pi\pi$ amplitude, we can predict the cross-sections for $\gamma\gamma \to \pi\pi$ where the only free parameter F_γ is related to the strength of the direct coupling $\sigma \to \gamma\gamma$. The fit of the model involving both the $\pi\pi$ rescattering and direct meson couplings to the data from Crystal Ball ($\pi^0\pi^0$) [390] and MARK-II ($\pi^+\pi^-$) [391] collaborations is shown in Figs. 8.9 and 8.10 for the cases with and without direct contribution. A simultaneous best fit of the data in the region below 700 MeV is obtained for $F_\gamma \simeq -0.070$, at the minimum $\chi^2 \simeq 25.6$ (19 data points) in Fig. 8.10(b).[13] However, a fit of the neutral channel alone gives a much better fit: $F_\gamma \simeq -0.090$, for a $\chi^2 = 3.38$ (eight data points) in Fig. 8.9(b). In the charged channel, this is due to the deviations of the fit at high mass; the large systematic errors and the absence of data points in the region 0.40–0.55 GeV do not permit a good understanding of this channel. One can note that the $I = 2$ S-wave amplitude which reproduces with high accuracy the CGL points [388] improves the agreement with the data in the neutral case (Fig. 8.9) but gives negligible contribution for the charged pion. For definiteness, we consider as a final estimate the mean value of the previous numbers:

$$F_\gamma \simeq -(0.080 \pm 0.014), \tag{8.141}$$

where we consider that the uncertainty in the fitting procedure 0.010 is expressed by this range. We have added linearly the systematics of the method to be about 5%, which we have estimated from the deviation of the mass and width determinations in Eq. (8.140) from the ones [388] in Eq. (8.125), as we have used their parametrization for fixing the parameters of the model. The corresponding values of the residues in units of $\alpha \times 10^{-3}$ GeV are:

$$g_\gamma^{\rm dir} \simeq (7 \pm 1) + {\rm i}\ (31 \pm 4), \quad g_\gamma^{\rm resc} \simeq (90 + 5) + {\rm i}\ (116 \pm 6). \tag{8.142}$$

Using Eqs. (8.134), (8.140), and (8.141) and $\Gamma_{\sigma\to\pi\pi} = 2{\rm Im}M_\sigma$, we can deduce the "partial" $\gamma\gamma$ widths at the complex pole:

$$\Gamma^{\rm dir}_{\sigma\to\gamma\gamma} \simeq (0.13 \pm 0.05)\ {\rm keV}, \quad \Gamma^{\rm resc}_{\sigma\to\gamma\gamma} \simeq (2.7 \pm 0.4)\ {\rm keV}, \tag{8.143}$$

and the total $\gamma\gamma$ width (direct + rescattering):

$$\Gamma^{\rm tot}_{\sigma\to\gamma\gamma} \simeq (3.9 \pm 0.6)\ {\rm keV}. \tag{8.144}$$

Improvements of these estimates need more precise data below 700 MeV and an extension of the analysis to higher energies.

[13]In the earlier analysis [378], a value F_γ of $\mathcal{O}(1)$ has been found, which corresponds to the DM2 data in Ref. [392] which are about twice the one of MARKII [391], i.e., twice the Born contribution.

- **Comparisons with existing results**

◇ **Direct $\gamma\gamma$ width**

Our result for the direct coupling is comparable with our previous value [393] obtained without using the shape function $f_0(s)$. In a parallel analysis [394], a K-matrix model analogous to Ref. [378] has been applied in a larger energy range, including the recent BELLE data [395], still assuming elementary pion and kaon exchange; a smaller direct coupling was obtained, but without error analysis.

◇ **Total $\gamma\gamma$ width**

– Our result is compatible with the range of values (1.2–3.2) keV obtained in Ref. [393] using a Breit–Wigner parametrization of the $\gamma\gamma \to \pi^0\pi^0$ data. It is also in the range of values based on dispersion relations (see introduction), though a direct comparison is difficult to do as these analyses are extended to higher energy region, and we have an additional determination of the direct term. In our analysis below 700 MeV, we have also neglected the ω and axial-vector exchange in agreement with Ref. [378]. Other calculations find contributions of 2.5% [396] or 16% [397] at 400 MeV. One can note that the effect of the direct term in our Fig. 8.9 distorts the shape of the unitarized Born curve and then permits a good fit of all points below 700 MeV, which is not the case of the one in Ref. [397]. Then, if our fit was restricted to lower masses to suppress these other exchanges, no different results are to be expected.

- **"On-shell" σ mass and widths**

For a more appropriate comparison with QSSR predictions, one can introduce the mass and widths of the "visible meson" on the real axis. This can be either the Breit–Wigner mass and width in Eq. (8.124) or/and the "on shell" mass and widths (see, e.g., Ref. [398]).

◇ **Mass of the σ**

A Breit–Wigner parametrization of the data leads to the value given in Eq. (8.124): $M_\sigma^{\rm bw} \approx \Gamma_\sigma^{\rm bw} \approx 1$ GeV, which can be compared with the "on-shell mass" $M_\sigma^{\rm os}$ in the model [378] where the amplitude is purely imaginary at the phase 90°:

$$\text{Re}\mathcal{D}((M_\sigma^{\text{os}})^2) = 0 \implies M_\sigma^{\text{os}} \approx 0.92 \text{ GeV}. \tag{8.145}$$

◇ Hadronic width

In the same way as for the mass, one can define an "on-shell width" [see Eqs. (8.128) and (8.129)] evaluated at $s = (M_\sigma^{\text{os}})^2$:

$$M_\sigma^{\text{os}}\Gamma_\sigma^{\text{os}} \simeq \frac{\text{Im }\mathcal{D}}{-\text{Re }\mathcal{D}'} \Longrightarrow \Gamma_{\sigma\to\pi\pi}^{\text{os}} \approx 1.02 \text{ GeV}, \tag{8.146}$$

to be compared with the Breit–Wigner width in Eq. (8.124).

◇ Direct $\gamma\gamma$ width

As there are no separated S-wave cross-sections, we use here only the estimates based on the analytic model [378] where we proceed as in the previous section, but we evaluate the parameters at $s_0 = (M_\sigma^{\text{os}})^2$. Then, we obtain from Eq. (8.134):

$$\Gamma_{\sigma\to\gamma\gamma}^{\text{os,dir}} \approx (1.0 \pm 0.4) \text{ keV}. \tag{8.147}$$

As the model is extrapolated here towards energies beyond its validity, we expect that the previous results are a crude approximation and will only serve as a guideline.

• QSSR/LET results

QCD spectral sum rules (QSSR) and low-energy theorems (LET) have given quantitative estimates for the mass and decay properties of the gluonic bound states. In particular, in a combined analysis of subtracted and unsubtracted sum rules, a low mass for the bare unmixed gluonium state is obtained [126, 129, 132]:

$$M_{\sigma B} \simeq (0.95 \sim 1.10)\text{GeV}, \tag{8.148}$$

besides the heavier state around $M_G \simeq (1.5-1.6)$ GeV. These masses are similar to the unquenched and quenched lattice results, respectively, quoted above. The hadronic width in this approach is obtained rather large:

$$\Gamma_{\sigma_B \to \pi^+\pi^-} \simeq 0.7\,\text{GeV}, \tag{8.149}$$

whereas the radiative decay width is found to be small:

$$\Gamma_{\sigma_B \to \gamma\gamma} \simeq (0.2 \sim 0.6)\,\text{keV}. \tag{8.150}$$

Indeed, this width is considerably smaller than the value of about 5 keV, which is expected for a state $S_2 \equiv (\bar{u}u + \bar{d}d)$ with mass of 1 GeV [124]. On the other hand, QSSR predicts, for a four-quark state having the same mass of 1 GeV, a $\gamma\gamma$ width of about 0.4 eV [124].

• Comparison with QSSR/LET results

One can note that the QSSR and LET predictions for the mass and widths of a low mass gluonium σ_B in Eqs. (8.148)–(8.150) are in remarkable agreement with the results in Eqs. (8.145)–(8.147) for an on-shell resonance. The direct $\gamma\gamma$ coupling, which can reveal the photon coupling to the intrinsic quark ($\bar{q}q$, $4q, \ldots$) or gluon ($gg, \ldots$) constituent structure of the resonance, can be related to the QSSR/LET evaluations of its width through quark or gluon loops as in the case of the quark triangle for the pion or $f_2(1270)$. Our $\sigma \to \gamma\gamma$ width in Eq. (8.147) (also the pole value in Eq. (8.143) are consistent with the small value predicted in Eq. (8.150). This "overall agreement" favours a large gluon component in the $\sigma/f_0(600)$ wave function. It disfavours, in particular, a $\bar{q}q$ interpretation because of the larger radiative width to be expected (∼5 keV). On the other hand, a four-quark interpretation would require a smaller radiative width and is disfavoured as well by our result in Eq. (8.143) (at the level of 2.5 σ).

8.9.2 Gluonium nature of the $\sigma/f_0(600)$ from $g_{\sigma K\bar{K}}$

• Fit of the $\pi\pi \to \pi\pi/\bar{K}K$ data

In Ref. [399], we extract the K^+K^- couplings of the isoscalar scalar mesons $\sigma/f_0(600)$ and $f_0(980)$ from $\pi\pi \to \pi\pi/K\bar{K}$ scatterings using the previous analytic K-matrix model [56, 377–380].

Using the normalization:

$$\Gamma_{\sigma_B \to \pi^+\pi^-} \equiv \frac{|g_{\sigma_B \pi^+\pi^-}|^2}{16\pi M_{\sigma_B}} \sqrt{1 - \frac{4m_\pi^2}{M_{\sigma_B}^2}}, \tag{8.151}$$

and the data from [382, 400–402], we obtain the results in Table 8.3. We compare these results with the ones from some other models in Table 8.3.

In particular, we rederive the results from a coupled channel model ($\pi\pi$ and $\bar{K}K$) using a unitary S-matrix approach [402]. The results of the fits have been improved by using the method in GKPY [406] based on the dispersive analysis of experimental data where the authors have shown that theoretical constraints given by forward dispersion relations (FDR), sum rules and by once and twice subtracted dispersion relation (GKPY and Roy's equations, respectively) allow us to determine the $\pi\pi$ scattering amplitudes consistent with analyticity, unitarity and crossing symmetry.

Table 8.3. Modulus of the $\pi^+\pi^-$ and K^+K^- complex couplings in GeV of σ and $f_0(980)$ from S- and K-matrix models for $\pi\pi \to \pi\pi/K\bar{K}$ scatterings compared with the ones from ϕ and J/ψ decays. $r_{S\pi K} \equiv |g_{SK^+K^-}|/|g_{S\pi^+\pi^-}|$: $S \equiv \sigma,\ f_0$.

Processes	$\|g_{\sigma\pi^+\pi^-}\|$	$r_{\sigma\pi K}$	$\|g_{f_0\pi^+\pi^-}\|$	$r_{f_0\pi K}$	Models
This work					
$\pi\pi \to \pi\pi/K\bar{K}$	2.90(10)	0.52(17)	1.42(6)	2.20(14)	[377]
	2.03(3)	0.65(18)	0.97(6)	1.71(14)	[402]
Average	*2.47(45)*	*0.59(19)*	*1.20(23)*	*1.96(29)*	
Others					
$\pi\pi \to \pi\pi/K\bar{K}$	2.5	0.62	1.55	1.20	[403]
$\phi \to \sigma/f_0(980)\ \gamma$	–	0.67	–	–	[404]
$\psi \to \phi\ \pi\pi/K\bar{K}$	–	–	2.35	1.80	[405]
Global Average	*2.5*	*0.6*	*1.7*	*1.7*	

Forward dispersion relations (calculated at $t = 0$) are used in Ref. [406] for three isospin combinations of $\pi\pi$ amplitudes: for $\pi^0\pi^+$, $\pi^0\pi^0$ and for *the amplitude with isospin 1 in the t-channel.*

In the case of non-forward dispersion relation (Roy's and GKPY equations) derived with imposed crossing symmetry condition, a difference between "output" and "input" partial wave amplitudes is minimized together with that for FDR.

- **Comparison with QSSR/LET results**

In the previous section, we found that the large $\pi\pi$ width of the σ and its coupling to $\gamma\gamma$ favour its interpretation as a gluonium/glueball state.

The previous finding that its coupling to $\bar{K}K$ is almost equal to the one to $\pi\pi$ reinforces this interpretation where from LET one expects, up to SU(3) symmetry breakings, that [126, 132]:

$$g_{\sigma\pi\pi} \simeq g_{\sigma\bar{K}K}. \tag{8.152}$$

This is not the case of a $\pi\pi$ molecule or a four-quark $(\bar{q}\bar{q})(qq)$ states with $q \equiv u, d$ where one expects $g_{\sigma\bar{K}K}$ to be relatively small.

8.9.3 The σ and $f_0(980)$ from $Ke4 \oplus \pi\pi$ scatterings

In Ref. [407],we systematically reconsider, within an improved "analytic K-matrix model" [377], the extraction of the $\sigma \equiv f_0(600)$ and $f_0(980)$ masses, widths and hadronic couplings using new $Ke4 \equiv K \to \pi\pi e\nu_e$ data on $\pi\pi$ phase shift below 390 MeV from NA48/2 [408] and different sets of $\pi\pi \to \pi\pi/K\bar{K}$ scattering data. The analysis in Ref. [377] from 400–700 MeV is extended until 1.4 GeV.

In this work, we discuss the approach for the case of

– *1 channel* $\oplus$ *0 "bare" resonance:* $\lambda\phi^4$ *model,*
– *1 channel* $\oplus$ *1 "bare" resonance (K-matrix pole),*
– *2 channels* $\oplus$ *2 "bare" resonances,*

and we restrict to the $SU(3)$ symmetric shape function $f_0(s)$.

- **Elastic $\pi\pi$ scattering**

We shall use the new precise data from NA48/2 on $Ke4 \equiv K \to \pi\pi e\nu_e$ for the $\pi\pi$-phase shift below 390 MeV [408] and use from 400 to 900 MeV the CERN-Munich [381] and Hyams *et al.* [382] $\pi\pi$-phase from $\pi\pi \to \pi\pi$ which agree with each other above 400 MeV. These data are shown in Fig. 8.11 together with the fit for different number of "bare" resonances.

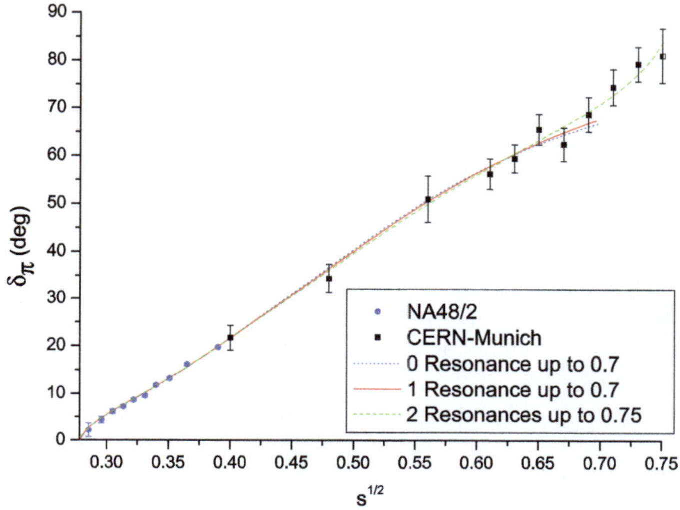

Figure 8.11. Central values of the best fit of the $\pi\pi$ phase shift below $\sqrt{s} = 0.75$ GeV for elastic $\pi\pi \to \pi\pi$ scattering: blue (dotted) line for 0 "bare" resonance with $\chi^2_{\rm min}$ reached at $\sqrt{s} = 0.7\,$GeV; red (continuous) line for 1 "bare" resonance and $\chi^2_{\rm min}$ reached at $\sqrt{s} = 0.7\,$GeV; green (dashed) line for 2 "bare" resonances and $\chi^2_{\rm min}$ reached at $\sqrt{s} = 0.75$ GeV.

Noting at the first step that the fitted value of the Adler zero is quite bad, we have fixed its value at the theoretical one:

$$s_{A\pi} = \frac{m_\pi^2}{2} = 0.0094 \text{ GeV}^2. \tag{8.153}$$

The results are shown in Table 8.4 where one can note that the results are stable against the choice of the number of resonances. We have used the fit program "MINUIT."

• Inelastic $\pi\pi/\bar{K}K$ scatterings

We show in Fig. 8.12 the fit of the $\pi\pi$ phase δ_π, the inelasticity η and the sum of π and K phase $\delta_{\pi K}$.

In so doing, we take in Table 8.5 three representative sets of $\pi\pi \to \pi\pi/\bar{K}K$ data in the existing literature.

Table 8.4. Values in MeV of the physical poles for each number of "bare" resonances. An average is given in the last column. The corresponding values of the K-matrix model and "bare" parameters can be found in Ref. [307]. The coupling is normalized as in Eq. (8.151).

Output	0 res.	1 res.	2 res.	Average
M_σ	468(181)	456(19)	448(18)	*452(13)*
$\Gamma_\sigma/2$	261(211)	265(18)	260(19)	*259(16)*
$\lvert g_{\sigma\pi^+\pi^-}\rvert$	2.58(1.31)	2.72(16)	2.58(14)	*2.64(10)*

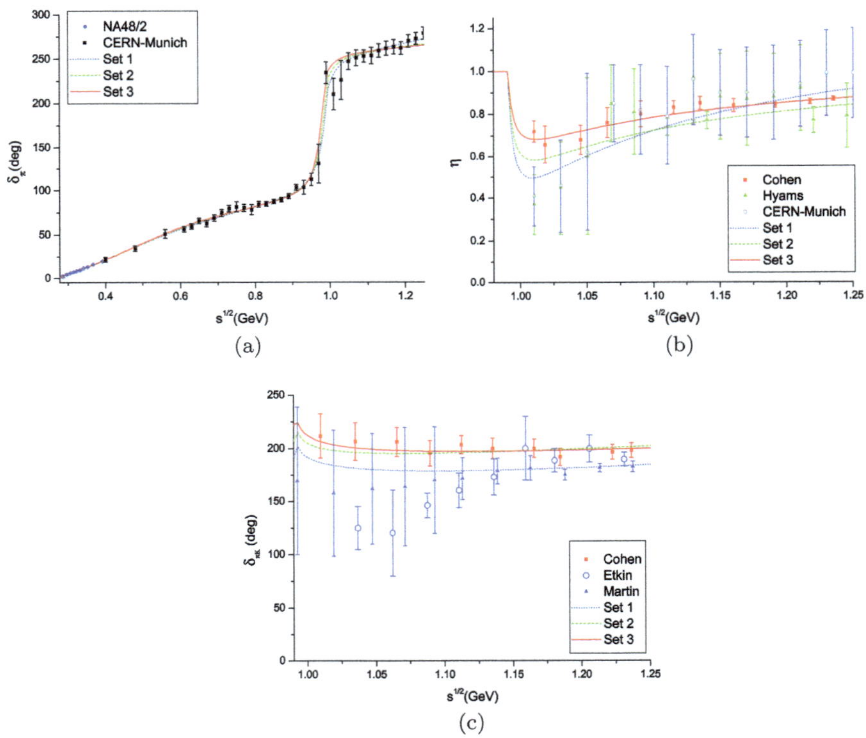

Figure 8.12. (a) Fit of the $\pi\pi$ phase δ_π versus $\sqrt{s}$: set 1 (blue: dotted); set 2 (green: dashed); set 3 (red: continuous); (b) fit of the inelasticity η; (c) fit of the sum of π and K phase $\delta_{\pi K}$.

◇ The $\sigma/f_0(600)$ meson

The average value of M_σ is in perfect agreement with the mean value from elastic $\pi\pi \to \pi\pi$ scattering. The average of the two leads to our

Table 8.5. Different data used for each different *sets* for determining the "bare" parameters in Table 4 of Ref. [407]. δ_π is the $\pi\pi$ phase shift, η is the inelasticity and $\delta_{\pi K}$ is the sum of the π and K phases.

Input	Set 1	Set 2	Set 3
δ_π	[381, 382, 408]	[381, 382, 408]	[381, 382, 408]
η	[381]	[382]	[400]
$\delta_{\pi K}$	[401]	[400]	[400]

Table 8.6. σ and $f_0(980)$ meson parameters from $\pi\pi \to \pi\pi/\bar{K}K$ scattering. The values of the "bare parameters" used to get these values can be found in Table 4 of Ref. [407]. An average is given in the last column.

Outputs	Set 1	Set 2	Set 3	Average
M_σ	435(74)	452(72)	457(76)	*448(43)*
$\Gamma_\sigma/2$	271(92)	266(65)	263(72)	*266(43)*
$\lvert g_{\sigma\pi^+\pi^-}\rvert$	2.72(78)	2.74(61)	2.73(61)	*2.73(38)*
$\lvert g_{\sigma K^+K^-}\rvert$	1.83(86)	0.80(55)	0.99(68)	*1.06(38)*
M_f	989(80)	982(47)	976(60)	*981(34)*
$\Gamma_f/2$	20(32)	18(16)	18(18)	*18(11)*
$\lvert g_{f\pi^+\pi^-}\rvert$	1.33(72)	1.22(60)	1.12(31)	*1.17(26)*
$\lvert g_{fK^+K^-}\rvert$	3.21(1.70)	2.98(70)	3.06(1.07)	*3.03(55)*

final estimate:

$$M_\sigma[\text{MeV}] = 452(12) - \text{i}\ 260(15). \tag{8.154}$$

Averaging the results in Tables 8.4 and 8.6, we deduce:

$$|g_{\sigma\pi^+\pi^-}| = 2.65(10)\ \text{GeV}, \quad r_{\sigma\pi K} \equiv \frac{|g_{\sigma K^+K^-}|}{|g_{\sigma\pi^+\pi^-}|} = 0.37(6), \tag{8.155}$$

which improves and confirms our previous rough findings in Ref. [399] and which is comparable with some other determinations given in Table 8.3 from Ref. [399].

Again, the sizeable coupling of σ to $\bar{K}K$ disfavours the usual $\pi\pi$ molecule and four-quark assignment of σ, where this coupling is expected to be negligible.

For better comparison of the results obtained in the complex plane with the theoretical predictions obtained in the real axis, it is more appropriate to introduce like in Ref. [377] the on-shell meson [398] masses and hadronic widths, where the amplitude is purely imaginary at the phase 90°:

$$\mathrm{Re}\mathcal{D}((M_\sigma^{\mathrm{os}})^2) = 0 \Longrightarrow M_\sigma^{\mathrm{os}} \simeq 0.9 \text{ GeV}. \tag{8.156}$$

In the same way as for the mass, one can also define an "on-shell width" [377]:

$$M_\sigma^{\mathrm{os}}\Gamma_\sigma^{\mathrm{os}} \simeq \frac{\mathrm{Im}\ \mathcal{D}}{\mathrm{Re}\ \mathcal{D}'} \Longrightarrow \Gamma^{\mathrm{os}}_{\sigma\to\pi^+\pi^-} \simeq 0.7 \text{ GeV}, \tag{8.157}$$

which are comparable with the Breit–Wigner mass and width [382–384]:

$$M_{BW} \simeq \Gamma_{BW} \simeq 1 \text{ GeV}. \tag{8.158}$$

These values lead to the on-shell coupling:

$$|g^{\mathrm{os}}_{\sigma\pi^+\pi^-}| \simeq 6 \text{ GeV}. \tag{8.159}$$

◇ The $f_0(980)$ meson

– For the $f_0(980)$, we obtain the mean value:

$$M_f[\mathrm{MeV}] = 981(34) - \mathrm{i}\ 18(11), \tag{8.160}$$

which is comparable with the PDG range of values [33]:

$$M_f[\mathrm{MeV}] = 980(10) - \mathrm{i}(20 \sim 50). \tag{8.161}$$

From Table 8.6, we also find:

$$|g_{f\pi^+\pi^-}| = 1.12(31) \text{ GeV}, \quad r_{f\pi K} \equiv \frac{|g_{fK^+K^-}|}{|g_{f\pi^+\pi^-}|} = 2.59(1.34) \tag{8.162}$$

in agreement with the determinations in the existing literature (see Table 8.3). *The large value of this ratio of coupling and the relative narrowness of the $f_0(980)$ width (compared to, e.g., the ρ-meson) does not favour the pure $(\bar{u}u+\bar{d}d)$ content of the $f_0(980)$ where $r_{f\pi K}$ is expected to be about 1/2 and the width of about 120 MeV* [57,126]. *This feature has been used as an indication of the four-quark nature of $f_0(980)$ (see, e.g., Ref.* [409]*) or alternatively of its large gluonium component via a maximal mixing with a $\bar{q}q$ state (see, e.g., Refs.* [125, 126]*).*

- **Comments**
  - ◇ The existence of a low mass glueball is necessary for a consistency between the subtracted and unsubtracted QSSR [132]. The mass and couplings of σ found in the previous analysis indicate that σ is a solid gluonium/glueball candidate beside $G(1.6)$ at higher mass.
  - ◇ The $S_2 \equiv \bar{u}u + \bar{d}d$ state is narrower and much higher in mass than the complex σ pole often identified with a $\bar{q}q$ state in the current literature [57,126].
  - ◇ The $f_0(980)$ cannot be a pure $(\bar{u}u + \bar{d}d)$ state due to the large ratio of its $\bar{K}K$ over its $\bar{\pi}\pi$ couplings $r_{f\pi K}$ (Eq. (8.162)) and to its relatively (compared to the ρ-meson) small $\pi\pi$ width. It cannot also be a pure $\bar{s}s$ or $\bar{K}K$ molecule due to its non-negligible width into $\pi\pi$.
  - ◇ A large gluonium component eventually mixed with a $\bar{q}q$ state in the σ and $f_0(980)$ wave functions [55,125,126,132,377,386,399] can be advocated for evading these previous difficulties.

8.9.4 σ and $f_0(980)$ substructures from $\gamma\gamma \to \pi\pi$

In Ref. [410], we pursue the analysis of the substructure of σ and $f_0(980)$ from $\gamma\gamma \to \pi\pi$, $J/\psi, \phi \to \gamma S$ and $D_s \to Sl\bar{\nu}_l$ processes.

- **$\gamma\gamma \to \pi\pi$**

In this paper, we extend the previous analysis by including $\pi\pi$ and $\bar{K}K$ loops and work in the region just above the $\bar{K}K$ threshold (the minimal χ^2/ndf of our fit is obtained for $\sqrt{s} = 1.09$ GeV), where the $\sigma(600)$, $f_0(980)$ contributions are dominant.

◇ **$I = 0$ D-wave mass and hadronic couplings**

In addition to the S-wave contribution with the masses and hadronic couplings obtained in Eqs. (8.154) and (8.155) and Eqs. (8.160) and (8.162) used here as inputs, we shall add the contribution of the D-wave ($I = 0$) (assumed to be dominated by $f_2(1270)$) into the

$\pi\pi \to \pi\pi/\bar{K}K$ scattering amplitudes. The fits of the phase and inelasticity are shown in Fig. 8.13 from which we deduce:

$$M_{f_2} = 1.27{-}i0.07\,\text{GeV}, \quad g_{f_2\pi\pi} = 0.48{-}i0.03, \quad g_{f_2\bar{K}K} = 0.28{-}i0.06. \tag{8.163}$$

Our results reproduce the PDG values quite well [33].

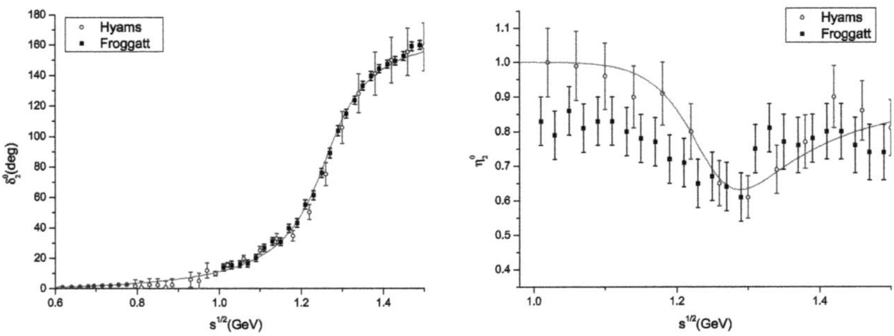

(a) (b)

Figure 8.13. (a) $I = 0$ D-wave $\pi\pi$ phase; (b) inelasticity. Experimental data are taken from Refs. [382, 411].

◇ K-matrix model analysis of $\gamma\gamma \to \pi\pi$ below 1.09 GeV

In this region, the contributions of vector (V) and axial-vector (A) can also be important in the t-channel exchange (Born approximation). We add the pion and kaon loops (P) for the unitarized part of the amplitudes (see Fig. 8.14) within the model of Ref. [378].

◇ Fitting $\gamma\gamma \to \pi^0\pi^0$ just above the $\bar{K}K$ threshold

In the fitting procedure, we fix the value of the helicity $\lambda = 2$ component of $f_2(1270)$ using the $f_2\,\gamma\gamma$ width given by PDG [33] and neglect the $\lambda = 0$ one. We use as inputs Set 2 and Set 3 hadronic parameters obtained in Ref. [407] using the largest range of hadronic data. Then, we perform a fit of the four direct couplings from the $\gamma\gamma \to \pi^0\pi^0$ total cross-section up to $\sqrt{s} \approx 1$ GeV.[14] We move $\sqrt{s}$ around 1 GeV and look for the minimum χ^2/ndf for the total cross-section, which is

[14]We choose not to fit much above 1 GeV in order to avoid the possible contribution of an eventual $f_0(1370)$ and minimize the contribution of the D-wave.

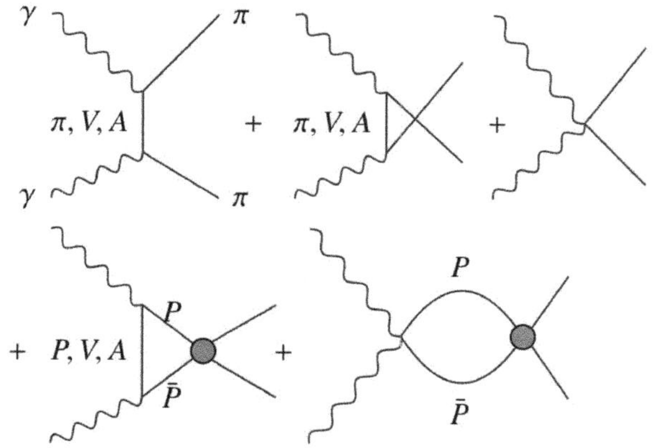

Figure 8.14. Born and unitarized amplitudes T_P^u: $P \equiv \pi,\ K;\ V \equiv \rho,\ \omega;\ A \equiv b_1,\ h_1,\ a_1$.

obtained at $\sqrt{s} = 1.09\,\text{GeV}$, where $\chi^2/ndf = 39.5/41 = 0.96$ for, e.g., Set 3 of the hadronic parameters, as shown in Fig. 8.15(a). In Fig. 8.15(b), we compare the $\gamma\gamma \to \pi^0\pi^0$ differential cross-section at $\sqrt{s} = 0.97$ GeV for the sum of all partial waves with the sum of the S + D-waves and with the Belle data. The results of the fit are given in Table 8.8 (the results for the bare parameters can be found in Ref. [410]). A comparison with some other determinations is given in Table 8.7.

◇ Comments on the results

The results are summarized in Tables 8.7 and 8.8.

– *The direct part of the $\gamma\gamma$ widths:* one can note that the K exchange and K-loop tend to decrease the direct width of the σ meson which is compensated by the V+A contributions, such that at the end, the final result is compatible with the one 0.13 keV obtained in the case: 1 resonance $\oplus$ 1 channel obtained below 0.7 GeV [377]. We consider as a final result the average from Set 2, Set 3 and from the one resonance $\oplus$ one channel analysis below 0.7 GeV of Ref. [377] for σ and the average from Set 2 and Set 3 for $f_0(980)$ (see Table 8.7):

$$\Gamma_\sigma^{\text{dir}} = 0.16(4)\ \text{keV}, \quad \Gamma_{f_0}^{\text{dir}} \simeq 0.28(1)\ \text{keV}. \tag{8.164}$$

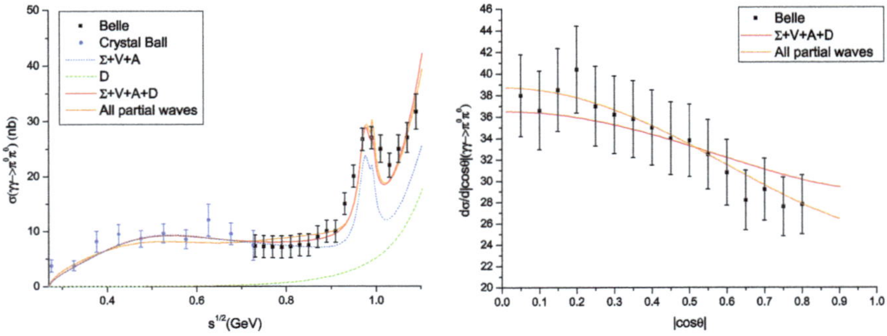

(a) (b)

Figure 8.15. (a) Fit of the $\gamma\gamma \to \pi^0\pi^0$ total cross-section ($\sqrt{s} \leq 1.09$ GeV and $|\cos\theta| \leq 0.8$) for Set 3 of hadronic parameters [407], where $\chi^2/ndf = 39.5/41 = 0.96$ is minimal. The data come from Crystal Ball (blue full circle) and from Belle (black full square); dotted blue (S-channel contribution); dashed green (D-wave contribution); continuous red (total contribution); dotted-dashed salmon (sum of all partial waves); (b) comparison of the effects of the sum of all partial waves (dashed-dotted salmon) with sum of S+D waves (continuous red) and the Belle data on the $\gamma\gamma \to \pi^0\pi^0$ differential cross-section at $\sqrt{s} = 0.97$ GeV.

Table 8.7. Summary of the fitted values of the $\gamma\gamma$ decay width (in unit of keV using Set 2 and Set 3 of hadronic parameters from Ref. [407] and comparison with some other determinations and the PDG [33] value. The total cross-section has been fitted until $\sqrt{s} = 1.09$ GeV (minimum value of χ^2/ndf) and the differential cross-section until $\sqrt{s} = 0.97$ GeV. The best $\chi^2/ndf = 0.96$ is obtained from Set 3.

	Set 2	Set 3	[377]	[412]	[413]	[413]	[403]	[414]	[415]	PDG [33]
$\sqrt{s}$	1.09	1.09	0.8	0.8	1.44	1.44	1.4		1.44	
$r_{0/2}$	0.0	0.0			0.13	0.26	0.15			
Γ_σ^{dir}	0.16	0.20	0.13						0.010	
Γ_σ^{resc}	1.53	1.43	2.70							
Γ_σ^{tot}	2.67	2.67	3.90	1.7	3.1	2.4	2.1	1.2		
$\Gamma_{f_0}^{dir}$	0.29	0.27							0.015	
$\Gamma_{f_0}^{resc}$	0.90	0.81								
$\Gamma_{f_0}^{tot}$	0.17	0.15			0.42	0.10	0.13			0.29 ± 0.08

Table 8.8. The physical couplings (in units of $\alpha \times 10^{-3}$ GeV) and the two photon widths (in units of keV) of σ and $f_0(980)$. Each contributions of different mesons exchanged in the t-channel and in the loops are shown explicitly. D corresponds to the D-wave contribution.

Fit	π	$\Sigma \equiv \pi{+}K$	$\Sigma{+}V$	$\Sigma{+}A$	$\Sigma{+}V{+}A$	$\Sigma{+}D$	$\Sigma{+}V{+}A{+}D$
Set 2							
$\Gamma_\sigma^{\rm dir}$	0.27	0.06	0.11	0.07	0.13	0.07	0.16
$\Gamma_\sigma^{\rm resc}$	1.60	1.71	3.97	2.05	1.53	1.71	1.53
$\Gamma_\sigma^{\rm tot}$	2.90	2.36	4.66	2.51	2.50	2.43	2.67
$\Gamma_{f_0}^{\rm dir}$	0.82	0.002	0.63	0.04	0.31	0.002	0.29
$\Gamma_{f_0}^{\rm resc}$	0.02	0.48	0.13	1.67	0.90	0.48	0.90
$\Gamma_{f_0}^{\rm tot}$	0.68	0.43	0.20	2.19	0.16	0.42	0.17
Set 3							
$\Gamma_\sigma^{\rm dir}$	0.13	0.10	0.11	0.19	0.15	0.14	0.20
$\Gamma_\sigma^{\rm resc}$	1.36	1.44	4.22	1.94	1.43	1.44	1.43
$\Gamma_\sigma^{\rm tot}$	2.34	2.35	4.70	2.32	2.49	2.43	2.67
$\Gamma_{f_0}^{\rm dir}$	0.61	0.01	0.48	0.03	0.23	0.01	0.27
$\Gamma_{f_0}^{\rm resc}$	0.01	0.49	0.14	1.51	0.81	0.49	0.81
$\Gamma_{f_0}^{\rm tot}$	0.53	0.39	0.10	1.93	0.19	0.36	0.15

For a consistent comparison with some other theoretical estimates (QSSR, etc.) obtained in the real axis, we translate these widths from the residues to the one evaluated at the on-shell σ mass or Breit–Wigner mass defined in Refs. [377,398] when the amplitude is purely imaginary at the phase 90°:

$$\text{Re}\ \mathcal{D}(M_\sigma^2) = 0 \quad \longrightarrow \quad M_\sigma^{os} \simeq 0.92\ \text{GeV}. \tag{8.165}$$

In this way, we obtain:

$$\Gamma_\sigma^{\rm dir}|_{\text{on-shell}} = \ 1.2(3)\ \text{keV}, \quad \Gamma_{f_0}^{\rm dir}|_{\text{on-shell}} \approx \Gamma_{f_0}^{\rm dir}, \tag{8.166}$$

which are similar with the results obtained by using a Breit–Wigner parametrization of the data [377].

– *The rescattering part* of the $\gamma\gamma$ widths are (in units of keV):

$$\Gamma_\sigma^{\rm resc} = 1.89(81)\ \text{keV}, \quad \Gamma_{f_0}^{\rm resc} = 0.85(5)\ \text{keV}, \tag{8.167}$$

where we take the average from Set 2, Set 3 and from the one resonance $\oplus$ one channel analysis below 0.7 GeV of Ref. [377] for σ and the average from Set 2 and Set 3 for $f_0(980)$ (see Table 8.7). One can note that in both cases, the rescatterings are relatively large, indicating the important rôle of meson loop contributions in the $\gamma\gamma$ widths of the scalar mesons. The large couplings of scalar to meson loops are often interpreted in the current literature as being related to their four-quark or/and molecule structure. However, these large couplings to $\pi\pi$ and $\bar{K}K$ are also expected if σ and f_0 have large gluon component and violate OZI rule in their hadronic decays [57, 126, 132]. In the case of the σ meson, one can note the large effect due to vector mesons which is partly compensated by the one of the axial-vector mesons. Compared with the result of 2.7 keV from the analysis below 0.7 GeV with a pion loop [377], one can note that $\Gamma_\sigma^{\rm resc}$ has been affected by the presence of $f_0(980)$ when doing the fitting procedure. In the case of $f_0(980)$, the effect of K and $V + A$ are quite important.

– *The total* $\equiv$ *direct + rescattering part* of the $\gamma\gamma$ widths are (in units of keV):

$$\Gamma_\sigma^{\rm tot} = 3.08(82)\ {\rm keV}, \quad \Gamma_{f_0}^{\rm tot} = 0.16(1)\ {\rm keV}. \qquad (8.168)$$

◇ Comparison with some other results

– *Dispersion relations*

A comparison of the total width in Eq. (8.168) with the results obtained using dispersion relations [403, 412–414][15] is shown in Table 8.7. The results can only be compared for the total width, as the authors have not performed the (model-dependent) separation of the direct and rescattering processes done in this approach. Our results for σ are in better agreement with the ones from [377,403,413]. The one of f_0 is in between the results of Refs. [403, 413] and of the PDG value [33]. However, the PDG value seems to be more consistent with our value of the direct width which might be identified with the one using a Breit–Wigner parametrization of the data.

[15]Some calculations based on the π and K loops can also be found in Ref. [416].

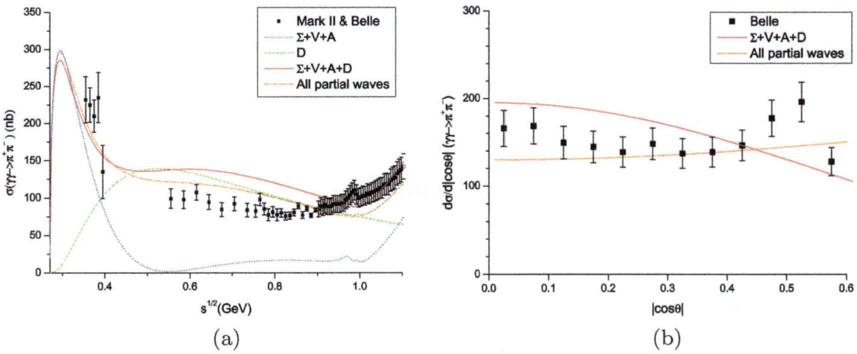

Figure 8.16. (a) Predictions of the $\gamma\gamma \to \pi^+\pi^-$ total cross-section ($\sqrt{s} \leq$ 1.09 GeV and $|\cos\theta| \leq 0.6$) using the fitted parameters from $\gamma\gamma \to \pi^0\pi^0$. The data come from Mark II and Belle: dotted blue (S-channel contribution including direct couplings); dashed green (D-wave contribution); continuous red (S+D partial waves); dot-dashed salmon (sum of all partial waves); (b) Comparison of the effects of the sum of all partial waves (dashed-dotted salmon) with sum of S + D-waves (continuous red) and the Belle data on the differential $\gamma\gamma \to \pi^+\pi^-$ cross-section at $\sqrt{s} = 0.95$ GeV.

– *Model of Ref.* [415]

This approach presents some similarities with ours where the direct and rescattering processes can also be separated. However, we differ in the treatment of the hadronic $\pi\pi \to \pi\pi/\bar{K}K$ process as a (*a priori ad hoc*) background phase has been explicitly introduced which multiplies the resonance contributions to the amplitude in this approach. Besides this difference, the authors also have a different (philosophical) view of the problem, where they claim that considering the residues of the σ pole for the hadronic amplitude does not give a clear understanding of the nature of σ, while, in our paper, we, on the contrary, claim that the residues in the complex plane can clarify its nature.

We note that the authors use the bare couplings for predicting the $\gamma\gamma$ width of the resonances while we transform the bare couplings (which are real numbers) to the residues in the complex plane by analytically continuing these results to the second sheet. In fact, we expect that working with the bare couplings for the wide complex σ meson cannot (*a priori*) be a good approximation.

We also note that the authors only include the S- and D-waves and neglect the contributions due to the vector and axial-vector mesons (V + A) which is a bad approximation for the direct width predictions (compare the last two columns of Table 8.8) though the numerical fits of the total cross-section are similar in the two cases (see Fig. 8.15) because of the dominant contribution of the rescattering amplitude which remains almost constant.

$-\gamma\gamma \to \pi^+\pi^-/K^+K^-$data

We use the previous fitted values of the parameters from $\gamma\gamma \to \pi^0\pi^0$ for predicting the $\gamma\gamma \to \pi^+\pi^-$ process. The results for the total and differential cross-sections are shown in Fig. 8.16, where one can note that the prediction is not good between 0.5 and 0.9 GeV if one only retains the S- and D-waves in the partial waves of the Born term. A similar observation can be done for the process $\gamma\gamma \to K^+K^-$, where the Born contributions due to the S- and D-waves are also very important and dominates over the scalar meson contributions, as can be shown in Fig. 8.10 of Ref. [417], which we have checked. We improve our analysis by adding all higher partial waves in the Born term and by including the direct terms of the S-waves. One can see in Fig. 8.16 that the predictions using the parameters from $\gamma\gamma \to \pi^0\pi^0$ are quite good compared with the data without introducing any *ad hoc* form factors.

8.9.5 Gluonium production from $J/\psi, \phi$ radiative decays

The previous possible gluonium assignment of σ and $f_0(980)$ can be tested from the J/ψ and ϕ radiative decays. This can be done following the work of Ref. [128], using dispersion relation techniques and the Euler–Heisenberg effective Lagrangian, where the gluonic part of the amplitude can be converted into the non-perturbative matrix element $\langle 0|\alpha_s G^2|S\rangle \sim M_S^2 f_S$, where the decay constant f_S ($S \equiv \sigma_B$, G) can be obtained from QSSR which predicts two scalar gluonia ($\sigma_B(1)$ coupled strongly to $\pi\pi, \bar{K}K$ due to OZI violations and the G(1.5–1.6) coupled to $\eta'\eta', \eta\eta'$ obtained using quenched lattice) [57,126,132].

In this way, one obtains:

$$\Gamma(J/\psi \to S\gamma) \simeq \frac{\alpha^3\pi}{\beta_1^2 656100}\left(\frac{M_{J/\psi}}{M_c}\right)^4\left(\frac{M_S}{M_c}\right)^4 \times \frac{\left(1-M_S^2/M_{J/\psi}^2\right)^3}{\Gamma\left(J/\psi \to e^+e^-\right)} f_S^2, \quad (8.169)$$

where $\beta_1 = -1/2(11-2n_f/3)$ for $SU(n)_f$ is the first coefficient of the β function, M_c is the constituent charm quark mass which we take to be about $M_{J/\psi}/2$. Using the decay constant $f_S \equiv f_{\sigma_B} \simeq (1.4 \sim 1.0)$ GeV for $M_{\sigma_B} \approx (0.75-1)$ GeV [57, 126, 132], one predicts:

$$B(J/\psi \to \sigma_B\gamma) \times B(\sigma_B \to \text{all}) \simeq (0.4 \sim 1.0) \times 10^{-3}, \quad (8.170)$$

which one can compare with [33]: $B(J/\psi \to \eta'\gamma) \simeq (4.7 \pm 0.27) \times 10^{-3}$ and $B(J/\psi \to f_2\gamma) \simeq (1.43 \pm 0.11) \times 10^{-3}$. Applying this result to the case of the $G(1.5\text{–}1.6)$ glueball of higher mass, one obtains [57, 126, 132]:

$$B(J/\psi \to G\gamma) \simeq (5.0 \pm 3.8) \times 10^{-4}, \quad (8.171)$$

where we have used $f_G \simeq (0.39 \pm 0.15)$ GeV. This prediction, which needs to be improved, is in agreement with the experimental lower bound of $(5.7 \pm 0.8) \times 10^{-4}$ [33].

Extrapolating this analysis to the case of the ϕ-meson, one obtains by using the strange quark constituent mass $M_s \simeq M_\phi/2$:

$$B(\phi \to S\gamma) \simeq 1.2 \times 10^{-4}, \quad (8.172)$$

which, despite the crude approximation used, compares quite will with the data [33]: $B(\phi \to \pi^+\pi^-\gamma) \simeq (0.41 \pm 0.1) \times 10^{-4}$ and $B(\phi \to f_0(980)\gamma) \simeq (3.22 \pm 0.19) \times 10^{-4}$.

8.9.6 Gluonium production from D_s semi-leptonic decays

An analogous analysis has been done for D_s semileptonic decays, where one expects that the gluonium production will be of similar strength as the one for a $\bar{q}q$ state [210]:

$$\frac{\Gamma[D_s \to \sigma_B(gg)l\nu]}{\Gamma[D_s \to S_2(\bar{q}q)l\nu]} \approx \frac{1}{|f_+(0)|^2}\left(\frac{f_{\sigma_B}}{M_c}\right)^2 \simeq \mathcal{O}(1), \tag{8.173}$$

for $f_{\sigma_B} \approx 1$ GeV, where $|f_+(0)| \simeq 0.5$ [211] is the form factor associated to the $\bar{q}q$ semileptonic production. Some other hadronic processes like $J/\psi \to \omega, \phi + \pi\pi/\bar{K}K$ (see, e.g., Ref. [380]) and $D_{(s)}, B_{(s)} \to 3\pi, \ldots$ (see, e.g., Ref. [418]) decays can also be studied, but the mechanism for the decays are expected to be more complex than the one of radiative and semi-leptonic processes discussed previously. However, these analyzes emphasize the important role of the kaon loops contributions, which in our approach are due to the rescattering contributions.

8.9.7 Some possible substructures of σ and $f_0(980)$

Our previous works have been motived for shedding light on the possible substructure of $\sigma/f_0(600)$ and $f_0(980)$ by analyzing their masses, hadronic and $\gamma\gamma$ widths, where the value of direct $\gamma\gamma$ width is of special interest for a comparison with theoretical calculations based on quark and/or gluon loops. However, for a direct comparison with theoretical calculations obtained in the real axis (QCD spectral sum rules, low-energy theorem, etc.), one should convert the complex pole mass and width to the on-shell or Breit-Wigner one given in Eqs. (8.156), (8.157), (8.160) and (8.162).

• σ meson

In our previous analysis of the hadronic couplings [399, 407] from $\pi\pi \to \pi\pi$, $\bar{K}K$ processes, we have noted that because of the large size of its coupling to $\pi\pi$ and $\bar{K}K$ $[|g_{\sigma K^+K^-}|/|g_{\sigma\pi^+\pi^-}| = 0.37(6)]$ (see also Refs. [378,402,403,419]), σ cannot be mainly a $\pi\pi$ molecule or/and a four-quark state components, while its broad width cannot be explained by a large $\bar{q}q$ component. These observations go in line to a large gluon component expected from a low-energy theorem analysis in [57,126,132].[16] The averaged result of the $\gamma\gamma$ direct width of about 1.2(3) keV (see Eq. (8.166)), when runned at the on-shell σ mass, can indicate an eventual large gluon component of the σ.[17] Moreover, our result seems to rule out a large four-quark component which has a too small direct $\gamma\gamma$ width of about some few eV [124] from QSSR or from some σ-like models [422].

• $f_0(980)$

The direct $\gamma\gamma$ width of 0.27 keV for $f_0(980)$ in Eq. (8.164) is neither compatible with a large four-quark component (too small) nor with a large S_2 component (too big). A pure $\bar{s}s$ component where a $\gamma\gamma$ width is expected to be about 0.4 keV [57, 126, 423] is not however favoured because of the non-zero coupling of $f_0(980)$ to $\pi\pi$ $[|g_{fK^+K^-}|/|g_{f\pi^+\pi^-}| = 2.59(1.34)]$ obtained in Refs. [399, 407].

• Conclusions

At first sight, a $\bar{q}q$-gluonium mixing scheme like the one proposed below 1 GeV in Refs. [125, 126] might be appropriate for describing σ and $f_0(980)$ obtained from our fits of $\pi\pi \to \pi\pi/\bar{K}K$ and $\gamma\gamma \to \pi\pi$ scatterings. The values of their "direct widths" favour a large gluon content for the σ meson but are not decisive for explaining the substructure of the $f_0(980)$ meson. However, the large values of the rescattering widths, due to meson loops because of the large couplings of σ and $f_0(980)$ to $\pi\pi$ or/and $\bar{K}K$, can also be obtained if they are gluonia states but not necessarily if they are four-quark (diquark–antidiquark) or molecule states as currently claimed in the

---

[16] A large gluon component of the σ meson with a Breit–Wigner mass and width of about 1 GeV has been also advocated in Ref. [386]. A large gluon component for σ and $f_0(980)$ also emerges from a calculation of the gluonia spectrum using a Nambu–Jona–Lasinio model [420].

[17] An isoscalar $S_2 \equiv (1/\sqrt{2})\,(\bar{u}u + \bar{d}d)$ state is expected to have a $\gamma\gamma$ width of about 4 keV [57, 126] from QCD spectral sum rules (QSSR) [3, 6, 7] and quark model [421], while a pure gluonium state σ_B with $M_{\sigma_B} = 1$ GeV has a width of about (0.2–0.6) keV [126, 132] from some low-energy theorems.

existing literature. The agreement of our predictions for a gluonium production through ϕ radiative decays with the data seems to support some large gluon component for σ and to a lesser extent for $f_0(980)$. This test can be pursued in the analysis of J/ψ radiative and D_s semi-leptonic decays.

8.10 Montpellier–Saskatchewan–Tanà Collaboration

This collaboration involves Siyuan Li and Tom Steele (Saskatchewan University) (see Fig. 8.17) and Davidson Rabetiarivony (iHEP-MAD) (see Fig. 7.30) and concerns some works on the gluonia/glueball spectra and couplings from QCD spectral sum rules (QSSR).

Figure 8.17. S. Li and T. Steele (Saskatchewan University).

8.10.1 2^{++} Tensor di-gluonium from LSR at NLO

In Ref. [424], we evaluate the next-to-leading (NLO) corrections to the perturbative (PT) and $\langle \alpha_s G^2 \rangle$ condensate and the LO constant term of the $\langle G^3 \rangle$ contributions to the 2^{++} tensor di-gluonium two-point correlator. Using these results into the inverse Laplace transform sum rule (LSR) moments and their ratio, we estimate the mass and coupling of the lowest ground state. We obtain $M_T = 3028(287)$ MeV and the renormalization group invariant (RGI) coupling $\hat{f}_T = 224(33)$ MeV within a vacuum saturation estimate of the $D = 8$ dimension gluon condensates ($k_G = 1$). We study the effect of k_G on the result and find $M_T = 3188(337)$ MeV and $\hat{f}_T = 245(32)$ MeV for $k_G = (3 \pm 2)$. Our result does not favour the pure gluonia/glueball nature of the observed $f_2(2010, 2300, 2340)$ states.

• **The Laplace transform sum rules**

We shall be concerned with the two-point function[18]:

$$\psi_T^{\mu\nu\rho\sigma}(q^2) \equiv i \int d^4x\, e^{iqx} \langle 0|\theta_G^{\mu\nu}(x) \left(\theta_G^{\rho\sigma}\right)^\dagger (0)|0\rangle$$
$$= \left(P^{\mu\nu\rho\sigma} \equiv \eta^{\mu\rho}\eta^{\nu\sigma} + \eta^{\mu\sigma}\eta^{\nu\rho} - \frac{2}{n-1}\eta^{\mu\nu}\eta^{\rho\sigma} \right) \psi_T(q^2), \tag{8.174}$$

built from the gluon component of the energy–momentum tensor[19]:

$$\theta_G^{\mu\nu} = \alpha_s \Big[- G_\alpha^{\mu,a} G_a^{\nu\alpha} + \frac{1}{4} g^{\mu\nu} G_{\alpha\beta}^a G_a^{\alpha\beta} \Big], \tag{8.175}$$

with:

$$\eta^{\mu\nu} \equiv g^{\mu\nu} - q^\mu q^\nu / q^2 \;:\quad P_{\mu\nu\rho\sigma} P^{\mu\nu\rho\sigma} = 2(n^2 - n - 2), \tag{8.176}$$

where $n = 4 + 2\epsilon$ is the space–time dimension used for dimensional regularization and renormalization. To LO and up to dimension $D = 8$ gluon condensates, the QCD expression is [81]:

$$\psi_T|_{LO}(q^2 \equiv -Q^2) = a_s^2 \left[-\frac{Q^4}{20} \log \frac{Q^2}{\nu^2} + \frac{5}{3}\pi^3 \alpha_s \langle 2O_1 - O_2 \rangle \right], \tag{8.177}$$

where $a_s \equiv \alpha_s/\pi$ and:

$$O_1 = \left(f_{abc} G_{\mu\alpha} G_{\nu\alpha}\right)^2, \quad O_2 = \left(f_{abc} G_{\mu\nu} G_{\alpha\beta}\right)^2. \tag{8.178}$$

Using the vacuum saturation hypothesis ($k_g = 1$), it reads [81]:

$$\langle 2O_1 - O_2 \rangle \simeq -k_G \left(\frac{3}{16} \right) \langle G^2 \rangle^2. \tag{8.179}$$

---

[18] For relations among different form factors, see, e.g., Ref. [235].

[19] Note the extra-factor α_s compared to the current used in Ref. [126] and references quoted therein.

We shall test the effect of this assumption by taking:

$$k_G \neq 1 \tag{8.180}$$

for an eventual violation of the factorization assumption like the one found for the $D = 6$ four-quark condensates (see, e.g., Refs. [6, 7]) and new papers in Chapter 9.

• PT expression of the two-point function up to NLO

◇ Lowest order (LO) contribution

This comes from the diagram in Fig. 8.18 and reads:

$$\psi_T^{pert}|_{LO} = -a_s^2 \frac{Q^4}{20} \log \frac{Q^2}{\nu^2}. \tag{8.181}$$

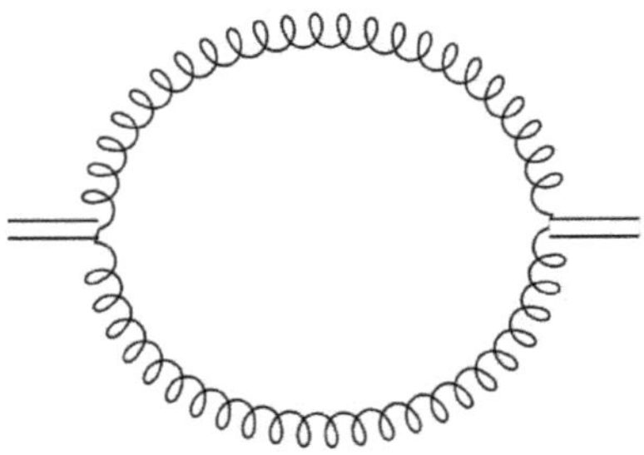

Figure 8.18. LO perturbative contribution to $\psi_T(Q^2)$.

◇ Next-to-Leading (NLO) contribution

We use two approaches to perform the calculation:

A diagrammatic renormalization [425] which requires an isolation of the subdivergences arising from the one-loop subdiagram(s) of an individual bare NLO diagram (see, e.g., Ref. [426]). Counterterm diagrams generated from the subdivergences are then calculated and subtracted from the bare diagram to obtain the renormalized diagram.

The conventional approach using standard Feynman diagrams techniques and the renormalization of the gluonic current using the renormalization constant obtained in Ref. [134] for the current $\theta_g^{\mu\nu}/\alpha_s$:

$$Z_\psi = 1 - \left(\frac{n_f}{3}\right)\frac{a_s}{2\epsilon}, \tag{8.182}$$

for $n = 4 + 2\epsilon$ dimensions, to which corresponds the anomalous dimension:

$$\gamma_\psi \equiv \left(\gamma_1 = \frac{n_f}{3}\right) a_s + \cdots . \tag{8.183}$$

The renormalized two-point function for n_f flavours reads:

$$\psi_T^{\rm pert}|_{NLO}^R(Q^2) = \psi_T^{\rm pert}|_{LO}\left[1 + a_s\left(\frac{n_f}{6}\log\frac{Q^2}{\nu^2} - \frac{101 n_f + 150}{90}\right)\right], \tag{8.184}$$

where $\psi_T^{\rm pert}|_{LO}$ can be deduced from Eq. (8.177). One can note that for gluodynamics ($n_f = 0$), we recover the earlier result of Ref. [427].

• Dimension-four gluon $\langle\alpha_s G^a_{\mu\nu} G^{\mu\nu}_a\rangle$ condensate

One can note from Eq. (8.177) that, unlike the case of scalar and pseudoscalar gluonia [81], the contributions of the gluon condensates are only due to the $D = 8$ dimension.

◇ LO contribution

From the diagram in Fig. 8.19, we have checked that to LO the leading log term does not contribute to the two-point function. The LO contribution comes from the constant term:

$$\psi_T^{G^2}|_{LO} = \frac{\alpha_s}{6}\langle\alpha_s G^2\rangle, \tag{8.185}$$

where we have used two different approaches (plane wave and conventional one using the projection in Eq. (8.174). The non-zero value of this constant term raises the question of the validity of the null result obtained in Ref. [128] based on instantons for dual/antidual background fields. However, this term is harmless in the LSR analysis as it will disappear when one takes the different derivatives of the two-point functions.

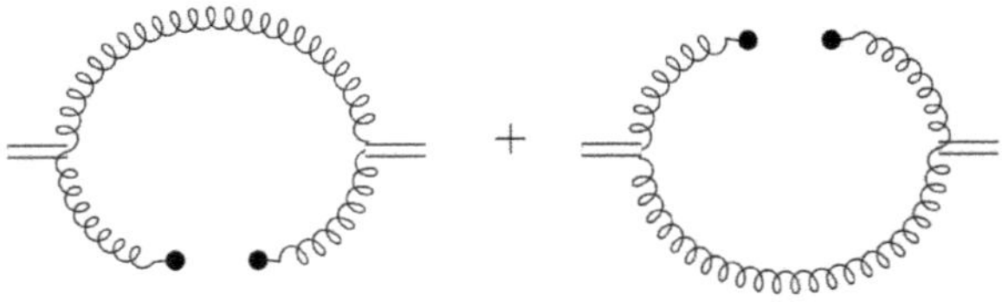

Figure 8.19. LO $\alpha_s G^2$ contribution to $\psi_T(Q^2)$.

◇ NLO contribution

The leading-log, contribution at NLO, can be derived from the renormalization group equation (RGE). Using the fact that $\langle \alpha_s G^2 \rangle$ obeys the RGE [428] (see different applications in Section 4.4 of Ref. [5])[20]:

$$\left\{ -\frac{\partial}{\partial t} + \beta(a_s)\, a_s \frac{\partial}{\partial a_s} - 2\gamma_\psi^\theta \right\} \psi_T^{G^2} = 0, \tag{8.186}$$

where $t \equiv (1/2)\log(Q^2/\nu^2)$ and γ_ψ^θ is the anomalous dimension defined in Eq. (8.183). Writing the α_s expansion as:

$$\psi_T^{G^2} = \left(g_0 \alpha_s^2 + 2g_1 \alpha_s^2 a_s\, t + g_1' \alpha_s^2 a_s - 2\gamma_\psi^\theta \alpha_s^2 + \cdots \right) \langle G^2 \rangle, \tag{8.187}$$

and considering that $\langle \alpha_s G^2 \rangle$ is a constant, one deduces:

$$g_1 = g_0 \left(\frac{\beta_1}{2} - \gamma_1^\theta \right) = \frac{1}{24}\left(11 + \frac{2n_f}{3} \right) \quad \text{with} \quad g_0 = \frac{1}{6}. \tag{8.188}$$

• Dimension 6 $g^3 \langle f_{abc} G^a_{\mu\nu} G^b_{\nu\rho} G^c_{\rho\mu} \rangle$ gluon condensate

◇ LO contribution

We calculate the coefficients of the $\langle gG^3 \rangle$ contribution using the conventional approach and the projection in Eq. (8.176). We show the different contributions in Fig. 8.20 where the total sum is zero (log coefficient and constant term) in agreement with the result of Ref. [128].

---

[20] We take into account the overall α_s factor appearing in the definition of the energy–momentum tensor current.

◇ Check of the result

We recompute the G^3 coefficient of the scalar gluonium two-point function using the same method. We recover the result of Ref. [81] which is an indirect test of our result.

• QCD expression of the LSR at NLO

We shall be concerned with the following inverse Laplace transform moments and their ratio [3, 5, 64, 69, 70]:

$$\mathcal{L}^c_{0,1}(\tau,\mu) = \int_{t>}^{t_c} dt\; t^{(0,1)} e^{-t\tau} \frac{1}{\pi} \mathrm{Im}\, \tilde{\psi}_T(t,\nu),$$

$$\mathcal{R}^c_{10}(\tau) \equiv \frac{\mathcal{L}^c_1}{\mathcal{L}^c_0} = \frac{\int_{t>}^{t_c} dt\; e^{-t\tau} t\, \mathrm{Im}\, \tilde{\psi}_T(t,\nu)}{\int_{t>}^{t_c} dt\; e^{-t\tau} \mathrm{Im}\, \tilde{\psi}_T(t,\nu)}. \tag{8.189}$$

To get the lowest moment $\mathcal{L}^c_0$, we take the third derivative of the two-point function which is superconvergent, while for the $\mathcal{L}^c_1$ moment, we take the fourth derivative of $Q^2\tilde{\psi}_T(Q^2)$. The NLO QCD expressions

Diagrams	$\psi_T(Q^2)$
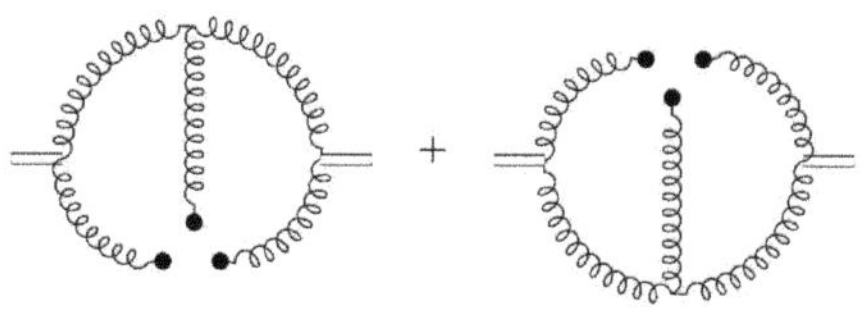	$= -\frac{1}{12Q^2}\langle gG^3\rangle$
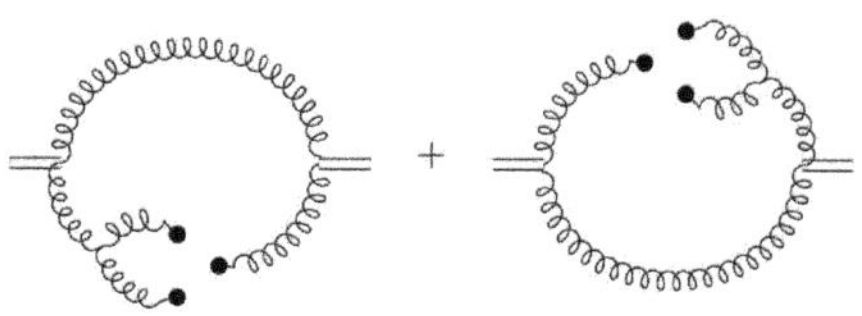	$= 0$
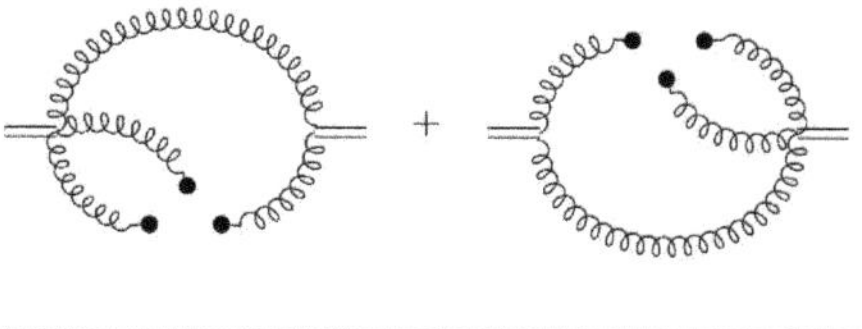 + Sym	$= +\frac{1}{12Q^2}\langle gG^3\rangle$

Figure 8.20. LO $g\langle G^3\rangle$ contribution to $\psi_T(Q^2)/\alpha_s^2$.

of the moments for $n_f = 3$ flavours are:

$$\mathcal{L}_0^c = \frac{\tau^{-3}}{10\pi^2}\left\{\left[1 - a_s\left(\frac{53}{15} + \gamma_E\right)\right]\rho_2^c - \frac{65\pi}{12}\langle\alpha_s G^2\rangle\tau^2\rho_0^c\right.$$
$$\left. -\frac{\pi^2}{a_s}\left(\frac{25}{8}\right)k_G\langle\alpha_s G^2\rangle^2\tau^4\right\}, \tag{8.190}$$

and:

$$\mathcal{L}_1^c = \frac{3\tau^{-4}}{10\pi^2}\left\{\left[1 - a_s\left(\frac{16}{5} + \gamma_E\right)\right]\rho_3^c - \frac{65\pi}{36}\langle\alpha_s G^2\rangle\tau^2\rho_1^c\right.$$
$$\left. +\frac{\pi^2}{a_s}\left(\frac{25}{24}\right)k_G\langle\alpha_s G^2\rangle^2\tau^4\right\}, \tag{8.191}$$

from which one can deduce the ratio $\mathcal{R}_{10}^c(\tau)$. $\gamma_E = 0.5772\cdots$ is the Euler constant and:

$$\rho_n^c = 1 - e^{-t_c\tau}\left(1 + (t_c\tau) + \cdots + \frac{(t_c\tau)^n}{n!}\right). \tag{8.192}$$

• Parametrization of the spectral function

To a first approximation, we parametrize the spectral function using the minimal duality ansatz (MDA):

$$\frac{1}{\pi}\mathrm{Im}\,\tilde{\psi}(t) = f_T^2 M_T^4 \delta(t - M_T^2) + \theta\,(t\_t_c)\text{ “QCD continuum”}, \tag{8.193}$$

where we assume that the QCD expression of the spectral function above the continuum threshold t_c smears all radial excitation contributions. f_T is normalized as $f_\pi = 132$ MeV. Within the MDA parametrization, one has:

$$\mathcal{R}_{10}^c \simeq M_T^2. \tag{8.194}$$

• Extraction of the ground state mass and coupling at LO

We show in Fig. 8.21 the behaviour of the mass from the ratio of moments $\mathcal{R}_{10}^c$ versus τ and t_c assuming a factorization of the $D = 8$

condensates ($k_G = 1$). The stabilities in τ are reached for the set $(\tau, t_c) = (0.18, 4.5)$ to $(0.68,\ 12)$ (GeV^{-2}, GeV^2). We deduce at LO:

$$M_T = 2091(234)_{t_c}(24)_{G^2}\text{MeV} \quad \rightarrow \quad t_c \simeq \left(6.5^{+5.5}_{-2.0}\right)\ \text{GeV}^2, \qquad (8.195)$$

and an upper bound from the positivity of the ratio of moments:

$$M_T \leq 2347\ \text{MeV}, \quad f_T \leq 174\ \text{MeV}. \qquad (8.196)$$

We have used:

$$\Lambda = 340(28)\,\text{MeV}, \quad \langle \alpha_s G^2 \rangle = (6.49 \pm 0.35)10^{-2}\ \text{GeV}^4, \qquad (8.197)$$

and the parametrization of the $D = 8$ gluon condensates given in Eq. (8.179).

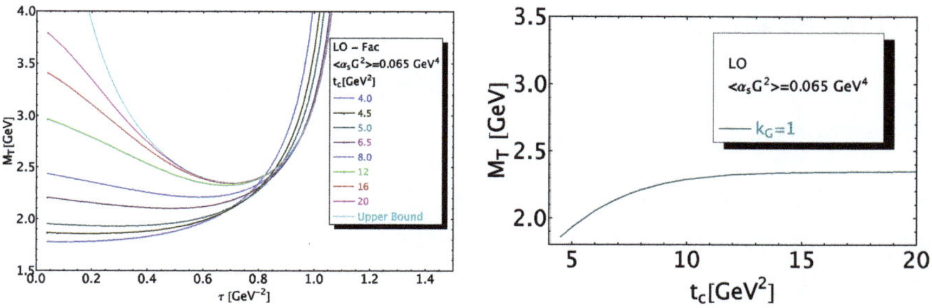

(a) (b)

Figure 8.21. Behaviour of the 2^{++} tensor di-gluonium mass from the ratio of moments $\mathcal{R}^c_{10}$ versus (a) τ for different values of t_c at LO; (b) t_c at the τ minimum assuming a factorization of the $D = 8$ condensates ($k_G = 1$).

We show in Fig. 8.22 the τ-behaviour of the coupling from the lowest moment at LO. One obtains the optimal value:

$$f_T = 156(9)_{t_c}(0.4)_{G^2}(22)_{M_T}\ \text{MeV} \quad \rightarrow \quad t_c \simeq \left(7.0^{+5.0}_{-0.5}\right)\ \text{GeV}^2. \qquad (8.198)$$

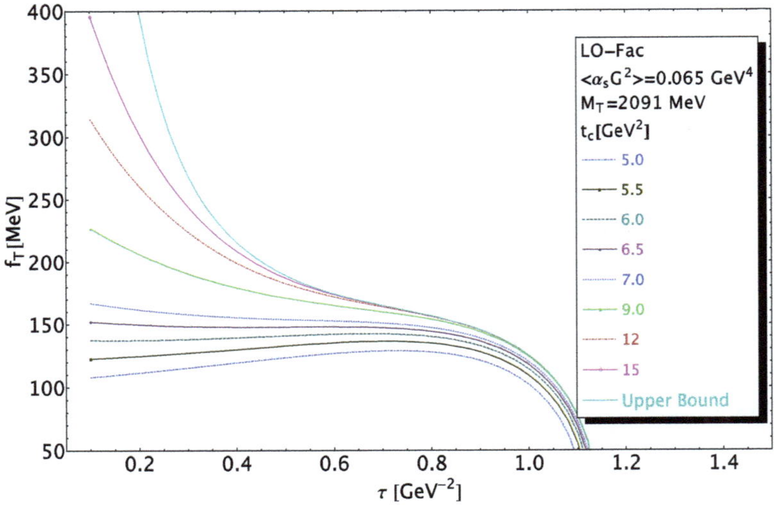

Figure 8.22. Behaviour of the 2^{++} tensor di-gluonium coupling from the moment $\mathcal{L}_0^c$ versus τ for different values of t_c at LO assuming a factorization of the $D = 8$ condensates ($k_G = 1$).

• Comparison with some other LSR results at LO

◇ These results agree within the errors with the ones:

$$f_T|_{LO} = 113(20)\ \text{MeV}, \quad M_T|_{LO} = 2.0(1)\ \text{GeV},$$
$$M_T|_{LO} \leq 2.7(4)\ \text{GeV}, \tag{8.199}$$

obtained at slightly low value of $t_c \simeq 5.5$ GeV^2 in Ref. [126]. The large error obtained here is due to the most conservative choice of the t_c-range.

◇ In Ref. [429], the result:

$$M_T = 1.86^{+014}_{-0.17}\ \text{GeV} \tag{8.200}$$

has been obtained to LO PT but including the NLO $\langle \alpha_s G^2 \rangle$ term and the LO constant term of the $\langle gG^3 \rangle$ condensates. Unfortunately, our results do not agree with the coefficients of these condensates. The difference of these coefficients may come from the different current used in Ref. [429].

◇ Result within instanton liquid model is about 1525 MeV [430], which is much lower than our above result.

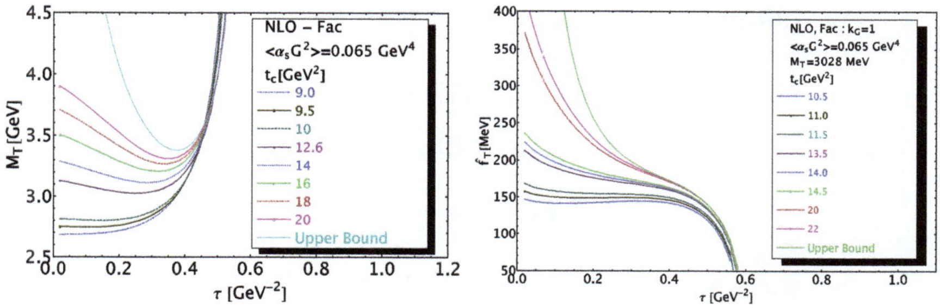

(a) (b)

Figure 8.23. Behaviour of the 2^{++} tensor di-gluonium (a) mass; (b) coupling at NLO from the ratio of moments $\mathcal{R}^c_{10}$ versus τ for different values of t_c assuming a factorization of the $D = 8$ condensates ($k_G = 1$).

• Mass and coupling at NLO assuming factorization

We introduce the renormalization group invariant (RGI) coupling $\hat{f}_T$ which is related to the running coupling $f_T(\nu)$ associated to the two-point correlator $\tilde{\psi}_T$ as:

$$f_T(\nu) = \frac{\hat{f}_T}{\left(\log \frac{\nu}{\Lambda}\right)^{\frac{\gamma_1}{-2\beta_1}}} \; : \; \gamma_1 = \frac{n_f}{3}. \qquad (8.201)$$

Doing a similar analysis in the LO case, we show the result in Fig. 8.23(a), assuming a factorization of the $D = 8$ condensates.

◇ We note that the optimal result is obtained at lower values of τ and higher values of t_c for the set (τ, t_c) equal to (0.2, 9.5) and (0.36, 20) (GeV^{-2}, GeV^2). We deduce to NLO:

$$M_T = 3028(429) \text{ MeV} \quad \multimap\!\!\to \; t_c \simeq \left(12.6^{+7.4}_{-3.1}\right) \text{ GeV}^2. \qquad (8.202)$$

◇ From Fig. 8.23(a), one can also deduce the optimal upper bound from the positivity of the ratio of moments. We obtain to NLO:

$$M_T \leq 3376(377) \text{ MeV}. \qquad (8.203)$$

◇ The analysis of the coupling is shown in Fig. 8.23(b). The curves present a minimum at 148 MeV and inflexion point at 193 MeV for the set (τ, t_c) equal to (0.2, 11) and (0.3, 22) (GeV^{-2}, GeV^2).

One obtains to NLO:

$$\hat{f}_T = 167(25)_{t_c}(1)_\tau(1.5)_\Lambda(1)_{G^2}(25)_{M_T}$$
$$(16)_{PT}(11)_{PT_{G2}}(0)_{OPE}\ \text{MeV},$$
$$= 167(40)\ \text{MeV} \quad \multimap\!\!\!\to t_c \simeq \left(14^{+8}_{-3}\right)\ \text{GeV}^2, \qquad (8.204)$$

and the optimal upper bound at the inflexion point $\tau \simeq 0.3$ GeV^{-2}:

$$\hat{f}_T \leq 186(1)_\Lambda(2)_{G^2}(32)_{M_T}(2)_{PT}(11)_{PT_{G2}}(0)_{OPE}\ \text{MeV}$$
$$\leq 186(34)\ \text{MeV}, \qquad (8.205)$$

where we have used the mass value in Eq. (8.202).

• Final results at NLO beyond factorization

For definiteness, we assume the conservative value:

$$k_G = 3 \pm 2, \qquad (8.206)$$

where the lowest one corresponds to factorization and the highest one to the same violation of factorization as the one of the four-quark condensate.

◇ Mass

In this way, we quote the final value:

$$M_T = 3188(291)_{t_c}(34)_\Lambda(47)_{G^2}(159)_{k_G}(157)_{PT}(162)_{PT_{G2}}(2)_{OPE}\ \text{MeV}$$
$$= 3188(405)\text{MeV} \quad \multimap\!\!\!\to t_c,$$
$$\simeq \left(14.1^{+7.9}_{-3.1}\right)\ \text{GeV}^2. \qquad (8.207)$$

The related absolute upper bound is:

$$M_T \leq 3580(17)_\Lambda(45)_{G^2}(168)_{k_G}(210)_{PT}(198)_{PT_{G2}}(8)_{OPE}\ \text{MeV},$$
$$\leq 3580(338)\ \text{MeV}. \qquad (8.208)$$

◇ Coupling

One obtains for the set of (τ, t_c): (0.12, 12) and (0.30, 22) in (GeV^{-2}, GeV^2) the values of 156.6 and 173.6 MeV which give:

$$\begin{aligned}\hat{f}_T &= 164(8)_{t_c}(3)_\Lambda(1.3)_{G^2}(3)_{k_G}(20)_{M_T},\\ &\quad (16)_{PT}(5)_{PT_{G2}}(0)_{OPE}\ \text{MeV},\\ &= 164(28)\ \text{MeV}, \quad \not\leftrightarrow t_c,\\ &\simeq \left(12.5^{+9.5}_{-0.5}\right)\ \text{GeV}^2, \end{aligned} \tag{8.209}$$

and the optimal upper bound at the inflexion point $\tau \simeq 0.3$ GeV^{-2}:

$$\begin{aligned}\hat{f}_T|_{NLO} &\leq 203(5)_\Lambda(2)_{G^2}(5)_{k_G}(28)_{M_T},\\ &\quad (19)_{PT}(11)_{PT_{G2}}(0)_{OPE}\ \text{MeV},\\ &\leq 203(36)\ \text{MeV}. \end{aligned} \tag{8.210}$$

- **Comments**

  ◇ Assuming factorization of the $D = 8$ condensate, we have seen that the inclusion of the PT α_s corrections has moved the mass from 2091(235) MeV to 3028(428) MeV (NLO) but only affects the coupling slightly.

  ◇ Our final result for the ground state mass is in line with the ones from some other approaches [431] and ADS/QCD [432], where its mass is expected to be above 2 GeV. It is slightly higher than recent lattice calculations in the range (2.27–2.67) GeV [433–435].

  ◇ However, it does not favour the interpretation of the observed $f_2(2010, 2300, 2340)$ states as a pure gluonia/glueball candidates (see, e.g., Ref. [436]). Moreover, we do not expect that an eventual meson–gluonium mixing will affect our result as this mixing is expected to be small ($\theta \simeq -10^0$) [125].

Chapter 9

Reviews of Author's Recent Works

In this chapter, I review some recent results on the determinations of some properties of hadrons ($\bar{Q}q$ heavy-light mesons decay constants and spectra, light 1^{-+} exotics, light scalar quarlonia, gluonia and molecule states)[1] and QCD parameters (light and heavy quark masses, condensates and α_s, topological and conformal charges and their slopes, proton spin) using QCD spectral sum rules (QSSR) approach. Some recent determinations of the Standard Model high-precision quantities such as the lepton anomalies $a_l \equiv (1/2)(g-2)_l$ and the hadronic contributions to the QED coupling $\Delta\alpha$ are also reviewed.

9.1 Light Quark Masses from Pseudoscalar Sum Rules

In Ref. [437], we improve our determinations of the sum of light quark masses discussed in the previous sections. A review of the status of the determinations of the light quark masses from QCD spectral sum rules prior to 2005 can be found in Ref. [358].

First, we use ratios of the inverse Laplace transform sum rules within stability criteria for the subtraction point μ in addition to the ones of the usual τ spectral sum rule variable and continuum

[1] A comprehensive review of the masses and couplings of heavy four-quark and molecule states is postponed in a future publication.

threshold t_c to extract the $\pi(1300)$ and $K(1460)$ decay constants to order α_s^4 of perturbative QCD by including power corrections up to dimension six condensates.

With these inputs, we extract, *in a model-independent way*, the sum of the renormalization group invariant (RGI) quark masses $(\hat{m}_u+\hat{m}_q) : q \equiv d, s$ and the corresponding running masses $(\overline{m}_u+\overline{m}_q)$ evaluated at 2 GeV. By giving the value of the ratio m_u/m_d, we deduce the running quark masses $\overline{m}_{u,d,s}$ and the scale-independent mass ratios: $2m_s/(m_u + m_d)$ and m_s/m_d.

Using the positivity of the spectral function, we also deduce new lower bounds on these running masses from the inverse Laplace transform sum rules. Our results are summarized in Table 9.3.

9.1.1 Abstract

Using ratios of the inverse Laplace transform sum rules within stability criteria for the subtraction point μ in addition to the ones of the usual τ spectral sum rule variable and continuum threshold t_c, we extract the $\pi(1300)$ and $K(1460)$ decay constants to order α_s^4 of perturbative QCD by including power corrections up to dimension six condensates, tachyonic gluon mass, instanton and finite width corrections. Using these inputs with enlarged generous errors, we extract, in model-independent and conservative ways, the sum of the scale-independent renormalization group invariant (RGI) quark masses $(\hat{m}_u + \hat{m}_q)$: $q \equiv d, s$ and the corresponding running masses $(\overline{m}_u + \overline{m}_q)$ evaluated at 2 GeV. By giving the value of the ratio mu/md, we deduce the running quark masses $\overline{m}_{u,d,s}$ and condensate $\langle\bar{u}u\rangle$ and the scale-independent mass ratios: $2m_s/(m_u + m_d)$ and m_s/m_d. Using the positivity of the QCD continuum contribution to the spectral function, we also deduce, from the inverse Laplace transform sum rules, for the first time to order α_s^4, new lower bounds on the RGI masses which are translated into the running masses at 2 GeV and into upper bounds on the running quark condensate $\langle\bar{u}u\rangle$. Our results summarized in Table 9.3 and compared with our previous results and with recent lattice averages suggest that precise phenomenological determinations of the sum of light quark masses require improved experimental measurements of the $\pi(1300)$ and $K(1460)$ hadronic widths and/or decay constants which are the dominant sources of errors in the analysis.

9.1.2 The pseudoscalar Laplace sum rule

- **The form of the sum rules**

We shall be concerned with the two-point correlator:

$$\psi_5^P(q^2) = i\int d^4x\ e^{iq\cdot x}\langle 0|TJ_5^P(x)J_5^P(0)^\dagger|0\rangle, \tag{9.1}$$

where $J_5^P(x)$ is the local pseudoscalar current:

$$J_5^P(x) \equiv (m_u + m_q)\bar{u}(i\gamma_5)q, \quad q = d,\ s;\ P = \pi,\ K. \tag{9.2}$$

The associated Laplace sum rules (LSR) $\mathcal{L}_5^P(\tau)$ and its ratio $\mathcal{R}_5^P(\tau)$ read [3][2]

$$\mathcal{L}_5^P(\tau,\mu) = \int_{(m_u+m_q)^2}^{t_c} dt\ e^{-t\tau}\frac{1}{\pi}\mathrm{Im}\psi_5^P(t,\mu), \tag{9.3}$$

$$\mathcal{R}_5^P(\tau,\mu) = \frac{\int_{(m_u+m_q)^2}^{t_c} dt\ t\ e^{-t\tau}\frac{1}{\pi}\mathrm{Im}\psi_5^P(t,\mu)}{\int_{(m_u+m_q)^2}^{t_c} dt\ e^{-t\tau}\frac{1}{\pi}\mathrm{Im}\psi_5^P(t,\mu)}, \tag{9.4}$$

where μ is the subtraction point which appears in the approximate QCD series. The ratio of sum rules $\mathcal{R}_5^P(\tau,\mu)$ is useful here for extracting the contribution of the radial excitation P' to the spectral function, while the Laplace sum rule $\mathcal{L}_5^P(\tau,\mu)$ will be used for determining the sum of light quark masses.

- **The QCD expression within the SVZ expansion**

As mentioned earlier, the perturbative expression of the two-point correlator $\psi_5^P(q^2)$ is known up to order α_s^4 from successive works [34, 438–441]. For convenience of the reader, we give the following numerical expression[3]:

---

[2] A quantum mechanics interpretation of these Laplace sum rules has been given in Ref. [70].

[3] In the following, we shall not expand the QCD expression $1/(1 + ka_s + \cdots + (np)\tau + \cdots)$ as: $1 - ka_s + k^2a_s^2 + \cdots - (np)\tau + (np)^2\tau^2 + \cdots$, but keep its non-expanded form.

$$\mathcal{L}_5^P(\tau) = \frac{3}{8\pi^2}(\overline{m}_u + \overline{m}_q)^2\tau^{-2}$$
$$\left[1 + \sum_{n=1,4} \delta_n^{(0)} a_s^n - 2m_q^2\tau\left(1 + \sum_{n=1,2} \delta_n^{(2)} a_s^n\right) + \tau^2\delta^{(4)} + \tau^3\delta^{(6)}\right], \tag{9.5}$$

where $\overline{m}_q$ is the running quark mass evaluated at the scale μ. From the analytic expression compiled in Ref. [442], we derive the following numerical PT corrections:

$$\begin{aligned}
\delta_1^{(0)} &= 4.82107 - 2l_\mu, \\
\delta_2^{(0)} &= 21.976 - 28.0729l_\mu + \frac{17}{4}l_\mu^2, \\
\delta_3^{(0)} &= 53.1386 - 677.987l_\mu + 102.82l_\mu^2 - \frac{221}{24}l_\mu^3, \\
\delta_4^{(0)} &= -31.6283 + 756701l_\mu + 1231.57l_\mu^2 - 321.968l_\mu^3 + \frac{7735}{384}l_\mu^4, \\
\delta_1^{(2)} &= 7.64213 - 4l_\mu, \\
\delta_2^{(2)} &= 51.0915 - 62.93l_\mu + \frac{25}{2}l_\mu^2,
\end{aligned} \tag{9.6}$$

where $a_s \equiv \alpha_s/\pi$; $l_\mu \equiv -\log(\tau\mu^2)$. The non-perturbative corrections are combinations of RGI quantities defined in Refs. [79, 113]:

$$\begin{aligned}
\overline{m_q\langle\bar{q}q\rangle} &= m_q\langle\bar{q}q\rangle + \frac{3}{7\pi^2}m_q^4\left(\frac{1}{a_s} - \frac{53}{24}\right), \\
\overline{\langle\alpha_s G^2\rangle} &= \langle\alpha_s G^2\rangle\left(1 + \frac{16}{9}a_s\right) - \frac{16}{9}\alpha_s\left(1 + \frac{91}{24}a_s\right)m_q\langle\bar{q}q\rangle \\
&\quad -\frac{1}{3\pi}\left(1 + \frac{4}{3}a_s\right)m_q^4.
\end{aligned} \tag{9.7}$$

In terms of these quantities, they read [6, 7, 34, 157, 443]:

$$\delta^{(4)} = \frac{4\pi^2}{3}\left(\delta_q^{(4)} + \delta_g^{(4)}\right), \tag{9.8}$$

with:

$$\delta_q^{(4)} = -2m_q\langle\bar{u}u\rangle\left[1+a_s\left(5.821-2l_\mu\right)\right] + \overline{m_q\langle\bar{q}q\rangle}$$
$$\times\left[1+a_s\left(5.266--2l_\mu\right)\right] - \frac{3}{7\pi^2}m_q^4\left(\frac{1}{a_s}+2.998-\frac{15}{4}l_\mu\right),$$
$$\delta_g^{(4)} = \frac{1}{4\pi}\overline{\langle\alpha_s G^2\rangle}\left[1+a_s\left(4.877-2l_\mu\right)\right]. \qquad (9.9)$$

The contribution of the $d=6$ condensate is:

$$\delta^{(6)} = -\frac{4\pi^2}{3}\left[m_q\langle\bar{u}Gu\rangle + \frac{32}{27}\pi\rho\alpha_s\left(\langle\bar{u}u\rangle^2+\langle\bar{q}q\rangle^2-9\langle\bar{u}u\rangle\langle\bar{q}q\rangle\right)\right], \qquad (9.10)$$

where $\langle\bar{u}Gu\rangle \equiv \langle\bar{u}(\lambda^a/2)G_a^{\mu\nu}\sigma_{\mu\nu}u\rangle \equiv M_0^2\langle\bar{u}u\rangle$ with $M_0^2 = (0.8\pm 0.2)$ GeV^2 [67, 297–301] is the quark-gluon mixed condensate and $\rho = (2.6\pm 0.6)$ [67, 115, 237, 297, 298] indicates the violation of the vacuum saturation assumption of the four-quark operators.

- **Some contributions beyond the SVZ expansion**

◇ **The tachyonic gluon mass contribution**

The tachyonic gluon mass λ of dimension two has been introduced in Refs. [232,235] for parametrizing the uncalculated higher order terms of the PT series [234]. It appears naturally in most holographic QCD models [228]. Its contribution reads [235]:

$$\mathcal{L}_5^P(\tau)|^{\text{tach}} = -\frac{3}{2\pi^2}(\overline{m}_u+\overline{m}_q)^2 a_s\lambda^2\tau^{-1}. \qquad (9.11)$$

Its value has been estimated from e^+e^- [236, 237] and τ-decay [360] data:

$$a_s\lambda^2 = -(0.07\pm 0.03)\ \text{GeV}^2. \qquad (9.12)$$

In Ref. [234], it has been demonstrated that this contribution is dual to non-calculated terms of the PT series where, up to now, the coeffecients behave like a geometric series. Then, the more one adds terms in the PT series, the size of the tachyonic gluon mass contribution decreases.

◇ The instanton contribution

The inclusion of this contribution into the operator product expansion (OPE) is not clear and controversial [130, 194, 446]. In addition, an analogous contribution might lead to some contradiction to the OPE in the scalar channel [240]. Therefore, we shall consider the sum rule including the instanton contribution as an alternative approach. For our purpose, we parametrize this contribution as in Refs. [130, 194], where its corresponding contribution to the Laplace sum rule reads:

$$\mathcal{L}_5^P(\tau)|^{\text{inst}} = \frac{3}{8\pi^2}(m_u + m_q)^2 \tau^{-3} \rho_c^2 e^{-r_c} \big[K_0(r_c) + K_1(r_c)\big], \quad (9.13)$$

where K_i is the Bessel–Mac-Donald function, $r_c \equiv \rho_c^2/(2\tau)$ and $\rho_c = (1.89 \pm 0.11)$ GeV^{-1} [290, 291] is the instanton radius.

Table 9.1. Input parameters in addition to the ones given in Table 7.10. Some other predictions and related references can be found in Ref. [33]. The error on $\Gamma_{K'}$ is a guessed estimate.

Parameters	Values	References
$\Lambda(n_f = 3)$	(353 ± 15) MeV	[33, 288, 289, 444]
$\hat{m}_s$	(0.114 ± 0.021) GeV	[6, 7, 123, 289, 294, 295]
$\kappa \equiv \langle \bar{s}s\rangle / \langle \bar{u}u\rangle$	(0.74 ± 0.06)	[445]
$a_s\lambda^2$	$(7 \pm 3) \times 10^{-2}$ GeV^2	[236, 237, 360]
ρ_c	(1.89 ± 0.11) GeV^{-1}	[290, 291]
$\Gamma_{\pi'}$	(0.6 ± 0.2) GeV	[33]
$\Gamma_{K'}$	(0.25 ± 0.05) GeV	[33]

9.1.3 The QCD input parameters

We shall work in the analysis with the input parameters given in Table 9.1. $\hat{m}_q$ and $\hat{\mu}_q$ are RGI invariant mass and condensates which are related to the corresponding running parameters as [29]:

$$\overline{m}_q(\tau) = \frac{\hat{m}_q}{(-\log\sqrt{\tau}\Lambda)^{2/-\beta_1}}(1 + \rho_m),$$

$$\langle \bar{q}q\rangle(\tau) = -\hat{\mu}_q^3 \left(-\log\sqrt{\tau}\Lambda\right)^{2/-\beta_1} / (1 + \rho_m), \quad (9.14)$$

where $\beta_1 = -(1/2)(11 - 2n_f/3)$ is the first coefficient of the QCD β-function for n_f-flavours. ρ_m is the QCD correction which reads to N4LO accuracy for $n_f = 3$ [6, 156]:

$$\rho_m = 0.8951a_s + 1.3715a_s^2 + 0.1478a_s^3, \tag{9.15}$$

where $a_s \equiv \alpha_s/\pi$ is the QCD running coupling. The value of $\hat{m}_s$ quoted here will serve as an initial value for the m_s corrections in the PT expression of the kaon correlator. It will be re-extracted by iteration in the estimate of m_s. The value of the gluon condensate used here comes from recent charmonium sum rules [290, 291].

9.1.4 A Laplace sum rule estimate of the decay constant $f_{\pi'}$

• The spectral function

We shall parametrize the spectral function as:

$$\frac{1}{\pi}\,\mathrm{Im}\psi_5^{\pi}(t) \simeq \sum_{\pi,\pi'} 2f_P^2 m_P^4 \delta(t - m_P^2) + \text{“QCD cont.”}\theta(t - t_c), \tag{9.16}$$

where the higher states (π'',...) contributions are smeared by the "QCD continuum" coming from the discontinuity of the QCD diagrams and starting from a constant threshold t_c. f_P is the well-known decay constant:

$$\langle 0|J_5^P(x)|P\rangle = \sqrt{2}m_P^2 f_P, \tag{9.17}$$

normalized here as $f_\pi = (93.2 \pm 0.2)$ MeV and $f_K \simeq (1.20 \pm 0.01)f_\pi$ [449]. We improve the $\pi' \equiv \pi(1300)$ contribution by taking into account the finite width correction by replacing the delta function with a Breit–Wigner shape:

$$\pi\delta(t - m_{\pi'}^2) \to \mathrm{BW}(t) = \frac{m_{\pi'}\Gamma_{\pi'}}{(t - m_{\pi'}^2)^2 + m_{\pi'}^2\Gamma_{\pi'}^2}. \tag{9.18}$$

Defined in this way, the π' can be considered as an "effective resonance" parametrizing the higher state contributions not smeared by the QCD continuum and may take into account some possible interference between the $\pi(1300)$ and $\pi(1800)$ contributions.

• The LSR analysis

In order to eliminate the unknown value of the sum of light quark masses $(m_u + m_d)$, it is convenient to work with the ratio of finite energy Laplace sum rules $\mathcal{R}_5^{\pi}(\tau, \mu)$ defined in Eq. (9.4). In so doing, we define the quantity:

$$r_\pi \equiv \frac{M_{\pi'}^4 f_{\pi'}^2}{m_\pi^4 f_\pi^2} = \frac{\mathcal{R}_5^{\pi}|_{\text{qcd}} - m_\pi^2}{BWI_1 - \mathcal{R}_5^{\pi}|_{\text{qcd}} BWI_0} e^{-m_\pi^2 \tau}, \tag{9.19}$$

which quantifies the relative weight between the π' and the pion contribution into the spectral function. $\mathcal{R}_5^{\pi}|_{\text{qcd}}$ is the QCD expression of the FESR in Eq. (9.4), where we have parametrised the spectral function by a step function corresponding to the perturbative expression for massless quarks from the threshold t_c. $\text{BW}I_n$ is the Breit–Wigner integral:

$$\text{BW}I_n \equiv \frac{1}{\pi} \int_{9m_\pi^2}^{t_c} dt\ t^n e^{t\tau} \text{BW}(t): \quad n = 0, 1. \tag{9.20}$$

With the set of input parameters in Table 9.1, we show in Fig. 9.1(a) the τ-behaviour of r_π at a given value of $\mu = 1.53$ GeV. We extract the optimal result at the value of $t_c = 2$ GeV^2 where both a minimum in the change of t_c is obtained and where a large range of τ-stability is reached. In Fig. 9.1(b), we study the influence of the choice of μ on this result. Our final optimal result corresponds to the values of $\tau \simeq (0.50–0.55)$ GeV^{-2} and the inflexion point $\mu = 1.53$ GeV at which we deduce:

$$\begin{aligned} r_\pi^{\text{SVZ}} &= 4.63(10)_\Lambda(1)_{\lambda^2}(19)_{\bar{u}u}(14)_{G^2}(0)_{\bar{u}Gu}(9)_\rho(181)_{\Gamma_\pi} \\ &= 4.63 \pm 1.83, \end{aligned} \tag{9.21}$$

where the main error comes from the experimental width of the $\pi(1300)$ which needs to be improved.

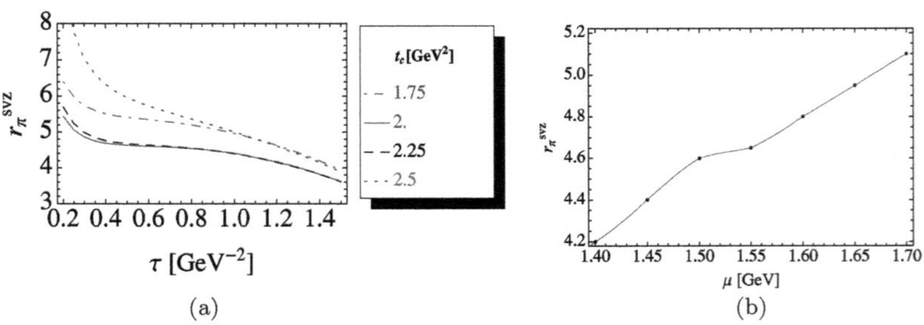

Figure 9.1. (a) τ-behaviour of r_π for $\mu = 1.53$ GeV and for different values of t_c within the SVZ expansion. (b) μ-behaviour of the optimal value of r_π deduced from (a).

• Convergence of the PT series and the OPE within SVZ

Here, we study the convergence of the PT series in the more general case where μ is arbitrary and not necessarily correlated to the value of τ. The particular case $\mu = \tau^{-1/2}$ often used in the literature will also be discussed in the following paragraph.

◇ We study in Fig. 9.2(a), the convergence of the PT QCD series at the value of the subtraction scale $\mu = 1.53$ GeV where a μ stability is obtained (inflexion point in Fig. 9.1(b)) for the ratio $\mathcal{R}_5^{\pi}(\tau, \mu)$ used for the estimate of r_π (lower family of curves). One can note that for $\tau \approx (0.5–0.6)$ GeV^{-2} where a τ stability (inflexion point) and t_c stability (minimum in t_c) are obtained (Fig. 9.1(a)), the $\alpha_s, \alpha_s^2, \alpha_s^3$ and α_s^4 effects are, respectively −7, −9, −9.5, and −7.2% of the preceding PT series: LO, up to NLO, up to N2LO and up to N3LO. The convergence of the PT series is not quite good, but each correction to r_π is reasonably small. The convergence of the PT series is much better than in the case (often used in the literature): $\mu = \tau^{-1/2}$ where the τ stability is obtained at larger value of $\tau = 1.9$ GeV^{-2} (Fig. 9.5(a)) which we shall discuss later.

◇ We show in Fig. 9.2(b) the convergence of the power corrections for r_π (lower family of curves). We see that the $d = 2,\ 4,\ 6$ dimension operator effects are −1.8, −6.8 and −4.6% of the preceding sum of contributions, indicating a good convergence and relatively small corrections.

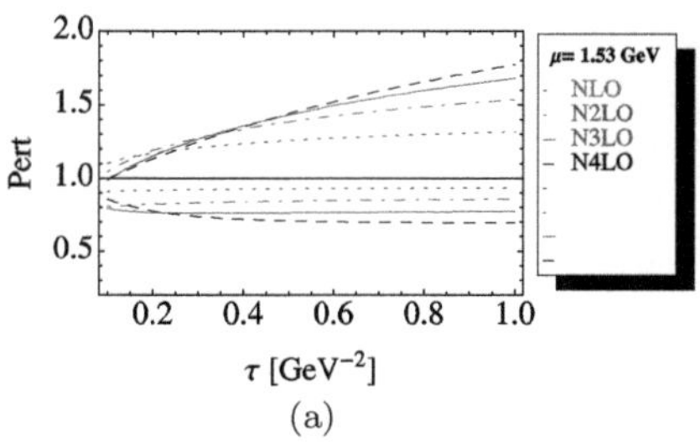

(a)

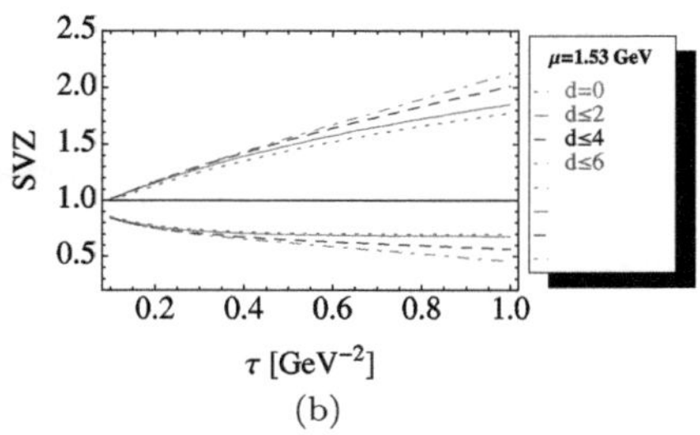

(b)

Figure 9.2. (a) τ-behaviour of the PT series of $\sqrt{\mathcal{L}_5^{\pi}}(\tau, \mu)$ (upper group of curves) and of $\mathcal{R}_5^{\pi}(\tau, \mu)$ (lower group of curves) appropriately normalised to 1 for $\tau = 0$ and using $\mu = 1.53$ GeV. (b) the same as (a) but for the power corrections within the SVZ expansion.

- **Contribution beyond the SVZ expansion**

◇ Tachyonic gluon mass to r_π

Using the expression in Eq. (9.11), one finds that it decreases the value of r_π by about 0.1 which is relatively negligible. By duality with the unknown higher order terms of the PT series [234], this small effect of the tachyonic gluon mass is in line with the good convergence of the PT series obtained previously. It may also indicate that the QCD long PT series can only bring small systematic uncertainties.

◇ Duality Violation to r_π

Some eventual additional contribution from duality violation (DV) [447] may also be considered. However, in the best measured case of V–A spectral function entering τ-decay, the parameters of the model are weakly constrained rendering the model less predictive. An attempt to estimate such a contribution in this channel shows that the effect is at the level of 0.2% of the LO perturbative contribution [289, 448] which is about the same strength as the one of the tachyonic gluon. In addition, the two contributions tend to compensate each other [289]. Extrapolating this result to the case of pseudoscalar channel, we then expect that the DV contribution is also negligible like the one of tachyonic gluon.

◇ r_π from instanton sum rule at arbitrary μ

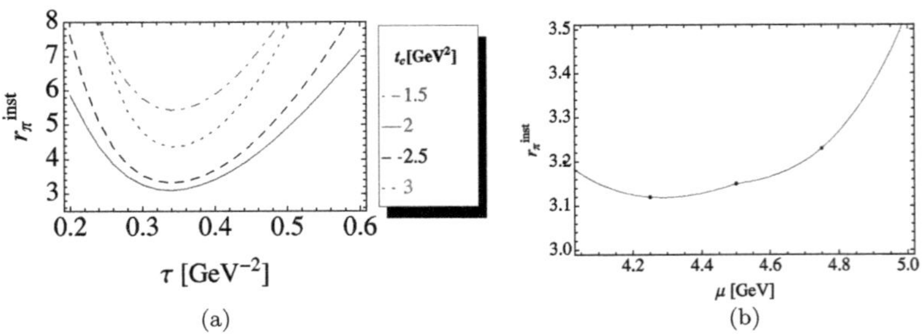

Figure 9.3. (a) τ-behaviour of r_π for $\mu = 4.25$ GeV and for different values of t_c from the instanton sum rule. (b) μ-behaviour of the optimal value of r_π deduced from (a).

We include the instanton contribution into the OPE using the expression given in Eq. (9.13). The variations of r_π versus τ and t_c for different values of μ are similar to the one in Fig. 9.3(a) which are

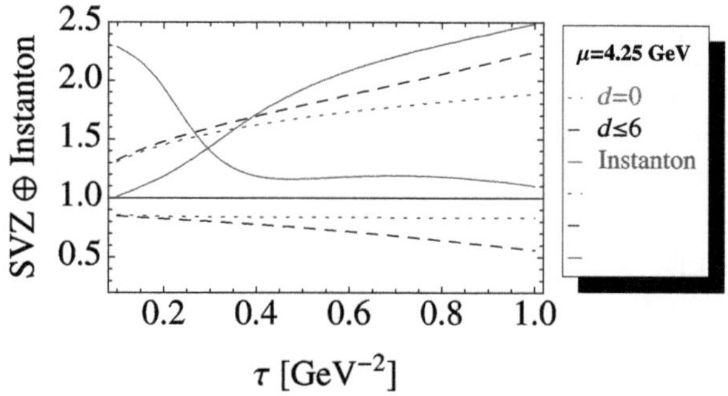

Figure 9.4. τ-behaviour of the Npert corrections including the instanton contribution to $\sqrt{\mathcal{L}_5^\pi}(\tau,\mu)$ (upper group of curves) and of $\mathcal{R}_5^\pi(\tau,\mu)$ (lower group of curves) appropriately normalised to 1 for $\tau = 0$ and using $\mu = 4.25$ GeV.

also qualitatively similar to the ones discussed without the instanton contribution. The optimal result is obtained for $\tau \approx (0.3\text{--}0.4)$ GeV^{-2} and $t_c \simeq 2$ GeV^2 (minimum in t_c). If we take $\mu = 1.53$ GeV where the optimal result in Eq. (9.21) is obtained, one would get $r_\pi^{\text{inst}} \simeq 1.62$. At this value of μ and for $\tau \simeq 0.3$ GeV^{-2}, where the optimal result is obtained, the sum of the $d \leq 6$ condensates is –6% of the PT contributions, while the one due to the instanton is +80% of the $PT \oplus d \leq 6$ contributions which are relatively large corrections.[4] One can improve the convergence of the OPE by working with larger value of μ. Then, we study in Fig. 9.3(b) the μ-dependence of r_π where a minimum in τ (respectively t_c) is obtained for various values of μ at (0.3–0.4) GeV^{-2} (respectively 2 GeV^2) like in Fig. 9.3(a). A μ-stability is obtained at about (4–4.75) GeV with a minimum at about μ= 4.25 GeV. At this scale, we show in Fig. 9.4 the convergence of the OPE for the ratio of moments (lowest families of curves). The instanton contribution is about 50% of the $PT \oplus d \leq 6$ contributions which is a more reasonable correction. Then, we deduce the optimal value of r_π at this value μ=4.25 GeV:

$$\begin{aligned} r_\pi^{\text{inst}} &= 3.10(2)_\Lambda(1)_{\lambda^2}(1)_{\bar{u}u}(3)_{G^2}(0)_{\bar{u}Gu}(1)_\rho(35)_{\rho_c}(114)_{\Gamma_\pi} \\ &= 3.10 \pm 1.19, \end{aligned} \tag{9.22}$$

which is in better agreement with the one obtained by using the SVZ expansion.

[4]The shape of the curves for $\mu = 1.53$ GeV is similar to the one in Fig. 9.3(b).

• r_π from the Laplace sum rule at $\mu = \tau^{-1/2}$

We complete your analysis in the case where the subtraction constant μ is equal to the sum rule variable $1/\sqrt{\tau}$. This case is interesting as it does not possess the $\text{Log}^n \mu^2\tau$ terms appearing in the PT series which have large coefficients and which are now absorbed into the running of $\alpha_s(\tau)$ from the renormalization group equation. This case has been largely used in the literature (for reviews, see, e.g., Refs. [6, 7, 41]). The analysis is very similar to the previous case. In Fig. 9.5, we show the τ-behaviour of the results in the case of the SVZ expansion and SVZ $\oplus$ instanton contribution where in both cases a minimum in t_c is obtained at $t_c = 2$ GeV^2. We obtain:

$$
\begin{aligned}
r_\pi^{\text{svz}} &= 5.70(88)_\Lambda(11)_{\lambda^2}(86)_{\bar{u}u}(63)_{G^2}(2)_{\bar{u}Gu}(113)_\rho(90)_{\Gamma_\pi} \\
&= 5.70 \pm 2.01 \qquad (9.23)
\end{aligned}
$$

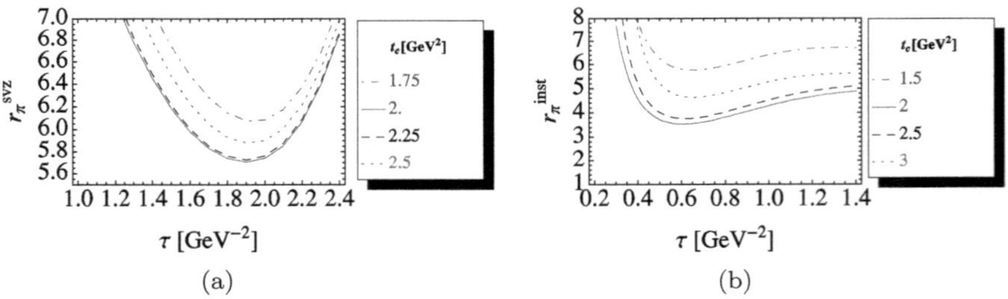

Figure 9.5. (a) τ-behaviour of r_π for the case $\mu = \tau^{-1/2}$ in the case of the SVZ expansion; (b) the same as in (a) but for SVZ $\oplus$ instanton contribution.

and:

$$
\begin{aligned}
r_\pi^{\text{inst}} &= 3.54(80)_\Lambda(1)_{\lambda^2}(6)_{\bar{u}u}(11)_{G^2}(0)_{\bar{u}Gu}(8)_\rho(119)_{\Gamma_\pi}(30)_{\rho_c} \\
&= 3.54 \pm 1.48. \qquad (9.24)
\end{aligned}
$$

• Final results

One can remark a nice agreement within the errors between the different results in Eqs. (9.21)–(9.24). We consider as a final estimate the means of these different results which correspond to:

$$
\begin{aligned}
\langle r_\pi^{\text{svz}} \rangle &= 5.12 \pm 1.36 \longrightarrow \frac{f_{\pi'}}{f_\pi} = (2.58 \pm 0.35)\,10^{-2}, \\
\langle r_\pi^{\text{inst}} \rangle &= 3.27 \pm 0.95 \longrightarrow \frac{f_{\pi'}}{f_\pi} = (2.06 \pm 0.30)\,10^{-2}, \qquad (9.25)
\end{aligned}
$$

where we have separated the determinations from the SVZ and SVZ $\oplus$ instanton sum rules.

• Comparison with some existing predictions

In Fig. 9.6, we compare the above two results, with the existing ones in the current literature, NPT83 [91], SN02 [6], BPR95 [450], KM02 [451] for the quantity:

$$L_\pi(\tau) \equiv r_\pi BWI_0, \tag{9.26}$$

involved in the Laplace sum rule estimate of (m_u+m_d) which we shall discuss in the following sections. Here, BWI_0 defined in Eq. (9.20) is the integrated spectral function entering into the lowest moment Laplace sum rule $\mathcal{L}_5^\pi(\tau)$.

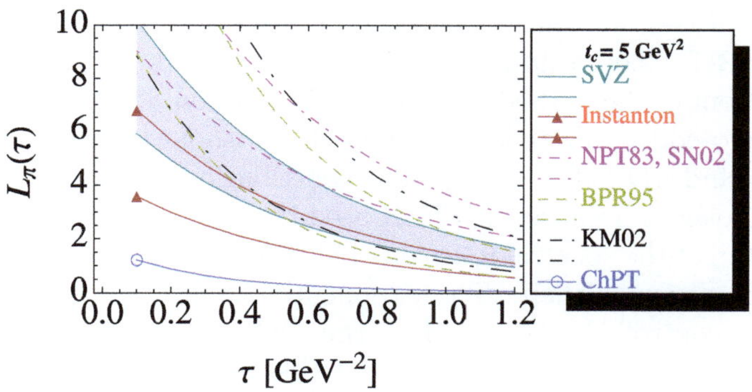

Figure 9.6. Comparison of some other determinations of r_π for a given value of $t_c = 5$ GeV^2 which corresponds to the optimal value of $(m_u + m_d)$. The blue continuous line with a circle is the ChPT prediction without a resonance. The results of NPT83 [91] and SN02 [6] are within a narrow width approximation. The errors due to the experimental width of $\pi(1300)$ have been introduced in the result of BPR95 [450].

For this comparison, we have used the following:

◇ $r_\pi = (9.5 \pm 2.5)$ and consistently the NWA for the results in NPT83 and SN02 from Ref. [6] (see also Ref. [439]).

◇ For KM02, we add coherently the $\pi(1300)$ and $\pi(1800)$ contributions which may be an overestimate as they may have a destructive interference like in Ref. [450]. We use the decay constants $f_\pi(1300) = (2.2 \pm 0.57)$ MeV and $f_\pi(1800) = (1.36 \pm 0.21)$

MeV obtained in Ref. [451] and consistently a Breit–Wigner parametrization of the spectral function. We have not introduced the (large) error due to the width of the $\pi(1300)$ because it is not clear if this error is already included in the quoted value of the decay constants. Therefore, we may have underestimated the errors for this comparison.

- ◇ For BPR95, we add, into their parametrization, the error due to the $\pi(1300)$ width which is not included in their original work.
- ◇ The CHPT parametrization from Ref. [450] without any resonance is also given in Fig. 9.6.

One can see that there is (within the errors) a complete agreement between the different determinations though the previous results tend to slightly overestimate the value of r_π at smaller values of τ which would imply a slightly larger value of $(m_u + m_d)$. The results obtained in Refs. [450, 451] are model-dependent as they depend on the way of treating the $\pi(1800)$ contribution into the spectral function. In Ref. [450], a destructive interference of about 10% suppression in front of the $\pi(1800)$ propagator is adopted, while in Ref. [451], it is supposed that there is a coherent sum without interference of the $\pi(1300)$ and $\pi(1800)$. The $\pi(1800)$ contribution is relevant in these two approaches because the FESR used there is sensitive to the high-energy tail of the spectral function. This is not the case of the Laplace sum rule discussed here. The exponential weight safely suppresses this less controlled region, while the QCD continuum introduced above the $\pi(1300)$ mass is expected (like in some other channels) to smear the contributions of higher mass states, namely, the one of $\pi(1800)$.

From the previous comparison, we note the following:

- ◇ The prediction from the SVZ expansion agrees (within the errors) with all existing predictions shown in Fig. 9.6, while the one including the instanton tends to underestimate the $\pi(1300)$ contribution to the spectral function.
- ◇ The tachyonic gluon mass contribution decreases r_π by about 0.1, which is negligible.
- ◇ The DV contribution is also negligible.

Hereafter, we shall show only present the detailed estimates within the standard SVZ expansion. The ones due to tachyonic gluon mass and duality violation will be neglected. The one due to instanton can be found in detail in the original paper [437].The final results due to this contribution will only be given for a comparison.

9.1.5 Estimate of $(m_u + m_d)$ within the SVZ expansion

• **The Laplace sum rule analysis at $\mu = 1.53$ GeV**

We find it convenient to extract the RGI scale-independent mass defined in Eq. (9.14):

$$\hat{m}_{ud} \equiv \frac{1}{2}(\hat{m}_u + \hat{m}_d) \tag{9.27}$$

from the Laplace sum rule $\mathcal{L}_5^{\pi}(\mu, \tau)$ in Eq. (9.3). The QCD expression of $\mathcal{L}_5^{\pi}(\mu, \tau)$ is given in Eq. (9.5). We shall use into the spectral function, parametrized as in Eq. (9.65), the value of r_π obtained in Eq. (9.25). In Fig. 9.7, we study the τ-dependence of $\hat{m}_{ud}$ for different values of t_c and for a given value of the subtraction point $\mu = 1.53$ GeV, where the optimal value of r_π has been obtained. One can find from this figure that the largest τ-stability range is obtained for $t_c = (5 \pm 0.5)$ GeV^2. Using the values of the parameters in Table 9.1, we extract the optimal value of the sum of the RGI u, d quark masses at the τ- minimum of 0.5 GeV^{-2}:

$$\begin{aligned}\hat{m}_{ud}^{\rm SVZ} &= 4.47(17)_{t_c}(12)_\Lambda(6)_{\lambda^2}(1)_{\bar{u}u}(4)_{G^2}(0)_{\bar{u}Gu}(1)_\rho(13)_{\Gamma_\pi}(39)_{r_\pi}\ \text{MeV}\\ &= (4.47 \pm 0.47)\ \text{MeV},\end{aligned} \tag{9.28}$$

where the error due to t_c corresponds to the localisation of the minimum, while moving τ around the minimum leads to a negligible uncertainty.

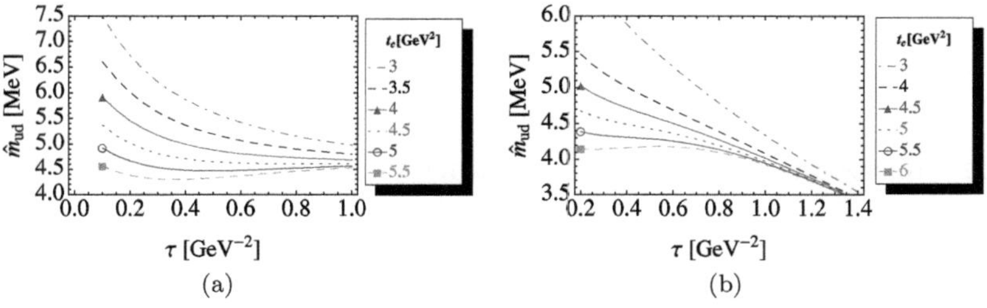

Figure 9.7. τ- and t_c-dependence of $\hat{m}_{ud}$ from the Laplace sum in Eq. (9.3): (a) at the subtraction scale $\mu = 1.53$ GeV; (b) for $\mu = \tau^{-1/2}$.

- **Convergence of the QCD series**

Like in the case of r_π, we study the convergence of the PT series in the more general case where μ is arbitrary and not necessarily correlated to the value of τ. The particular case $\mu = \tau^{-1/2}$ often used in the literature is also discussed.

- We shall study the different contributions of the truncated PT series to $\sqrt{\mathcal{L}_5^\pi}(\mu,\tau)$ for $\mu = 1.53$ GeV where a μ stability is obtained (inflexion point in Fig. 9.1(b)). The relative strengths of each truncated contribution are given in Fig. 9.2(a) (upper family of curves). One can deduce that for $\tau \simeq 0.4$ GeV^{-2} where the τ-stability is obtained (Fig. 9.7), the $\alpha_s, \alpha_s^2, \alpha_s^3$ and α_s^4 effects are, respectively, +27, +15.5, +4.6 and +2% of the preceding PT series: LO, NLO, N2LO and N3LO, which indicate a much better convergence than the case (often used in the literature): $\mu = \tau^{-1/2}$ where the τ-stability is obtained at larger value of $\tau = 0.6$ GeV^{-2} (Fig. 9.7).
- We show in Fig. 9.2(b) the convergence of the power corrections for $\sqrt{\mathcal{L}_5^\pi}(\mu,\tau)$ for $\mu = 1.53$ GeV (upper family of curves). We see that the $d = 2, 4, 6$ contributions are +3.4, +3.1 and +1.3% of the preceding sum of contributions (PT, $PT \oplus d = 2$, $PT \oplus d = 2{+}4$), indicating a good convergence and a relatively small correction.

- **$\hat{m}_{ud}$ from the Laplace sum rule at $\mu = \tau^{-1/2}$**

As mentioned previously, this sum rule has been widely used in the current literature for extracting m_{ud}. We shall use it here as another method for determining m_{ud}. The analysis is similar to the one for arbitrary μ. We show the τ-dependence for different t_c in Fig. 9.7, where one observes τ-and t_c stabilities for $\tau = 0.6$ GeV$^{-1/2}$ and $t_c = (5.5 \pm 0.5)$ GeV^2. At such values, we extract the optimal result:

$$\begin{aligned}\hat{m}_{ud} &= 4.30(12)_{t_c}(11)_\Lambda(5)_{\lambda^2}(1)_{\bar{u}u}(3)_{G^2}(0)_{\bar{u}Gu}(1)_\rho(14)_{\Gamma_\pi}(39)_{r_\pi}\ \text{MeV}\\ &= (4.30 \pm 0.45)\ \text{MeV}.\end{aligned} \tag{9.29}$$

One can note that the couple of values (τ, t_c) differ for r_π and m_{ud}.

- **Tachyonic gluon mass contribution to $\hat{m}_{ud}$**

If we do not include the tachyonic gluon mass into the SVZ expansion, the value of $\hat{m}_{ud}$ obtained in Eqs. (9.28) and (9.29) would increase by

0.14 MeV which is relatively negligible, confirming again the good convergence of the PT series if the duality between the tachyonic gluon mass and the not yet calculated higher order PT corrections are used [234].

• Final estimate of $\hat{m}_{ud}$ within the SVZ expansion

We consider, as a *final estimate* of $\hat{m}_{ud}$ within the SVZ expansion, the mean value of the results in Eqs. (9.28) and (9.29):

$$\langle \hat{m}_{ud}^{\rm SVZ} \rangle = (4.38 \pm 0.33)\ {\rm MeV} \quad \longrightarrow \quad \overline{m}_{ud}^{\rm SVZ} = 3.77(28)\ {\rm MeV}, \quad (9.30)$$

where the error is the mean of the sum of quadratic errors (the overlap of the two determinations leads to about the same value) from which one deduces the running masses at order α_s^4 evaluated at 2 GeV.

9.1.6 m_{ud} from the instanton Laplace sum rules

For optimizing the instanton contribution, we work at the same subtraction point $\mu = 4.25$ GeV where $r_\pi^{\rm inst}$ has been obtained. We repeat the previous analysis by taking into account the instanton contribution. Its contribution to $\sqrt{\mathcal{L}_5^\pi}(\mu,\tau)$ compared to the OPE up to $d = 6$ condensates is shown in Fig. 9.4 (upper family of curves) for $\mu = 4.25$ GeV where the estimate of r_π is optimized in this case (see Fig. 9.3). The instanton contribution is about $+14\%$ of the perturbative $\oplus\ d \leq 6$ condensate ones at $\tau = 0.4$ GeV^{-2} where the instanton sum rule is optimized. The behaviour of $\hat{m}_{ud}$ versus τ and for different values of t_c is given in Fig. 9.8. t_c-stability is almost reached for $t_c \simeq (5.0 \pm 0.5)$ GeV^2 and for $\tau \simeq 0.7$ GeV^{-2}. We deduce, respectively, from the sum rule subtracted at $\mu = 4.25$ GeV and $\mu = \tau^{-1/2}$:

$$\begin{aligned}
\hat{m}_{ud}^{\rm inst}|_{4.25} &= 3.82(14)_{t_c}(6)_\Lambda(2)_{\lambda^2}(0)_{\bar{u}u}(1)_{G^2}(0)_{\bar{u}Gu} \\
&\quad \times (0)_\rho(8)_{\rho_c}(10)_{\Gamma_\pi}(30)_{r_\pi}\ {\rm MeV}, \\
&= (3.82 \pm 0.36)\ {\rm MeV}, \\
\hat{m}_{ud}^{\rm inst}|_{\tau^{-1/2}} &= 3.04(6)_{t_c}(9)_\Lambda(3)_{\lambda^2}(1)_{\bar{u}u}(3)_{G^2}(0)_{\bar{u}Gu} \\
&\quad \times (0)_\rho(3)_{\rho_c}(7)_{\Gamma_\pi}(19)_{r_\pi}\ {\rm MeV}, \\
&= (3.04 \pm 0.24)\ {\rm MeV}, \qquad (9.31)
\end{aligned}$$

from which we deduce the mean value:

$$\langle \hat{m}_{ud}^{\rm inst} \rangle \simeq (3.28 \pm 0.22)\ {\rm MeV} \qquad \longrightarrow \qquad \overline{m}_{ud}^{\rm inst} = 2.82(18)\ {\rm MeV}, \tag{9.32}$$

which we consider as a determination of $\hat{m}_{ud}$ if the SVZ $\oplus$ instanton sum rules are used. This value is lower than the one within the standard SVZ expansion. We have not taken the mean value of the two results taking into account the controversial contribution of the instanton into the pseudoscalar sum rule [194, 240, 258, 446].

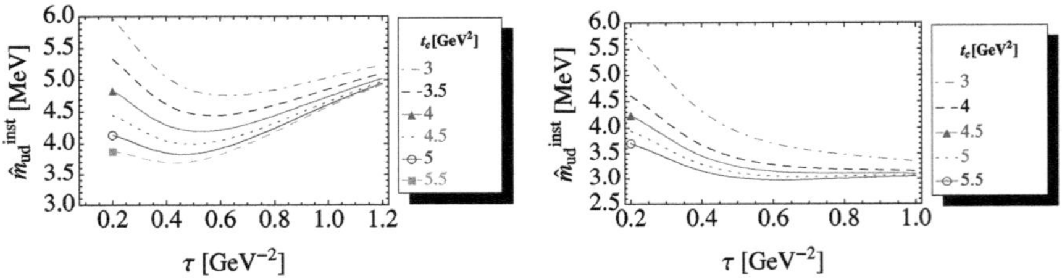

(a) (b)

Figure 9.8. (a) τ-behaviour of m_{ud} in the case $\mu = 4.25$ GeV; (b) the same as in (a) but in the case $\mu = \tau^{-1/2}$.

Using the mean of the range of different results quoted in PDG13 [33] for the ratio:

$$\frac{m_u}{m_d} = 0.50(3), \tag{9.33}$$

which (*a priori*) does not favour the solution $m_u = 0$, one can deduce the value of the u and d running quark masses at 2 GeV in units of MeV:

$$\overline{m}_u^{\rm svz} = 2.51(21), \qquad \overline{m}_u^{\rm inst} = 1.88(14),$$
$$\overline{m}_d^{\rm svz} = 5.02(42), \qquad \overline{m}_d^{\rm inst} = 3.76(28), \tag{9.34}$$

where one can note that the sum rule within the SVZ $\oplus$ instanton contribution tends to give lower values of the light quark masses. We summarize our results in Table 9.3.

9.1.7 The kaon channel

• Laplace sum rule estimate of $f_{K'}$

Using the same method as in the case of π', we shall estimate the $K' \equiv K(1460)$ decay constant through:

$$r_K \equiv \frac{M_{K'}^4 f_{K'}^2}{m_K^4 f_K^2}. \tag{9.35}$$

◇ r_K within the SVZ expansion at arbitrary μ

We show its τ behaviour for a given μ in Fig. 9.9(a) and its μ behaviour in Fig. 9.9(b) at the optimal values of τ. The analysis

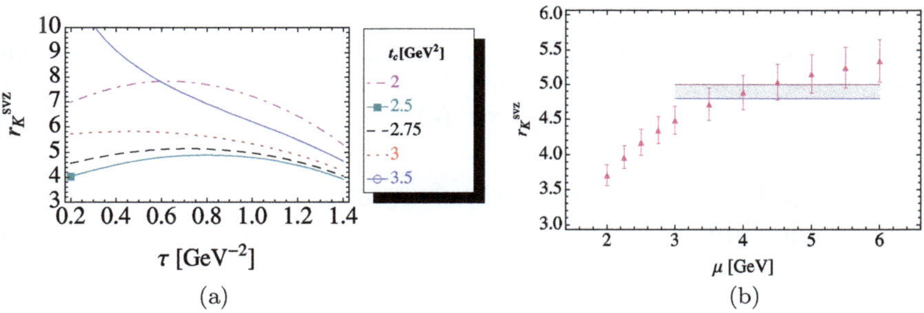

Figure 9.9. (a) τ-behaviour of r_K for a given value $\mu = 4$ GeV of subtraction point and for different values of t_c. (b) μ-behaviour of the optimal value of r_K deduced from (a). The error corresponds to $t_c = (2.5 \pm 0.25)$ GeV^2. The dashed region is the mean value obtained from $\mu = (3–6)$ GeV, where r_K is almost stable.

for $\mu = \tau^{-1/2}$ shown in Ref. [437] is not conclusive. Our final result from arbitrary μ is:

$$\begin{aligned} r_K^{\text{svz}} &= 4.79(8)_\mu(2)_\Lambda(4)_{\lambda^2}(20)_{\bar{u}u}(12)_{G^2}(5)_{\bar{u}Gu}(20)_\rho \\ &\quad (49)_{\Gamma_K}(11)_{m_s}(8)_{\bar{s}s/\bar{u}u}, \\ &= 4.79 \pm 0.60 \longrightarrow \frac{f_{K'}}{f_K} = (25.0 \pm 1.5)10^{-2}. \end{aligned} \tag{9.36}$$

One can remark that $r_\pi \approx r_K$ as might be expected from chiral symmetry arguments. One can note that the tachyonic gluon mass contribution included here has decreased the value of r_K by a small amount of 0.10.

◇ r_K from instanton sum rules for arbitrary μ

The analysis for $\mu = \tau^{-1/2}$ also does not lead to a conclusive result like in the case of the usual SVZ expansion. Then, we shall also not consider it here and study the sum rule for arbitrary value of μ.

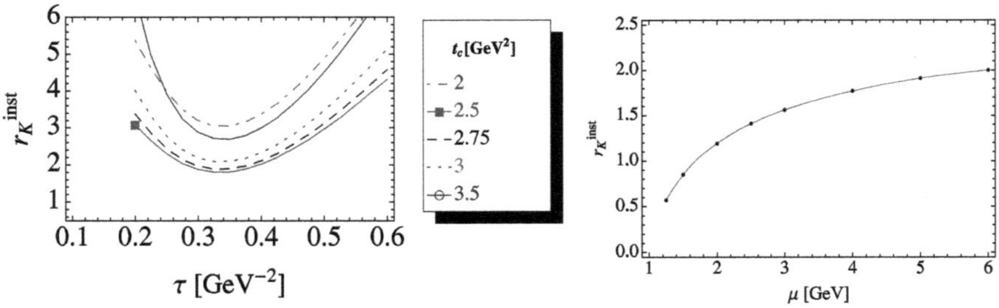

(a) (b)

Figure 9.10. (a) τ-behaviour of r_K for a given value $\mu = 4$ GeV of subtraction point and for different values of t_c from the instanton sum rule; (b) μ-behaviour of the optimal value of r_K deduced from (a).

We show in Fig. 9.10(a) the τ-behaviour of r_K for a given value $\mu = 4$ GeV of the subtraction point and for different values of t_c from the instanton sum rule. One has a τ-minimum for $\tau \simeq 0.3$ GeV^{-2}, while a t_c-minimum is obtained for $t_c = 2.5$ GeV^2. At this τ-scale, the instanton contribution to $\mathcal{R}_5^K(\tau, \mu)$ is about 60% of the perturbative $+d \leq 6$ condensate contributions.

We show in Fig. 9.10(b) the μ-behaviour of the optimal value of r_K deduced from Fig. 9.10(a). One can note like in the previous analysis without instanton that the curve is (almost) stable for $\mu \geq 3$ GeV. We use as a final result the mean obtained for μ in the range 3–6 GeV:

$$r_K^{\rm inst} \simeq 1.8(1)_\mu (3.8)_{\rm svz} (2.4)_{\rho_c} \quad \longrightarrow \quad \frac{f_{K'}}{f_K} = (15.3 \pm 2.0) 10^{-2}. \quad (9.37)$$

◇ Comparison with some other predictions

We compare in Fig. 9.11 our results from Eqs. (9.36) and (9.37) for

$$L_K(\tau) \equiv r_K BW I_0, \quad (9.38)$$

with the existing ones in the current literature (NPT83 [91], SN02 [6], KM02 [451] and DPS98 [452]). The results of NPT83 [91] and SN02 [6] are obtained within a narrow width approximation. The ones of KM02 [451] and DPS98 [452] include finite width correction. There is fair agreement between different determinations with the exception of the one from Ref. [452] which is relatively high (central value shown in Fig. 9.11). This high value may be either due to the coherent sum and equal coupling of the $K(1460)$ and $K(1800)$ contribution assumed in the amplitude or due to an overall normalization factor.[5] We also see in Fig. 9.11 that the instanton sum rule estimate is relatively small compared with the one from the sum rule within the SVZ expansion and with some other determinations. As we shall see later, this low value of the $K(1460)$ contribution will imply a smaller value of m_s from the instanton sum rule estimate.

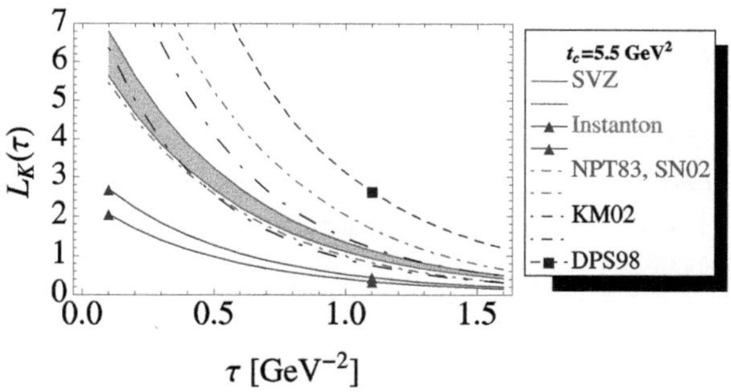

Figure 9.11. Comparison of our determination of r_K from SVZ and instanton sum rules with the ones in the current literature: NPT83 [91], SN02 [6], KM02 [451] and DPS98 [452]). We use $t_c = 5.5$ GeV^2.

• Laplace sum rule estimate of $m_{us} \equiv (m_u + m_s)$

Defining:

$$m_{us} = (m_u + m_s), \tag{9.39}$$

we now turn to the estimate of the RGI $\hat{m}_{us}$ and running $\overline{m}_{us}$ sum of masses.

[5]Note that instead of Ref. [452] in the kaon channel, a destructive interference has been assumed by Ref. [450] in the pion channel. Our estimate in the pion channel agrees with Ref. [450].

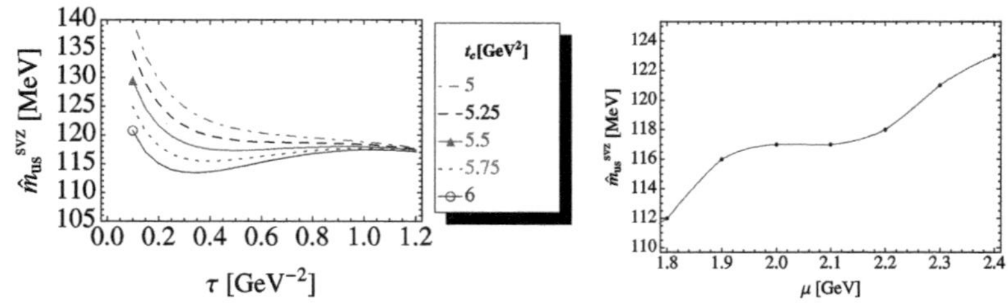

(a) (b)

Figure 9.12. (a) τ-behaviour of $\hat{m}_{us}$ for a given value $\mu = 2.1$ GeV of subtraction point and for different values of t_c; (b) μ-behaviour of the optimal value of $\hat{m}_{us}$ deduced from (a).

◇ $\hat{m}_{us}$ within the SVZ expansion for arbitrary μ

We show in Fig. 9.12(a) the τ-behaviour of $\hat{m}_{us}$ for a given value $\mu = 2.1$ GeV of subtraction point and for different values of t_c where we have used the value of r_K in Eq. (9.36). The largest range of τ-stability of about (0.4–0.6) GeV^{-2} is reached at $t_c \simeq (5.25-5.75)$ GeV^2. We show in Fig. 9.12(b) the μ-behaviour of the optimal value of $\hat{m}_{us}$ deduced from Fig. 9.12(a). We obtain a stability at $\mu = 2.1$ GeV at which corresponds the value:

$$\begin{aligned}\hat{m}_{us}^{\rm SVZ} &= 117.3(16)_{t_c}(30)_{\Lambda}(13)_{\lambda^2}(4)_{\bar{u}u}(10)_{G^2}(2)_{\bar{u}Gu}(4)_{\rho}\\ &\quad (30)_{r_K}(4)_{\Gamma_K}(2)_{m_s}(2)_{\bar{s}s/\bar{u}u}\ \text{MeV},\\ &= (117.3 \pm 4.9)\ \text{MeV}. \end{aligned} \tag{9.40}$$

◇ $\hat{m}_{us}$ within the SVZ expansion for $\mu = \tau^{-1/2}$

We redo the previous analysis but for $\mu = \tau^{-1/2}$. This is shown in Fig. 9.13, where one can see that a τ-stability (inflexion point here) is obtained for $t_c \geq 5.75$ GeV^2 and where the result is almost t_c-stable. We obtain for $t_c = (6 \pm 0.25)$ GeV^2 and $\tau = 0.4$ GeV$^{-1/2}$:

$$\begin{aligned}\hat{m}_{us}^{\rm SVZ} &= 110.5(10)_{\tau}(18)_{t_c}(27)_{\Lambda}(5)_{\lambda^2}(3)_{\bar{u}u}(7)_{G^2}(1)_{\bar{u}Gu}(2)_{\rho}\\ &\quad (32)_{r_K}(9)_{\Gamma_K}(1)_{m_s}(1)_{\bar{s}s/\bar{u}u}\ \text{MeV},\\ &= (110.5 \pm 4.8)\ \text{MeV}. \end{aligned} \tag{9.41}$$

Note that in both cases ($\mu = 2.1$ GeV and $\tau^{-1/2}$), the tachyonic gluon mass decreases by about 3 MeV the value of $\hat{m}_{us}$ which is relatively small but indicates, by duality with the uncalculated higher order terms [234], that the QCD PT series converge quite well.

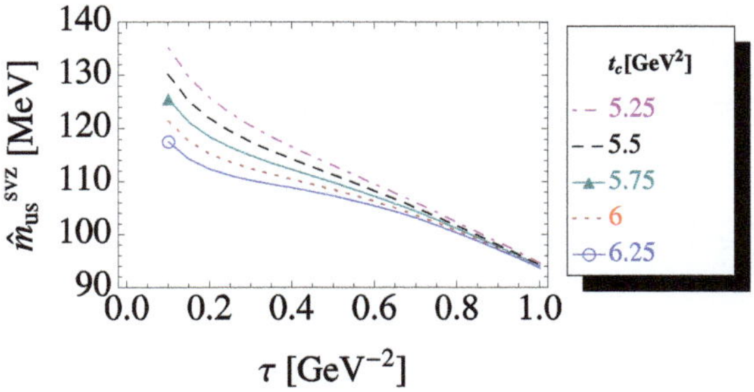

Figure 9.13. τ-behaviour of $\hat{m}_{us}$ from the SVZ expansion for $\mu = \tau^{-1/2}$ and for different values of t_c.

◇ $\hat{m}_{us}$ from the instanton sum rule at arbitrary μ

We show in Fig. 9.14 the τ-behaviour of $\hat{m}_{us}$ from the instanton sum rule at different values of t_c and for a given value $\mu = 2$ GeV. The largest τ-stability is obtained for $t_c \simeq (3.5 \pm 0.5)$ GeV^2. We take the value at the τ-minimum of about 0.6 GeV^2. As shown in Fig. 9.14, the results increase smoothly for $\mu \geq 3$ GeV. Therefore, we take for definiteness the mean of the results for $\mu = (3–6)$ GeV:

$$\begin{aligned}\hat{m}_{us}^{\text{inst}} &= 105.3(22)_{t_c}(17)_{\rho_c}(44)_{\text{svz}} \text{ MeV},\\ &= (105.3 \pm 5.2) \text{ MeV},\end{aligned} \tag{9.42}$$

where the value corresponds approximately to $\mu = 4.25$ GeV.

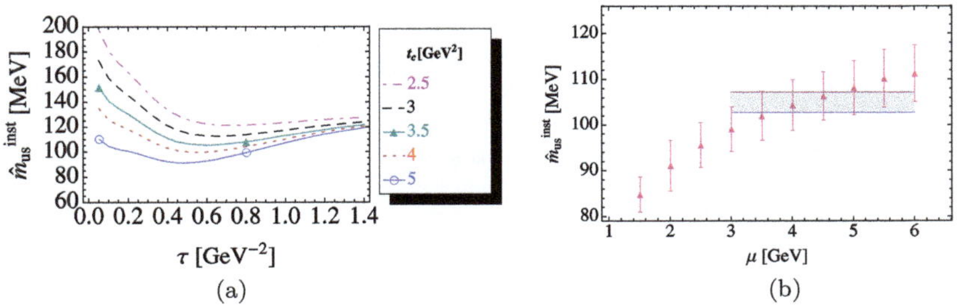

Figure 9.14. (a) τ-behaviour of $\hat{m}_{us}$ from the instanton sum rule for a given value $\mu = 2$ GeV of the subtraction point and for different values of t_c; (b) μ-behaviour of the optimal results. The error is due to t_c. The dashed region corresponds to the mean value for $\mu = (3–6)$ GeV.

◇ $\hat{m}_{us}$ from the instanton sum rule at $\mu = \tau^{-1/2}$

The analysis is shown in Fig. 9.15. Our optimal results correspond to $\tau = (0.8 \pm 0.5)$ GeV^{-2} and $t_c = (3.5 \pm 0.5)$ GeV^2. We deduce:

$$\begin{aligned}\hat{m}_{us}^{\text{inst}} &= 77.7(7)_\tau(25)_{t_c}(24)_\Lambda(8)_{\lambda^2}(3)_{\bar{u}u}(9)_{G^2}(1)_{\bar{u}Gu}(3)_\rho \\ &\quad (23)_{r_K}(2)_{\Gamma_K}(3)_{m_s}(2)_{\bar{s}s/\bar{u}u}\ \text{MeV}, \\ &= (77.7 \pm 4.4)\ \text{MeV}. \end{aligned} \tag{9.43}$$

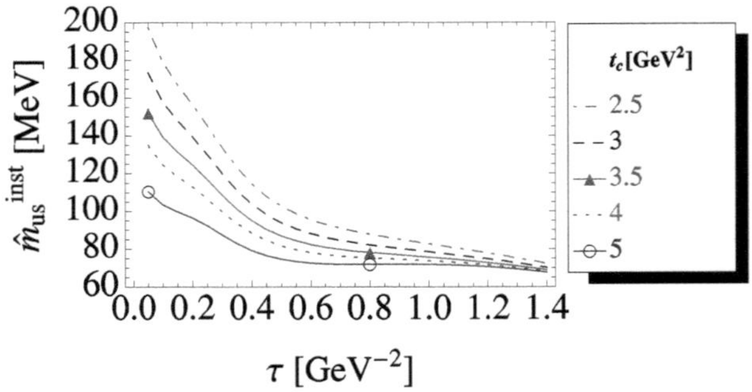

Figure 9.15. τ-behaviour of $\hat{m}_{us}$ from the instanton sum rule for $\mu = \tau^{-1/2}$ and for different values of t_c.

◇ Final value of $\hat{m}_{us}$ and $\overline{m}_{us}$

By combining the previous determinations from Eqs. (9.40) and (9.41) and from Eqs. (9.42) and (9.43), one can deduce the mean final estimate in units of MeV:

$$\hat{m}_{us}^{\text{svz}} = 113.8(34), \qquad \hat{m}_{us}^{\text{inst}} = 89.2(34), \tag{9.44}$$

and the corresponding running mass evaluated at 2 GeV:

$$\overline{m}_{us}^{\text{svz}} = 98.0(29), \qquad \overline{m}_{us}^{\text{inst}} = 76.8(29). \tag{9.45}$$

Using as input the values of $\overline{m}_u$ given in Eq. (9.34), one can deduce:

$$\overline{m}_{s}^{\text{svz}} = 95.5(29), \qquad \overline{m}_{s}^{\text{inst}} = 74.9(29). \tag{9.46}$$

Combining this result with the value of $\overline{m}_{ud}$ in Eq. (9.30), one predicts the scale-independent mass ratios:

$$\left(\frac{m_s}{m_{ud}}\right)^{\text{svz}} = 25.3 \pm 1.1, \qquad \left(\frac{m_s}{m_{ud}}\right)^{\text{inst}} = 26.6 \pm 0.7, \tag{9.47}$$

and:

$$\left(\frac{m_s}{m_d}\right)^{\text{svz}} = 19.0 \pm 1.0, \qquad \left(\frac{m_s}{m_d}\right)^{\text{inst}} = 19.9 \pm 0.7. \tag{9.48}$$

These results are summarized in Table 9.3.

9.1.8 Lower bounds on $(\overline{m}_u + \overline{m}_q)$ from Laplace sum rule

Lower bounds on quark masses have been first derived in Refs. [34, 69] and improved later on in Ref. [73] using a finite number of Q^2-derivatives of the two-point function. In Ref. [7], the α_s^3 corrections to the result of Ref. [73] have been included leading to the (improved) lower bounds:

$$\begin{aligned} \overline{m}_{ud} &\equiv \frac{1}{2}(\overline{m}_u + \overline{m}_d) \geq (3.0 \pm 0.5)\ \text{MeV}, \\ \overline{m}_{us} &\equiv (\overline{m}_u + \overline{m}_s) \geq (82.7 \pm 13.3)\ \text{MeV}. \end{aligned} \tag{9.49}$$

In the present work, we shall use the positivity of the spectral function into the Laplace sum rule $\mathcal{L}_5^P(\tau, \mu)$ defined in Eq. (9.3) for extracting a lower bound on the sum of light quark RGI masses $(\hat{m}_u + \hat{m}_q)$.

• Bounds from Laplace sum rules at $\mu = \tau^{-1/2}$

We study the lower bounds obtained from sum rules within the SVZ expansion (Fig. 9.16(a)) and the ones where the instanton contribution is added into the OPE (Fig. 9.16(b)). Similar curves are obtained in the s-quark channel. For $\mu = \tau^{-1/2}$, we obtain[6] in units of MeV:

$$
\begin{aligned}
\hat{m}_{ud}^{\rm svz} &\geq 2.84(14)_\Lambda(4)_{\lambda^2}(4)_{\bar{u}u}(8)_{G^2}(0)_{\bar{u}Gu}(5)_\rho \\
&\geq (2.84 \pm 0.18), \\
\hat{m}_{us}^{\rm svz} &\geq 75(3)_\Lambda(1)_{\lambda^2}(0.7)_{\bar{u}u}(1.5)_{G^2}(0)_{\bar{u}Gu}(0)_\rho \\
&\quad (0)_{m_s}(0.1)_{\bar{s}s/\bar{u}u}, \\
&\geq (75.0 \pm 3.6), \qquad (9.50)
\end{aligned}
$$

and:

$$
\hat{m}_{ud}^{\rm inst} \geq 2.52(16)_{\rm svz}(0)_{\rho_c}, \qquad \hat{m}_{us}^{\rm inst} \geq 65.2(3.1)_{\rm svz}(0.4)_{\rho_c}. \qquad (9.51)
$$

One can translate the previous bounds on the RGI masses into the ones for the running masses evaluated at 2 GeV in units of MeV:

$$
\begin{aligned}
\overline{m}_{ud}^{\rm svz} &\geq 2.66(12), \qquad \overline{m}_{us}^{\rm svz} \geq 69.6(2.4), \\
\overline{m}_{ud}^{\rm inst} &\geq 2.29(10), \qquad \overline{m}_{us}^{\rm inst} \geq 59.9(2.1). \qquad (9.52)
\end{aligned}
$$

Using the value of the ratio m_u/m_d in Eq. (9.33), one can deduce from the bound on m_{ud} in units of MeV:

$$
\begin{aligned}
\overline{m}_{u}^{\rm svz} &\geq 1.77(11), \qquad \overline{m}_{u}^{\rm inst} \geq 1.53(9), \\
\overline{m}_{d}^{\rm svz} &\geq 3.54(21), \qquad \overline{m}_{u}^{\rm inst} \geq 3.06(18). \qquad (9.53)
\end{aligned}
$$

Using the value of $\overline{m}_u$ in Eq. (9.34), one can deduce from the bound on m_{us} in units of MeV:

$$
\overline{m}_{s}^{\rm svz} \geq 67.1(24)), \qquad \overline{m}_{s}^{\rm inst} \geq 58.0(21). \qquad (9.54)
$$

These bounds are interesting though weaker than the ones in Eq. (9.49). The results are summarized in Table 9.2.

[6]The sum rules with an arbitrary μ do not present a τ-stability and will not be considered here.

Table 9.2. Values of the set of external parameters (μ, τ, t_c) obtained at the stability points corresponding to the optimal values of r_P and m_q. μ in GeV, τ in GeV^{-2} and t_c in GeV^2. The indices SVZ and I correspond to the SVZ and SVZ $\oplus$ instanton expansions.

Observables	μ	τ	t_c	$\tau = \mu^{-2}$	t_c
Pion channel					
r_π^{SVZ}	1.5	0.6	2	1.8	2
r_π^I	3–6	0.4	2	0.6	2
m_{ud}^{SVZ}	1.5	0.4	5	0.5	5
m_{ud}^I	3–6	0.7	5	0.7	4
Kaon channel					
r_K^{SVZ}	3–6	0.8	2	1.8	2
r_K^I	3–6	0.3	2	0.6	2
m_{us}^{SVZ}	2	0.5	5.5	0.4	6
m_{us}^I	3–6	0.6	3.5	0.8	4

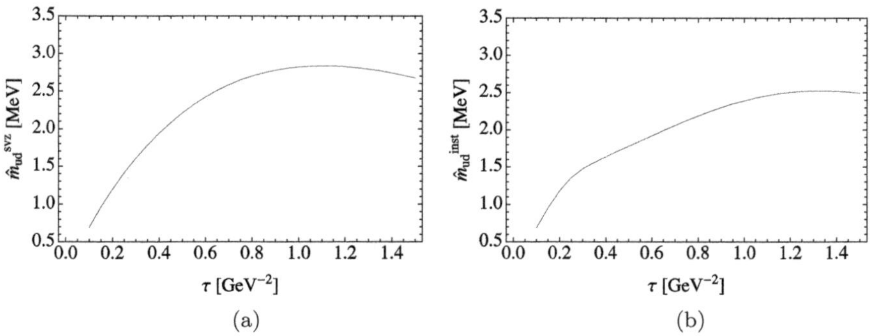

Figure 9.16. (a) τ-behaviour of the lower bound of $\hat{m}_{ud}$ from the sum rule within the SVZ expansion for $\mu = \tau^{-1/2}$; (b) the same as in (a) but for the instanton sum rule.

9.1.9 Summary and conclusions

◇ We have re-estimated the $\pi(1300)$ and $K(1460)$ decay constants using pseudoscalar Laplace sum rules which we have compared with some existing ones in the literature. We have used these results for improving the determinations of (m_u+m_q): $q \equiv d, s$ from these channels. Our results obtained from the set of parameters are summarized in Table 9.2.

◇ The results using the SVZ expansion without the instanton contribution agree within the errors with our previous determinations from the pseudoscalar and nucleon sum rules [6,294,295] and the ones from e^+e^- and τ-decay data [294,295,359,360]. Therefore, this comparison tends to favour the results from the SVZ expansion without instanton. Another point which does not also favour the instanton sum rule is its underestimate of the strength of the spectral integral as compared to some alternative estimates shown in Figs. 9.6 and 9.11.

Table 9.3. Summary of the main results of this work. For deriving the values and bounds of $\overline{m}_{u,d}$, we have used $m_u/m_d = 0.50 \pm 0.03$ deduced from the compilation of PDG13 [33]. The estimated value of $\overline{m}_u$ has been used for an estimate and for giving a bound on $\overline{m}_s$. The running masses $\overline{m}_q$ evaluated at 2 GeV are in units of MeV.

Parameters	SVZ	SVZ $\oplus$ instanton	Equations
Estimated values			
$f_{\pi'}/f_\pi$	$(2.58 \pm 0.35)10^{-2}$	$(2.06 \pm 0.30)10^{-2}$	(9.25)
$f_{K'}/f_K$	$(25.0 \pm 1.5)10^{-2}$	$(15.3 \pm 2.0)10^{-2}$	(9.36, 9.37)
$\overline{m}_{ud}$	3.77 ± 0.28	2.82 ± 0.18	(9.30)
$\overline{m}_u$	2.51 ± 0.21	1.88 ± 0.14	(9.34)
$\overline{m}_d$	5.02 ± 0.42	3.76 ± 0.28	(9.34)
$\overline{m}_{us}$	98.0 ± 2.9	76.8 ± 2.9	(9.45)
$\overline{m}_s$	95.5 ± 2.9	74.9 ± 2.9	(9.46)
m_s/m_{ud}	25.3 ± 1.1	26.6 ± 0.7	(9.47)
m_s/m_d	19.0 ± 1.0	19.9 ± 0.7	(9.48)
Lower bounds			
$\overline{m}_{ud}$	2.66 ± 0.12	2.29 ± 0.10	(9.52)
$\overline{m}_{us}$	69.6 ± 2.4	59.9 ± 2.1	(9.52)
$\overline{m}_u$	1.77 ± 0.11	1.53 ± 0.09	(9.53)
$\overline{m}_d$	3.54 ± 0.21	3.06 ± 0.18	(9.53)
$\overline{m}_s$	67.1 ± 2.4	58.0 ± 2.1	(9.54)

◇ An inclusion of the tachyonic gluon mass into the SVZ expansion showing that its effect is relatively small (it decreases r_π and r_K by 0.1 and $\hat{m}_{ud}$ (respectively $\hat{m}_{us}$) by 0.13 (respectively 3) MeV) which is reassuring as, by duality, the tachyonic gluon mass contribution is expected to provide an estimate of the uncalculated higher order terms of the QCD PT series [234].

9.2 Heavy–Light Meson Sum Rules

In this section, we report on the different works [455–457] reviewed in Ref. [51] and on the more recent works in Refs. [453, 454] on the determination of the heavy-light mesons decay constants and masses from QCD spectral sum rules (QSSR) and on the estimate of the c, b running quark masses.

9.2.1 Introduction

The decay constants $f_{P/S}, f_{V/A}$ of (pseudo)scalar and (axial)vector mesons are of prime interest for understanding the realizations of chiral symmetry in QCD and for controlling the meson (semi-)leptonic decay widths, hadronic couplings and form factors. In addition to the well-known values of f_π=130.4(2) MeV and f_K=156.1(9) MeV which control the light flavour chiral symmetries [449], it is also desirable to extract the ones of the heavy–light charm and bottom quark systems with high accuracy.

We shall be concerned with the two-point correlators:

$$\psi_{P/S}(q^2) = i \int d^4x \; e^{iq.x} \langle 0|T J_{P/S}(x) J_{P/S}(0)^\dagger|0\rangle,$$

$$\begin{aligned}\Pi^{\mu\nu}_{V/A}(q^2) &= i \int d^4x \; e^{iq.x} \langle 0|T J^\mu_{V/A}(x) J^\nu_{V/A}(0)^\dagger|0\rangle,\\ &= -\left(g^{\mu\nu} - \frac{q^\mu q^\nu}{q^2}\right)\Pi^T_{V/A}(q^2) + \frac{q_\mu q_\nu}{q^2}\Pi^L_{V/A}(q^2)\end{aligned} \quad (9.55)$$

associated to the vector/axial-vector: $J^\mu_{V/A}(x) \equiv \bar{q}(\gamma_\mu/\gamma_\mu\gamma_5)Q$ and pseudo/scalar current $J_{P/S}(x) \equiv (m_q + / - M_Q)\bar{q}(i\gamma_5/1)Q$ currents; $q \equiv d, c$; $Q \equiv c, b$; $P \equiv D, B, B_c$; $S \equiv D_0^*, B_0^*$, $V \equiv D^*, B^*$, $A \equiv D_1, B_1$ and their s-quark analogue.

The decay constants $f_{P/S}, f_{V/A}$ related to the leptonic widths $\Gamma[P/S \to l^+\nu_l]$ and $\Gamma[V/A \to l^+l^-]$ are normalized as:

$$\langle 0|J_{P/S}(x)|P\rangle = f_{P/S}M^2_{P/S};\ \langle 0|J^{\mu}_{V/A}(x)|V/A\rangle = f_{V/A}M_V V/A\epsilon^{\mu}, \tag{9.56}$$

where ϵ_μ is the vector polarization.

$\Pi^T_{V/A}(q^2)$ has more power of q^2 than the transverse two-point function used in the current literature for $m_q = m_Q$ in order to avoid mass singularities at $q^2 = 0$ when m_q goes to zero. $\Pi^{\mu\nu}_{V/A}(q^2)$ (vector/axial) and $\psi^{S/P}(q^2)$ (scalar/pseudoscalar) correlators are related to each other through the Ward identities:

$$q_\mu q_\nu \Pi^{\mu\nu}_{V/A}(q^2) = \psi_{S/P}(q^2) - \psi_{S/P}(0), \tag{9.57}$$

where the (non)perturbative parts of $\psi_{S/P}(0)$ are known (see, e.g., Ref. [6]) and play an important rôle for absorbing mass singularities which appear during the evaluation of the PT two-point function.

For extracting the decay constants, we shall work with the well-established (inverse) Laplace sum rules [we use the terminology (inverse) Laplace instead of Borel sum rule as it has been demonstrated in Ref. [69] that its QCD radiative corrections satisfy the (inverse) Laplace transform properties][7]:

$$\mathcal{L}_{P/S,V/A}(\tau,\mu) = \int_{(m_q+M_Q)^2}^{t_c} dt\ e^{-t\tau}\frac{1}{\pi}\text{Im}\ \left[\psi_{P/S}, \Pi^T_{V/A}\right](t,\mu). \tag{9.59}$$

For improving the extraction of the decay constants $f_{D^*_{(s)}}$ and $f_{B^*_{(s)}}$, we shall also work with the ratio:

$$\mathcal{R}_{V/A,P} \equiv \frac{\mathcal{L}_{V,A}(\tau_{V/A})}{\mathcal{L}_P(\tau_P)}, \tag{9.60}$$

in order to minimize the systematics of the approach, the effects of heavy quark masses and the continuum threshold uncertainties

---

[7] One can also work with moment sum rules:

$$\mathcal{M}^{(n)}_{\bar{q}b}(\mu) = \int_{(m_q+M_b)^2}^{t_c} \frac{dt}{t^{n+2}}\ \frac{1}{\pi}\text{Im}\psi_{\bar{q}b}(t,\mu), \tag{9.58}$$

like in Ref. [455] or with τ-decay like finite energy sum rules [74] inspired from τ-decay [79], but these different sum rules give approximately the same results as the one from (inverse) Laplace sum rules.

which are one of the main sources of errors in the determinations of the individual decay constants. $\tau_{V/A}$ and τ_P denote the values of the sum rule τ-variable at which each individual sum rule is optimized (minimum or inflexion point). In general, $\tau_{V/A} \neq \tau_P$, as we shall see later, which requires some care for a precise determination of the ratio of decay constants. This ratio of sum rule led to a successful prediction of the SU(3) breaking ratio f_{P_s}/f_P [167] such that, from it, one expects to extract precise values of the ratio $f_{V/A}/f_P$.

9.2.2 QCD expression of the two-point correlators

The QCD expression of the Laplace sum rule $\mathcal{L}_{P/S}(\tau,\mu)$ in the (pseudo)scalar channel can be found in Refs. [301, 455, 456] for full QCD, including N2LO perturbative QCD corrections and contributions of non-perturbative condensates up to the complete $d = 6$ dimension condensates.[8] The one of the vector channel has been given within the same approximation in Ref. [457]. Some comments are in order:

- The expressions of NLO PT in Refs. [458, 459], of N2LO PT in Ref. [460], of the non-perturbative in Refs. [157,372] and the light quark mass corrections in Refs. [157,371,458] have been used.
- The N3LO PT contributions have been estimated by assuming the geometric growth of the series [234] which is dual to the effect of a $1/q^2$ term [232, 235].
- The PT expressions have been originally obtained using a pole quark mass which is transformed into the $\overline{MS}$-scheme by using the relation between the running $\bar{m}_Q(\mu)$ and pole mass M_Q in the $\overline{MS}$-scheme to order α_s^2 [154, 156, 158–160, 461].
- The LO contributions in α_s up to the $d = 4$ gluon: $\langle\alpha_s G^2\rangle \equiv \langle\alpha_s G^a_{\mu\nu}G_a^{\mu\nu}\rangle$, $d = 5$ mixed quark-gluon: $\langle\bar{q}Gq\rangle \equiv \langle\bar{q}g\sigma^{\mu\nu}(\lambda_a/2)G^a_{\mu\nu}q\rangle = M_0^2\langle\bar{q}q\rangle$ and $d = 6$ four-quark condensates,

$$\begin{aligned}\langle\bar{d}jd\rangle &\equiv \left\langle \bar{d}g\gamma_\mu D^\mu G_{\mu\nu}\frac{\lambda_a}{2}d\right\rangle,\\ &= g^2\left\langle \bar{d}\gamma_\mu\frac{\lambda_a}{2}d\sum_q \bar{q}\gamma_\mu\frac{\lambda_a}{2}q\right\rangle \simeq -\frac{16}{9}(\pi\alpha_s)\,\rho\langle\bar{d}d\rangle^2, \quad (9.61)\end{aligned}$$

[8] Note an unfortunate misprint of $1/\pi$ in front of $\mathrm{Im}\psi(t)$ in Ref. [455].

(after the use of the equation of motion) condensates, have been obtained originally by NSV2Z [372]. $\rho \simeq 3\text{–}4$ measures the deviation from the vacuum saturation estimate of the $d = 6$ four-quark condensates [115, 289, 297].

- The contribution of the $d = 6$ gluon condensates:

$$\langle g^3 G^3 \rangle \equiv \langle g^3 f_{abc} G^a_{\mu\nu} G^b_{\nu\rho} G^c_{\rho\mu} \rangle,$$

$$\langle j^2 \rangle \equiv g^2 \langle (D_\mu G^a_{\nu\mu})^2 \rangle = g^4 \left\langle \left(\sum_q \bar{q} \gamma_\nu \frac{\lambda^a}{2} q \right)^2 \right\rangle,$$

$$\simeq -\frac{64}{3} (\pi \alpha_s)^2 \rho \langle \bar{d} d \rangle^2, \tag{9.62}$$

after the use of the equation of motion, have been deduced from Ref. [157] (Eqs. II.4.28 and Table II.8).

- One can note that the gluon condensates $\langle \alpha_s G^2 \rangle$ and $\langle G^3 \rangle$ contributions flip sign from the pseudoscalar to the vector correlators, while there is an extra $m_Q M_0^2 \langle \bar{d} d \rangle$ term with a positive contribution in the pseudoscalar channel from the mixed condensate. Similar observation arises for the signs of the chiral quark and mixed quark gluon condensates from the pseudoscalar to the scalar and from the vector to the axial correlators. We shall see in Fig. 9.17 that these different signs transform, e.g., the minimum in τ for the pseudoscalar into an inflexion point for the vector sum rule.
- It is clear that, for some non-perturbative terms which are known to leading order of perturbation theory, one can use either the running or the pole mass. However, we shall see that this distinction does not affect, in a visible way, the present result, within the accuracy of our estimate, as the non-perturbative contributions are relatively small though vital in the analysis.

9.2.3 QCD input parameters

The QCD parameters which shall appear in the following analysis will be the charm and bottom quark masses $m_{c,b}$ (we shall neglect the light quark masses $q \equiv u, d$), the light quark condensate $\langle \bar{q} q \rangle$, the gluon condensates $\langle \alpha_s G^2 \rangle$ and $\langle g^3 G^3 \rangle$ the mixed condensate $\langle \bar{q} G q \rangle$ defined previously and the four-quark condensate $\rho \alpha_s \langle \bar{q} q \rangle^2$. Their values are given in Tables 7.10 and 9.1.

- We shall work with the running light quark condensates and masses. They read:

$$\langle\bar{q}q\rangle(\tau) = -\hat{\mu}_q^3(-\beta_1 a_s)^{2/\beta_1}/C(a_s),$$
$$\langle\bar{q}Gq\rangle(\tau) = -M_0^2\hat{\mu}_q^3(-\beta_1 a_s)^{1/3\beta_1}/C(a_s),$$
$$\overline{m}_s(\tau) = \hat{m}_s/\left(-\log\sqrt{\tau}\Lambda\right)^{2/-\beta_1}C(a_s), \tag{9.63}$$

  where $\beta_1 = -(1/2)(11 - 2n_f/3)$ is the first coefficient of the β function for n_f flavours, $a_s \equiv \alpha_s(\tau)/\pi$ and $\hat{\mu}_q$ is the spontaneous RGI light quark condensate [29]. The QCD correction factor $C(a_s)$ in the previous expressions is numerically [156]:

$$C(a_s) = 1 + 1.1755a_s + 1.5008a_s^2 + \cdots : n_f = 5, \tag{9.64}$$

  which shows a good convergence.
- The value of the running $\langle\bar{q}q\rangle$ condensate is deduced from the well-known GMOR relation: $(m_u+m_d)\langle\bar{u}u+\bar{d}d\rangle = -m_\pi^2 f_\pi^2$, where $f_\pi =$ 130.4(2) MeV [449] and the value of $(\overline{m}_u + \overline{m}_d)(2) = (7.9 \pm 0.6)$ MeV obtained in Ref. [74] which agrees with the PDG in Ref. [33] and lattice averages in Ref. [462]. Then, we deduce the RGI light quark spontaneous mass $\hat{\mu}_q$ given in Table 7.10.
- For the heavy quarks, we shall use the running mass and the corresponding value of α_s evaluated at the scale μ.
- To be conservative, we have enlarged the original errors of some parameters (gluon condensate, SU(3) breaking, etc.) by a factor 2–3. We do not consider the lattice result [463] for κ which is in conflict with SR one from various channels and ChPT.
- We shall see that the effects of the gluon and four-quark condensates on the values of the decay constants are small though they play a crucial role in the stability analysis.

9.2.4 Spectral function and Stability criteria

• Minimal Duality Ansatz (MDA)

We shall use the minimal duality ansatz (MDA) for parametrizing the spectral function:

$$\frac{1}{\pi}\,\mathrm{Im}\psi_{P/S}(t) \simeq f_P^2 M_{P/S}^4 \delta(t - M_P^2) + \text{“QCD cont.”}\theta(t - t_c^P),$$
$$\frac{1}{\pi}\,\mathrm{Im}\Pi_V^T(t) \simeq f_V^2 M_V^2 \delta(t - M_V^2) + \text{“QCD cont.”}\theta(t - t_c^V),$$
$$(9.65)$$

where $f_{P,V}$ are the decay constants defined in Eq. (9.56) and the higher state contributions are smeared by the “QCD continuum” coming from the discontinuity of the QCD diagrams and starting from a constant threshold t_c^P, t_c^V which is independent on the subtraction point μ in this standard minimal model.

- **Stability criteria and optimization**

We shall use τ-, t_c- and μ-stability criteria for extracting our optimal results.

9.2.5 The decay constants f_D and f_{D^*}

- **Direct determinations from $\mathcal{L}_{P,V}(\tau, \mu)$**

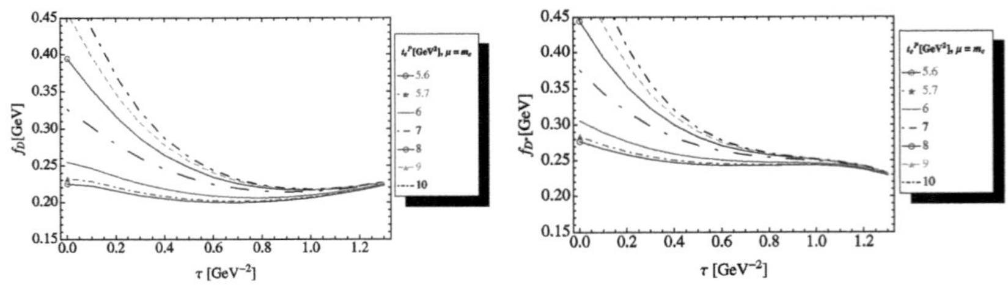

(a) (b)

Figure 9.17. (a) τ-behaviour of f_D from $\mathcal{L}_P$ for different values of t_c at a given value of the subtraction point $\mu = m_c$; (b) the same as in (a) but for f_{D^*} from $\mathcal{L}_V$.

We start by showing in Fig. 9.17 the τ-behaviour of the decay constants f_{D^*} and f_D at a given value of the subtraction point $\mu = m_c$ for different values of the continuum threshold t_c. We have assumed that $(t_c^{D^*})^{1/2} - (t_c^D)^{1/2} \simeq M_{D^*} - M_D = 140.6$ MeV for the vector and pseudoscalar channels where the QCD expressions are truncated at the same order of PT and NP series. At this value of μ, we deduce the optimal value versus τ (minimum or inflexion point for $\tau \simeq 0.8$–1 GeV^{-2}) and t_c (5.6–8 GeV^2). Next, we study the μ

variation of these results, which we show in Fig. 9.18. We deduce the *mean result* from [455]:

$$f_D = 204(6)\ \text{MeV}, \tag{9.66}$$

where the largest error for each data point comes from t_c.[9] The value of f_{D^*} at the minimum in $\mu = (1.5 \pm 0.1)$ GeV is [457]:

$$f_{D^*} = 253.5(11.5)_{t_c}(5.7)_{\tau}(13)_{\text{svz}}(1)_{\mu} = 253.5(18.3)\ \text{MeV}, \tag{9.67}$$

with:

$$(13)_{\text{svz}} = (0.5)_{\alpha_s}(12.3)_{\alpha_s^3}(0.6)_{m_c}(3.8)_{\langle\bar{d}d\rangle}(1.8)_{\langle\alpha_s G^2\rangle}$$
$$(1.4)_{\langle\bar{d}Gd\rangle}(0.4)_{\langle g^3G^3\rangle}(0.4)_{\langle\bar{d}d\rangle^2}, \tag{9.68}$$

where the SVZ-OPE error is dominated by the estimate of the α_s^3 corrections (95%) and where again the error due to μ has been mutiplied by a factor 2 for a more conservative error. One should note that the accurate value of f_D comes from the mean of different data shown in Fig. 9.18, while the error from f_{D^*} is taken from the one at the minimum for $\mu \simeq 1.45$ GeV.

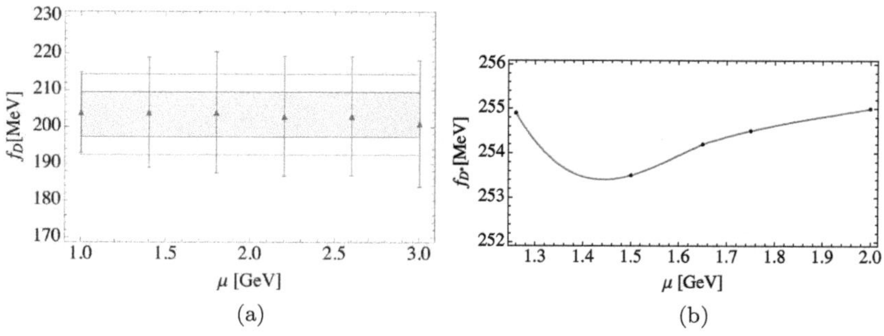

Figure 9.18. (a) f_D from LSR versus μ and for $\hat{m}_c$=1467 MeV. The filled (grey) region is the average with the corresponding averaged errors. The dashed horizontal lines denote the ones where the errors come from the best determination; (b) f_{D^*} for different values of μ.

[9]One should note that the extraction of the charm quark running mass by requiring that the sum rules should reproduce the D-meson mass is obtained in the same range of values of the previous set of stability parameters [455]. The same results are obtained for the other mesons. This fact increases the confidence on the predictions of the unknown decay constants.

• The ratio f_{D^*}/f_D and improved determination of f_{D^*}

One can note in Fig. 9.17 that working directly with the ratio in Eq. (9.60) by taking the same value $\tau_V = \tau_P$ is inaccurate as the two sum rules $\mathcal{L}_V(\tau)$ and $\mathcal{L}_P(\tau)$ are not optimized at the same value of τ (minimum for f_D and inflexion point for f_{D^*}). Therefore, for a given value of t_c, we take separately the value of each sum rule at the corresponding value of τ where they present minimum and/or inflexion point and then take their ratio. For a given μ, the optimal result corresponds to the mean obtained in range of values of t_c where one starts to have a τ-stability ($t_c \simeq 5.6{-}5.7$ GeV^2 for $\tau \simeq 0.6$ GeV^{-2}) and a t_c-stability ($t_c \simeq 9.5 \sim 10.5$ GeV^2 for $\tau \simeq 0.8$ GeV^{-2}). In Fig. 9.19, we look for the μ-stability of the previous optimal ratio f_{D^*}/f_D in the set (τ, t_c). We obtain, at the minimum $\mu = (1.5 \pm 0.1)$ GeV:

$$f_{D^*}/f_D = 1.218(6)_{t_c}(27)_\tau(23)_{\rm svz}(4)_\mu, = 1.218(36), \tag{9.69}$$

with:

$$(23)_{\rm svz} = (7)_{\alpha_s}(2)_{\alpha_s^3}(3)_{m_c}(0)_{\langle\bar dd\rangle}(18)_{\langle\alpha_s G^2\rangle}(12)_{\langle\bar dGd\rangle}(0)_{\langle g^3G^3\rangle}(1)_{\langle\bar dd\rangle^2}. \tag{9.70}$$

The error from μ has been multiplied by 2 to be conservative. The largest error comes from τ which is due to the inaccurate localization of the inflexion point. The one due to t_c is smaller as expected in the ratio which is not the case for the direct extraction of the decay constants. The ones due to $\langle\alpha_s G^2\rangle$ and $\langle\bar dGd\rangle$ are large due to the opposite sign of their contributions in the vector and pseudoscalar channels which add when taking the ratio. Using the value $f_D = 204(6)$ MeV in Eq. (9.66) and the ratio in Eq. (9.69), we deduce the improved value:

$$f_{D^*} = 248.5(10.4)\ \text{MeV}, \tag{9.71}$$

where the errors have been added quadratically. Taking the mean of the two results in Eqs. (9.67) and (9.71), one obtains:

$$\langle f_{D^*}\rangle = 249.7(10.5)(1.2)_{\rm syst} = 250(11)\ \text{MeV}, \tag{9.72}$$

where the first error comes from the most precise determination and the second one from the distance of the mean value to it.

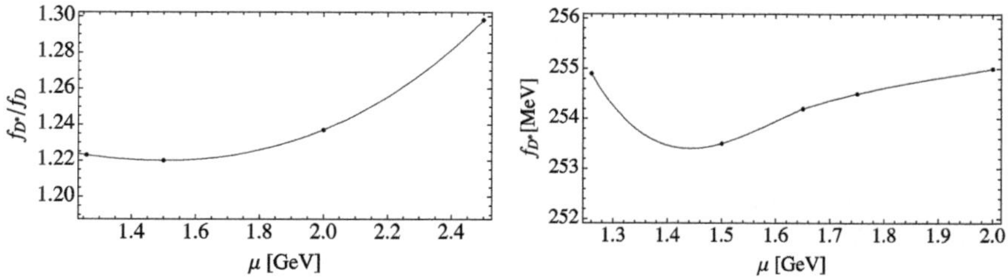

Figure 9.19. (a) f_{D^*}/f_D versus the subtraction point μ; (b) the same as (a) but for f_{D^*}.

• Upper bound on f_{D^*}

We derive an upper bound on f_{D^*} by considering the positivity of the QCD continuum contribution to the spectral function and by taking the limit where $t_c \to \infty$ in Eq. (9.59) which corresponds to a full saturation of the spectral function by the lowest ground-state contribution. The result of the analysis versus the change of τ for a given value of $\mu = 1.5$ GeV is given in Fig. 9.20(a), where one can observe like in the previous analysis the presence of a τ-inflexion point. We also show the good convergence of the PT series by comparing the result at N2LO and the one including an estimate of the N3LO term based on the geometric growth of the PT coefficients. We show in Fig. 9.20(b) the variation of the optimal bound versus the subtraction point μ, where we find a region of μ stability from 1.5 to 2 GeV. We obtain:

$$f_{D^*} \leq 267(10)_\tau(14)_{\rm svz}(0)\mu \text{ MeV}, \tag{9.73}$$

with:

$$\begin{aligned}(14)_{\rm svz} = {}& (1.4)_{\alpha_s}(13)_{\alpha_s^3}(1.3)_{m_c}(3)_{\langle\bar dd\rangle}\\ & (3)_{\langle\alpha_s G^2\rangle}(2.5)_{\langle\bar dGd\rangle}(0)_{\langle g^3G^3\rangle}(0.7)_{\langle\bar dd\rangle^2}.\end{aligned} \tag{9.74}$$

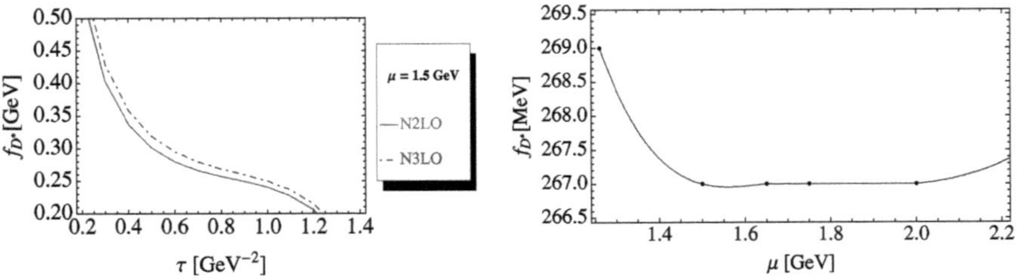

(a) (b)

Figure 9.20. Upper bounds on f_{D^*}: (a) at different values of τ for a given value μ =1.5 GeV of the subtraction point μ; one can note a good convergence of the PT series by comparing the calculated N2LO and estimated N3LO terms; (b) versus the values of the subtraction point μ.

Alternatively, we combine the upper bound $f_D \leq 218.4(1.4)$ MeV obtained in Refs. [455, 456] with the ratio in Eq. (9.69) and deduce:

$$f_{D^*} \leq 266(8) \text{ MeV}, \tag{9.75}$$

where we have added the errors quadratically. The good agreement of the results in Eqs. (9.73) and (9.75) indicates the self-consistency of the approaches. This bound is relatively strong compared to the estimate in Eq. (9.72). A comparison of our results with the ones from some other sources is shown in Table 9.4.

9.2.6 The decay constants f_B and f_{B^*} from SR in full QCD

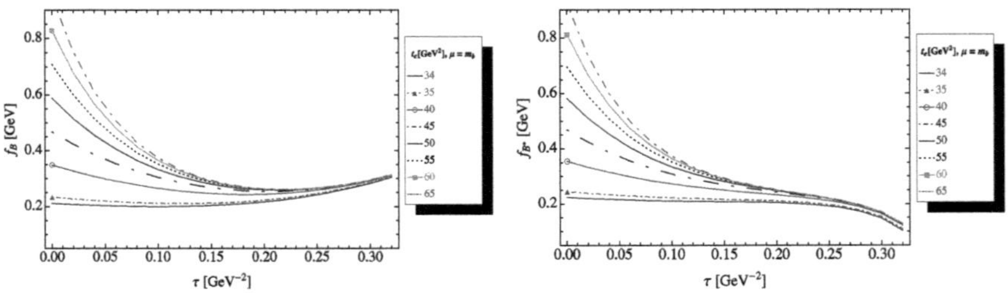

(a) (b)

Figure 9.21. (a) τ-behaviour of f_B from $\mathcal{L}_P$ for different t_c at $\mu = m_b$; (b) the same as in (a) but for f_{B^*} from $\mathcal{L}_V$.

• Direct estimate of f_B and f_{B^*}

We extend the analysis to the case of the B^* meson. We use the set of parameters in Table 1. The τ-behaviours of f_B and f_B^* in Fig. 9.21 have a shape similar to the case of f_D and f_{D^*}. We estimate f_{B^*} from Fig. 9.21(b) where the τ-stability is reached from $t_c = 34$ GeV^2, while the t_c-one starts from $t_c = (55-60)$ GeV^2. We show the μ-behaviour of the optimal result in Fig. 9.22. At the inflexion point $\mu = (5.5\pm0.5)$ GeV, we deduce:

$$\begin{aligned} f_{B^*} &= 239(38)_{t_c}(1)_\tau(2.7)_{\rm svz}(1.4)_\mu = 239(38)\ \text{MeV}, \\ \text{with :}\quad & (2.7)_{\rm svz} = (0.7)_{\alpha_s}(2)_{\alpha_s^3}(0.4)_{m_b}(1.6)_{\langle\bar dd\rangle}, \\ & (0.4)_{\langle\alpha_s G^2\rangle}(0.3)_{\langle\bar dGd\rangle}(0)_{\langle g^3G^3\rangle}(0)_{\langle\bar dd\rangle^2}. \end{aligned} \tag{9.76}$$

• The ratio f_{B^*}/f_B

We extract directly the ratio f_{B^*}/f_B from the ratio of sum rules. We show in Fig. 9.23(a) its τ-behaviour for different values of t_c for $\mu = 5.5$ GeV from which we deduce as optimal value the mean of the τ-minima obtained from $t_c = 34$–60 GeV^2. We show in Fig. 9.23(b) the μ-behaviour of these optimal results where we find a minimum in μ around 3.8–4.5 GeV and a slight inflexion point around 5.5 GeV. We consider as a final result the mean of the ones from these two regions of μ:

$$f_{B^*}/f_B = 1.016(12)_{t_c}(1)_\tau(9)_{\rm svz}(6)_\mu = 1.016(16), \tag{9.77}$$

with:

$$(9)_{\rm svz} = (3)_{\alpha_s}(6)_{\alpha_s^3}(3)_{m_b}(4)_{\langle\bar dd\rangle}(3)_{\langle\alpha_sG^2\rangle}.(1)_{\langle\bar dGd\rangle}(1)_{\langle g^3G^3\rangle}(1)_{\langle\bar dd\rangle^2}. \tag{9.78}$$

Combining the results in Eq. (9.77) with the value $f_B = 206(7)$ MeV obtained in [455, 456], we deduce:

$$f_{B^*} = 209(8)\ \text{MeV}, \tag{9.79}$$

which is more accurate than the direct determination in Eq. (9.76) and where the main error comes from the one of f_B extracted in Refs. [455, 456]. We consider the result in Eq. (9.79) which is

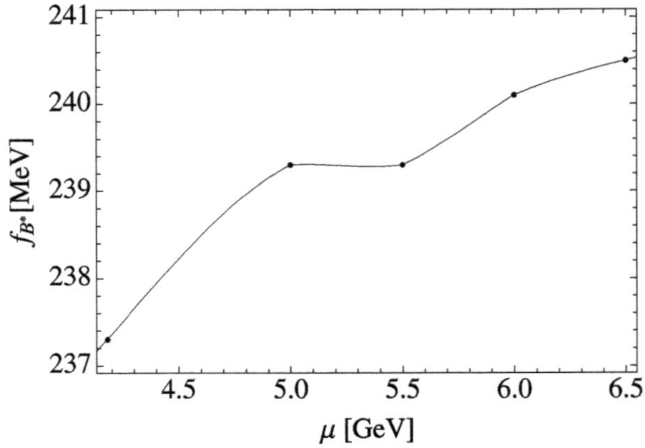

Figure 9.22. Values of f_{B^*} at different values of the subtraction point μ.

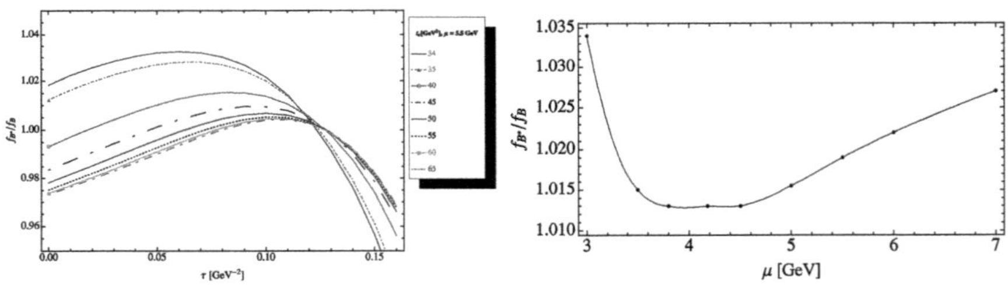

(a) (b)

Figure 9.23. f_{B^*}/f_B: (a) τ-behaviour for different t_c at $\mu = 5.5$ GeV; (b) μ-behaviour.

also the mean of the results in Eqs. (9.79) and (9.76) as our final determination.

• Upper bound on f_{B^*}

Like in the case of the D^* meson, we extract directly an upper bound on f_{B^*} by using the positivity of the QCD continuum to the spectral function. We show the τ- and the μ-behaviours of the optimal bound in Fig. 9.24. We deduce, at $\mu = (6 \pm 0.5)$ GeV:

$$f_{B^*} \leq 295(14)_\tau(4)_{\text{svz}}(10)_\mu = 295(18) \text{ MeV}. \tag{9.80}$$

We consider the previous values of f_{D^*} and f_{B^*} as improvement of our earlier results in Refs. [7, 168].

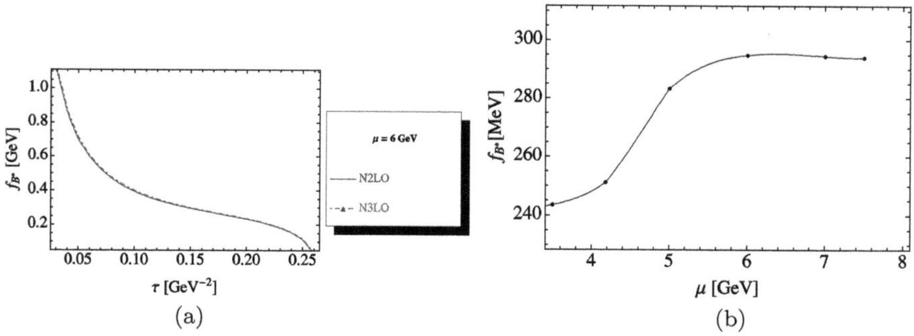

Figure 9.24. Upper bound of f_{B^*} from $\mathcal{L}_P$: (a) τ-behaviour at $\mu = 6$ GeV for N2LO and N3LO truncation of the PT series; (b) μ-behaviour.

9.2.7 f_{B^*}/f_B from HQET sum rules

This ratio is under good control from heavy quark effective theory (HQET) sum rules. It reads in the large mass limit [180, 220, 468]:

$$\begin{aligned} R_b &\equiv \frac{f_{B^*}\sqrt{M_{B^*}}}{f_B\sqrt{M_B}} = R_{PT} + \frac{R_{NPT}}{M_b}, \\ R_{\rm PT} &= 1 - \frac{2}{3}a_s - 6.56a_s^2 - 84.12a_s^3 = 0.892(13), \\ R_{\rm NPT} &= \frac{2}{3}\bar{\Lambda}\left(1 + \frac{5}{3}a_s\right) - 4G_\Sigma\left(1 + \frac{3}{2}a_s\right), \end{aligned} \tag{9.81}$$

with $a_s \equiv (\alpha_s/\pi)(M_b) = 0.070(1)$, $\bar{\Lambda} \equiv M_B - M_b = (499 \pm 60)$ MeV, where we have used the pole mass $M_b = 4.810(60)$ MeV to order α_s^2 deduced from the running mass $\bar{m}_b(m_b) = 4.177(11)$ MeV [290, 291] and we have retained the larger error from PDG [33]. We have estimated the error in the PT series by assuming a geometric growth of the coefficients [234]:

$$G_\Sigma(M_b) \equiv \hat{G}_\Sigma[\alpha_s(M_b)]^{-3/2\beta_1} + \frac{8}{27}\bar{\Lambda}, \tag{9.82}$$

where the invariant quantity $\hat{G}_\Sigma \simeq -(0.20 \pm 0.01)$ GeV has been extracted inside the t-stability region of Fig. 7 from Ref. [220]. We obtain:

$$R_b = 0.938(13)_{PT}(5)_{NP}(6)_{M_b,\bar{\Lambda}} = 0.938(15). \tag{9.83}$$

To convert the above HQET result to the one of the full theory at finite quark and meson masses, we have to include the normalization factor $(M_B/M_b)^2$ relating the pseudoscalar to the universal HQET correlators according to the definition in Eq. (9.56). Then, we deduce (see, e.g., Refs. [220, 456, 470]):

$$\frac{f_{B^*}}{f_B} = \left(\frac{M_B}{M_b}\right)\sqrt{\frac{M_B}{M_{B^*}}}R_b = 1.025(13)_{M_b}(15)_{R_b}, \tag{9.84}$$

in fair agreement with our result in Eq. (9.77) and the ones in [220, 464, 469] but higher than the ones in Refs. [471, 490]. To make a direct comparison of our results with the lattice calculations, it is desirable to have a direct lattice calculation from the pseudoscalar correlator built from the current in Eq. (9.56). The result of [490] is difficult to interpret as they use a non-standard threshold of the QCD continuum. However, their requirement of maximal stability for $\tau \leq 0.15$ GeV^{-2} is outside the "standard sum rule (SR) window" obtained around 0.2–0.3 GeV^{-2} (minimum or inflexion point in our figures) where the lowest resonance dominates the sum rule. Taking literally their set $(t_c, \tau) = (33$ GeV^2, 0.05 GeV$^{-2})$ where their "duality" is obtained (Fig. 3 of Ref. [490]) into our Fig. 9.23, one obtains $f_{B^*}/f_B = 0.985$ more comparable with their result of 0.944, but meaningless from the SR point of view as it comes from a region dominated by the QCD continuum.

9.2.8 SU(3) breaking for $f_{D_s^*}$ and $f_{D_s^*}/f_{D^*}$

We pursue the same analysis for studying the $SU(3)$ breaking for $f_{D_s^*}$ and the ratio $f_{D_s^*}/f_{D^*}$. We work with the complete massive ($m_s \neq 0$) LO expression of the PT spectral function obtained in Ref. [29] and the massless ($m_s = 0$) expression known to N2LO used in the previous sections. We include the NLO PT corrections due to linear terms in m_s obtained in Ref. [469]. We show the τ-behaviour of the results in Fig. 9.25(a) for a given $\mu = 1.5$ GeV and different t_c. We study their μ-dependence in Fig. 9.25(b) where a nice μ stability is reached for $\mu \simeq 1.4\text{–}1.5$ GeV. We have used $(t_c^{D_s^*})^{1/2} - (t_c^{D^*})^{1/2} \approx M_{D_s^*} - M_{D^*} = 102$ MeV. Taking the conservative result ranging from the beginning of τ-stability ($t_c \simeq 5.7$ GeV^2) until the beginning of

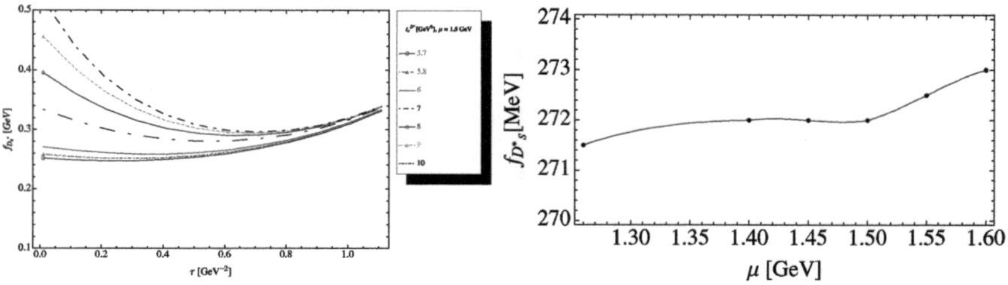

(a) (b)

Figure 9.25. $f_{D_s^*}$ from $\mathcal{L}_V$: (a) τ-behaviour for different t_c at μ =1.5 GeV; (b) μ-behaviour.

t_c-stability of about (9–10) GeV^2, we obtain, at μ=1.5 GeV:

$$f_{D_s^*} = 272(24)_{t_c}(2)_\tau(18)_{\rm svz}(2)_\mu = 272(30)\ \text{MeV},$$

with:

$$\begin{aligned}(18)_{\rm svz} &= (0)_{\alpha_s}(14)_{\alpha_s^3}(1)_{m_c}(2)_{\langle\bar dd\rangle}(1.5)_{\langle\alpha_s G^2\rangle}\\ &(0.8)_{\langle\bar dGd\rangle}(0)_{\langle g^3G^3\rangle}(0)_{\langle\bar dd\rangle^2}(0.3)_{m_s}(2)\kappa.\end{aligned} \tag{9.85}$$

Taking the PT linear term in m_s at lowest order and $t_c = 7.4$ GeV^2, we obtain $f_{D_s^*} = 291$ MeV in agreement with the one 293 MeV of Ref. [469] obtained in this way. The inclusion of the complete LO term decreases this result by about 5 MeV, while the inclusion of the NLO PT $SU(3)$ breaking terms increases the result by about the same amount. Combining the result in Eq. (9.85) with the one in Eq. (9.72), we deduce the ratio:

$$f_{D_s^*}/f_{D^*} = 1.090(70), \tag{9.86}$$

where we have added the relative errors quadratically. Alternatively, we extract directly the previous ratio using the ratio of sum rules. We show the results in Fig. 9.26(a) versus τ and for different values of t_c at $\mu = 1.5$ GeV. τ-stabilities occur from $\tau \simeq 1$ to 1.5 GeV^{-2}. We also show in Fig. 9.26(b) the μ-behaviour of the results where a good stability in μ is observed for $\mu \simeq (1.4–1.5)$ GeV in the same way as for $f_{D_s^*}$. We deduce

$$f_{D_s^*}/f_{D^*} = 1.073(1)_{t_c}(16)_\tau(2)_\mu(50)_{\rm svz} = 1.073(52),$$

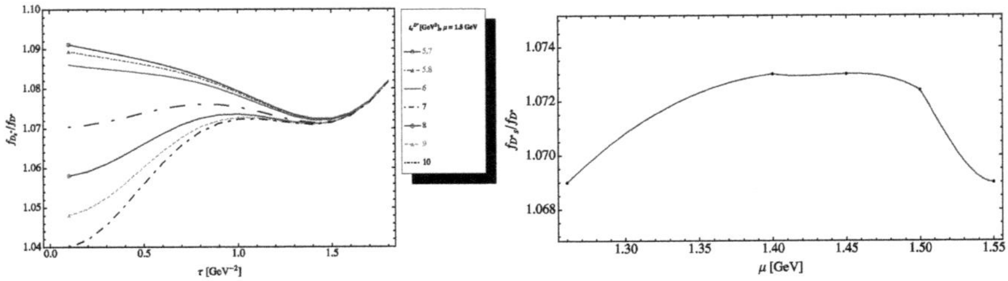

(a) (b)

Figure 9.26. $f_{D_s^*}/f_{D^*}$: (a) τ-behaviour for different t_c and for $\mu = 1.5$ GeV; (b) μ-behaviour.

with:

$$(50)_{\rm SVZ} = (1)_{\alpha_s}(45)_{\alpha_s^3}(0)_{m_c}(2)_{\langle\bar{d}d\rangle}(16)_{\langle\alpha_s G^2\rangle}$$
$$(3)_{\langle\bar{d}Gd\rangle}(1)_{\langle g^3G^3\rangle}(2)_{\langle\bar{d}d\rangle^2}(4)_{m_s}(13)\kappa, \tag{9.87}$$

where, for asymmetric errors, we have taken the mean of the two extremal values. The error associated to τ take into account the fact that, for some values of t_c, the τ-minima for f_{D^*} and $f_{D_s^*}$ do not coincide. Comparing Eqs. (9.86) and (9.87), one can see the advantage of a direct extraction from the ratio of moments due to the cancellation of systematic errors in the analysis. Taking the mean of Eqs. (9.86) and (9.87), we deduce:

$$f_{D_s^*}/f_{D^*} = 1.08(6)(1)_{\rm syst}, \tag{9.88}$$

where the first error comes from the most precise determination and the second one from the distance of the mean value to the central value of this precise determination. Using Eq. (9.88), f_{D^*} in Eq. (9.72) and its upper bound in Eq. (9.75), we predict in MeV:

$$f_{D_s^*} = 270(19), \quad f_{D_s^*} \leq 287(8.6)(16) = 287(18). \tag{9.89}$$

Future experimental measurements of f_{D^*} and $f_{D_s^*}$ though most probably quite difficult should provide a decisive selection of these existing theoretical predictions.

9.2.9 SU(3) breaking for $f_{B_s^*}$ and $f_{B_s^*}/f_{B^*}$

We extend the analysis done for D_s^* to the case of the B_s^*-meson. We show, in Fig. 9.27(a), the τ-behaviour of the ratio $f_{B_s^*}/f_{B^*}$ at μ=5

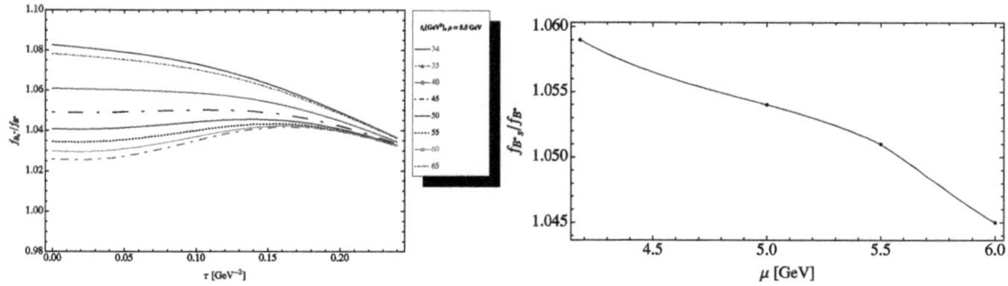

(a) (b)

Figure 9.27. $f_{B_s^*}/f_{B^*}$: (a) τ-and t_c-behaviours at a given $\mu = 5$ GeV; (b) μ-behaviour.

GeV and for different values of t_c where the τ-stability starts from $t_c = 40$ GeV^2, while the t_c one is reached for $t_c \simeq (60-65)$ GeV^2. Our optimal result is taken in this range of t_c. We study the μ-behaviour in Fig. 9.27(b) where an inflexion point is obtained for $\mu = (5 \pm .5)$ GeV. At this point, we obtain:

$$f_{B_s^*}/f_{B^*} = 1.054(8)_{t_c}(0)_\tau(6)_\mu(4.6)_{\rm SVZ} = 1.054(11),$$

with:

$$(4.6)_{\rm SVZ} = (2)_{\alpha_s}(2.5)_{\alpha_s^3}(0)_{m_b}(2)_{\langle\bar{d}d\rangle}(1.5)_{\langle\alpha_s G^2\rangle}$$
$$(0)_{\langle\bar{d}Gd\rangle}(0)_{\langle g^3G^3\rangle}(0)_{\langle\bar{d}d\rangle^2}(0)_{m_s}(2)\kappa. \qquad (9.90)$$

We show, in Fig. 9.28(a), the τ-behaviour of the result for $f_{B_s^*}$ at $\mu = 5.5$ GeV and for different values of t_c. For $f_{B_s^*}$, τ-stability starts from $t_c \simeq 34$ GeV^2, while t_c-stability is reached from $t_c \simeq (50-65)$ GeV^2. We show in Fig. 9.28(b) the μ-behaviour of these optimal

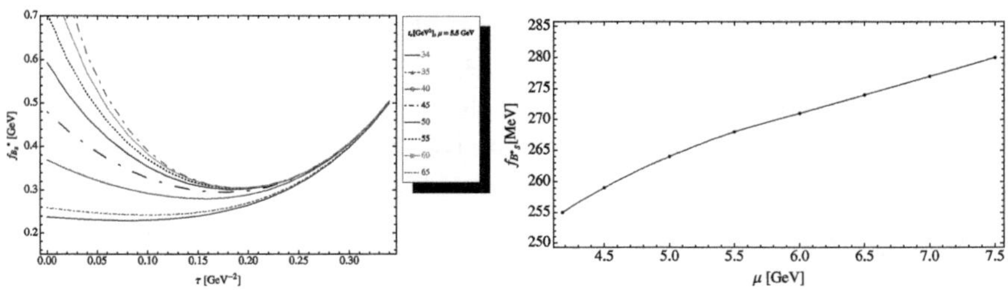

(a) (b)

Figure 9.28. $f_{B_s^*}$: (a) τ-behaviour for different t_c at a given $\mu = 6$ GeV; (b) μ-behaviour.

results. One can note a slight inflexion point at μ=6 GeV which is about the one (5.0–5.5) GeV obtained previously for the ratio $f_{B_s^*}/f_{B^*}$. At this value of μ, we obtain:

$$\begin{aligned} f_{B_s^*} &= 271(39)_{t_c}(0)_\tau(3)_{\rm svz}(6)_\mu = 271(40)\ \text{MeV}, \\ \text{with :}\quad & (3)_{\rm svz} = (1)_{\alpha_s}(1.5)_{\alpha_s^3}(0.5)_{m_b}(2)_{\langle\bar dd\rangle}(0.5)_{\langle\alpha_s G^2\rangle} \\ & (0.5)_{\langle\bar dGd\rangle}(0)_{\langle g^3G^3\rangle}(0.5)_{\langle\bar dd\rangle^2}(0)_{m_s}(1)_\kappa. \end{aligned} \tag{9.91}$$

Combining this result with the one of f_{B^*} in Eq. (9.76) obtained within the same approach and conditions, we deduce the ratio:

$$f_{B_s^*}/f_{B^*} = 1.13(25), \tag{9.92}$$

where the large error is due to the determinations of each absolute value of the decay constants. We take as a final value of the ratio $f_{B_s^*}/f_{B^*}$ the most precise determination in Eq. (9.90). Combining this result with the final value of f_{B^*} in Eq. (9.79) and with the upper bound in Eq. (9.80), we deduce our final estimate:

$$f_{B_s^*} = 220(9)\ \text{MeV}, \qquad f_{B_s^*} \leq 311(19)\ \text{MeV}. \tag{9.93}$$

9.2.10 The scalar meson decay constants $f_{D^*_{0(0s)}}$ and $f_{B^*_{0(0s)}}$

• Due to large discrepancies of the $f_{D_0^*}$ values among different SR [307, 472, 473] and between lattice [475, 476] results (see Table 9.4), we update our previous estimates in Refs. [7, 168, 301]. We use the same QCD inputs (N2LO for PT within the $\overline{MS}$-scheme $\oplus$ $d \leq 6$ condensates) and the same criteria as for $f_{D,B}$ [455] for obtaining the optimal results. The difference between the scalar and pseudoscalar SR is the overall factor $(M_Q \pm m_s)^2$ and the signs of the $\langle\bar qq\rangle$ and mixed condensate effects for $m_s = 0$. We show $f_{D_0^*}$ versus τ in Figs. 9.29(a) and (b) where the τ-stability about (0.4−0.6) GeV^{-2} of the SR for $\mu = \tau^{-1/2}$ constrains t_c inside (7−9) GeV^2 (Fig. 9.29(a)).[10] In Fig. 9.29(b), we show the SR for a given $\mu = 2$ GeV and t_c = (7−11) GeV^2-stability. The effect of the truncation of

[10] Larger t_c corresponds to an apparent small τ-stability outside the SR window and leads to a larger $f_{D_0^*}$ (Fig. 9.29(b)).

the PT series is analyzed by using instead the invariant quark mass in Eq. (9.63) and the QCD scale Λ. We show in Fig. 9.29(c) the results versus μ and using either the running (blue open circle data points) or the invariant (red triangle data) mass. Taking their average, we deduce the value in Table 9.4 where 95% of the error comes from the above range of t_c-values. The error becomes 12 instead of 7 if one considers the most precise determination. Upper bound is derived in Fig. 9.30 using the positivity of the spectral function. We extract from the SR at $\mu = \sqrt{1/\tau}$ the SU(3) breaking ratio $f_{D^*_{0s}}/f_{D^*_0}$ with the result in Table 9.4. The error of the ratio comes mainly from the $SU(3)$ breaking terms. These results confirm and improve the previous one in Ref. [301].

• Doing similar analysis for $f_{B^*_0}$ and $f_{B^*_{0s}}$, we note that only the SR subtracted at $\mu = \tau^{-1/2}$ (Fig. 9.29(d)) exhibits τ-stability for $t_c \simeq$ (35–38) GeV^2. We use $M_{B^*_0} \simeq M_{B^*_{0s}} = 5733$ MeV, assuming that $M_{B^*_0} - M_B \simeq M_{D^*_0} - M_D = 448$ MeV supported by the SR estimate 413(210) MeV in Refs. [7, 168]. We deduce the results in Table 9.4, where like for $f_{D^*_0}$, the errors come mainly from t_c and the $SU(3)$ breaking parameters. These results improve and confirm previous estimates in Refs. [7, 168].

• $f_{D^*_0}$ agrees with the (inaccurate) phenomenological value 206(120) MeV from $B \to D^{**}\pi$ [475]. The agreement for $f_{B^*_0}$ but not $f_{D^*_0}$ with the SR one in Ref. [307] is puzzling and could be due to a different treatment of the QCD input and evolution parameters which are sensitive at low scale. A comparison with some other determinations are given in Table 9.4. A future precise measurement of $f_{D^*_0}$ will select among the theoretical estimates.

9.2.11 The axial meson decay constants f_{D_1} and f_{B_1}

In the zero light quark mass limit, the axial-vector correlator can be deduced from the vector one by changing the sign of the chiral condensate contributions. Using the expression known to N2LO $\oplus$ $d \leq 6$ condensates, we extract directly the ratio f_{D_1}/f_D and f_{B_1}/f_B by taking each value of f_{D_1} and f_{B_1} at a τ-minimum for a given (μ, t_c) (Figs. 9.31(a) and (b)) and the corresponding value of f_D (Fig. 9.17) and f_B (Fig. 9.21) at the same of set (μ, t_c). We use $M_{B_1} - M_{B^*} \simeq M_{B^*_0} - M_B$=448 MeV as indicated by the result 417(212) MeV in

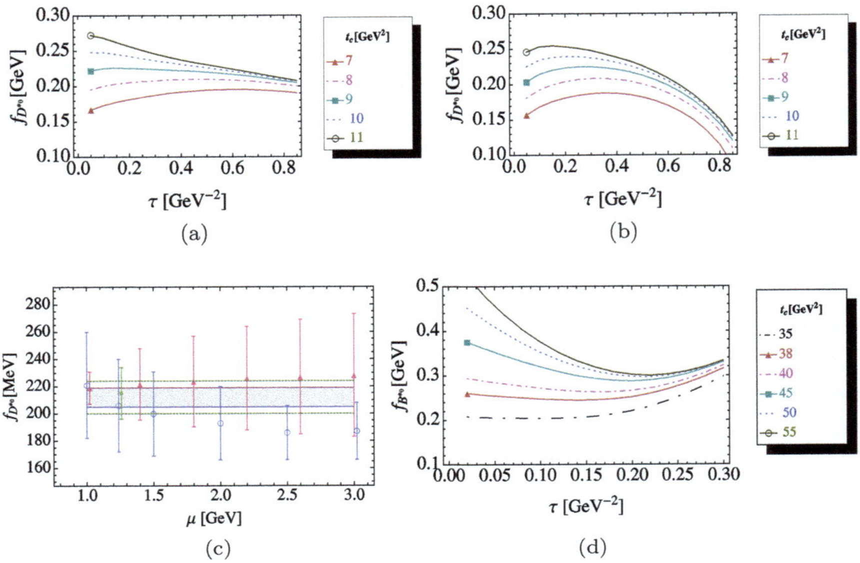

Figure 9.29. (a) $f_{D_0^*}$ versus τ and t_c for $\mu = \tau^{-1/2}$; (b) the same as (a) but for $\mu = 2$ GeV; (c) optimal value of $f_{D_0^*}$ versus μ: blue open circles are data from SR with running mass. Red triangles are from SR using invariant mass. The green open circle comes from the SR for $\mu = \tau^{-1/2}$. The grey region is the average with averaged errors. The green dotted horizontal lines correspond to the error coming from the best determination; (d) $f_{B_0^*}$ versus τ and t_c for $\mu = 1/\sqrt{\tau}$. A similar behaviour is obtained for $f_{B_{0s}^*}$.

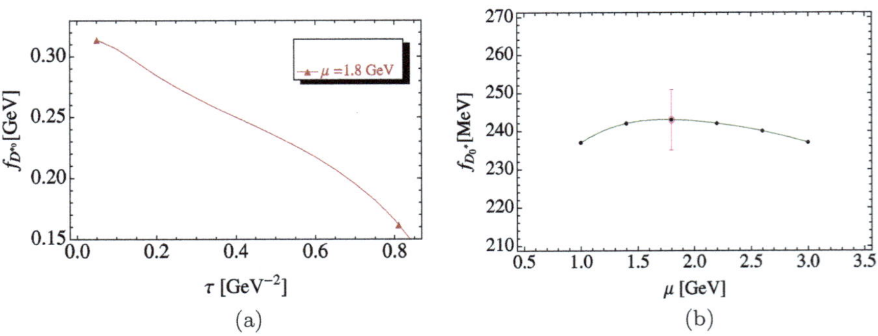

Figure 9.30. Upper bound of $f_{D_0^*}$: (a) versus τ for $\mu = 1.8$ GeV; (b) for different μ. The red open circle is the optimal value.

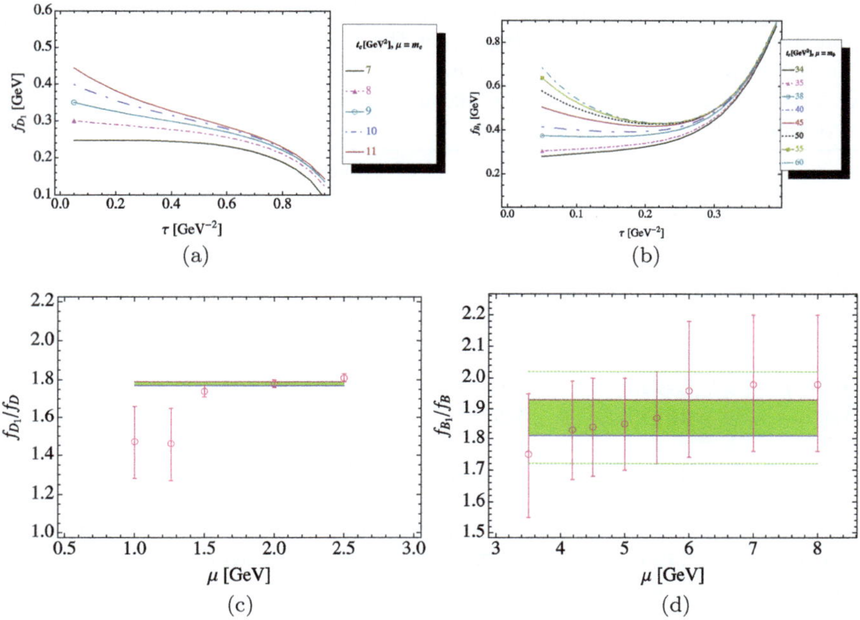

Figure 9.31. (a) f_{D_1} versus τ and t_c for $\mu = m_c$; (b) the same as (a) but for f_{B_1}; (c) optimal value of f_{D_1} for different μ: the green dashed line denotes the average; the green region denotes the value if one takes the errors from the weighted average; (d) The same as (c) but for f_{B_1}. The horizontal green lines correspond to the error from the best determination.

Refs. [7, 168]. The optimal ratios obtained in this way are given for different μ in Figs. 9.31(c) and (d). The average given in Table 9.4 comes from the dashed coloured regions.

9.2.12 The decay constants f_{B_c} and $f_{B_c^*}$

We complete the analysis by the estimate and bound of the decay constant f_{B_c} of the $B_c(6277)$ meson $\bar{b}c$ bound state and of $f_{B_c^*}$ of its vector partner B_c^*, where the light quarks d and s are replaced by the heavy quark c.

Our analysis will be very similar to the one in Refs. [168, 217, 221], but we shall use the running c and b quark masses instead of the pole masses and we shall include N2LO radiative corrections.

Table 9.4. Pseudoscalar $D_{(s)}, B_{(s)}$ meson decay constants. NS2R refers to non-standard sum rules not using either the minimal QCD continuum ansatz nor the exponential/moments weights. The errors of the global average are rescaled by 1.2 as in Ref. [449] for accounting slight tensions between different determinations.

			Sources	Refs.
f_D [MeV]	f_{D_s} [MeV]	f_{D_s}/f_D		
204(6)	246(6)	1.21(4)	SR	[455]
203(23)	235(24)	1.15(4)	SR	[167, 465]
$\leq 218(2)$	$\leq 254(2)$	–	SR	[455]
≤ 279	$\leq 321(12)$	–	SR	[465]
≤ 230	≤ 270	–	SR	[467]
201(13)	238(18)	1.15(5)	SR	[469]
208(10)	240(10)	1.15(6)	SR	[307]
195(20)	–	–	HQET-SR	[470]
204.0(4.6)	**243.2(4.9)**	**1.170(23)**	**SR Average**	
209.2(3.3)	249.3(1.8)	1.187(2)	Latt average $2 \oplus 1$	[449, 462]
212.1(1.2)	248.8(1.3)	1.172(3)	Latt average $2 \oplus 1 \oplus 1$	[478, 479]
211.8(1.1)	**249.0(1.1)**	**1.173(3)**	**Latt final average**	
206(9)	245(16)	1.19(3)	NS2R	[480]
177(21)	205(22)	1.16(3)	NS2R	[481]
211.2(1.1)	**249.1(1.1)**	**1.173(3)**	**Global average**	
203.7(4.7)	**257.8(4.1)**	–	**Data**	[449]
f_B [MeV]	f_{B_s} [MeV]	f_{B_s}/f_B		
206(7)	234(5)	1.14(3)	SR	[455]
203(23)	236(30)	1.16(5)	SR	[167, 465]
$\leq 235(4)$	$\leq 251(6)$	–	SR	[455]
197(23)	232(25)	–	SR	[466]
194(15)	231(16)	1.19(10)	SR	[307]
207(13)	242(15)	1.17(4)	SR	[469]
199(29)	–	–	HQET-SR	[456]
206(20)	–	–	HQET-SR	[470]
204.0(5.1)	**234.5(4.4)**	**1.154(21)**	**SR Average**	
190(4)	227(2)	1.207(12)	Latt average $2 \oplus 1$	[421, 462]
186(4)	227(4)	1.216(8)	Latt average $2 \oplus 1 \oplus 1$	[482, 483]
188(3)	**227(2)**	**1.213(7)**	**Latt final average**	
186(14)	222(12)	1.19(5)	NS2R	[484]
192(15)	228(20)	1.18(2)	NS2R	[485]
191.9(2.5)	**228.1(1.8)**	**1.204(6)**	**Global average**	
196(24)	–	–	**Data**	[449]

Table 9.5. Scalar $D^*_{0(s)}, B^*_{0(s)}$ meson decay constants. Values marked with † are not included in the average. The one with * has been rescaled by 1.5 for accounting a slight tension with our estimate.

			Sources	Ref.
$f_{D^*_0}$ [MeV]	$f_{D^*_{0s}}$ [MeV]	$f_{D^*_{0s}}/f_{D^*_0}$		
122(43)†–360(90)†	–	–	Latt.	[475, 476]
128(13)†–373(19)†	–	–	SR	[307, 472, 473]
217(24)	202(23)	0.93(2)	SR	[301]
221(12)	202(19)	0.912(23)	SR	New
$\leq$ 243(8)	$\leq$ 222(6)	–	SR	New
220(11)	**202(15)**	**0.922(15)**	**SR Average**	
$f_{B^*_0}$ [MeV]	$f_{B^*_{0s}}$ [MeV]	$f_{B^*_{0s}}/f_{B^*_0}$		
263(51)	–	–	SR	[7, 168]
281(14)	274(20)*	–	SR	[307]
271(26)	233(21)	0.865(54)	SR	New
$\leq$ 302(8)	$\leq$ 261(17)	–	SR	New
278(12)	**255(15)**	–	**SR Average**	

• The dynamics of B_c and B_c^* is expected to be different from the B and B^* one because, by using the heavy quark mass expansion, the heavy quark $\langle\bar{c}c\rangle$ and quark-gluon mixed $\langle\bar{c}Gc\rangle$ condensates defined in Section 2 behave as [217]:

$$\langle\bar{c}c\rangle = -\frac{1}{12\pi m_c}\langle\alpha_s G^2\rangle - \frac{\langle g^3G^3\rangle}{1440\pi^2 m_c^3},$$
$$\langle\bar{c}Gc\rangle = \frac{m_c}{\pi}\left(\log\frac{m_c}{\mu}\right)\langle\alpha_s G^2\rangle - \frac{\langle g^3G^3\rangle}{48\pi^2 m_c}. \tag{9.94}$$

These behaviours are in contrast with the ones of the light quark $\langle\bar{q}q\rangle$ and mixed quark-gluon $\langle\bar{q}Gq\rangle$ condensates [6, 7].

• The complete expression of the perturbative NLO pseudoscalar spectral function has been obtained in Ref. [157] and explicitly written in Ref. [217], where we transform the pole to the running masses. We add to this expression the N2LO result in Ref. [460] for $m_c = 0$. We consider as a source of errors the N3LO contribution

Table 9.6. Vector D^*, B^* and axial D_1, B_1 meson decay constants. The $n_f = 2$ lattice and global average errors marked with * is rescaled by 1.5 for accounting slight tensions between different estimates.

f_{D^*} [MeV]	f_{D^*}/f_D	f_{B^*} [MeV]	f_{B^*}/f_B	Sources	Ref.
250(11)	1.218(36)	209(8)	1.016(16)	SR	[457]
242(16)	1.20(8)	210(11)	1.02(5)	SR	[469]
263(21)	–	213(18)	–	SR	[307]
–	–	–	1.025(16)	HQET-SR	New
$\leq$ 266(8)	–	$\leq$ 295(18)	–	SR	[457]
$\leq$ 297	–	$\leq$ 261	–	SR	[469]
250(8)	**1.215(30)**	**210(6)**	**1.020(11)**	**SR Average**	
278(24)*	1.208(27)	–	1.051(17)	Latt. $n_f = 2$	[464]
–	–	175(6)	0.941(26)	Latt. NR	[471]
–	–	–	**1.018(14)**	**Latt Average**	
252(22)	1.22(8)	182(14)	0.944(21)	NS2R	[490]
253(7)	**1.212(20)**	**192(6)***	**1.008(12)***	**Global Average**	
f_{D_1} [MeV]	f_{D_1}/f_D	f_{B_1} [MeV]	f_{B_1}/f_B		
363(11)	1.78(1)	385(18)	1.87(6)	SR	New
$\leq$ 388(5)	–	$\leq$ 440(16)	–	SR	New
333(20)	–	335(18)	–	SR	[307]
356(16)*	–	**360(19)***	**SR Average**		

from an assumed geometric growth of the PT series [234] which mimics the $1/q^2$-term parametrizing the large order terms of PT series [232, 235].

• The Wilson coefficients of the non-perturbative $\langle \alpha_s G^2 \rangle$ and $\langle g^3 G^3 \rangle$ contributions are also given in Ref. [217].

• Like in the previous cases, we study the SR versus τ and t_c, as shown in Fig. 9.32. One can note that the non-perturbative contributions are small (about 1–2 MeV), indicating that the dynamics of the B_c meson is dominated by the perturbative contributions. This feature might explain the success of the non-relativistic potential models for describing the B_c-like hadrons [217]. The optimal result is obtained from $t_c = 44$ GeV^2 (beginning of τ-stability) until $t_c = (50–60)$ GeV^2 (beginning of t_c-stability). We show in Fig. 9.33 the μ-behaviour of the results, where there is an inflexion point for

Table 9.7. $SU(3)$ breakings for the vector D_s^* and B_s^* decay constants.

$f_{D_s^*}$ [MeV]	$f_{D_s^*}/f_{D^*}$	$f_{B_s^*}$ [MeV]	$f_{B_s^*}/f_{B^*}$	Sources	Ref.
270(29)	1.08(6)	220(14)	1.054(11)	SR	[457]
293(17)	1.21(5)	210(11)	1.20(4)	SR	[469]
308(21)	–	255(19)	–	SR	[307]
$\leq$ 287(18)	–	$\leq$ 317(17)	–	SR	[457]
$\leq$ 347	–	$\leq$ 296	–	SR	[469]
290(11)	**1.16(4)**	**221(7)**	**1.064(10)**	**SR Average**	
306(27)	1.21(6)	214(19)	–	NS2R	[490]
311(14)*	1.16(6)	–	–	Latt. $n_f = 2$	[464]
274(6)	–	213(7)	–	Latt. $n_f = 2 \oplus 1$ & NR	[471, 488]
282.5(4.8)	**1.17(3)**	**216.9(4.6)**	**1.064(11)**	**Global Average**	

Table 9.8. Pseudoscalar B_c and vector B_c^* meson decay constants. The $n_f = 2$ lattice error marked by * is rescaled by 1.5. The same for NS2R for accounting the tension with other estimates.

f_{B_c} [MeV]		$f_{B_c^*}$ [MeV]	$f_{B_c^*}/f_{B_c}$	Sources	Ref.
436(40)	–	–	–	SR	[457]
$\leq$ 466(16)	–	–	–	SR	[457]
503(171)	–	–	–	Pot. Mod.	[217]
528(29)*	–	–	–	NS2R	[487]
427(6)	–	–	–	Latt. NR	[486]
434(23)*	–	422(20)*	0.988(27)	Latt. $n_f = 2$	[471]
–		453(20)	1.043(45)	SR	New
–		$\leq$ 505(6)	–	SR	New
432(6)	–	**438(14)**	**1.003(23)**	**Global Average**	

$\mu = (7.5 \pm 0.5)$ GeV for the estimate (Fig. 9.33(a)) and the upper bound (Fig. 9.33(b)). At these optimal points, we deduce:

$$f_{B_c} = 436(38)_{tc}(2)_{\alpha_s}(2)_{\alpha_s^3}(7)_{m_c}(6)_{\mu} = 436(40) \text{ MeV},$$
$$f_{B_c} \leq 466(9)_{\alpha_s}(2)_{\alpha_s^3}(12)_{m_c}(8)_{\mu} = 466(16) \text{ MeV}. \qquad (9.95)$$

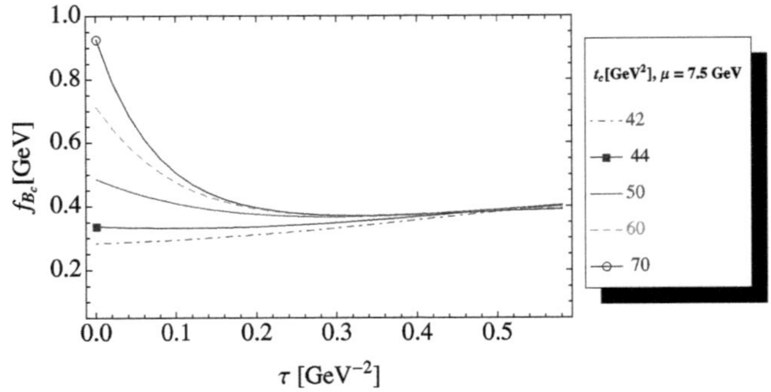

Figure 9.32. τ-behaviour of f_{B_c} from $\mathcal{L}_P$ for different t_c and for $\mu = 7.5$ GeV.

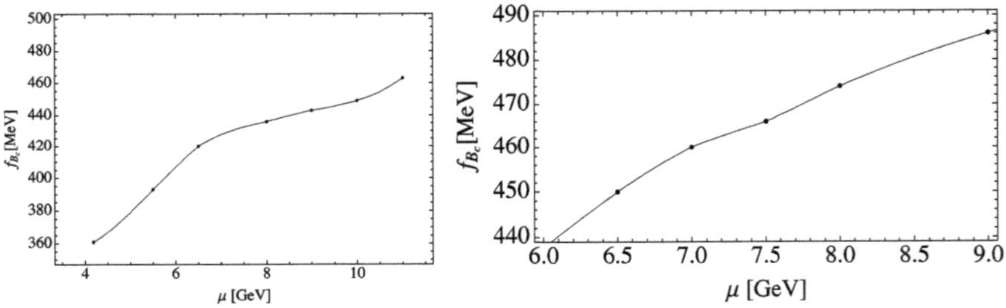

(a) (b)

Figure 9.33. f_{B_c} at different values of the subtraction point μ: (a) estimate; (b) upper bound.

We may consider these results as a confirmation and improvement of the earlier ones in Refs. [7, 168, 217, 221]. The results in Eq. (9.95) will restrict the wide range of f_{B_c} given in the current literature and may be used for extracting the CKM angle V_{cb} from the $B_c \to \tau\nu_\tau$ leptonic width.

• We do a similar analysis for the B_c^* vector meson where we replace the strange by the charm quark mass in the B_s^* SR expression.[11] We use the rescaled $M_{B_c^*} = 6350(20)$ MeV from potential model [217] using the experimental value of M_{B_c}. The different steps are shown in Fig. 9.34. We obtain the results in Table 9.4 for $\tau \simeq 0.15$ GeV^{-2}.

[11] A complete expression of the spectral function to NLO is given in Ref. [62], but it has singularity for $m_c \to 0$ and needs to be checked.

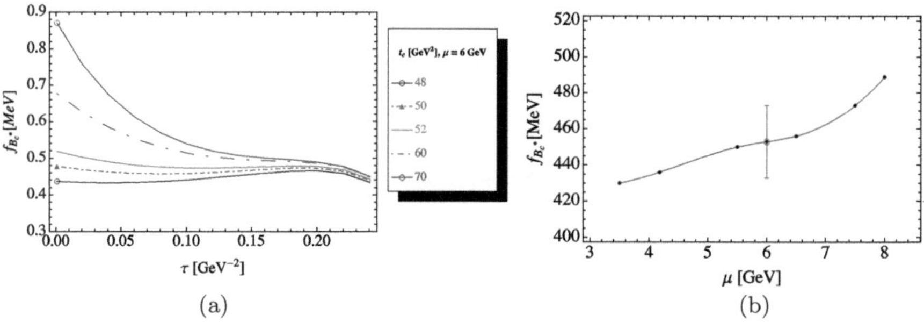

Figure 9.34. $f_{B_c^*}$: (a) versus τ for different t_c and at a given μ; (b) optimal values at $\tau \simeq 0.1$ GeV^{-2} versus μ. The open red circle denotes the optimal value at the inflexion point.

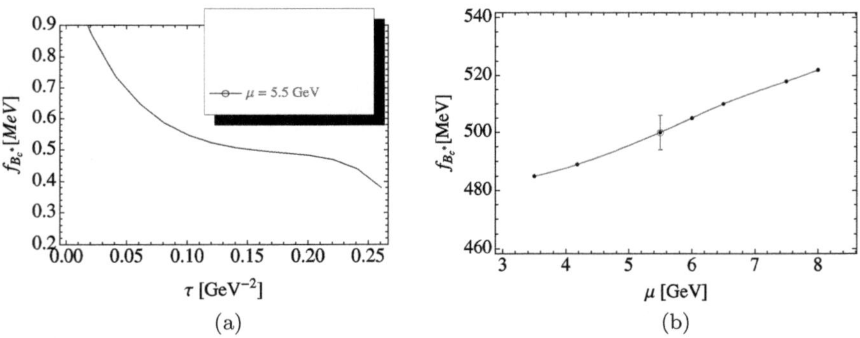

Figure 9.35. $f_{B_c^*}$ bound: (a) versus τ for different t_c and at a given μ; (b) optimal values at $\tau \simeq 0.14$ GeV^{-2} versus μ. The open red circle denotes the optimal bound at the inflexion point.

9.2.13 Correlated values of $\overline{m}_c(\overline{m}_c)$ and $\overline{m}_b(\overline{m}_b)$ from M_{B_c}

In Ref. [453], we study the correlation between $\overline{m}_c(\overline{m}_c)$ and $\overline{m}_b(\overline{m}_b)$ from the measured mass of the B_c meson. The explicit QCD expression of the corresponding correlator is given in the original paper [217] and reported in [453]. The expression of the spectral function obtained on-shell is converted into the $\overline{MS}$ one through the relation between the pole mass and running one [154–156, 158–160]. We use the ratio of Laplace sum rule (LSR) defined as:

$$\mathcal{L}_n^c(\tau, \mu) = \int_{(m_c+m_b)^2}^{t_c} dt\, t^n\, e^{-t\tau} \frac{1}{\pi} \text{Im}\, \psi_5(t, \mu), \quad \mathcal{R}_n^c(\tau) = \frac{\mathcal{L}_{n+1}^c}{\mathcal{L}_n^c}, \qquad (9.96)$$

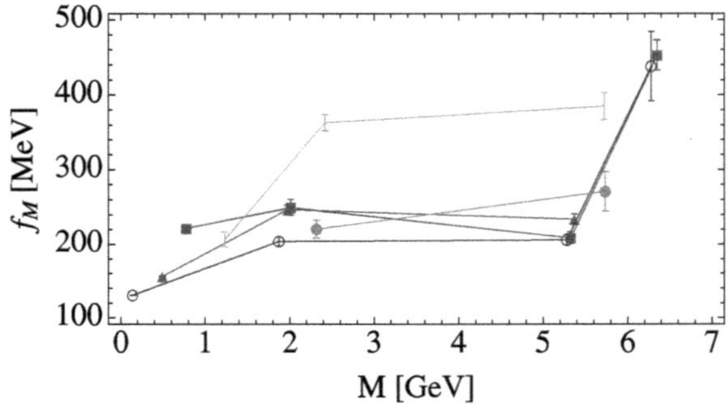

Figure 9.36. Behaviour of the meson decay constants versus the meson masses: green open circle: pseudoscalar π, D, B, B_c; purple triangle: $SU(3)$ breaking: K, D_s, B_s; blue box: vector ρ, D^*, B^*, B_c^*; brown full circle: scalar D_0^*, B_0^*; light green: axial A_1, D_1, B_1.

where τ is the LSR variable, n is the degree of moments, t_c is the threshold of the "QCD continuum" which parametrizes, from the discontinuity of the Feynman diagrams, the spectral function $\text{Im}\psi_5(t, m_Q^2, \mu)$, where $\psi_5(t, m_Q^2, \mu)$ is the (pseudo)scalar correlator. We use the minimal duality ansatz (MDA) to get M_{B_c}.

• τ- and t_c-stabilities

At the first step, we study the τ-, t_c- and μ-stabilities of the sum rule for given values of $\overline{m}_c(\overline{m}_c)$ and $\overline{m}_b(\overline{m}_b)$ from Table 7.10 and Ref. [33].

We show in Fig. 9.37(a) the τ and t_c behaviour of M_{B_c}. The central values of $\overline{m}_c(\overline{m}_c)$=1264 MeV and $\overline{m}_b(\overline{m}_b)$=4188 MeV given in Table 7.10 have been used. We see that the inflexion points at $\tau \simeq (0.30 \sim 0.32)$ GeV^{-2} appear for $t_c \geq 52$ GeV^2.

We show in Fig. 9.37(b) the behaviour of M_{B_c} versus t_c, which is very stable. For definiteness, we take t_c in the range 52–79 GeV^2 where we have a slight maximum at $t_c \simeq 60$ GeV^2. At this range of t_c values, one can easily check that the QCD continuum contribution to the sum rule is (almost) negligible.

The smallest value of $\sqrt{t_c}$ is around the $B_c(2S)$ mass of 6872(1.5) MeV recently discovered by CMS [477] but does not necessarily coincide with it as the QCD continuum is expected to smear all higher state contributions to the spectral function. Instead, its value is related by duality to the ground state parameters as discussed in Ref. [521] from an FESR analysis of the ρ-meson channel.

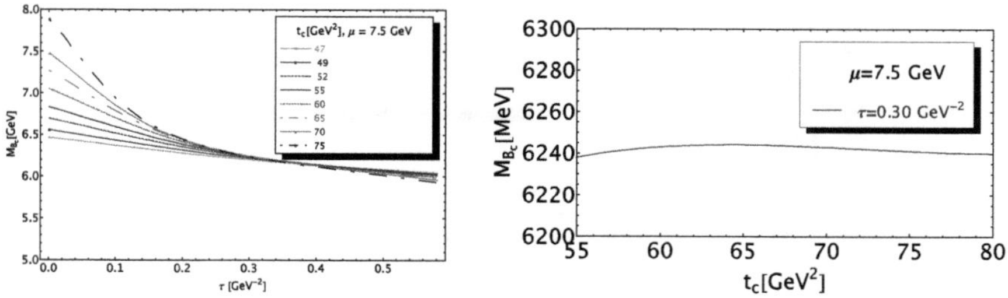

(a) (b)

Figure 9.37. M_{B_c} for given values of $\overline{m}_{c,b}(\overline{m}_{c,b})$ in Table 7.10 as function (a) of τ for different values of t_c, for μ=7.5 GeV and (b) of t_c for μ=7.5 GeV and for the range of τ-stability values.

To gain more insights on this contribution, we show in Fig. 9.38 the ratio of the continuum over the lowest ground state contribution as predicted by QCD:

$$r_c \equiv \frac{\int_{t_c}^{\infty} dt e^{-t\tau} \Delta \mathrm{Im}\psi_5^{\mathrm{cont}}}{\int_{(m_c+m_b)^2}^{t_c} dt e^{-t\tau} \Delta \mathrm{Im}\psi_5^{B_c}}, \tag{9.97}$$

where one can indeed see that the QCD continuum to the moment sum rule $\mathcal{L}_0^c$ is negligible in this range of t_c values. This contribution is even less in the ratio of moments $\mathcal{R}_0^c$ used to get M_{B_c}.

- **μ stability**

Given, e.g., the central value of $\overline{m}_b(\overline{m}_b) = 4188$ MeV from Table 7.10 and using $\tau = 0.32$ GeV^{-2} and $t_c = 60$ GeV^2, we show in Fig. 9.39 the correlated values of $\overline{m}_c(\overline{m}_c)$ at different values of μ needed for reproducing M_{B_c}. We obtain an inflexion point at:

$$\mu = (7.5 \pm 0.1) \text{ GeV}, \tag{9.98}$$

which we shall use in the following. This value agrees with the one $\mu = (7.5 \pm 0.5)$ GeV quoted in Ref. [51] using different ways.

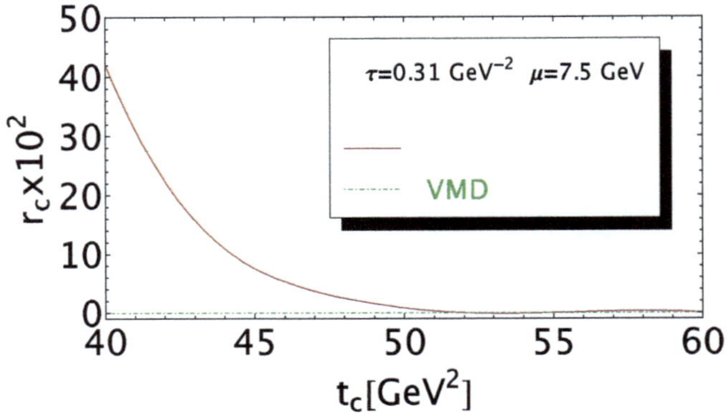

Figure 9.38. Ratio r_c of the continuum over the lowest ground state contribution as function of t_c at the corresponding τ-stability points for $\mu = 7.5$ GeV and for given values of $\overline{m}_{c,b}(\overline{m}_{c,b})$ in Table 7.10. The dashed-dotted line is the contribution for a vector meson dominance (VDM) assumption of the spectral function.

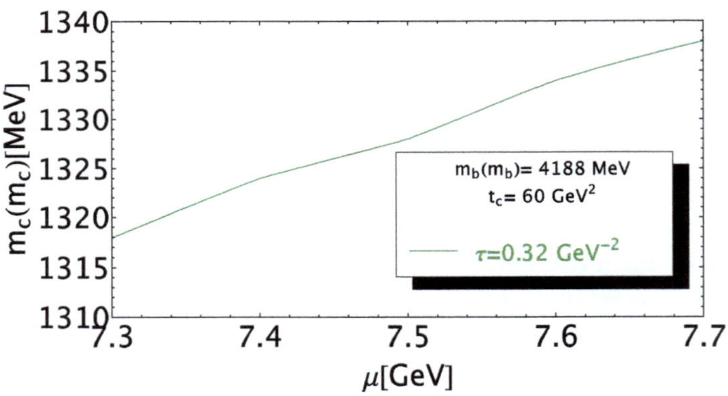

Figure 9.39. $\overline{m}_c(\overline{m}_c)$ as function of μ for $\tau \simeq 0.32$ GeV^2 and for the central value of $\overline{m}_b(\overline{m}_b)$ given in Table 7.10.

• Extracting the set $(\overline{m}_c, \overline{m}_b)$

In the following, we study the correlation between $\overline{m}_c$ and $\overline{m}_b$ needed for reproducing the experimental mass [33]:

$$M_{B_c}^{exp} = 6275.6(1.1) \text{ MeV}, \tag{9.99}$$

from the ratio $\mathcal{R}_0^c$ of Laplace sum rules defined in Eq. (9.96).

– Allowing $\overline{m}_c(\overline{m}_c)$ to move in the range:

$$\overline{m}_c(\overline{m}_c) \simeq (1252\text{--}1282)\ \text{MeV}, \tag{9.100}$$

from the J/ψ and $M_{\chi_{0c}-M_{\eta c}}$ mass splitting sum rules, we show in Fig. 9.40, the predictions for M_{B_c} as a function of $\overline{m}_b(\overline{m}_b)$. The band is the error induced by the choice of the stability points $\tau = (0.30\text{--}0.32)\ \text{GeV}^{-2}$ which is about (12–13) MeV, while the error due to some other parameters are negligible. Then, we deduce:

$$\overline{m}_b(\overline{m}_b) = (4195 - 4245)\ \text{MeV}, \tag{9.101}$$

which leads to the correlated set of values in units of MeV:

$$[\overline{m}_b(\overline{m}_b), \overline{m}_c(\overline{m}_c)] = [4195, 1282] ...\ [4245, 1252]. \tag{9.102}$$

This result shows that a small value of m_c is correlated to a large value of m_b and vice versa.

– Scrutinizing Fig. 9.40, one can see that, at fixed m_b, e.g., 4220 MeV, M_{B_c} increases with the m_c values given in the legend (vertical line), as intuitively expected. On the other hand fixing the value of m_c at the one in the legend, say 1282 MeV, one can see (straight line with a positive slope) that M_{B_c} increases when m_b increases on the m_b-axis, as also intuitively expected.

– Considering that the values of $\overline{m}_b(\overline{m}_b)$ are inside the range:

$$\overline{m}_b(\overline{m}_b) = (4176\text{--}4209)\ \text{MeV}, \tag{9.103}$$

allowed from the Υ sum rules, as given in Table 9.9, we can deduce from Fig. 9.40 stronger constraints on $\overline{m}_b(\overline{m}_b)$:

$$\begin{aligned} \overline{m}_b(\overline{m}_b) &= (4195 - 4209)\ \text{MeV}, \\ &= 4202(7)\ \text{MeV}, \end{aligned} \tag{9.104}$$

where the error is similar to the accurate value from the Υ sum rule in Table 9.9. This is due to the small intersection region of the results from the J/ψ, Υ and the B_c sum rules. With this range of values, we deduce:

$$\begin{aligned} \overline{m}_c(\overline{m}_c) &= 1286(8)_{fig.4}(14)_{\tau}(1)_{t_c}\ \text{MeV}, \\ &= 1286(16)\ \text{MeV}, \end{aligned} \tag{9.105}$$

where the errors due to some other parameters and to the truncation of the PT series are negligible. The subscript $fig.4$ indicates that the error comes from the intersection region between the Υ and B_c sum rules in Fig. 9.40.

We consider the values in Eqs. (9.104) and (9.105) as our final determinations of $\overline{m}_b(\overline{m}_b)$ and $\overline{m}_c(\overline{m}_c)$ from M_{B_c} and combined constraints from the J/ψ and Υ sum rules.

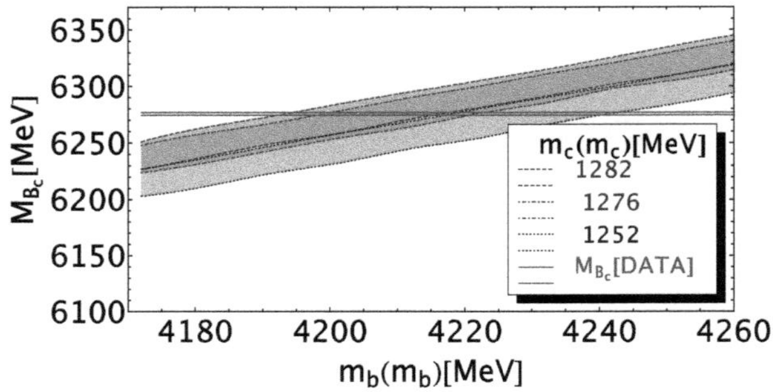

Figure 9.40. M_{B_c} as function of $\overline{m}_b(\overline{m}_b)$ for different values of $\overline{m}_c(\overline{m}_c)$ for $\mu = 7.5$ GeV and for the range of τ-stability values $\tau = (0.30–0.32)$ GeV^{-2}.

9.2.14 Revisiting f_{B_c}

Using the previous correlated values of $(\overline{m}_c, \overline{m}_b)$, we reconsider the estimate of f_{B_c} recently done in Ref. [51].

• τ- and t_c-stabilities

We show the τ-behaviour of f_{B_c} in Fig. 9.41(a) for different values of t_c for $\mu = 7.5$ GeV and for $[\overline{m}_c(\overline{m}_c), \overline{m}_b(\overline{m}_b)] = [1264, 4188]$ MeV from Table 7.10. We start to have τ-stability (minimum) for the set $[t_c, \tau]$=[47 GeV^2, 0.22 GeV^{-2}] and t_c-stability for the set [60 GeV^2, (0.30–0.32) GeV^{-2}]. One can note that the τ-stability starts earlier for smaller t_c-value than to the case of the ratio of moments used in the preceding sections. To be conservative, we shall consider the value of f_{B_c} obtained inside this larger range of t_c-values and take as a final value its mean. In this range, the value of f_{B_c} increases by about 15 MeV. We study the influence of μ on f_{B_c} in Fig. 9.41(b) given the values of τ and $m_{b,c}$. We see a clear stability for $\mu = (7.2–7.5)$ GeV which is consistent with the one for M_{B_c} and with the one obtained in Ref. [51] indicating the self-consistency of the analysis.

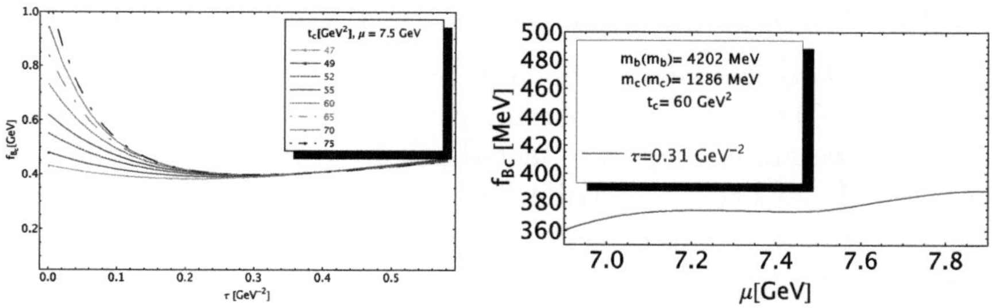

(a) (b)

Figure 9.41. f_{B_c} given $[\overline{m}_c(\overline{m}_c), \overline{m}_b(\overline{m}_b)] = [1264, 4188]$ MeV from Table 7.10: (a) versus τ given $\mu = 7.5$ GeV; (b) as function of μ for $\tau \simeq 0.31$ GeV^{-2}.

• Results [453]

The result of the analysis in units of MeV is:

$$\begin{aligned} f_{B_c} &= 368(1)_\tau(8)_{t_c}(7)_{m_c}(5)_{m_b}(1)_{\alpha_s}(1)_\mu 9)_{N2LO}(10)_{\text{HO}}, \\ &= 368(18)\ \text{MeV}, \end{aligned} \tag{9.106}$$

if one uses the mass values obtained in Eqs. (9.104) and (9.105), while it is:

$$\begin{aligned} f_{B_c} &= 381(1)_\tau(8)_{t_c}(3)_{m_c}(5)_{m_b}(1)_{\alpha_s}(1)_\mu(9)_{N2LO}(10)_{\text{HO}}, \\ &= 381(17)\ \text{MeV}, \end{aligned} \tag{9.107}$$

if one uses the tentative mass averages given in Table 9.9. We take as a final result the mean of the two determinations:

$$f_{B_c} = 371(17)\ \text{MeV}. \tag{9.108}$$

This result improves the previous one $f_{B_c} = 436(40)$ MeV obtained recently in Ref. [51] and the earlier results in Ref. [217, 221]. It confirms and improves the one $f_{B_c} = 388(29)$ MeV averaged from Moments and LSR in Ref. [217] where the values of the pole masses have been used. However, it disagrees with some results, including the lattice one reviewed in Table 6 of Ref. [51b] (Table 9.8 in this section). New estimates from the lattice approach is needed for clarifying the issue. Comments related to some of the previous works have been already addressed in Ref. [51] and can be consulted there.

9.2.15 Attempted upper bound for $f_{B_c(2S)}$

We attempt to give an upper bound to $f_{B_c(2S)}$ by using a "two resonances + QCD continuum" parametrization of the spectral function. However, we are aware of the fact that due to the exponential suppression of the $B_c(2S)$ contribution compared to $B_c(1S)$ and of its eventual smaller coupling as expected for the observed radial excitations in some other channels, we may not extract with good precision the $B_c(2S)$ decay constant from this approach. Instead by using the positivity of the QCD continuum contribution for $t_c \geq 47$ GeV^2 just above $M^2_{B_c(2S)}$, one obtains the semi-analytic expression from $\mathcal{L}^c_0$:

$$\rho_{B_c} \equiv \left(\frac{f_{B_c(2S)}}{f_{B_c}}\right)^2 \left(\frac{M_{B_c(2S)}}{M_{B_c}}\right)^4 \mathrm{e}^{-\left(M^2_{B_c(2S)} - M^2_{B_c}\right)\tau} \leq 3.6\%, \quad (9.109)$$

at the τ-stability of about 0.22 GeV^{-2}, as can be deduced from Fig. 9.38. Using the previous value of f_{B_c} in Eq. (9.95), we deduce:

$$f_{B_c(2S)} \leq (139 \pm 6)\ \text{MeV}, \quad (9.110)$$

indicating that the radial excitation couples weaker to the corresponding quark current than the ground state meson. This feature has been already observed experimentally in the case of light (π, ρ) and heavy (ψ, Υ) mesons.

9.2.16 Spectra and decay constants of B_c-like and B_0^* Mesons

• 0^{++} Scalar channel

◇ τ- and t_c-stabilities

In the first step, fixing the value of $\mu = 7.5$ GeV which we shall justify later and which is the central value obtained in Refs. [51,454], we show in Fig. 9.42 the τ-behaviour of $f_{B_{c^*0}}$ and $M_{B_{c^*0}}$ for different values of t_c where the central values of $\overline{m}_{c,b}(\overline{m}_{c,b})$ given in Table 9.1 have been used. We see that $f_{B_{c^*0}}$ but not $M_{B_{c^*0}}$ presents inflexion points at $\tau \simeq (0.11–0.12)$ GeV^{-2}, which appear for $t_c \geq 55$ GeV^2. We shall use these inflexion points to fix the values of $M_{B_{c^*0}}$.

We study the t_c-behaviour of of $f_{B_{c^*0}}$ and $M_{B_{c^*0}}$ in Fig. 9.43, where we see that $f_{B_{c^*0}}$ starts to stabilize from $t_c = 70$ GeV^2.

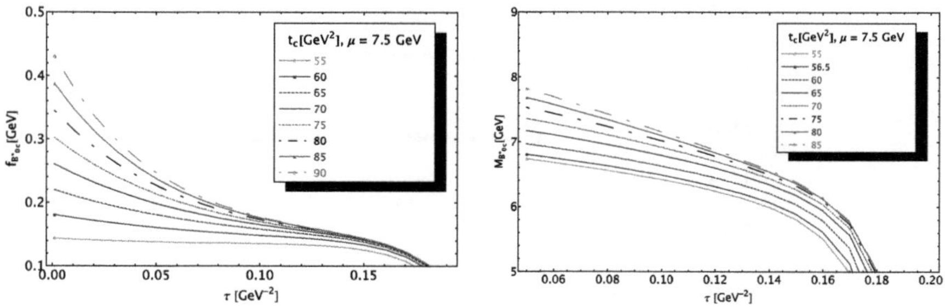

(a) (b)

Figure 9.42. $f_{B^*_{0c}}$ and $M_{B^*_{0c}}$ as function of τ for different values of t_c for μ=7.5 GeV and for values of $\overline{m}_{c,b}(\overline{m}_{c,b})$ given in Table 9.1.

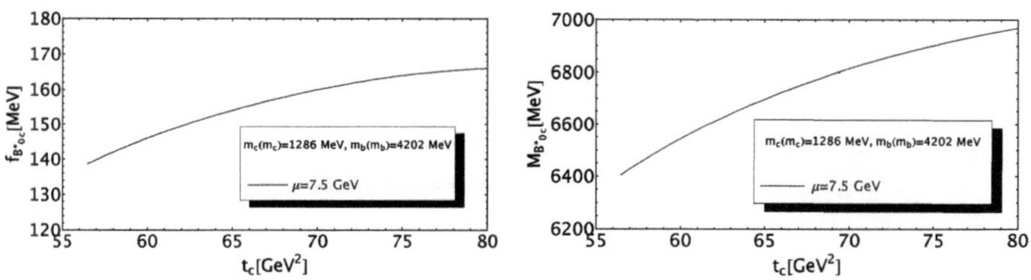

Figure 9.43. $f_{B_{c^*0}}$ and $M_{B_{c^*0}}$ as function of t_c for μ=7.5 GeV and for $\tau \simeq 0.11$ GeV^{-2}.

◇ μ-stability

Fixing $t_c = 70$ GeV^2 and $\tau = 0.11$ GeV^{-2}, we show in Fig. 9.44 the μ behaviour of $f_{B_{c^*0}}$, where we note an inflexion point at:

$$\mu = (7.5 \pm 0.5)\ \text{GeV}, \tag{9.111}$$

in agreement with the one quoted in Refs. [51, 454] using different ways and/or from different channels. The μ-behaviour of $M_{B_{c^*0}}$ is not shown as it is almost constant in this range of t_c-values. To have more insights on the QCD continuum contribution, we show in Fig. 9.45 the ratio of the continuum over the lowest ground state contribution as predicted by QCD:

$$r_{B_{c^*0}} \equiv \frac{\int_{t_c}^{\infty} dt e^{-t\tau} \text{Im}\psi_S^{\text{cont}}}{\int_{(m_c+m_b)^2}^{t_c} dt e^{-t\tau} \text{Im}\psi_S^{B_{c^*0}}}, \tag{9.112}$$

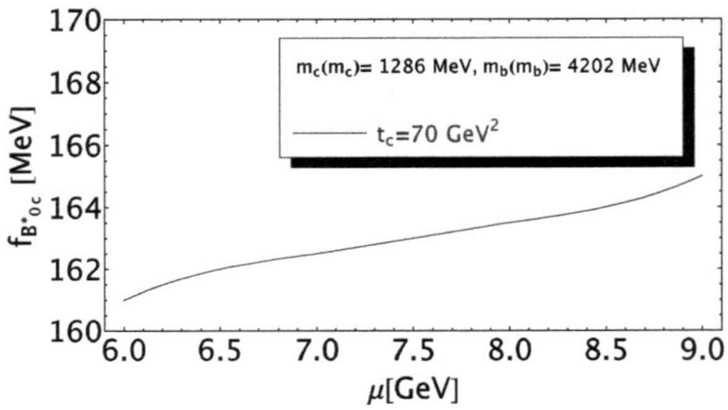

Figure 9.44. $f_{B_{c^*0}}$ as function of μ for $\tau \simeq 0.11$ GeV^{-2} at given t_c.

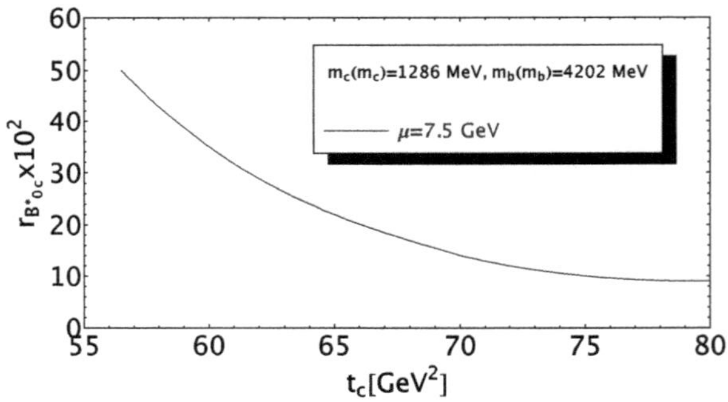

Figure 9.45. Ratio $r_{B_{c^*0}}$ of the continuum over the lowest ground state contribution as function of t_c at the corresponding τ-inflexion point for $\mu = 7.5$ GeV.

The curve started from $t_c = 56.5$ GeV^2 where the QCD continuum contribution to the spectral function is half of the resonance contribution. One can also note from Fig. 9.42 that the τ-stability is reached from this value.

◇ Predictions for $M_{B_{c^*0}}$ and $f_{B_{c^*0}}$

– From the previous analysis and taking the large range of t_c from 56.5 GeV^2 where the τ-stability starts and where the QCD continuum is less than 50% of the resonance one to t_c=75 GeV^2 where the t_c-stability is (almost) reached, at which the lowest resonance dominates the spectral function (meson dominance model), we obtain, for $\tau \simeq$

0.11 GeV^{-2}, the conservative range of predictions in units of MeV :

$$M_{B_{c^*0}} \simeq (6400 - 6965) \text{ and } f_{B_{c^*0}} \simeq (135 - 168). \qquad (9.113)$$

– To improve these results, we request that the QCD continuum contribution to the spectral function is less than (20–25)% of the resonance one. In this way, the t_c-values is restricted to be (70 ± 5) GeV^2. Then, we deduce the improved predictions for $\tau \simeq (0.10\text{–}0.12)$ GeV^{-2} in units of MeV:

$$M_{B_{c^*0}} \simeq 6689(146)_{t_c}(112)_\tau(0)_\mu(11)_{m_{b,c}}(15)_{\alpha_s}(19)_{G^2}(67)_{\text{syst}},$$
$$f_{B_{c^*0}} \simeq 155(15)_{M_{B_{c^*0}}}(4)_{t_c}(5)_\tau(0.5)_\mu(3)_{m_{b,c}}(5)_{\alpha_s}(0)_{G^2}. \qquad (9.114)$$

– We test the accuracy of the approximate expression expanded up to order m_c^2 by taking the example of the pseudoscalar channel where the complete expression of the non-perturbative contribution is used. We note that the approximate result overestimates the mass prediction by about 1.01%. We take into account this effect by adding to the prediction in Eq. (9.114) a systematic error of about 1% to the mass prediction and dividing by 1.01 the estimate from the analysis.[12]

– Examining the analytical form of the (pseudo)scalar sum rules, one can deduce the approximate LO relation:

$$f_{B_c} \simeq f_{B_{c^*0}} \left(\frac{\overline{m}_b + \overline{m}_c}{\overline{m}_b - \overline{m}_c}\right)\left(\frac{M_{B_{c^*0}}}{M_{B_c}}\right)^2 \times \rho \approx (348\text{–}468) \text{ MeV}, \quad (9.115)$$

where $\rho \equiv \text{Exp}[(M^2_{B_c} - M^2_{B_{c^*0}})\tau/2]$ and $\tau \approx (0.1\text{–}0.2)$ GeV^{-2}. We have evaluated the running mass at $\mu = 7.5$ GeV. The result is comparable to the more involved estimate $f_{B_c} = 371(17)$ MeV in Ref. [454].

• Vector $B_c^*(1^{--})$ and axial-vector $B_{c1}(1^{++})$ channels

We extend the analysis to the vector $B_c^*(1^{--})$ and axial-vector $B_{c1}(1^{++})$ mesons. In both cases, the μ-stability is the same as

[12]Here and in the following, the quoted results for the masses already take into account this systematic effects.

in Eq. (9.111). For the B_c^* [respectively B_{c1}], the τ- and t_c-stabilities are obtained for the set $(0.1{-}0.2, 52{-}65)$ [respectively $(0.09{-}0.1, 65{-}75)$] (GeV^{-2}, GeV^2). We obtain:

$$
\begin{aligned}
M_{B_c^*} &\simeq 6451(52)_{t_c}(1)_\tau(1)_\mu(11)_{m_{b,c}}(7)_{\alpha_s}(17)_{G^2}(65)_{\text{syst}},\\
f_{B_c^*} &\simeq 442(41)_{M_{B_c^*}}(11)_{t_c}(1)_\mu(1)_\tau(6)_{m_{b,c}}(7)_{\alpha_s}(4)_{G^2},
\end{aligned}
\tag{9.116}
$$

and:

$$
\begin{aligned}
M_{B_{c1}} &\simeq 6794(68)_{t_c}(44)_\tau(32)_\mu(16)_{m_{b,c}}(60)_{\alpha_s}(11)_{G^2}(68)_{\text{syst}},\\
f_{B_{c1}} &\simeq 274(19)_{M_{B_{c1}}}(10)_{t_c}(1)_\mu(7)_\tau(1)_{m_{b,c}}(3)_{\alpha_s}(1)_{G^2}.
\end{aligned}
\tag{9.117}
$$

• Mass splittings from heavy quark symmetry (HQS)

To improve the predictions on the meson masses, we shall use the properties of heavy quark symmetry (HQS) [180, 493]. In so doing, we confront the observed values of the $D_{(s)}, B_{(s)}$-like mass splittings to the LO expectations of HQS in the heavy quark $(1/M_Q)$ inverse mass expansion, where $Q \equiv c, b$. Then, we extrapolate this result for predicting the B_c-like meson masses.

◇ Hyperfine splittings

From spin symmetry, one expects that, to LO, the hyperfine splittings are independent of the flavour of the "brown muck" [180, 493] which is realized experimentally [33]. In units of GeV^2, one has indeed:

$$
\begin{aligned}
M_{B^*}^2 - M_B^2 &= 0.488 \approx M_{D^*}^2 - M_D^2 = 0.543,\\
M_{B_s^*}^2 - M_{B_s}^2 &= 0.518 \approx M_{D_s^*}^2 - M_{D_s}^2 = 0.588,
\end{aligned}
\tag{9.118}
$$

with negligible errors. These results indicate that the $1/M_Q$ corrections to LO are quite small. We shall use the B meson results, where the $1/M_b$ corrections are smaller than the one of the D-mesons. Extrapolating to the B_c-like mesons and using M_{B_c}= 6274.9(0.8) MeV, one can deduce:

$$
M_{B_c^*} \simeq (6315 \pm 1)\ \text{MeV}. \tag{9.119}
$$

◇ **Heavy flavour independence of excitation energies**

– One also expects from HQS that the excitation energies for states with different quantum numbers of the light degrees of freedom are heavy flavour-independent [180, 493], which is approximately reproduced by the data [33]. In units of MeV, we have:

$$\begin{aligned} M_{B_1} - M_B &= 447(1) \approx M_{D_1} - M_D = 551(1), \\ M_{B_{s1}} - M_{B_s} &= 462(1) \approx M_{D_{s1}} - M_{D_s} = 491(1), \end{aligned} \quad (9.120)$$

and:

$$\begin{aligned} M_{B_2} - M_B &= 458(1) \approx M_{D_2} - M_D = 596(1), \\ M_{B_{s2}} - M_{B_s} &= 473(0) \approx M_{D_{s2}} - M_{D_s} = 601(1). \end{aligned} \quad (9.121)$$

The approximate equalities between the B and D mass splittings again indicate that the $1/M_Q$ corrections to the LO relations are negligible. Extrapolating the values for the B to the B_c-like mesons, we deduce in units of MeV:

$$M_{B_{c1}} = (6730 \pm 8) \quad \text{and} \quad M_{B_{c2}} = (6741 \pm 8). \quad (9.122)$$

– For estimating the scalar meson mass $M_{B^*_{0,c}}$, we assume the flavour independence (within the errors) of the mass splitting of chiral multiplets as given by the sum rule results [6, 168, 301]:

$$M_{B^*_0} - M_B \simeq 422(196) \text{ MeV} \approx M_{D^*_{s0}} - M_{D_s}, \quad (9.123)$$

and the data [33]:

$$M_{D^*_0} - M_D \simeq 448(29) \text{ MeV}. \quad (9.124)$$

Using the previous experimental value, we deduce [453]:

$$M_{B^*_0} \simeq 5733 \text{ MeV and } M_{B^*_{c0}} \simeq 6723 \text{ MeV}. \quad (9.125)$$

The value of $M_{B^*_0}$ improves previous LSR estimate in Refs. [5,168]. It suggests that the experimental candidate $B^*_J(5732)$ can be fairly identified with a 0^{++} B-like meson. The decay constant of B^*_0 has been already estimated in Refs. [5,51,168] within LSR. It is quoted in Table 3 of Ref. [51] (see Table 9.5) and agrees with the one in Ref. [307].

Table 9.9. Values of the masses from LSR and HQS compared with lattice and Potential Model (PM) results.

Channel	LSR	HQS	Lattice [494]	PM [495]
Masses				
$B_c^*(1^{--})$	6451(86)	6315(1)	6331(7)	6330(20) [217]
$B_{0c}^*(0^{++})$	6689(198)	6723(29)	6712(19)	6693
$B_1(1^{++})$	6794(128)	6730(8)	6736(18)	6731
$B_{c2}(2^{++})$	–	6741(8)	–	7007
$B_0^*(0^{++})$	5701(196) [301]	5733		

• Summary and comparison with some other estimates

◇ Spectra

- We collect the results for the masses obtained in the previous sections in Table 9.9 which we compare with some recent lattice calculations and Potential Model (PM) results.
- We note that, within the errors, the results from different approaches agree with each other except the one for the 2^{++} meson. Note that the original numbers from Ref. [495] are quoted without any errors, but we expect that the PM results are known within 20 MeV error as estimated in Ref. [217].

Table 9.10. Values of the decay constants f_H in units of MeV using as input the values of the masses from LSR and HQS quoted in Table 9.9.

Masses	$B_c(0^{--})$	$B_c^*(1^{--})$	$B_{0c}^*(0^{++})$	$B_{1c}(1^{++})$	$B_0^*(0^{++})$
LSR	371(17) [454]	442(44)	155(17)	274(23)	–
HQS	–	387(15)	158(9)	266(14)	271(26) [51]

- One should mention that the errors from LSR are relatively large, which are mainly due to the large, range of t_c-values. The predictions can only be improved after a complete measurement of the spectral functions which is out of reach at present.
- The quoted errors from HQS come only from the data. We are aware that some systematic errors not included here are present

using the HQS results to LO, but the agreement of these results with the accurate data may indicate that these corrections are small. The inclusion of such HQS higher order corrections is beyond the scope of this paper.

◇ **Decay constants**

- The new predictions for the decay constants are collected in Tables 9.5 to 9.8. One should note that the values of the decay constants are largely affected by the value of the masses.
- We have also reported in Table 9.9 the predictions using the relatively precise predictions on the masses from HQS.
- The difference of the LSR and HQS results for f_H is due to the LO factor $\text{Exp}[M_H^2\tau/2]/M_H^2$ entering in the LSR expression of f_H.

• **Summary and conclusions**

◇ We have reviewed our recent determinations of the heavy–light pseudoscalar [455,456], vector [457] mesons decay constants from the standard Laplace and Moment sum rules (SR) introduced by SVZ [3]. Motivated by different tensions in the literature, we have completed the review by presenting new determinations of the vector f_{B^*}/f_B, scalar $f_{D_0^*,B_0^*}$, axial f_{D_1,B_1} and vector $f_{B_c^*}$ meson decay constants.
Our results based on stability criteria are summarized in the tables and compared with some other recent sum rules and lattice results. Our quoted errors are mainly due to the choice of the QCD continuum threshold and to the estimate of the α_s^3 PT series. This latter is not considered in some other SR results.

◇ Then, we have attempted to give a global average of sum rules and lattice results (results from some other approaches can, e.g., be found in Ref. [489]), which can be used for further phenomenological applications. To take into account the slight tensions among different determinations, we have rescaled some errors.

◇ One can see in these tables that there are fair agreements among estimates from different standard sum rule (SR) approaches. The slight difference is mainly due to a different appreciation of the optimal results from the choice of the QCD continuum threshold, subtraction constant μ and of the set of input QCD parameters. The quoted upper bounds come from the positivity of the spectral functions.

◇ A slight tension exists between SR, NS2R and lattice results for f_{B^*}/f_B [490]. To gain insight into this problem, we have rederived this ratio from HQET sum rules. There is also a tension for f_{B_c} from NS2R [487] (see comments in Ref. [457]). The N2SR results do not affect significantly the SR average and they are considered in the global average.

◇ Large tensions exist in the scalar sector among various SR and between lattice results for $f_{D_0^*}$. Our updated results given in Table 9.4 using the running $\overline{MS}$ charm mass and SR stability criteria confirm our previous findings [301].

◇ The results indicate a good realization of heavy quark symmetry for the B and B^* mesons ($f_B \approx f_{B^*}$) as expected from HQET [180] but signal large charm quark mass and radiative QCD corrections for the D and D^* mesons ($f_D \approx f_B, ..$) which are known since a long time (see, e.g., among others Refs. [161, 492]).

◇ The $SU(3)$ breaking is typically 20% for $f_{D_s}/f_D \approx f_{B_s}/f_B$ and 10% for $f_{D_s^*}/f_{D^*} \approx f_{B_s^*}/f_{B^*}$ which are almost constant as they behave like m_s/ω_c [167], where $\omega_c \equiv \sqrt{t_c} - M_b \approx M'_B - M_B$ is independent of M_b. The reverse effect for $f_{D_s^*}/f_{D^*} \approx f_{B_s^*}/f_{B^*} < 1$ comes mainly from the overall $(1 - m_s/m_Q)$ factor not compensated by $M_{D_{0s}^*/B_{0s}^*} \simeq M_{D_0^*/B_0^*}$ [167].

◇ It is informative to show the behaviour of our predictions of the pseudoscalar and vector meson decay constants versus the corresponding meson masses in Fig. 9.36. We use $f_\rho = (221.6 \pm 1.0)$ MeV from its electronic width [33] and $f_{A_1} = 207$ MeV from SR analysis [7]. One can note similar M_Q behaviours of these couplings where the $1/\sqrt{M_Q}$ HQET relations [180, 491] are not satisfied due to large α_s and $1/M_Q$ corrections, as noted earlier [161, 163, 164, 492]. One can note small $SU(3)$ but large $SU(4)$ breaking for B_c, B_c^*. Chiral symmetries are badly broken between the vector D^*, B^* and axial D_1, B_1 meson couplings.

◇ We do not find any deviation of these Standard Model results from the present data.

9.3 Gluon Condensates and $\overline{m}_{c,b}$ from Moments to Order α_s^2 and $\langle G^3\rangle$

9.3.1 Moments and their ratios

In Ref. [290], QCD moments and their ratios have been used to extract the charm and bottom quark running masses and the gluon condensates $\langle\alpha_s G^2\rangle$ and $\langle g^3 f_{abc} G^3\rangle$.

Here, we shall be concerned with the two-point correlator of a heavy quark Q:

$$i\int d^4x\, e^{-\mathrm{iqx}}\langle 0|\mathcal{T} J_Q^\mu(x)\left(J_Q^\nu(0)\right)^\dagger|0\rangle \equiv -\left(g^{\mu\nu}q^2 - q^\mu q^\nu\right)\Pi_Q(q^2), \tag{9.126}$$

where $J_Q^\mu = \bar{Q}\gamma^\mu Q$ is the heavy quark neutral vector current. Im $\Pi_c(s)$ can be related to the charmonium leptonic widths and masses. In a narrow width approximation (NWA),

$$\mathcal{R}_c(t) \equiv 4\pi\mathrm{Im}\Pi_c(t+i\epsilon) = \frac{\pi N_c}{Q_c^2\alpha^2}\sum M_\psi\Gamma_{\psi\to e^+e^-}\delta((t-M_\psi^2), \tag{9.127}$$

where $N_c = 3$, M_ψ and $\Gamma_{\psi\to e^+e^-}$ are the mass and leptonic width of the J/ψ mesons, $Q_c = 2/3$ is the charm electric charge in units of e, and $\alpha = 1/133.6$ is the running electromagnetic coupling evaluated at M_ψ^2. We shall use the experimental values of the J/ψ parameters compiled in Table 9.11.

Table 9.11. Masses and electronic widths of the J/ψ family from PDG 08 [33].

Name	Mass [MeV]	$\Gamma_{J/\psi\to e^+e^-}$ [keV]
$J/\psi(1S)$	3096.916(11)	5.55(14)
$\psi(2S)$	3686.093(34)	2.33(7)
$\psi(3770)$	3775.2(1.7)	0.259(16)
$\psi(4040)$	4039(1)	0.86(7)
$\psi(4160)$	4153(3)	0.83(7)
$\psi(4415)$	4421(4)	0.58(7)

We find that it is convenient to introduce the instanton radius ρ_c:

$$\frac{\langle g^3 f_{abc} G^3 \rangle}{\langle \alpha_s G^2 \rangle} = \frac{4}{5} \frac{12\pi}{\rho_c^2}. \tag{9.128}$$

The value of ρ_c ranges from 1/3 fm = 1.5 GeV^{-1} [130], 0.5 fm = 2.5 GeV^{-1} [496] to 0.9 fm = 4.5 GeV^{-1} [3]. As $\langle g^3 f_{abc} G^3 \rangle$ contributes like $1/\rho_c^2$ in the OPE analysis, a more precise value of ρ_c is crucial for checking the convergence of the OPE. The estimate of ρ_c from charmonium sum rules is feasible as the light quark condensate $m_q\langle \bar{q}q \rangle$ contributes, to higher loop order, which are chiral suppressed and are negligible, while the heavy quark condensate contribution can be absorbed into the gluon one, as mentioned earlier.

Different forms of QSSR exist in the literature [5–7, 42]. We shall work here with the moments:

$$\mathcal{M}_n\left(-q^2 \equiv Q^2\right) = \int_{4m_Q^2}^{\infty} dt \frac{\mathcal{R}_c(t, m_c^2)}{(t+Q^2)^{n+1}}, \tag{9.129}$$

and more likely with their ratios:

$$r_{n/n+1}(Q^2) = \frac{\mathcal{M}_n}{\mathcal{M}_{n+1}}, \qquad r_{n/n+2}(Q^2) = \frac{\mathcal{M}_n}{\mathcal{M}_{n+2}}, \tag{9.130}$$

where the experimental sides are more precise than the absolute moments $\mathcal{M}_n$. Also, in the ratios, partial cancellations of different perturbative as well as non-perturbative terms occur, which render the QCD approximation more precise than in the absolute moments.

The QCD sides of the sum rules are known in the literature since the original works of SVZ [3]. Their expressions at the subtraction scale $\nu^2 = m_Q^2$ are given explicitly numerically in the Appendix of Ref. [293] to 3-loops ($\mathcal{O}(\alpha_s^2)$) accuracy in terms of the running heavy quark mass[13] using the pQCD results of Ref. [497], while the $Q^2 = 0$ moments to 4-loops ($\mathcal{O}(\alpha_s^3)$) are given in Ref. [498] using the pQCD results in Refs. [499, 500].[14] Among the different moments given in

[13]We shall use these expressions in our analysis and we shall correct our final results on the quark masses by adding an estimate of the 4-loops ($\mathcal{O}(\alpha_s^3)$) contributions.

[14]Some Pade approximants are given in Ref. [501].

Ref. [293], we shall select three moments where both the $(\alpha_s)^n$ $(n = 1,\ 2)$, the gluon condensate contributions and the effects of the higher resonances plus the QCD continuum are relatively small but not negligible. These conditions can be simultaneously satisfied by the moments[15]:

$$\begin{aligned} &\mathcal{M}_{2,3,4} && \text{for } Q^2 = 0, \\ &\mathcal{M}_{8,9,10} && \text{for } Q^2 = 4m_Q^2, \\ &\mathcal{M}_{13,14,15} && \text{for } Q^2 = 8m_Q^2. \end{aligned} \tag{9.131}$$

One may also work with more moments, but these will not bring newer information. Lower moments are more sensitive to the experimental errors and the QCD continuum, while higher moments are more sensitive to higher dimension condensates which are not under good control.[16] Moreover, one can also note from the QCD expressions given in Ref. [293] that for n larger than in our previous selected choice, the signs of the pQCD corrections start to change compared to the original ones of the two-point correlator. Such a change may introduce some systematical difficulties inherent to the approach for controlling the size of higher order terms.[17]

We shall work with the ratios of moments:

$$\begin{aligned} &r_{2/3} && \text{and} && r_{2/4} && \text{for } Q^2 = 0, \\ &r_{8/9} && \text{and} && r_{8/10} && \text{for } Q^2 = 4m_c^2, \\ &r_{13/14} && \text{and} && r_{13/15} && \text{for } Q^2 = 8m_c^2, \end{aligned} \tag{9.132}$$

and use as inputs, in this first step:

$$\overline{m}_c(m_c) = 1.26(3) \text{ GeV}, \tag{9.133}$$

as given by different approaches using charmonium moment sum rules [3,6,7,33,41,62,158–160,293,498,505] which we shall re-estimate

---

[15]However, one should note that the accuracy of the $Q^2 = 0$ moments is less than that of the $Q^2 \neq 0$ moments, while one cannot use higher moments due to the bad convergence of the OPE in this case.

[16]The contributions of the dimension 8 condensates have been evaluated in Ref. [502] and can be sizeable if one assumes factorization, which might not be applied here [338, 339].

[17]In Ref. [503], some low-energy gluon contributions to order α_s^3 to the correlator can invalidate the uses of the $Q^2 = 0$-moments for $n > 4$ (see, however, Ref. [504]).

Table 9.12. QCD expressions of the moments and their ratios normalized to the lowest order terms and given to order α_s^2 and including the dimension 4 and 6 gluon condensates derived from Ref. [293]; $d_4 \equiv \langle\alpha_s G^2\rangle/(4m_c^2)^2 \simeq (1.49\pm0.50)\times10^{-3}$ if we use $m_c = 1.261$ GeV from Table 7.10 and $\langle\alpha_s G^2\rangle = 0.06$ GeV^4 in Eq. (9.305); $\rho_{64} \equiv \langle g^3 f_{abc}G^3\rangle/(\langle\alpha_s G^2\rangle 4m_c^2) \simeq (4.88\pm2.05)$ is the ratio between the dimension 6 and 4 gluon condensate contributions if we use the numerical value in Eq. (9.136); $a_s \equiv \alpha_s/\pi$.

Mom	QCD expression
Q^2=0	
$\mathcal{M}_2$	$1 + 2.427a_s + 6.110a_s^2 - d_4\,(18.61 - 0.83\rho_{64})$
$\mathcal{M}_3$	$1 + 1.917a_s + 6.115a_s^2 - d_4\,(45.71 - 4.00\rho_{64})$
$\mathcal{M}_4$	$1 + 1.100a_s + 4.402a_s^2 - d_4\,(90.21 - 12.76\rho_{64})$
$r_{2/3}$	$1 + 0.510a_s - 0.983a_s^2 + d_4\,(27.10 - 3.17\rho_{64})$
$r_{2/4}^{1/2}$	$1 + 0.664a_s - 0.095a_s^2 + d_4\,(35.80 - 5.97\rho_{64})$
$Q^2 = 4m_c^2$	
$\mathcal{M}_8$	$1 + 1.118a_s + 4.253a_s^2 - d_4\,(77.51 - 5.02\rho_{64})$
$\mathcal{M}_9$	$1 + 0.601a_s + 2.700a_s^2 - d_4\,(104.86 - 9.20\rho_{64})$
$\mathcal{M}_{10}$	$1 + 0.045a_s + 1.136a_s^2 - d_4\,(137.83 - 15.63\rho_{64})$
$r_{8/9}$	$1 + 0.517a_s + 1.242a_s^2 + d_4\,(27.35 - 4.18\rho_{64})$
$r_{8/10}^{1/2}$	$1 + 0.537a_s + 1.390a_s^2 + d_4\,(30.16 - 5.31\rho_{64})$
$Q^2 = 8m_c^2$	
$\mathcal{M}_{13}$	$1 + 0.776a_s + 3.061a_s^2 - d_4\,(90.37 - 5.19\rho_{64})$
$\mathcal{M}_{14}$	$1 + 0.412a_s + 1.909a_s^2 - d_4\,(109.75 - 7.89\rho_{64})$
$\mathcal{M}_{15}$	$1 + 0.031a_s + 0.770a_s^2 - d_4\,(137.72 - 11.53\rho_{64})$
$r_{13/14}$	$1 + 0.364a_s + 1.002a_s^2 + d_4\,(19.38 - 2.70\rho_{64})$
$r_{13/15}^{1/2}$	$1 + 0.373a_s + 1.065a_s^2 + d_4\,(23.68 - 3.17\rho_{64})$

later on. We shall also use:

$$\alpha_s(M_\tau) = 0.3249(80) \longrightarrow \alpha_s(m_c)|_{n_f=4} = 0.408(14) \qquad (9.134)$$

from τ-decay [289], a value which agrees with the central value of the world average [33, 444] when run until M_Z.

The QCD expressions of the moments and their ratios are given in Table 9.12.

9.3.2 Instanton radius ρ_c from ratios of moments

We shall work with the ratios of moments in Eq. (9.132). We parametrize the spectral function by a sum of the six J/ψ-like narrow resonances below 4.6 GeV[18] and use its pQCD expression from $\sqrt{t} = (4.6 \pm 0.1)$ GeV. We extract the value of ρ_c for a large range of $\langle \alpha_s G^2 \rangle$.

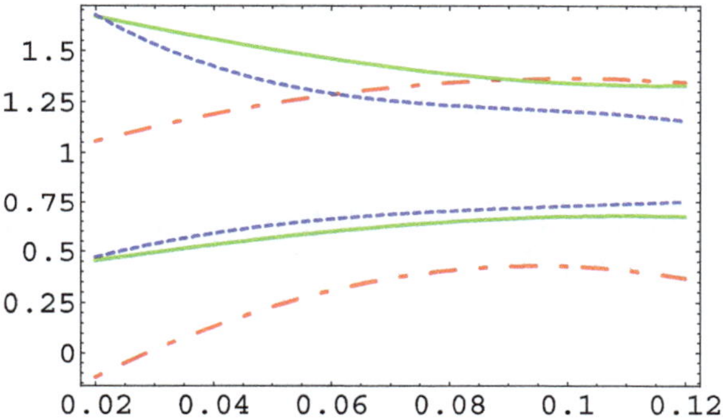

Figure 9.46. Determinations of ρ_c in GeV^{-1} versus $\langle \alpha_s G^2 \rangle$ in GeV^4 from different sets of moments: $Q^2 = 0$ (red: dashed-dotted) ; $Q^2 = 4m_c^2$ (green: continuous); $Q^2 = 8m_c^2$ (blue: dotted).

Table 9.13. Ratio ρ_c of $\langle g^3 f_{abc} G^3 \rangle / \langle \alpha_s G^2 \rangle$ as defined in Eq. (9.128), and value of $\overline{m}_c(m_c)$ from charmonium moments known to 3-loops. The errors in ρ_c come from the choice of moments and from the error on m_c given in Eq. (9.133). The ones due to α_s, $\langle \alpha_s G^2 \rangle$ and to the data on the J/ψ family are negligible. The value of m_c is taken at $\langle \alpha_s G^2 \rangle = 0.06(2)$ GeV^4, as given in Eq. (9.305). The errors on m_c come, respectively, from the ones of ρ_c and $\langle \alpha_s G^2 \rangle$. The ones due to α_s and the data are about 1 MeV, each which are negligible.

Mom	ρ_c [GeV^{-1}]	$m_c(m_c)$ [MeV]
Q^2=0: $r_{2/3}$, $r_{2/4}$	0.800(490)	1234(34)(8)
Q^2=$4m_c^2$: $r_{8/9}$, $r_{8/10}$	1.025(425)	1265(29)(9)
Q^2=$8m_c^2$: $r_{13/14}$, $r_{13/15}$	1.025(275)	1268(18)(7)
Average	0.98(21)	1261(15)

[18]One can improve this parametrization by taking into account finite width corrections using BES data [507], but these corrections will be negligible in the moments which we shall use.

One can see in Fig. 9.46 that the results are very stable for all moments for $\langle\alpha_s G^2\rangle \geq 0.06$ GeV^4, from which we deduce the value of ρ_c given in Table 9.13. The errors in ρ_c come, respectively, from the values of m_c and the choice of the moments at given Q^2. These errors are included in the regions given in Fig. 9.46. The one due to α_s and experiments are negligible. Our final averaged result is:

$$\rho_c = 0.98(21)\ \text{GeV}^{-1} \quad \text{for} \quad \langle\alpha_s G^2\rangle \geq 0.06\ \text{GeV}^4, \tag{9.135}$$

which we consider as an improvement of the different estimates based on instanton liquid models [3, 130, 496] recalled in the introduction. However, our result agrees within the error with the one in Ref. [130] but is smaller by a factor of about 4 than the SVZ estimate [3]. Using Eq. (9.128), our result for ρ_c corresponds to:

$$\langle g^3 f_{abc} G^3\rangle = (31 \pm 13)\ \text{GeV}^2\langle\alpha_s G^2\rangle, \tag{9.136}$$

indicating that it is much bigger than usually assumed in the literature. It is also smaller than the lattice result in $SU(2)$ pure Yang–Mills [508] and than a rough estimate extended to $SU(3)$ with dynamical fermions [509] using the result in Refs. [508, 510–512]. One should note that this value of the gluon condensate has been extracted by assuming that the OPE including the dimension 6 gluon condensate gives a good description of the experimental data which implicitly assumes that the contributions of higher dimension condensates are negligible in the analysis. In other words, one may also interpret this value as the one of an *effective gluon condensate* which parametrizes all higher dimension condensates contributing to the OPE.

9.3.3 $\overline{m}_c(\overline{m}_c)$ from ratios of moments

For extracting $\overline{m}_c(\overline{m}_c)$, we equate the QCD and experimental sides of the ratios of moments $r_{n/n+1}(Q^2)$ and $r_{n/n+2}(Q^2)$ and solve exactly numerically these equations using the Mathematica subroutine FindRoot. Though the equations look simple for the $Q^2 = 0$-moment:

$$F_{\text{th}}(x) \equiv ax + b + \frac{c}{x} + \frac{d}{x^2} + \frac{e}{x^3} = F_{\text{exp}}(x), \tag{9.137}$$

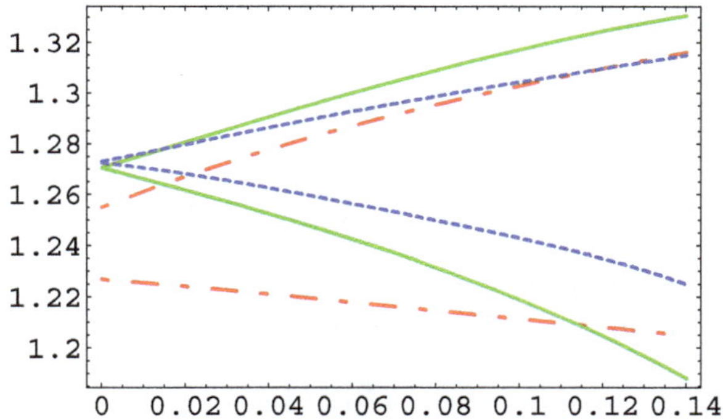

Figure 9.47. Determinations of $\overline{m}_c(\overline{m}_c)$ in GeV versus $\langle\alpha_s G^2\rangle$ in GeV^4 from different sets of moments: $Q^2 = 0$ (red: dashed-dotted); $Q^2 = 4m_c^2$ (green: continuous); $Q^2 = 8m_c^2$ (blue: dotted).

$x \equiv m_c^2$, they are highly non-trivial for $Q^2 = 4m_c^2, 8m_c^2$ and for higher values of n due to the appearance of the term:

$$F_{\rm exp}(x) \sim \sum_\psi \frac{g_\psi^2}{(M_\psi^2 + Q^2)^{n+1}}, \tag{9.138}$$

in the experimental side of the ratio of moments. We show in Fig. 9.47 the different solutions of $\overline{m}_c(m_c)$ versus $\langle\alpha_s G^2\rangle$ for each ratio $r_{n/n+1}(Q^2)$ and $r_{n/n+2}(Q^2)$ of Q^2-moments. One can note that common solutions of different ratios moments occur in the range of values:

$$\langle\alpha_s G^2\rangle \geq 0.02 \text{ GeV}^4, \tag{9.139}$$

indicating that the central values of $\overline{m}_c(m_c)$ are not very sensitive to the one of $\langle\alpha_s G^2\rangle$. This feature confirms the unconclusive range of values obtained for $\langle\alpha_s G^2\rangle$ in Ref. [293]. However, zero and negative values of $\langle\alpha_s G^2\rangle$ as obtained from some analysis of τ-decays [506][19] are excluded by our present result and by the one in Ref. [293]. The slight difference between [293] and our analysis is that we put

[19]One should note that due to the kinematical structure of the original τ-decay width [79], the gluon condensate contribution acquires there an extra α_s coefficient compared with the one of the two-point correlator which suppresses its contribution and can render inaccurate its extraction from this observable.

implicitly the previously determined value of $\langle g^3 f_{abc} G^3 \rangle$ and its correlation with $\langle \alpha_s G^2 \rangle$ in the extraction of $\overline{m}_c(m_c)$. This fact explains the increase of the errors for increasing values of $\langle \alpha_s G^2 \rangle$ in our analysis. In order to improve the determination of m_c, we use the values:

$$\langle \alpha_s G^2 \rangle = 6(2) \times 10^{-2}\ \text{GeV}^4, \tag{9.140}$$

obtained by enlarging the error of the average value 0.06(1) GeV^4 from the heavy quarkonia mass splittings [223]:

$$\langle \alpha_s G^2 \rangle = 7.5(2.5) \times 10^{-2}\ \text{GeV}^4, \tag{9.141}$$

and from $e^+e^- \to I = 1$ hadrons sum rules [236, 237],

$$\langle \alpha_s G^2 \rangle = 6.1(0.7) \times 10^{-2}\ \text{GeV}^4. \tag{9.142}$$

These previous values agree with the one about 0.069 GeV^4 obtained from $SU(3)$ lattice with dynamical fermions [512]. Using this value, we show in Table 9.9 the value of $\overline{m}_c(m_c)$ from each set of ratios of moments, from which we can deduce the mean value:

$$\overline{m}_c(\overline{m}_c)|_{\text{3-loops}} = 1261(15)\ \text{MeV}, \tag{9.143}$$

obtained from a 3-loop ($\mathcal{O}(\alpha_s^2)$) expression of the ratios of moments. This value is more weighted by the one from the $Q^2 = 8m_c^2$ ratios of moments which give the most precise predictions.

9.3.4 Comments on the results and error estimates

• $Q^2 = 0$ moments

We note that our result $\overline{m}_c(m_c) = 1234(35)$ MeV from the $Q^2 = 0$ moments agrees within the error with the 4-loops recent estimate 1279(13) MeV in Ref. [498] based on the lowest $n = 1$ moment. Our result is less precise due mainly to the errors induced by the presence of $\langle g^3 f_{abc} G^3 \rangle$ and its correlated $\langle \alpha_s G^2 \rangle$ condensate in our analysis and to the error induced by the choice of the ratio of moments, as can be seen in Fig. 9.47.

From the QCD expressions given in Table 9.12:

– One should first note that each perturbative and non-perturbative corrections tend to partially cancel out in the ratios of moments, which render the QCD PT series and OPE more convergent for the ratios than for the corresponding individual moments.

- Using the previous values of the QCD parameters and for definiteness $\langle\alpha_s G^2\rangle = 0.06$ GeV^4, one finds that the contribution of $\langle\alpha_s G^2\rangle$ in $\mathcal{M}_2(0)$ is about -2.8% which is comparable with the one -1.6% from $-0.23\alpha_s^3$ [498], while, for $\mathcal{M}_3(0)$, it is -6.8% which is about the one 10% from α_s^2. For $\mathcal{M}_2(0)$, the $\langle g^3 f_{abc} G^3\rangle$ contribution is about 0.6% which is about 1/3 of the α_s^3 one, while for $\mathcal{M}_3(0)$, it is about 3% compared with 2% from $0.299\alpha_s^3$ and with 10% from α_s^2.
- Finally, the leading experimental error due to the J/ψ leptonic widths, which gives a strong limitation to the accuracy of the low n moments (see, e.g., Ref. [498] and references therein), partially cancel out in the ratio of moments such that the experimental error induces only a negligible error of about 1 MeV in the determination of m_c.

- **$Q^2 \neq 0$ moments**

- Comparing, e.g., the QCD sides of $n = 2, Q^2 = 0$ $\mathcal{M}_2(0)$ with that of the $n = 8, Q^2 = 4m_c^2$ $\mathcal{M}_8(4m_c^2)$-moments, we find from Table 9.12 that the sum of the PT corrections up to order α_s^2 for $\mathcal{M}_8(4m_c^2)$ (21%) are about 1/2 of the ones for $\mathcal{M}_2(0)$ (42%). Moreover, though the size of the sum of the NP terms increases from -2.2% for $\mathcal{M}_2(0)$ to -9.6% for $\mathcal{M}_8(4m_c^2)$, one can see that the ratio between $\langle g^3 f_{abc} G^3\rangle$ over the $\langle\alpha_s G^2\rangle$ contributions are almost unchanged of about 20%, indicating the good convergence of the OPE in the analysis.
- One can also deduce from Table 9.12 that the PT and NP QCD corrections are much lower for the ratios of moments. The sum of PT corrections is typically 6.5%, while the NP ones are 2%. A convergence of the OPE is still observed though the ratio between $\langle g^3 f_{abc} G^3\rangle$ over the $\langle\alpha_s G^2\rangle$ contributions is larger ($0.6 \sim 0.8$) for the ratios of moments than for the moments (0.3–0.4).
- Our results from these $Q^2 \neq 0$ ratios of moments given in Table 9.3 agree with the ones 1275(15) MeV obtained in Ref. [293] though less accurate due to the effect of the error on $\langle g^3 f_{abc} G^3\rangle$ included in our analysis.

- **Error due to the subtraction point ν**

Our previous results in Eq. (9.143) have been obtained at the subtraction point $\nu^2 = m_c^2$. If the whole series is known, the results should be independent of ν. The knowledge of the PT series to 4-loops decreases the sensitivity of the results on ν. To order α_s^2 where the moments have been evaluated, one can introduce this ν-dependence through the replacement (see, e.g., Refs. [6, 41]):

$$\alpha_s(m_c) \to \alpha_s(\nu) \times \left(1 - \beta_1 \frac{\alpha_s(\nu)}{\pi} \log \frac{\nu}{m_c}\right), \tag{9.144}$$

where $\beta_1 = -(1/2)(11 - 2n_f/3)$ for n_f-flavours. Taking $0.5 \leq \nu^2/m_c^2 \leq 2$, we deduce from the $Q^2 = 8m_c^2$ ratio of moments:

$$\delta_{m_c}|_\nu = \pm 6 \text{ MeV}. \tag{9.145}$$

- **Shift due to Coulombic corrections**

We can approximately estimate this effect by working with the resummed Coulombic expression of the spectral function [513]:

$$\mathcal{R}_c|_{\text{Coul}} \simeq \frac{3}{2} v \frac{x}{1 - e^{-x}}, \tag{9.146}$$

where $x \equiv \pi C_F \alpha_s / v$, $C_F = 4/3$ and $v = \sqrt{1 - 4m_Q^2/t}$. We compare the value of the ratio of moments using this perturbative expression for the spectral function with the one obtained from PT theory including radiative corrections. In the case $Q^2 = 8m_c^2$ and $n = 14$, where the most precise result is obtained, the corrections induced by the Coulombic contributions to the value of m_c are negligible[20]:

$$\delta_{m_c}|_{\text{Coul}} = -(0.4 \pm 0.4) \text{ MeV}, \tag{9.147}$$

where we have assumed that our determination is known within 100% error.

[20]Some further arguments justifying the smallness of these contributions can be found in Ref. [293].

• Estimate of the $\mathcal{O}(\alpha_s^3)$ and higher order corrections

Observing that the coefficients of the PT corrections for the moment decrease when n increases and do not flip sign compared with the lowest moments (see Table 9.12) and assuming that the ratio of the 3-loop over the 4-loop coefficients are approximatively the same for each moments, we can write to 4-loops:

$$\begin{aligned}\mathcal{M}_{13}(8m_c^2) &\sim 1 + 0.78a_s + 3.06a_s^2 - 5.6a_s^3,\\ \mathcal{M}_{14}(8m_c^2) &\sim 1 + 0.41a_s + 1.91a_s^2 - 3.5a_s^3,\\ \mathcal{M}_{15}(8m_c^2) &\sim 1 + 0.03a_s + 0.77a_s^2 - 1.4a_s^3,\end{aligned} \tag{9.148}$$

where $a_s \equiv \alpha_s/\pi$. We have used the 4-loops coefficient -5.6 obtained in Refs. [499, 500] for low $n = 1$, $Q^2 = 0$ moment, which we expect to be an overestimate of the coefficient of $\mathcal{M}_{13}(8m_c^2)$. These expressions lead to an α_s^2 corrections of about 2% for both $r_{13/14}(8m_c^2)$ and $r_{13/15}(8m_c^2)$ and an α_s^3 correction of about -0.3% and -0.9%. Then, taking the average of the two corrections, we may expect that the α_s^3 corrections can provide a maximal shift of the charm quark mass of about -0.3% leading to:

$$\delta_{m_c}|_{4-\text{loops}} \simeq \pm(2 \times 4) \text{ MeV}, \tag{9.149}$$

a range of values expected from some alternative estimates [498]. The factor 2 in front assumes the estimate of higher order PT ($\mathcal{O}(\alpha_s^n)$: $n \geq 4$) contributions or by duality the $1/s$ corrections due to the tachyonic gluon mass λ^2 [235].

• Final value of $\overline{m}_c(\overline{m}_c)$

Adding the previous estimates of new sources of contributions and errors into the 3-loops result in Eq. (9.143), we obtain to $\mathcal{O}(\alpha_s^2)$:

$$\overline{m}_c(\overline{m}_c)|_{\text{3-loops}} = 1261(18) \text{ MeV}. \tag{9.150}$$

This result is comparable with the existing ones obtained from moment sum rules in the literature [3, 6, 7, 33, 41, 62, 158–160, 293, 498, 505].

9.3.5 $\overline{m}_b(\overline{m}_b)$ from ratios of moments

We extend the previous analysis of the charmonium system to bottomium. In the following, we shall use the value:

$$\alpha_s(m_b)|_{n_f=5} = 0.219(4), \tag{9.151}$$

Table 9.14. Masses and electronic widths of the Υ-family from PDG 08 [33].

Name	Mass [MeV]	$\Gamma_{\Upsilon\to e^+e^-}$ [keV]
$\Upsilon(1S)$	9460.30(26)	1.340(18)
$\Upsilon(2S)$	10023.26(31)	0.612(11)
$\Upsilon(3S)$	10355.2(5)	0.443(8)
$\Upsilon(4S)$	10579.4(1.2)	0.272(29)
$\Upsilon(10860)$	10865(8)	0.31(7)
$\Upsilon(11020)$	11019(8)	0.13(3)

deduced from $\alpha_s(m_\tau)$ in Eq. (9.134). We shall use as experimental inputs the Υ-family parameters in Table 9.14 using NWA and parametrize the spectral function above $\sqrt{t} = (11.098 \pm 0.079)$ GeV by its pQCD expression (QCD continuum), where the error in the continuum threshold is given by the total width of the $\Upsilon(11020)$. Using the previous moments, the dominant contributions will come from the two lowest ground states, while finite width corrections will not be observable. We show in Table 9.15 the results from different moments known to 3-loops.

• Error due to some other sources

Like in the case of charmonium, we estimate the different sources of errors. The one due to the choice of the subtraction point is:

$$\delta_{m_b}|_\nu = \pm 6 \text{ MeV}. \tag{9.152}$$

The shift due to the Coulombic corrections is:

$$\delta_{m_b}|_{\rm Coul} \simeq \pm 6 \text{ MeV}, \tag{9.153}$$

Table 9.15. Value of $\overline{m}_b(m_b)$ from bottomium moments known to 3-loops. The errors on m_b come, respectively, from the choice of the moments, α_s, the data on the Υ-family and the choice of the QCD continuum threshold. The ones due to the gluon condensates are negligible here.

Mom	$\overline{m}_b(\overline{m}_b)$ [MeV]
Q^2=0: $r_{2/3}$, $r_{2/4}$	4160(4)(2)(3)(3)
Q^2=4m_b^2 : $r_{8/9}$, $r_{8/10}$	4177(2)(3)(3)(6)
Q^2=8m_b^2: $r_{13/14}$, $r_{13/15}$	4183(2)(4)(2)(6)
Average	4173(4)

which is about the same value as the one obtained in Ref. [505]. The one due to the estimate of α_s^3 PT corrections is:

$$\delta_{m_b}|_{\text{4-loops}} \simeq \pm(2 \times 2) \text{ MeV}, \tag{9.154}$$

where the factor 2 is assumed to include higher order or/and λ^2-tachyonic gluon mass corrections.

- **Final value of $\overline{m}_b(\overline{m}_b)$ to $\mathcal{O}(\alpha_s^2)$**

Adding these new corrections to the one in Table 9.15, we deduce to 4-loops accuracy:

$$\overline{m}_b(\overline{m}_b)|_{\text{3-loops}} = 4173(10) \text{ MeV}, \tag{9.155}$$

which is relatively more precise than that of m_c as the non-perturbative contributions are much smaller here, while α_s is evaluated at a higher scale.

9.4 Gluon Condensates and $\overline{m}_{c,b}$ from Moments to order α_s^3 and $\langle G^4\rangle$

We extend the previous analysis by including the α_s^3 PT and $\langle G^4\rangle$ dimension $D = 8$ gluon condensate corrections [291].

9.4.1 Test and parametrization of the spectral function

We use a narrow width approximation (NWA) to parametrize the J/ψ family contributions $J/\psi(1S)$ to $\psi(4415)$ to the spectral function.

We shall parametrize the contributions from $\sqrt{t_c} \geq (4.6 \pm 0.1)$ GeV using either of the following:

- *Model 1:* The approximate PT QCD expression of the spectral function to order α_s^2 up to order $(m_c^2/t)^6$ is given in Ref. [440] and the α_s^3 contribution from non-singlet contribution up to order $(m_c^2/t)^2$ in Ref. [514].
- *Model 2:* The asymptotic PT expression of the spectral function known to order α_s^3 where the quark mass corrections are neglected.[21]
- *Model 3:* Fits of different data above the $\psi(2S)$ mass: The most recent fit is done in Ref. [514], where a comparison of results from different fitting procedures can be found.

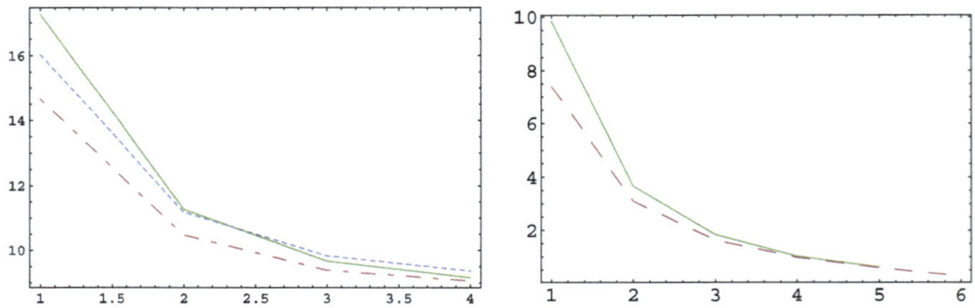

(a) (b)

Figure 9.48. Behaviour of moments $\mathcal{M}_n(Q^2)$ in units of $(4\overline{m}_c^2)^n \times 10^{n+1}$ versus n for different models of the continuum as defined in Section 9.4.1: (a) $\mathcal{M}_n(0)$: Model 1: green (continuous), Model 2: red (dot-dashed), Model 3: blue (dot); (b) the same as Fig. 9.38(a) but for $\mathcal{M}_n(4\overline{m}_c^2)$ for Models 1 and 2.

We test the model dependence of the experimental side of the moments using the previous models for parametrizing the continuum (high-energy) contribution to the spectral function. The analysis is

[21]Original papers are given in Refs. [317–321] of the book in Ref. [6].

shown in Fig. 9.48(a) for the moments $\mathcal{M}_n(0)$ using Models 1–3 for different values of n and in Fig. 9.48(b) for the moments $\mathcal{M}_n(4\overline{m}_c^2)$ using Models 1 and 2. One can deduce that this model dependence can be avoided when working with values of $n \geq 3, 4$. One can also note that for $\mathcal{M}_1(0)$, the continuum (high-energy) contribution to the moments is about (40–50)% of the total contribution, which indicates that the resulting value of m_c from the low moments $n \leq (2-3)$ will depend strongly on the appreciation of this high-energy behaviour which is not measured accurately.

9.4.2 QCD expression of the moment sum rules

- The expression of the moments, up to order α_s^2 for PT series and up to the dimension $D = 6$ condensates, is the same as in the previous section.
- To order α_s^3, the three lowest $\mathcal{M}_1(0)$ [499] and $\mathcal{M}_{2,3}(0)$ moments [515] are known analytically. Semi-analytic expressions of higher moments $\mathcal{M}_n(0)$ using Padé approximants [501] and Mellin–Barnes transform [516] are also available.
- The $D = 8$ condensates are given explicitly in Ref. [291] while they have been estimated in Ref. [502] using factorization (fac $= 1$) and in Ref. [338] using a modified factorisation (fac $= 0.5$). Their contributions have been calculated in Ref. [502] to lowest order. The corresponding expressions of $\mathcal{M}_n(Q^2)$ have been given in Refs. [62, 502]. In general, one can form eight operators for the $D = 8$ gluon condensates:

$$\begin{aligned}
O_1 &= \langle \mathbf{Tr}\ G^2\ \mathbf{Tr}\ G^2 \rangle, & O_2 &= \langle \mathbf{Tr}\ G_{\nu\mu} G^{\rho\mu}\ \mathbf{Tr}\ G_{\nu\tau} G^{\rho\tau} \rangle, \\
O_3 &= \langle \mathbf{Tr}\ G_{\nu\mu} G^{\tau\rho}\ \mathbf{Tr}\ G_{\nu\mu} G^{\tau\rho} \rangle, & O_4 &= \langle \mathbf{Tr}\ G_{\nu\mu} G^{\tau\rho}\ \mathbf{Tr}\ G^{\nu}_{\tau} G^{\mu}_{\rho} \rangle, \\
O_5 &= \langle \mathbf{Tr}\ G_{\nu\mu} G^{\mu\rho}\ G_{\rho\tau} G^{\tau\nu} \rangle, & O_6 &= \langle \mathbf{Tr}\ G_{\nu\mu} G^{\nu\mu} G^{\tau\rho} G_{\tau\rho} \rangle, \\
O_7 &= \langle \mathbf{Tr}\ G_{\nu\mu} G^{\nu\rho}\ G_{\mu\tau} G^{\rho\tau} \rangle, & O_8 &= \langle \mathbf{Tr}\ G_{\nu\mu} G^{\rho\tau} G^{\nu\mu} G^{\rho\tau} \rangle.
\end{aligned} \tag{9.156}$$

Using the symmetry properties of the colour indices and an explicit evaluation of the trace, one can show that one has only six independent operators and the relation for $N = 3$ colours [338]:

$$O_5 + 2O_7 = O_2 + \frac{1}{2}O_4, \qquad O_8 + 2O_6 = O_3 + \frac{1}{2}O_1. \tag{9.157}$$

Normalized to $\langle G^2\rangle^2$, the use of the vacuum saturation in the large N-limit gives:

$$O_1 = \tfrac{1}{4}\left(1 + \tfrac{1}{3}\tfrac{1}{N^2-1}\right), \qquad O_2 = \frac{1}{4}\left(\frac{1}{4} + \frac{1}{3}\frac{1}{N^2-1}\right),$$

$$O_3 = \tfrac{1}{4}\left(\tfrac{1}{6} + \tfrac{7}{6}\tfrac{1}{N^2-1}\right), \qquad O_4 = \frac{1}{4}\left(\frac{1}{12} + \frac{1}{2}\frac{1}{N^2-1}\right),$$

$$O_5 = \tfrac{1}{4N}\left(\tfrac{1}{2} - \tfrac{1}{12}\tfrac{1}{N^2-1}\right), \qquad O_6 = \frac{1}{4N}\left(\frac{7}{6} - \frac{1}{6}\frac{1}{N^2-1}\right),$$

$$O_7 = \tfrac{1}{4N}\left(\tfrac{1}{3} - \tfrac{1}{4}\tfrac{2}{N^2-1}\right), \qquad O_8 = \frac{1}{4N}\left(\frac{1}{3} - \frac{1}{N^2-1}\right), \tag{9.158}$$

which indicates that only the first four operators are leading in $1/N$, and the previous constraints in Eq. (9.157) are not satisfied for large N. Moreover, the $1/N^2$ corrections to these leading term are also large for $N = 3$ in the case of O_3 and O_4, and raise some doubts on the validity of the $1/N$-approximation. Therefore, a modified factorization has been proposed in Ref. [338], where the $D = 8$ gluon condensates have been expressed in terms of O_2 which is not constrained. Normalized to $\langle G^2\rangle^2$, one has:

$$O_1 = 3O_6 = \tfrac{1}{4}, \qquad O_3 = 2O_4 = -\frac{1}{16} + 2O_2$$

$$O_5 = O_7 = -\tfrac{1}{192} + \tfrac{1}{2}O_2, \qquad O_8 = -\frac{5}{48} + 2O_2. \tag{9.159}$$

Reference [338] estimates O_2 using either its large N or its factorization value. Noting that the dominant contribution to the sum rule is due to O_5. Reference [338] notes that the factorization proposed in Ref. [502] overestimates the $D = 8$ gluon condensate contributions.

• From the different expressions of the PT series given in Ref. [293], we observe that, unlike $\mathcal{M}_n(Q^2 = 0)$, where the coefficients increase approximately like n for large n (the same feature occurs for the α_s^3 term given in Refs. [501, 515, 516]), the ones of $\mathcal{M}_n(Q^2 \neq 0)$ remains (within a factor 2) almost constant though change sign from low to high moments. Therefore, we estimate the coefficient of the $\mathcal{O}(\alpha_s^3)$ term of the moments $\mathcal{M}_n(Q^2 \neq 0)$ to be about:

$$c_3|_{Q^2\neq 0,n} \simeq \pm c_3|_{Q^2=0,n=1} \simeq \pm 5.6, \tag{9.160}$$

which is larger than the estimate used in Ref. [290], where it has been assumed that the ratio of the α_s^2 over the α_s^3 coefficients are approximatively the same for each moment.

- We shall use the input values [236, 237, 289, 290]:

$$\begin{aligned} \overline{m}_c(\overline{m}_c) &= 1261(18)\ \text{MeV}, \\ \alpha_s(m_c)|_{n_f=4} &= 0.408(14)\ \text{from}\ \tau-\text{decays}, \\ \alpha_s\langle\bar{\psi}\psi\rangle^2 &= 4.5\times 10^{-4}\ \text{GeV}^6\ \text{from}\ \text{e}^+\text{e}^-. \end{aligned} \tag{9.161}$$

The error in the value of α_s is the distance between its value and the world average [33, 444].

9.4.3 Hunting $\langle\alpha_s G^2\rangle$ from higher moments $\mathcal{M}_n(Q^2)$

As mentioned in the introduction, the gluon condensate plays a key rôle in QCD gluon dynamics like the quark condensate $\langle\bar{\psi}\psi\rangle$ for chiral symmetry breaking. We have also mentioned that there is a spread of predictions of its value in the literature. The extraction of $\langle\alpha_s G^2\rangle$ in this paper is closed to the one using charmonium sum rules in the early literatures, which follow the pioneer work of SVZ [3].

- In our analysis, we shall work with higher moments which are more sensitive to $\langle\alpha_s G^2\rangle$ but limit ourselves to the ones where the higher dimension six and eight condensates remain still small corrections such that the OPE remains valid. This compromise choice eliminates the use of higher $Q^2 = 0$ moments where their convergence has been the subject of hot debate in the past [127, 502]. Instead, the $Q^2 \neq 0$ moments converge faster [62], which allow us to work with higher n-values. In the following, we shall work with $\mathcal{M}_n(Q^2 = 0)$ for $n \leq 5$, $\mathcal{M}_n(Q^2 = 4\overline{m}_c^2)$ for $n \leq 11\text{–}12$ and $\mathcal{M}_n(Q^2 = 8\overline{m}_c^2)$ for $n \leq 20$, where the OPE still makes sense when using the values of the vacuum condensates given in the literature [6].

• We extract $\langle\alpha_s G^2\rangle$ using its correlations with the $D = 6$ and 8 condensates introduced above. We allow the instanton radius ρ_c which correlates $\langle\alpha_s G^2\rangle$ and $\langle g^3 f_{abc} G^3\rangle$ to move from 1 to 5 GeV^{-1}, where the latter would be the value given by a dilute gas instanton model estimate [3]. We shall also use the relation of $\langle\alpha_s G^2\rangle$ and $\langle g^3 f_{abc} G^3\rangle$ with the $D = 8$ condensates if one assumes a factorization hypothesis [502] or its modified form [338].

• Note that, unlike [293], we fix $\overline{m}_c$, which is, at present, known with good accuracy in order to give stronger constraints on the value of $\langle\alpha_s G^2\rangle$. We show the results as function of the number n of derivatives for $Q^2 = 4\overline{m}_c^2$ and $8\overline{m}_c^2$ and for different values of the QCD input parameters.

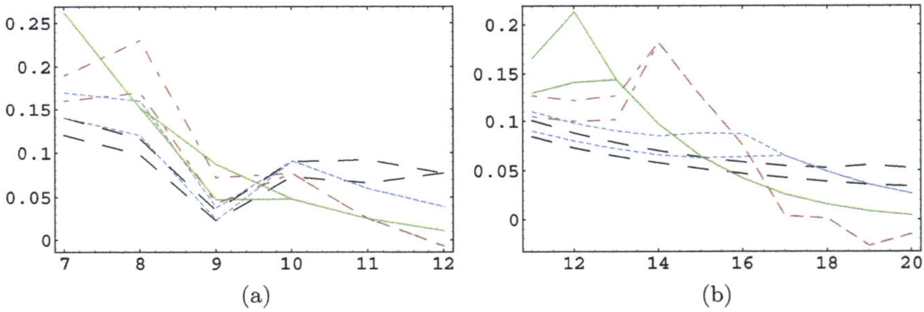

Figure 9.49. Behaviour of $\langle\alpha_s G^2\rangle$ in units of GeV^4 versus n from $\mathcal{M}_n(Q^2)$ moments for different values of the QCD inputs. We use the central value of α_s given in Eq. (9.161). The region between the same curves correspond to the values of c_3 from Eq. (9.160): (a) $\mathcal{M}_n(4m_c^2)$: $\rho_c = 1$ GeV^{-1}, fac = 1: green (continuous); $\rho_c = 5$ GeV^{-1}, fac = 1: blue (dot); $\rho_c = 1$ GeV^{-1}, fac = 0.5: red (dot-dashed); $\rho_c = 5$ GeV^{-1}, fac = 0.5: black (dashed); (b) the same as Fig. 9.49(a) but for $\mathcal{M}_n(8m_c^2)$.

One can note from Fig. 9.49 that the effect of α_s^3 is relatively small. Much more stable values of $\langle\alpha_s G^2\rangle$ correspond to the case of a modified factorisation of $\langle G^4\rangle$ which sounds better found from the analysis of Ref. [338] based on the $1/N$ approach. Taking into account these remarks, we deduce in units of GeV^4:

$$\begin{aligned}\langle\alpha_s G^2\rangle &= (4.8 - 9.2)\times 10^{-2} \quad \text{from } \mathcal{M}_n(4m_c^2),\\ &\quad (5.6 - 8.3)\times 10^{-2} \quad \text{from } \mathcal{M}_n(8m_c^2), \qquad (9.162)\end{aligned}$$

where we have used the *Mathematica Package FindRoots*, which we shall also use later on for deriving all the results in this paper.

• We consider as a final result the most precise determination from $\mathcal{M}_n(8m_c^2)$ which can be written as:

$$\langle \alpha_s G^2 \rangle = (7.0 \pm 1.3) \times 10^{-2}\ \text{GeV}^4. \tag{9.163}$$

This result goes in line with different claims [6,7,64,70,223–225,236, 237,502,522] that the SVZ value given in Eq. (9.201) underestimates the value of the gluon condensate.[22] This result agrees quite well with the one derived from the charmonium and bottomium mass splittings using double ratio of sum rules (DRSR) [223]:

$$\langle \alpha_s G^2 \rangle = (7.5 \pm 2.5) \times 10^{-2}\ \text{GeV}^4, \tag{9.164}$$

and from τ-like sum rule for $e^+e^- \to I = 1$ hadrons data [236, 237]:

$$\langle \alpha_s G^2 \rangle = 6.1(0.7) \times 10^{-2}\ \text{GeV}^4. \tag{9.165}$$

Our result is more precise than the one in Ref. [293], using some particular choice of moments, as, here, we have fixed the value of $\overline{m}_c$, while in Ref. [293], a two-parameter fit ($\overline{m}_c, \langle \alpha_s G^2 \rangle$) has been performed. Indeed, using as input the value of $\overline{m}_c(\overline{m}_c)$ in Eq. (9.161), one would deduce from the different figures given in Ref. [293]:

$$\langle \alpha_s G^2 \rangle \simeq (3.5 - 7.5) \times 10^{-2}\ \text{GeV}^4, \tag{9.166}$$

obtained to order α_s^2 and without the inclusion of $\langle G^4 \rangle$. This range of values is in agreement with the one in Eq. (9.162) but less precise.

9.4.4 Re-extraction of $\langle g^3 f_{abc} G^3 \rangle$ and factorisation test of $\langle G^4 \rangle$

Using the previous new information, we re-extract the value of $\langle g^3 f_{abc} G^3 \rangle$ firstly obtained in Ref. [290] using the sum rules approach. As remarked in Ref. [290], the moment $\mathcal{M}_n(8m_c^2)$ can provide the most accurate value of $\langle g^3 f_{abc} G^3 \rangle$. We plot in Fig. 9.50 the value of the instanton radius ρ_c defined in Eq. (9.128) versus the number of moments for given values of $\overline{m}_c$, α_s, $\langle \alpha_s G^2 \rangle$ and the factorisation factor fac of the $\langle G^4 \rangle$ condensates. We have only shown in Fig. 9.50

[22] A compilation of different determinations can be found in Table 9.9 of Ref. [223] and in Ref. [6] (see reprinted papers in Chapters 51 and 52).

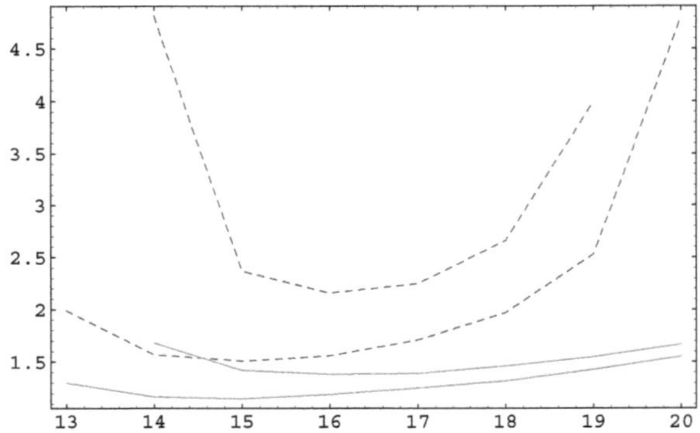

Figure 9.50. Behaviour of the instanton radius ρ_c in units of GeV^{-1} versus n from $\mathcal{M}_n(8m_c^2)$ moments for the central values of $\overline{m}_c(\overline{m}_c)$ and α_s given in Eq. (9.161). The curves correspond to fac=0.5 for the factorisation of $\langle G^4\rangle$ and $\langle\alpha_s G^2\rangle = 0.070$ GeV^4. The region between the same curves correspond to the values of c_3 from Eq. (9.160): dashed (red) with $\langle G^4\rangle$ and green (continuous) without $\langle G^4\rangle$.

the curves for fac = 0.5 because the one for fac = 1 gives unrealistic values of ρ_c. This can be an indirect indication that the value fac = 1 is less favoured, a result which supports the $1/N$ result in Ref. [338]. A similar feature is also signalled when extracting $\langle\alpha_s G^2\rangle$ because for fac = 0.5, larger stabilities versus the change of n (see Fig. 9.49) are obtained. At the minimas of the curves in Fig. 9.50, we deduce the optimal value of ρ_c in GeV^{-1} when the effect of $\langle G^4\rangle$ is included:

$$\rho_c = 1.84 \pm 0.24_{\alpha_s} \pm 0.33_{\alpha_s^3} \pm 0.27_{G^2}, \tag{9.167}$$

which after adding the errors quadratically gives[23]:

$$\rho_c = (1.84 \pm 0.49)\ \text{GeV}^{-1} \quad \longrightarrow \quad \frac{\langle g^3 f_{abc} G^3\rangle}{\langle\alpha_s G^2\rangle} = (8.8 \pm 2.3)\ \text{GeV}^2. \tag{9.168}$$

We consider this result as improvement of the previous result quoted in Eq. (9.136), which has been affected by the presence of $\langle G^4\rangle$ in the OPE (see the two continuous (green) curves in Fig. 9.46 when $\langle G^4\rangle$

[23]The error in Ref. [291] has been multiplied by a conservative factor 2.

is not included[24]) as (*a priori*) expected. This value of $\langle g^3 f_{abc} G^3 \rangle$ is in the range of lattice calculations in pure $SU(2)$ Yang–Mills [508], though an eventual future result for $SU(3)$ is desirable.

9.4.5 $\overline{m}_c(\overline{m}_c)$ from low moments $\mathcal{M}_{n\leq 5}(0)$

$Q^2 = 0$ moments possess several drawbacks (convergence of the OPE, large PT α_s corrections, etc.) while the lowest moment $\mathcal{M}_1(0)$ is largely affected by the non-accurate higher energy data.

To order α_s^3, the PT series normalized to $(4\overline{m}_c^2)^n$ read in our normalization as follows:

$$\begin{aligned}
\mathcal{M}_1^{PT}(0) &= 0.80(1 + 2.39a_s + 2.38a_s^2 - 5.64a_s^3),\\
\mathcal{M}_2^{PT}(0) &= 0.3429(1 + 2.43a_s + 6.11a_s^2 - 7.64a_s^3),\\
\mathcal{M}_3^{PT}(0) &= 0.2032(1 + 1.92a_s + 6.12a_s^2 - 10.48a_s^3),\\
\mathcal{M}_4^{PT}(0) &= 0.1385(1 + 1.10a_s + 4.40a_s^2 - 18.13a_s^3),\\
\mathcal{M}_5^{PT}(0) &= 0.1023(1 + 0.08a_s + 2.16a_s^2 - 27.4a_s^3),
\end{aligned} \tag{9.169}$$

where one can note that the coefficient of a_s^3 grows with the order n of the moments, but the coefficient of α_s decreases.

- The $D = 4$ contribution including the α_s corrections normalized to the lowest order PT moments and without the overall factor $\langle a_s G^2 \rangle/(4\overline{m}_c^2)^2$ read as follows:

$$\begin{aligned}
\mathcal{M}_1^{D=4}(0) &= -15.04(1 + 2.48a_s),\\
\mathcal{M}_2^{D=4}(0) &= -58.49(1 + 1.05a_s),\\
\mathcal{M}_3^{D=4}(0) &= -143.6(1 - 0.48a_s),\\
\mathcal{M}_4^{D=4}(0) &= -283.4(1 - 2.11a_s),\\
\mathcal{M}_5^{D=4}(0) &= -491.3(1 - 3.80a_s),
\end{aligned} \tag{9.170}$$

where one should note that one cannot go beyond $n = 5$ because the α_s correction to the $D = 4$ contribution is larger than 49%, indicating the divergence of the QCD series, as also emphasized by Ref. [293].

---

[24]These values agree with the one obtained in Ref. [290] using some judicious choice of the ratios of moments $r_{13/14}$ and $r_{14/15}$.

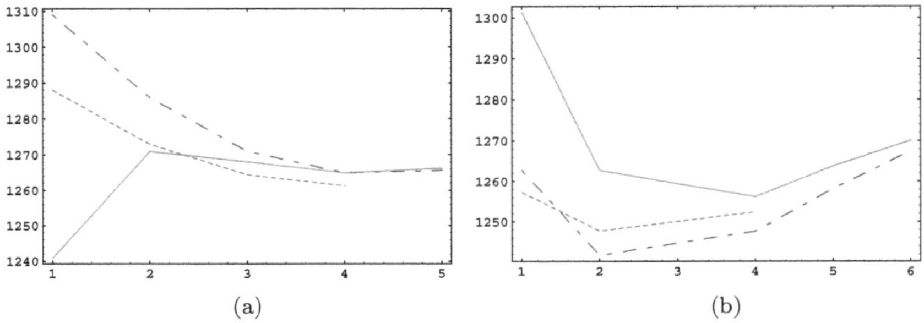

Figure 9.51. (a) Behaviour of $\overline{m}_c(\overline{m}_c)$ in units of MeV versus n from $\mathcal{M}_n(0)$ moments for the central value of $\langle\alpha_s G^2\rangle$ in Eq. (9.163) and α_s given in Eq. (9.134). We use fac = 0.5 for the factorisation of $\langle G^4\rangle$. The curves correspond to different models for the continuum defined in Section 9.4.1: Model 1: green (continued); Model 2: red (dot-dashed); Model 3 (data fit): blue (dot). (b) Behaviour of $\overline{m}_c(\overline{m}_c)$ versus different ratios of $\mathcal{M}_n(0)$ moments in units of MeV. The inputs and legends are the same as in Fig. 9.51(a)). In the n-axis: $1 \equiv r_{1/2}$, $2 \equiv r_{2/3}$, $3 \equiv r_{2/4}$, $4 \equiv r_{3/4}$, $5 \equiv r_{3/5}$ and $6 \equiv r_{4/5}$.

- Then, we limit ourselves to use the relatively low moments $\mathcal{M}_{n\leq 5}(0)$ for extracting the running mass $\overline{m}_c(\overline{m}_c)$ within fixed order PT series and for a given set of NP parameters determined in the previous section. We show the results from the moments in Fig 9.51(a) and the one from the ratios in Fig. 9.51(b).

As expected, the result for $n \leq 2$ is sensitive to the model for the continuum which contributes for 40% to the moments.

The result only stabilizes versus the variation of n for $n \geq 3$–5, where an optimal result can be taken. For definiteness, we take $n \simeq 4$, where all *continuum models* give consistent predictions. Then, we deduce, from Fig. 9.51(a), in units of MeV:

$$\overline{m}_c(\overline{m}_c)|_0^4 = 1263.7(1.3)_{\text{cont}}(3.5)_{\alpha_s}(4.9)_{\alpha_s^4}$$
$$(3.9)_\mu(4.4)_{G^2}(4.7)_{G^3}(3.1)_{G^4}(1.7)_{\text{exp}}, \qquad (9.171)$$

where the central value is the average from different continuum models. It leads to the result in Table 9.16.

- The first error in Eq. (9.171) is due to the different models for the continuum and the second one to the value of α_s given in Eq. (9.134). The third error is an estimate of higher order terms of PT assumed to be equal to the contribution of the α_s^3 one, while the fourth error is an estimate of the effect of the subtraction point ν by varying it from m_c to M_τ and using the substitution in Eq. (9.144). The fifth and sixth errors are due, respectively, to the ones of the gluon condensates $\langle\alpha_s G^2\rangle$ and $\langle g^3 f_{abc} G^3\rangle$ estimated previously. The seventh error is due to the $\langle G^4\rangle$ condensates, allowing it to move from fac=0.5 [338] (favoured value from our preceding fit) to fac=1 [502]. The last error is due to the experimental J/ψ widths given in Table 9.11.

• We consider as a final value the one obtained from $\mathcal{M}_4(0)$ where both PT corrections are still small for the unit and dimension 4 operators. Indeed, for the unit operator, the dominant correction is due to α_s, which is about 14% for $\mathcal{M}_4(0)$ or for m_c^2 [25] and which is about half of the one of $\mathcal{M}_1(0)$. Then, we may expect that the error induced by the organization of the PT series (fixed order, contour improved, etc.) discussed in Ref. [514] is smaller for $\mathcal{M}_4(0)$ though the PT series converges faster for $\mathcal{M}_1(0)$ as on can note in Eq. (9.169).

• The result from the ratios of moments in Fig. 9.51(b) is not very conclusive as the model dependence of the result starts to disappear from the ratio of moments $r_{3/5}$, but for these ratios, the result increases with n. Then, we shall not retain the results from the ratios of moments for the charmonium channel.

9.4.6 $\overline{m}_c(\overline{m}_c)$ from higher $\mathcal{M}_n(Q^2)$ moments

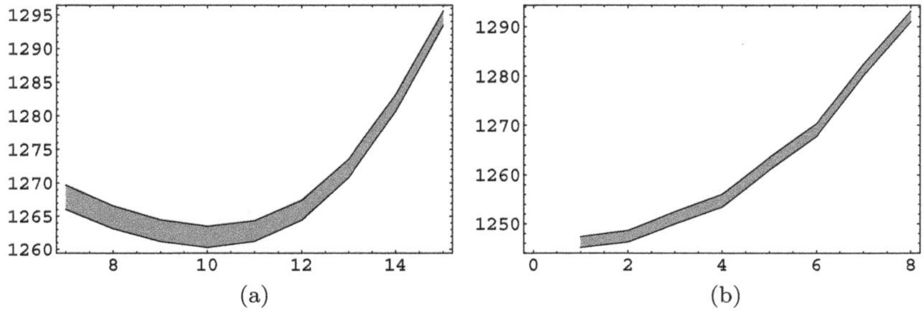

Figure 9.52. (a) Behaviour of $\overline{m}_c(\overline{m}_c)$ in units of MeV versus n from $\mathcal{M}_n(4\overline{m}_c^2)$ moments and for the central value of $\langle\alpha_s G^2\rangle$ in Eq. (9.163). We use fac=0.5 for the factorisation of $\langle G^4\rangle$ and Model 2 for the continuum. The coloured region corresponds to the range of c_3 values given in Eq. (9.160). (b) Behaviour of $\overline{m}_c(\overline{m}_c)$ in units of MeV versus different ratios of moments $r_{n/n+1}(4\overline{m}_c^2)$. The inputs and legends are the same as in Fig. 9.52(a). In the n-axis, $1 \equiv r_{7/9}$, $2 \equiv r_{8/9}$, $3 \equiv r_{8/10}$, $4 \equiv r_{9/10}$, etc.

[25] The α_s^2 (resp α_s^3) are relatively small, i.e., 7.4% (respectively 3.9%).

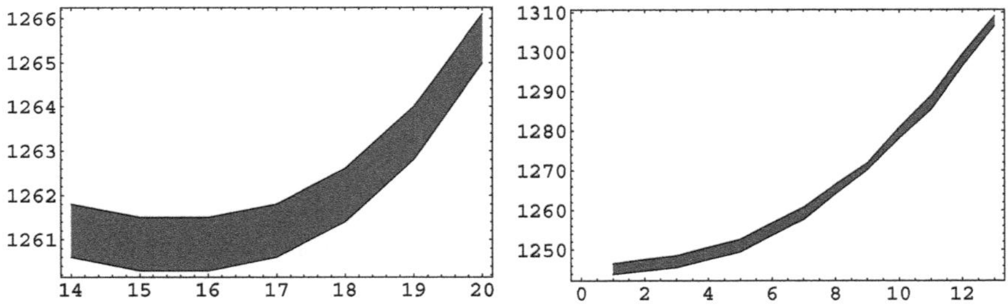

(a) (b)

Figure 9.53. (a) Behaviour of $\overline{m}_c(\overline{m}_c)$ in units of MeV versus n from $\mathcal{M}_n(8\overline{m}_c^{\,2})$ moments for the central value of $\langle\alpha_s G^2\rangle$ in Eq. (9.163) and α_s given in Eq. (9.134). We use fac $= 0.5$ for the factorisation of $\langle G^4\rangle$ and Model 2 for the continuum. The coloured region corresponds to the range of c_3 values given in Eq. (9.160). (b) Behaviour of $\overline{m}_c(\overline{m}_c)$ versus different ratios of moments $\mathcal{M}_n(8\overline{m}_c^{\,2})$ in units of MeV. The inputs and legends are the same as in Fig. 9.53(a). In the n-axis, $1 \equiv r_{13/14}$, $2 \equiv r_{13/15}$, $3 \equiv r_{14/15}$, $4 \equiv r_{14/16}$, $5 \equiv r_{15/16}$, etc.

• In the following, we shall extract $\overline{m}_c(\overline{m}_c)$ from higher $\mathcal{M}_n(4m_c^2)$ and $\mathcal{M}_n(8m_c^2)$ moments. We show the results of the analysis, respectively, from the moments in Figs. 9.52(a) and 9.53(a) and from the ratios of moments in Figs 9.52(b) and 9.53(b). One can note that in both cases, the results from the moments present minimas versus n.

• The minimum is obtained from $\mathcal{M}_{10}(4m_c^2)$ and $\mathcal{M}_{15}(8m_c^2)$, which give in units of MeV:

$$\begin{aligned}
\overline{m}_c(\overline{m}_c)|^{10}_{4m_c^2} &= 1261.9(0.7)_{\alpha_s}(1.6)_{\alpha_s^3}(1.6)_{\alpha_s^{n\geq 4}}(0.4)_\mu \\
&\quad (1.1)_{G^2}(1.0)_{G^3}(1.7)_{G^4}(3.0)_{\rm exp}, \\
\overline{m}_c(\overline{m}_c)|^{15}_{8m_c^2} &= 1260.9(0.5)_{\alpha_s}(1.6)_{\alpha_s^3}(1.6)_{\alpha_s^{n\geq 4}}(0.4)_\mu \\
&\quad (1.0)_{G^2}(0.7)_{G^3}(1.3)_{G^4}(2.6)_{\rm exp},
\end{aligned} \tag{9.172}$$

which lead to the result in Table 9.16. The different sources of errors are the same as the ones discussed in Eq. (9.171). The one from α_s^3 here is due to the distance of the average of the α_s^3 contribution to the $\pm$ assumed value of the coefficient in Eq. (9.160). We have estimated the error due to the unknown α_s^n $(n \geq 4)$ to be equal to that of α_s^3. Equation (9.172) leads to the result in Table 9.16.

• One can also see in Figs. 9.52(b) and 9.53(b) that the results from the ratios of moments increase with n. Though the outputs obtained from the ratios of optimal moments are consistent with the ones from these moments and with the ones obtained in Ref. [290] where a judicious choice (small PT corrections) of these ratios have been used, we shall not consider these numbers in the final results because of the absence of stabilities or minimas versus the variation of n.

Table 9.16. Value of $\overline{m}_c(\overline{m}_c)$ from charmonium moments known to α_s^3 for $Q^2 = 0$ with an estimate of the α_s^3 contribution for $Q^2 \neq 0$.

Moments	$\overline{m}_c(\overline{m}_c)$ [MeV]
Q^2=0: $\mathcal{M}^4$	1263.7(10.3)
Q^2=$4m_c^2$: $\mathcal{M}^{10}$	1261.9(4.5)
Q^2=$8m_c^2$: $\mathcal{M}^{15}$	1260.9(4.0)

9.4.7 Final value of $\overline{m}_c(\overline{m}_c)$ and Coulombic corrections

Like in the previous section [290], we approximately estimate the Coulombic correction by working with the resummed expression of the spectral function [513]. We obtain:

$$\delta_{m_c}|_{\text{Coul}} = \pm 16 \text{ MeV}. \tag{9.173}$$

One can see from Table 9.16 that the estimate from different forms of the moments are consistent each other. We shall consider as a final estimate the most precise one from $\mathcal{M}^{15}(8m_c^2)$, where the Coulombic correction obtained previously is also small. Adding this correction, we obtain:

$$\overline{m}_c(\overline{m}_c) = 1261(16) \text{ MeV}, \tag{9.174}$$

in excellent agreement with the one [290]:

$$\overline{m}_c(\overline{m}_c) = 1261(18) \text{ MeV}, \tag{9.175}$$

obtained from a judicious choice of ratios of high moments having small PT and NP corrections. The previous results also improve earlier results obtained by the author to lower orders in this channel [158–160]. The two determinations lead to the mean value:

$$\langle\overline{m}_c(\overline{m}_c)\rangle = 1261(12) \text{ MeV}, \tag{9.176}$$

9.4.8 Determination of $\overline{m}_b(\overline{m}_b)$

We extend the previous analysis to the bottomium systems. In the following, we shall use the value:

$$\alpha_s(m_b)|_{n_f=5} = 0.219(4), \tag{9.177}$$

deduced from $\alpha_s(m_\tau)$ in Eq. (9.161). The spectral function is parametrized by using NWA for the Υ-family from $\Upsilon(1S)$ to $\Upsilon(11020)$. Above, $\sqrt{t} = (11.098 \pm 0.079)$ GeV by its pQCD expression (Model 2), where the error in the continuum threshold is given by the total width of the $\Upsilon(11020)$. We shall work with higher moments in order to minimize the contributions of the QCD continuum. We use moments known to order α_s^3 for $Q^2 = 0$, while for $Q^2 \neq 0$, we have added the estimate of the α_s^3 contribution given in Eq. (9.160).

• Results from $\mathcal{M}_n(0)$

We show the results from $\mathcal{M}_n(0)$ in Fig. 9.54, where one can note that the result is (almost) stable versus the variation of n for $n \simeq 3$–7 while for the ratios of moments, the stability is reached from $r_{4/5}$. At these values, the contribution of the QCD continuum is less than 29% of the total which is much less than the one for $n = 1$, where it is about 66%. This feature raises serious doubts on the accurate value of m_b from this low moment $\mathcal{M}_1(0)$ given in the literature [498,499,515] due to the inaccuracy of the data in this high-energy region. From the moments, we obtain in units of MeV:

$$\begin{aligned}
\overline{m}_b(\overline{m}_b)|_0^4 &= 4169.6(1.9)_{\alpha_s}(4.1)_{\alpha_s^{n\geq 4}}(2.7)_\mu(1.1)_{G^2}(1.2)_{G^3} \\
&\quad (1.2)_{G^4}(10.6)_{\rm exp}, \\
\overline{m}_b(\overline{m}_b)|_0^{4/5} &= 4166.3(4)_{\alpha_s}(7.1)_{\alpha_s^{n\geq 4}}(16.6)_\mu(0.6)_{G^2}(0.4)_{G^3} \\
&\quad (0.4)_{G^4}(1.9)_{\rm exp},
\end{aligned} \tag{9.178}$$

giving the results in Table 9.17. One can note that, at the optimal choice $r_{4/5}(0)$, PT corrections are large, which induce larger PT errors than in the case of $\mathcal{M}_4(0)$. The different sources of errors are similar to the case of charmonium.

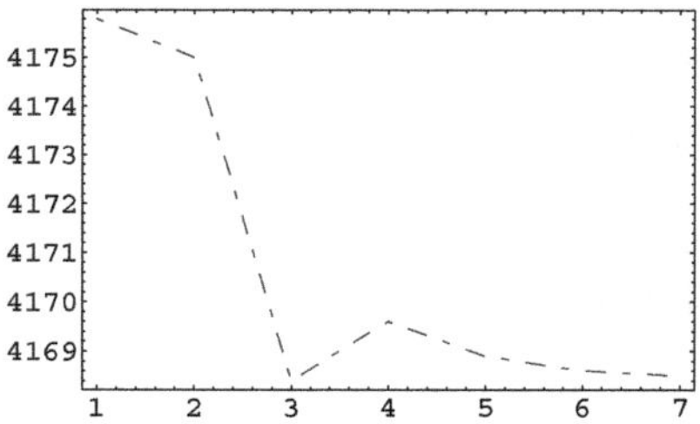

(a)

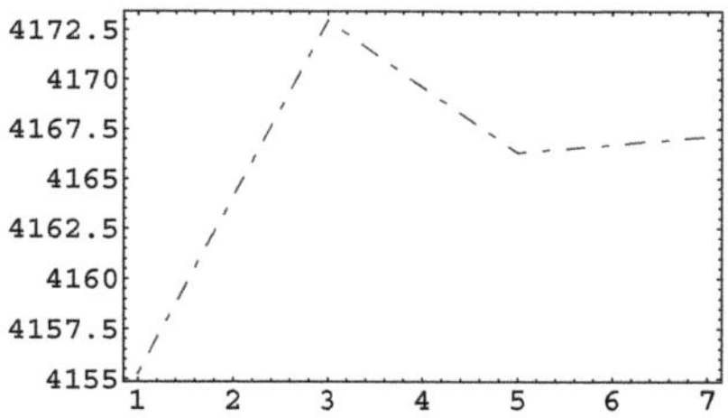

(b)

Figure 9.54. (a) Behaviour of $\overline{m}_b(\overline{m}_b)$ in units of MeV versus n from $\mathcal{M}_n(0)$ moments for the central value of $\langle\alpha_s G^2\rangle$ in Eq. (9.163) and α_s given in Eq. (9.134). We use fac=0.5 for the factorisation of $\langle G^4\rangle$ and Model 2 for the high-energy part of the spectral function. (b) Behaviour of $\overline{m}_b(\overline{m}_b)$ versus different ratios of moments $\mathcal{M}_n(0)$ in units of MeV. The inputs and legends are the same as in Fig. 9.54(a). In the n-axis, $1 \equiv r_{2/3}$, $2 \equiv r_{2/4}$, $3 \equiv r_{3/4}$, $4 \equiv r_{3/5}$, $5 \equiv r_{4/5}$, etc.

• Results from $\mathcal{M}_n(4m_b^2)$

The results from $\mathcal{M}_n(4m_b^2)$ are shown in Fig 9.55, where a stability versus the variation of n is obtained for $n = 14$, while for the ratios of moments, it is reached for $r_{10/11}(4m_b^2)$ and $r_{10/12}(4m_b^2)$. In both cases, the errors due to the NP contributions and induced by the $\pm$ sign for the estimate of the α_s^3 coefficient and of the higher order $\alpha_s^{n\geq 4}$ are tiny (≤ 0.4 MeV) and can be neglected. We obtain in units of MeV:

$$\overline{m}_b(\overline{m}_b)|_{4m_b^2}^{14} = 4174.2(0.6)_{\alpha_s}(2.6)_\mu(5.1)_{\rm exp},$$
$$\overline{m}_b(\overline{m}_b)|_{4m_b^2}^{10/11} = 4178.6(4.2)_{\alpha_s}(10.8)_\mu(3.6)_{\rm exp}, \qquad (9.179)$$

from which we deduce the result in Table 9.17.

• Results from $\mathcal{M}_n(8m_b^2)$

The results from $\mathcal{M}_n(8m_b^2)$ and from the ratios of moments are shown in Fig 9.56, where stabilities versus the n-variations are, respectively, obtained for $n = 9-11$ and for $r_{15/17}$, $r_{16/17}$. Non-perturbative corrections and the one due to the $\pm$ sign of the α_s^3 coefficient are also negligible (≤ 0.3 MeV). We obtain in units of MeV:

$$\overline{m}_b(\overline{m}_b)|_{8m_b^2}^{10} = 4175.1(0.5)_{\alpha_s}(5.1)_{\alpha_s^3}(0.5)_{\alpha_s^{n\geq 4}}(1.9)_{\mu}(10)_{\rm exp},$$
$$\overline{m}_b(\overline{m}_b)|_{8m_b^2}^{16/17} = 4170.9(1.6)_{\alpha_s}(9.5)_{\alpha_s^3}(4.1)_{\alpha_s^{n\geq 4}}(7.2)_{\mu}(3.6)_{\rm exp}. \quad (9.180)$$

Then, we deduce the result in Table 9.17.

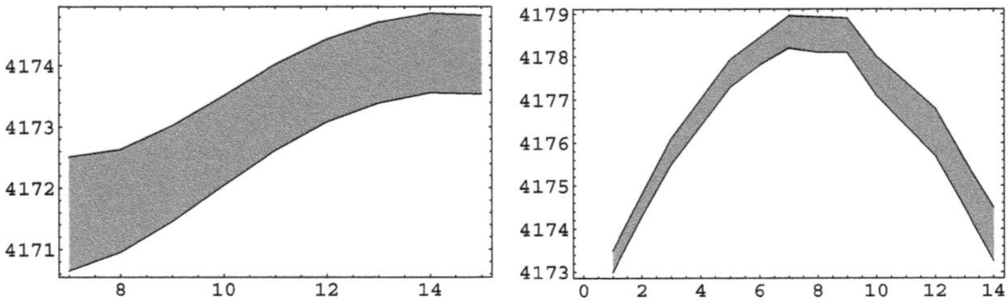

(a) (b)

Figure 9.55. (a) Behaviour of $\overline{m}_b(\overline{m}_b)$ in units of MeV versus n from $\mathcal{M}_n(4\overline{m}_b^2)$ moments for the central value of $\langle\alpha_s G^2\rangle$ in Eq. (9.163) and α_s given in Eq. (9.134). We use fac = 0.5 for the factorisation of $\langle G^4\rangle$ and Model 2 for the high-energy part of the spectral function. The coloured region corresponds to the range of c_3 values given in Eq. (9.160). (b) Behaviour of $\overline{m}_b(\overline{m}_b)$ versus different ratios of moments $\mathcal{M}_n(4\overline{m}_b^2)$ in units of MeV. The inputs and legends are the same as in Fig. 9.55(a). In the n-axis, $1 \equiv r_{7/8}$, $2 \equiv r_{7/9}$, $3 \equiv r_{8/9}$, $4 \equiv r_{8/10}$, etc.

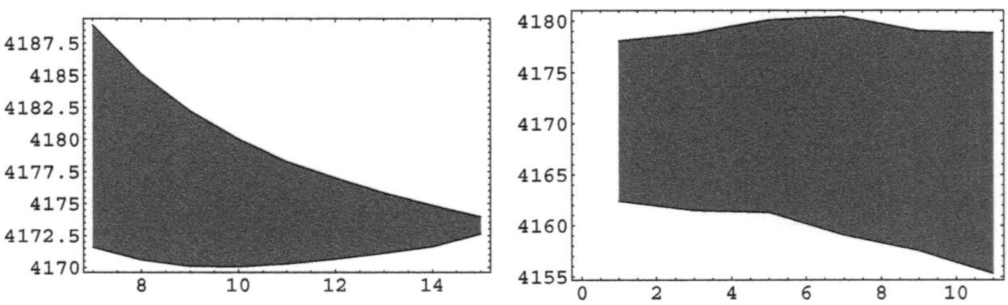

(a) (b)

Figure 9.56. (a) Behaviour of $\overline{m}_b(\overline{m}_b)$ in units of MeV versus n from $\mathcal{M}_n(8\overline{m}_b^2)$ moments for the central value of $\langle\alpha_s G^2\rangle$ in Eq. (9.163) and α_s given in Eq. (9.134). We use fac = 0.5 for the factorisation of $\langle G^4\rangle$. Model 2 for the high-energy part of the spectral function is used. The coloured region corresponds to the range of c_3 values given in Eq. (9.160). (b) Behaviour of $\overline{m}_b(\overline{m}_b)$ versus different ratios of moments $\mathcal{M}_n(8\overline{m}_b^2)$ in units of MeV. The inputs and legends are the same as in Fig. 9.56(a). In the n-axis, $1 \equiv r_{13/14}$, $2 \equiv r_{13/15}$, $4 \equiv r_{14/15}$, etc.

Table 9.17. Value of $\overline{m}_b(\overline{m}_b)$ from bottomium moments known to α_s^3 for $Q^2 = 0$ and with an estimate of the α_s^3 contribution for $Q^2 \neq 0$.

Q^2	Moments	$\overline{m}_b(\overline{m}_b)$ [MeV]	Ratio	$\overline{m}_b(\overline{m}_b)$ [MeV]
0	$\mathcal{M}^4$	4169.6(11.3)	$r_{4/5}$	4166.4(18.6)
$4m_b^2$	$\mathcal{M}^{14}$	4174.2(5.8)	$r_{10/11}$	4178.6(12.1)
$8m_b^2$	$\mathcal{M}^{10}$	4175.1(5.5)	$r_{16/17}$	4170.9(12.7)

9.4.9 Final value of $\overline{m}_b(\overline{m}_b)$ and Coulombic corrections

Here, we analyze the Coulombic corrections like in the case of charm. The ones for the moments are relatively large which are, respectively, 1.7%, 1.1% and 4% for $\mathcal{M}^{10}(8m_b^2)$, $\mathcal{M}^{14}(4m_b^2)$ and $\mathcal{M}^4(0)$. The ones for the ratios of moments $r_{16/17}(8m_b^2)$, $r_{10/11}(4m_b^2)$ and $r_{4/5}(0)$ are, respectively, 1.2, 2. 1 and 3.6 per mil, which are about one order of magnitude smaller. Among these different determinations, the one from $r_{16/17}(8m_b^2)$ is the most precise. We consider it as our best final result:

$$\overline{m}_b(\overline{m}_b) = 4171(14)\ \text{MeV}. \tag{9.181}$$

It is informative to compare the previous result with the one in Ref. [290] (see Table 9.15) using some judicious choices of the ratios of moments having the smallest PT corrections and where the $\langle G^4 \rangle$ contribution has not been included. Adding the errors ± 6 MeV due to the Coulombic, ± 6 MeV due to the subtraction point and ± 4 MeV due to the α_s^3 contributions, the average from Table 9.15 becomes:

$$\overline{m}_b(\overline{m}_b) = 4173(10)\ \text{MeV}, \tag{9.182}$$

which is in excellent agreement with the one obtained in Eq. (9.181).

9.5 Gluon Condensates and $\overline{m}_b(\overline{m}_b)$ from LSR

In Ref. [291], we pursue the estimate of the gluon condensates $\langle g^3 f_{abc} G^3 \rangle$ and the running b-quark mass but using the QCD-exponential or Laplace sum rules (LSR):

$$\mathcal{L}_p(\tau) = \int_{4m_Q^2}^{\infty} dt\ t^p\ e^{-t\tau} R_{e^+e^-}(t, m_c^2) \quad \text{for } p = 0, \tag{9.183}$$

and its ratio:

$$\mathcal{R}_0(\tau) \equiv -\frac{d}{d\tau}\log[\mathcal{L}_0] = \frac{\mathcal{L}_1}{\mathcal{L}_0}, \qquad (9.184)$$

when higher order PT and NP terms are included into their QCD expressions; $R_{e^+e^-}$ is the ratio of the total cross-section of $\sigma(e^+e^- \to \text{hadrons})$ over $\sigma(e^+e^- \to \mu^+\mu^-)$, which is normalized asymptotically to one; τ is the sum rule variable. Using the minimal duality ansatz (MDA) for parametrizing the spectral function, one can transform the previous sum rule into a finite energy (FESR) one:

$$\mathcal{L}_0^c(\tau) = \int_{4m_Q^2}^{t_c} dt e^{-t\tau} R_{e^+e^-}(t, m_c^2), \quad \mathcal{R}_0^c(\tau) = \frac{\mathcal{L}_1^c}{\mathcal{L}_0^c} \simeq M_R^2, \quad (9.185)$$

where M_R is the lowest resonance mass.

9.5.1 Tests of the convergence of the sum rules

To illustrate the analysis, we shall use as QCD input parameters the values of the gluon condensate:

$$\langle \alpha_s G^2 \rangle = (0.07 \pm 0.01)\,\text{GeV}^4,$$
$$\langle g^3 f_{abc} G^3 \rangle = (8.8 \pm 5.5)\,\text{GeV}^2 \times \langle \alpha_s G^2 \rangle, \qquad (9.186)$$

and the quark masses (average of the two consistent determinations in Refs. [290, 291]):

$$\overline{m}_c(\overline{m}_c) = 1262(17)\,\text{MeV}, \quad \overline{m}_b(\overline{m}_b) = 4172(12)\,\text{MeV}. \qquad (9.187)$$

• We show in Fig. 9.57 the semi-numerical PT expression of the moment in the $\overline{MS}$ scheme for different truncation of the PT series and for $\tau = 1.35\,\text{GeV}^{-2}$ where the sum rule stabilizes (see Fig. 9.58):

$$\mathcal{L}_0(\overline{m}_c^2)|_{PT} \simeq 1 + 22.6a_s + 85.2a_s^2 + 152.8a_s^3. \qquad (9.188)$$

One can note that the PT corrections are large.

• For the ratio of moments normalized to the lowest order expression, we obtain, within the same approximation and for $\tau = 1.35\,\text{GeV}^{-2}$:

$$\sqrt{\mathcal{R}_0}(\overline{m}_c^2)|_{PT} \simeq 1 - 0.12a_s - 0.17a_s^2 + 0.57a_s^3, \qquad (9.189)$$

where in this case the PT corrections are much smaller and the series converge quite well.

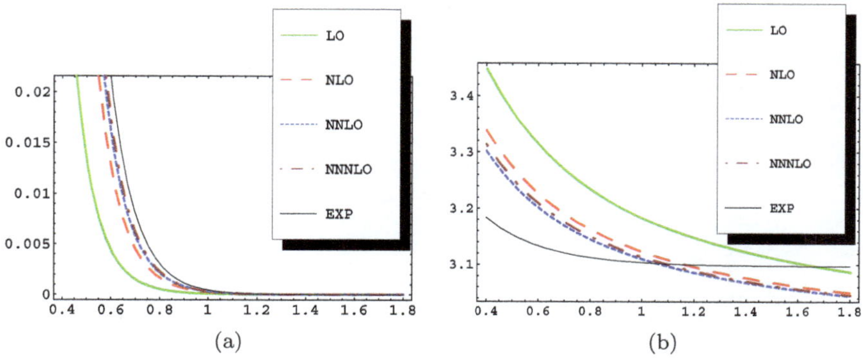

Figure 9.57. (a) Behaviour of the moment $\mathcal{L}_0^c(\overline{m}_c^2)$ in GeV^2 versus τ in GeV^{-2} at different orders of perturbation theory: LO: green (thick continuous); NLO: red (long dashed); NNLO: blue (short dashed); NNNLO: red-wine (dot-dashed). The black (continuous) curve comes from the central values of the data including the QCD continuum; (b) the same as in (a) but for the ratio of moments $\sqrt{\mathcal{R}_0^c}(\overline{m}_c^2)$.

9.5.2 $\langle g^3 f_{abc} G^3 \rangle$ versus $\langle \alpha_s G^2 \rangle$ from charmonium

In this section, our aim is to improve the estimate of the triple gluon condensate $\langle g^3 f_{abc} G^3 \rangle$ from the ratio of the exponential sum rules using its correlation with the gluon condensate $\langle \alpha_s G^2 \rangle$ by giving as input the precise value of the charm quark mass $\overline{m}_c(m_c)$ in Eq. (9.187) obtained in Refs. [290, 291] from moment sum rules.

• We show in Fig. 9.58 the τ-behaviour of $\sqrt{\mathcal{R}_0^c}(\overline{m}_c^2)$ by using the central values of the QCD parameters in Eq. (9.187) and the best central value of the instanton radius $\rho_c = 1.89$ GeV^{-1} obtained in Eq. (9.168) by adjusting the minimum of $\sqrt{\mathcal{R}_0^c}(\overline{m}_c^2)$ to the experimental data. One can note that at the minimum, one has:

$$\sqrt{\mathcal{R}_0^c}(\overline{m}_c^2)|_{\min} = M_{J/\psi}, \tag{9.190}$$

indicating that the higher state contributions to the ratio of moments are negligible at this value of τ.

• We show in Fig. 9.46 the correlation between ρ_c and $\langle \alpha_s G^2 \rangle$, from which we can find a slight stability (inflexion point) for the range:

$$\langle \alpha_s G^2 \rangle = (7.5 \pm 2.0) \times 10^{-2}\,\text{GeV}^4, \tag{9.191}$$

which is comparable with the one in Eq. (9.186) obtained from the Q^2-moments though less accurate here. In this region, we can deduce

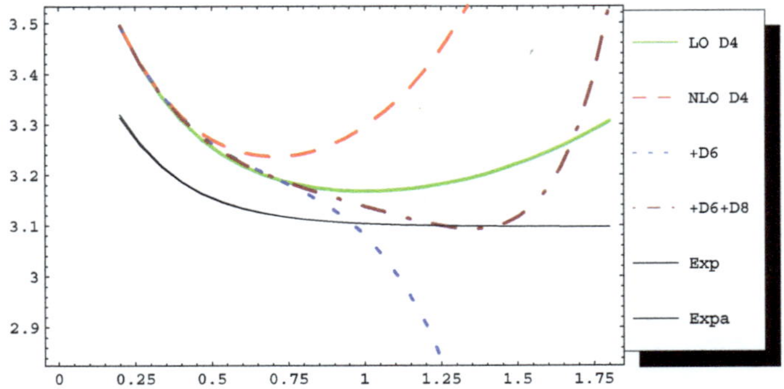

Figure 9.58. Behaviour of the ratio of moments $\sqrt{\mathcal{R}_0^c(\overline{m}_c^2)}$ in GeV versus τ in GeV^{-2}. The two black curves almost degenerate are the data including error bars. The green (thick continuous) denotes the PT contribution, including the $D = 4$ condensate to LO. The long dashed (red) curve denotes the contribution including the α_s correction to the $D = 4$ contribution. The short dashed (blue) curve denotes the QCD expression including the $D = 6$ condensate and the dot-dashed (red-wine) denotes the QCD expression, including the $D = 8$ contribution.

in GeV^{-1}[26]:

$$\rho_c = 1.89\ (10)_{G^2}\ (1)_{\alpha_s}\ (5)_{\nu}\ (1)_{\tau}\ = 1.89 \pm 0.11\,\mathrm{GeV}^{-1}, \qquad (9.192)$$

where the error comes, respectively, from $\langle\alpha_s G^2\rangle$, α_s, the choice of subtraction point ν, the localisation of the minimum: $\tau \simeq (1.25\text{–}1.45)\,\mathrm{GeV}^{-2}$. We have estimated the effect of the subtraction point ν.

9.5.3 $\overline{m}_b(\overline{m}_b)$ from bottomium

Due to the smaller values of $\alpha_s(m_b)$ and of τ at which the ratio of sum rules stabilizes (see Fig. 9.59), the convergence of the QCD PT and NP series is faster here than for the case of charmonium. More explicitly, for $\tau = 0.2\,\mathrm{GeV}^{-2}$, the PT expression of the ratio of sum

[26]If we have used the value of $\langle\alpha_s G^2\rangle$ in Eq. (9.186) from the moments, we would have obtained a more precise value of ρ_c.

rules reads numerically as:

$$\sqrt{\mathcal{R}_0}(\overline{m}_b^2)|_{PT} \simeq 1 - 0.14a_s - 0.35a_s^2 - 0.76a_s^3, \qquad (9.193)$$

where we have used the QCD inputs in Eqs. (9.168), (9.186) and (9.187). We shall use the value of α_s deduced from the one from τ-decay [289]:

$$\alpha_s(m_b) = 0.219(4). \qquad (9.194)$$

Using $\langle \alpha_s G^2 \rangle$ from Eq. (9.186), the improved value of the triple gluon condensate in Eq. (9.192) from exponential charmonium sum rule and the modified factorized value of the $D = 8$ dimension condensates, we reconsider here the determination of $\overline{m}_b(\overline{m}_b)$ from bottomium by adjusting the minimum (inflexion point) of the QCD value of $\sqrt{\mathcal{R}_0^b(\overline{m}_b^2)}$ with the one from the data. This is achieved by the value of the running mass in GeV:

$$\overline{m}_b(\overline{m}_b) = 4212(18)_\tau(0.2)_{\alpha_s}(25)_\nu(1.5)_{G^2+G^3+G^4}\,(8)_{\rm Coul}, \qquad (9.195)$$

corresponding to $\tau = (0.20 \pm 0.06)\,\text{GeV}^{-2}$ where the inflexion point of the theoretical curve meets the experimental one (see Fig. 9.59).

- We note that the sum of the NP condensate contributions is about 1.9 per mil of m_b which is quite small.

- The error due to the $\alpha_s^{n\geq 4}$ is obtained by taking the $\pm$ sign in the coefficient of α_s^3. The one due to ν is induced by moving it in the range $(\overline{m}_b \pm 1)\,\text{GeV}$. The one due to G^4 is the deviation from the modified to the usual factorization. The one from Coulombic corrections has been estimated like in Ref. [291] from the resummed expression of the spectral function. Then, we deduce the final value:

$$\overline{m}_b(\overline{m}_b) = 4212(32)\,\text{MeV}, \qquad (9.196)$$

which is slightly higher and less accurate than the one from the Q^2-moments given in Eq. (9.187) [290, 291], but the two results are consistent within the errors.

- By comparing the two results from the Q^2-moments and exponential sum rules, we are tempted to take the average of the results in Eqs. (9.187) and (9.104) as a final result from the sum rules approaches:

$$\overline{m}_b(\overline{m}_b) = 4177(11)\,\text{MeV}. \qquad (9.197)$$

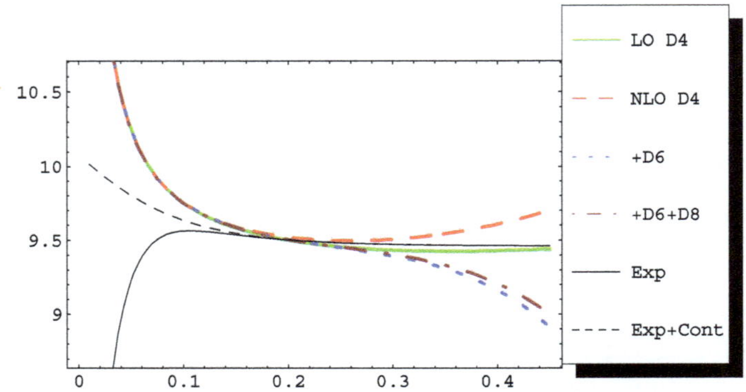

Figure 9.59. Behaviour of the ratio of moments $\sqrt{\mathcal{R}_0^b(\overline{m}_b^2)}$ in GeV versus τ in GeV^{-2} and for $\overline{m}_b(\overline{m}_b) = 4212$ MeV. The same legend as in Fig. 9.58 for the QCD contributions. The black continuous (respectively short dashed) curves are the experimental contribution including (respectively without) the QCD continuum.

9.5.4 Final values of $\overline{m}_{c,b}(\overline{m}_{c,b})$ and condensates from J/ψ and Υ sum rules

• The final values of $\overline{m}_{c,b}(\overline{m}_{c,b})$ from J/ψ and Υ sum rules are quoted in Eqs. (9.176) and (9.197) to order α^3 and including NP corrections up to dimension $D = 8$:

$$\overline{m}_c(\overline{m}_c) = 1261(12)\,\text{MeV}, \quad \overline{m}_b(\overline{m}_b) = 4177(11)\,\text{MeV}. \tag{9.198}$$

• The best value of the ratio of gluon condensates comes from the LSR analysis in Eq. (9.192) quantified by the instanton radius ρ_c defined in Eq. (9.128):

$$\frac{\langle g^3 f_{abc} G^3\rangle}{\langle \alpha_s G^2\rangle} = \frac{4}{5}\frac{12\pi}{\rho_c^2} = (8.4 \pm 0.9)\,\text{GeV}^2, \tag{9.199}$$

which is much larger than the one from instanton liquid model [3, 130].

• The best value of gluon condensate $\langle \alpha_s G^2\rangle$ comes from the high-moments analysis at $Q^2 = 8m_c^2$ (Eq. (9.163)):

$$\langle \alpha_s G^2\rangle = (7.0 \pm 1.3) \times 10^{-2}\,\text{GeV}^4. \tag{9.200}$$

9.6 QCD Parameters Correlations from Quarkonia

In Ref. [517], we study (for the first time) the correlations between the QCD coupling α_s, the gluon condensate $\langle\alpha_s G^2\rangle$, and the c, b-quark running masses $\overline{m}_{c,b}$ in the $\overline{MS}$-scheme from a global analysis of the (axial-)vector and (pseudo)scalar charmonium and bottomium spectra using optimized ratios of Laplace sum rules (LSR) where PT @N2LO, N3LO, $\langle\alpha_s G^2\rangle$ @NLO and LO $D = 6$–8 dimensions non-perturbative condensate corrections are included.

This analysis solves the apparent discrepancies of different sum rule results quoted in the literature. Il also provides improved estimates of the gluon condensate $\langle\alpha_s G^2\rangle$. New estimates of α_s from mass splittings are obtained.

9.6.1 Review of the determination of $\langle\alpha_s G^2\rangle$

According to different analyses, gluon condensates introduced by SVZ [3, 40] play an important rôle in gluodynamics and in the QCD spectral sum rules analysis where they enter as high-dimension operators in the OPE of the hadronic correlators [3, 6, 7, 41–44] and especially in the case of the heavy quarkonia and gluonia/glueball channels. We review in Table 9.18 the determinations of $\langle\alpha_s G^2\rangle$ from different sources prior to 2017. The average value given in the table can be compared with the SVZ original one:

$$\langle\alpha_s G^2\rangle \simeq 0.04\,\text{GeV}^4, \tag{9.201}$$

extracted (for the first time) from charmonium sum rules [3].

Results from τ-decay [525–528] are not stable. We shall come back to this point in the following section.

Earlier lattice calculations indicate a non-zero positive value of $\langle\alpha_s G^2\rangle$ [508, 510–512], while recent estimates in Table 9.18 give positive values about 2–7 times higher than the phenomenological estimates. However, the subtraction of the perturbative contribution in the lattice analysis which is scheme-dependent is not yet well understood [534] and does not permit a precise determination of the gluon condensates.

Table 9.18. Selected determinations of $\langle \alpha_s G^2 \rangle$ from charmonium, bottomium and light quark systems. The numbers marked with * are not included in the average. This average does not take into account the new results of this paper. Estimates from variants of the SVZ sum rules using some weight functions are not considered here. The ones from high moments of τ decays and from the lattices are only mentioned for comparisons.

Sources	$\langle \alpha_s G^2 \rangle \times 10^2$ [GeV^4]	References
Charmonium		
$q^2 = 0$-moments	4 ± 2	SVZ 79 [3] (guessed error)
$q^2 \neq 0$-moments	5.3 ± 1.2	RRY 81-85 [62]
–	9.2 ± 3.4	Miller-Olsson 82 [518]
–	$\approx 6.6^*$	Broadhurst *et al.* 94 [334]
–	2.8 ± 2.2	Ioffe-Zyablyuk 07 [293, 519]
–	7.0 ± 1.3	Narison 12a [292]
Exponential	12 ± 2	Bell-Bertlmann 82 [64, 70, 224]
–	17.5 ± 4.5	Marrow *et al.* 87 [225]
–	7.5 ± 2.0	Narison 12b [291]
Exponential $M_\psi - M_{\eta_c}$	10 ± 4	Narison 96 [223, 520]
Bottomium		
Exponential $M_{\chi_b} - M_\Upsilon$	6.5 ± 2.5	Narison 96 [223, 520]
Non-rel. moments	5.5 ± 3	Yndurain 99 [303]
$e^+e^- \to$ **I=1 Hadrons**		
Exponential	$0.9 - 6.6^*$	Eidelman *et al.* 79 [71]
Ratio of Exponential	4 ± 1	Launer *et al.* 84 [115]
FESR	13 ± 6	Bertlmann *et al.* 88 [521, 522]
Infinite norm	$1 - 30^*$	Causse-Mennessier [523]
τ-like decay	7 ± 1	Narison 95 [236, 237]
τ**-decay**		
Axial spectral function	6.9 ± 2.6	Dominguez-Sola 88 [524]
Sum Rule Average	$\mathbf{6.25 \pm 0.45}$	**Prior 2017**
τ**-decay with high moments**		
ALEPH collaboration	6.3 ± 1.2	Duflot 95 [525]
CLEO II collaboration	2.4 ± 1.0	Duflot 95 [525]
OPAL collaboration	$-0.9 - +4$	Ackerstaff *et al.* 99 [526]
ALEPH collaboration	$-5 - +6$	Schael *et al.* 05 [527]
ALEPH collaboration	$-12 - 0.6$	Davier *et al.* 14 [528]
Lattice		
$O(\alpha_s^{12})$	≈ 13	Rakow 05 [529–531]
$O(\alpha_s^{35})$	≈ 27	Bali-Pineda 15 [532, 533]
Average plaquette	≈ 44	Lee 14 [534]

9.6.2 The form of sum rules

In this section, we choose to work with the finite energy version of the QCD Laplace sum rule (LSR) [3, 64, 69, 70] and their ratios:

$$\mathcal{L}_n^c(\tau) = \int_{4m_Q^2}^{t_c} dt\, t^n e^{-t\tau} \mathrm{Im}\Pi_{V(A)}(t), \quad \mathcal{R}_n^c(\tau) = \frac{\mathcal{L}_{n+1}^c}{\mathcal{L}_n^c}, \tag{9.202}$$

where τ is the LSR variable, t_c is the threshold of the "QCD continuum" which parametrizes, from the discontinuity of the Feynman diagrams, the spectral function $\mathrm{Im}\Pi_{V(A)}(t, m_Q^2, \mu)$ associated to the transverse part $\Pi_{V(A)}(q^2, m_Q^2, \mu)$ of the two-point correlator:

$$\begin{aligned}\Pi^{\mu\nu}_{V(A)}(q^2) &\equiv i\int d^4x e^{-\mathrm{iqx}}\langle 0|\mathcal{T} J^{\mu}_{V(A)}(x)\left(J^{\nu}_{V(A)}(0)\right)^{\dagger}|0\rangle,\\ &= -\left(g^{\mu\nu}q^2 - q^{\mu}q^{\nu}\right)\Pi_{V(A)}(q^2) + q^{\mu}q^{\nu}\Pi^{(0)}_{V(A)}(q^2),\end{aligned} \tag{9.203}$$

where $J^{\mu}_{V(A)}(x) = \bar{Q}\gamma^{\mu}(\gamma_5)Q(x)$ is the heavy quark local vector (axial-vector) current. In the (pseudo)scalar channel associated to the local current $J_{S(P)} = \bar{Q}i(\gamma_5)Q(x)$, we work with the correlator:

$$\Psi_{S(P)}(q^2) = i\int d^4x\, e^{-iqx}\langle 0|\mathcal{T} J_{S(P)}(x)\left(J_{S(P)}(0)\right)^{\dagger}|0\rangle, \tag{9.204}$$

which is related to the longitudinal part $\Pi^{(0)}_{V(A)}(q^2)$ of the (axial-) vector one through the Ward identity [6, 7, 34]:

$$q^2\Pi^{(0)}_{A(V)}(q^2) = \Psi_{P(S)}(q^2) - \Psi_{P(S)}(0). \tag{9.205}$$

Working with $\Psi_{P(S)}(q^2)$ is safe as $\Psi_{P(S)}(0)$ should affect the Q^2-moments and the exponential sum rules derived from $\Pi^{(0)}_{A(V)}(q^2)$, which is not accounted for in, e.g., Refs. [62, 225, 535].

- **Parametrisation of the spectral function**

$\mathrm{Im}\Pi_V(t)$ is related to the ratio $R_{e^+e^-}$ of the total cross-section of $\sigma(e^+e^- \to \text{hadrons})$ over $\sigma(e^+e^- \to \mu^+\mu^-)$ through the optical theorem. It can be expressed in terms of the leptonic widths and meson masses as in previous sections.

In the case of the axial-vector and (pseudo)scalar channels where there are no complete data, we use the duality ansatz:

$$\mathrm{Im}\{\Pi(t);\,\Psi(t)\} \simeq f_H^2 M_H^{\{0;2\}}\delta(t-M_H^2)+\Theta(t-t_c)\text{“}QCD\text{ continuum”}, \tag{9.206}$$

where M_H and f_H are the lowest ground state mass and coupling analogue to f_ρ and f_π. This implies that

$$\mathcal{R}_n^c \equiv \mathcal{R} \simeq M_H^2, \tag{9.207}$$

indicating that the ratio of moments appears to be a useful tool for extracting the masses of hadrons [6,7]. We shall work with the lowest ratio of moments $\mathcal{R}_0^c$.

• QCD Perturbative expressions @N2LO

The perturbative QCD expression of the vector channel has been given in the previous section.

The one of the axial-vector current has been obtained in Refs. [62, 157, 536]. To order α_s^2 (N2LO), the spectral functions are usually parametrized as:

$$R^{(2)} \equiv C_F^2 R_A^{(2)}+C_A C_F R_{NA}^{(2)}+C_F T_Q n_l R_l^{(2)}+C_F T_Q(R_F^{(2)}+R_S^{(2)}+R_G^{(2)}), \tag{9.208}$$

which are, respectively, the abelian (A), non-abelian (NA), massless (l) and heavy (F) internal quark loops, singlet (S) and double bubble gluon (G) contributions. $C_F = 4/3$, $C_A = 3, T_Q = 1/2$ are usual SU(3) group factors and n_l is the number of light quarks. We use the (approximate) but complete result in the on-shell scheme given by Ref. [497] for the abelian and non-abelian contributions. The one from light quarks comes from Refs. [537–539]. The one from heavy fermion internal loop comes from Ref. [540] for the vector current, while the one from the axial current is (to our knowledge) not available. The singlet one due to double triangle loop comes from Ref. [541]. The one from the gluonic double bubble reconstructed from massless fermions comes from Refs. [537,538,540].

In order to use the previous expression of the spectral function, we transform the pole mass into the running one [154, 155, 158, 159]:

$$M_Q = \overline{m}_Q(\mu)\left[1 + \frac{4}{3}a_s + (16.2163 - 1.0414n_l)a_s^2 \right.$$
$$+ \ln\left(a_s + (8.8472 - 0.3611n_l)a_s^2\right)$$
$$\left. + \ln^2\left(1.7917 - 0.0833n_l\right)a_s^2 + \cdots\right], \tag{9.209}$$

for n_l light flavours, where μ is the arbitrary subtraction point and $a_s \equiv \alpha_s/\pi$, $\ln \equiv \log\left(\mu/M_Q\right)^2$.

• QCD non-perturbative expressions

Using the OPE à la SVZ, the non-perturbative contributions to the two-point correlator can be parametrized by the sum of higher dimension condensates:

$$\mathrm{Im}\Pi(t) = \sum C_{2n}(t, m^2, \mu^2))\langle O_{2n}\rangle : n = 1, 2, \ldots, \tag{9.210}$$

where C_{2n} are Wilson coefficients calculable perturbatively and $\langle O_{2n}\rangle$ are non-perturbative condensates. In the exponential sum rules, the order parameter is the sum rule variable τ, while for the heavy quark systems, the relevant condensate contributions at leading order in α_s are the gluon condensate $\langle\alpha_s G^2\rangle$ of dimension four [3], the dimension six gluon $\langle g^3 f_{abc} G^3\rangle$ and light four-quark $\alpha_s\langle\bar{u}u\rangle^2$ condensates [502]. The condensates of dimension eight entering in the sum rules are of seven types [502]. They can be expressed in different basis depending on how each condensate is estimated (vacuum saturation [502] or modified vacuum saturation [338]). Our estimate of these D = 8 condensates is the same as in Ref. [292]. For the vector channel, we use the analytic expressions of the different condensate contributions given by Bertlmann [224]. We shall not include the eventual $D = 2$ cooperator induced by a tachyonic gluon mass [232, 235] as it is dual to the contribution of large order terms [234], which we estimate using a geometric growth of the PT series.

• **Initial QCD input parameters**

In the first iteration, we shall use the following QCD input parameters:

$$\alpha_s(M_\tau) = 0.325^{+0.008}_{-0.016}, \quad \langle\alpha_s G^2\rangle = (0.07 \pm 0.04)\,\text{GeV}^4,$$
$$\overline{m}_c(\overline{m}_c) = (1261 \pm 17)\,\text{MeV}, \overline{m}_b(\overline{m}_b) = (4177 \pm 11)\,\text{MeV}. \tag{9.211}$$

The central value of α_s comes from τ-decay. The value of $\langle\alpha_s G^2\rangle$ almost covers the range from different determinations mentioned in Table 9.18 and reviewed in Refs. [6, 7, 223, 520]. We shall use the ratio of condensates given in Eq. (9.192). For the light four-quark condensate, we shall use the value given in Table 7.10:

$$\alpha_s\langle\bar{u}u\rangle^2 = (5.8 \pm 1.8) \times 10^{-4}\,\text{GeV}^6. \tag{9.212}$$

9.6.3 Charmonium ratio of moments $\mathcal{R}_{J/\psi(\chi_{c1})}$

• **Tests of convergence of the PT series**

We show in Fig. 9.60 the behaviour of the perturbative contribution to $\mathcal{R}_{J/\psi(\chi_{c1})}$ for different truncation of the PT series where. We fix $\sqrt{t_c} = 4.6\,\text{GeV}$. One can note the importance of the N2LO contribution which is dominated by the abelian and non-abelian contributions which go towards the good direction of the values of the experimental masses.

• **Tests of convergence of the OPE**

We show in Fig. 9.61 the behaviour of $\mathcal{R}_{J/\psi(\chi_{c1})}$ for different dimension condensates D in the OPE in terms of the sum rule variable τ. One can note that the presence of condensates is vital for having τ-stabilities which are not there for the PT-terms alone. The τ-stability is reached for $\tau \simeq 0.6\,\text{GeV}^{-2}$. At a given order of the PT series, the contributions of the $D = 8$ condensates are negligible at the τ-stability region, while the $D = 6$ contribution goes again to the right track compared with the data.

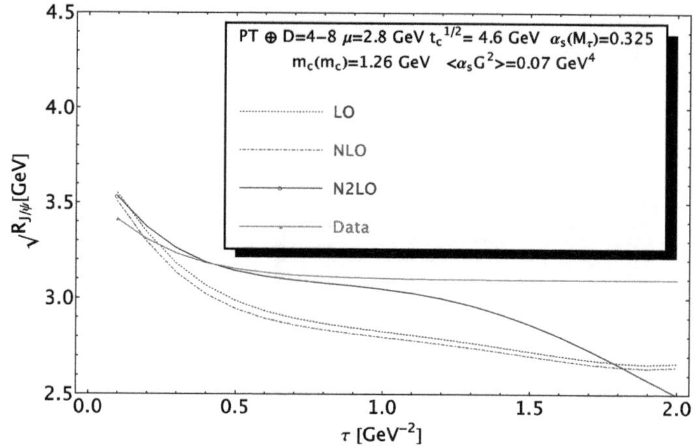

(a)

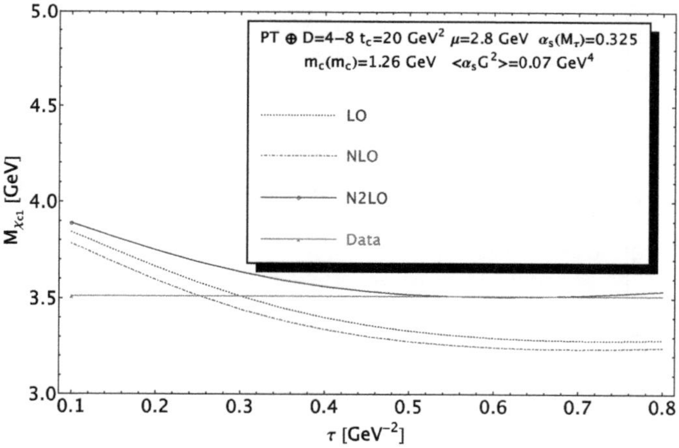

(b)

Figure 9.60. Behaviour of the ratio of moments $\mathcal{R}$ versus τ in GeV^{-2} at different orders of perturbation theory. The input and the meaning of each curve are given in the legends: (a) J/ψ; (b) χ_{c1}.

• Continuum threshold t_c-stability for $\mathcal{R}_{\chi_{c1}}$

We show the analysis in Fig. 9.62, where the curves correspond to different t_c-values. We find nice t_c-stabilities where we take the value:

$$t_c \simeq (17 \sim 22)\,\text{GeV}^2, \tag{9.213}$$

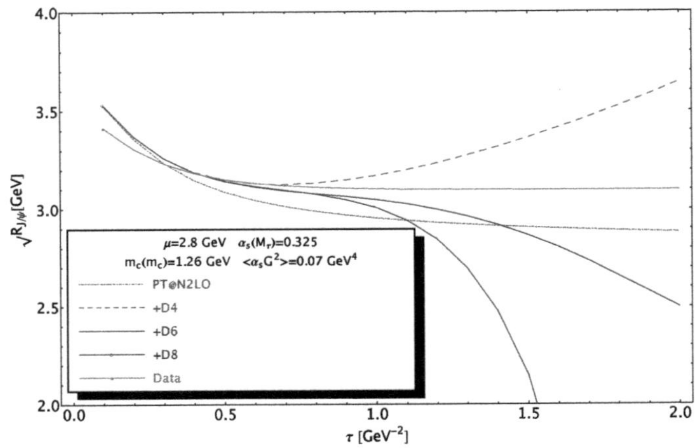

(a)

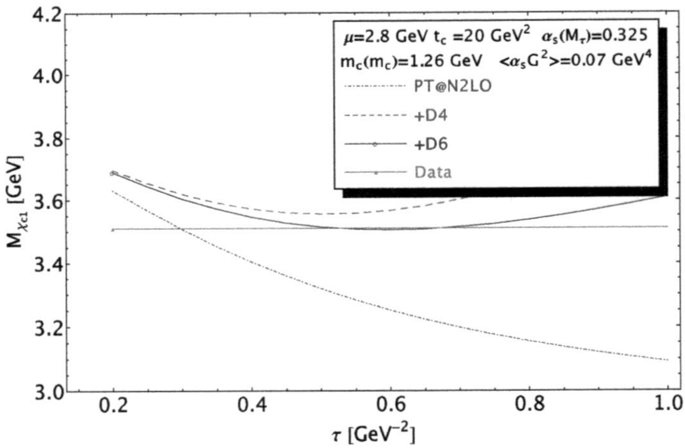

(b)

Figure 9.61. The same as in Fig. 9.57 but for different truncation of the OPE: (a) J/ψ; (b) χ_{c1}.

where the lowest value corresponds to the phenomenological estimate $M_{\chi_{c1}}(2P) - M_{\chi_{c1}}(1P) \approx M_{\psi}(2S) - M_{\psi}(1S)$, while the higher one corresponds to the beginning of t_c-stability. This range of t_c-values induces an error of about 8 MeV in the meson mass determination.

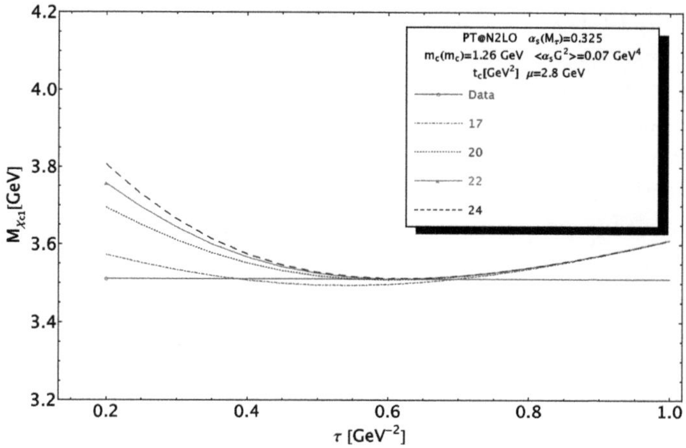

Figure 9.62. Behaviour of the ratio of moments $\mathcal{R}_{\chi_{c1}}$ versus τ in GeV^{-2}. The input and the meaning of each curve are given in the legend.

• Subtraction point μ-stability

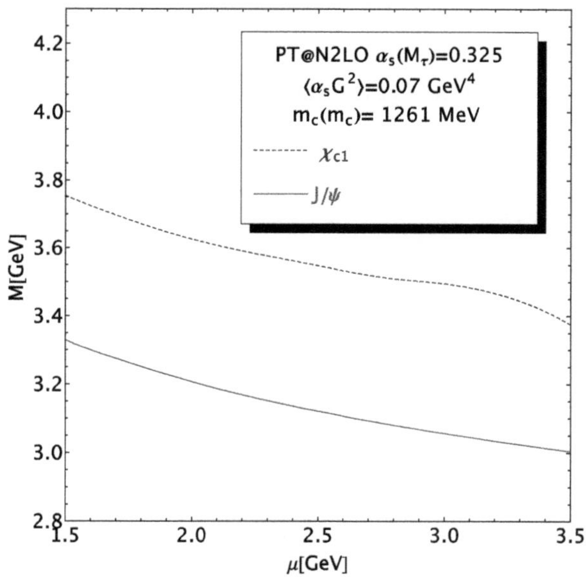

Figure 9.63. Behaviour of the ratio of moments $\mathcal{R}_{J/\psi}$ and $\mathcal{R}_{\chi_{c1}}$ versus μ for $t_c = 20\,\text{GeV}^2$. The inputs and the meaning of each curve are given in the legends.

The subtraction point μ is an arbitrary parameter. It is popularly taken between 1/2 and two times an "*ad hoc*" choice of scale. However, the physical observables should not be quite sensitive to μ even for a truncated PT series. In the following, like in the previous case of external (unphysical) variable, we shall fix its value by looking for a μ-stability point, if it exists, at which the observable will be evaluated. This procedure has been used recently for improving the LSR predictions on molecules and four-quark charmonium and bottoming states (see Chapter 7). Taking here the example of the ratios of moments, we show in Fig. 9.63 their μ- dependence. We note that $\mathcal{R}_\psi$ is a smooth decreasing function of μ, while $\mathcal{R}_{\chi_{c1}}$ presents a slight stability at:

$$\mu = (2.8 \sim 2.9)\,\text{GeV}, \tag{9.214}$$

at which we shall evaluate the two ratios of moments. This value is about the same as the one obtained from the analysis of the molecules/four-quark states (see Chapter 7). One can note that at a such higher scale, one has a better convergence of the $\alpha_s(\mu)$ PT series.

• Correlations of the QCD parameters

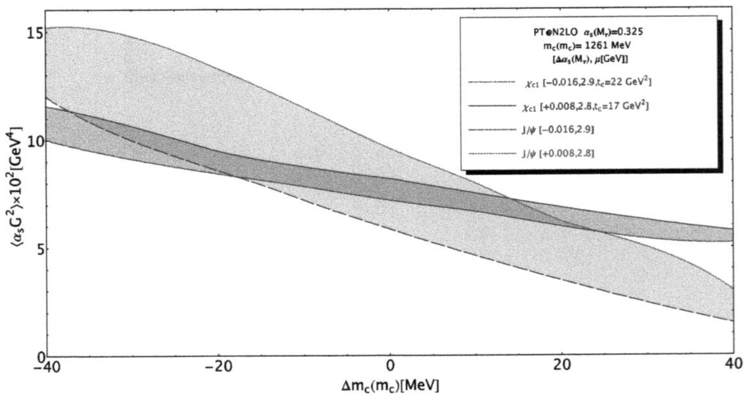

Figure 9.64. Correlation between $\langle\alpha_s G^2\rangle$ and $\overline{m}_c(\overline{m}_c)$ for the range of α_s values given in Eq. (9.211) and for μ given in Eq. (9.230).

Once these preliminaries are fixed, we are now ready to study the correlation between α_s, the gluon condensate $\langle\alpha_s G^2\rangle$, and the c-quark running masses $\overline{m}_c(\overline{m}_c)$. In so doing, we request that the $\sqrt{\mathcal{R}_{J/\psi}}$ sum rule reproduces within (2–3) MeV accuracy the experimental measurement, while the χ_{c1} mass is reproduced within (8–10) MeV which is the error induced by the choice of t_c in Eq. (9.213). The results of the analysis are obtained at the τ-stability points which are about 1.1 (respectively 0.6) GeV^{-2} for the J/ψ (respectively χ_{c1}) channels. They are shown in Fig. 9.64 for the two values of μ given in Eq. (9.230). One can note that $\langle\alpha_s G^2\rangle$ decreases smoother from the χ_{c1} (grey region) than from the J/ψ sum rule when $\overline{m}_c$ increases. In the J/ψ sum rule, it moves from 0.15 to 0.02 GeV^4 for $\overline{m}_c(\overline{m}_c)$, varying from 1221 to 1301 MeV. This feature may explain the apparent discrepancy of the results reviewed in the introduction from this channel.

One should note that the results from the J/ψ sum rules are quite sensitive to the choice of the subtraction point (no μ-stability) which then does not permit accurate determinations of $\langle\alpha_s G^2\rangle$ and $\overline{m}_c(\overline{m}_c)$ from LSR as discussed previously. Some accurate results reported in the literature for an "*ad hoc*" choice of μ may be largely affected by the μ variation.

One can also see from Fig. 9.64 that within the lone J/ψ sum rule, the values of $\langle\alpha_s G^2\rangle$ and $\overline{m}_c(\overline{m}_c)$ cannot be strongly constrained.

Once the constraint from the χ_{c1} sum rule is introduced, one obtains a much better selection. Taking as a conservative result the range covered by the change of μ in Eq. (9.214), one deduces:

$$\langle\alpha_s G^2\rangle = (8.5 \pm 3.0) \times 10^{-2}\,\text{GeV}^4, \quad \overline{m}_c(\overline{m}_c) = (1256 \pm 30)\,\text{MeV}. \tag{9.215}$$

This result remains the same when including N3LO PT and @NLO gluon condensate contributions. These results agree within the errors within our recent estimates of $\langle\alpha_s G^2\rangle$ and $\overline{m}_c(\overline{m}_c)$ discussed in previous sections. Taking the weighted average of different sum rule determinations given in Table 9.18 with the new result in Eq. (9.215), we obtain the *sum rule average*:

$$\langle\alpha_s G^2\rangle|_{\text{average}} = (6.30 \pm 0.45) \times 10^{-2}\,\text{GeV}^4. \tag{9.216}$$

9.6.4 Bottomium ratios of moments $\mathcal{R}_{\Upsilon(\chi_{b1})}$

The analysis is very similar to the previous $J/\psi/\chi_{c1}$ sum rules.

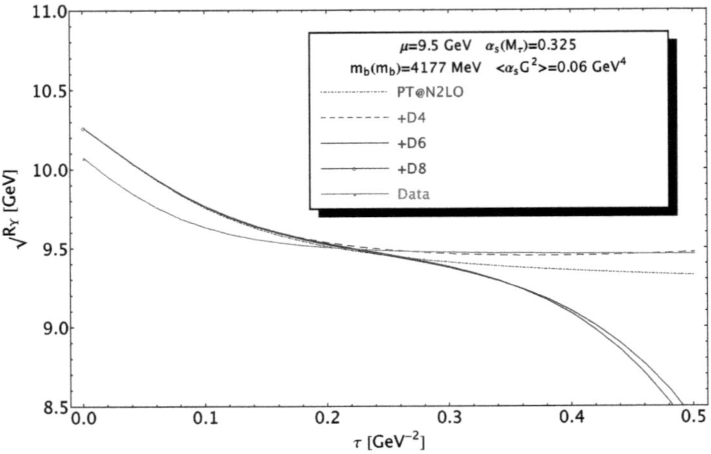

Figure 9.65. Behaviour of the ratio of moments $\mathcal{R}_\Upsilon$ versus τ in GeV^{-2} for different truncation of the OPE. The input and the meaning of each curve are given in the legend.

We show in Fig. 9.65 the τ-behaviour of $\mathcal{R}_\Upsilon$ for different truncation of the OPE. A nice stability which coincides with M_Υ is obtained for $\tau \simeq 0.2\,\mathrm{GeV}^{-2}$. t_c-stability is obtained for $\sqrt{t_c} \simeq 11\,\mathrm{GeV}$.

The ratio $\mathcal{R}_{\Upsilon(\chi_{b1})}$ versus τ is shown in Fig. 9.66 for fixed $\mu = 9.5\,\mathrm{GeV}$ and for different values of t_c. The τ-stability is reached for $\tau \simeq 0.28\,\mathrm{GeV}^{-2}$. The sum rule predicts at the minimum:

$$M_{\chi_{b1}} \simeq 9677(26)_{t_c}(8)_{\alpha_s}(11)_{G^2}(9)_{m_b}(99)_\mu\,\mathrm{MeV}, \tag{9.217}$$

which is (within the error) about 100 MeV lower than the experimental mass $M^{\mathrm{exp}}_{\chi_{b1}} = 9893\,\mathrm{MeV}$.

μ-stability is obtained for:

$$\mu \simeq (9\text{–}10)\,\mathrm{GeV}, \tag{9.218}$$

which is about the same as the one obtained in the four-quark meson analysis (see Chapter 7). It may indicate that this value of μ is universal for the bottomium-like states.

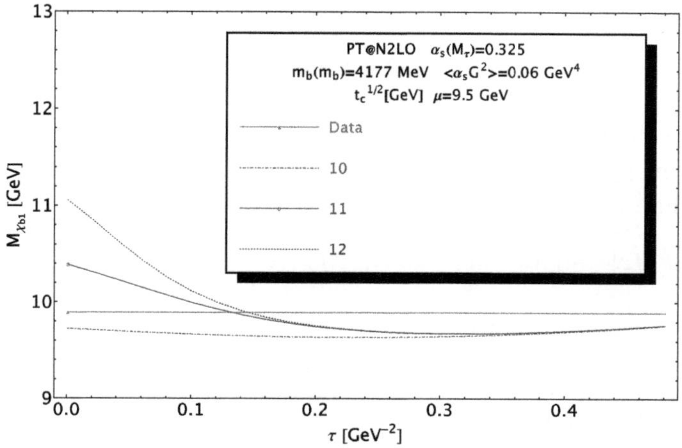

Figure 9.66. Behaviour of the ratio of moments $\mathcal{R}_{\chi_{b1}}$ versus τ for different values of t_c. The input and the meaning of each curve are given in the legend.

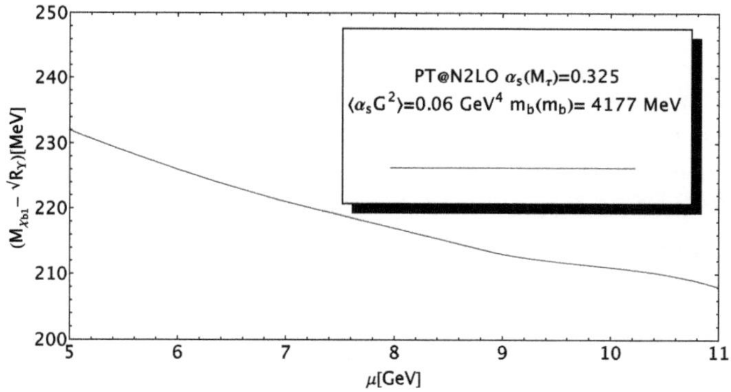

Figure 9.67. Behaviour of $M_{\chi_{b1}} - \sqrt{\mathcal{R}_\Upsilon}$ versus μ.

• Correlation between $\alpha_s(\mu)$ and $\overline{m}_b(\overline{m}_b)$ from $\mathcal{R}_\Upsilon$

From the previous analysis, one can note that the χ_{b1} channel cannot help from a precise study of the correlation between α_s and $\overline{m}_b(\overline{m}_b)$. We show in Fig. 9.68 the result of the analysis from the Υ channel by requiring that the experimental value of $\sqrt{\mathcal{R}_\Upsilon}$ is reproduced within (1−2) MeV accuracy. First, one can note that the error due to the gluon condensate with the value given in Eq. (9.215) is negligible. Given the range of α_s quoted in Eq. (9.211), one can deduce the prediction:

$$\overline{m}_b(\overline{m}_b) = 4192(15)(8)_{\text{Coul}}\,\text{MeV}, \tag{9.219}$$

where we have added in Eq. (9.219) an error of about 8 MeV from Coulombic corrections as estimated in Refs. [290, 291] (previous sections). The result in Eq. (9.219) corresponds to:

$$\alpha_s(M_\tau) = 0.321(12) \longrightarrow \alpha_s(M_Z) = 0.1186(15)(3), \tag{9.220}$$

given by the range in Eq. (9.211). The running from M_τ to M_Z due to the choice of the thresholds induces the last error (3). The result in Eq. (9.219) is consistent with the ones from moments sum rules quoted in Eq. (9.211) and with [291]:

$$\overline{m}_b(\overline{m}_b) = 4212(32)\,\text{MeV}, \tag{9.221}$$

from LSR with RG resummed PT expressions. Taking the average of our three determinations, we obtain the final estimate:

$$\overline{m}_b(\overline{m}_b)|_{\text{average}} = (4184 \pm 11)\,\text{MeV}, \tag{9.222}$$

where the errors come from the most precise determination. Due to the large errors induced by the subtraction scale as shown in Fig. 9.68, one cannot accurately extract the value of α_s given the present value of $\overline{m}_b$.

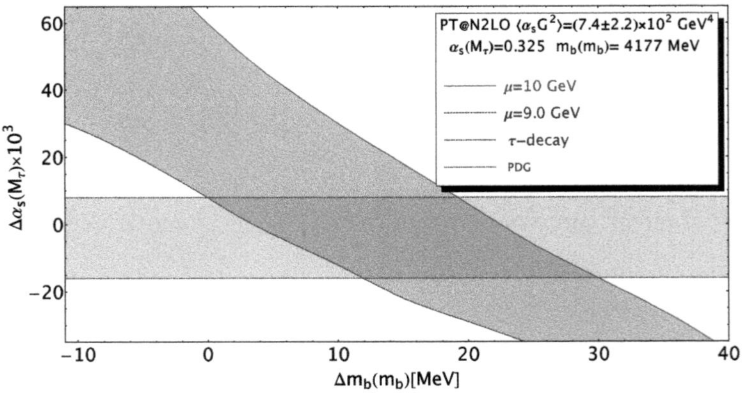

Figure 9.68. Behaviour of $\Delta\alpha_s(M_\tau)$ versus $\overline{m}_b(\overline{m}_b)$ from the ratio of moments $\mathcal{R}_\Upsilon$. The horizontal band corresponds to the range of α_s value given in Eq. (9.211). The input and the meaning of each curve are given in the legend.

9.6.5 (Pseudo)scalar charmonium

In these channels, we shall work with the ratio of sum rules associated to the two-point correlator $\Psi_{P(S)}(q^2)$ defined in Eq. (9.204) which is not affected by $\Psi_{P(S)}(0)$. We shall use the PT expression known as @N2LO [537–541], the contribution of the gluon condensates of dimensions four and six to LO [502].

• η_c and χ_{c0} masses

The η_c sum rule shows a smooth decreasing function of μ but does not present a μ-stability. Then, we choose the value of μ given in Eq. (9.214) for evaluating it. We show in Fig. 9.69 the τ-behaviour of the η_c-mass for different values of t_c which we take from 10 GeV^2 [around the mass squared of the $\eta_c(2P)$ and $\eta_c(3P)$] until 13 GeV^2 (t_c-stability). Similar analysis is done for χ_{c0} associated to the scalar current $\bar{Q}(i)Q$ which is shown in Fig. 9.70, where we take $t_c \simeq (16 \sim 24)$ GeV^2. Using the averaged values of $\langle\alpha_s G^2\rangle$ and $\overline{m}_c(\overline{m}_c)$ in Eqs. 9.216 and (9.225), we deduce the optimal result in units of MeV:

$$
\begin{aligned}
M_{\eta_c} &= 2979(5)_\mu(11)_{t_c}(11)_{\alpha_s}(30)_{m_c}(10)_{G^2}, \\
M_{\chi_{c0}} &= 3411(1)_\mu(17)_{t_c}(26)_{\alpha_s}(30)_{m_c}(20)_{G^2},
\end{aligned} \tag{9.223}
$$

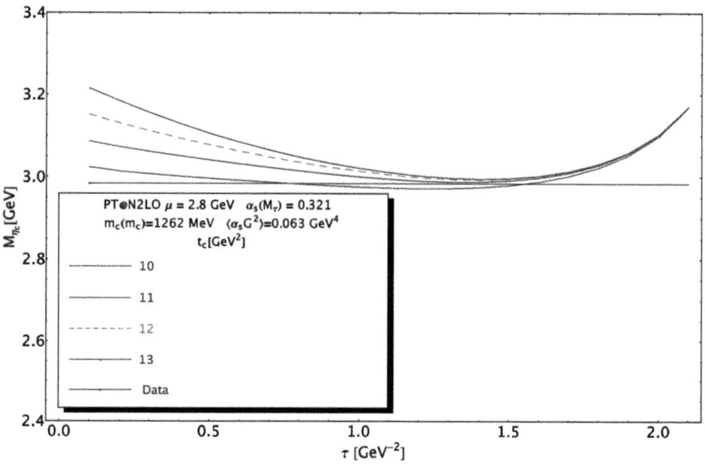

Figure 9.69. Behaviour of M_{η_c} versus τ for different values of t_c.

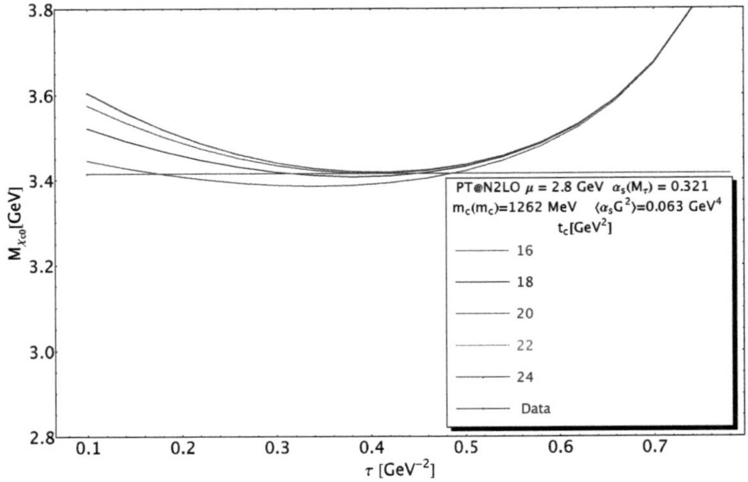

Figure 9.70. Behaviour of $M_{\chi_{c0}}$ versus τ for different values of t_c.

in good agreement within the errors with the experimental masses: $M_{\eta_c} = 2984\,\text{MeV}$ and $M_{\chi_{c0}} = 3415\,\text{MeV}$ but not accurate enough for extracting with precision the QCD parameters.

• Correlation between $\overline{m}_c(\overline{m}_c)$ and $\langle \alpha_s G^2 \rangle$

We study the correlation between $\overline{m}_c(\overline{m}_c)$ and $\langle \alpha_s G^2 \rangle$ by requiring that the sum rules reproduce the masses of η_c and χ_{c0} within the error induced by the choice of t_c, respectively, 11 and 17 MeV. We show the result of the analysis in Fig. 9.71, keeping only the strongest constraint from M_{η_c}. We deduce:

$$\overline{m}_c(\overline{m}_c) = 1266(16)\,\text{MeV}, \tag{9.224}$$

in good agreement with the one in Ref. [542] from pseudoscalar moments. We combine our determinations in Eqs. (9.215) and (9.224) with the two determinations [290, 292] from vector moments sum rules quoted in Eq. (9.211). As a final result, we quote the average from exponential and moment sum rules from a global fit of the quarkonia spectra:

$$\overline{m}_c(\overline{m}_c)|_{\text{average}} = (1263 \pm 14)\,\text{MeV}, \tag{9.225}$$

where we have retained the error from the most precise prediction rather than from the weighted average. It is remarkable that this value agrees with the original SVZ estimate [3] of the Euclidean mass.

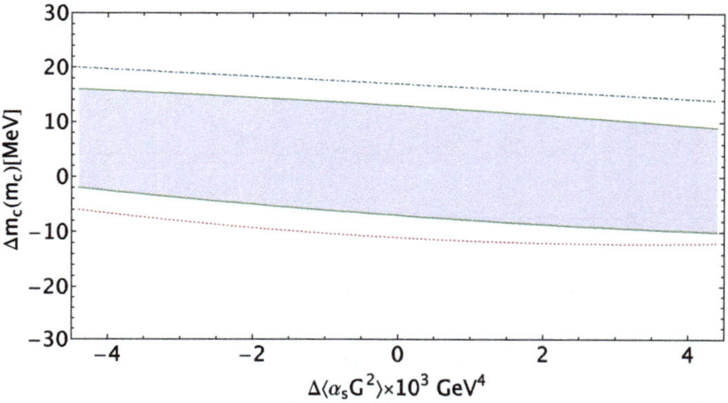

Figure 9.71. Behaviour of $\Delta\overline{m}_c(\overline{m}_c)$ versus $\Delta\langle\alpha_s G^2\rangle$ from M_{η_c}. The dashed region corresponds to $\Delta\alpha_s = 0$ and for different values of $t_c \simeq 10-13\,\mathrm{GeV}^2$. The two extremal lines correspond to $\Delta\alpha_s \pm 12$. We use the range of α_s in Eq. (9.211) and $\langle\alpha_s G^2\rangle$ in Eq. (9.216).

9.6.6 (Pseudo)scalar bottomium

- **η_b and χ_{b0} masses**

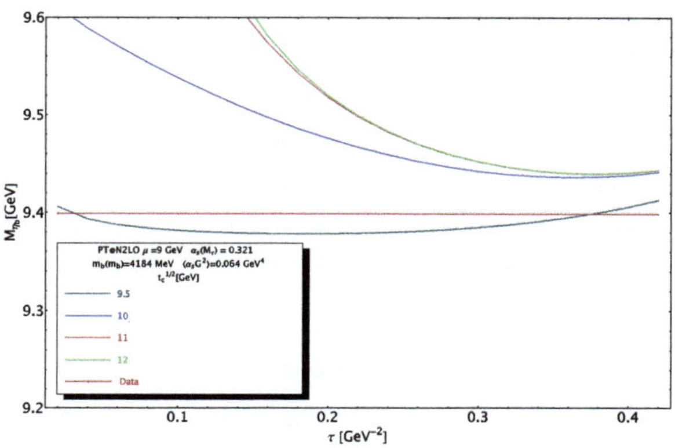

Figure 9.72. Behaviour of M_{η_b} versus τ for different values of t_c.

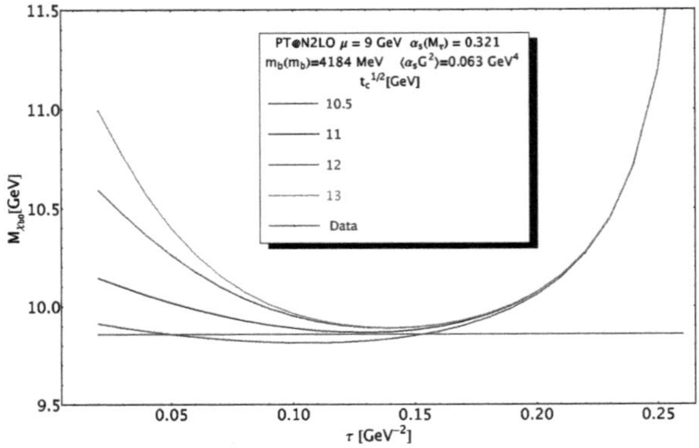

Figure 9.73. Behaviour of $M_{\chi_{b0}}$ versus τ for different values of t_c.

The masses of $\eta_b(0^{-+})$ and $\chi_{b0}(0^{++})$ are extracted in a similar way using the value of μ in Eq. (9.218) and the parameters in Eqs. (9.216) and (9.222). We take the range $\sqrt{t_c} = (9.5-12)$ [respectively $(10.5-13)$] GeV for the η_b [respectively χ_{b0}] channels, as shown in Figs. 9.72 and 9.73, from which we deduce, in units of MeV:

$$M_{\eta_b} = 9394(16)_\mu(30)_{t_c}(7)_{\alpha_s}(16)_{m_b}(8)_{G^2},$$
$$M_{\chi_{b0}} = 9844(7)_\mu(35)_{t_c}(6)_{\alpha_s}(17)_{m_b}(29)_{G^2}, \qquad (9.226)$$

in good agreement with the data $M_{\eta_b} = 9399\,\text{MeV}$ and $M_{\chi_{b0}} = 9859\,\text{MeV}$.

• Correlation between $\overline{m}_b(\overline{m}_b)$ and $\langle\alpha_s G^2\rangle$

The analysis done for charmonium is repeated here and we request that the sum rule reproduces the η_b and χ_{b0} masses with the error induced by the choice of t_c. Unfortunately, this constraint is too weak and leads to $\overline{m}_b(\overline{m}_b)$ with an accuracy of about 40 MeV which is less interesting than the estimate from the vector channel in Eq. (9.219).

9.6.7 α_s and $\langle\alpha_s G^2\rangle$ from $M_{\chi_{0c(0b)}} - M_{\eta_{c(b)}}$

As the sum rules reproduce quite well the absolute masses of the (pseudo)scalar states, we can confidently use their mass splittings for extracting α_s and $\langle\alpha_s G^2\rangle$. We shall not work with the double ratio of LSR [6, 7, 167, 168, 171, 236, 263, 445] as each sum rule does not optimize at the same points. We check that, in the mass difference, the effect of the choice of the continuum threshold is reduced and induces an error from 6 to 14 MeV instead of 11 to 35 MeV in the absolute value of the masses. The effect due to $\overline{m}_{c,b}$ in Eqs. (9.225) and (9.222) and to μ in Eqs. (9.230) and (9.218) induce, respectively, an error of about (1–2) MeV and 8 MeV. The largest effects are due to the changes of α_s and $\langle \alpha_s G^2 \rangle$. We show their correlations in Fig. 9.74, where we have run the value of α_s from $\mu = 2.85$ GeV to M_τ in the charm channel and from $\mu = 9.5$ GeV to M_τ in the bottom one, where the values of μ correspond to the scales at which the sum rules have been evaluated. We have requested that the method reproduces within the errors the experimental mass splittings by about 2–3 MeV. With the central values given in Eqs. (9.216) and (9.220), the allowed region leads to the predictions:

$$\begin{aligned} &\alpha_s(M_\tau) = 0.318(15) \longrightarrow \alpha_s(M_Z) = 0.1183(19)(3), \\ &\langle \alpha_s G^2 \rangle = (6.34 \pm 0.39) \times 10^{-2}\,\text{GeV}^4. \end{aligned} \tag{9.227}$$

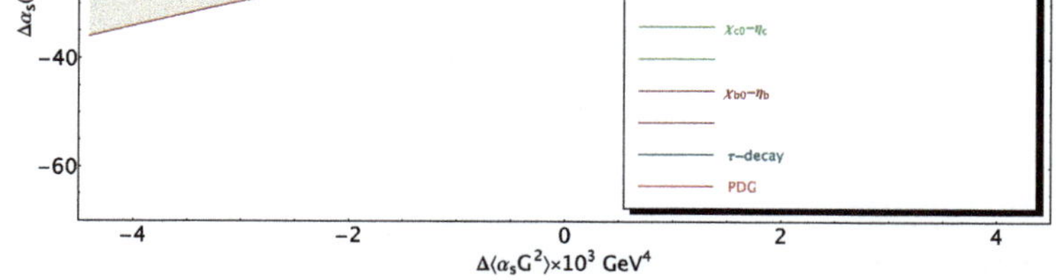

Figure 9.74. Correlation between α_s and $\langle \alpha_s G^2 \rangle$ by requiring that the sum rules reproduce the (pseudo)scalar mass splittings.

• Gluon condensate $\langle \alpha_s G^2 \rangle$

Adding into the analysis the range of input α_s values given in Eq. (9.211) (light grey horizontal band in Fig. 9.74), one can deduce stronger constraints on the value of $\langle\alpha_s G^2\rangle$:

$$\langle\alpha_s G^2\rangle = (6.39 \pm 0.35) \times 10^{-2}\,\text{GeV}^4. \tag{9.228}$$

Combining the previous values in Eqs. (9.215), (9.227) and (9.228) with the ones in Table 9.18, one obtains the *new sum rule average*:

$$\langle\alpha_s G^2\rangle|_{\text{average}} = (6.35 \pm 0.35) \times 10^{-2}\,\text{GeV}^4, \tag{9.229}$$

where we have retained the error from the most precise determination in Eq. (9.228) instead of the weighted error of 0.23. This result definitely rules out some eventual lower and negative values quoted in Table 9.18.

QCD coupling $\alpha_s(\mu)$

For the range of $\langle\alpha_s G^2\rangle$ values allowed by different analysis (x-axis) and requiring that the sum rule reproduces the experimental mass splittings $M_{\chi_{0c}} - M_{\eta_c}$ by about (2–3) MeV, one obtains the grey band limited by the two green (continuous) curves in Fig. 9.74, leading to:

$$\begin{aligned} &\alpha_s(2.85) = 0.262(9) \rightsquigarrow \alpha_s(M_\tau) = 0.318(15) \\ &\rightsquigarrow \alpha_s(M_Z) = 0.1183(19)(3). \end{aligned} \tag{9.230}$$

In the same way, the $M_{\chi_{0b}} - M_{\eta_b}$ bottomium sum rule evaluated at the optimization scale $\mu_b = 9\,\text{GeV}$ gives (sand colour band limited by two dotted red curves):

$$\begin{aligned} &\alpha_s(9.50) = 0.180(8) \rightsquigarrow \alpha_s(M_\tau) = 0.312(27) \\ &\rightsquigarrow \alpha_s(M_Z) = 0.1175(32)(3). \end{aligned} \tag{9.231}$$

• Comparison with the world average

These values of $\alpha_s(\mu)$ estimated at different μ-scales are shown in Fig. 9.67 where they are compared with the running of the world average $\alpha_s(M_Z) = 0.1181(11)$ [33,444]. We have added, in the figure, your previous estimate of $\alpha_s(M_\tau)$ [289] obtained from the original τ-decay rate (lowest moment) [79]:

$$\alpha_s(M_\tau) = 0.325(8), \tag{9.232}$$

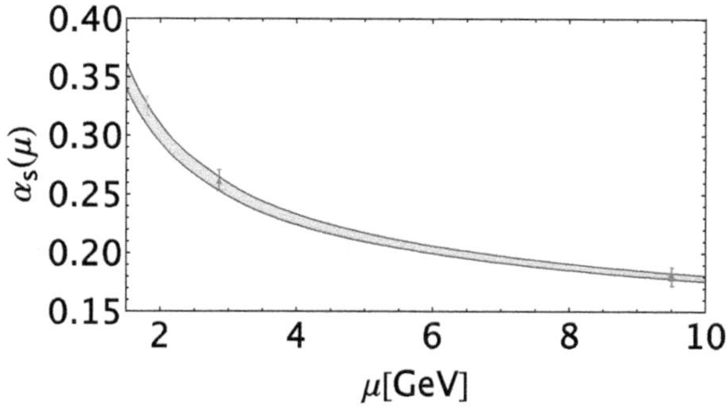

Figure 9.75. Comparison with the running of the world average $\alpha_s(M_Z) = 0.1181(11)$ [33,444] (grey band limited by the two green curves) of our predictions at three different scales: $\mu_\tau = M_\tau$ for the original τ-decay width [289] (open circle), $\mu_c = 2.85\,\text{GeV}$ for $M_{\chi_{c0}} - M_{\eta_c}$ (full triangle) and $\mu_b = 9.5\,\text{GeV}$ for $M_{\chi_{b0}} - M_{\eta_b}$ (full square) [517].

where one should note that non-perturbative corrections beyond the standard OPE (tachyonic gluon mass and duality violations) do not affect sensibly the above value of $\alpha_s(M_\tau)$ as indicated by the coincidence of the central value with the recent one from high moments [288].

Our most precise prediction for α_s from the heavy-quarkonia mass splittings comes from the (pseudo)scalar charmonium one in Eq. (9.230) which corresponds to:

$$\alpha_s(M_Z) = 0.1183(19)(3), \tag{9.233}$$

which agrees with the world average: $\alpha_s(M_Z) = 0.1181(11)$ [33,444].

9.7 QCD Parameters from $e^+e^- \to$ hadrons

In this section, we review the recent determinations of the QCD condensates and α_s from $e^+e^- \to$ hadrons where the original paper [18] has been summarized in Ref. [19] and updated in Ref. [20]. In so doing, we use the ratio of Laplace sum rules and some τ-decay-like moments.

Then, we obtain the values of condensates up to dimension $D = 20$ in Table 9.20 which confirms a large violation (about a factor 5.8) of the factorization of the four-quark condensates and the value of the gluon one $\langle \alpha_s G^2 \rangle$ from some other sources. Up to $D = 20$, I do not observe any factorial nor exponential growth of the size of these power corrections.

Using these new values of the $D = 6, 8$ power corrections to extract α_s from the Braaten–Narison–Pich (BNP) [79] lowest moment, to order α_s^4, I find within the SVZ expansion: $\alpha_s(M_\tau)|^{\rm SVZ}_{e^+e^-} = 0.3081(49)_{\rm fit}(71)_{\alpha_s^5}$ [respectively $0.3260(47)_{\rm fit}$ $(62)_{\alpha_s^5}$], implying $\alpha_s(M_Z)|^{\rm SVZ}_{e^+e^-} = 0.1170(6)(3)_{\rm evol}$ [respectively $0.1192(6)(3)_{\rm evol}$] for fixed order (FO) [respectively contour improved (CI)] PT series. They lead to the mean: $\alpha_s(M_\tau)|^{\rm SVZ}_{e^+e^-} = 0.3179(58)_{\rm fit}$ $(81)_{\rm syst}$ and $\alpha_s(M_Z)|^{\rm SVZ}_{e^+e^-} = 0.1182(12)(3)_{\rm evol}$ where the systematic error(syst) takes into account the discrepancy between the FO and CI results.

Using again the lowest BNP moment, we obtain from the vector (V) component of τ-decay data: $\alpha_s(M_\tau)|^{\rm SVZ}_{\tau,V} = 0.3128(19)_{\rm fit}(77)_{\alpha_s^5}$ [respectively $0.3291(25)_{\rm fit}(65)_{\alpha_s^5}$], implying $\alpha_s(M_Z)|^{\rm SVZ}_{\tau,V} = 0.1176(10)(3)_{\rm evol}$ [respectively $0.1196(8)(3)_{\rm evol}$] for FO [respectively CI] PT series, giving the mean $\alpha_s(M_\tau)|^{\rm SVZ}_{\tau,V} = 0.3219(52)(91_{\rm syst}$, leading to $\alpha_s(M_Z)|^{\rm SVZ}_{\tau,V} = 0.1187(13)(3)_{\rm evol}$.

The average of the two determinations from e^+e^- and τ-decay data is $\langle\alpha_s(M_\tau)\rangle = 0.3198(72)$, which implies $\langle\alpha_s(M_Z)\rangle = 0.1185(9)(3)_{\rm evol}$.

Some (eventual) contributions beyond the SVZ expansion ($1/Q^2$, instantons and duality violation) are discussed in Sections 10 and 11 which are expected to be relatively small.

9.7.1 The $I = 1$ isovector two-point function

We shall be concerned with the two-point correlator:

$$\Pi_H^{\mu\nu}(q^2) = i\int d^4x e^{-iqx}\langle 0|\mathcal{T} J_H^\mu(x)\,(J_H^\nu(0))^\dagger\,|0\rangle$$
$$= -(g^{\mu\nu}q^2 - q^\mu q^\nu)\Pi_H(q^2) \tag{9.234}$$

built from the T-product of the bilinear $I = 1$ vector current:

$$J_H^\mu(x) = \frac{1}{2}[\,:\bar\psi_u\gamma^\mu\psi_u - \bar\psi_d\gamma^\mu\psi_d:\,]. \tag{9.235}$$

It obeys the dispersion relation:

$$\Pi_H(q^2) = \int_{t>}^{\infty} \frac{dt}{t - q^2 - i\epsilon} \frac{1}{\pi} \, \mathrm{Im}\Pi_H(t) + \cdots, \tag{9.236}$$

where $\cdots$ are subtraction constants polynomial in q^2 and $t>$ is the hadronic threshold. From the optical theorem, the spectral function $\mathrm{Im}\Pi_H(t)$ can be related to the $e^+e^- \to$ hadrons total cross-section as:

$$R_{ee}^{I=1} \equiv \frac{\sigma(e^+e^- \to \text{hadrons})}{\sigma(e^+e^- \to \mu^+\mu^-)} = \left(\frac{3}{2}\right) 8\pi \, \mathrm{Im}\Pi_H(t). \tag{9.237}$$

9.7.2 Data handling

Data handling has been discussed in details in Ref. [18] and will not be repeated here. We use the compilation of PDG22 [33] which take into account the correlations among different data, while we use the new CMD3 [543] data for the pion form factor below 1 GeV. In this paper, we shall be concerned with the data below 1.875 GeV.

◇ Along this paper, we shall use the Mathematica program FindFit with optimized χ^2 and the program NSolve for our numerical analysis.

◇ To be conservative, we shall fit separately the ensemble of high data points and low data points. Our final result will be the mean of these two extremal values.

◇ From $2m_\pi$ to 0.993 GeV, we shall use a simple Breit–Wigner parametrization of the pion form factor after subtracting the $\omega(782)$ contribution using a narrow width approximation (NWA). Using as input the PDG value:

$$\Gamma(\rho \to e^+e^-) = (7.0.3 \pm 0.04) \, \mathrm{keV}, \tag{9.238}$$

we deduce in units of MeV:

$$M_\rho = (755.605 \pm 4.05), \quad \Gamma_\rho^{\mathrm{tot}} = (131.98 \pm 0.06), \tag{9.239}$$

which leads to the fit in Fig. 9.76.

◇ From 0.993 to 1.55 GeV, we fit the data using a simple interpolation program with polynomials. We subtract the 3π backgrounds by

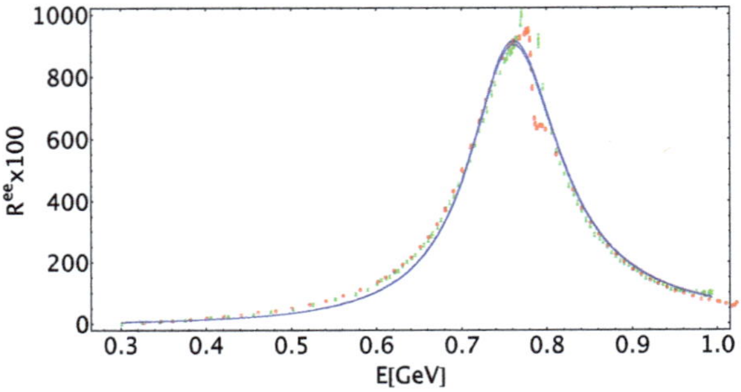

Figure 9.76. Fit of the data without ω using a minimal Breit–Wigner parametrization of the PDG [33] compilation (green triangle). A comparison with the new CMD3 [543] data is given (open red circle).

using the $SU(2)$ relation between the isoscalar and isovector states (a suppression 1/9 factor). We neglect the $\bar{K}K$ contributions from isoscalar sources which, in addition to the $SU(3)$ suppression factor, is also suppressed by phase space. The fit is shown in Fig. 9.77.

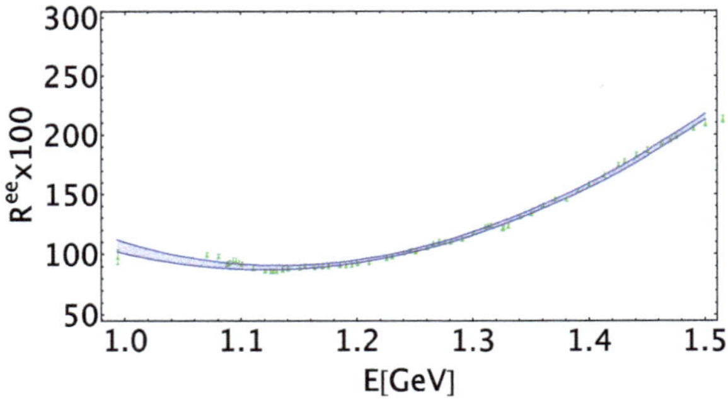

Figure 9.77. Fit of the data using a cubic polynomial interpolation formula.

◇ From 1.5 to 1.875 GeV, we use a Breit–Wigner fit for the ρ' meson and obtain:

$$M_{\rho'} = 1.6\,\text{GeV},$$
$$\Gamma(\rho' \to e^+e^-) = [10.5\,(\text{resp.}\,9.73)]\,\text{keV},$$
$$\Gamma^{\text{tot}}_{\rho'} = [720\,(\text{resp.}\,694)]\,\text{MeV}, \tag{9.240}$$

from the high (respectively low) data points. The fit is shown in Fig. 9.78.

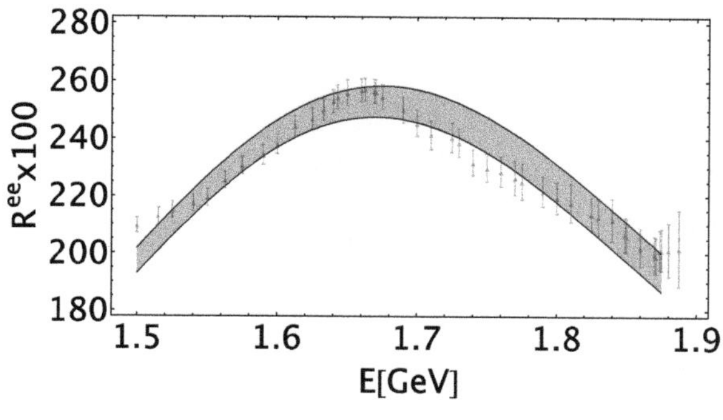

Figure 9.78. Fit of the data using a minimal Breit–Wigner parametrization.

9.7.3 The Laplace sum rules (LSR) and their ratios

• Form of the LSR and their ratio

In Ref. [115] and SN [18], the dimension $D = 4, 6$ and 8 condensates appearing in the OPE of the two-point vector correlator have been estimated using the ratio of Laplace sum rule moments [3, 69, 70]:

$$\mathcal{R}^c_{10}(\tau) \equiv \frac{\mathcal{L}^c_1}{\mathcal{L}^c_0} = \frac{\int_{t>}^{t_c} dt\ e^{-t\tau} t\, R^{I=1}_{ee}(t,\mu)}{\int_{t>}^{t_c} dt\ e^{-t\tau} R^{I=1}_{ee}(t,\mu)}, \tag{9.241}$$

where τ is the LSR variable and $t >$ is the hadronic threshold. Here, t_c is the threshold of the "QCD continuum" which parametrizes, from the discontinuity of the Feynman diagrams, the spectral function $\text{Im}\,\Pi_H(t, m_q^2, \mu^2)$. m_q is the quark mass and μ is an arbitrary subtraction point. The exponential weight enhances the contribution of the low-energy region of the spectral function accessible experimentally. We shall see in Eq. (9.243) that it introduces a factorial factor in front of each condensate which accelerates the convergence of the OPE.

• The PT QCD expression of the LSR

To order α_s^4, the perturbative (PT) expression of the lowest moment reads [546]:

$$\mathcal{L}^{\rm PT}_0(\tau) = \frac{3}{2}\tau^{-1}\Big[1 + a_s + 2.93856\, a_s^2 + 6.2985\, a_s^3 + 22.2233\, a_s^4\Big]. \tag{9.242}$$

Then, taking its derivative in τ, one gets $\mathcal{L}_1(\tau)$ and then their ratio $\mathcal{R}_{10}(\tau)$.

• The NPT QCD expression of the LSR

From Eq. (9.298), one can deduce the lowest moment LSR:

$$\mathcal{L}_0^{NPT}(\tau) = \frac{3}{2}\tau^{-1}\sum_D \frac{d_D}{(D/2-1)!}\tau^{D/2}, \tag{9.243}$$

from which one can deduce $\mathcal{L}_1^{NPT}$ and $\mathcal{R}_{10}$, where $d_D \equiv C_D\langle O_D\rangle$.

9.7.4 Fixing the value of t_c

• t_c from the moment $\mathcal{L}_0^{ee}$

We request that the data and the perturbative QCD expression of the moment $\mathcal{L}_0^{ee}$ coincides for the sum rule variable $\tau \leq 0.7\,\text{GeV}^{-2}$ where the non-perturbative effects to the QCD expression are expected to be negligible while the difference between the QCD contribution with the data one without a QCD continuum remains positive. Using the Mathematica optimized χ^2 fit program FindFit, we obtain to order α_s^4 accuracy:

$$t_c = (4.74 \sim 4.93)\,\text{GeV}^2, \tag{9.244}$$

where $t_c = 4.74$ (respectively 4.93) GeV^2 corresponds to the fit of the low (respectively high) set of values of the data points. The result of the analysis is shown in Fig. 9.79. We have improved the fit for the low data points by adjusting manually the value of t_c. Then, we arrive at the conservative range of values:

$$t_c = (4 \sim 5)\,\text{GeV}^2, \tag{9.245}$$

which we shall use for the forthcoming analysis.

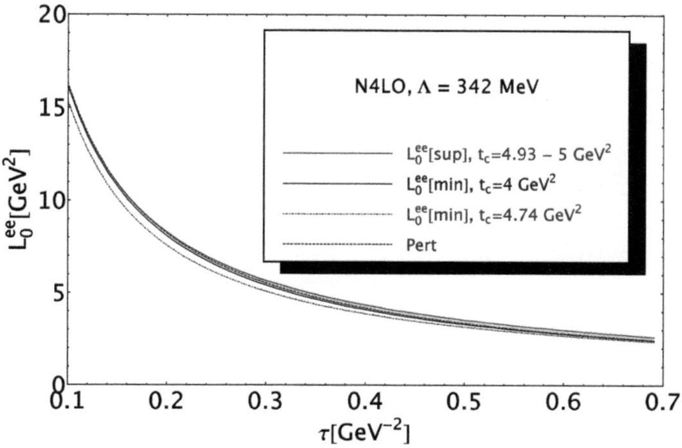

Figure 9.79. Comparison of the QCD expression of $\mathcal{L}_0^{ee}$ with the data for the obtained value of t_c given in Eq. (9.245).

.

• Comments and comparison with FESR

It is informative to compare this result with the one required by FESR for a duality between the experimental and QCD side of the sum rule [521, 522]:

$$F_0(t_c) \equiv \int_0^{t_c} dt R_{ee}^{I=1}(t)$$
$$= \frac{3}{2}t_c\left[1 + a_s(t_c) + a_s^2(t_c)\left(1.6398 - \frac{\beta_1}{2}\right) + \cdots\right], \tag{9.246}$$

where $\cdots$ comes from a numerical integration of the integral at order α_s^3. The analysis for the ratio:

$$R_0^{\text{fesr}} \equiv \frac{F_0^{\text{data}}}{F_0^{\text{qcd}}} \tag{9.247}$$

is shown in Fig. 9.80 from which one can deduce to order α_s^3 the duality region:

$$t_c^{\text{fesr}} = (5.1 \sim 5.6)\,\text{GeV}^2, \tag{9.248}$$

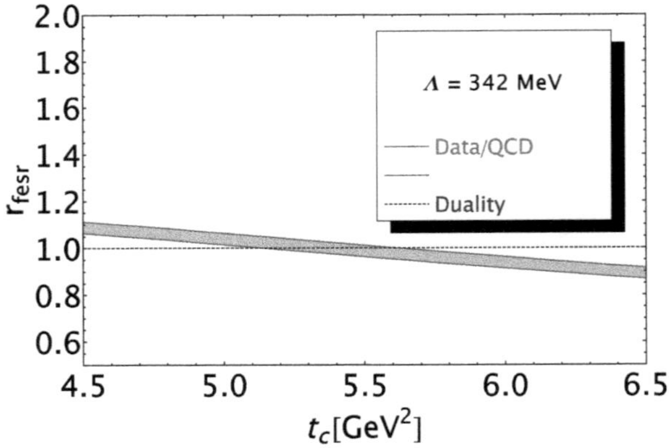

Figure 9.80. Variation of the ratio in Eq. (9.247) versus t_c.

The value in Eq. (9.248) is higher than the value around 2.1 GeV^2 obtained in Refs. [521, 522]. We interpret the origin of this discrepancy as due to the effect of the old data and its parametrization in the ρ'-region. This value does not also justify the choice $t_c = 1.55$ GeV^2 used in Ref. [544] for the unpinched moment which is equivalent to the lowest degree of FESR used here.

9.7.5 QCD condensates from the ratio of LSR

We attempt to extract the dimension 4, 6 and 8 condensates using different strategies.

• d_4 and d_6 condensates

Using a two-parameter fit, we attempt to extract in Fig. 9.81 the d_4 gluon $\langle\alpha_s G^2\rangle$ (green curves) and d_6 four-quark condensates (red curves). The analysis does not indicate τ-stability but reproduces the results of Ref. [115] at $\tau \simeq 1.9\,\text{GeV}^{-2}$.

• d_6 and d_8 condensates to order α_s^4

Instead, fixing the value of $\langle\alpha_s G^2\rangle = (6.49 \pm 0.35) \times 10^{-2}\,\text{GeV}^4$ from heavy quarkonia [517], we estimate the d_6 and d_8 condensates to order α_s^4. The analysis is shown in Fig. 9.82. We obtain

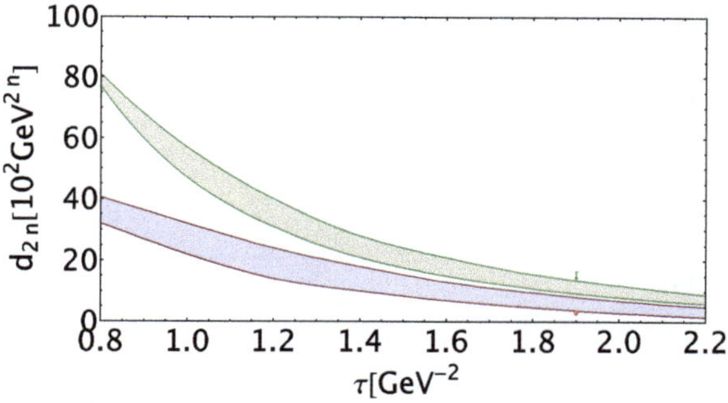

Figure 9.81. Two-parameter fit of $\langle \alpha_s G^2 \rangle$ and $-d_6$ from the ratio of moments $\mathcal{R}_{10}$.

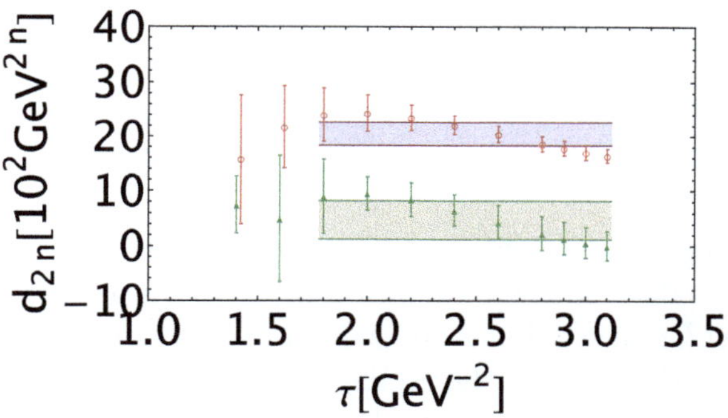

Figure 9.82. Two-parameter fit of $-d_6$ and $+d_8$ to order α_s^4 for a given value of $\langle \alpha_s G^2 \rangle$ from Eq. (9.254). The red (respectively green) points are the values of $-d_6$ (respectively d_8).

for $\tau \simeq (2 \sim 3)\,\text{GeV}^{-2}$:

$$d_6 = -(20.5 \pm 2.2) \times 10^{-2}\,\text{GeV}^6,$$
$$d_8 = (4.7 \pm 3.5) \times 10^{-2}\,\text{GeV}^8. \tag{9.249}$$

The value of d_6 remains unchanged from α_s^2 to α_s^4 which is not the case of d_8 which has decreased by about a factor 2.5! Then, we try to improve the determination of d_8 by using d_6 as input in addition to the previous input parameters. The analysis shown in Fig. 9.83 indicates a much better stability though we do not (unfortunately) gain much on the accuracy.

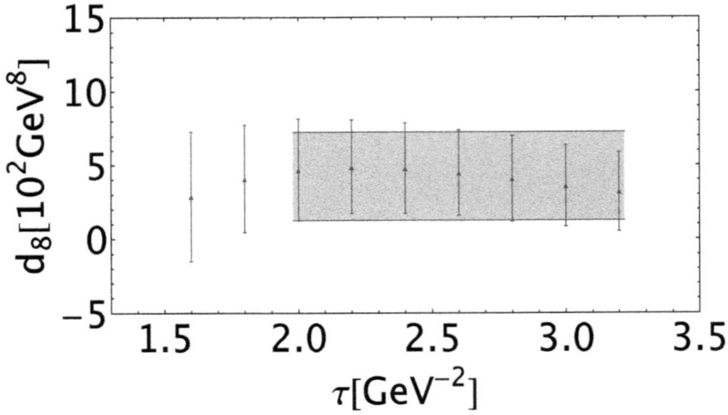

Figure 9.83. One-parameter fit of d_8 to order α_s^4 for a given value of $\langle\alpha_s G^2\rangle$ from Eq. (9.254) and d_6 from Eq. (9.249).

We obtain

$$d_8 = (4.3 \pm 3.0) \times 10^{-2}\,\text{GeV}^8, \tag{9.250}$$

which we consider as a final result for d_8. However, this low value of d_8 compared to the one obtained at order α_s^2 is a good news for QCD sum rules users. It (*a posteriori*) justifies the neglect of high-dimension condensates in the analysis which gives a good description of different hadron parameters when only the dimension $D \leq 6$ condensates are retained in the OPE.

From the value of d_6 in Eq. (9.249), one can deduce the value of the four-quark condensate to order α_s^4:

$$\rho\langle\bar{\psi}\psi\rangle^2 = (5.98 \pm 0.64) \times 10^{-4}\,\text{GeV}^6, \tag{9.251}$$

which (agreeably) confirms the value $(5.9 \pm 0.8) \times 10^{-4}\,\text{GeV}^6$ obtained from different determinations from light quark systems [6, 7, 47].

• A second attempt to extract $\langle\alpha_s G^2\rangle$ to order α_s^4

◇ $\langle\alpha_s G^2\rangle$ and d_6 from a two-parameter fit

First, we repeat the analysis done previously and neglect the $d = 8$ condensates. One can note (see Fig. 9.84) that the presence of the α_s^4

term leads in this case to τ-stability (minimum) at $\tau \simeq 2.4\,\text{GeV}^{-2}$ from which we extract the optimal values:

$$\langle\alpha_s G^2\rangle = (5.9 \pm 2.6) \times 10^{-2}\,\text{GeV}^4,$$
$$d_6 = -(17.1 \pm 4.1) \times 10^{-2}\,\text{GeV}^6, \tag{9.252}$$

where the main error is due to the data fitting procedure. These values agree with the determination of $\langle\alpha_s G^2\rangle$ from heavy quarkonia in Eq. (9.254) and with the one of d_6 in Eq. (9.249) but inaccurate.

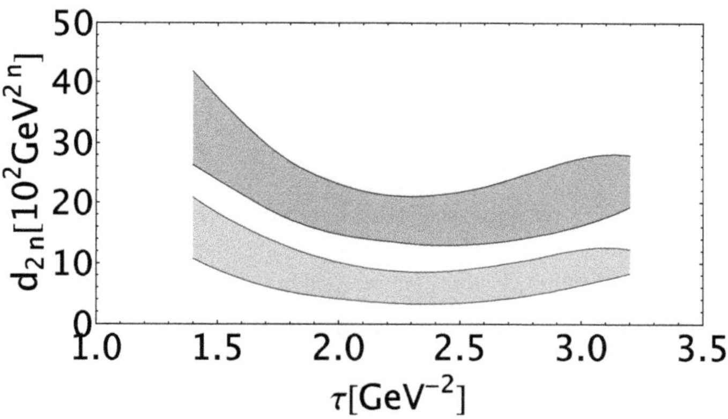

Figure 9.84. Two-parameter fit of $\langle\alpha_s G^2\rangle$ (lower curves) and d_6 (upper curves) to order α_s^4.

◇ $\langle\alpha_s G^2\rangle$ from a one-parameter fit

To improve the determination of $\langle\alpha_s G^2\rangle$, we use as input the value of d_6 in Eq. (9.249) and include d_8 from Eq. (9.250). The analysis is shown in Fig. 9.85. We deduce, at the stability point $\tau \simeq 1.5\,\text{GeV}^{-2}$:

$$\langle\alpha_s G^2\rangle = (6.12 \pm 0.61) \times 10^{-2}\,\text{GeV}^4, \tag{9.253}$$

which improves the result in Eq. (9.252). Combining this value with the average in Eq. (9.229), we deduce the QCD spectral sum rules (QSSR) average:

$$\langle\alpha_s G^2\rangle = (6.29 \pm 0.30) \times 10^{-2}\,\text{GeV}^4. \tag{9.254}$$

- **Comparison with some other approaches**

– *LNT* [115]: We consider the previous results as an improvement of the ones of LNT (note a slightly different normalization):

$$d_6|_{lnt} = -(12.7 \pm 4.2) \times 10^{-2}\,\text{GeV}^6,$$
$$d_8|_{lnt} = (27 \pm 24) \times 10^{-2}\,\text{GeV}^8, \tag{9.255}$$

where LNT have used the same ratio of moments $\mathcal{R}_{10}$ with a different strategy and older data. The lower value of d_6 is correlated to the low value of $\langle\alpha_s G^2\rangle$ obtained by LNT (see Fig. 9.81).

– *FESR [521, 522]*: These results are consistent in sign and in magnitude with the FESR ones:

$$\langle\alpha_s G^2\rangle|_{\text{fesr}} \simeq (7 \sim 18) \times 10^{-2}\ \text{GeV}^4,$$
$$d_6|_{\text{fesr}} = -(0.33 - 0.55)\ \text{GeV}^6. \tag{9.256}$$

The inaccuracy of the FESR results is related to their strong t_c-dependence and the inaccuracy of the data at higher energy.

– *τ-decay moments*: LSR and FESR results do not favour some results with negative signs obtained from different moments of τ-decay (see Table 9.19).

Table 9.19. Values of the QCD condensates of dimension $D = 2n$ in units of $10^{-2}\,\text{GeV}^{2n}$ from τ-like decays [545].

$\langle\alpha_s G^2\rangle$	$-d_6$	d_8	PT	Refs.
0.67 ± 0.89	15.2 ± 2.2	22.3 ± 2.5	FO	ALEPH [247]
5.34 ± 3.64	14.2 ± 3.5	21.3 ± 2.5	FO	OPAL [248]
$0.8^{+0.7}_{-1.4}$	32^{+8}_{-5}	50^{+4}_{-7}	FO	[545]
0.31 ± 2.45	13.5 ± 1.8	20.0 ± 1.6	CI	OPAL [248]
-1.57 ± 0.94	14.7 ± 1.1	20.4 ± 1.3	CI	ALEPH [506]
$-0.8^{+0.7}_{-0.7}$	35 ± 3	49^{+4}_{-5}	CI	[545]

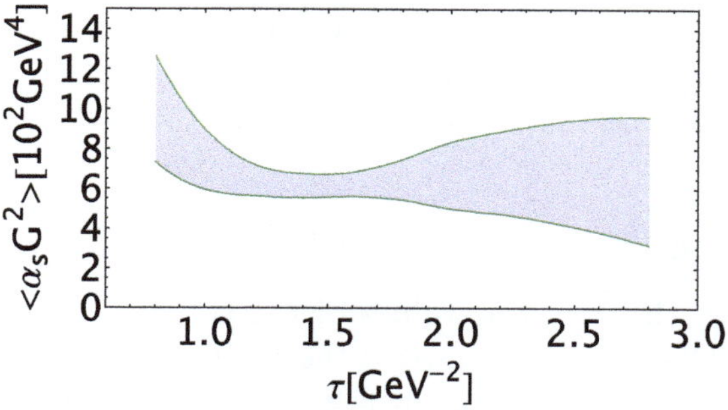

Figure 9.85. One-parameter fit of $\langle \alpha_s G^2 \rangle$ to order α_s^4 using d_6 in Eq. (9.249) and d_8 from Eq. (9.250).

• **Correlated values of d_6 and d_8 versus $\langle \alpha_s G^2 \rangle$**

We show in Fig. 9.86 the behaviour of d_6 and d_8 versus $\langle \alpha_s G^2 \rangle$. One can note the following:

- The value of the four-quark condensate estimated from factorization is inconsistent with the SVZ value of the gluon condensate.
- The ratio:

$$r_{46} \equiv \frac{\rho \langle \bar{\psi}\psi \rangle^2}{\langle \alpha_s G^2 \rangle} \approx 0.9 \times 10^{-2}\,\mathrm{GeV}^2 \tag{9.257}$$

is almost constant. This result has also been obtained from different approaches independently on the absolute size of the extracted

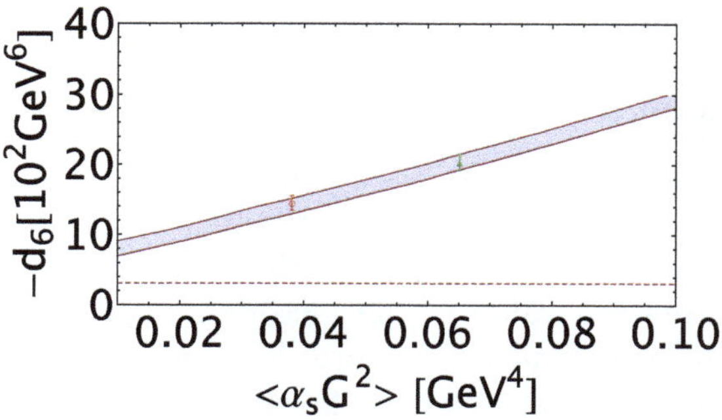

Figure 9.86. Correlated values of d_6 and d_8 versus $\langle \alpha_s G^2 \rangle$. The dashed horizontal line is the value of d_6 estimated from factorization of the four-quark condensate. The red (respectively oliva) point corresponds to the value of $\langle \alpha_s G^2 \rangle$ given by SVZ and by Eq. (9.254).

condensates and on the kind of methods as discussed in Ref. [237]. Moreover, we also note that some τ-like moments results given in Table 9.19 do not fulfill such a relation.

• Confronting theory and experiment for $\mathcal{R}_{10}^{ee}$ and $\mathcal{L}_{0}^{ee}$

Using as input the previous values of the condensates and the value of α_s from PDG [33], we compare the QCD and experimental sides of $\mathcal{R}_{10}$ and $\mathcal{L}_0$ which we show in Figs. 9.87 and 9.88.

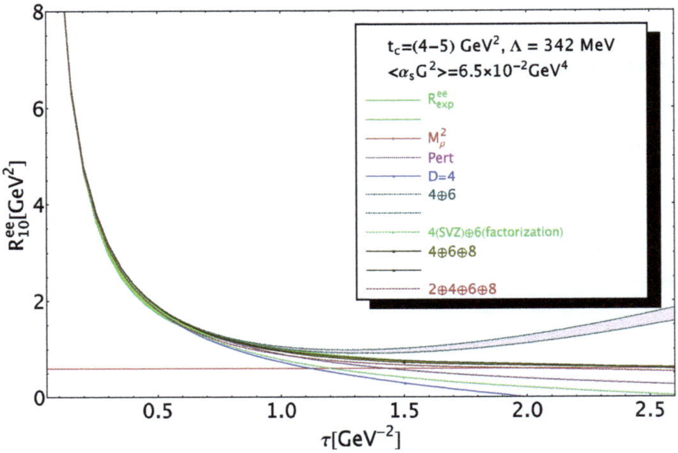

Figure 9.87. τ-behaviour of the ratio of moments $\mathcal{R}_{10}^{ee}$ for different truncation of the OPE.

◇ Ratio of moments $\mathcal{R}_{10}$

– The use of the standard SVZ value of the gluon condensate ⊕ the value of the four-quark condensate estimated using the factorization assumption underestimates the data. However, the simultaneous use of the SVZ standard value ⊕ the factorization (combination often used in the QCD sum rules literature) is inconsistent from the result of simultaneous two-parameter fit shown in the previous section and in Fig. 9.86.

– The inclusion of the $D = 8$ condensate enlarges the region of agreement between the QCD prediction and the experiment until $\tau = 2.5$ GeV^{-2} where the ρ-meson mass is reached and where, exceptionally, the PT series for the spectral function still make sense.

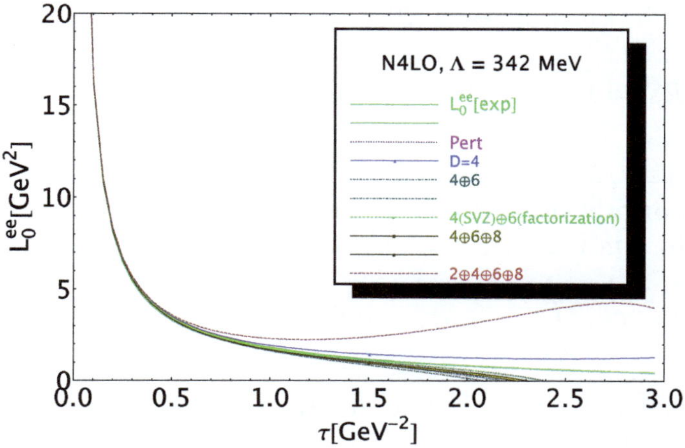

Figure 9.88. τ-behaviour of the lowest moment $\mathcal{L}_0$ for different truncation of the OPE.

– We complete the analysis by adding the contribution of the dimension two tachyonic gluon mass beyond the SVZ expansion [235]. We see that it tends to decrease the agreement with the data, but the effect is (almost) negligible within our precision in the high-energy region.

◇ Lowest moment $\mathcal{L}_0^{ee}$

We have tried to extract these parameters using $\mathcal{L}_0^{ee}$, but the results are unconclusive.

– Contrary to the case of $\mathcal{R}_{10}^{ee}$, we note an important sensitivity of the results on the truncation of the PT series. This problem might be evaded if one uses the π^2 resummation of the higher order terms where the PT series converges faster [546].

– However, we also remark that the determination from the low and high sets of data points varies in a large range, leading to a very inaccurate result.

9.7.6 The τ-decay-like moment sum rules

• The BNP lowest moment

The lowest τ-like decay moment applied to $e^+e^- \to$ hadrons data have been initially used in Refs. [237, 332] to extract α_s and the

QCD condensates. It reads [79]:

$$\mathcal{R}_0^{ee} = 12\pi \int_{t>}^{x_0} dt\,(1-x_0)^2(1+2x_0)\,2\,R_{ee}^{I=1}(x_0), \tag{9.258}$$

where $x_0 \equiv t/M_0^2$ with M_0 is the τ-like lepton mass. The moment can be conveniently reformulated in terms of the Cauchy integral along the contour in Fig. 9.89 à la Shankar [31][27]:

$$\mathcal{R}_0^{ee} = 6i\,\pi \int_{|x_0|=1} dx_0(1-x_0)^2(1+2x_0)2\,R_{ee}^{I=1}(x_0), \tag{9.259}$$

with a radius $|x_0| = 1$. To order α_s^4, the PT corrections to the lowest BNP moment reads [79, 438, 441, 547]:

$$\begin{aligned}\delta_0^{(0)} &= a_s + 5.2026\,a_s^2 + 26.368\,a_s^3 + 127.085\,a_s^4 + \mathcal{O}(a_s^5) \quad \text{(FO)},\\ &= 1.364\,a_s + 2.54\,a_s^2 + 9.71\,a_s^3 + 64.29\,a_s^4 + \mathcal{O}(a_s^5) \quad \text{(CI)},\end{aligned} \tag{9.260}$$

for fixed order (FO) and contour improved (CI) PT series.

Unlike different finite energy sum rule (FESR) used in the current literature [521, 522] and its pinched versions [448], the main advantage of the τ-decay-like moment is the presence of the weight factor $(1-x_0)^2$ which suppresses the contribution near the real axis where the data are inaccurate. This approach has led to an accurate determination of α_s at the τ mass.

- **The τ-like high moments**

Stimulated by the original work of Ref. [547], different forms of high moments have been proposed in the literature for extracting simultaneously α_s and the QCD condensates [247, 248, 506, 545]. In this paper, we shall be concerned with the moments of the form:

$$\mathcal{R}_n^{ee} = 6i\,\pi \int_{|x_0|=1} dx_0\,(1-x_0)^2(1+2x_0)(x_0^n)\,2\,R_{ee}^{I=1}(x_0), \tag{9.261}$$

where $n = 1$–6.

[27]Some other applications in QCD can, e.g., be found in Ref. [29].

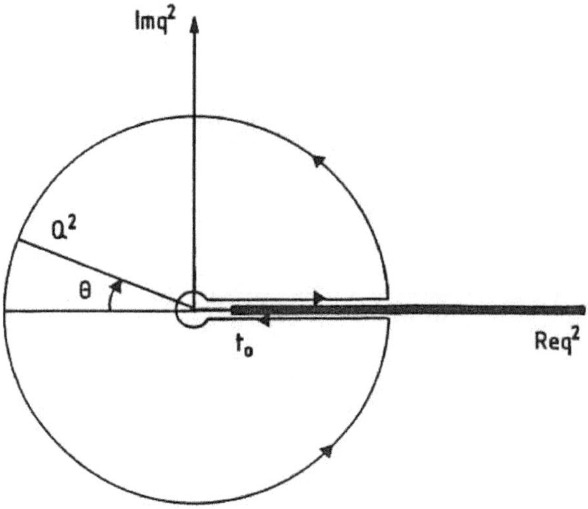

Figure 9.89. Cauchy contour integral in the complex $Q^2 \equiv |q^2|$-plane.

◇ PT corrections to the τ-like high moments at order α_s^4

The perturbative contributions read to order α_s^4 for fixed order (FO) perturbation series:

$$
\begin{aligned}
\delta_1^{(0)} &= a_s + 3.7776\, a_s^2 + 6.8810\, a_s^3 - 66.7235\, a_s^4 + \mathcal{O}(a_s^5),\\
\delta_2^{(0)} &= a_s + 3.2151\, a_s^2 + 1.0687\, a_s^3 - 98.2293\, a_s^4 + \mathcal{O}(a_s^5),\\
\delta_3^{(0)} &= a_s + 2.8990\, a_s^2 - 1.7710\, a_s^3 - 108.034\, a_s^4 + \mathcal{O}(a_s^5),\\
\delta_4^{(0)} &= a_s + 2.6928\, a_s^2 - 3.4664\, a_s^3 - 111.868\, a_s^4 + \mathcal{O}(a_s^5),\\
\delta_5^{(0)} &= a_s + 2.4365\, a_s^2 - 5.4048\, a_s^3 - 119.087\, a_s^4 + \mathcal{O}(a_s^5),\\
\delta_6^{(0)} &= a_s + 2.5464\, a_s^2 - 4.5966\, a_s^3 - 108.552\, a_s^4 + \mathcal{O}(a_s^5).
\end{aligned}
\tag{9.262}
$$

◇ τ-like high moments with NPT corrections at LO

To lowest order in α_s (LO) and in the massless quark limit $m_{u,d} = 0$, the non-perturbative (NPT) corrections to the different moments give the following:

$$\mathcal{R}_0^{ee} = \frac{3}{2}\left[1 + \delta_0^{(0)} - 6\frac{d_6}{M_0^6} - 4\frac{d_8}{M_0^8} + \sum_D \delta_0^{(D)}\right]:$$

$$\delta_0^{(D)} = 0 \quad (for\ D \geq 10),$$

$$\mathcal{R}_1^{ee} = \frac{9}{20}\left[1 + \delta_1^{(0)} - \frac{20}{3}\frac{d_4}{M_0^4} + 20\frac{d_8}{M_0^8} + \frac{40}{3}\frac{d_{10}}{M_0^{10}} + \sum_D \delta_1^{(D)}\right]:$$

$$\delta_1^{(D)} = 0 \quad (for\ D \geq 12),$$

$$\mathcal{R}_2^{ee} = \frac{1}{5}\left[1 + \delta_2^{(0)} + 15\frac{d_6}{M_0^6} - 45\frac{d_{10}}{M_0^{10}} - 30\frac{d_{12}}{M_0^{12}} + \sum_D \delta_2^{(D)}\right]:$$

$$\delta_2^{(D)} = 0 \quad (for\ D \geq 14),$$

$$\mathcal{R}_3^{ee} = \frac{3}{28}\left[1 + \delta_3^{(0)} - 28\frac{d_8}{M_0^8} + 84\frac{d_{12}}{M_0^{12}} + 56\frac{d_{14}}{M_0^{14}} + \sum_D \delta_3^{(D)}\right]:$$

$$\delta_3^{(D)} = 0 \quad (for\ D \geq 16),$$

$$\mathcal{R}_4^{ee} = \frac{9}{140}\left[1 + \delta_4^{(0)} + \frac{140}{3}\frac{d_{10}}{M_0^{10}} - 140\frac{d_{14}}{M_0^{14}} - \frac{280}{3}\frac{d_{16}}{M_0^{16}} + \sum_D \delta_4^{(D)}\right]:$$

$$\delta_4^{(D)} = 0 \quad (for\ D \geq 18),$$

$$\mathcal{R}_5^{ee} = \frac{1}{24}\left[1 + \delta_5^{(0)} - 72\frac{d_{12}}{M_0^{12}} - 216\frac{d_{16}}{M_0^{16}} - 144\frac{d_{18}}{M_0^{18}} + \sum_D \delta_5^{(D)}\right]:$$

$$\delta_5^{(D)} = 0 \quad (for\ D \geq 20),$$

$$\mathcal{R}_6^{ee} = \frac{1}{35}\left[1 + \delta_6^{(0)} + 75\frac{d_{14}}{M_0^{14}} - 225\frac{d_{18}}{M_0^{18}} - 150\frac{d_{20}}{M_0^{20}} + \sum_D \delta_6^{(D)}\right]:$$

$$\delta_6^{(D)} = 0 \quad (for\ D \geq 22), \tag{9.263}$$

where $\delta_n^{(D)}$ is the contribution of the condensate of dimension D and n is the degree of the moment.

9.7.7 QCD condensates from τ-like high moments

In so doing, we fix the value of α_s from the world average [33]. Then, we are looking for stability in the change of M_0 for extracting the optimal values of the condensates.

• $\mathcal{R}_0^{ee}$ moment

– Using a two-parameter (d_6, d_8) fit in the lowest BNP moment, we do not find a conclusive result. We interpret this fact due to the opposite sign of the two contributions which tend to cancel out. However, this feature is welcome for determining α_s as the sum of the non-perturbative contributions are small, but may indicate that it is not a good place for accurately determining the condensates.

– We impose more constraints by using a one-parameter fit with an initial value of the one:

$$d_6 = -(20.5 \pm 2) \times 10^{-2}\,\text{GeV}^6, \tag{9.264}$$

obtained recently in SN23 [18, 19] from the ratio of LSR into $\mathcal{R}_0^{ee}$ in an attempt to fix d_8. The resulting value of d_8 does not show any M_0 stability as also observed in SN23. Then, we conclude that the analysis from $\mathcal{R}_0^{ee}$ is also not conclusive.

• $\mathcal{R}_1^{ee}$ moment

In the following, we shall only work with a one-parameter fit d_0 and use as initial values of $\langle\alpha_s G^2\rangle$ in Eq. (9.254) obtained from heavy quarks mass splittings and some other sources and the value of d_8 from SN23:

$$d_8 = (4.7 \pm 3.5) \times 10^{-2}\,\text{GeV}^8. \tag{9.265}$$

We extract d_{10}. Then, we repeat the analysis by fixing d_{10} and d_6, and deduce d_8.

• $\mathcal{R}_2^{ee}$ moment

We use the value of d_{10} into $\mathcal{R}_2^{ee}$ and extract d_{12}. Using this value of d_{12}, we reiterate the analysis to extract d_{10} and continue the iterations.

• **Higher moments $n = 2$–6**

We repeat the procedure for higher moments and extract d_{14} to d_{20}. The quoted values are obtained without any iterations as they are already quite precise. However, the quoted error does not include the systematic ones which may increase with the degree of moments.

• **Conclusions from τ-like moments**

The values of d_8 to d_{20} from the τ-like moments are shown in Figs. 9.83 and 9.91, where we note nice stabilities of the results versus M_0. The optimal values are quoted in Table 9.20.

9.7.8 d_4 and d_6 from the ratio of LSR

Once we obtain the previous values of the $d = 8$ to $d = 20$ condensates, we inject them to the ratio $\mathcal{R}_{10}$ of LSR for re-extracting d_6 and d_4. To strongly constrain the parameters, we continue to use a one-parameter fit. Fixing again $\langle\alpha_s G^2\rangle$ as in Eq. (9.254), we re-extract d_6 by including into the QCD expression of $\mathcal{R}_{10}$ the contributions of condensates up to d_{20}. The analysis is shown in Fig. 9.92. The results are summarized in Table 9.20 and compared with the other estimates quoted in Table 9.21. One can note a good agreement with some published results, but the values obtained in the present work are more accurate.

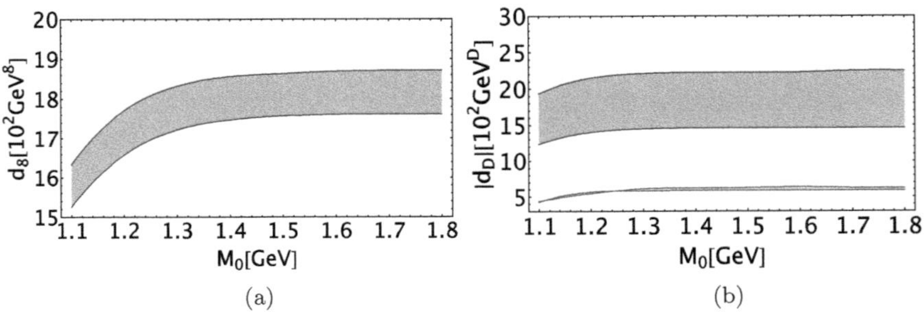

Figure 9.90. (a) Value of the condensate d_8 from $\mathcal{R}_1^{ee}$ versus M_0; (b) values of the condensates d_{10} and d_{12} from $\mathcal{R}_1^{ee}$ and $\mathcal{R}_2^{ee}$ versus M_0.

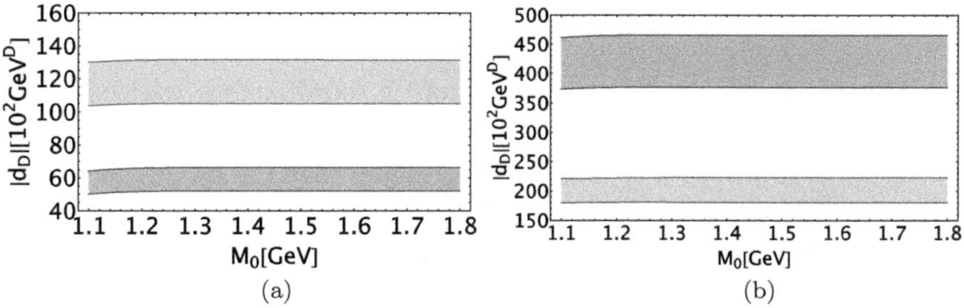

Figure 9.91. (a) Value of the condensate d_{14} and d_{16} from $\mathcal{R}_3^{ee}$ and $\mathcal{R}_4^{ee}$ versus M_0; (b) value of the condensates d_{18} and d_{20} from $\mathcal{R}_5^{ee}$ and $\mathcal{R}_6^{ee}$ versus M_0.

Table 9.20. Values of the QCD condensates of dimension D in units of 10^{-2} GeV^D from the present analysis.

$\langle\alpha_s G^2\rangle$	$-d_6$	d_8	$-d_{10}$	$-d_{12}$	d_{14}	$-d_{16}$	$-d_{18}$	d_{20}
7.8 ± 3.5	23.6 ± 3.7	18.2 ± 0.6	6.1 ± 0.2	18.4 ± 3.8	59.2 ± 7.1	118.3 ± 13.2	202.0 ± 20.7	421.3 ± 44.3

Table 9.21. Values of the QCD condensates in units of 10^{-2} GeV^D from some other τ-moments at fixed order (FO) of the PT series.

$\langle\alpha_s G^2\rangle$	$-d_6$	d_8	$-d_{10}$	$-d_{12}$	Refs.
0.67 ± 0.89	15.2 ± 2.2	22.3 ± 2.5			ALEPH [247]
5.34 ± 3.64	14.2 ± 3.5	21.3 ± 2.5			OPAL [248]
$3.5^{+2.2}_{-3.8}$	$19.7^{+11.8}_{-7.9}$	$23.7^{+11.8}_{-15.8}$	11.8 ± 19.7	7.9 ± 19.7	($d_{14,16} = 0$) Pich-Rodriguez [545]

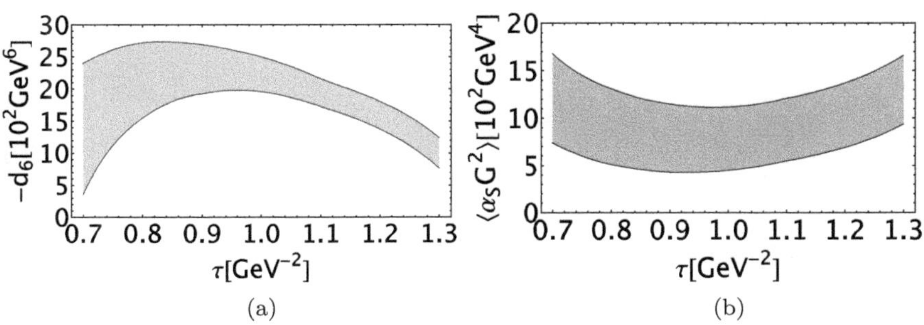

Figure 9.92. (a) Value of the condensate d_6 from $\mathcal{R}_{10}^{ee}$ versus M_0; (b) value of the condensate d_4 from $\mathcal{R}_{10}^{ee}$ versus M_0.

9.7.9 Comments on the results

◇ The value of $\langle\alpha_s G^2\rangle$ obtained here is compatible with the one in Eq. (9.254) but less accurate. However, we consider this result as an improvement of the ones from some other τ-decay moments given in Table 9.21 using FO perturbative series. It disfavours some negative values obtained from CI perturbative series. This value is also consistent with the one $(8 \pm 4) \times 10^{-2}\,\text{GeV}^4$ obtained from a matching of the short and long distance contributions to the Adler function at large N_c [548].

◇ The value of d_6 is in fair agreement with the one in Eq. (9.264) from ratio of LSR obtained in SN23 [18,19] where the OPE has been truncated at $D = 8$. The inclusion of higher dimension condensates has increased slightly $|d_6|$ by about 13.6%. The ratio $\langle\alpha_s G^2\rangle/\rho\alpha_s\langle\bar\psi\psi\rangle^2 = 95.4$ is in the range of the ones obtained using different methods quoted in Ref. [237].

◇ The value of d_8 is larger than the one in Eq. (9.265) from ratio of LSR. As the extraction of d_8 from LSR in SN23 has been done by neglecting all higher dimension condensates included in the present analysis, we interpret the value of d_8 obtained in SN23 as an effective condensate including all high dimension ones. To check this argument, we evaluate the NP contributions to $\mathcal{R}^{ee}_{10}$ at the typical optimization sum rule scale $\tau \approx 1\ \text{GeV}^{-2}$. We neglect all higher dimension condensates and find:

$$\sum_4^8 \delta^{NP} = 5.1 \times 10^{-2},$$

$$\sum_4^6 \delta^{NP} = -1.0 \times 10^{-2} \to d_8 \approx 6.1 \times 10^{-2}\,\text{GeV}^8, \quad (9.266)$$

in perfect agreement with the fitted value in Eq. (9.265).

◇ Using the previous values of the condensates, we check the convergence of the OPE. We obtain, at $\tau \simeq 1\,\text{GeV}^{-2}$:

$$\sum_{10}^{20} \delta^{NP} = -0.4 \times 10^{-2}, \quad (9.267)$$

which is negligible. This feature justifies (*a posteriori*) the phenomenological success of the SVZ sum rules by truncating the OPE up to dimension $D = 6, 8$ condensates.

◇ From Table 9.20, we observe that there is no signal of a factorial or/and exponential growth of the size of the condensates up to $D = 20$, where the OPE is truncated. Similar observation can be made in the $V - A$ channel for the OPE up to $D = 18$ [343, 549].

9.7.10 α_s from the lowest BNP moment

Armed with these improved values of the QCD condensates, we extract α_s from the lowest BNP moment $\mathcal{R}_0^{ee}$. For convenience, we rescale at M_τ the value of α_s obtained for different values of M_0.

- **Standard SVZ expansion**

◇ Fit of the data points

The analysis is shown in Fig. 9.93 (lower set of points) for (FO) and (CI) perturbative series to order α_s^4. We observe a stability for M_0 between 1.5 and 1.7 GeV. A least square fit of the data points gives:

$$\begin{aligned} \alpha_s(M_\tau) &= 0.3081(49)_{\text{fit}} \quad \text{(FO)}, \\ &= 0.3260(47)_{\text{fit}} \quad \text{(CI)}. \end{aligned} \tag{9.268}$$

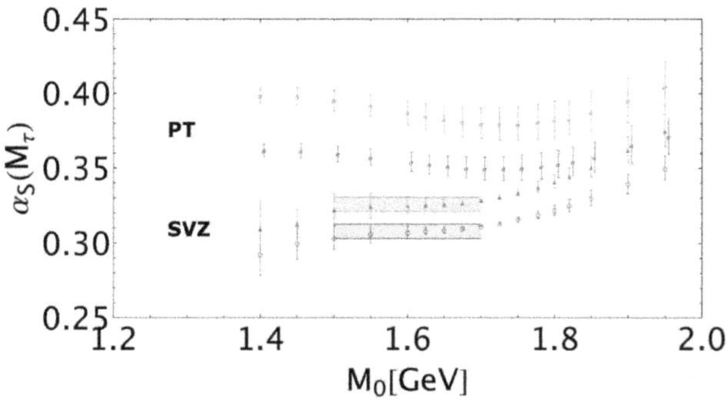

Figure 9.93. (a) Value of $\alpha_s(M_\tau)$ from the BNP lowest moment for different values of M_0. The red (green) points correspond to (FO) [respectively (CI)] PT series. The lower curves correspond to the SVZ expansion. The upper ones correspond to the case of zero value of the QCD condensates.

◇ Estimate of the α_s^5 corrections

Noting that the coefficient of α_s for the Adler D-function grows geometrically [234]:

$$D(Q^2) = \sum_n a_s^n c_n\colon \quad c_0 = c_1 = 1, \quad c_2 = 1.656,$$
$$c_3 = 6.37, \quad c_4 = 49.09, \tag{9.269}$$

we deduce:

$$c_4 \approx c_3^2 \longrightarrow 2\ c_5 \simeq (c_3/c_2)\, c_4^2 \simeq 228. \tag{9.270}$$

This value of c_5 is comparable with some other estimates from FAC: $\simeq 275$ [441], conformal mappings: $= 255$ [550], Padé approximants: $= (277\pm51)$ [551] and a linear behaviour of FO coefficients: $\simeq 283$ [552]. Writing the QCD expression of the moment as:

$$\mathcal{R}_0^{ee} = \sum_n a_s^n (g_n + c_n), \tag{9.271}$$

where g_n comes from a RG resummation [547, 553], one obtains:

$$\begin{aligned} g_5 &= -780 \longrightarrow 2\ \Delta\alpha_s^5(M_\tau) \simeq \pm 71 \times 10^{-4} \quad \text{(FO)},\\ &= 0 \longrightarrow 2\ \Delta\alpha_s^5(M_\tau) \simeq \pm 62 \times 10^{-4} \quad \text{(CI)}. \end{aligned} \tag{9.272}$$

◇ Final value of α_s using the SVZ expansion

Adding the different corrections, we deduce:

$$\begin{aligned} \alpha_s(M_\tau)|_{e^+e^-}^{\rm SVZ} &= 0.3081(49)_{\rm fit}(71)_{\alpha_s^5} \longrightarrow \alpha_s(M_Z)|_{e^+e^-}^{\rm SVZ}\\ &= 0.1170(6)(3)_{\rm evol} \quad \text{(FO)}\\ &= 0.3260(47)_{\rm fit}(62)_{\alpha_s^5} \longrightarrow \alpha_s(M_Z)|_{e^+e^-}^{\rm SVZ}\\ &= 0.1192(6)(3)_{\rm evol} \quad \text{(CI)}, \end{aligned} \tag{9.273}$$

where the sum of non-perturbative corrections at the τ-mass is given in Table 9.22. We note that the increase of $|d_6|$ has slightly decreased the value of α_s compared to the one in Ref. [18].

• QCD models without power corrections

Here, we analyze a QCD model without power corrections. We show the result in Fig. 9.93 (upper set of curves). We obtain at the minimum $M_0 \simeq 1.75\,\text{GeV}$ of the curves:

$$\begin{aligned}
\alpha_s(M_\tau)|^{\rm PT}_{e^+e^-} &= 0.3579(115)_{\rm fit}(34)_{\alpha_s^5} \longrightarrow \alpha_s(M_Z)|^{\rm PT}_{e^+e^-} \\
&= 0.1228(12)(3)_{\rm evol} \quad ({\rm FO}), \\
&= 0.3855(117)_{\rm fit}(83)_{\alpha_s^5} \longrightarrow \alpha_s(M_Z)|^{\rm PT}_{e^+e^-} \\
&= 0.1255(13)(3)_{\rm evol} \quad ({\rm CI}),
\end{aligned} \tag{9.274}$$

This result is about 4.6 σ higher than the PD23 weighted average [33]:

$$\langle \alpha_s(M_Z) \rangle = 0.1178(5), \tag{9.275}$$

and disfavours some QCD models without power corrections.

• Update of α_s from the lowest BNP moment of τ-decay

Within the new set of non-perturbative contributions, we update the determination of α_s from the vector (V) component of τ-decay. The QCD expression of the lowest moment reads [79]:

$$\mathcal{R}^{\rm qcd}_{\tau,V} = \frac{3}{2}\,|V_{ud}|^2 S_{ew}\left[1 + \delta'_{ew} + \delta_0^{(0)} + \delta_{ud}^{(D=4)} - 6\frac{d_6}{M_\tau^6} - 4\frac{d_8}{M_\tau^8}\right], \tag{9.276}$$

where $\delta_0^{(0)}$ is the PT corrections and $\delta_{ud}^{(D=4)}$ is the dimension four contributions given in Eq. (3.11) of Ref. [79]. The values of d_6 and d_8 condensate contributions are given in Table 9.20, while the electroweak parameters are [79]:

$$|V_{ud}| = 0.97425(22), \quad S_{ew} = 1.0194, \quad \delta'_{ew} = 0.0010. \tag{9.277}$$

Using the updated ALEPH data [506]:

$$\mathcal{R}^{\rm exp}_{\tau,V} = 1.782 \pm 0.009, \tag{9.278}$$

we obtain, from the lowest moment:

$$\begin{aligned}\alpha_s(M_\tau)|_{\tau,V} &= 0.3128(19)_{\rm fit}(77)_{\alpha_s^5} \longrightarrow \alpha_s(M_Z)|_{\tau,V}\\ &= 0.1176(10)(3)_{\rm evol} \quad ({\rm FO}),\\ &= 0.3291(25)_{\rm fit}(65)_{\alpha_s^5} \longrightarrow \alpha_s(M_Z)|_{\tau,V}\\ &= 0.1196(8)(3)_{\rm evol} \quad ({\rm CI}),\end{aligned} \tag{9.279}$$

- **Average of α_s from e^+e^- and τ-decay**

One can note an excellent agreement between the value of $\alpha_s(M_\tau)$ from e^+e^- in Eq. (9.273) and the one from the vector component of τ-decay in Eq. (9.279). We consider as a final value of α_s from the present analysis based on the lowest τ-decay-like moment the average of the results in Eqs. (9.273) and (9.279) which is:

$$\begin{aligned}\langle\alpha_s(M_\tau)\rangle &= 0.3094(58) \longrightarrow \langle\alpha_s(M_Z)\rangle = 0.1172(8)(3)_{\rm evol}\\ &= 0.3273(52) \longrightarrow \qquad\qquad\quad = 0.1194(6)(3)_{\rm evol}\end{aligned} \tag{9.280}$$

This mean value is in excellent agreement with the latest PDG average given in Eq. (9.275).

- **Comparison with α_s from τ-decay within the SVZ expansion**

Estimates of α_s using the vector component of τ-decay data within different τ-decay moments are available in the literature [247, 248, 506, 545], as shown in Table 9.22. One can note that the central value of $\alpha_s(M_\tau)$ from $e^+e^- \to$ hadrons data using the lowest BNP moment is slightly lower than the one from τ-decay data but agrees within the errors. We also note that our value of $\delta^{(NP)}$ at the τ-mass agrees with the OPAL one but higher than the ones using ALEPH data.

Table 9.22. $\alpha_s(M_\tau)$ within the SVZ expansion from the vector (V) component of τ-decay data using some high-τ-moments. For a better comparison with our result where the condensates have been estimated within (FO), we quote the value of $\delta^{(NP)}$ corresponding to (FO) at M_τ. We also show our prediction for the V component of τ-decay using d_6 and d_8 in Table 9.20.

	This work					
	e^+e^-	τ-decay	ALEPH [247]	OPAL [248]	PR [545]	ALEPH [506]
FO	0.3081(86)	0.3128(79)	0.3200(220)	0.3230(160)	0.3200(150)	–
CI	0.3260(78)	0.3291(70)	0.3400(230)	0.3470(220)	0.3370(200)	0.3460(110)
$\delta^{(NP)} \times 10^2$	3.8 ± 0.8	3.7 ± 1.0	2.0 ± 0.3	3.6 ± 0.4	1.7 ± 0.3	2.0 ± 0.3

9.7.11 The quest of $1/Q^2$ and tachyonic gluon mass beyond the SVZ expansion

• The quest of $1/Q^2$

A priori, this question is irrelevant as one does not expect to have a dimension-two condensate $\langle A^2 \rangle$ (A_μ is the gluon field) due to gauge invariance (see, however, Ref. [227]). However, this term is present in some holographic models [228] and in a lattice calculation of the QCD potential [229] and gluon condensates [230,231]. It has also been proposed [232, 233] for a phenomenological parametrization of UV renormalon as an alternative to the large β-approximation. The large β-approximation is not yet fully justified due to the non-observation of the factorial growth and alternate signs of the calculated coefficients of the Adler D-function and some other QCD observables known to order α_s^4. Instead, these calculated coefficients grow as a geometric sum [234].

• Tachyonic gluon mass λ^2

◇ In Refs. [235, 554], a systematic way to account for such a $1/Q^2$ term is the introduction of a tachyonic gluon mass squared λ^2. An estimate of this mass using $e^+e^- \to$ hadrons data and some other data lead to [235–238]:

$$a_s\lambda^2 = -(7 \pm 3) \times 10^{-2}\,\mathrm{GeV}^2, \tag{9.281}$$

which we re-estimate here from the ratio of LSR $\mathcal{R}_{10}^{ee}$. The result is shown in Fig. 9.94, from which we deduce to order α_s^4 at the stability regions:

$$a_s\lambda^2 = -(6.7 \pm 1.5) \times 10^{-2}\,\mathrm{GeV}^2, \tag{9.282}$$

in perfect agreement with the previous result.

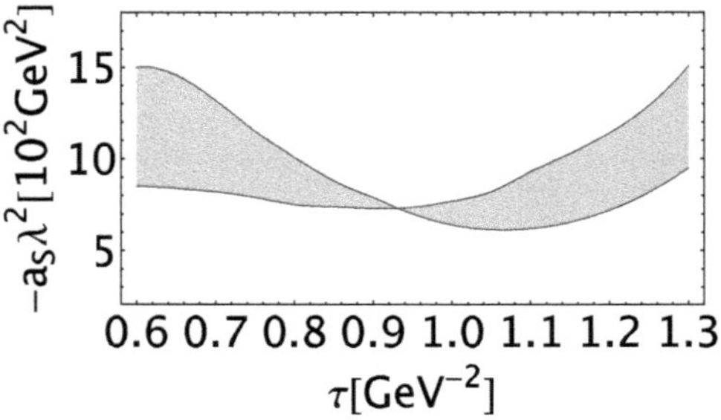

Figure 9.94. The tachyonic gluon mass squared $a_s\lambda^2$ versus M_0.

◇ Such a value of λ^2 resolves the hierarchy of the π, ρ mesons and glueball sum rule scales [235]. However, a direct inclusion of this effect to the τ-decay moment is quite delicate. It has been argued in Ref. [235] that this term is dual to the large order terms of the PT series and decreases when more terms in the PT series are added. To check this argument, we study the variation of λ^2 versus the number of α_s term added in the PT series of $\mathcal{R}_{10}$. We show the analysis in Fig. 9.95(a) where we, indeed, see that $|\lambda^2|$ decreases when more α_s^n terms are added in the PT series even at low values of n. A similar feature is observed at large orders in the calculation of the gluon condensate $\langle\alpha_s G^2\rangle$ on the lattice [230, 231]. We have also checked that the change of d_6 does not almost affect the size of λ^2, while we show in Fig. 9.95(b) the effect of $\langle\alpha_s G^2\rangle$. From the above results, we consider that the λ^2 effect to the BNP moment is:

$$\delta_{\lambda^2}^{(2)} \leq 4\%, \tag{9.283}$$

of the PT lowest order one which is much less than the size of sum of the α_s corrections.

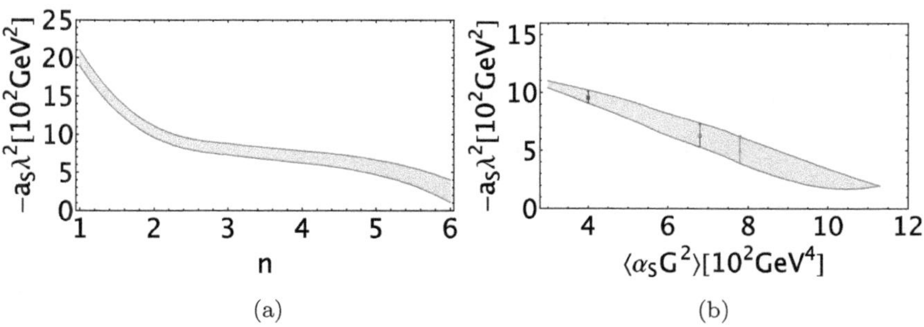

Figure 9.95. (a) $a_s\lambda^2$ for different truncations of the PT series; (b) $a_s\lambda^2$ for different values of the gluon condensate $\langle\alpha_s G^2\rangle$: The blue point denotes the SVZ value of $\langle\alpha_s G^2\rangle$, the red point the value in Eq. (9.254), and the green point the value quoted in Table 9.20.

◇ Assuming that our estimate of α_s^5 correction to the τ-decay moment is a good approximation of the contribution of higher order terms and using the duality property with λ^2 advocated by Ref. [234], we do not add this λ^2 correction in $\mathcal{R}_0^{ee}$ to avoid a double counting.

9.7.12 Some other contributions beyond the SVZ expansion

• Small size instantons

◇ Reference [446] found that these corrections can be parametrized as:

$$\delta^{\text{inst}} \simeq \left(\frac{3.64\,\Lambda}{M_\tau}\right)^9 \left(\frac{\hat{m}_u\hat{m}_d\hat{m}_s}{M_\tau^3}\right) \simeq 1.6\times 10^{-8}, \tag{9.284}$$

where [74] $\hat{m}_u = 3.1(2)$ MeV, $\hat{m}_d = 6.1(4)$ MeV and $\hat{m}_s = (114.3 \pm 6.4)$ MeV are the renormalization group invariant light quark masses and we use $\Lambda = 0.34$ GeV. This contribution is completely negligible.

◇ In Ref. [196], the instanton effect is introduced via the constituent quark mass:

$$m_{\text{const}} = m_{\text{current}} - \frac{2}{3}\pi^2\langle\bar{\psi}\psi\rangle\rho^2, \tag{9.285}$$

where ρ is the instanton size. The authors conclude that at the τ-mass, the effect behaves as $1/M_\tau^{18}$ and is

$$\delta^{\text{inst}} \simeq 0.03 - 0.05, \tag{9.286}$$

which is comparable with the $D = 6$ dimension condensate contribution. However, one should note that this effect depends crucially on the value of the instanton size ρ, which is not under good control.

◇ Reference [357] has tested the validity of Ref. [196] prediction by noting that Ref. [196] result contributes to the $V - A$ τ-decay rate as:

$$\delta^{\rm inst}_{V-A} \simeq 0.09\text{--}0.15, \tag{9.287}$$

while the ALEPH data [247] only allow the range of values:

$$\delta^{\rm inst}_{V-A} \simeq (1 \pm 3) \times 10^{-3}. \tag{9.288}$$

Using the (accepted) relation:

$$\delta^{\rm inst}_{V-A} \simeq \frac{\delta^{\rm inst}_{V-A}}{20}, \tag{9.289}$$

one concludes that:

$$\delta^{\rm inst}_{V} \simeq (0.5 \pm 1.5) \times 10^{-3}, \tag{9.290}$$

which is less than 5% of the d_6 contribution to the $\mathcal{R}^{ee}_\tau$ decay rate. Moreover, the result of Ref. [196] is much larger than the direct fit of the dimension $D = 18$ term in the OPE quoted in Table 9.20.

• Duality violation

◇ This effect is maintained by Refs. [448, 544]. It corresponds to an additional QCD contribution to the spectral function introduced by hand as:

$$\Delta {\rm Im}\, \Pi(t)|_{DV} \sim e^{-(\delta+\gamma t)} \sin(\alpha + \beta t), \tag{9.291}$$

above a certain threshold: $t_c \equiv \hat{s}_0 \simeq 1.5\,{\rm GeV}^2$, where δ, γ, α, β are free unknown fitted parameters.

◇ The eventual presence of such a term in QCD [556] and in some toy model [557] has been discussed at large N_c with an infinite number of narrow resonances in order to modelize the data in the Minkowski space (see also Ref. [555] for QCD in two dimensions). *This oscillating term in the Minkowski space is expected to be dual to an asymptotic exponential behaviour of power corrections in the Euclidean region* [556].

◇ The value of $\hat{s}_0 \equiv t_c$ can be fixed either from LSR of from the lowest FESR moment (see Section 9.7.4).

◇ Using a ρ-meson dominance, one obtains [18, 521]:

$$\hat{s}_0 = 1.5 \text{ GeV}^2, \tag{9.292}$$

◇ Using the $e^+e^- \to$ hadrons data, until $\sqrt{t_c} \simeq 1.875\,\text{GeV}$, the duality condition requires:

$$\hat{s}_0 \equiv t_c = (5.1 \sim 5.6) \text{ GeV}^2, \tag{9.293}$$

as shown in Fig. 9.80 from SN23 [18], where we plot the ratio:

$$r^{\text{fesr}} \equiv \frac{F_0^{\text{data}}}{F_0^{\text{qcd}}}, \tag{9.294}$$

where $r^{\text{fesr}} = 1$ corresponds to the duality region.

This value does not support the one $1.6\,\text{GeV}^2$ obtained in Ref. [544] using the $e^+e^- \to$ hadrons data (note that in Ref. [522] a value around $2.1\,\text{GeV}^2$ has been obtained). However, one may (intuitively) expect a value of t_c above the data region.

◇ One can inspect that with such a value of $\hat{s}_0$, the additional contribution due to duality violation becomes unobservable when one uses complete data thanks to the exponential weight for its parametrization.

◇ A critical analysis of the effects of DV for values of $\hat{s}_0$ around $1.5\,\text{GeV}^2$ on the determination of the QCD condensates and α_s using τ-decay ALEPH data is explicitly discussed in Ref. [558].

◇ One should also note that power corrections in the Euclidean space estimated previously in Table 9.20 up to $D = 20$ do not show any exponential behaviour which disfavours by *a duality analytic continuation* the presence of this oscillating DV term in the Minkowski space [556].

9.8 QCD Condensates and α_s from τ-Decays

In this section, we review the paper in Ref. [559] where we pursue the previous analysis from $e^+e^- \to$ hadrons and vector (V) component of τ-decays discussed in the previous section by studying here the axial-vector (A) and V–A components of τ-decays.

9.8.1 Abstract

We improve the determinations of the QCD condensates within the SVZ expansion in the axial-vector (A) channel using the ratio of Laplace sum rule (LSR) $\mathcal{R}^A_{10}(\tau)$ within stability criteria and τ-like higher moments $\mathcal{R}_{n,A}$ within stability for arbitrary τ-mass squared s_0. We find the same violation of the factorization by a factor 6 of the four-quark condensate as from $e^+e^- \to$ hadrons data. One can note a systematic alternate sign and no exponential growth of the size of these condensates. Then, we extract α_s from the lowest τ-decay-like moment. We obtain to order α_s^4 the conservative value from the s_0-stability until M_τ^2: $\alpha_s(M_\tau)|_A = 0.3178(66)$ (FO) and 0.3380 (44) (CI), leading to $\alpha_s(M_Z)|_A = 0.1182(8)_{\rm fit}(3)_{\rm evol.}$ (FO) and $0.1206(5)_{\rm fit}(3)_{\rm evol.}$ (CI). We extend the analysis to the V–A channel and find: $\alpha_s(M_\tau)|_{V-A} = 0.3135(83)$ (FO) and 0.3322 (81) (CI), leading to $\alpha_s(M_Z)|_{V-A} = 0.1177(10)_{\rm fit}(3)_{\rm evol.}$ (FO) and $0.1200(9)_{\rm fit}(3)_{\rm evol.}$ (CI). We observe that in different channels ($e^+e^- \to$ hadrons, A, V–A), the extraction of $\alpha_s(M_\tau)$ at the observed τ-mass leads to an overestimate of its value. Our determinations from these different channels lead to the mean: $\alpha_s(M_\tau) = 0.3140(44)$ (FO) and 0.3346 (35) (CI), leading to $\alpha_s(M_Z) = 0.1178(6)_{\rm fit}(3)_{\rm evol.}$ (FO) and $0.1202(4)_{\rm fit}(3)_{\rm evol.}$ (CI). Comparisons with some other results are done.

9.8.2 The axial-vector (A) two-point function

• The two-point function

We shall be concerned with the two-point correlator:

$$\Pi^{\mu\nu}_{V(A)}(q^2) = i\int d^4x\; e^{-iqx}\langle 0|\mathcal{T} J^\mu_{V(A)}(x)\left(J^\nu_{V(A)}(0)\right)^\dagger|0\rangle$$
$$= -(g^{\mu\nu}q^2 - q^\mu q^\nu)\Pi^{(1)}_{V(A)}(q^2) + q^\mu q^\nu \Pi^{(0)}_{V(A)}(q^2) \tag{9.295}$$

built from the T-product of the bilinear axial-vector current of u, d quark fields:

$$J^\mu_{V(A)}(x) =: \bar\psi_u\gamma^\mu(\gamma_5)\psi_d. \tag{9.296}$$

The upper indices (0) and (1) correspond to the spin of the associated hadrons. The two-point function obeys the dispersion relation:

$$\Pi_{V(A)}(q^2) = \int_{t>}^{\infty} \frac{dt}{t - q^2 - i\epsilon} \frac{1}{\pi} \mathrm{Im}\Pi_{V(A)}(t) + \cdots, \tag{9.297}$$

where $\cdots$ are subtraction constants polynomial in q^2 and $t>$ is the hadronic threshold.

◇ QCD expression of the two-point function

Within the SVZ expansion [3], the two-point function can be expressed in terms of the sum of higher and higher quark and gluon condensates:

$$\begin{aligned} 4\pi^2 \Pi_H(-Q^2, m_q^2, \mu) &= \sum_{D=0,2,4,\ldots} \frac{C_{D,H}(Q^2, m_q^2, \mu)\langle O_{D,H}(\mu)\rangle}{(Q^2)^{D/2}} \\ &\equiv \sum_{D=0,2,4,\ldots} \frac{d_{D,H}}{(Q^2)^{D/2}}, \end{aligned} \tag{9.298}$$

where $H \equiv V(A)$, μ is the subtraction scale which separates the long (condensates) and short (Wilson coefficients) distance dynamics and m_q is the quark mass.

◇ In the chiral limit [79, 331, 332]:

$$\begin{aligned} d_{2,A} &= d_{2,V} = 0, \\ d_{4,A} &= d_{4,V} = \frac{\pi}{3}\langle \alpha_s G^2\rangle \left(1 + \frac{7}{6} a_s\right), \\ d_{6,A} &= -\left(\frac{11}{7}\right) d_{6,V} = \frac{1408}{81}\pi^3 \rho \alpha_s \langle \bar{\psi}\psi\rangle^2, \\ d_{8,A} &\approx d_{8,V} \approx -\frac{39}{162}\pi \langle \alpha_s G^2\rangle^2, \end{aligned} \tag{9.299}$$

where ρ measures the deviation from the factorization of the four-quark condensates and the last equality for d_8 is based on vacuum saturation estimate of some classes of computed dimension-eight diagrams.

◇ The perturbative expression of the spectral function is known to order α_s^4 [438, 441]. It reads for three flavours:

$$4\pi \, \mathrm{Im}\Pi_H(t) = 1 + a_s + 1.6398a_s^2 - 10.2839a_s^3 - 106.8798a_s^4 + \mathcal{O}(a_s^5), \tag{9.300}$$

where:

$$a_s \equiv \frac{\alpha_s}{\pi} = \frac{2}{-\beta_1 \mathrm{Log}(t/\Lambda^2)} + \cdots, \tag{9.301}$$

where $-\beta_1 = (1/2)(11 - 2n_f/3)$ is the first coefficient of the β-function and n_f is the number of quark flavours; $\cdots$ stands for higher order terms which can, e.g., be found in Ref. [7]. We shall use the value $\Lambda = (342 \pm 8)$ MeV for $n_f = 3$: from the PDG average [33].

◇ Spectral function from the data

We shall use the recent ALEPH data [528] in Fig. 9.96 for the spectral function $a_1(s)$.

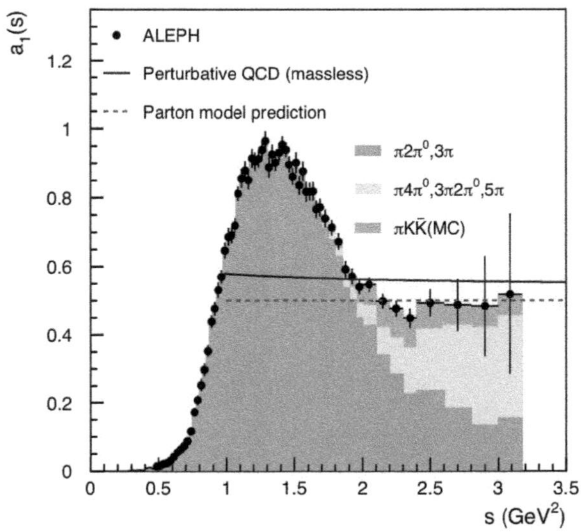

Figure 9.96. ALEPH data of the axial-vector spectral function.

Like in the case of the vector spectral function from $e^+e^- \to$ hadrons [18], we subdivide the region from 3π threshold to $M_\tau^2 = 3.16$

GeV^2 into different subregions in s (units in $[\text{GeV}^2]$):

$$s = [0.4, 0.8], \quad [0.8, 1.29], \quad [1.29, 1.42], \quad [1.42, 2.35], \quad [2.35, 3.16] \tag{9.302}$$

and fit the data with third-order polynomials using the optimized Mathematica program FindFit except for $[1.29, 1.42]$ where a second-order polynomial is used. We show the different fits in Fig. 9.97.

9.8.3 The ratio of Laplace sum rule (LSR) moments

Like in the case of vector channel, we shall use here the ratio of LSR moments [3, 69, 70]:

$$\mathcal{R}_{10}^A(\tau) \equiv \frac{\mathcal{L}_1^c}{\mathcal{L}_0^c} = \frac{\int_{t>}^{t_c} dt\ e^{-t\tau} t\, \frac{1}{\pi}\, \text{Im}\Pi_H(t, \mu^2, m_q^2)}{\int_{t>}^{t_c} dt\ e^{-t\tau}\, \frac{1}{\pi}\, \text{Im}\Pi_H(t, \mu^2, m_q^2)}, \tag{9.303}$$

where τ is the LSR variable and $t>$ is the hadronic threshold. Here, t_c is the threshold of the "QCD continuum" which parametrizes, from the discontinuity of the Feynman diagrams, the spectral function $\text{Im}\,\Pi_H(t, m_q^2, \mu^2)$. m_q is the quark mass and μ is an arbitrary subtraction point.

• QCD expression of the LSR moments

To order α_s^4, the perturbative (PT) expression of the lowest moment reads [18]:

$$\mathcal{L}_0^{\text{PT}}(\tau) = \frac{3}{2}\tau^{-1}\left[1 + a_s + 2.93856\, a_s^2 + 6.2985\, a_s^3 + 22.2233\, a_s^4\right]. \tag{9.304}$$

Then, taking its derivative in τ, one gets $\mathcal{L}_1(\tau)$ and then their ratio $\mathcal{R}_{10}(\tau)$.

From Eq. (9.298), one can deduce the non-perturbative contribution to the lowest LSR moment (see Eq. (9.243)), from which one can deduce $\mathcal{L}_1^{NPT}$ and $\mathcal{R}_{10}^A$.

• $d_{6,A}$ and $d_{8,A}$ from $\mathcal{R}_{10}^A$

We show in Fig. 9.98 the τ-behaviour of $\mathcal{R}_{10}^A$ for values of t_c around the physical τ-lepton mass squared where the effect of t_c is negligible.

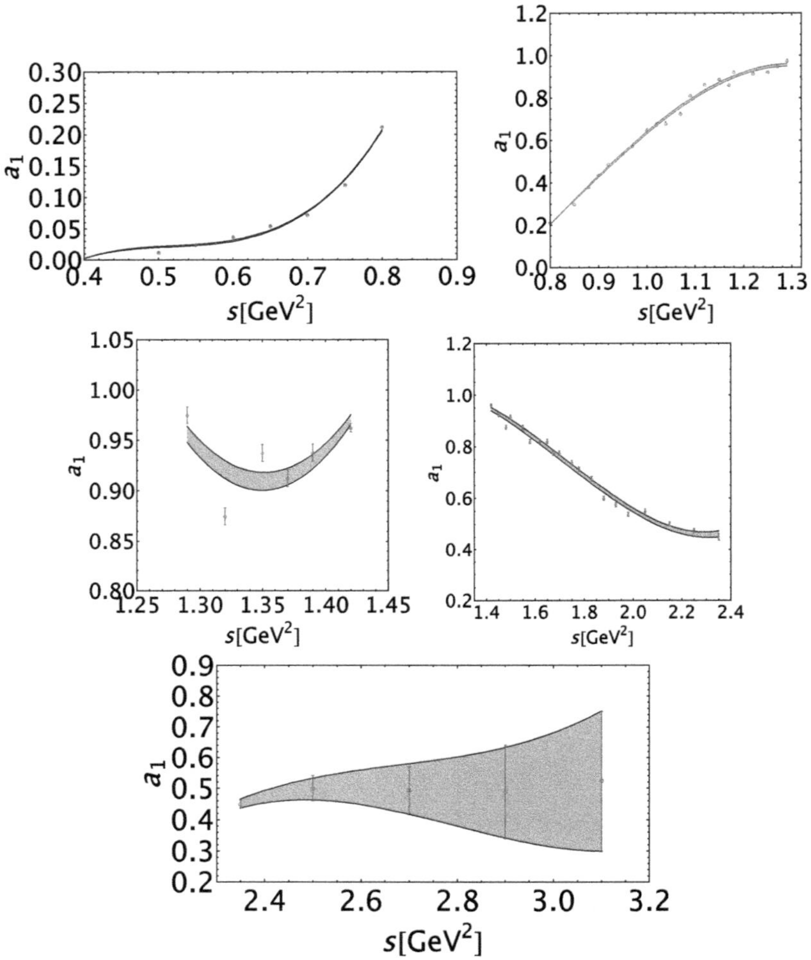

Figure 9.97. Fit of the data using a second- and third-order polynomial fits.

For determining $d_{6,A}$ and $d_{8,A}$, we use as input the more precise value of $\langle\alpha_s G^2\rangle$ from the heavy quark mass splittings and some other sum rules [290–292, 517]:

$$\langle\alpha_s G^2\rangle = (6.39 \pm 0.35) \times 10^{-2}\,\mathrm{GeV}^4. \tag{9.305}$$

and perform a two-parameter fit ($d_{6,A}$ and $d_{8,A}$) by confronting the phenomenological and QCD side of $\mathcal{R}^A_{10}$ for different values of $\wedge$. The QCD coupling α_s is evaluated at the LSR sum rule scale τ where we use $\wedge = (342 \pm 8)$ MeV for $n_f = 3$ flavours deduced from the PDG world average. The results of the analysis are shown in Fig. 9.98.

There is an inflexion point around $\tau \simeq 2.5\,\text{GeV}^{-2}$ at which we extract the optimal values of the condensates. One should note that at this scale the OPE converges quite well in the vector channel [18]. Then, we may expect that for the axial-vector the same feature occurs. We obtain the optimal result:

$$
\begin{aligned}
d_{6,A} &= (33.5 \pm 3.0 \pm 2.7) \times 10^{-2}\,\text{GeV}^6, \\
d_{8,A} &= -(47.2 \pm 2.8 \pm 3.2) \times 10^{-2}\,\text{GeV}^8,
\end{aligned}
\tag{9.306}
$$

where the first error comes from the fitting procedure and the second one from the localization of $\tau \simeq (2.5 \pm 0.1)\,\text{GeV}^{-2}$.

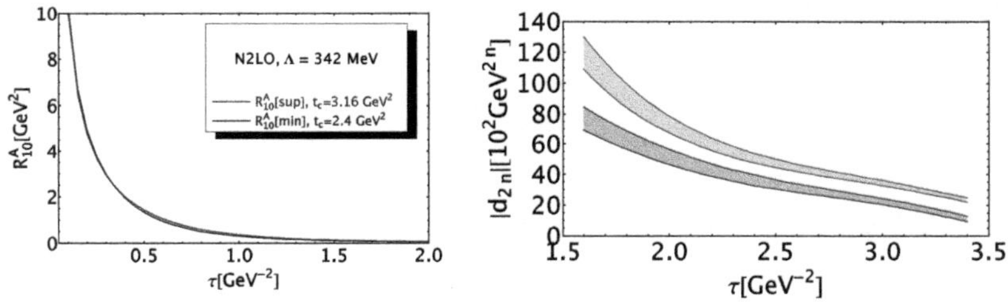

(a) (b)

Figure 9.98. (a) R_{10}^A versus the LSR variable τ; (b) $d_{6,A}$ and $d_{8,A}$ from $\mathcal{R}_{10}^A$.

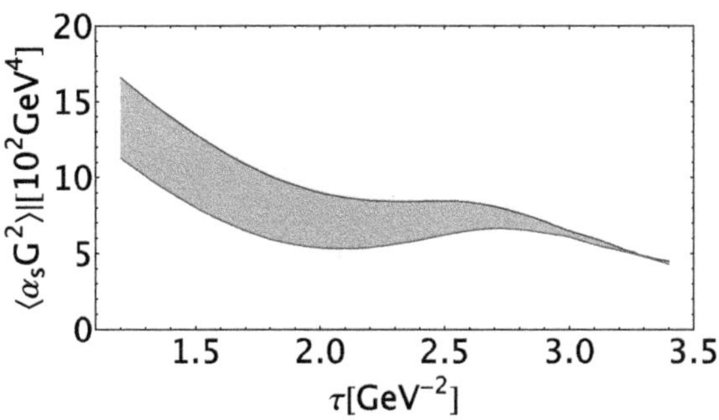

Figure 9.99. $\langle \alpha_s G^2 \rangle$ versus the LSR variable τ.

- **$\langle \alpha_s G^2 \rangle$ from $\mathcal{R}_{10}^A$**

We use the previous values of $d_{6,A}$ and $d_{8,A}$ into $\mathcal{R}_{10}^A$ and we re-extract $\langle \alpha_s G^2 \rangle$ using a one-parameter fit. The analysis is shown in

Fig. 9.99. We obtain:

$$\langle \alpha_s G^2 \rangle = (6.9 \pm 1.5) \times 10^{-2}\,\text{GeV}^4, \tag{9.307}$$

in good agreement with the one in Eq. (9.305) used previously as input. This result also indicates the self-consistency of the set of condensates entering in the analysis.

9.8.4 The lowest BNP moment $\mathcal{R}_{0,A}$

The QCD expression of $\mathcal{R}_{0,A}$ including the dimension six ($d_{6,A}$) and eight ($d_{8,A}$) condensates within the SVZ expansion [3] can be deduced from Ref. [79]. To simplify for the reader, we give its expression in the chiral limit[28]:

$$\mathcal{R}_{0,A} = \frac{N_c}{2}|V_{ud}|^2\, S_{ew} \left\{ 1 + \delta'_{ew} + \delta_A^{(0)} + \sum_{D=1,2,\ldots} \delta_A^{(2D)} \right], \tag{9.308}$$

where the electroweak factors and corrections are:

$$V_{ud} = 0.97418, \quad S_{ew} = 1.019, \quad \delta'_{ew} = 0.0010. \tag{9.309}$$

The QCD corrections copied from BNP are as follows:

◇ Perturbative corrections to order α_s^4:

$$\begin{aligned} \delta_A^{(0)}|_{FO} &= a_s + 5.2023\, a_s^2 + 26.366\, a_s^3 + 127.079\, a_s^4, \\ \delta_A^{(0)}|_{CI} &= 1.364\, a_s + 2.54\, a_s^2 + 9.71\, a_s^3 + 64.29\, a_s^4, \end{aligned} \tag{9.310}$$

for fixed order (FO) [79] and contour improved (CI) [547].

Observing that the PT series grows geometrically [234] from the calculated coefficients in different channels, we estimate the a_s^5 coefficient to be [18]

$$\delta_5^{FO} \approx \pm 552\, a_s^5, \quad \delta_5^{CI} \approx \pm 228\, a_s^5, \tag{9.311}$$

which one can consider either to be the error due to the unknown higher order terms of the series or (more optimistically) to be the estimate of the uncalculated α_s^5 coefficient.

---

[28] We use the complete expression including quark masses and α_s corrections in the numerical analysis.

◇ Power corrections up to $d_{8,A}$:
They read as follows:

$$\delta_A^{(2)} = -8(1+\frac{16}{3}a_s)\frac{(m_u^2+m_d^2)}{M_0^2} - 4\left(1+\frac{25}{3}a_s\right)\frac{(m_u m_d)}{M_0^2},$$

$$\delta_A^{(2)}|_{tach} = -2\times 1.05\frac{a_s\lambda^2}{M_0^2},$$

$$\delta_A^{(4)} = \frac{11\pi}{4}a_s^2\frac{\langle\alpha_s G^2\rangle}{M_0^4} + 32\pi^2\left(1+\frac{63}{16}a_s^2\right)\frac{(m_u+m_d)\langle\bar\psi_u\psi_u\rangle}{M_0^4} - 8\pi^2\sum_k\frac{m_k\langle\bar\psi_k\psi_k\rangle}{M_0^4},$$

$$\delta_A^{(6)} = -6\frac{d_{6,A}}{M_0^6},\quad \delta_A^{(8)} = -4\frac{d_{8,A}}{M_0^8}, \tag{9.312}$$

where $s_0 \equiv M_0^2$, while $d_{6,A}$ and $d_{8,A}$ have been defined in Eq. (9.299). We have assumed $\langle\bar\psi_u\psi_u\rangle = \langle\bar\psi_d\psi_d\rangle$. We shall not include the $D = 2$ contribution due to an eventual tachyonic gluon mass within the standard OPE. Using duality [234], this term can be included in the estimate of the non-calculated higher order terms of the PT series discussed previously. Instanton contributions are expected to have higher dimensions and their contributions can be safely neglected [18].

- $d_{6,A}$ **and** $d_{8,A}$ **condensates from** $\mathcal{R}_{0,A}$

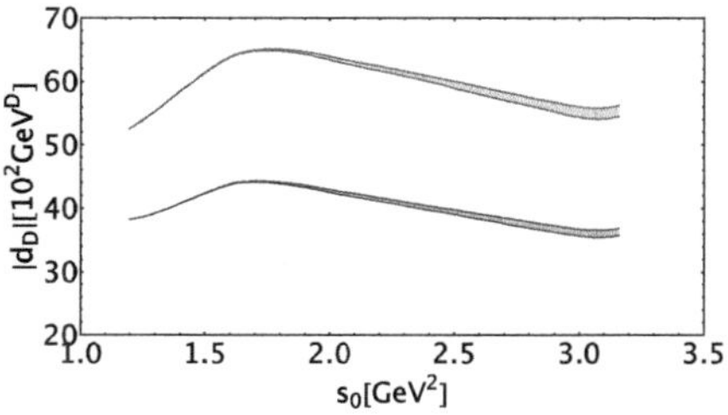

(a)

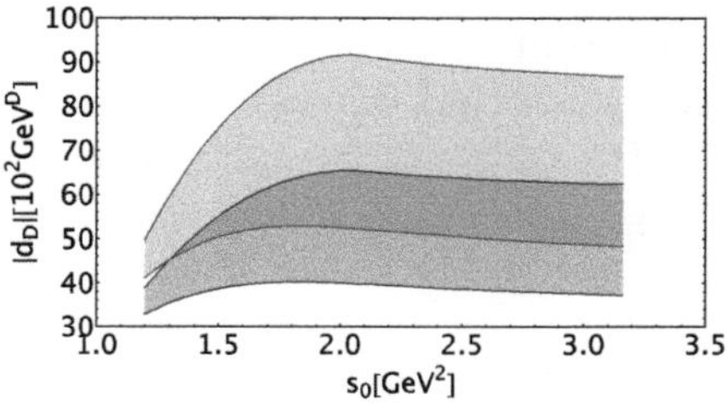

(b)

Figure 9.100. (a) $|d_{6,A}|$ (lowest curve) and $|d_{8,A}|$ (highest curve) versus M_0^2; (b) similar to (a) but for $|d_{8,A}|$ and $|d_{10,A}|$.

We use a two-parameter fit to extract the values of these condensates and consider the PT series up to order α_s^4. The results versus the hypothetical τ-mass squared $s_0 = M_0^2$ are shown in Fig. 9.100.

We consider as a reliable value the one from $s_0 = 1.93$ GeV^2 beyond the peak of the A_1 meson. We note a stability (minimum) around 3 GeV^2 just below the physical τ mass which we consider as our optimal value:

$$\begin{aligned} d_{6,A} &= (36.2 \pm 0.5) \times 10^{-2}\,\text{GeV}^6, \\ d_{8,A} &= -(55.2 \pm 0.9) \times 10^{-2}\,\text{GeV}^8 \quad (\text{FO}), \\ d_{6,A} &= (33.1 \pm 0.5) \times 10^{-2}\,\text{GeV}^6, \\ d_{8,A} &= -(50.6 \pm 0.8) \times 10^{-2}\,\text{GeV}^8 \quad (\text{CI}). \end{aligned} \tag{9.313}$$

One can note from the analysis that the absolute values of the condensates are slightly higher for FO than for CI. To be conservative, we take the arithmetic average of the FO and CI values and add as a systematic the largest distance between the mean and the individual value:

$$d_{6,A} = (34.6 \pm 1.8) \times 10^{-2}\,\text{GeV}^6, \quad d_{8,A} = -(52.9 \pm 2.4) \times 10^{-2}\,\text{GeV}^8. \tag{9.314}$$

9.8.5 $d_{8,A}$ and $d_{10,A}$ from $\mathcal{R}_{1,A}$

The expression of $\mathcal{R}_{1,A}$ is similar to $\mathcal{R}_{1,V}$ given in Eq. (20) of Ref. [18]. We show the result of the analysis in Fig. 9.100 using FO PT series. One obtains:

$$d_{8,A} = -(51.4 \pm 11.0) \times 10^{-2}\,\text{GeV}^6, \quad d_{10,A} = (70.1 \pm 21.6) \times 10^{-2}\,\text{GeV}^8. \tag{9.315}$$

The results are stable versus s_0 but less accurate than in the case of $\mathcal{R}_{0,A}$ such that one cannot differentiate an FO from CI truncation of the PT series. Then, for higher moments, we shall only consider FO PT series.

9.8.6 Final values of $d_{6,A}$ and $d_{8,A}$

As a final value of $d_{6,A}$, we take the mean of the values in Eqs. (9.306) and (9.314). We obtain:

$$d_{6,A} = (34.4 \pm 1.7) \times 10^{-2}\,\text{GeV}^6, \tag{9.316}$$

while for $d_{8,A}$, we take the mean of the values obtained in Eqs. (9.306), (9.314) and (9.315). We obtain:

$$d_{8,A} = -(51.51 \pm 2.08) \times 10^{-2}\,\text{GeV}^8. \tag{9.317}$$

◇ We note that the relation:

$$d_{6,A} \simeq -(11/7)\,d_{6,V} \tag{9.318}$$

is quite well satisfied within the errors. This result also suggests a violation of the four-quark condensate vacuum saturation (see Eq. (9.299)) similar to the one found from $e^+e^- \to$ hadrons data [18–20]:

$$\rho\alpha_s\langle\bar\psi\psi\rangle^2 = (6.38 \pm 0.30) \times 10^{-4}\,\text{GeV}^6 \quad \longrightarrow 2 \quad \rho \simeq (6.38 \pm 0.30). \tag{9.319}$$

9.8.7 Determination of $\alpha_s(M_\tau)$

We use the previous values of $d_{6,A}$ and $d_{8,A}$ together with the one of $\langle\alpha_s G^2\rangle$ in Eq. (9.305) as inputs in the lowest BNP moment $\mathcal{R}_{0,A}$ in order to determine $\alpha_s(M_\tau)$. We show in Fig. 9.101 the behaviour of $\alpha_s(M_\tau)$ versus a hypothetical τ mass squared $M_0^2 \equiv s_0$. One can note an inflexion point in the region $2.5^{+0.10}_{-0.15}$ GeV^2 at which we extract the optimal result. The conservative result from $s_0 = 2.1$ GeV^2 to M_τ^2 (see Fig. 9.101) is:

$$\begin{aligned}\alpha_s(M_\tau)|_A &= 0.3178(10)(65) \;\longrightarrow\; \alpha_s(M_Z)|_A = 0.1182(8)(3)_{\text{evol}}\ (\text{FO}),\\ &= 0.3380(10)(43) \;\longrightarrow\; \alpha_s(M_Z)|_A = 0.1206(5)(3)_{\text{evol}}\,(\text{CI}).\end{aligned} \tag{9.320}$$

The first error in $\alpha_s(M_\tau)|_A$ comes from the fitting procedure. The second one comes from an estimate of the α_s^5 contribution from Ref. [18]. At the scale s_0 =2.5 GeV^2, the sum of non-perturbative contributions to the moment normalized to the parton model is:

$$\delta_{NP,A} \simeq -(7.9 \pm 1.1) \times 10^{-2}. \tag{9.321}$$

One can note that extracting $\alpha_s(M_\tau)|_A$ at the observed τ-mass tends to overestimate the result:

$$\alpha_s(M_\tau)|_A = 0.3352(40) \quad (\text{FO}), \quad 0.3592(47) \quad (\text{CI}). \tag{9.322}$$

These values extracted at M_τ agree with the ones from Ref. [545] obtained at the same scale. The same feature has been observed in the case of $e^+e^- \to$ hadrons.

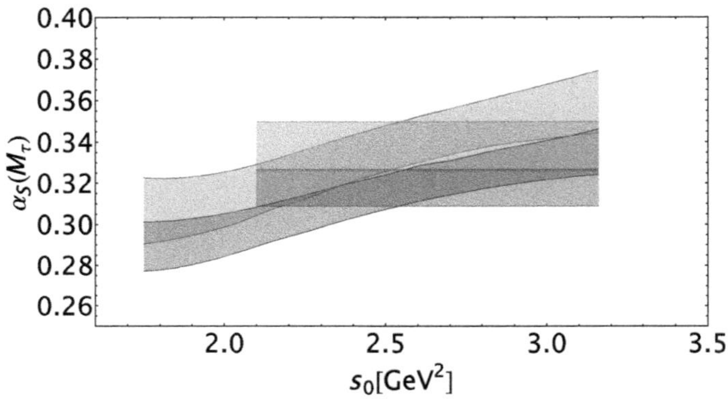

Figure 9.101. $\alpha_s(M_\tau)$ versus the hypothetical τ-mass squared s_0.

9.8.8 Comparison with previous results

We compare our results with the ones from e^+e^- and τ-decay vector channel [18–20] and with the results obtained by different authors in the axial-vector channel:

◇ Our values of $d_{6,A}$ and $d_{8,A}$ are in good agreement within the errors with the ones of Ref. [545] but about two times larger than the ones of Refs. [247, 528].

◇ The value of $d_{8,A}$ suggests that the assumption in Eq. (9.299):

$$d_{8,A} \approx d_{8,V} \tag{9.323}$$

is not satisfied by the fitted values given in Table 9.23.

Table 9.23. Values of the QCD condensates from some other τ-moments at fixed order (FO) PT series and of $\alpha_s(M_\tau)$ for FO and contour improved (CI) PT series.

Channel	d_6	$-d_8$	$\alpha_s(M_\tau)$ FO	$\alpha_s(M_\tau)$ CI	Reference
e^+e^-	$-(26.3 \pm 3.7)$	$-(18.2 \pm 0.6)$	0.3081(86)	0.3260(78)	[18, 20]
V			0.3129(79)	0.3291(70)	[18, 20]
A	34.4 ± 1.7	51.5 ± 2.1	0.3157(65)	0.3368(45)	This work
A	43.4 ± 13.8	59.2 ± 19.7	0.3390(180)	0.3640(230)	[545]
A	19.7 ± 1.0	27 ± 1.2	–	0.3350(120)	[247]
A	9.6 ± 3.3	9.0 ± 5.0	0.3230(160)	0.3470(230)	[248]

9.8.9 High-dimension condensates

To determine the high-dimension condensates, we use the analogue of the moments given in Eq. (19) of Ref. [20] for the vector channel.

• $d_{10,A}$ and $d_{12,A}$ condensates from $\mathcal{R}_{2,A}$

We use a two-parameter fit to extract $(d_{10,A}, d_{12,A})$ from $\mathcal{R}_{2,A}$ and $(d_{12,A}, d_{14,A})$ from $\mathcal{R}_{3,A}$. The s_0 behaviour of the results is given in Fig. 9.102.

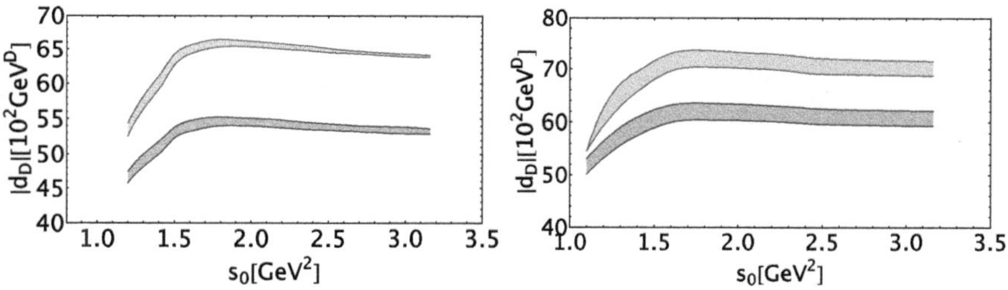

(a) (b)

Figure 9.102. (a) $|d_{10,A}|$ (lowest curve) and $|d_{12,A}|$ (highest curve) versus M_0^2; (b) similar to (a) but for $|d_{12,A}|$ and $|d_{14,A}|$.

We deduce from $\mathcal{R}_{2,A}$:

$$d_{10,A} = (53.9 \pm 0.7) \times 10^{-2}\,\text{GeV}^{10},$$
$$d_{12,A} = -(65. \pm 1.) \times 10^{-2}\,\text{GeV}^{12}, \tag{9.324}$$

inside the stability region 1.93 GeV^2 to M_τ^2. We take as a final value of $d_{10,A}$ the mean from $\mathcal{R}_{1,A}$ in Eq. (9.315) and the one from $\mathcal{R}_{2,A}$ in Eq. (9.324):

$$d_{10,A} = (53.9 \pm 0.7) \times 10^{-2}\,\text{GeV}^{10}. \tag{9.325}$$

• $d_{2n,A}$ and $d_{2(n+1),A}$ condensates

We do the same procedure as previously for higher dimension condensates $n \geq 6$. The s_0 behaviours of the condensates are shown in Figs. 9.103 and 9.104. The results are summarized in Table 9.24.

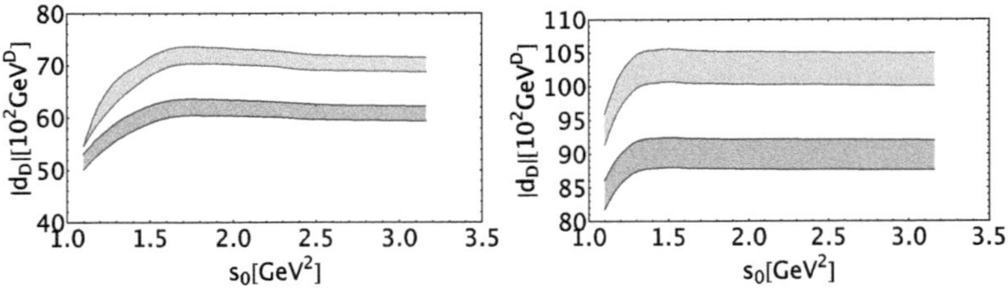

(a) (b)

Figure 9.103. (a) $|d_{14,A}|$ (lowest curve) and $|d_{16,A}|$ (highest curve) versus M_0^2; (b) similar to (a) but for $|d_{16,A}|$ and $|d_{18,A}|$.

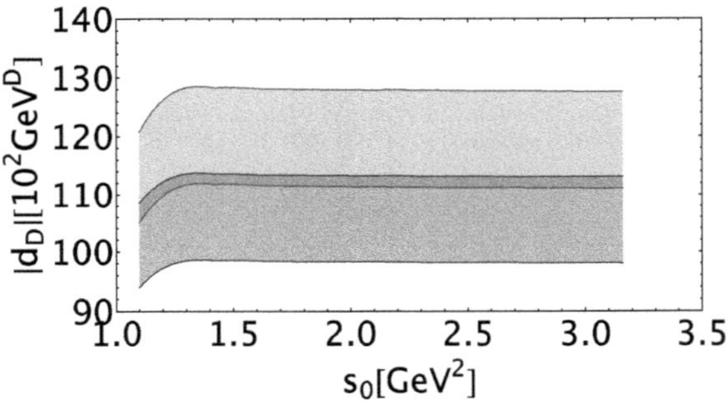

Figure 9.104. $|d_{18,A}|$ (lowest curve) and $|d_{12,A}|$ (highest curve) versus M_0^2.

◇ One can note that the condensates in the A channel has alternate sign.

◇ Their size is almost constant and more accurate than previous determinations in the literature. There is not also any sign of an exponential growth. This feature in the Euclidean region does not favour a sizeable duality violation in the time-like region [556].

Table 9.24. Values of the QCD condensates of dimension D in units of 10^{-2} GeV^D from this work and some other estimates.

$d_{6,A}$	$-d_{8,A}$	$d_{10,A}$	$-d_{12,A}$	$d_{14,A}$	$-d_{16,A}$	$d_{18,A}$	$-d_{20,A}$	Reference
34.4 ± 1.7	51.5 ± 2.1	53.9 ± 0.7	63.3 ± 1.8	77.0 ± 6.0	93.1 ± 4.0	104.3 ± 7.7	119.7 ± 8.2	This work
43.4 ± 13.8	59.2 ± 19.7	63.2 ± 33.6	43.4 ± 31.6					[545]
19.7 ± 1.0	27 ± 1.2							[247]
9.6 ± 3.3	9.0 ± 5.0							[248]

9.8.10 The V-A channel

• The two-point function

It corresponds to the $V - A$ quark current:

$$J^\mu_{V-A}(x) =: \bar{\psi}_u \gamma^\mu (1 - \gamma_5)\psi_d. \tag{9.326}$$

In the chiral limit, the QCD expression of the corresponding two-point function is similar to the one in Eq. (9.299), where

$$\begin{aligned} d_{6,V-A} &= \frac{1}{2}(d_{6,V} + d_{6,A}) = -\frac{1}{2}\left(\frac{4}{7}\right) d_{6,V}, \\ &= \frac{512}{27}\pi^3 \rho \alpha_s \langle\bar{\psi}\psi\rangle^2 = (37.5 \pm 1.9) \times 10^{-2}\,\text{GeV}^6, \\ d_{8,V-A} &= \frac{1}{2}(d_{8,V} + d_{8,A}) \simeq -(14.5 \pm 2.2) \times 10^{-2}\,\text{GeV}^8. \end{aligned} \tag{9.327}$$

• Data handling of the spectral function

The spectral function has been measured by ALEPH [247] which we show in Fig. 9.105. In order to fit the data, we proceed like in the

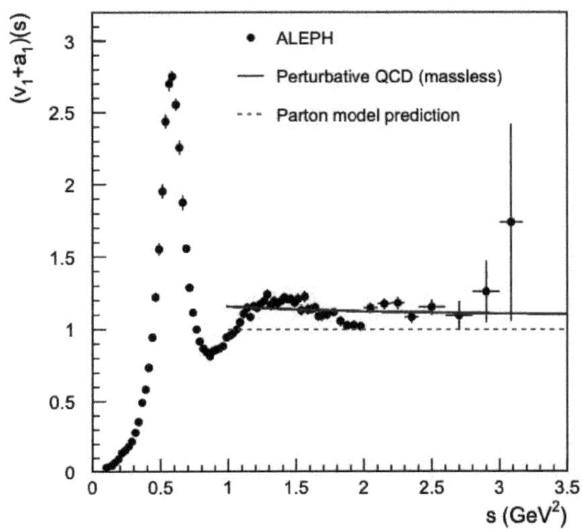

Figure 9.105. V–A spectral function measured by ALEPH.

case of the axial-vector channel by subdividing the region into five subregions:

$$[4m_\pi^2, 0.585],\ [0.585, 0.85],\ [0.85, 1.45],\ [1.45, 1.975],\ [1.975, M_\tau^2], \tag{9.328}$$

in units of GeV^2. We use third-order polynomials. The results of the fits are given in Fig. 9.106.

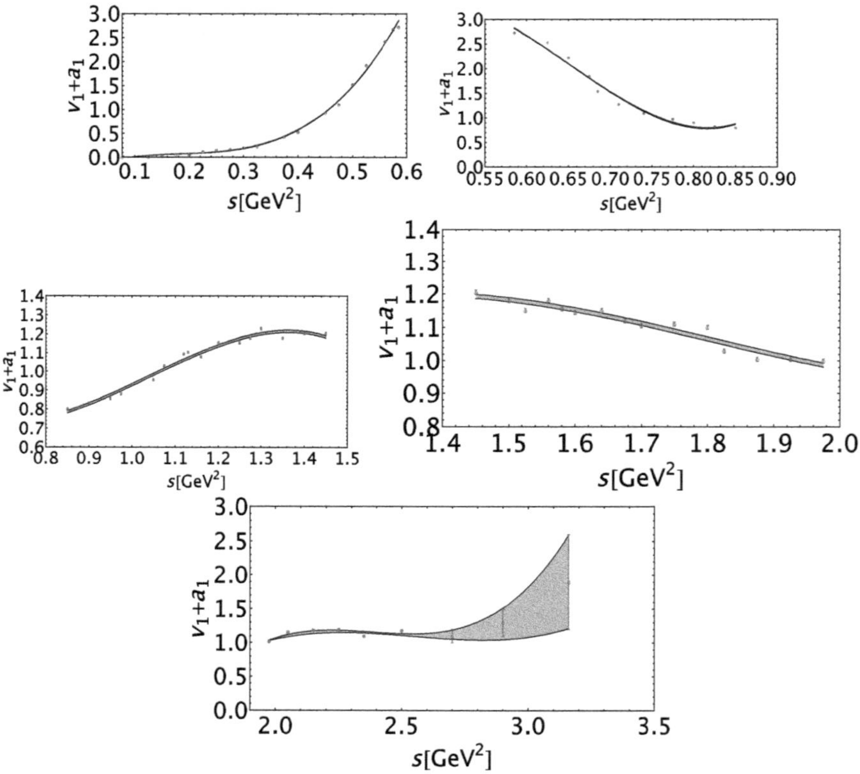

Figure 9.106. Fit of the data using second- and third-order polynomial fits.

• The lowest BNP moment $\mathcal{R}_{0,V-A}$

In the chiral limit, their expression is similar to the previous ones for the axial-vector current modulo an overall factor 2 and the change into the condensate contributions $d_{D,V-A}$ of dimension D. We show

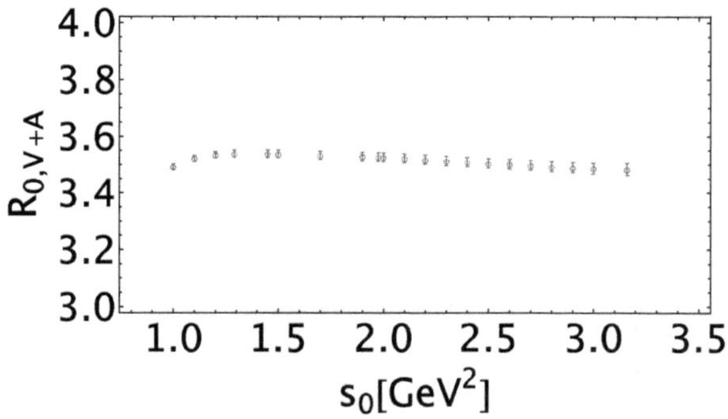

Figure 9.107. s_0-behaviour of the experimental moment $\mathcal{R}_{0,V-A}$.

the s_0-behaviour of the moment in Fig. 9.107. For $s_0 = M_\tau^2$, we obtain:

$$\mathcal{R}_{0,V-A} = 3.484 \pm 0.022, \tag{9.329}$$

to be compared with the ALEPH data [528]:

$$\mathcal{R}_{0,V-A}|_{\text{Aleph}} = 3.475 \pm 0.011, \tag{9.330}$$

where our error is larger, which may come due to the fact that we have separately fitted the upper and lower values of the data.

- **The $d_{6,V-A}$ and $d_{8,V-A}$ condensates**

We shall use the values of the corresponding condensates from the $e^+e^- \to$ hadrons in Refs. [18–20] and the ones from the axial-vector channel obtained in the previous section. They read as follows:

$$d_{6,V-A} \equiv \frac{1}{2}\,(d_{6,V} + d_{6,A}) = +(3.6 \pm 0.5) \times 10^{-2}\,\text{GeV}^6,$$
$$d_{8,V-A} \equiv \frac{1}{2}\,(d_{8,V} + d_{8,A}) = -(14.5 \pm 2.2) \times 10^{-2}\,\text{GeV}^8, \tag{9.331}$$

where the error is the largest relative % error from each channel. We note (as can be found in Ref. [79]) that for $\mathcal{R}_{0,V-A}(s_0)$, the contribution of $\langle \alpha_s G^2 \rangle$ is α_s suppressed, while the one of $d_{6,V-A}$ is smaller than in the individual V and A channels. Then, we expect that V–A is a *golden channel* for extracting α_s.

• **α_s from $\mathcal{R}_{0,V-A}(s_0)$**

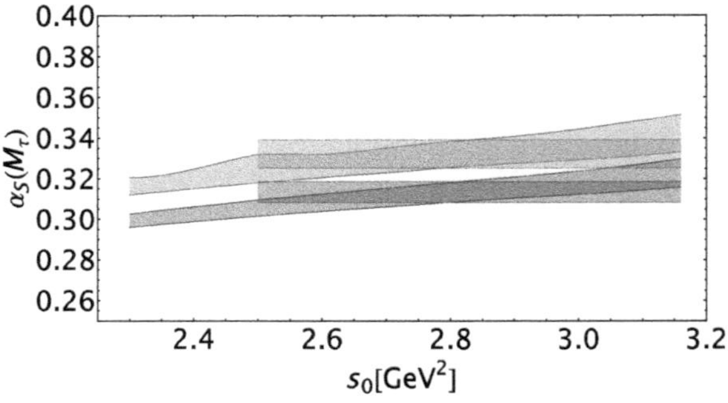

Figure 9.108. $\alpha_s(M_\tau)$ versus a hypothetical τ-mass squared s_0. The upper curves correspond to CI perturbative series and the lower ones to FO. The horizontal lines come from a least-square fit of the data in the optimal region $s_0 \simeq (2.3 \sim 2.9)$ GeV^2.

Using the value of $\langle\alpha_s G^2\rangle$ in Eq. (9.305) and the previous values of $d_{6,V-A}$ and $d_{8,V-A}$ in Eq. (9.331), we extract the value of $\alpha_s(M_\tau)$ as a function of s_0 (see Fig. 9.108) from $\mathcal{R}_{0,V-A}(s_0)$. We note a stable result in the region $s_0 \simeq (2.5 \sim 2.6)$ GeV^2 though not quite convincing. Then, we consider as a conservative value the one obtained from a least-square fit of the values inside the region $[2,5,M_\tau^2]$. The optimal result corresponds to $s_0 = 2.8$ GeV^2 (see Fig. 9.108):

$$\begin{aligned}\alpha_s(M_\tau)|_{V-A} &= 0.3135(51)(65) \longrightarrow \alpha_s(M_Z)|_{V-A},\\ &= 0.1177(10)(3)_{\text{evol}} \text{ (FO)},\\ &= 0.3322(69)(43) \longrightarrow \alpha_s(M_Z)|_{V-A},\\ &= 0.1200(9)(3)_{\text{evol}} \text{ (CI)}.\end{aligned} \tag{9.332}$$

The first error in $\alpha_s(M_\tau)|_{V-A}$ comes from the fitting procedure. The second one comes from an estimate of the α_s^5 contribution from Ref. [18]. At this scale, the sum of non-perturbative contributions to the moment normalized to the parton model is:

$$\delta_{NP,V-A} \simeq +(2.7 \pm 1.1) \times 10^{-4}, \tag{9.333}$$

which is completely negligible. One can note from Fig. 9.108 that extracting $\alpha_s(M_\tau)|_{V-A}$ at the observed M_τ-mass tends to overestimate its value:

$$\alpha_s(M_\tau)|_{V-A} = 0.3227(69)(65) \quad (\text{FO}), \quad 0.3423(92)(43) \quad (\text{CI}). \tag{9.334}$$

• Comparison with some previous results

Table 9.25. Values of the QCD condensates from $\mathcal{R}_{0,e^+e^-}$ and $\mathcal{R}_{0,A}$ at fixed order (FO) PT series and $\alpha_s(M_\tau)$ for FO and contour improved (CI) PT series. The condensates are in units of 10^{-2} GeV^D.

$d_{6,V-A}$	$-d_{8,V-A}$	$\alpha_s(M_\tau)$ FO	$\alpha_s(M_\tau)$ CI	s_0 [GeV]^2	Reference
3.6 ± 0.5	14.5 ± 2.2	0.3135(83)	0.3322(81)	$2.5 \to M_\tau^2$	This work
3.6 ± 0.5	14.5 ± 2.2	0.3227(95)	0.3423(102)	M_τ^2	This work
$5.1^{+5.5}_{-3.1}$	3.2 ± 2.2	$0.3170^{+0.0100}_{-0.0050}$	$0.3360^{+0.0110}_{-0.0090}$	M_τ^2 ($D \leq 10$)	Table 7 [545]
24^{+24}_{-16}	32^{+25}_{-32}	$0.3290^{+0.0120}_{-0.0110}$	$0.3490^{+0.0160}_{-0.0140}$	M_τ^2 ($D \leq 12$)	Table 8 [545]
2.4 ± 0.8	3.2 ± 0.8	–	0.3410(78)	M_τ^2	[247, 528]
1.5 ± 4.8	3.7 ± 9.0	0.3240(145)	0.3480(212)	M_τ^2	[248]

In Table 9.25, we compare our results with some other determinations [247, 248, 545][29] obtained at the scale $s_0 = M_\tau^2$.

◇ There is quite a good agreement for $d_{6,V-A}$ but not for $d_{6,V-A}$ from different determinations.
◇ The central values of the condensates given in Table 7 of Ref. [545] using a truncation of the OPE up to $D = 10$ are systematically smaller than the ones in their Table 8 using the OPE up to $D = 12$ though they agree within the errors.
◇ Extracting α_s at M_τ, there is a good agreement among different determinations where the values are slightly higher than the ones from the optimal region given in the first row (see Fig. 9.108).

[29]Some estimates including renormalon within a large β approximation (respectively duality violation) effects can be, e.g., found in Refs. [560] (respectively [448]).

9.8.11 Mean value of α_s from e^+e^- and τ-decays

Using the result from $e^+e^- \to$ hadrons and from the A and V–A τ-decay channels, we deduce the mean:

$$\begin{aligned}\langle\alpha_s(M_\tau)\rangle &= 0.3128(51) \longrightarrow \langle\alpha_s(M_Z)\rangle = 0.1176(7)(3)_{\text{evol}}\\ &= 0.3330(57) \longrightarrow \qquad\qquad\quad = 0.1201(7)(3)_{\text{evol}}. \quad (9.335)\end{aligned}$$

The result from the vector component of τ-decays quoted in Table 9.22 is not included as it has been obtained from the data measured at the physical τ-mass which does not give the optimal result.

9.8.12 Summary

- We have improved the determinations of the QCD condensates in the axial-vector channel using a ratio of LSR and higher BNP-like moments. Our results are summarized in Table 9.24. We observe alternate signs, an almost constant value of their size. The absence of an exponential behaviour in the Euclidean region may not favour a duality violation in the time-like region [556].
- We have used as inputs the value of $\langle\alpha_s G^2\rangle$ better determined from the heavy quark channels and the ones of the previous condensates $d_{6,A}$ and $d_{8,A}$ to extract $\alpha_s(M_\tau)$ from the lowest BNP-moment $\mathcal{R}_{0,A}(s_0)$. Our conservative result in Eq. (9.320) is obtained from $s_0 \simeq 2.1$ GeV^2 to M_τ^2 where we note from Fig. 9.101 that extracting $\alpha_s(M_\tau)$ at the observed τ-mass leads to an overestimate of its value.
- We combine the previous values of $d_{6,A}$ and $d_{8,A}$ with the ones of $d_{6,V}$ and $d_{8,V}$ from $e^+e^- \to$ hadrons [18–20] into the lowest moment $\mathcal{R}_{0,V-A}(s_0)$ in the V–A channel. Then, we extract the conservative value of $\alpha_s(M_\tau)$ given in Eq. (9.334) at $s_0 = [2.5$ GeV$^2 \to M_\tau^2]$. Like in the case of the $e^+e^- \to$ hadrons and axial-vector channel, we also note that extracting $\alpha_s(M_\tau)$ at M_τ leads to an overestimate.
- We have not considered some eventual effects beyond the SVZ expansion listed below:
  - ◇ According to Ref. [234], the effect of a tachyonic gluon mass [235] for parametrizing phenomenologically the UV renormalon effect is equivalent by duality to the contribution of the

uncalculated higher order PT terms. We assume that these terms are well approximated by the estimate of the α_s^5 term done in the paper using the observation that the coefficients of the calculated PT terms behave as a geometric sum.

- ◇ The observation that the calculated PT terms behave as a geometric sum and that no signal of alternate sign of these PT terms (*a priori*) favours the motivation of a large β-approximation for the estimate of the UV renormalon.
- ◇ The non-observation of an exponential behaviour of the non-perturbative condensate effects in the Euclidean region may indicate that duality violation in the time-like region [448] may not be sizeable [556].
- ◇ Instanton effects act as high-dimension operators and their effects have been shown to be negligible in the vector channel [18–20]. We expect similar features for the A and V–A channels.

9.9 Hadronic Contributions to a_l and $\Delta\alpha$

In this section, we review the re-estimate of the lowest order hadronic vacuum polarization contributions $a_l|_{\rm l.o}^{\rm hvp}$: $l \equiv \mu, \tau$ to the anomalous magnetic moments of the μ and τ leptons in Refs. [18–20], where the new CMD3 data [543] and the PDG compilation [33] have been used. Using the new experimental data on a_μ [322], one deduces $\Delta a_\mu \equiv a_\mu^{\rm exp} - a_\mu^{\rm th} = (143 \pm 42_{\rm th} \pm 22_{\rm exp}) \times 10^{-11}$ indicating about 3σ discrepancy between the SM predictions and experiment. We also present an estimate of $a_\tau|_{\rm l.o}^{\rm hvp}$ and the hadronic contribution to $\Delta\alpha$. The results are compiled in Table 9.26.

9.9.1 $a_\mu|_{\rm l.o}^{\rm hvp}$ from $e^+e^- \to$ hadrons data

We complete the analysis by updating our previous determination of the lowest order hadronic contributions to the vacuum polarization of a_μ [561]. This analysis will also serve as a test of our parametrization of the $I = 1$ part of the spectral function used in previous sections.

The lowest order hadronic contributions to the vacuum polarization can be obtained from the well-known dispersion relation given in Eq. (3.2).

• $a_\mu|_{\rm l.o}^{\rm hvp}$ from $e^+e^- \to I = 1$ hadrons below 1.875 GeV

This analysis is an update of the one in Ref. [561] and will also serve as a test of our parametrization of the data, especially for the ρ-meson, where we have only used a minimal Breit–Wigner for the pion form factor. We obtain the contributions given in Table 9.26 for the three subdivision regions of the data below 1.875 GeV, as shown in Figs. 9.76–9.78.

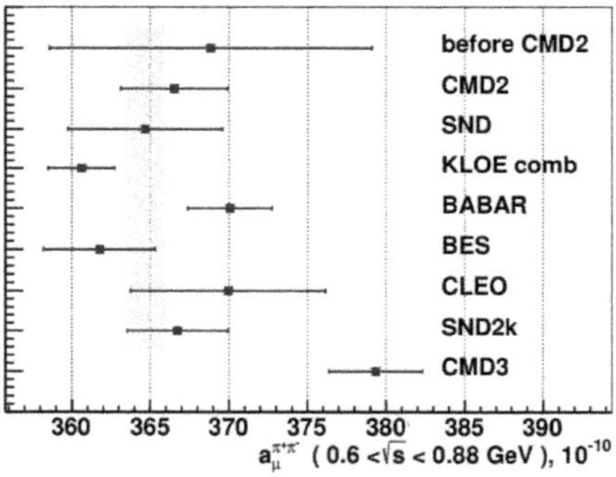

Experiment	$a_\mu^{\pi^+\pi^-,LO}$, 10^{-10}
before CMD2	368.8 ± 10.3
CMD2	366.5 ± 3.4
SND	364.7 ± 4.9
KLOE	360.6 ± 2.1
BABAR	370.1 ± 2.7
BES	361.8 ± 3.6
CLEO	370.0 ± 6.2
SND2k	366.7 ± 3.2
CMD3	379.3 ± 3.0

Figure 9.109. Comparison of the ρ-meson contribution to a_μ for $0.6 \le \sqrt{s} \le 0.88$ GeV from Ref. [543].

◇ Comparison with some others for $0.60 \le \sqrt{t} \le 0.88$ GeV

We compare our determinations from the ρ-meson using a minimal Breit–Wigner parametrization of the pion form factor for $0.6 \le \sqrt{t} \le 0.88$ GeV:

$$a_\mu^\rho|_{\rm l.o}^{\rm hvp}[0.6 \to 0.88] = (370.5 \pm 2.1) \times 10^{-10}. \tag{9.336}$$

Despite our minimal parametrization of the pion form factor, our result is in good agreement with the previous ones in Fig. 9.109 using a more involved parametrization. We interpret our slightly lower value compared to CMD3 [543] from the fact that the minimal fit cannot accommodate accurately the data in this region.

Table 9.26. $a_\tau|_{\rm l.o}^{\rm hvp}$, $a_\mu|_{\rm l.o}^{\rm hvp}$ and $\alpha(M_z^2)$ within our parametrization of the compiled PDG [33] ⊕ the recent CMD3 [543].

$\sqrt{t}$ [GeV]	$a_\mu\|_{\rm l.o}^{\rm hvp} \times 10^{11}$	$a_\tau\|_{\rm l.o}^{\rm hvp} \times 10^{9}$	$\Delta\alpha_{\rm had}^{(5)}(M_Z^2) \times 10^5$
Light $I=1$			
$\rho(2m_\pi \to 0.50)$	489.4 ± 3.1	85.1 ± 0.6	12.38 ± 0.08
$\rho(0.50 \to 0.60)$	524.5 ± 13.7	135.9 ± 3.5	22.84 ± 0.59
$\rho(0.60 \to 0.776)$	2712.0 ± 30.2	943.7 ± 10.3	182.87 ± 1.97
$\rho(0.776 \to 0.993)$	1297.3 ± 9.8	1797.3 ± 20.3	117.82 ± 0.98
$0.993 \to 1.5$	354.4 ± 6.7	228.2 ± 4.1	67.3 ± 1.14
$1.5 \to 1.875$	237.6 ± 5.7	206.8 ± 4.9	80.11 ± 1.9
Total Light I=1 (≤ 1.875)	*5615.2 ± 36.0*	*2453.3 ± 11.9*	*483.3 ± 4.4*
Light $I=0$			
ω (NWA)	417.1 ± 13.7	163.3 ± 5.3	33.6 ± 1.1
ϕ (NWA)	389.6 ± 4.6	20.6 ± 0.2	51.2 ± 0.6
$0.993 \to 1.5$	44.3 ± 0.8	28.5 ± 0.5	8.4 ± 0.1
$\omega(1650)$(BW)	24.3 ± 0.1	16.7 ± 0.1	5.2 ± 0.1
$\phi(1680)$(BW)	1.8 ± 0.9	1.3 ± 0.6	0.4 ± 0.2
Total Light I=0 (≤ 1.875)	*877.1 ± 14.5*	*230.4 ± 5.3*	*99.4 ± 1.3*
Light I=0 ⊕ 1			
$1.875 \to 1.913$	14.7 ± 0.7	14.4 ± 0.7	6.3 ± 0.3
$1.913 \to 1.96$	17.6 ± 0.6	17.5 ± 0.6	7.9 ± 0.3
$1.96 \to 2$	14.3 ± 0.5	14.5 ± 0.5	6.7 ± 0.2
$2 \to 3.68$: QCD (u, d, s)	247.2 ± 0.3	308.3 ± 0.5	202.8 ± 0.5
Total Light I=0 ⊕ 1 ($1.875 \to 3.68$)	*293.8 ± 1.1*	*354.7 ± 1.2*	*223.7 ± 0.7*
Total Light I=0 ⊕ 1 ($2m_\pi \to 3.68$)	6786.1 ± 38.8	3038.4 ± 24.5	806.4 ± 4.6
Charmonium			
$J/\psi(1S)$ (NWA)	65.1 ± 1.2	92.7 ± 1.8	73.5 ± 1.4
$\psi(2S)$ (NWA)	16.4 ± 0.6	26.0 ± 0.9	26.1 ± 0.8
$\psi(3773)$ (NWA)	1.7 ± 0.1	2.7 ± 0.2	2.9 ± 0.2
Total $J/\psi(NWA)$	*83.2 ± 1.4*	*121.4 ± 2.0*	*102.5 ± 1.6*
$3.69 \to 3.86$	11.4 ± 1.0	18.3 ± 1.6	19.0 ± 1.6
$3.86 \to 4.094$	16.6 ± 0.5	27.5 ± 0.8	30.9 ± 0.9
$4.094 \to 4.18$	6.6 ± 0.2	11.2 ± 0.3	13.2 ± 0.4

(*Continued*)

Table 9.26. (*Continued*)

$\sqrt{t}$ [GeV]	$a_\mu\vert^{\rm hvp}_{\rm l.o} \times 10^{11}$	$a_\tau\vert^{\rm hvp}_{\rm l.o} \times 10^9$	$\Delta\alpha^{(5)}_{\rm had}(M_Z^2) \times 10^5$
$4.18 \to 4.292$	6.5 ± 0.2	11.2 ± 0.4	13.7 ± 0.5
$4.292 \to 4.54$	11.8 ± 0.6	20.7 ± 0.8	26.8 ± 1.1
$4.54 \to 10.50$: QCD (u, d, s, c)	92.0 ± 0.0	186.2 ± 0.0	$458.7.3 \pm 0.1$
Total Charmonium (3.69 → 10.50)	*145.1 ± 1.3*	*275.6 ± 2.0*	*564.9 ± 2.2*
Total Charmonium	228.1 ± 1.9	396.5 ± 2.8	664.8 ± 2.7
Bottomium			
$\Upsilon(1S)$ (NWA)	0.54 ± 0.02	1.25 ± 0.07	5.65 ± 0.29
$\Upsilon(2S)$ (NWA)	0.22 ± 0.02	0.51 ± 0.06	2.54 ± 0.29
$\Upsilon(3S)$ (NWA)	0.14 ± 0.02	0.33 ± 0.04	1.77 ± 0.23
$\Upsilon(4S)$ (NWA)	0.10 ± 0.01	0.23 ± 0.03	1.26 ± 0.16
$\Upsilon(10.86 \oplus 11)$ (NWA)	0.1 ± 0.06	0.20 ± 0.06	1.67 ± 0.39
Total Bottomium (NWA)	*1.0 ± 0.1*	*2.3 ± 0.1*	*11.2 ± 0.5*
Z − pole	–	–	29.2 [562]
$10.59 \to 2m_t$: QCD (u, d, s, c, b)	22.4 ± 0.3	57.5 ± 0.1	1282.9 ± 1.2
Total Bottomium	23.4 ± 0.3	59.8 ± 0.1	1323.3 ± 1.3
$2m_t \to \infty$: QCD (u, d, s, c, b, t)	0.03	0.08	−28.2
Total sum	**7036.5 ± 38.9**	**3494.8 ± 24.7**	**2766.3 ± 4.5**

◇ Improving the ρ meson contribution below 0.993 GeV

Though our minimal BW parametrization is in agreement with previous determination given in Fig. 9.109, we can see in Fig. 9.76 that there are regions which need to be improved. This is necessary for the high-precision determination of a_μ. Note that, within the modest accuracy of the results in previous sections, this improvement only affects slightly these results. In so doing, we subdivide the region below 0.993 GeV into six subregions in units of GeV:

$$[2m_\pi \to 0.5],\ [0.5 \to 0.6],\ [0.6 \to 0.776],\ [0.776 \to 0.786],\ [0.786 \to 0.810],\ [0.810 \to .993]. \tag{9.337}$$

We use the tail of the Breit–Wigner for $[2m_\pi \to 0.5]$ as shown in Fig. 9.76 and different polynomials within the FindFit Mathematica program with optimised χ^2 to fit the other regions from $0.5 \to 0.993$ GeV. We remind that the new CMD3 data [543] are 2–3 times more precise than the one compiled by PDG [43], but the central values are approximately similar except in the low energy region where the CMD3 data are slightly higher. We show the analysis in Figs. 9.110 and 9.111. Note that, in the region 0.776–0.86 GeV, we have only considered the data from CMD3 (red). The effect of this region to a_μ from CMD3 data is $(206.5 \pm 1.7) \times 10^{-11}$, while the one using the BW in Fig. 9.76 gives $(205.7 \pm 1.03) \times 10^{-11}$ such that the difference by using the two data is negligible. The different contributions are given in Table 9.26. From the fit improvement, the result in Eq. (9.336) becomes:

$$a_\mu^\rho|_{1.o}^{hvp}[0.6 \to 0.88] = (377.4 \pm 3.1) \times 10^{-10}. \quad (9.338)$$

It is also informative to give our result in the so-called intermediate region $[0.6 \to 0.993]$ GeV:

$$a_\mu^\rho|_{1.o}^{hvp}[0.6 \to 0.993] = (400.9 \pm 3.2) \times 10^{-10}. \quad (9.339)$$

◇ $a_\mu^{I=1}|_{1.o}^{hvp}$ from 0.993 to 1.875 GeV

The fits of these regions have been done in the previous sections. The result is given in Table 9.26.

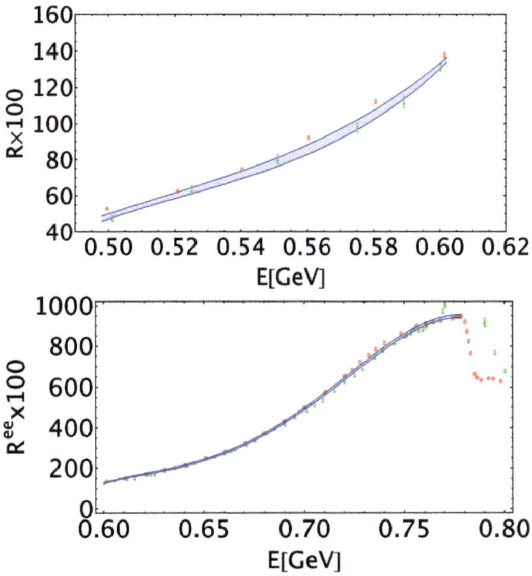

Figure 9.110. Fit of the data in the region 0.5–0.776 GeV.

• a_μ from $e^+e^- \to I = 0$ hadrons below 1.875 GeV

◇ ω and ϕ meson contributions

We estimate the ω and ϕ meson contributions using a narrow width approximation (NWA). It reads [561]:

$$a_\mu^{I=0}\Big|_{\rm l.o}^{\rm hvp} = \frac{3}{\pi}\sum_{R=\omega,\phi}\frac{\Gamma_R^{ee}}{M_R}K_\mu(M_R^2). \tag{9.340}$$

Using [33]:

$$\begin{aligned}\Gamma_\omega^{ee} &= (64 \pm 1.9 \pm 0.96)\times 10^{-2}\ \text{keV},\\ \Gamma_\phi^{ee} &= (126.6 \pm 1.4 \pm 0.4)\times 10^{-2}\ \text{keV},\\ M_\omega &= (782.66 \pm 0.13)\ \text{MeV}, \quad M_\phi = (1019.461 \pm 0.016)\ \text{MeV},\end{aligned} \tag{9.341}$$

the result is given in Table 9.26.

◇ $a_\mu^{I=0}\big|_{\rm l.o}^{\rm hvp}$ from .993 to 1.5 GeV

We add the $I = 0$ contribution subtracted for the $I = 1$ analysis in the previous sections by using the $SU(3)$ relation. We obtain the value in Table 9.26.

◇ $a_\mu^{I=0}\big|_{\rm l.o}^{\rm hvp}$ from 1.5 to 1.875 GeV

We consider the contributions of the $\omega(1650)$ and $\phi(1680)$. We use the parameters deduced from PDG22 [33]:

$$\begin{aligned}\omega(1650)\colon M_\omega &= (1670 \pm 30)\,\text{MeV}, \quad \Gamma_\omega^{ee} = 1.35 \pm 0.14\,\text{keV},\\ \Gamma_\omega^h &= 315 \pm 35\,\text{MeV},\\ \phi(1680)\colon M_\phi &= (1680 \pm 20)\,\text{MeV}, \quad \Gamma_\phi^{ee} = 0.18 \pm 0.06\,\text{keV},\\ \Gamma_\psi^h &= 150 \pm 50\,\text{MeV}.\end{aligned} \tag{9.342}$$

Using a Breit–Wigner (BW) parametrization, we obtain the value in Table 9.26.

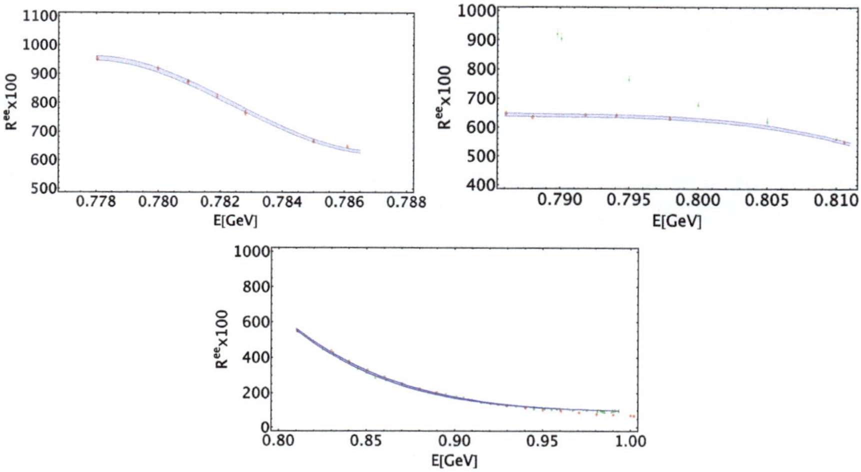

Figure 9.111. Fit of the data in the region 0.776 to 0.993 GeV.

• $a_\mu|^{\rm hvp}_{\rm l.o}$ from 1.875 to 2 GeV

To get this contribution, we divide the data into three regions and fit with polynomials using the optimized χ^2 Mathematica program FindFit. The fits of the data are shown in Fig. 9.112.

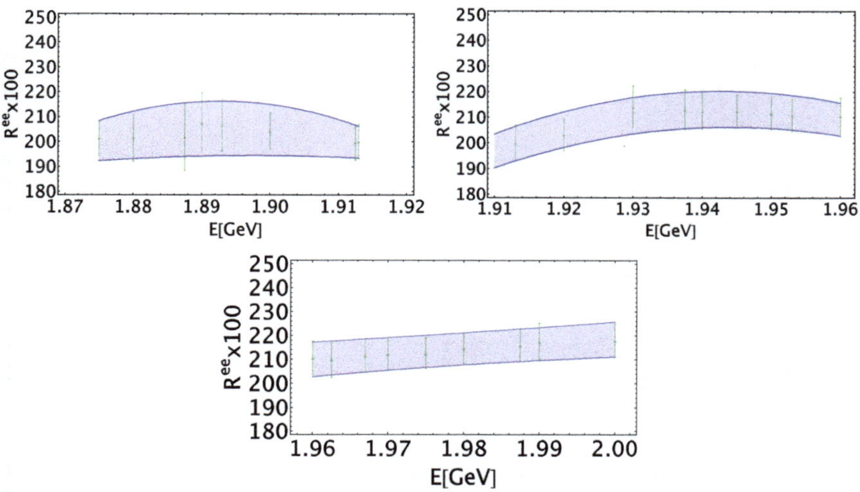

Figure 9.112. Fit of the data in the region 1.875 to 2 GeV.

• $a_\mu|_{\rm l.o}^{\rm hvp}$ from 2 to 3.68 GeV

The data in this region are well fitted by the QCD expression of the spectral function for three flavours as one can see from Fig. 9.113 given by PDG. To the massless PT expression known to order α_s^4, we include the quark and gluon condensates of dimensions $D = 4, 6$ (see Eqs. (9.300)–(9.264)). We add the quadratic m_s^2-corrections to order α_s^3 and the quartic m_s^4-mass corrections to order α_s^2 [156]. These quark mass corrections read in the $\overline{MS}$-scheme:

$$
\begin{aligned}
d_{2m} &= -\frac{\overline{m_s}^2}{t}\left(12\,a_s + 104.833\,a_s^2 + 541.735\,a_s^3\right),\\
d_{4m} &= \frac{\overline{m_s}^4}{t^2}\Bigg[-6\left(1+\frac{11}{3}a_s\right) + a_s^2\Bigg[n_f\left(\frac{\rm L_m}{3} - 1.841\right)\\
&\quad - \frac{11}{2}{\rm L_m} + 136.693 + 12 - 0.475 - {\rm L_m}\Bigg]\Bigg],
\end{aligned}
\tag{9.343}
$$

where $\overline{m}$ is the running quark mass and $L_m \equiv \log(\overline{m}_s^2/t)$. In the numerical analysis, we shall also introduce the RGI quark mass defined as [29]

$$
\overline{m_s}(\nu) = \hat{m}_s\left(-\beta_1 a_s(\nu)\right)^{-\gamma_1\beta_1}\Big[1 + \mathcal{O}(a_s)\Big],
\tag{9.344}
$$

where $\gamma_1 = 2$ is the quark mass anomalous dimension and $\beta_1 = -(11-2n_f/3)/2$ the first coefficient of the β-function. The expression

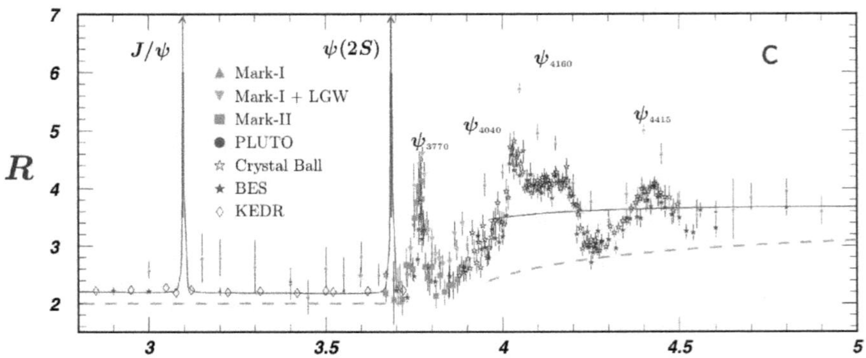

Figure 9.113. $e^+e^- \to$ hadrons data in the charmonium region from 2 to 5 GeV from PDG [33].

to order a_s^2 which we shall use can be found, e.g., in Ref. [7]. We shall use $\hat{m}_s = 116(6)$ MeV [47].

• $a_\mu|_{\rm l.o}^{\rm hvp}$ from charmonium

◇ The $J/\psi(1S), \psi(2S)$ and $\psi(3773)$ contributions

We estimate their contributions using an NWA and the values of the masses (in units of MeV) and widths (in units of keV except $\Gamma_\psi^h(3773)$) from PDG [33]:

$$
\begin{aligned}
\psi(1S)\colon M_\psi &= (3069.9 \pm 0.006), \quad \Gamma_\psi^{ee} = (5.53 \pm 0.10),\\
\Gamma_\psi^h &= (92.6 \pm 1.0),\\
\psi(2S)\colon M_\psi &= (3681.1 \pm 0.06), \quad \Gamma_\psi^{ee} = (2.33 \pm 0.08),\\
\Gamma_\psi^h &= (294 \pm 8),\\
\psi(3773)\colon M_\psi &= (3773.7 \pm 0.4), \quad \Gamma_\psi^{ee} = (0.26 \pm 0.02),\\
\Gamma_\psi^h &= (27.2 \pm 1.0)\,\text{MeV}.
\end{aligned}
\tag{9.345}
$$

The result is given in Table 9.26.

◇ Region from 3.68 to 4.55 GeV

We divide this region into five subregions, as shown in Fig. 9.114. The region between 3.68 and 3.86 is better fitted using a Breit–Wigner, while in the others, we use polynomials. The different contributions are shown in Table 9.26.

◇ Region from 4.55 to 10.50 GeV

As shown in Figs. 9.113 and 9.115, the data without resonance peaks are well fitted by QCD for four flavours. We add to the previous QCD expressions the charm contributions with m_c^2 and m_c^4 mass corrections, where $\overline{m}_c(\overline{m}_c) = 1266(6)$ MeV [47]. The contribution is given in Table 9.26.

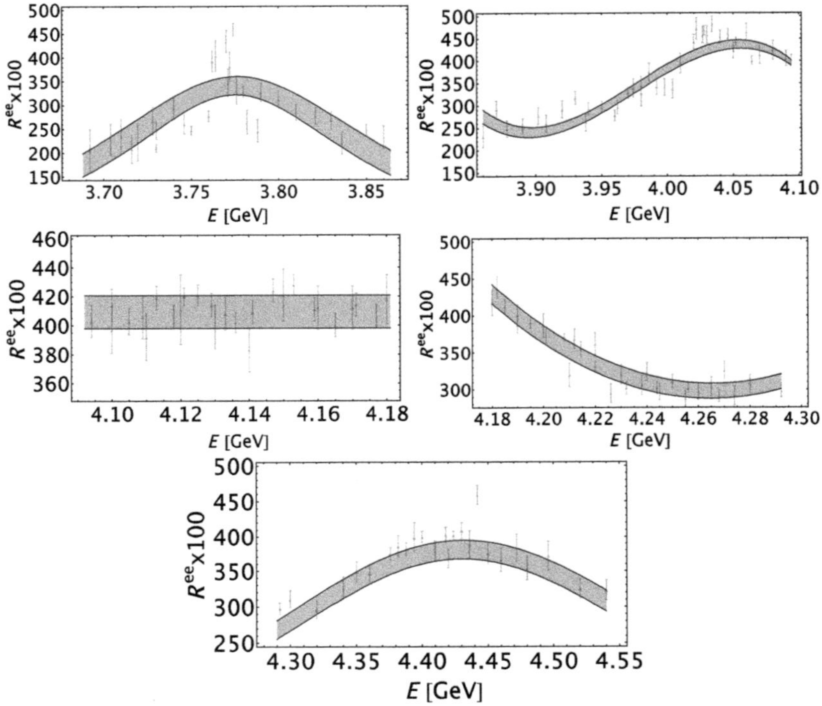

Figure 9.114. Fits of the $e^+e^- \to$ hadrons data in the charmonium region from 3.68 to 4.55 GeV from PDG [33].

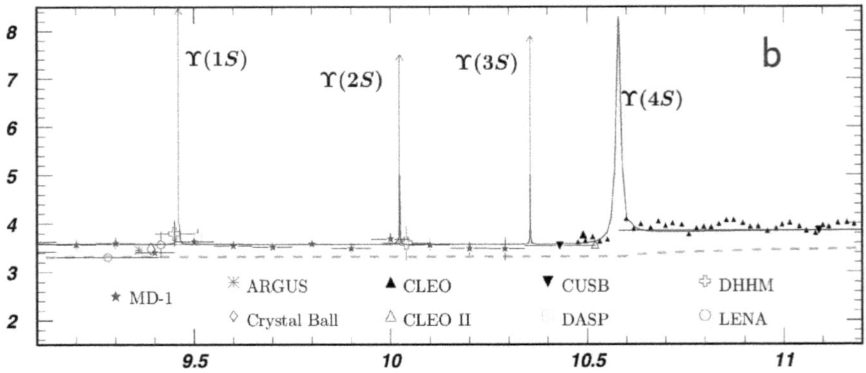

Figure 9.115. $e^+e^- \to$ hadrons data in the bottomium region from 9 to 10.5 GeV from PDG [33].

• $a_\mu|_{\rm l.o}^{\rm hvp}$ from bottomium

⋄ $\Upsilon(1S \to 11.02)$ contributions

We use an NWA to estimate their contributions with:

$$\begin{aligned}
\Upsilon(1S)\text{: } M_\Upsilon &= (9460.30 \pm 0.26), \quad \Gamma_\Upsilon^{ee} = (1.286 \pm 0.067),\\
\Gamma_\Upsilon^{h} &= (954.02 \pm 1.25),\\
\Upsilon(2S)\text{: } M_\Upsilon &= (10023.26 \pm 0.31), \quad \Gamma_\Upsilon^{ee} = (0.611 \pm 0.071),\\
\Gamma_\Upsilon^{h} &= (31.98 \pm 2.63),\\
\Upsilon(3S)\text{: } M_\Upsilon &= (10355.2 \pm 0.5), \quad \Gamma_\Upsilon^{ee} = (0.443 \pm 0.057),\\
\Gamma_\Upsilon^{h} &= (20.32 \pm 1.85),\\
\Upsilon(4S)\text{: } M_\Upsilon &= (10579.4 \pm 1.2), \quad \Gamma_\Upsilon^{ee} = (0.322 \pm 0.042),\\
\Gamma_\Upsilon^{h} &= (20.5 \pm 2.5).
\end{aligned} \tag{9.346}$$

⋄ QCD continuum contribution from 10.59 GeV to $2m_t$

We add the b-quark contribution to the previous QCD expression where b-quark mass corrections to order $a_s^3\overline{m}_b^2/t$ and $a_s^2\overline{m}_b^4/t^2$ are included. We use $\overline{m}_b(\overline{m}_b) = 4202(8)$ MeV [47]. For the analysis, we consider the region from 10.59 MeV to $2m_t$ just after $\Upsilon(4S)$ where the QCD continuum is expected to smear $\Upsilon(10860, 11020)$ and some eventual higher resonances. The result is given in Table 9.26.

⋄ QCD continuum contribution from $2m_t \to \infty$

Due to the heaviness of the top quark mass, we shall use the approximate Schwinger formula near the $\bar{t}t$ threshold for a much better description of the spectral function up to order α_s due to the top quark:

$$R_t^{ee} = \frac{4}{3}v\frac{(3-v^2)}{2}\left[1 + \frac{4}{3}\alpha_s f(v)\right], \tag{9.347}$$

with:

$$f(v) = \frac{\pi}{2v} - \frac{(3+v)}{4}\left(\frac{\pi}{2} - \frac{3}{4\pi}\right) : \quad v = \left(1 - \frac{m_t^2}{t}\right)^{1/2}. \tag{9.348}$$

Here, m_t is the on-shell top quark mass which we fix to be [33][30]:

$$\overline{m}_t(\overline{m}_t^2) = (172.7 \pm 0.3)\,\text{GeV} \qquad (9.349)$$

from some direct measurements. We add to this expression the one due to α_s^2 and α_s^3 given in Eq. (9.343) within the $\overline{MS}$-scheme. For this $\overline{MS}$-scheme expression in terms of running mass, we need the value of the RGI top mass:

$$\hat{m}_t = (254 \pm 0.4)\,\text{GeV} \qquad (9.350)$$

deduced from Eq. (9.349). Adding the above expressions to the ones in the previous sections, we obtain the result given in Table 9.26.

9.9.2 Conclusion for $a_\mu|_{\text{l.o}}^{\text{hvp}}$ and extension to $a_\tau|_{\text{l.o}}^{\text{hvp}}$

- **Results for $a_\mu|_{\text{l.o}}^{\text{hvp}}$**

We have used the sum exclusive $e^+e^- \to$ hadrons data compiled by PDG $\oplus$ some resonances to extract the lowest order vacuum polarization contribution to the muon anomaly. We find from Table 9.26:

$$a_\mu|_{\text{l.o}}^{\text{hvp}} = (7036.5 \pm 38.9) \times 10^{-11}, \qquad (9.351)$$

where the largest contribution and error come (as expected) from the ρ-meson low-energy one. It is amazing that the central value remains stable when comparing it with our old results in Refs. [9, 10, 561, 570] though the accuracy has increased by a huge factor of about 24 (!) thanks to the experimental efforts for improving the data during about half century! The total sum is slightly larger than the recent data-based determinations in Refs. [563, 564],[31] but in better agreement with a recent analysis of τ-decay data [565] and with some recent lattice results [21, 566, 567].

[30]We should note that the definition of the top quark mass from different experiments is still ambiguous.

[31]Some earlier references are quoted in these papers and in Ref. [561].

In Ref. [21], we have updated the previous estimate and found a slightly larger value though consistent within the errors:

$$a_\mu|_{l.o}^{hvp} = (7043 \pm 37) \times 10^{-11}. \tag{9.352}$$

Adding the contributions of the $\pi^0\gamma$ and $\eta\gamma$ not considered in Refs. [18–20], the result becomes:

$$a_\mu|_{l.o}^{hvp} = (7094 \pm 37) \times 10^{-11}. \tag{9.353}$$

Using this value into Table 8 of Ref. [568] and in Table 1 of Ref. [21] where some other sources of contributions are reviewed (see also Refs. [22]), we deduce the total hadronic contributions, from phenomenology, including Ligh-by-Light scattering and Higher Order α_s contributions [21]:

$$a_\mu^h|_{pheno} = (7112 \pm 38) \times 10^{-11}, \tag{9.354}$$

leading to the phenomenological total estimate:

$$a_\mu^{pheno} = (116591985.4 \pm 38.5) \times 10^{-11}. \tag{9.355}$$

Using the recent experimental average [23]:

$$\langle a_\mu^{exp} \rangle = (116592071.5 \pm 14.5) \times 10^{-11}, \tag{9.356}$$

we deduce:

$$\Delta a_\mu^{pheno} \equiv \langle a_\mu^{exp} \rangle - a_\mu^{pheno} = (86 \pm 41) \times 10^{-11}, \tag{9.357}$$

which signals a 2.1σ deviation from the SM prediction. If, instead, we attempt to average this value with the one from the most precise lattice calculation [24]:

$$\Delta a_\mu^{lattice} = (90 \pm 56) \times 10^{-11}, \tag{9.358}$$

we obtain:

$$\langle \Delta a_\mu^{th} \rangle = (87 \pm 33) \times 10^{-11}. \tag{9.359}$$

which is a deviation of about 2.6σ from the SM prediction. If it persists, it may signal the presence of new physics beyond the Standard Model and stimulates the needs of a much better precision for measuring the $e^+e^- \to$ Hadrons total cross-section and for the lattice calculations.

- **Some comments on the determination of $a_\mu|_{\mathrm{l.o}}^{\mathrm{hvp}}$**

⋄ Instead of our simple Breit–Wigner parametrization of pion form factor done in previous sections, we have subdivided the region below 0.993 GeV to have a better fit of the data which is necessary for the extraction of the lepton anomalies. We have compared our result in this region with the ones from different experimental groups quoted in Fig. 9.109.

⋄ Our analysis differs from the most recent ones in Refs. [563, 564] as we fit the sum of exclusive modes except the ρ-meson. For the narrow resonances, we use the NWA.

⋄ We note that compared to the existing analysis, we have also studied in detail the contributions from the heavy quark sectors, taking into account all possible resonances and, in particular, carefully analyzed the charmonium region.

⋄ We have also carefully parametrized the QCD continuum contributions, taking into account higher order PT quark mass corrections and the non-perturbative ones.

⋄ Comparing our results with the most recent ones in Refs. [563,564], we found that in the low-energy region below 1.875 GeV, our result is higher by about $(100\text{–}136) \times 10^{-11}$ than the ones in these references. This can be due to the pion form factor where we use the new data of CMD3 $\oplus$ PDG, which leads to higher value of a_μ than some other determinations, as shown in Fig. 9.109 for, e.g., the region $0.60 \leq \sqrt{t} \leq 0.88$ GeV.

⋄ In the high-energy region $\sqrt{t} \geq 1.875$ GeV, our result is $(545.3 \pm 2.2) \times 10^{-11}$ which is about the same as the one $(535.5 \pm 7.0) \times 10^{-11}$ of Ref. [563] but smaller by about 38×10^{-11} than $(583.6 \pm 3.3) \times 10^{-11}$ in Ref. [564]. This difference is mainly due to the choice of the QCD continuum threshold $\sqrt{t_c} = 1.8$ GeV in Ref. [564] which is lower than the one found from the asymptotic coincidence of the two sides of $\mathcal{L}_0$ Laplace moment in Eq. (9.245) and requested by duality from FESR in Eq. (9.248).

9.9.3 Extension to the determination of $a_\tau|_{\text{l.o}}^{\text{hvp}}$

◇ The muon analysis has been extended to the case of the τ-lepton by simply changing the lepton mass. The total sum is from Table 9.26:

$$a_\tau|_{\text{l.o}}^{\text{hvp}} = (3494.8 \pm 24.7) \times 10^{-9}, \tag{9.360}$$

which we consider as an improvement of the *pioneer determination* of this quantity and of a_τ in Ref. [570].

◇ One can note that the relative weight of the light quarks over the heavy ones moves from 23 to 6 from μ to τ, indicating that a measurement of the τ-anomaly will probe higher energy region thanks to the behaviour of the QED kernel function given in Eq. (3.2).

◇ Our result is slightly lower than the one of Ref. [563] where the origin may come from the $I = 0$ light meson region.

9.9.4 Determination of $\Delta\alpha^{(5)}_{\text{had}}(M_Z^2)$

We conclude the paper by updating our previous determination of $\Delta\alpha^{(5)}_{\text{had}}(M_Z^2)$ in Ref. [572]. The hadronic contribution to this quantity can be expressed as:

$$\Delta\alpha^{(5)}_{\text{had}}(M_Z^2) = -\left(\frac{\alpha}{3\pi}\right) M_Z^2 \int_{4m_\pi^2}^{\infty} \frac{R_{ee}(t)}{t(t - M_Z^2)}, \tag{9.361}$$

where R_{ee} is the ratio of the $e^+e^- \to$ Hadrons over the $e^+e^- \to \mu^+\mu^-$ total cross-sections. The results from different regions are shown in Table 9.26 where at the Z_0 pole, we take the principal value of the integral which we take as:

$$\int_{4m_\pi^2}^{(M_Z-\Gamma_Z/2)^2} dt\, f(t) + \int_{(M_Z+\Gamma_Z/2)^2}^{\infty} dt\, f(t), \tag{9.362}$$

where $\Gamma = 2.5$ GeV is the total hadronic Z-width. We add to the QCD continuum contribution the one of the Z-pole estimated to be [562]:

$$\Delta\alpha^{(5)}_{\rm had}|_{M_Z} = 29.2 \times 10^{-5}. \quad (9.363)$$

Then, we obtain the total sum:

$$\Delta\alpha^{(5)}_{\rm had}(M_Z^2) = (2766.3 \pm 4.5) \times 10^{-5}. \quad (9.364)$$

This value is comparable with the ones in the literature reviewed, e.g., in Ref. [571]. It improves and confirms our previous determination in Ref. [572].

9.10 Light 1^{-+} Exotic Mesons

The properties of the light hybrid states have been discussed in Sections 6.14 and 7.6.1 and more recently in Ref. [573] where these previous results have been revised and confronted with the ones from four-quark and molecule states. In this recent paper, the exotic nature of the 1^{-+} exotic states has been discussed.

9.10.1 Abstract

We systematically re-examine the extraction of the masses and couplings of the 1^{-+} hybrid, four-quark and molecule mesons from QCD spectral sum rules (QSSR). To NLO for the perturbative and power corrections, the hybrid mass is $M_H = 1.81(6)$ GeV and $M_H \leq 2.2(2)$ GeV from the positivity of the spectral function. In the same way, but to LO, the four-quark state mass is $M_{4q} = 1.70(4)$ GeV and $M_{4q} \leq 2.4(1)$ GeV, while the molecule mass is about 1.3(1) GeV. The observed $\pi_1(1400)$ and $\pi_1(1600)$ might be explained by a two-component mixing with the set of input masses $(1.2-1.3; 1.70-1.74)$ GeV and with a mixing angle $\theta \simeq -(11.7 \pm 2.2)^\circ$, which slightly favours a molecule/four-quark mixing, and which eventually suggests that the $\pi_1(2015)$ is mostly a hybrid meson. Isospin and non-exotic partners of the previous states and some of their radial excitations are also expected to be found in the energy region around 2 GeV. Further tests of this phenomenological scenario are required.

9.10.2 The hybrid meson

• The two-point correlator

We shall be concerned with the two-point correlator:

$$\Pi^{\mu\nu}_{V/A}(q^2) \equiv i \int d^4x\, e^{iqx} \langle 0|\mathcal{T}\mathcal{O}^{\mu}_{V/A}(x) \left(\mathcal{O}^{\nu}_{V/A}(0)\right)^{\dagger} |0\rangle$$
$$= -\left(g^{\mu\nu}q^2 - q^{\mu}q^{\nu}\right)\Pi^{(1)}_{V/A}(q^2) + q^{\mu}q^{\nu}\Pi^{(0)}_{V/A}(q^2), \qquad (9.365)$$

built from the hadronic local currents $\mathcal{O}^{V/A}_{\mu}(x)$:

$$\mathcal{O}^{\mu}_{V}(x) \equiv: g\bar{\psi}_i\lambda_a\gamma_{\nu}\psi_j G^{\mu\nu}_a, \quad \mathcal{O}^{\mu}_{A}(x) \equiv: g\bar{\psi}_i\lambda_a\gamma_{\nu}\gamma_5\psi_j G^{\mu\nu}_a, \qquad (9.366)$$

which select the specific quantum numbers of the hybrid mesons; A and V refer, respectively, to the vector and axial-vector currents. The invariant $\Pi^{(1)}$ and $\Pi^{(0)}$ refer to the spin one and zero mesons. The anomalous dimension of the current is [201]:

$$\gamma \equiv \nu\frac{\mathrm{d}}{\mathrm{d}\nu}\mathcal{O}^{\mu}_{V} = \left[\gamma_1 \equiv -\frac{16}{9}\right]\frac{\alpha_s}{\pi}\mathcal{O}^{\mu}_{V}. \qquad (9.367)$$

One exploits, in the sum rule approaches, the analyticity property of the correlator which obeys the well-known Källen–Lehmann dispersion relation:

$$\Pi^{(1,0)}_{V/A}(q^2) = \int_0^{\infty} \frac{dt}{t - q^2 - i\epsilon}\,\frac{1}{\pi}\,\mathrm{Im}\,\Pi^{(1,0)}_{V/A} + \cdots, \qquad (9.368)$$

where $\cdots$ represent subtraction terms which are polynomials in the q^2-variable.

The QCD expression of the two-point function including NLO perturbative and the four-quark condensate contributions is given in Ref. [573]. We use the input parameters given in Table 7.10 except the value of the three-gluon condensates where we have used the old value from instanton liquid model not yet updated in the time where the paper has been written:

$$g^3\langle G^3\rangle \simeq (1.2\pm 0.2)\ \mathrm{GeV}^2\langle \alpha_s \mathrm{G}^2\rangle. \tag{9.369}$$

We have also used a less accurate value of the two-gluon condensate:

$$\langle \alpha_s G^2\rangle \simeq (6.8\pm 1.3)10^{-2}\ \mathrm{GeV}^4. \tag{9.370}$$

We have included the effect of the tachyonic gluon mass:

$$\frac{\alpha_s}{\pi}\lambda^2 \simeq -(0.07\pm 0.03)\ \mathrm{GeV}^2. \tag{9.371}$$

- **The spectral function**

In the absence of the complete data, the spectral function is often parametrized using the "naïve" duality ansatz:

$$\frac{1}{\pi}\ \mathrm{Im}\Pi^{(1,0)}_{V/A}(t) \simeq 2M_H^4 f_H^2 \delta(t - M_H^2) \oplus \text{"QCD continuum"} \times \theta(\mathrm{t}-\mathrm{t_c}), \tag{9.372}$$

- **The sum rules**

We shall work with the following Laplace sum rule (LSR) and its ratio [3, 69, 70]:

$$\mathcal{L}_n^{(1,0)}(\tau) = \int_0^\infty dt\ t^n\ \exp(-t\tau)\ \frac{1}{\pi}\ \mathrm{Im}\Pi^{(1,0)}_{V/A}(t),$$
$$\mathcal{R}_n(\tau) \equiv -\frac{d}{d\tau}\log\mathcal{L}_n, \quad (n \geq 0). \tag{9.373}$$

The ratio of the sum rules is a useful quantity to work with in the determination of the resonance mass, as it is equal to the meson mass squared in the usual duality ansatz parametrization. As one can note, there are "*a priori*" two free external parameters (τ, t_c) in the analysis. The optimized result will be (in principle) insensitive to their variations. In some cases, the t_c-stability is not reached due to the too naïve parametrization of the spectral function. In order to restore the t_c-stability of the results, one can fix the t_c-values with the help of FESR (local duality) [521, 522]:

$$\mathcal{M}_n^{(1,0)} = \int_0^{t_c} dt\ t^n\ \frac{1}{\pi}\ \mathrm{Im}\Pi_{V/A}^{(1,0)}(t),$$

$$R_n(t_c) \equiv \frac{\int_0^{t_c} dt\ t^{n+1}\ \frac{1}{\pi}\ \mathrm{Im}\Pi_{V/A}^{(1,0)}(t)}{\int_0^{t_c} dt\ t^n\ \frac{1}{\pi}\ \mathrm{Im}\Pi_{V/A}^{(1,0)}(t)} \simeq M_H^2. \qquad (9.374)$$

• Upper bound on $M_H(1^{-+})$ from LSR

Table 9.27. Value of the upper bound on M_H in GeV at the minima of $\mathcal{R}_0(\tau)$ and at its intersection with $\mathcal{R}_1(\tau)$ for two extremal values of the sets of power corrections in units of GeV^d (d is their dimension).

Power Corrections	$(\alpha_s/\pi)\lambda^2$	$\alpha_s\langle G^2\rangle$	$\alpha_s\langle\bar\psi\psi\rangle^2 \times 10^4$	M_{π_1}
Set 1	−0.04	0.055	4.8	≤ 2.0
Set 2	−0.10	0.081	4.2	≤ 2.4

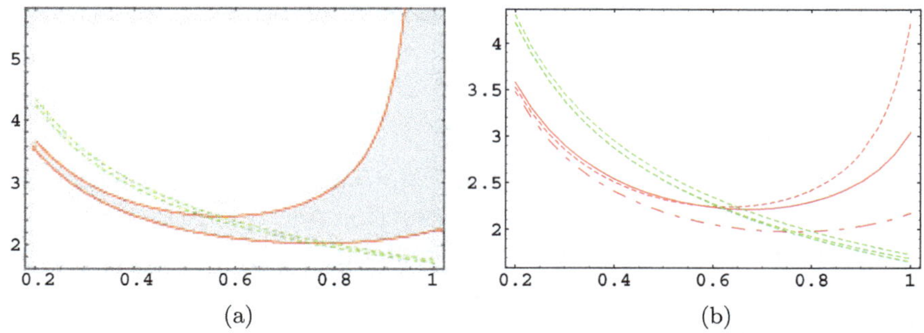

Figure 9.116. (a) Domain of upper bound on M_H in GeV versus the LSR variable τ in GeV^{-2} for two extremal values of the sets of power corrections in Table 9.27 from $\mathcal{R}_0(\tau)$ (red: continuous) and $\mathcal{R}_1(\tau)$ (green: dashed); (b) Effect of radiative corrections on the central value of the bound of M_H from $\mathcal{R}_0(\tau)$ (red: LO (dashed-dotted); NLO condensate+LO perturbative (dashed); total NLO (continuous); green curves are from $\mathcal{R}_1(\tau)$).

One can note that the prediction increases by 100 MeV each when $|\lambda^2|$ and $\alpha_s\langle G^2\rangle$ decrease from their central value to the one allowed in the range given in Section 2, while it increases by 25 MeV when $10^4\alpha_s\langle\bar{\psi}\psi\rangle^2$ increases from 4.5 to 4.8. The effects of the variation of Λ and M_0^2 within the range in Section 2 are invisible. At the stability points (in the sum rule variable τ) of the continuous (red) curves which are also the intersections with the dashed (green) curves, one can deduce:

$$M_H(1^{-+}) \leq (2.2 \pm 0.2)\ \text{GeV}. \tag{9.375}$$

This value confirms and improves previous results in Refs. [199, 201, 575]. The result (2.2 ± 0.2) GeV from lattice with two dynamical quarks though still consistent with this bound is on its border. We show in Fig. 9.116 the effects of radiative corrections on the results. One can see that the ones due to the condensates give larger effects and increase the bound from 2.0 (LO) to 2.25 GeV, while the ones due to the perturbative terms decrease the mass prediction by 0.05 MeV, where the α_s-correction and the one induced by the anomalous dimension act in the opposite directions. The ratio $\mathcal{R}_1$ is almost unaffected by the radiative corrections (dotted (green) curves on top of each other).

• QSSR predictions of f_H and M_H for 1^{-+}

Following Ref. [576], we introduce the RGI coupling $\hat{f}_H$ defined as:

$$f_H(\nu) = \frac{\hat{f}_H}{(\log \nu/\Lambda)^{\gamma_1/-\beta_1}}, \tag{9.376}$$

with γ_1 the anomalous dimension in Eq. (9.367) and $-\beta_1 = 1/2(11 - 2n_f/3)$ the first coefficient of the β function for n_f flavours.

Using FESR, we show the predictions for decay constant Fig. 9.117(a) and the mass of the 1^{-+} in Fig. 9.117(b). Using the stability criterion on the variation of t_c, and the intersection of the predictions of different moments, and the sets of power corrections in Table 9.27, one obtains the optimal value for 1^{-+}:

$$\begin{aligned} \hat{f}_H &= 9.6(1.4)\ \text{MeV}: t_c = (4.4 \sim 5.2)\ \text{GeV}^2, \\ M_H &= 1.80(10)\ \text{GeV}: t_c = (2.8 \sim 5.0)\ \text{GeV}^2. \end{aligned} \tag{9.377}$$

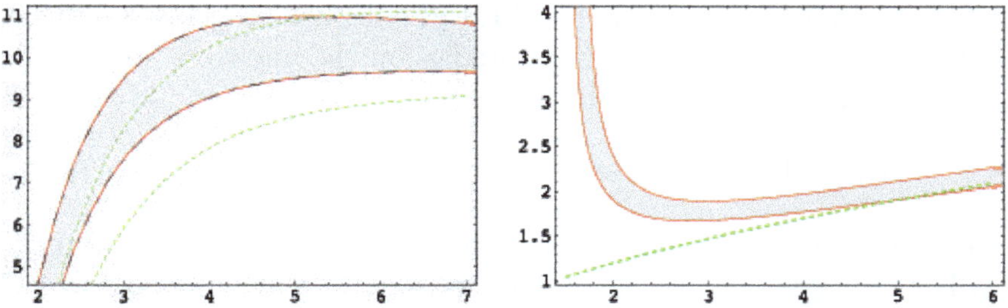

(a) (b)

Figure 9.117. (a) Domain of FESR predictions versus the QCD continuum threshold t_c using the set of power correction in Table 9.27 for f_H from $\mathcal{M}_0$ (red: continuous) and $\mathcal{M}_1$ (black: dashed (on top of the red curve)) and $\mathcal{M}_2$ (green: dotted); (b) The same as for f_H but for M_H: from $R_0(t_c)$ (red: continuous); from $R_1(t_c)$ (green: dotted).

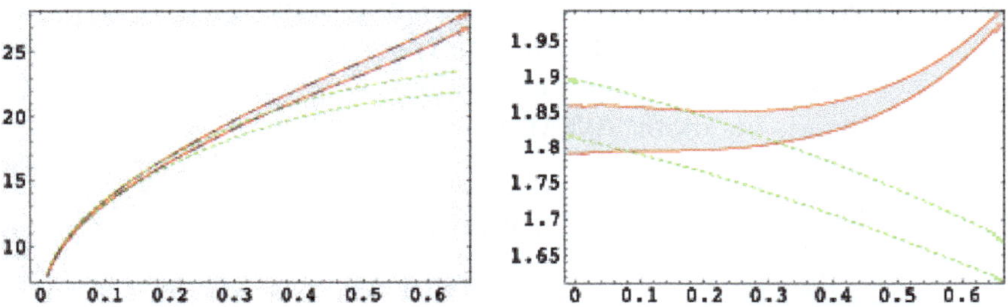

(a) (b)

Figure 9.118. Predictions of the LSR in the 1^{-+} hybrid channel using the central values of the QCD input parameters and two values of $t_c = 4.6$ (beginning of the τ-stability) and 5 GeV^2 fixed from the previous FESR results: (a) RG invariant coupling constant $\hat{f}_H$ from $\mathcal{L}_0$ (red: continuous), from $\mathcal{L}_1$ (black: dashed-dotted (on top of the red curve)), from $\mathcal{L}_1$ (green: dashed); (b) M_H from $\mathcal{R}_0(\tau)$ (red: continuous), from $\mathcal{R}_1(\tau)$ (green: dotted).

We show in Fig. 9.118 the predictions of the LSR in the 1^{-+} channel using the central values of the QCD input parameters and using two values of $t_c = 4.6$ (beginning of the τ-stability) and 5 GeV^2 fixed from the previous FESR results. The prediction for the decay constant is not conclusive as there is not a stability in the LSR variable τ. However, one can see from Fig. 9.118 that the different predictions interact for:

$$\hat{f}_H = (10 \sim 22) \text{ MeV}, \tag{9.378}$$

which is consistent with the previous FESR result but less accurate. Therefore, we consider as a final prediction the one from FESR. For the mass prediction, the LSR gives:

$$M_H = 1.81(3)(7) \text{ GeV} : t_c = (4.6 \sim 5.0) \text{ GeV}^2, \tag{9.379}$$

where the first and second errors come, respectively, from t_c and the QCD input parameters. We take the (naïve arithmetic) average of the FESR and LSR results and we take the quadratic average of the errors. Then, we obtain:

$$\langle M_H \rangle = 1.81(6) \text{ GeV}. \tag{9.380}$$

The central value of the result in Eq. (9.380) has increased by 100–200 MeV compared to the previous QSSR NLO perturbative results quoted in Ref. [573]. The difference with the one in Refs. [201, 574] is due to the inclusion of radiative corrections for both perturbative and power corrections here, which increase the mass prediction by about 200 MeV.

9.10.3 Masses of the 1^{-+} four-quark states

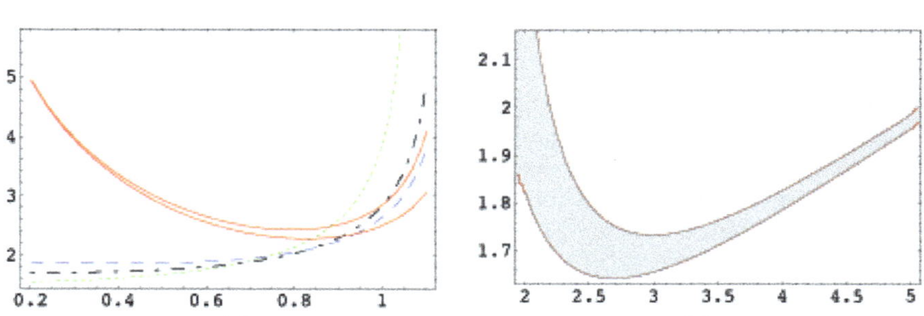

(a) (b)

Figure 9.119. Predictions of the four-quark state mass in GeV using the current in Eq. (9.382): (a) LSR: domain spanned by the upper bound versus τ using the largest range spanned by the QCD parameters given in Table 7.10 (red: continuous curve); mass using the central values of the QCD parameters for $t_c = 3\, GeV^2$ (green: dashed), $t_c = 4\, GeV^2$ (black: dotted-dashed), $t_c = 5\, GeV^2$ (blue: dashed); (b) FESR versus t_c: region spanned by the largest range of QCD parameters.

We reconsider the QSSR analysis in Ref. [577] using the previous set of QCD parameters where the values of the gluon and four-quark condensates are about a factor 2 higher. We shall introduce the renormalization group invariant $\langle\bar\psi\psi\rangle$ condensate [29]:

$$\hat\mu_i^3 = \frac{\langle\bar\psi_i\psi_i\rangle(\nu)}{(\log\nu/\Lambda)^{2/-\beta_1}}, \tag{9.381}$$

with [74]: $\hat\mu_u = -(263\pm7)$ MeV. We neglect the small Q^2-dependence of $\alpha_s\langle\bar\psi\psi\rangle$ and $\langle g\bar\psi G\psi\rangle$ as well as the anomalous dimension of the four-quark current, which should have small effect in the mass determination. The four-quark meson can be described by the diquark anti-diquark operators[32]:

$$\begin{aligned}\eta^M_{1\mu} &= u_a^T C\gamma_\mu d_b(\bar u_a C\bar d_b^T + \bar u_b C\bar d_a^T) + \cdots,\\ \eta^M_{3\mu} &= u_a^T C d_b(\bar u_a\gamma_\mu C\bar d_b^T - \bar u_b\gamma_\mu C\bar d_a^T) + \cdots.\end{aligned} \tag{9.382}$$

where $\cdots$ denotes an interchange between γ_u and 1 or γ_5. The QCD expression of the LSR of the corresponding correlator can be written to LO in α_s [577].

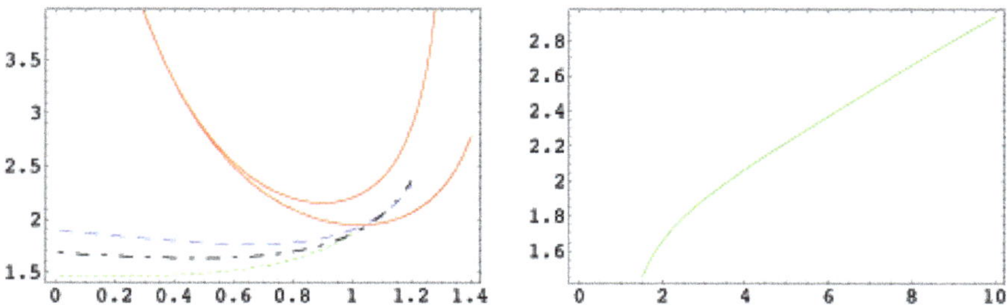

(a) (b)

Figure 9.120. Predictions of the four-quark state mass in GeV using the current in Eq. (9.385): (a) LSR: domain spanned by the upper bound versus τ using the largest range spanned by the QCD parameters given in Table 7.10 (red: continuous curve); mass using the central values of QCD parameters for $t_c = 3$ GeV^2 (green: dashed), $t_c = 4$ GeV^2 (black: dotted-dashed), $t_c = 5$ GeV^2 (blue: dashed); (b) FESR versus t_c using the central values of QCD parameters.

[32]In principle, these operators should mix under renormalizations [113] and one should build their renormalization group invariant (RGI) combination for describing the physical state.

The result of the analysis is shown in Fig. 9.119, from which we can deduce in units of GeV:

$$M_{4q1} \leq 2.4(1)\text{: LSR;} \quad M_{4q1} = 1.70(4)\text{: FESR,} \tag{9.383}$$

for $t_c \simeq (2.5{\sim}3)$ GeV^2.

Here, one can note that the LSR result does not have τ- and t_c-stabilities such that only an upper bound can be extracted from the positivity of the spectral function. This result is slightly higher than the one in Ref. [577] due to the consideration of the violation of factorization in the estimate of the high-dimension condensates here. We have repeated the analysis for $\eta^M_{3\mu}$. We show in Fig. 9.120(a) the result of the analysis from the LSR moment $\mathcal{R}_0(\tau)$ which presents τ-stability for $t_c \geq 4$ GeV^2 but no t_c-stability. In the expanded form, FESR $R_0(t_c)$ presents a zero for $2.5 \leq t_c \leq 7$ GeV^2 due to $\mathcal{M}_0(t_c)$ signalling large non-perturbative effects for this quantity. This zero shows up like a minimum in t_c in the non-expanded form of $R_0(t_c)$ given a mass $M \approx 1.2$ GeV, but at too low value of $t_c \approx 2.$ GeV^2, where the OPE does not converge.[33] At this t_c-values, the LSR also does not show a τ-stability. Therefore, we do not retain this result from our analysis. $R_1(t_c)$ moment has a much better behaviour but (unfortunately) does not show a t_c-stability (see Fig. 9.120(b)). The most conservative result from both LSR and FESR is:

$$1.64 \leq M_{4q3} \leq 2.1(1): \quad t_c \geq 4 \text{ GeV}^2, \tag{9.384}$$

where the upper bound comes from the positivity of the LSR moment.

Currents with higher number of γ matrices have been also considered in Ref. [577]:

$$\eta^M_{2\mu,4\mu} \equiv u_a^T C\gamma^\nu\gamma_5 d_b(\bar{u}_a C\sigma_{\mu\nu}\gamma_5\bar{d}_b^T \pm \bar{u}_b C\sigma_{\mu\nu}\gamma_5\bar{d}_a^T + \cdots), \tag{9.385}$$

where $\cdots$ is an interchange between γ^ν and $\sigma_{\mu\nu}$. The analysis of the corresponding correlators leads to similar results (within the errors) than the ones in Eq. (9.382). For definiteness, we fix the four-quark state mass to the value from Eq. (9.383):

$$M_{4q} = 1.70(4) \text{ GeV}. \tag{9.386}$$

[33]In Ref. [577], this minimum corresponds to a slightly higher value of the mass due to the different choice of the QCD set of parameters.

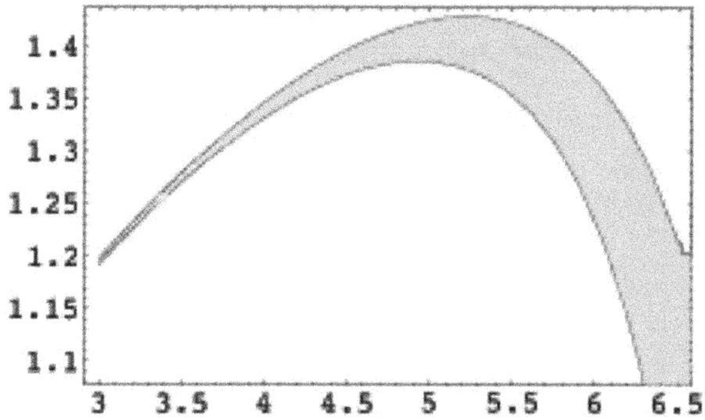

Figure 9.121. Mass of the molecule state associated to the current J_1^μ in Eq. (9.387) versus t_c using the largest range spanned by the QCD parameters given in Table 7.10 (red: continuous curve).

9.10.4 Masses of the 1^{-+} molecules

The possibility to form two 1^{-+} molecules with the two operators:

$$J_1^\mu \sim (\bar{\psi}\gamma_5\psi)(\bar{\psi}\gamma_5\gamma^\mu\psi),$$
$$J_2^{\mu\nu} \sim \epsilon^{\nu\nu\rho\sigma}\Big[(\bar{u}\gamma_5\gamma_\rho d)(\bar{d}\gamma^\sigma u) - (\bar{d}\gamma_5\gamma_\rho u)(\bar{u}\gamma^\sigma d)\Big], \qquad (9.387)$$

has been discussed in Ref. [578], where the meson associated to the first (respectively second) operator is expected to decay into $\eta\pi$, $\eta'\pi$ (respectively $\rho\pi$, $b_1\pi$). Using the QCD expression given in Ref. [578] without the one-direct instanton contribution which is largely affected by the uncertainty of the instanton density ρ appearing as ρ^6, we use the lowest FESR moment $R_0(t_c)$ for the analysis of the meson mass M_{mol1} associated to the current J_1^μ where a t_c-stability is obtained at $t_c \simeq 5$ GeV^2 (Fig. 9.121), while the LSR moments $\mathcal{R}_{0,1,2}$ do not give conclusive results. The mass M_{mol2} associated to the current J_2^μ increases with t_c for the FESR, while the LSR moments decrease slowly with the sum rule variable τ but present a zero around $\tau = 0.7{-}0.8$ GeV^{-2} (not shown in Fig. 9.122), while it presents t_c-stability around 3 GeV^2. The optimal results are:

$$M_{\text{mol1}} = 1.41(3)\ \text{GeV}, \quad M_{\text{mol2}} \simeq (1.2 \sim 1.4)\text{GeV}, \qquad (9.388)$$

which agree within the errors with the ones in Ref. [578]. However, we can consider that the two operators in Eq. (9.387) mix under renormalization such that only one physical state couples with the corresponding RGI operator, which we fix to be:

$$M_{\rm mol} = 1.3(1)\ {\rm GeV}. \tag{9.389}$$

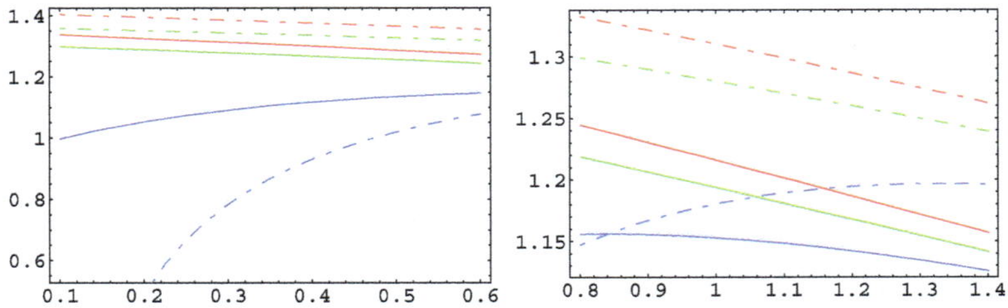

Figure 9.122. Mass of the molecule state associated to the current J_2^μ in Eq. (9.387) from LSR versus the sum rule variable τ= 0.1 to 0.6 GeV^{-2} and 0.8 to 1.4 GeV^{-2} and for different values of t_c using the central values of the QCD parameters given in Table 7.10. and for two LSR moments $\mathcal{R}_0$ (continuous curve) and $\mathcal{R}_1$ (dotted-dashed curve): t_c = 2.5 GeV^2 (green), t_c = 3 GeV^2 (red) and t_c = 3.5 GeV^2 (blue).

9.10.5 Nature of $\pi_1(1400)$, $\pi_1(1600)$ and $\pi_1(2015)$

From your previous analysis and for a further phenomelogical use, we shall consider one hybrid, one four-quark and one molecule states below 2 GeV, with the masses from Eqs. (9.380), (9.386) and (9.389) in GeV:

$$M_H = 1.81(6),\ \ M_{4q} = 1.70(4),\ \ M_{\rm mol} = 1.3(1). \tag{9.390}$$

$\pi_1(1400)$ is intriguing as it is seen to decay into $\eta\pi$ and $\eta'\pi$ but not into $\rho\pi$, $b_1\pi$ [33].[34] The latter decays being also expected for a hybrid state [574]. Looking at our *bare (unmixed)* mass predictions in Eq. (9.390), one may expect that the molecule state is the most probable candidate. We consider that the suppression of the $\rho\pi$ decay can be due to a mixing of this molecular state with the four-quark or/and hybrid states. We use a minimal two-component mixing:

$$\begin{aligned}\pi_1(1400) &= \cos\theta_{\rm mol}|{\rm mol}\rangle + \sin\theta_{\rm mol}|X\rangle,\\ \pi_1(1600) &= -\sin\theta_{\rm mol}|{\rm mol}\rangle + \cos\theta_{\rm mol}|X\rangle,\end{aligned} \tag{9.391}$$

[34] For a review on different experimental results and related problems, see, e.g., Ref. [436].

where X is a four-quark or hybrid state. The "best fit" is obtained for the sets:

$$(M_{\rm mol}, M_X) = (1.2\text{–}1.3;\ 1.70, 1.74)\ \text{GeV} \longrightarrow$$
$$\theta_{\rm mol} \simeq -(11.7 \pm 2.2)^0, \tag{9.392}$$

which slightly favours a molecule/four-quark mixing. This result may suggest that the $\pi_1(2015)$, quoted in the extended version of PDG [33] can have more hybrid state in its wave function. One also expects that the non-exotic meson 1^{--} as well as the isospin partners of these 1^{-+} exotics should be almost degenerate in masses with these exotic states, while some of their radial excitations have masses of the order of optimal continuum threshold:

$$M' \approx \sqrt{t_c} \simeq (1.7{\sim}2.2)\ \text{GeV}. \tag{9.393}$$

Further experimental and theoretical tests of this mixing scheme are required.

9.11 Light Scalar Quarkonia from LSR at HO

In this section, we review the paper in Ref. [579].

9.11.1 Abstract

We scrutinize and improve some determinations of the masses and couplings of light scalar quarkonia ($\bar{q}q$ and four-quark states) and present new results for the $\pi^+\pi^-, K^+K^-, \ldots$ molecules using QCD Laplace sum rule (LSR) truncated at the $D = 6$ dimension vacuum condensates. We pay special attention to the higher order perturbative (PT) corrections up to the (estimated) $\mathcal{O}(\alpha_s^5)$ which improve the quality of the analysis. We request that the optimal results obey the rigorous constraint: Resonance $\geq$ QCD continuum contributions ($R_{P/C} \geq 1$) in the LSR which excludes a Breit–Wigner/on-shell (not to be confused with a complex pole) scalar meson mass around (500–600) MeV obtained for values [$t_c \leq (1 - 1.5)$ GeV^2] of the QCD continuum threshold. Mass splittings due to $SU(3)$ breakings are small. We discuss the different assignments of the observed scalar mesons below 2 GeV in the conclusions where the $I = 0$ states are compared with the scalar gluonia. The results are compiled in Tables 9.30 and 9.31. *We hope that the systematic analysis done in this paper helps to clarify the complex spectra of the light scalar mesons.*

9.11.2 The light scalar-isoscalar ($\bar{u}u + \bar{d}d$) state

In QCD, the isoscalar states are associated to the trace of the energy–momentum tensor defined in the introduction. We shall separate the quark part of θ_μ^μ into two currents:

$$J_2 = \frac{m_q}{\sqrt{2}}\left(\bar{u}u + \bar{d}d\right) \quad \text{and} \quad J_3 = m_s\bar{s}s, \tag{9.394}$$

where $m_q = (m_u + m_d)/2$. In this section, we shall first analyze the correlator:

$$\psi_{\bar{q}q}(q^2) = i\int d^4x\, e^{iqx}\,\langle 0|\mathcal{T} J_2(x) J_2^\dagger(0)|0\rangle. \tag{9.395}$$

In this paper, we shall improve the estimate done in Refs. [7, 62, 85, 124, 125, 580] by adding higher order PT corrections. The mass corrections and non-perturbative contributions up to $D = 6$ are listed in Eqs. (53.24)–(53.31) of Ref. [6] and in Refs. [85, 443, 581].

• Estimate of the mass and coupling

One can estimate the coupling from the low moment $\mathcal{L}_0$ and the mass from the ratio of moments $\mathcal{R}_{10}$. The LSR expression of the moment reads to N3LO of PT series and to leading order of the NPT contributions truncated at $D = 6$:

$$\begin{aligned}\mathcal{L}_0^c(\tau) &\equiv \int_{(m_u+m_d)^2}^{t_c} dt\, e^{-t\tau}\,\frac{1}{\pi}\,\mathrm{Im}\,\psi_S(t)\\ &= \overline{m}_q^2\left\{\int_{(m_u+m_d)^2}^{t_c} dt\, e^{-t\tau}\,\frac{1}{\pi}\,\mathrm{Im}\,\psi_{\mathrm{Pert}}(t) + \frac{3}{8\pi^2}\tau^{-2}\sum_{n=1}^{3}\delta^{(2n)}\tau^n\right\},\end{aligned} \tag{9.396}$$

where $\text{Im}\,\psi_{\text{Pert}}(t)$ is known up to order α_s^4 [34, 439–441, 459]. For three flavours, it reads including the π^2-term from the analytic continuation:

$$\frac{1}{\pi}\,\text{Im}\psi_\pi(t)|_{\text{Pert}}$$
$$= \frac{3}{8\pi^2}\,t\left(1+\frac{17}{3}a_s+31.863a_s^2+89.157a_s^3-536.749a_s^4+\mathcal{O}(a_s^5)\right). \tag{9.397}$$

The mass corrections to order α_s and non-perturbative corrections to lowest order read:

$$\delta^{(2)} = -2\Big[(1+7.64\,a_s)(\bar{m}_u^2+\bar{m}_d^2)+(1+11.64\,a_s)\,\bar{m}_u\bar{m}_d\Big],$$
$$\delta^{(4)} = \frac{8\pi^2}{3}\left[\left(m_u+\frac{m_d}{2}\right)\langle\bar{u}u\rangle+\left(m_d+\frac{m_u}{2}\right)\langle\bar{d}d\rangle\right]+\frac{\pi}{3}\langle\alpha_s G^2\rangle,$$
$$\delta^{(6)} = \frac{4\pi^2}{3}\Big[\left(m_u\langle\bar{d}Gd\rangle+m_d\langle\bar{u}Gu\rangle\right)$$
$$-\frac{32}{27}\pi\alpha_s\left(\langle\bar{u}u\rangle^2+\langle\bar{d}d\rangle^2+9\langle\bar{u}u\rangle\langle\bar{d}d\rangle\right)\Big]. \tag{9.398}$$

We extract the meson mass from the ratio $\mathcal{R}_{10}(\tau,t_c)$ of moments defined in Eq. (9.4). The analysis for the N3LO PT series approximation is shown in Fig. 9.123.

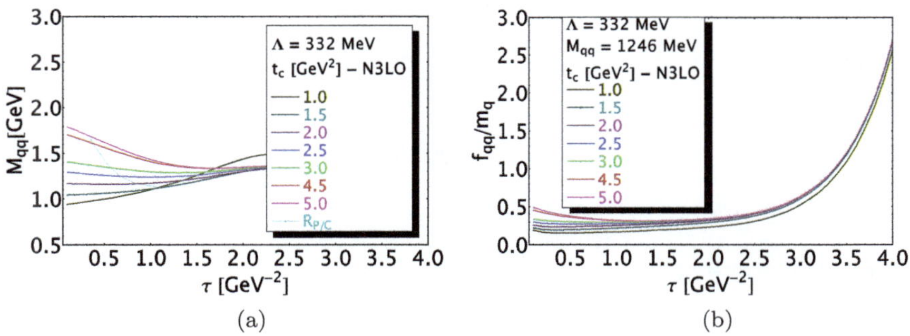

Figure 9.123. Behaviour of (a) mass and (b) coupling of the $\bar{q}q \equiv \frac{1}{\sqrt{2}}\left(\bar{u}u \pm \bar{d}d\right)$ meson states versus τ for different values of t_c.

• Optimization procedure

We shall base our analysis on the stability of the results versus the change of the external input parameters (sum rule variable τ and continuum threshold t_c [6, 7] to which we implement the (rigorous) condition [285]:

$$R_{P/C} \equiv \frac{\text{Lowest Pole}}{\text{QCD continuum}} = \frac{\int_{t_>}^{t_c} dt\ e^{-t\tau}\,\text{Im}\,\psi_S(t,\mu)}{\int_{t_c}^{\infty} dt\ e^{-t\tau}\text{Im}\,\psi_S(t,\mu)} \geq 1. \quad (9.399)$$

This condition is necessary for a reliable prediction of the resonance contribution which is disentangled from the QCD continuum one.

• Truncation of the PT series

In order to estimate the error due to the truncation of the PT series, we first assume that the next term of the series has a positive sign (alternate sign). We estimate its absolute size by observing that the coefficients of the series have a geometrical growth behaviour similar to the one found in Ref. [234] for the complete two-point correlator:

$$\frac{1}{\pi}\,\text{Im}\psi_{\bar{q}q}(t)|_{\text{Pert}} = \frac{3}{8\pi^2}\,t \times \left\{1 + c\,a_s + (c\,a_s)^2 + \frac{1}{2}\left[(c\,a_s)^3 - (c\,a_s)^4 + (c\,a_s)^5\right]\right\}, \quad (9.400)$$

where $c = 17/3$ and the last term is our prediction. Then, the estimated coefficients lead to:

$$\frac{1}{\pi}\,\text{Im}\psi_{\bar{q}q}(t)|_{\text{Pert}}^{\text{estimate}} = \frac{3}{8\pi^2}\,t\left(1 + \frac{17}{3}a_s + 32.1a_s^2 + 90.98a_s^3 - 515.56a_s^4 + 2921.52a_s^5\right), \quad (9.401)$$

where the coefficients up to order α_s^4 agree remarkably with the analytic ones given in Eq. (9.397). The estimated N5LO contribution is:

$$\frac{1}{\pi}\text{Im}\,\psi_{\bar{q}q}(t)|_{\text{N5LO}} = \left(\frac{3}{8\pi^2}\right)\,t \times 2921.51\,a_s^5, \quad (9.402)$$

which should be checked from a direct calculation. We truncate the PT series at N3LO for extracting the optimal result and consider that the error due to the truncation of the PT series comes from the sum N4LO $\oplus$ N5LO. Then, we obtain, at the optimal value of $\tau \simeq 1$ GeV^{-2}:

$$\Delta M_{\bar{q}q}|_{\text{Geom}} = \pm 1\,\text{MeV}, \quad \Delta f_{\bar{q}q}/m_q|_{\text{Geom}} = \pm 15 \times 10^{-3}. \tag{9.403}$$

An alternative way to estimate the error is to parametrize the contribution of the non-calculated order term by the tachyonic gluon mass λ proposed in Refs. [228, 235, 554] which is a phenomenological alternative to the large β approach on UV renormalon:

$$\frac{1}{\pi}\,\text{Im}\,\psi(t)|_{\text{Tach}} = -\left(\frac{3}{8\pi^2}\right) 4\, a_s\, \lambda^2, \tag{9.404}$$

with [235, 236, 238, 304]:

$$a_s\, \lambda^2 = -(7 \pm 3) \times 10^{-2}\ \text{GeV}^2. \tag{9.405}$$

This leads to:

$$\Delta M_{\bar{q}q}|_{\text{Tach}} = \pm 27\,\text{MeV}, \quad \Delta f_{\bar{q}q}/m_q|_{\text{Tach}} = \pm 24 \times 10^{-3}. \tag{9.406}$$

Though it has been shown in Ref. [234] that the two approaches are equivalent, we shall see in the following sections that the tachyonic gluon mass tends to overestimate the error in some cases. For definiteness, we shall keep, here and in the following sections, the most optimistic errors from the geometric growth of the α_s-coefficient.

One can note from Figs. 9.124 and 9.125 that the size of the coupling is strongly affected by the PT radiative corrections and starts to be stable from N2LO corrections.

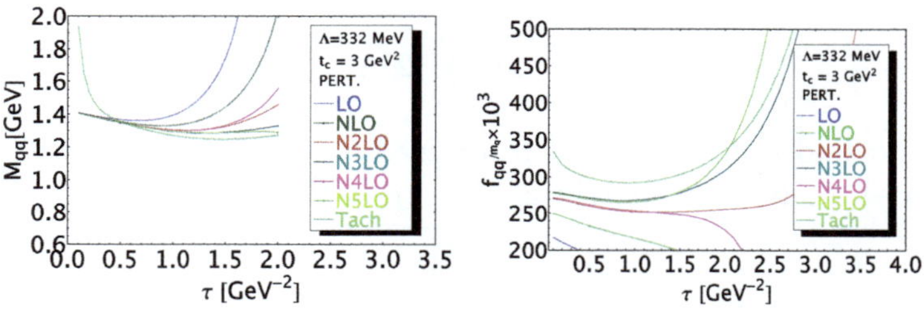

(a) (b)

Figure 9.124. Behaviour of (a) mass and (b) coupling of the $\bar{q}q \equiv \frac{1}{\sqrt{2}}(\bar{u}u \pm \bar{d}d)$ meson states for different truncation of the PT series for fixed value of $t_c = 3$ GeV^2 versus τ.

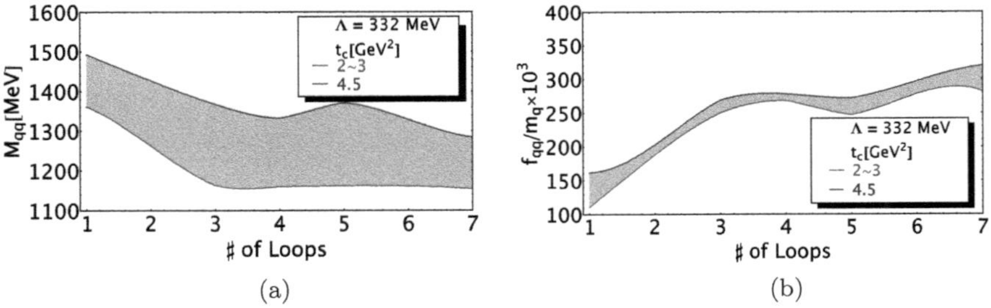

Figure 9.125. Behaviour of (a) mass and (b) coupling of the $\bar{q}q \equiv \frac{1}{\sqrt{2}}\left(\bar{u}u \pm \bar{d}d\right)$ meson states for different truncation of the PT series and for two extremal values of t_c where the τ-stability is reached.

• On the truncation of the OPE

We truncate the OPE by assuming that the next non-calculated term is of the form:

$$\Delta\text{OPE} = (\Lambda^2\tau) \times (D = 6\,\text{contributions}). \tag{9.407}$$

Adding these estimated contributions as some other sources of the errors in Table 9.28, we obtain the final result within an NWA:

$$M_{\bar{q}q} = 1246(95)\ \text{MeV}, \quad f_{\bar{q}q}/\bar{m}_q(\tau) = 274(43) \times 10^{-3}, \tag{9.408}$$

where the mass $\bar{m}_q(\tau)$ is evaluated at $\tau \simeq 0.9$ GeV^{-2} at which the optimal result has been extracted.

• Finite width correction

In the literature, one often identifies the lightest scalar meson with the broad σ found from $\pi\pi$ and $\gamma\gamma$ scattering:

$$M_\sigma^{\text{pole}}[\text{MeV}] = (455 \pm 12) - \text{i}\,(260 \pm 12), \tag{9.409}$$

while its hadronic couplings from $Ke4 \oplus \pi\pi$ scatterings within an improved analytic K-matrix model are [407]:

$$|g_{\sigma\pi^+\pi^-}| \simeq 2.65(10)\ \text{GeV}, \quad r_{\sigma\pi K} \simeq \frac{|g_{\sigma K^+K^-}|}{|g_{\sigma\pi^+\pi^-}|} \simeq 0.37(6). \tag{9.410}$$

In order to take into account the finite width correction to the previous result obtained using a narrow width approximation (NWA), we make the replacement:

$$\pi\delta(t - M_\sigma^2) \to BW(t) = \frac{M_\sigma \Gamma_\sigma}{(t - M_\sigma^2)^2 + M_\sigma^2 \Gamma_\sigma^2}, \tag{9.411}$$

in the parametrization of the spectral function. Then, we study the effect of the width to the ratio:

$$\left(M_{\bar{q}q}^{\rm BW}\right)^2 = \frac{\int_0^{t_c} dt\, t^2\, e^{-t\tau}\, BW(t)}{\int_0^{t_c} dt\, t\, e^{-t\tau} BW(t)}. \tag{9.412}$$

We use in the integral the mass from the NWA and the value $\tau = 1.05$ GeV^{-2} where the optimal value of $M_{\bar{q}q}|_{NWA}$ has been obtained for $t_c = 2.56$ GeV^2 corresponding to the mean value of the mass from the extremal values of t_c. We show the analysis in Fig. 9.126. From the range $\Gamma_{\pi\pi} = 120$ MeV predicted from vertex sum rule [7,125,582] to 520 MeV for the complex pole (see Eq. (9.409)) and 700 MeV for the on-shell/Breit–Wigner mass [410,523], one can see that the width decreases, respectively, slightly the mass by (in units of MeV):

$$\Delta M_{\bar{q}q}^{\rm BW} = -22|_{\rm vertex\,SR}, \quad -60|_{\rm pole}, \quad -70|_{\rm on\text{-}shell}, \tag{9.413}$$

which is within the errors of the LSR determinations. Using the width predicted from vertex sum rules, we deduce:

$$M_{\bar{q}q}|_{\rm vertex\,SR} = 1229(95)\ {\rm MeV}, \tag{9.414}$$

which can be compared with the on-shell mass in Eq. (9.473).

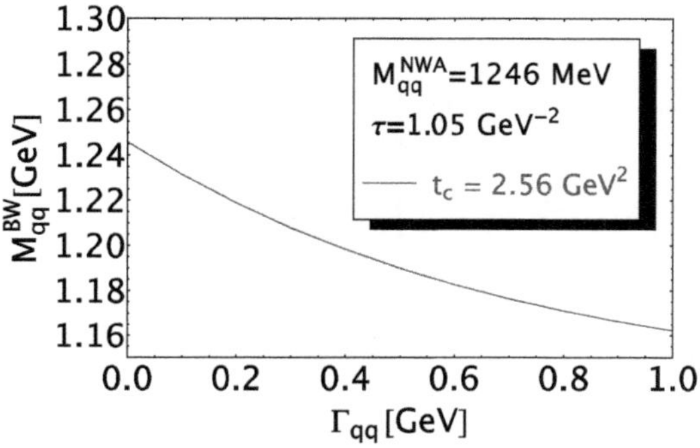

Figure 9.126. Analysis of the finite width effect on $M_{\bar{q}q}$ for $t_c = 2.56$ GeV^2 corresponding to the central value of $M_{\bar{q}q} = 1246$ MeV in an NWA at $\tau = 1.05$ GeV^{-2}.

• Factorization of the four-quark condensates

Another point which does not make a consensus in the literature is the vacuum saturation or factorization in order to estimate the size of the four-quark condensate:

$$\langle 0|\bar{q}q\bar{q}'q'|0\rangle \simeq \rho\langle 0|\bar{q}q|0\rangle\langle 0|\bar{q}'q'|0\rangle, \tag{9.415}$$

where ρ quantifies the deviation from the vacuum saturation. Several analysis of $e^+e^- \to$ hadrons and τ-decay data have shown that $\rho \simeq (3-6)$ [18, 19, 67, 115, 120, 289, 359, 522, 524] while the renormalization of the four-quark operators indicates that vacuum saturation is inconsistent with a renormalization group invariance of the four-quark operators [113]. In the following, we shall test the effect of this assumption on the mass and coupling predictions. The analysis is shown in Fig. 9.127. One can note that the τ-stability of the mass is less good than in the case of a violation of factorization, where it is an inflexion point here. It makes its localization less precise while the range of t_c-values is more restricted for $t_c = (3-4.5)$ GeV^2. For the mass, this leads to an error of 33 MeV from t_c and 77 MeV taking $\tau = 1.5$ GeV^{-2} and $\Delta\tau = 0.2$ GeV^{-2}. For the coupling, the minimum appears at $\tau \simeq 0.9$ and 1.2 GeV^{-2}. The errors due to t_c and τ on the coupling are 12×10^{-3} and 3×10^{-3}. Adding the other sources of errors from Table 9.28, one obtains:

$$M_{\bar{q}q} = 1131(90) \text{ MeV}, \quad f_{\bar{q}q}/\bar{m}_q(\tau) = 338(24)\times 10^{-3}, \tag{9.416}$$

where we have subtracted to the mass the width correction of –22 MeV.

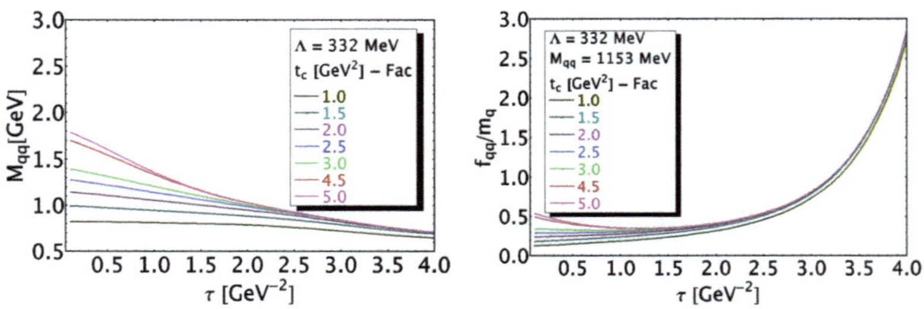

(a) (b)

Figure 9.127. Behaviour of (a) mass and (b) coupling of the $\bar{q}q \equiv \frac{1}{\sqrt{2}}(\bar{u}u \pm \bar{d}d)$ meson states versus τ for different values of t_c using a factorization of the four-quark condensate.

9.11.3 The $\bar{u}s$ and $\bar{s}s$ states

One extends the analysis to the current:

$$J_{\bar{u}s} \equiv \partial_\mu V^\mu_{\bar{u}s} = (m_u - m_s)\bar{u}s, \quad J_3 = m_s\,\bar{s}s. \tag{9.417}$$

We use the previous expressions in Eqs. (9.397) and (9.398) by replacing d by s for $\bar{u}s$ state and u, d by s for the $\bar{s}s$ states. The behaviours of the different curves are similar to the previous case of $\bar{q}q$ state and will not be repeated here. We just quote the results from Table 9.28:

$$M_{\bar{u}s} = 1276(61)\ \text{MeV}, \quad f_{\bar{u}s}/(\bar{m}_u - \bar{m}_s)(\tau) = 264(25)\times 10^{-3},$$
$$M_{\bar{s}s} = 1288(65)\ \text{MeV}, \quad f_{\bar{s}s}/(\bar{m}_s)(\tau) = 256(19)\times 10^{-3}. \tag{9.418}$$

Comparing these values with the one for the $\bar{u}u \pm \bar{d}d$ state, one can note that the SU(3) breakings are small. They shift the mass by about 28 MeV for $\bar{u}s$ and 40 MeV for $\bar{s}s$ states. The corresponding shift for the coupling is -10×10^{-3} (respectively -18×10^{-3}) for the $\bar{u}s$ (respectively $\bar{s}s$ state). These shifts are relatively tiny. The ratio of masses:

$$\frac{M_{\bar{s}s}}{M_{\bar{u}s}} \simeq 1.01 \tag{9.419}$$

is consistent with the direct determination (1.03 ± 0.02) in Ref. [582].

9.11.4 Comments on ordinary $\bar{q}q$ mesons

◇ We consider the results in our previous analysis as an update of the ones obtained earlier in Refs. [7, 580, 582].

◇ A comparison of these results with the on-shell mass for σ and $f_0/a_0(980)$ are more appropriate than with the residue at the complex plane. These results indicate that the lightest $\bar{q}q$ mesons are in the range:

$$M_{\bar{q}q'} \simeq (1040\sim 1353)\ \text{MeV} \qquad \text{for} \qquad q, q' \equiv u, d, s \tag{9.420}$$

within the accuracy of the LSR approach.

◇ This range of values is consistent with the on-shell masses of the observed mesons $\sigma/f_0(500)$, $f_0/a_0(980)$ and $K_0^*(1430)$. However, the estimated hadronic width of about 120 MeV does not favour a pure $\bar{q}q$ interpretation of the broad $\sigma/f_0(500)$ but may favour the meson–gluonium mixing scenario proposed in Refs. [125, 580, 582].

◇ The predicted value $M_{\bar{u}s} \simeq 1276(58)$ MeV is comparable with the one of the $K_0^*(1350)$, but its predicted total width from vertex sum rule is expected to be narrower than the experimental data as can be deduced from Ref. [582].

◇ The predicted value of the $\bar{s}s$ state mass is too low compared to the candidate $f_0(1710)$, while the predicted K^+K^- width [582] is about 1/2 of the observed one.

9.11.5 The $\sigma/f_0(500)$ as a dipion molecule

• **The $\pi^+\pi^-$ dipion field and its two-point function**

The interpolating current of the dipion state:

$$\mathcal{O}_{\pi^+\pi^-} = J_{\pi^+} \otimes J_{\pi^-}(x) \equiv (m_u + m_d)^2 \tilde{\mathcal{O}}_{\pi^+\pi^-}(x), \tag{9.421}$$

is the convolution of the two renormalization group invariant divergences of axial current (pion current):

$$J_{\pi^+}(x) = (m_u + m_d) : \bar{d}(i\gamma_5)u(x). \tag{9.422}$$

We do not consider the scalar $\bar{d}u$ current which cannot participate to leading order to the decay $\sigma \to \pi^+\pi^-$. A similar choice will be done for the K^+K^- and $\eta\pi^0$ molecules. For convenience, we shall omit the global factor $(m_u + m_d)$ of the pion current such that the corresponding two-point function has an anomalous dimension. We shall see that this procedure will not affect the mass but the coupling. Then, the σ two-point correlator is given by:

$$\psi_{\pi^+\pi^-}(q^2) = i \int d^4x \, e^{iqx} \mathcal{T} \left\langle 0 | \tilde{\mathcal{O}}_{\pi^+\pi^-}(x) \left(\tilde{\mathcal{O}}_{\pi^+\pi^-}(0) \right)^\dagger | 0 \right\rangle, \tag{9.423}$$

where $(1/\pi)\,\mathrm{Im}\,\psi_{\pi^+\pi^-}(t) \equiv \rho_{\pi^+\pi^-}(t)$ is

$$\begin{aligned}
\rho^{\text{pert}}_{\pi^+\pi^-} &= \frac{t^4}{5 \times 2^{14}\pi^6}, \\
\rho^{\langle \bar{q}q \rangle}_{\pi^+\pi^-} &= \frac{(m_d - 2m_u)\langle \bar{d}d \rangle + (m_u - 2m_d)\langle \bar{u}u \rangle}{2^8\pi^4} t^2, \\
\rho^{\langle G^2 \rangle}_{\pi^+\pi^-} &= \frac{\langle \alpha_s G^2 \rangle}{2^{10}\pi^5} t^2, \\
\rho^{\langle \bar{q}Gq \rangle}_{\pi^+\pi^-} &= \frac{(2m_d + 3m_u)\langle \bar{d}Gd \rangle + (2m_u + 3m_d)\langle \bar{u}Gu \rangle}{2^8\pi^4} t, \\
\rho^{\langle \bar{q}q \rangle^2}_{\pi^+\pi^-} &= \frac{\rho \langle \bar{u}u \rangle \langle \bar{d}d \rangle}{2^4\pi^2} t, \quad \rho^{\langle G^3 \rangle}_{\pi^+\pi^-} = \mathcal{O}(m_q^2 \langle G^3 \rangle),
\end{aligned} \tag{9.424}$$

where:

$$(m_u + m_d)\langle \bar{u}u + \bar{d}d\rangle = -2f_\pi^2 m_\pi^2 \tag{9.425}$$

and ρ indicates the deviation from the factorization of the four-quark condensate.

• Higher order PT QCD corrections

For this purpose, we use a factorization of the molecule spectral function which is given by the product of the two pseudoscalar ones. In this way, we obtain the convolution integral [183, 286]:

$$\frac{1}{\pi}\mathrm{Im}\,\psi_{\pi^+\pi^-}(t) = k_\pi \int_0^t dt_1 \int_0^{(\sqrt{t}-\sqrt{t_1})^2} dt_2\, \lambda^{1/2}\left(\frac{t_1}{t}+\frac{t_2}{t}-1\right)^2 \times\frac{1}{\pi}\mathrm{Im}\,\psi_{\pi^+}(t_1)\frac{1}{\pi}\mathrm{Im}\,\psi_{\pi^-}(t_2), \tag{9.426}$$

with the phase space factor:

$$\lambda = \left(1-\frac{\left(\sqrt{t_1}-\sqrt{t_2}\right)^2}{t}\right)\left(1-\frac{\left(\sqrt{t_1}+\sqrt{t_2}\right)^2}{t}\right), \tag{9.427}$$

where:

$$k_\pi = \frac{35}{208\pi^2} \tag{9.428}$$

is an appropriate normalisation factor. The convolution representation is expected to be valid for large N_c, while the non-factorized contribution is found to be small in the example of the $\bar{B}B$ system [183]. The expression of the pion spectral function is known up to order α_s^4, as defined in Eq. (9.397).

• Parametrization of the spectral function

The contribution of σ to the spectral function can be introduced within the minimal duality ansatz:

$$\frac{1}{\pi}\mathrm{Im}\,\psi_{\pi^+\pi^-}(t) = 2f_{\pi^+\pi^-}^2 M_{\pi^+\pi^-}^8\,\delta\left(t-M_{\pi^+\pi^-}^2\right) \oplus \theta(t-t_c)\text{“QCD continuum”} \tag{9.429}$$

with the normalization:

$$\langle 0|J_{\pi^+\pi-}|\pi^+\pi-\rangle = \sqrt{2}f_{\pi^+\pi-}M^4_{\pi^+\pi-}. \qquad (9.430)$$

At the first step, we shall use an NWA for σ and shall estimate later on the finite width correction.

• Mass and coupling of the $\pi^+\pi^-$ molecule

Using the previous QCD expressions, one can estimate the mass from the ratio of moments $\mathcal{R}_{10}$ and the coupling from the low moment $\mathcal{L}_0$. We show the analysis of the mass and the one of the coupling in Fig. 9.128 at N3LO of PT series, retaining the $D = 6$ condensate contributions.

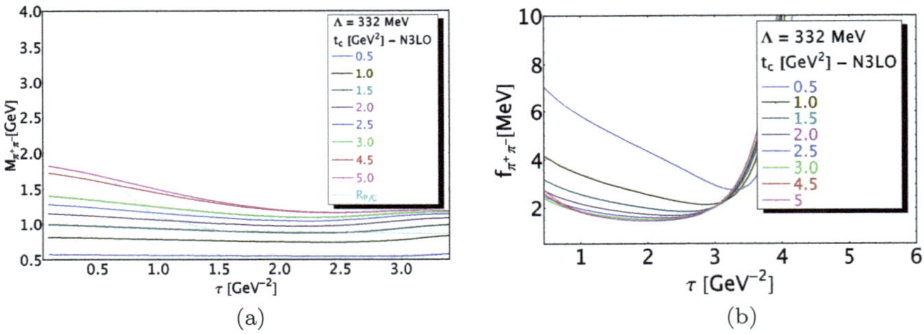

Figure 9.128. $\pi^+\pi^-$ molecule: (a) mass and (b) coupling versus τ for different values of t_c.

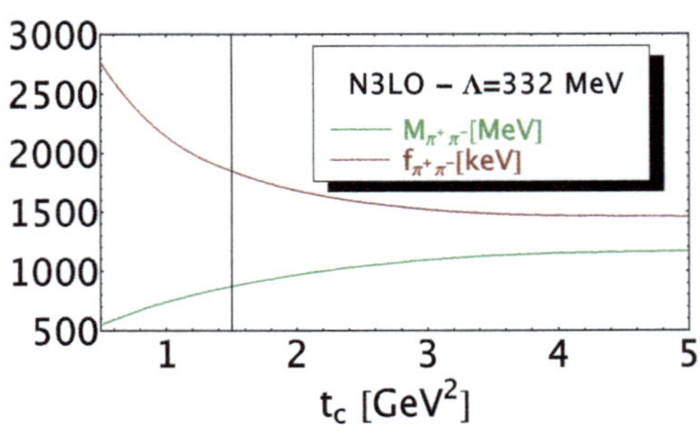

Figure 9.129. Behaviour of the optimal results versus t_c for the $\pi^+\pi^-$ molecule.

The τ-stability of the mass is obtained for $t_c \geq 0.5$ GeV^2, while the result reaches a t_c-stability above $t_c = 4.5$ GeV^2 (see Fig. 9.128).

The $R_{P/C}$ condition does not allow the region on the left of the $R_{P/C}$ curve.

We show in Fig. 9.129 the t_c-behaviour of the optimal results (minimum in τ). The vertical line $t_c = 1.5$ GeV^2 is the minimum value of t_c allowed by the $R_{P/C} \geq 1$ condition. Then, we obtain the optimal result for $t_c = (1.5-4.5)$ GeV^2 at N3LO within an NWA:

$$M_{\pi^+\pi^-} = 1017(144)_{t_c} \text{ MeV}, \quad f_{\pi^+\pi^-} = 1657(193)_{t_c} \text{ keV}, \quad (9.431)$$

where only the errors from t_c are quoted.

• Finite width corrections

We proceed as in the previous section for estimating the finite width corrections by doing the replacement in Eq. (9.411). Then, one obtains:

$$\left(M^{\rm BW}_{\pi^+\pi^-}\right)^2 \simeq \frac{\int_0^{t_c} dt\, t^5\, e^{-t\tau}\, \mathrm{BW}(t)}{\int_0^{t_c} dt\, t^4\, e^{-t\tau} \mathrm{BW}(t)}. \quad (9.432)$$

One can remark that the shape of the width correction differs completly from the one of the $\bar{q}q$ meson. This is due to the different powers of t entering in the spectral integral. From $\Gamma_{\pi\pi} = 0$ to 700 MeV (width of the on-shell mass), the mass increases by 154 MeV, leading to the final prediction:

$$M^{BW}_{\pi^+\pi^-} = 1171(159) \text{ MeV}. \quad (9.433)$$

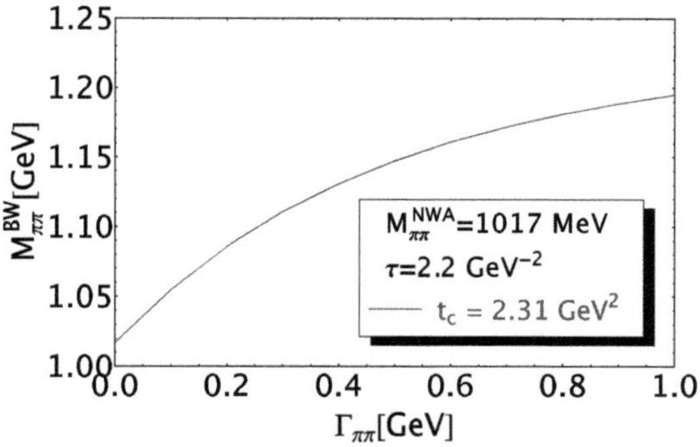

Figure 9.130. Analysis of the finite width effect on $M_{\bar{q}q}$ for $t_c = 2.38$ GeV^2 corresponding to the central value of $M_{\bar{\pi}^+\pi^-} = 1042$ MeV in an NWA at $\tau = 2$ GeV^{-2}.

9.11.6 The K^+K^- molecule

We shall work with the current:

$$\mathcal{O}_{K^+K^-}(x) = (\bar{s}\, i\,\gamma_5\, u) \otimes (\bar{u}\, i\,\gamma_5\, s)(x). \tag{9.434}$$

The expression of the corresponding spectral function can be deduced from the one in Eq. (9.424) by replacing the d by the s quark to which we add the LO m_s^2 correction:

$$\frac{1}{\pi}\mathrm{Im}\,\psi_{K^+K^-}(t)|_{m_s^2} = -\frac{m_s^2}{2^{11}\pi^6}t^3. \tag{9.435}$$

The analysis and shape of the different curves are very similar to the case of $\pi^+\pi^-$ and will not be shown. The different sources of the errors are given in Table 9.28. We obtain in the NWA:

$$M_{K^+K^-} = 1056(214)\ \mathrm{MeV}, \quad f_{K^+K^-} = 1380(255)\ \mathrm{keV}. \tag{9.436}$$

9.11.7 The $K^+\pi^-$ molecule

We shall work with the current:

$$\mathcal{O}_{K^+\pi^-}(x) = (\bar{s}\, i\,\gamma_5\, u) \otimes (\bar{u}\, i\,\gamma_5\, d)(x). \tag{9.437}$$

The expression of the corresponding spectral function reads to LO and up to dimension six:

$$\begin{aligned}
\rho^{\mathrm{pert}}_{K^+\pi^-} &= \frac{t^4}{5\times 2^{14}\pi^6} - \frac{m_s^2}{2^{12}\pi^6}t^3, \quad \rho^{\langle G^2\rangle}_{K^+\pi^-} = \frac{\langle\alpha_s G^2\rangle}{2^{10}\pi^5}\,t^2,\\
\rho^{\langle\bar{q}q\rangle}_{K^+\pi^-} &= \frac{\begin{array}{c}(m_d-2m_u)\,\langle\bar{d}d\rangle - 2(m_d+m_s-m_u)\langle\bar{u}u\rangle\\ +\,(m_s-2m_u)\,\langle\bar{s}s\rangle\end{array}}{2^9\pi^4}t^2,\\
\rho^{\langle\bar{q}Gq\rangle}_{K^+\pi^-} &= \frac{\begin{array}{c}(2m_d+3m_u)\langle\bar{d}Gd\rangle + (4m_u+3(m_d+m_s))\langle\bar{u}Gu\rangle\\ +\,(2m_s+3m_u)\langle\bar{s}Gs\rangle\end{array}}{2^9\pi^4}t,\\
\rho^{\langle\bar{q}q\rangle^2}_{K^+\pi^-} &= \frac{\rho\langle\bar{u}u\rangle(\langle\bar{d}d\rangle+\langle\bar{s}s\rangle)}{2^5\pi^2}\,t, \quad \rho^{\langle G^3\rangle}_{\pi^+\pi^-} = \mathcal{O}(m_q^2\,\langle G^3\rangle).
\end{aligned} \tag{9.438}$$

We include the PT corrections like in the preceding sections. The analysis and shape of different curves are similar to the previous case and will not be shown. The different sources of the errors are given in Table 9.28. We obtain, within an NWA for $\tau \simeq 2.5\ \text{GeV}^{-2}$ and $t_c \simeq (1.5\text{–}4.5)\ \text{GeV}^2$:

$$M_{K^+\pi^-} = 1035(134)\ \text{MeV}, \quad f_{K^+\pi^-} = 1504(275)\ \text{keV}. \tag{9.439}$$

Taking into account the correction to the NWA due to the experimental total width $\Gamma_{K\pi} \simeq (270 \pm 80)$ MeV [33], the mass result becomes:

$$M^{BW}_{K^+\pi^-} = 1110(135)\ \text{MeV}. \tag{9.440}$$

9.11.8 The $\eta\pi^0$ molecule

We shall work with the current:

$$\begin{aligned} \mathcal{O}_{\eta\pi^0}(x) = \frac{1}{\sqrt{6}}\Big[(\bar{u}\, i\gamma_5\, u) + (\bar{d}\, i\gamma_5\, d) - 2(\bar{s}\, i\gamma_5\, s)\Big] \\ \otimes \frac{1}{\sqrt{2}}\Big[(\bar{u}\, i\gamma_5\, u) - (\bar{d}\, i\gamma_5\, d)\Big]. \end{aligned} \tag{9.441}$$

The expression of the corresponding spectral function reads:

$$\begin{aligned} \rho^{\text{pert}}_{\eta\pi^0} &= \frac{7}{3^2 \times 2^{16}\pi^6} t^4 - \frac{m_s^2}{3 \times 2^{11}\pi^6} t^3, \quad \rho^{\langle G^2\rangle}_{\eta\pi^0} = \frac{11}{3}\frac{\langle \alpha_s G^2\rangle}{2^{12}\pi^5}\, t^2, \\ \rho^{\langle \bar{q}q\rangle}_{\eta\pi^0} &= \frac{19[\, m_d\langle \bar{d}d\rangle + \, m_u\langle \bar{u}u\rangle] + 24\, m_s\langle \bar{s}s\rangle}{3^2 \times 2^{10}\pi^4} t^2, \\ \rho^{\langle \bar{q}Gq\rangle}_{\eta\pi^0} &= \frac{121[m_d\langle \bar{d}Gd\rangle + m_u\langle \bar{u}Gu\rangle] + 120 m_s\langle sGs\rangle}{3^2 \times 2^{10}\pi^4}\, t, \\ \rho^{\langle \bar{q}q\rangle^2}_{\eta\pi^0} &= \rho\frac{7[\langle \bar{u}u\rangle^2 + \langle \bar{d}d\rangle^2] + 8\langle \bar{s}s\rangle^2}{3 \times 2^7\pi^2}\, t, \quad \rho^{\langle G^3\rangle}_{\eta\pi^0} = \mathcal{O}(m_q^2\, \langle G^3\rangle). \end{aligned} \tag{9.442}$$

We obtain for the mass and coupling:

$$M_{\eta\pi^0} = 1040(139)\ \text{MeV}, \quad f_{\eta\pi^0} = 1462(249)\ \text{keV}. \tag{9.443}$$

Table 9.28. Sources of errors and values of the scalar meson masses M_S and couplings f_S for the molecules and some four-quark configurations within an NWA. For $\pi^+\pi^-$, the PCAC relation $(m_u + m_d)\langle\bar\psi\psi\rangle = -f_\pi^2 m_\pi^2$ has been used for the estimate of the $m_\psi\langle\bar\psi\psi\rangle$ contribution (f_π=92.3 MeV). The error due to M_S on the coupling has been implicitly included in the one due to t_c. We estimate the error due to the truncation of PT series and the OPE as explained in the text. We take $|\Delta\tau| \simeq 0.2$ GeV^{-2}. Finite width corrections which are tiny are discussed in the text.

Currents	Δt_c	$\Delta\tau$	$\Delta\Lambda$	ΔPT	Δm_q	$\Delta\bar qq$	$\Delta\kappa$	ΔG^2	$\Delta\bar qGq$	ΔG^3	$\Delta\bar qq^2$	ΔOPE	Value
Ordinary $\bar qq$													
Masses [MeV]													
$\frac{1}{\sqrt2}(\bar uu \pm \bar dd)$	88	4.5	3.0	1.0	0	0	–	1.0	1.0	0	19	24.4	1246(94)
$\bar us$	43	4.3	6.0	1.0	3.4	3	13.4	1.1	4.2	0	23	25.3	1276(58)
$\bar ss$	51	2.8	3.5	0.9	7.6	4.2	21.8	1.3	4.9	0	16.5	17.7	1288(62)
(Couplings/m_q) $\times 10^3$													
$\frac{1}{\sqrt2}(\bar uu \pm \bar dd)$	38	1.1	3.0	1.5	0	0	–	2.8	0.0	0.0	1.6	0.7	274(38)
$\bar us$	23	1.2	2.4	2.1	0.5	0.2	0.8	3	0.15	0	1.2	0.5	264(24)
$\bar ss$	17	0.95	2.4	1.9	1.5	0.4	2.5	3.4	0.2	0	1.3	0.5	256(18)
Molecules													
Masses [MeV]													
$\pi^+\pi^-$	144	5	25	31	–	11	–	0.1	0.4	0	11.5	52	1017(159)
K^+K^-	165	3.9	18.4	115	1.1	11.8	7.4	0.6	3.5	0	13	67	1056(214)
$K^+\pi^-$	54	4.5	19.4	105	0.2	10.4	2.5	0.1	1.8	0	13.1	59	1035(134)
$\eta\pi$	57	4.5	15.8	109	0.2	10.8	2.8	0.2	1.8	0	10	60	1040(139)
Couplings [keV]													
$\pi^+\pi^-$	193	24.5	2.0	81	–	87	–	2.5	2.0	0.	70	140	1657(277)
K^+K^-	161	15	0.8	32	4	70	39	2.8	24	0	78	158	1380(255)
$K^+\pi^-$	170	16.5	0.2	56	3	73	16	1.9	13	0	78	176	1504(275)
$\eta\pi$	165	19	0.1	53	2	71	9.6	2.6	13.7	0	77	142	1462(249)

9.11.9 Comments on molecule states

◇ Within an NWA, the masses of the molecule states are about (230–250) MeV lower than the corresponding $\bar{q}q$ states. However, by including the width corrections which act with opposite signs in the two cases, the two predictions tend to meet around 1.1 GeV.

◇ It is amusing to observe that the previous value of the $\bar{q}q$ and molecule masses coincide with the one of the lowest scalar digluonium:

$$M_{\sigma_B}^{\text{glue}} = 1041(111)\ \text{MeV}, \tag{9.444}$$

obtained in Ref. [583], indicating that, at this stage, one cannot yet distinguish the $\bar{q}q$, $\pi\pi$ molecule and gluonium nature of σ.

◇ One can note that the SU(3) breakings to the mass values are small in these two configurations.

9.11.10 The pseudoscalar ⊕ scalar four-quark currents

The four-quark configuration is a "MESS" as many diquark–antidiquark currents can describe the four-quark scalar states [137, 138, 270, 577]. In general, the physical state should be their combination with arbitrary mixing parameters which is (almost) impossible to control. Among these different possibilities, we choose to work with the combination of scalar and pseudoscalar currents which we compare with the combination of vector and axial-vector ones.[35]

• The $\bar{u}\bar{d}ud$ current

We consider the current:

$$\mathcal{O}_{\bar{u}\bar{d}ud}^{S/P} = \epsilon_{abc}\epsilon_{dec}\Big[(\bar{u}_a\gamma_5 C\bar{d}_b^T)\otimes(u_d^T C\gamma_5 d_e) + r\,(\bar{u}_a C\bar{d}_b^T)\otimes(u_d^T C d_e)\Big], \tag{9.445}$$

where r is an arbitrary mixing parameter. In Ref. [137], only the scalar current has been considered ($r = 0$). In the configuration

[35] We shall comment on some choices given in the literature in the following sections.

proposed in Ref. [598], the value of r is $1/\sqrt{2}$. We shall work with the normalized current:

$$\tilde{\mathcal{O}}^{S/P}_{\bar{u}\bar{d}ud} \equiv \mathcal{O}^{S/P}_{\bar{u}\bar{d}ud}/(1+r^2)^{1/2} \quad \text{with:} \quad \langle 0|\tilde{\mathcal{O}}^{S/P}_{\bar{u}\bar{d}ud}|M_{\bar{u}\bar{d}ud}\rangle = \sqrt{2} f_{\bar{u}\bar{d}ud} M^4_{\bar{u}\bar{d}ud}. \tag{9.446}$$

The QCD expression of the spectral function normalized to $(1+r^2)$ reads at LO [36]:

$$\begin{aligned}
\rho^{\text{pert}}_{\bar{u}\bar{d}ud} &= \frac{t^4}{5\times 3\times 2^{12}\pi^6}, \quad \rho^{\langle G^2\rangle}_{\bar{u}\bar{d}ud} = \frac{\langle \alpha_s G^2\rangle}{3\times 2^9\pi^5}\, t^2,\\
\rho^{\langle \bar{q}q\rangle}_{\bar{u}\bar{d}ud} &= \frac{\begin{array}{c}[m_d\langle \bar{d}d\rangle + m_u\langle \bar{u}u\rangle](1+r^2)\\ -\,2[m_u\langle \bar{d}d\rangle + m_d\langle \bar{u}u\rangle](1-r^2)\end{array}}{3\times 2^6\pi^4(1+r^2)}\, t^2,\\
\rho^{\langle \bar{q}Gq\rangle}_{\bar{u}\bar{d}ud} &= \frac{\begin{array}{c}[m_d\langle \bar{d}Gd\rangle + m_u\langle \bar{u}Gu\rangle](1+r^2)\\ +\,6[m_u\langle \bar{d}Gd\rangle + m_d\langle \bar{u}Gu\rangle](1-r^2)\end{array}}{3\times 2^7\pi^4(1+r^2)}\, t,\\
\rho^{\langle \bar{q}q\rangle^2}_{\bar{u}\bar{d}ud} &= \frac{\rho\langle \bar{u}u\rangle\langle \bar{d}d\rangle}{3\times 2^2\pi^2}\frac{(1-r^2)}{(1+r^2)}\, t, \quad \rho^{\langle G^3\rangle}_{\bar{u}\bar{d}ud} = \mathcal{O}(m_q^2\,\langle g^3G^3\rangle),
\end{aligned} \tag{9.447}$$

where ρ indicates the deviation from the factorization of the four-quark condensate. The main difference with $\pi^+\pi^-$ molecule spectral function is the flip of signs of the $D=5,6$ condensate contributions for the pseudoscalar current.

◇ The analysis is shown in Figs. 9.131–9.133 for typical values of the mixing parameter r: 1, $1/\sqrt{2}$ and 0, where the OPE is truncated up to dimension $D=6$. The mass presents stability (plateau) at low values of t_c which becomes an inflexion point for $t_c \geq 3$ GeV^2, while the stability of the coupling depends on r.

We also see in these figures that the constraint $R_{P/C}\geq 1$ excludes values of $t_c \leq 1.5$ GeV^2 like in the previous sections.

[36]The current $\mathcal{O}^{P/S}_{\bar{u}\bar{d}ud} = (\bar{u}_a\gamma_5 C\bar{d}_b^T)\otimes(u_a^T C\gamma_5 d_b - u_b^T C\gamma_5 d_a) + k\,(\bar{u}_a C\bar{d}_b^T)\otimes(u_a^T C d_b - u_b^T C d_a)$ gives the same spectral function. For the symmetric current, $\mathcal{O}^P_{\bar{u}\bar{d}ud} = (\bar{u}_a C\bar{d}_b^T)\otimes(u_a^T C d_b)$, the contributions of G^2 and the gluon exchange for the mixed condensates are zero.

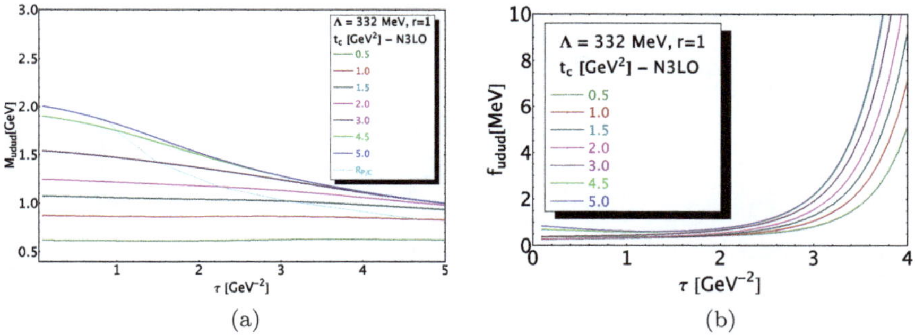

Figure 9.131. Behaviour of (a) mass and (b) coupling of the $\bar{u}\bar{d}ud$ four-quark states versus τ for different values of t_c and for $r = 1$.

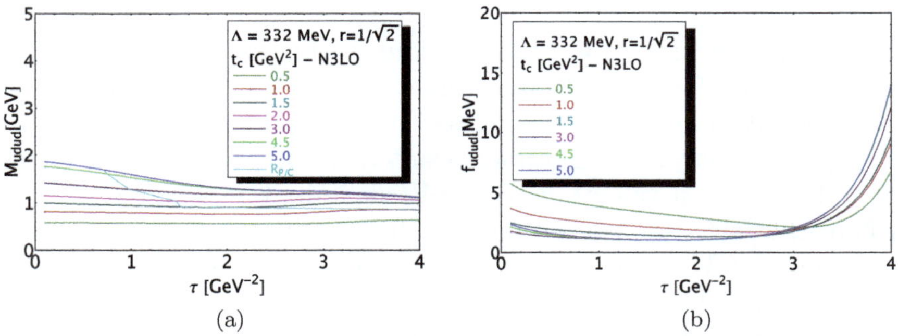

Figure 9.132. Behaviour of (a) mass and (b) coupling of the $\bar{u}\bar{d}ud$ four-quark states versus τ for different values of t_c and for $r = 1/\sqrt{2}$.

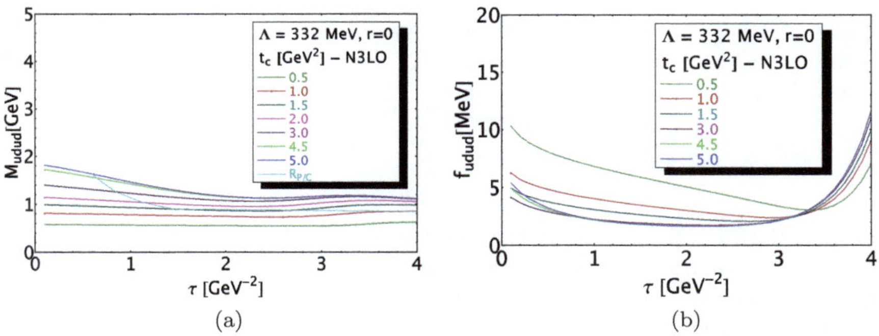

Figure 9.133. Behaviour of (a) mass and (b) coupling of the $\bar{u}\bar{d}ud$ four-quark states versus τ for different values of t_c and for $r = 0$.

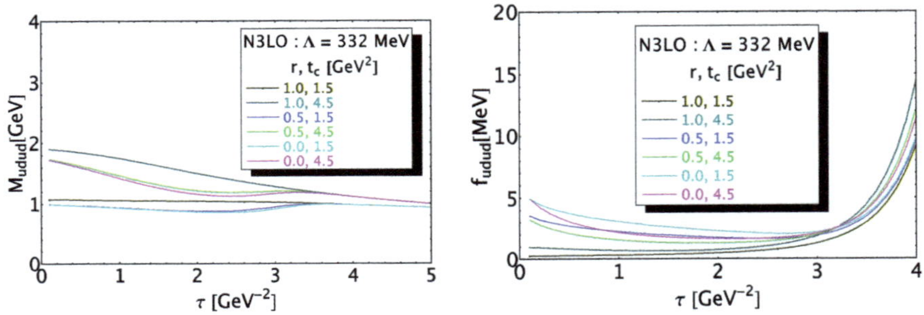

(a) (b)

Figure 9.134. Behaviour of (a) mass and (b) coupling of the $\bar{u}\bar{d}ud$ four-quark states versus τ for two extremal values of t_c allowed by the constraint: $R_{P/C} \geq 1$ and for $r = 0,\ 0.5,\ 1$.

◇ We show in Fig. 9.134 the behaviour of the optimal values of the mass versus τ for two extremal values of $t_c = 1.5$ and 4.5 GeV^2 and for different values of the mixing parameter r. In Fig. 9.135, we show the t_c-behaviour of the optimal result in τ for three typical values of r. The vertical value $t_c = 1.5$ GeV^2 delimits the region $R_{P/C} \geq 1$ where the left region is excluded. This result does not favour the low values of M_σ found from QCD spectral sum rules in the current literature which are obtained at low values of $t_c \simeq (0.5 \sim 1.0)$ GeV^2 corresponding to a (misleading) plateau in the sum rule variable τ.

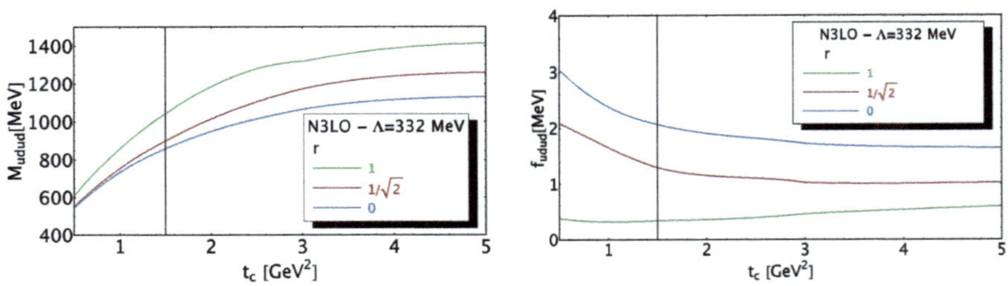

(a) (b)

Figure 9.135. Behaviour of the optimal values in τ for the (a) mass and (b) coupling of the $\bar{u}\bar{d}ud$ four-quark states versus t_c for three values of r (1, $1/\sqrt{2}$, 0) The right side of the vertical line $t_c = 1.5$ GeV^2 is the region allowed by the constraint: $R_{P/C} \geq 1$.

◇ Taking the lower values of $t_c = 1.5$ (respectively 2) GeV^2 for $r = 1/\sqrt{2}$, 1 (respectively 0) allowed by $R_{P/C} \geq 1$ and the higher value of $t_c = 4.5$ GeV^2, we show the r-behaviour of the optimal results in τ in Fig. 9.136. One can note that the value of the mass is not strongly affected by the choice of the mixing parameter which is not the case of the coupling.

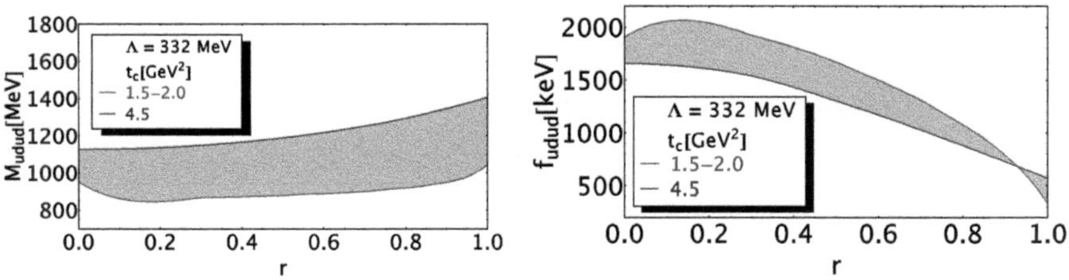

(a) (b)

Figure 9.136. Behaviour of the optimal values in τ for the (a) mass and (b) coupling of the $\bar{u}\bar{d}ud$ four-quark state versus the mixing parameter r for two extremal values of t_c allowed by the constraint: $R_{P/C} \geq 1$.

◇ We consider as a final value the one inside the region $t_c \simeq (1.5{-}2){-}4.5$ GeV^2 allowed by the $R_{P/C} \geq 1$ condition. We obtain

$$
\begin{aligned}
M_{\bar{u}\bar{d}ud} &= 1225(184)\ \text{MeV} \quad & f_{\bar{u}\bar{d}ud} &= 458(288)\ \text{keV} \quad & r &= 1,\\
&= 1078(186)\ \text{MeV} & &= 1160(357)\ \text{keV} & &= 1/\sqrt{2},\\
&= 1040(119)\ \text{MeV}, & &= 1160(540)\ \text{keV} & &= 0,
\end{aligned}
\tag{9.448}
$$

where the different sources of the errors are given in Table 9.28. We note that the errors are dominated by the one due to t_c, ΔPT and (to a lesser extent) by the truncation of the OPE. The one due to ΔPT is mainly due to the tachyonic gluon mass given in Eq. 9.404, while the one due to N5LO is relatively small. We have taken their mean.

- **The $\bar{u}\bar{d}us$ current**

We consider the current

$$
\mathcal{O}^{S/P}_{\bar{u}\bar{d}us} = \epsilon_{abc}\epsilon_{dec}\Big[(\bar{u}_a\gamma_5 C\bar{d}_b^T) \otimes (u_d^T C\gamma_5 s_e) + r\,(\bar{u}_a C\bar{d}_b^T) \otimes (u_d^T C s_e)\Big],
\tag{9.449}
$$

which is expected to have the quantum numbers of $K_0^*(1430)$. The corresponding spectral function normalized to $(1+r^2)$ reads

$$
\begin{aligned}
\rho^{\text{pert}}_{\bar u\bar d us} &= \frac{t^4}{5\times 3\times 2^{12}\pi^6} - \frac{m_s^2}{3\times 2^{10}\pi^6}t^3, \quad \rho^{\langle G^2\rangle}_{\bar u\bar d us} = \frac{\langle\alpha_s G^2\rangle}{3\times 2^9\pi^5}\,t^2,\\
\rho^{\langle\bar q q\rangle}_{\bar u\bar d us} &= \frac{\begin{array}{c}[2m_u\langle\bar u u\rangle + m_d\langle\bar d d\rangle + m_s\langle\bar s s\rangle] - 2[(m_d+m_s)\langle\bar u u\rangle\\ + m_u(\langle\bar s s\rangle + \langle\bar d d\rangle)](1-r^2)/(1+r^2)\end{array}}{3\times 2^7\pi^4}\,t^2,\\
\rho^{\langle\bar q Gq\rangle}_{\bar u\bar d us} &= \frac{\begin{array}{c}[2m_u\langle\bar u Gu\rangle + m_d\langle\bar d Gd\rangle + m_s\langle\bar s Gs\rangle] + 6[m_u(\langle\bar d Gd\rangle\\ + \langle\bar s Gs\rangle) + (m_d+m_s)\langle\bar u Gu\rangle](1-r^2)/(1+r^2)\end{array}}{3\times 2^8\pi^4}\,t,\\
\rho^{\langle\bar q q\rangle^2}_{\bar u\bar d us} &= \frac{\rho\langle\bar u u\rangle(\langle\bar d d\rangle + \langle\bar s s\rangle)}{3\times 2^3\pi^2}\frac{(1-r^2)}{(1+r^2)}\,t, \quad \rho^{\langle G^3\rangle}_{\bar u\bar d us} = \mathcal{O}(m_q^2\,\langle g^3G^3\rangle),
\end{aligned}
\tag{9.450}
$$

The analysis and shape of different curves are very similar to the case of $f_0/a_0(980)$ and will not be shown. The $R_{P/C} \geq 1$ condition also restricts the value of t_c to be in the range (2–4.5) GeV^2 for $r = 0$ as explicitly shown in Fig. 9.137(a) and (1.5–4.5) GeV^2 for $0 < r \leq 1$, as illustrated in Fig. 9.137(b) for $r = 1$.

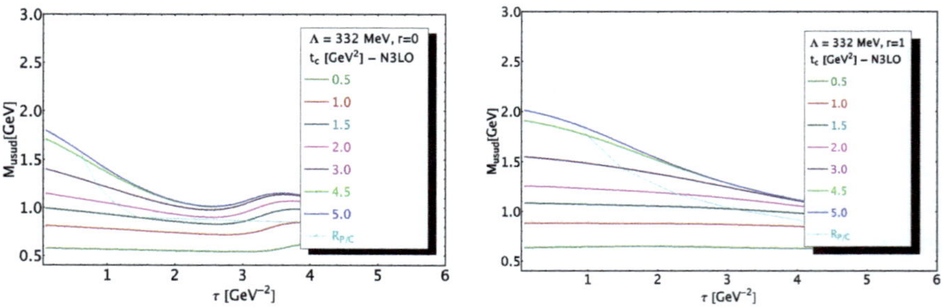

(a) (b)

Figure 9.137. Behaviour of the optimal values in τ for the mass of the $\bar u\bar s ud$ four-quark state: (a) $r = 0$; (b) $r = 1$.

We show in Fig. 9.138 the r-behaviour of the optimal results in τ for the previous range of t_c. We obtain for $\tau \simeq 2.6$, 2.4, 2.3 GeV^{-2}

for $r = 0,\ 1/\sqrt{2}, 1$:

$$\begin{aligned} M_{\bar u\bar s ud} &= 1215(157)\ \text{MeV} & f_{\bar u\bar s ud} &= 422(131)\ \text{keV} & r &= 1,\\ &= 997(147)\ \text{MeV} & &= 1835(382)\ \text{keV} & &= 1/\sqrt{2},\\ &= 957(123)\ \text{MeV}, & &= 2902(638)\ \text{keV}, & &= 0. \end{aligned} \tag{9.451}$$

Taking into account the correction to the NWA due to the experimental total width $\Gamma_{K\pi} \simeq (270 \pm 80)$ MeV [33], the mass result becomes:

$$\begin{aligned} M^{BW}_{\bar u\bar s ud} &= 1236(157)\ \text{MeV} & r &= 1,\\ &= 1086(147)\ \text{MeV} & &= 1/\sqrt{2},\\ &= 1071(123)\ \text{MeV}, & &= 0. \end{aligned} \tag{9.452}$$

• The $\bar u\bar s ds$ current

We consider the current:

$$\mathcal{O}^{S/P}_{\bar u\bar s ds} = \epsilon_{abc}\epsilon_{dec}\Big[(\bar u_a\gamma_5 C\bar s_b^T)\otimes(d_d^T C\gamma_5 s_e) + r\,(\bar u_a C\bar s_b^T)\otimes(d_d^T C s_e)\Big]. \tag{9.453}$$

Its QCD spectral function normalized to $(1+r^2)$ reads:

$$\begin{aligned}
\rho^{\text{pert}}_{\bar u\bar s ds} &= \frac{t^4}{5\times 3\times 2^{12}\pi^6} - \frac{m_s^2}{3\times 2^9\pi^6}t^3, \quad \rho^{\langle G^2\rangle}_{\bar u\bar s ds} = \frac{\langle\alpha_s G^2\rangle}{3\times 2^9\pi^5}\,t^2,\\
\rho^{\langle\bar qq\rangle}_{\bar u\bar s ds} &= \frac{\begin{array}{c}[m_d\langle\bar dd\rangle + m_u\langle\bar uu\rangle + 2m_s\langle\bar ss\rangle] - 2[m_s(\langle\bar dd\rangle + \langle\bar uu\rangle)\\ +(m_d+m_u)\langle\bar ss\rangle](1-r^2)/(1+r^2)\end{array}}{3\times 2^7\pi^4}\,t^2,\\
\rho^{\langle\bar qGq\rangle}_{\bar u\bar s ds} &= \frac{\begin{array}{c}[m_d\langle\bar dGd\rangle + m_u\langle\bar uGu\rangle + 2m_s\langle\bar sGs\rangle] + 6[m_s(\langle\bar dGd\rangle\\ +\langle\bar uGu\rangle) + (m_d+m_u)\langle\bar sGs\rangle](1-r^2)/(1+r^2)\end{array}}{3\times 2^8\pi^4}\,t,\\
\rho^{\langle\bar qq\rangle^2}_{\bar u\bar s ds} &= \frac{\rho\langle\bar ss\rangle(\langle\bar dd\rangle + \langle\bar uu\rangle)}{3\times 2^3\pi^2}\frac{(1-r^2)}{(1+r^2)}\,t, \quad \rho^{\langle G^3\rangle}_{\bar u\bar s ds} = \mathcal{O}(m_q^2\,\langle g^3G^3\rangle),
\end{aligned} \tag{9.454}$$

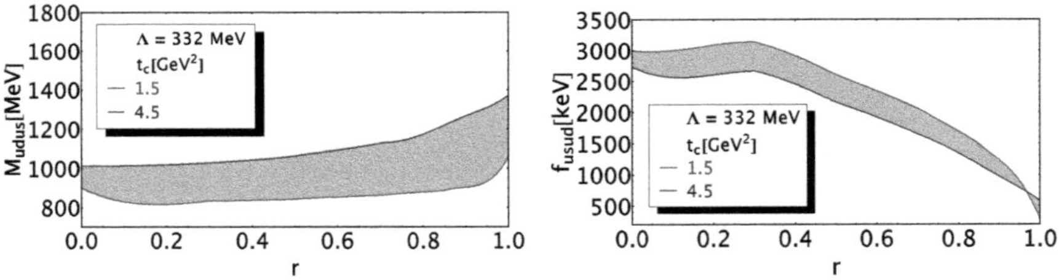

(a) (b)

Figure 9.138. Behaviour of the optimal values in τ for the (a) mass and (b) coupling of the $\bar{u}\bar{s}ud$ four-quark state versus the mixing parameter r for two extremal values of t_c allowed by the constraint: $R_{P/C} \geq 1$.

The analysis and shape of different curves are very similar to the case of the σ and will not be reported here. The $R_{P/C} \geq 1$ condition restricts the value of t_c to be in the range (2–4.5) GeV^2 for $r = 0$ and (1.5–4.5) GeV^2 for $0 < r \leq 1$. We show in Fig. 9.139 the r-behaviour of the optimal results in τ for the previous range of t_c. We obtain for $\tau \simeq 2.5,\ 2.4,\ 2.8$ GeV^{-2} for $r = 0,\ 1/\sqrt{2},\ 1$:

$$
\begin{aligned}
M_{\bar{u}\bar{s}ds} &= 1214(166)\ \text{MeV} & f_{\bar{u}\bar{s}ds} &= 413(153)\ \text{keV} & r &= 1,\\
&= 1012(170)\ \text{MeV} & &= 1683(342)\ \text{keV} & &= 1/\sqrt{2},\\
&= \ 971(112)\ \text{MeV}, & &= 2637(514)\ \text{keV}, & &= 0.
\end{aligned}
\tag{9.455}
$$

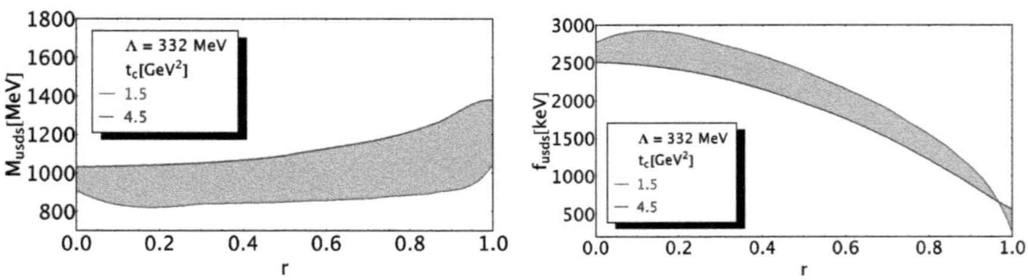

(a) (b)

Figure 9.139. Behaviour of the optimal values in τ for the (a) mass and (b) coupling of the $\bar{u}\bar{s}ds$ four-quark state versus the mixing parameter r for two extremal values of t_c allowed by the constraint: $R_{P/C} \geq 1$.

Table 9.29. The same caption as for Table 9.28 but for the four-quark states in the scalar $\oplus$ pseudoscalar configurations and for three typical values of the mixing parameter $r = 1,\ 1/\sqrt{2},\ 0$. Note that the lower value of t_c allowed by $R_{P/C} \geq 1$ is 2 GeV^2 for $r = 0$ and 1.5 GeV^2 for $r = 1/\sqrt{2},\ 1$.

Currents	Δt_c	$\Delta\tau$	$\Delta\Lambda$	ΔPT	Δm_q	$\Delta\bar{q}q$	$\Delta\kappa$	ΔG^2	$\Delta\bar{q}Gq$	ΔG^3	$\Delta\bar{q}q^2$	ΔOPE	Value
$\mathcal{O}^{S/P}_{\bar{u}\bar{d}ud}$													
Masses [MeV]													
1	181	13	3	28	0	0	–	1	0.2	0	0	–	1225(184)
$1/\sqrt{2}$	177	6.1	22	31	0	10.2	–	0.2	0.4	0	10.3	32	1078(184)
0	87	5.3	25	31	0	10.5	–	0	0	0	11.5	60	1040(114)
Couplings [keV]													
1	124	14	12	5	0	0	–	2.6	0	0	0	–	449(126)
$1/\sqrt{2}$	140	12	6.5	26	0	52	–	2.5	1.5	0	53	54	1160(170)
0	122	29	7.5	119	0	94	–	1.5	1.5	0	95	152	1780(266)
$\mathcal{O}^{S/P}_{\bar{u}\bar{s}ud}$													
Masses [MeV]													
1	155	26	2	3.5	0.7	1.2	1	0.8	1.7	0	0	1.9	1215(157)
$1/\sqrt{2}$	134	7	25	31	0.1	11	2.6	0.2	0.4	0	26	33	997(147)
0	56	8	21	62	0.5	7	2.7	0.2	0.3	0	23	84	957(123)

Couplings [keV]													
1	129	13	8.5	5	4	1.7	1.7	4	2.3	0	0.4	8.5	422(131)
$1/\sqrt{2}$	203	28	5.5	118	1	90	20.5	1.5	5.5	0	189	214	1835(382)
0	169	60	7.5	355	1.5	140	33.5	1	9.5	0	293	377	2902(638)
$\boldsymbol{\mathcal{O}^{S/P}_{\bar{u}\bar{s}ds}}$													
Masses [MeV]													
1	164	25	1	2.5	0.5	2.0	2.0	0.15	2.2	0	0	2	1214(166)
$1/\sqrt{2}$	153	5	25	30	0.2	11.5	6.5	0.4	2.1	0	24.3	56	1012(170)
0	60	8	27	37	0	9.5	5.6	0.15	1.1	0	22.4	79	971(112)
Couplings [keV]													
1	150	16	12	11	3.5	3.5	5	4	4.5	0	0	16	413(153)
$1/\sqrt{2}$	192	25	4	98	2.5	81	45.5	1.5	10	0	178	171	1683(342)
0	129	56	11	297	2.5	113	79	0.7	38	0	278	241	2637(514)

9.11.11 The vector ⊕ axial-vector currents

- **The $\bar{u}\bar{d}us$ current in QCD**

◇ We consider the current:

$$\mathcal{O}^{V/A}_{\bar{u}\bar{d}ud} = \frac{1}{\sqrt{2}}\Big[(\bar{u}_a\gamma_\mu\gamma_5 C\bar{d}_b^T)\otimes(u_a^T C\gamma^\mu\gamma_5 q_b - u_b^T C\gamma^\mu\gamma_5 q_a)$$
$$+ r\,(\bar{u}_a\gamma_\mu C\bar{d}_b^T)\otimes(u_a^T C\gamma^\mu q_b + u_b^T C\gamma^\mu q_a)\Big]. \qquad (9.456)$$

In Refs. [306, 577], the value $r = 1/\sqrt{2}$ has been taken, which corresponds to the choice 3_c and $\bar{6}_c$ in the colour representation (denoted A_6 and V_3 currents). In our analysis, we leave r as a free parameter inside the range 0–1.

◇ QCD lowest order expression up to $D = 6$

The expression of the spectral function normalized to $(1+2r^2)$ reads:

$$\rho^{\rm pert}_{\bar{u}\bar{d}us} = \frac{t^3}{5\times 3\times 2^{11}\pi^6}\Bigg[[t - 20(2m_u^2 + m_d^2 + m_s^2)] - \frac{(1-2r^2)}{(1+2r^2)}10m_u(m_d+m_s)\Bigg],$$

$$\rho^{\langle\bar{q}q\rangle}_{\bar{u}\bar{d}us} = \frac{\begin{array}{c}[(3+2r^2)m_u + 2m_d + (1-2r^2)m_s]\langle\bar{q}q\rangle\\ +[(1-2r^2)m_u + (1+2r^2)m_s]\langle\bar{s}s\rangle\end{array}}{3\times 2^6\pi^4(1+2r^2)}\,t^2,$$

$$\rho^{\langle G^2\rangle}_{\bar{u}\bar{d}us} = \frac{\langle\alpha_s G^2\rangle}{3\times 2^9\pi^5}\frac{(1+6r+5r^2)}{(1+2r^2)}\,t^2,$$

$$\rho^{\langle\bar{q}Gq\rangle}_{\bar{u}\bar{d}us} = -[\langle\bar{q}Gq\rangle[(5-36r-11r^2)m_u + 2(2-9r-2r^2)m_d$$
$$+3(1+r^2)m_s] + \langle\bar{s}Gs\rangle[(1-18r-7r^2)m_s + 3(1+r^2)m_u)]]$$
$$\times\frac{t}{3\times 2^8\pi^4(1+2r^2)},$$

$$\rho^{\langle\bar{q}q\rangle^2}_{\bar{u}\bar{d}us} = -\frac{\rho[\langle\bar{q}q\rangle^2 + \langle\bar{q}q\rangle\langle\bar{s}s\rangle]}{24\,\pi^2}\frac{(1-2r^2)}{(1+2r^2)}\,t,\quad \rho^{\langle G^3\rangle}_{\bar{u}\bar{d}us} = \mathcal{O}(m_q^2\,\langle g^3G^3\rangle),$$
$$(9.457)$$

where $\langle\bar{q}q\rangle \equiv \langle\bar{u}u\rangle \simeq \langle\bar{d}d\rangle$.

◇ Truncation of the OPE

We truncate the series at the dimension six condensate contributions due to the less controlled values of the higher dimension condensates and to the mixing of different operators which appear after the uses of the equation of motion [113]. We estimate the unknown higher dimension contributions (some classes of dimension-eight contributions are given in Ref. [577]) by using the scaling factor in Eq. (9.407).

◇ Approximate higher order PT QCD corrections

We introduce the higher order PT corrections from the convolution integral [183, 286]:

$$\frac{1}{\pi}\mathrm{Im}\,\psi_{\bar{u}\bar{d}ud}(t) = k_{V/A}\int_0^t dt_1 \int_0^{(\sqrt{t}-\sqrt{t_1})^2} dt_2\,\lambda^{1/2} \times\left[\left(\frac{t_1}{t}+\frac{t_2}{t}-1\right)^2+8\frac{t_1t_2}{t^2}\right] \times\frac{1}{\pi}\mathrm{Im}\,\Pi_{V/A}(t_1)\frac{1}{\pi}\mathrm{Im}\,\Pi_{V/A}(t_2), \tag{9.458}$$

where

$$k_{V/A} = \frac{1}{96\pi^2}, \tag{9.459}$$

is an appropriate normalisation factor. $\Pi_{V/A}(t)$ is the two-point function associated to the vector $V \equiv \bar{u}\gamma_\mu d$ and axial-vector $A \equiv \bar{u}\gamma_\mu\gamma_5 d$ currents. Its spectral function is known to order α_s^4 in the massless quark limit. It reads for three flavours and in the $\overline{MS}$-scheme[37]:

$$\frac{1}{\pi}\mathrm{Im}\,\Pi_{V/A}(t) = \frac{1}{4\pi^2}\left[1+a_s+1.623\,a_s^2-6.370\,a_s^3-106.8798\,a_s^4+1092\,a_s^5\right]. \tag{9.460}$$

We have estimated the last a_s^5 term, assuming a geometric growth of the a_s coefficient, while the alternate sign is fixed, assuming that the series reaches its asymptotics at this order.

[37]For reviews, see, e.g., Refs. [6, 584].

• **The $\bar{u}\bar{d}ud$ state**

The QCD expression can be deduced from Eq. (9.457) by replacing the quark s by d.

⋄ LSR analysis within an NWA

– We show the analysis in Figs. 9.140 and 9.141 for $r = 1$ and $1/\sqrt{2}$. The analysis for $0 \leq r < 1/\sqrt{2}$ is not conclusive within our truncation of the OPE as we do not have τ-stability. This feature is expected from the negative contribution of the dimension six condensate (see Eq. (9.457)).

– One can note from the figures that the stability is reached for large values of $\tau \simeq (2.3–3.7)$ GeV^{-2} for the coupling and (2.8–2.9) GeV^{-2} for the mass due to the relative small contributions of the $\langle\bar{q}q\rangle^2$ condensate, which vanish for $r = 1/\sqrt{2}$. This fact is reflected on the large errors due to the truncation of the OPE and the N5LO PT corrections, as given in Table 9.30. We obtain, within an NWA:

$$\begin{aligned} M^{V/A}_{\bar{u}\bar{d}ud} &= 831(141)\ \text{MeV} \qquad f_{\bar{u}\bar{d}ud} = 938(278)\ \text{keV} \qquad r = 1, \\ &= 941(193)\ \text{MeV}, \qquad\qquad\ = 601(66)\ \text{keV}, \qquad = 1/\sqrt{2}, \end{aligned} \tag{9.461}$$

for $t_c = (1 - 4.5)$ GeV^2 and $\tau \simeq (2.8–3.7)$ GeV^{-2}: $t_c = 1$ GeV^2 corresponds to the beginning of τ-stability allowed by $R_{P/C}$ and $t_c = 4.5$ GeV^2 is the beginning of t_c-stability.

⋄ Effects of the higher order PT corrections

– Our approximate NLO coefficient is 1.72 smaller than the exact NLO result of Ref. [306]. The difference is due to the fact that the exact result takes into account the contribution of the off-diagonal diagram and the non-leading term in $1/N_c$. We compare the effect of different truncation of the PT series in Fig. 9.142 for the case $r = 1/\sqrt{2}$ where we have fixed $t_c \simeq 1.7$ GeV^2 for approximately reproducing the central values of the mass and coupling obtained previously:

$$M^{V/A}_{\bar{u}\bar{d}ud}|_{1.7} \simeq 945 \text{ MeV}, \quad f^{V/A}_{\bar{u}\bar{d}ud}|_{1.7} \simeq \quad 601 \text{ keV}. \tag{9.462}$$

We find that the use of the factorized α_s correction underestimates the value of the mass by about 35 MeV compared to the exact result of 881 MeV. However, this shift is smaller than the error of the mass determination.

– We note, like in the previous cases, that the inclusion of higher order PT corrections improve the predictions of the LSR as the stability is obtained at lower values of τ. From NLO to N4LO, the value of the mass shifts by about 64 MeV. The effects of the α_s^n corrections to the coupling are marginal.

◇ Finite width corrections

We proceed as in the previous sections and find that the masses become:

$$\begin{aligned} M^{V/A}_{\bar{u}\bar{d}ud}|_{BW} &= 1014(141) \text{ MeV} \quad r = 1, \\ &= 1050(193) \text{ MeV}, \quad = 1/\sqrt{2}, \end{aligned} \tag{9.463}$$

where we have used as input the experimental width $\Gamma_\sigma^{\text{pole}}$ 520 MeV and $\Gamma_\sigma^{\text{BW}} = 700$ MeV.

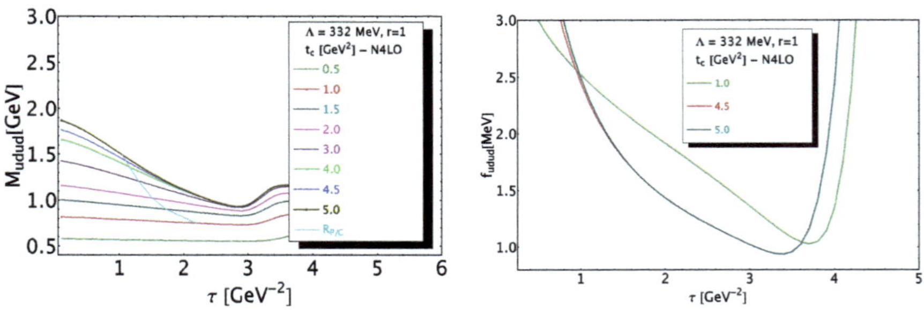

(a) (b)

Figure 9.140. Behaviour of (a) mass and (b) coupling of the $\bar{u}\bar{d}ud$ four-quark states versus τ for different values of t_c and for $r = 1$ in the case of vector $\oplus$ axial-vector current.

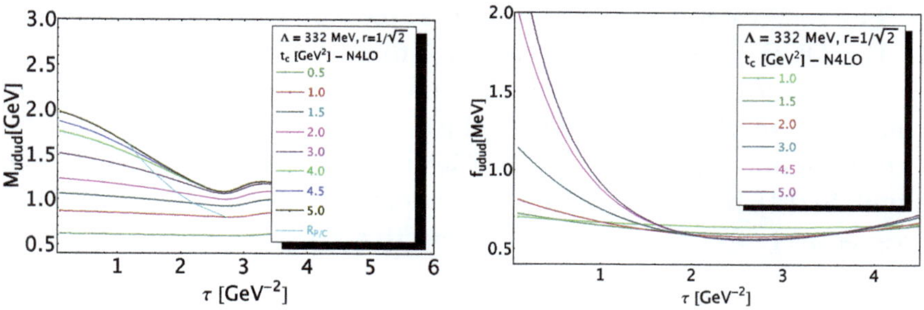

(a) (b)

Figure 9.141. Behaviour of (a) mass and (b) coupling of the $\bar{u}\bar{d}ud$ four-quark states versus τ for different values of t_c and for $r = 1/\sqrt{2}$ in the case of vector $\oplus$ axial-vector current.

• The $\bar{u}\bar{d}us$ state

◇ The QCD expression of the spectral function is given in Eq. (9.457). The analysis and shape of the curves are very similar to the case of the $\bar{u}\bar{d}ud$ state and will not be shown as well as the range of $t_c = (1-4.5)$ GeV^2 allowed by the $R_{P/C}$ condition. We obtain

$$\begin{aligned} M^{V/A}_{\bar{u}\bar{d}us} &= 834(141)\ \text{MeV} \qquad & f^{V/A}_{\bar{u}\bar{d}us} &= 1042(295)\ \text{keV} \qquad & r &= 1, \\ &= 978(185)\ \text{MeV}, & &= 408(57)\ \text{keV}, & &= 1/\sqrt{2}. \end{aligned} \tag{9.464}$$

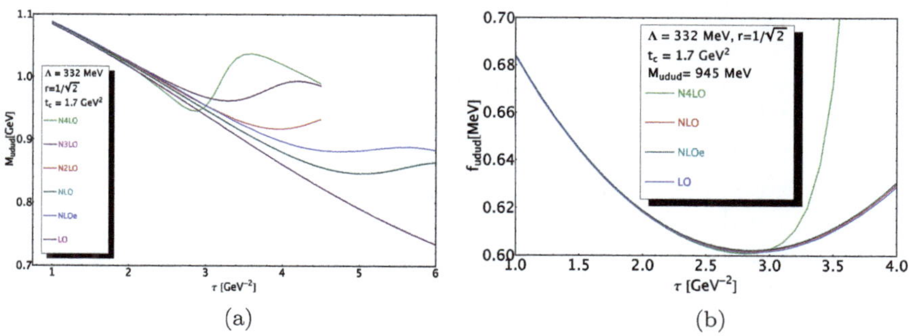

(a) (b)

Figure 9.142. Behaviour of (a) mass and (b) coupling of the $\bar{u}\bar{d}ud$ four-quark states for different truncation of the PT series versus τ for t_c =1 GeV^2, $r = 1/\sqrt{2}$ in the case of vector $\oplus$ axial-vector current. NLOe is the exact result from Ref. [306].

◇ The sources of errors are given in Table 9.30 where we have added to the QCD errors of $\bar{u}\bar{d}ud$ the ones due to $SU(3)$ breakings. One can note the relative small error due to t_c and τ for the coupling compared to the one of $\bar{u}\bar{d}ud$ which is due to the vicinity of the two minimum for $t_c = 1$ and 4.5 GeV^2, as shown in Fig. 9.143. The small $SU(3)$ breakings increase the central value of the mass by 3 (respectively 37) MeV for r=1 (respectively $1/\sqrt{2}$).

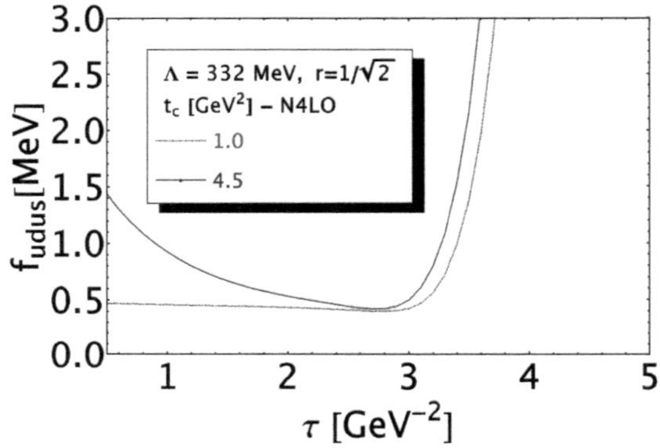

Figure 9.143. Behaviour of the coupling of the $\bar{u}\bar{d}us$ four-quark states versus τ for two extremal values of t_c and $r = 1/\sqrt{2}$ in the case of vector ⊕ axial-vector current.

• **The $\bar{u}\bar{s}ds$ state**

◇ The QCD expression of the spectral function normalized to $(1 + 2r^2)$ reads

$$\rho^{\text{pert}}_{\bar{u}\bar{s}ds} = \frac{t^3}{5 \times 3 \times 2^{11}\pi^6}\left[\left[t - 20(m_u^2 + m_d^2 + 2m_s^2)\right] - \frac{(1-2r^2)}{(1+2r^2)}10\, m_s(m_u + m_d)\right],$$

$$\rho^{\langle\bar{q}q\rangle}_{\bar{u}\bar{s}ds} = \frac{(m_u + m_d)\langle\bar{q}q\rangle + 2m_s\langle\bar{s}s\rangle + (1-2r^2)\left[(m_u+m_d)\langle\bar{s}s\rangle + 2m_s\langle\bar{q}q\rangle\right]}{3 \times 2^6\pi^4(1+2r^2)}t^2,$$

$$\rho^{\langle G^2\rangle}_{\bar{u}\bar{s}ds} = \frac{\langle\alpha_s G^2\rangle}{3 \times 2^9\pi^5}\frac{(1 + 6r + 5r^2)}{(1+2r^2)}t^2,$$

$$\rho_{\bar{u}\bar{s}ds}^{\langle\bar{q}Gq\rangle} = -\Big[\langle\bar{q}Gq\rangle\left[(m_u+m_d)(1-18r-7r^2)+6m_s(1+r^2)\right]$$
$$+\langle\bar{s}Gs\rangle\left[3(m_u+m_d)(1+r^2)+2m_s(1-18r-7r^2)\right]\Big]$$
$$\times\frac{t}{3\times 2^8\pi^4(1+2r^2)},$$
$$\rho_{\bar{u}\bar{s}ds}^{\langle\bar{q}q\rangle^2} = -\frac{\rho[\langle\bar{q}q\rangle\langle\bar{s}s\rangle]}{12\,\pi^2}\frac{(1-2r^2)}{(1+2r^2)}t, \quad \rho_{\bar{u}\bar{s}ds}^{\langle G^3\rangle} = \mathcal{O}(m_q^2\,\langle g^3G^3\rangle), \quad (9.465)$$

where $\langle\bar{q}q\rangle \equiv \langle\bar{u}u\rangle \simeq \langle\bar{d}d\rangle$.

◇ The analysis and shape of the curves are very similar to the case of the $\bar{u}\bar{d}ud$ state except that the $R_{P/C} \geq 1$ condition requires the range $t_c = (2-4.5)$ GeV^2, as shown in Fig. 9.144 for $r = 1/\sqrt{2}$:

$$M_{\bar{u}\bar{s}ds}^{V/A} = 840(140)\ \text{MeV} \qquad f_{\bar{u}\bar{s}ds} = 829(282)\ \text{keV} \qquad r = 1,$$
$$= 1282(145)\ \text{MeV}, \qquad = 354(69)\ \text{keV}, \qquad = 1/\sqrt{2}. \quad (9.466)$$

The sources of errors are given in Table 9.30 where we have added to the QCD errors of $\bar{u}\bar{d}ud$ the ones due to $SU(3)$ breakings which are negligible.

◇ One can note the relative small error due to t_c and τ for the coupling compared to the one of $\bar{u}\bar{d}ud$ which is due to the vicinity of the two minimum for $t_c = 2$ and 4.5 GeV^2, as shown in Fig. 9.144 for $r = 1/\sqrt{2}$. The shape of the coupling is similar to the case of the $\bar{u}\bar{d}ud$ state shown in Fig. 9.140(b).

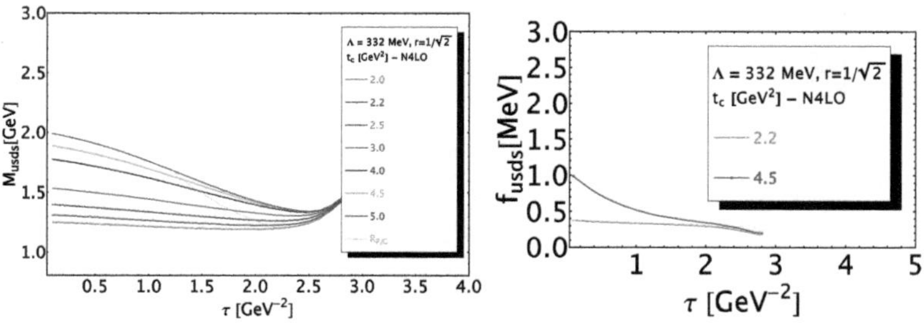

(a) (b)

Figure 9.144. Behaviour of (a) mass and (b) coupling of the $\bar{u}\bar{s}ds$ four-quark states versus τ for different values of t_c and for $r = 1/\sqrt{2}$ in the case of vector $\oplus$ axial-vector current.

Table 9.30. The same caption as for Table 9.28 but for the four-quark states in the vector $\oplus$ axial-vector configurations and for two typical values of the mixing parameter $r = 1$ and $1/\sqrt{2}$. Note that the lower value of t_c allowed by $R_{P/C} \geq 1$ is 1 GeV^2.

Currents	Δt_c	$\Delta\tau$	$\Delta\Lambda$	ΔPT	Δm_q	$\Delta\bar{q}q$	$\Delta\kappa$	ΔG^2	$\Delta\bar{q}Gq$	ΔG^3	$\Delta\bar{q}q^2$	ΔOPE	Value
$\mathcal{O}^{V/A}_{\bar{u}\bar{d}ud}$													
Masses [MeV]													
1	100	3.5	16	97	0	4.5	–	1.8	0	0	0	9.3	831(141)
$1/\sqrt{2}$	141	14	15.8	130	0	0.8	–	0.9	2.5	0	0	3.6	941(193)
Couplings [keV]													
1	45	16.5	4	269	0	19.5	–	16.5	2.9	0	2.1	45.5	983(278)
$1/\sqrt{2}$	40	1.5	0	20	0	1.6	–	48	5.6	0	0	7	601(66)
$\mathcal{O}^{V/A}_{\bar{u}\bar{s}ud}$													
Masses [MeV]													
1	99	18	16	97	0.1	4.5	1	1.8	0	0	0	9.3	834(141)
$1/\sqrt{2}$	129	14.5	15.8	130	0.1	0.8	10	0.9	2.5	0	0	3.6	978(185)
Couplings [keV]													
1	108	16.5	4	269	0.1	19.5	1.7	16.5	2.9	0	2.1	45.5	1042(295)
$1/\sqrt{2}$	11	2.2	0	20	0.1	1.6	19.5	48	5.6	0	0	7	408(57)
$\mathcal{O}^{V/A}_{\bar{u}\bar{s}ds}$													
Masses [MeV]													
1	97	23	16	97	0.1	4.5	3	1.8	0	0	0	9.3	840(140)
$1/\sqrt{2}$	57	21	15.8	130	0.1	0.8	10	0.9	2.5	0	0	3.6	1282(145)
Couplings [keV]													
1	63	21.5	4	269	0.3	19.5	4	16.5	2.9	0	2.1	45.5	829(282)
$1/\sqrt{2}$	40	13	0	20	0.15	1.6	16	48	5.6	0	0	7	354(69)

◇ The *anomalously* large value of the mass for $r = 1/\sqrt{2}$ is essentially due to the range of t_c-value required by the $R_{P/C}$ condition.

9.11.12 First radial excitations

In this section, we attempt to extract the masses and couplings of the first radial excitations for different assignments. We limit ourselves to the case of $SU(2)_F$ by observing that $SU(3)$ breakings shift only slightly the values of these observables and cannot be seen within the errors of the determination. In so doing, we shall subtract the contribution of the lowest ground states obtained in previous section from the sum rules and work with the first radial excitation $\oplus$ QCD continuum in a higher range of t_c-values. This procedure is more helpful than a direct extraction of two resonance parameters where extracting simultaneously the masses and decay constants of the ground state and first radial excitation from $\mathcal{R}_{10}$ is hopeless. One could also work with higher ratio of moments more sensitive to the higher states like in Ref. [583] for the case of multiple resonances. Unfortunately, within our truncation of the OPE up to $D = 6$ condensates, the next moment $\mathcal{R}_{21}$ cannot help as it does not present a τ-minimum but a slight inflexion point. Therefore, the obtained result is less accurate than the one from $\mathcal{R}_{10}$. We also emphasize that high moments which are more sensitive to higher radial excitations cannot help improve the mass and coupling of the ground state, which is one of the main goal of this work.

• The $\bar{u}d$ current

We show the result of the analysis in Fig. 9.145, where the contribution of the ground state has been subtracted from the sum rule. One obtains:

$$M^{(1)}_{\bar{u}d} = 1378(186)\ \text{MeV} \quad f^{(1)}_{\bar{u}d} = 212(38)\ \text{keV}. \tag{9.467}$$

Compared to the case of the ground state, the optimal value of the mass and coupling is obtained at lower values of τ as intuitively expected. The mass shift from the ground state is about 350 MeV which is relatively small compared to the case of the ρ-meson of 680 MeV.

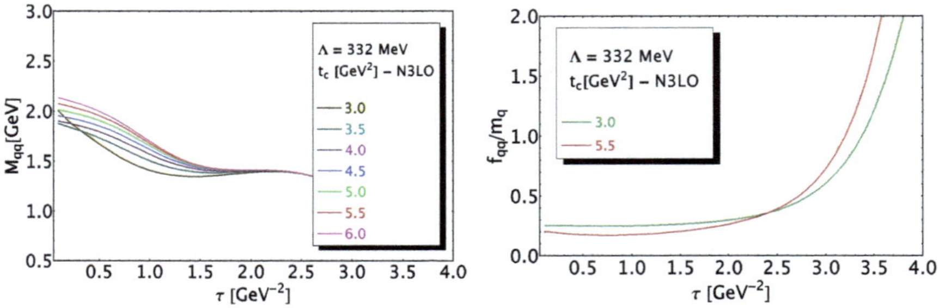

(a) (b)

Figure 9.145. Behaviour of the optimal values in τ for the (a) mass and (b) coupling of the $\bar{u}\bar{d}$ state versus τ for different values of t_c.

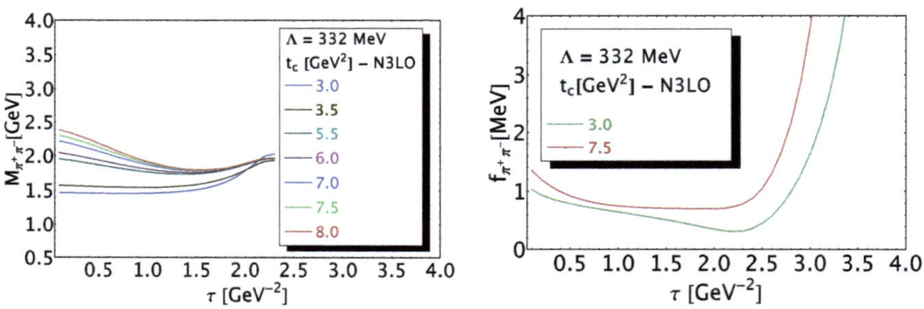

(a) (b)

Figure 9.146. Behaviour of the optimal values in τ for the (a) mass and (b) coupling of the $\pi^+\pi^-$ molecule state versus τ for different values of t_c.

• The $\pi^+\pi^-$ molecule current

The analysis is shown in Fig. 9.146. We deduce:

$$M^{(1)}_{\pi^+\pi^-} = 1621(514)\ \text{MeV}, \quad f^{(1)}_{\pi^+\pi^-} = \ 665(338)\ \text{keV}, \tag{9.468}$$

where one should note that the optimal value for $t_c = 3\ \text{GeV}^2$ is taken at the inflexion point $\tau = 1.1\ \text{GeV}^{-2}$.

• The $\bar{u}\bar{d}ud$ pseudo (scalar) four-quark current

We show the analysis for three typical values of the mixing parameter in Figs. 9.147–9.149. We obtain:

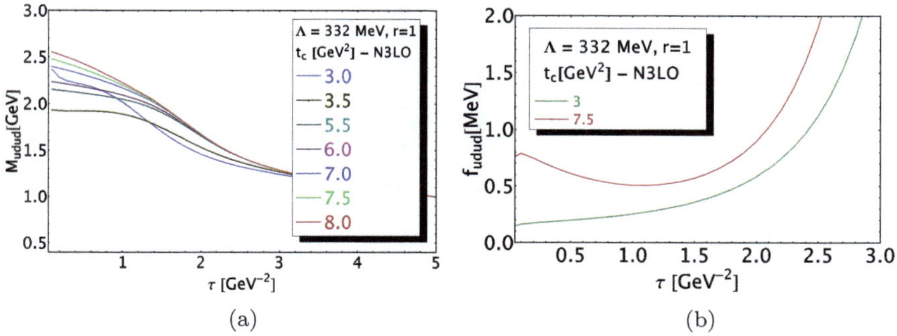

Figure 9.147. Behaviour of the optimal values in τ for the (a) mass and (b) coupling of the $\bar{u}\bar{d}ud$ first radial state versus τ for different values of t_c for $r = 1$ for the (pseudo)scalar currents.

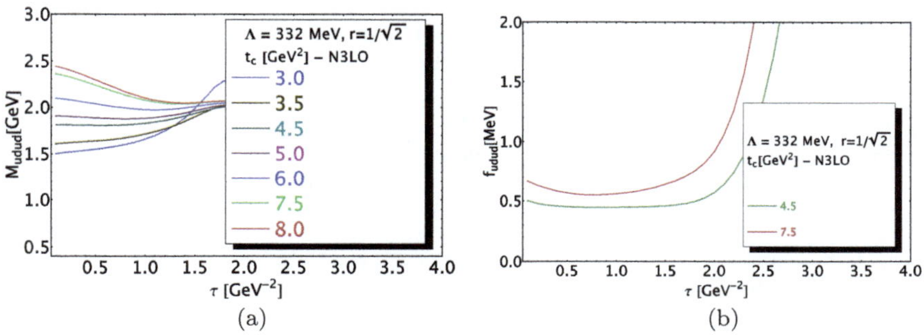

Figure 9.148. Behaviour of the optimal values in τ for the (a) mass and (b) coupling of the $\bar{u}\bar{d}ud$ first radial state versus τ for different values of t_c for $r = 1/\sqrt{2}$ for the (pseudo)scalar currents.

$$
\begin{aligned}
M_{\bar{u}\bar{d}ud}^{S/P(1)} &= 1670(168)\ \text{MeV} \quad & f_{\bar{u}\bar{d}ud}^{S/P(1)} &= 381(128)\ \text{keV} \quad & r &= 1, \\
&\ 1920(317)\ \text{MeV} & &= 498(97)\ \text{keV} & &= 1/\sqrt{2}, \\
&\ 1588(511)\ \text{MeV}, & &= 777(189)\ \text{keV}, & &= 0.
\end{aligned}
\tag{9.469}
$$

One should note, like in the case of $\pi^+\pi^-$, that the optimal value of the coupling for $r = 0$ and $t_c = 3\ \text{GeV}^2$ is taken at the inflexion point $\tau = 1.3\ \text{GeV}^{-2}$. One should also note that the relatively large value of the mass for $r = \sqrt{2}$ is due to the fact that we have taken the minimum value of $t_c = 4.5\ \text{GeV}^2$ where the coupling starts to have stability instead of $3\ \text{GeV}^2$ in the other channels $r = 0, 1$.

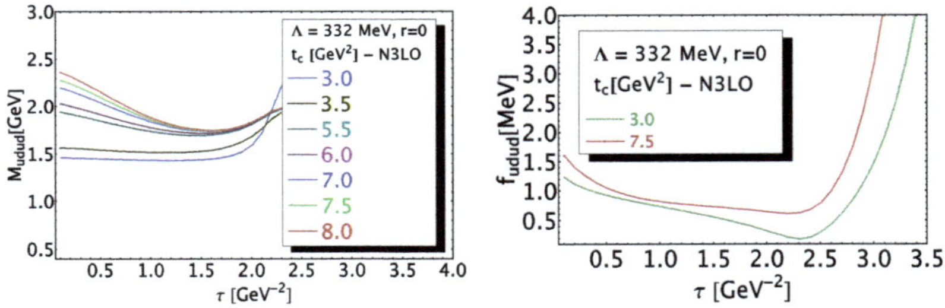

(a) (b)

Figure 9.149. Behaviour of the optimal values in τ for the (a) mass and (b) coupling of the $\bar{u}\bar{d}ud$ state versus τ for different values of t_c for $r = 0$ for the (pseudo)scalar currents.

• The $\bar{u}\bar{d}ud$ axial(vector) four-quark currents

We do a similar analysis for the (axial) vector currents. The analysis is shown in Figs. 9.150 and 9.151. We obtain:

$$\begin{aligned} M^{V/A(1)}_{\bar{u}\bar{d}ud} &= 931(193)\ \text{MeV}, & f^{V/A(1)}_{\bar{u}\bar{d}ud} &= 1137(334)\ \text{keV} & r &= 1, \\ & \ 1489(380)\ \text{MeV}, & &= 287(99)\ \text{keV}, & &= 1/\sqrt{2}, \end{aligned} \tag{9.470}$$

where the different sources of the errors can be found in Table 9.31. We note that the τ-stability starts earlier at $t_c = 1.5$ GeV^2 for the case $r = 1$ than the one $r = 1/\sqrt{2}$ at 3 GeV^2. This feature explains the low mass of the first radial excitation in this case in addition to the larger value of τ where the minimum is obtained (see Fig. 9.150). The optimal value of the coupling is taken at the inflexion point in

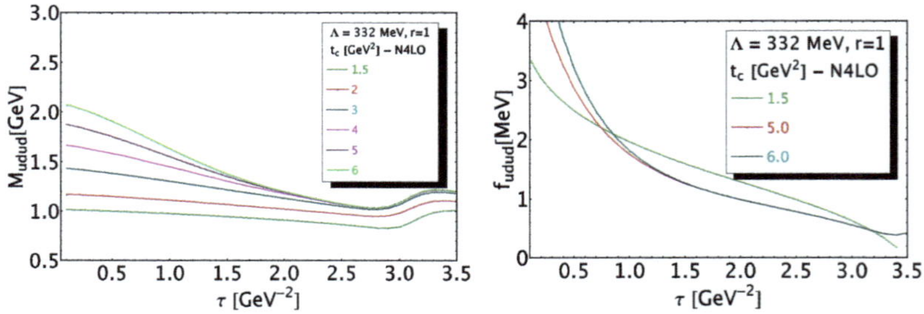

(a) (b)

Figure 9.150. Behaviour of the optimal values in τ for the (a) mass and (b) coupling of the first radial excitation of the $\bar{u}\bar{d}ud$ state versus τ for different values of t_c for $r = 1$ for the axial (vector) currents.

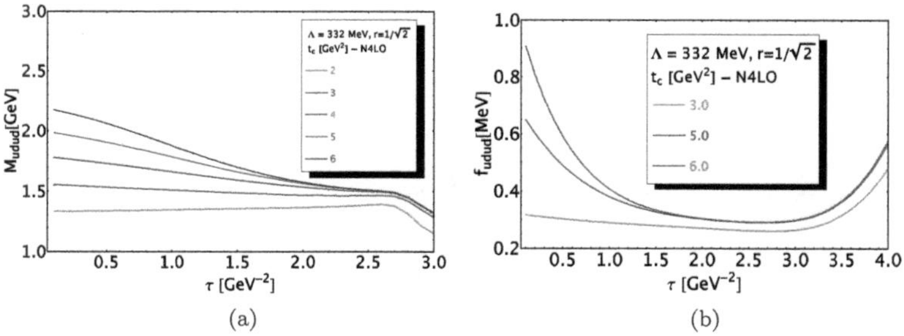

Figure 9.151. Behaviour of the optimal values in τ for the (a) mass and (b) coupling of the first radial excitation of the $\bar{u}\bar{d}ud$ state versus τ for different values of t_c for $r = 1/\sqrt{2}$ for the (axial) vector currents.

$\tau \simeq 2$ GeV^{-2} for $r = 1$ and at the minimum $\tau \simeq 2$ GeV^{-2} for $r = 1/\sqrt{2}$.

• Comments on the radial excitations

◇ Comparing the values of the radial excitation masses with the ones of the ground states, one can note that the mass splitting for the $\bar{q}q$ state of about 132 MeV is relatively low compared to the case of the ρ-meson of 680 MeV. On the contrary, the ones of the molecule and four-quark states are in the range of 450–840 MeV.

◇ We expect that the masses of the radial excitations with strange quarks are almost degenerated with the non-strange one due to the small $SU(3)$ breakings found for the lowest ground states. This range of mass values is comparable with the one obtained from light front holographic approach [585].

◇ The coupling of the first radial excitation is comparable with the one of the ground state for the $\bar{q}q$ state and $r = 1$ four-quark state but much smaller for the molecule and $r = 1/\sqrt{2}, 0$ four-quark states. Therefore, one may wonder if the one-resonance parametrization done for estimating the ground state mass can be affected by the presence of the nearby first radial excitation.

• Nearby radial excitations effects on the $\bar{u}d$ and $r = 1$ $\bar{u}\bar{d}ud$ ground states

◇ We reiterate the determination of the lowest ground state mass and coupling by including now into the spectral function the effect of

Table 9.31. The same caption as for Table 9.28 but for the first radial excitations for different assignments. $r = 1, 1/\sqrt{2}, 0$ are typical values of the four-quark mixing of currents.

Currents	Δt_c	$\Delta\tau$	$\Delta\Lambda$	ΔPT	Δm_q	$\Delta\bar{q}q$	ΔG^2	$\Delta\bar{q}Gq$	ΔG^3	$\Delta\bar{q}q^2$	ΔOPE	ΔM_S	Δf_S	Value
$\boldsymbol{\frac{1}{\sqrt{2}}(\bar{u}u+\bar{d}d)}$														
Masses [MeV]	33	7	36	28	0	0.5	4	1.5	0	104	88	101	51	1378(186)
Couplings [keV]	38	1.9	1.8	1.2	0	0.1	0.2	0.1	0	0.9	4	0.1	0.1	212(38)
$\boldsymbol{\pi^+\pi^-}$ molecule														
Masses [MeV]	167	9	21.5	9	0	113	1.2	1.2	0	109	126	322	303	1621(514)
Couplings [keV]	193	28	11.5	14	0	80	3.5	1.7	0	128	62	190	113	665(338)
$\boldsymbol{\mathcal{O}^{S/P}_{\bar{u}\bar{d}ud}}$ four-quark														
Masses [MeV]														
1	128	107	22.5	18.5	0	0.3	4	0.5	0	0.3	0.4	45	98	1670(201)
$1/\sqrt{2}$	115	4.3	9	17.5	0	67	2.3	0.8	0	69	53	237	136	1920(317)
0	155	7	31.5	0.2	0	110	1.25	1.25	0	153	108	242	361	1588(511)
Couplings [keV]														
1	93	35	46	54	0	0.05	2	0.15	0	0.05	0.05	35	16	381(128)
$1/\sqrt{2}$	51	4	2.3	5.8	0	21.4	1.25	0.15	0	25.3	10	69	28	498(97)
0	41	53	9.5	3.9	0	15	1	0.8	0	61	15	144	77	777(189)
$\boldsymbol{\mathcal{O}^{V/A}_{\bar{u}\bar{d}ud}}$														
Masses [MeV]														
1	102	29	30.5	154	0	15.5	2.4	1.7	0	6.5	33.4	45.5	34.4	931(193)
$1/\sqrt{2}$	25	95	9	57	0	4.5	53	14.5	0	0.4	15.5	329	141	1489(380)
Couplings [keV]														
1	149	105	40.5	6.5	0	88	13	2.5	0	86.5	125	170.5	129.5	1137(334)
$1/\sqrt{2}$	16	6.4	0.5	17	0	0.7	18	3.7	0	0	4.5	92.5	17	287(99)

the first radial excitation having the parameters given in Table 9.31 $\oplus$ the QCD continuum. In this case, we shall only retain the value of $t_c \simeq 2.5$ GeV^2 above the first radial excitation mass.

$\diamond$ We have shown in the previous analysis that the nearby first radial excitation only affects slightly the determination of the lowest ground state mass obtained from the minimal duality ansatz (MDA) one resonance $\oplus$ QCD continuum parametrization of the spectral function. The errors on the determination of the first radial excitations induce slightly larger errors on the ground state mass.

• Final predictions for the four-quark states

One can note from the previous analysis that the predictions of the four-quark states are a MESS and are not conclusive as there are too many freedom for fixing the masses and couplings (choice of the currents and of the mixing parameter). For definiteness, we take the mean of the results from different currents and mixing parameter inside the range $r = 0$ to 1. We deduce, as final predictions for four-quark ground states:

$$\begin{aligned} \bar{M}_{\bar{u}\bar{d}ud} &= 1009(114)\ \text{MeV}, & \bar{f}_{\bar{u}\bar{d}ud} &= 690(66)\ \text{keV}, \\ \bar{M}_{\bar{u}\bar{s}ud} &= 991(123)\ \text{MeV}, & \bar{f}_{\bar{u}\bar{s}ud} &= 444(55)\ \text{keV}, \\ \bar{M}_{\bar{u}\bar{s}ds} &= 1045(112)\ \text{MeV}, & \bar{f}_{\bar{u}\bar{s}ds} &= 457(55)\ \text{keV}, \end{aligned} \tag{9.471}$$

and for the 1st radial excitation:

$$\bar{M}^{(1)}_{\bar{u}\bar{d}ud} = 1409(112)\ \text{MeV}, \quad \bar{f}^{(1)}_{\bar{u}\bar{d}ud} = 449(69)\ \text{keV}, \tag{9.472}$$

where we have taken the error from the most precise determinations. Finite width corrections increase slightly these NWA masses by about (67–80) MeV which is inside the error bars of the determinations. The mass of the radial excitations including a strange quark is expected to be (almost) degenerated with the non-strange one due to the small $SU(3)$ breakings.

9.11.13 Summary and conclusions

We have systematically revisited the existing estimates of the masses and couplings of the scalar quarkonia (ordinary $\bar{q}q$ and four-quark) states and presented new results for $\pi^+\pi^-$-like ($\pi^+\pi^-, \bar{K}K, \bar{K}\pi$ and $\eta\pi$) molecule states. Within our choice of the currents for a given configuration, these states can be analyzed separately like in any QCD spectral sum rules approach. The parameters of these unmixed states can be used for a further analysis of their mixings, which is beyond the scope of this paper. The results are compiled in Table 9.28–9.31.

• QCD expressions

We have checked some existing expressions of the four-quark and molecule currents given in the literature.

◇ Molecule currents

We found several errors in the expression given in the pioneering work of Ref. [136], ($\langle\alpha_s G^2\rangle$, $\langle\bar{s}Gs\rangle$ and $\langle\bar{u}u\rangle^2 + \langle\bar{s}s\rangle^2$), using the pseudoscalar $\oplus$ scalar currents.

◇ Four-quark currents

– We agree with the expression given in Ref. [137] for the scalar four-quark current.

– We recover the different results given in Ref. [577] by putting to zero the $\langle\bar{s}Gs\rangle$ mixed condensate contributions absent in their QCD expressions! Then, we suspect that the strange quark may have been treated as a heavy quark along the calculation of Ref. [577].

• $SU(3)$ breakings

One can note from Tables 9.28 to 9.30 that the $SU(3)$ breakings are tiny (about some few tens of MeV). However, one can observe that for the S/P four-quark states, the central values go in the reverse direction.

• $\sigma/f_0(500)$

The $R_{P/C}$ condition in Eq. (9.399) excludes the values $M_\sigma \simeq$ (0.5–0.6) GeV obtained in the recent literature [138, 577] for four-quark state where its contribution does not exceed 60% of the QCD continuum one at the τ-stability point but, instead, favours the first estimate of about 1 GeV obtained in Refs. [7, 124, 136]. This 1 GeV mass obtained in the real axis from LSR can be identified with the on-shell/Breit–Wigner mass from fits of the $\pi\pi$ scattering data [56, 377, 410]:

$$(M_\sigma^{os}, \Gamma_\sigma^{os}) \simeq (920, 700) \text{ MeV} \tag{9.473}$$

evaluated at $s = (M_S^{os})^2 \; : \; S \equiv \sigma, f(980)$. Therefore, one can see from the previous tables that the molecules and mean of the four-quark (Eq. (9.471)) assignments provide predictions compatible with this on-shell mass definition of σ like in the case of the lightest scalar gluonium [132, 582, 583].

However, the additional constraint on the $\sigma\bar{K}K$ coupling from $\pi\pi \to \bar{K}K$ scattering data [399, 410] quoted in Eq. (9.410) does not favour the pure $\bar{u}\bar{d}ud$ and $\pi^+\pi^-$ assignment for σ often advocated in the literature. This is not the case of the light scalar gluonium which is expected to couple universally (up to $SU(3)$ breakings) to pair of pseudoscalar $\bar{q}q$ states [132, 582, 583] from the low-energy property of the energy–momentum tensor form factor.

• $f_0(980)$

The σ and $f_0(980)$ seems to emerge from a maximal meson–gluonium mixing [125] with $(M_{\bar{q}q}, \Gamma_{\pi\pi}) = (1229, 120)$ MeV from Eq. (9.414) and the light scalar gluonium mass $(M_G, \Gamma_G) = (1070, 890)$ from Ref. [583].

However, the result in Table 9.28 indicates that a K^+K^- molecule with $M_{K^+K^-} = 1056(214)$ MeV is compatible with $f_0(980)$, while the mean prediction of the different four-quark states leads to $M_{\bar{u}\bar{s}ds}$=1045(112) MeV (Eq. (9.471)).

• $a_0(980)$

One of the main motivations for introducing the four-quark/molecule assignment for $a_0(980)$ is its vicinity to $(\bar{u}u + \bar{d}d)$ and its strong coupling with $\bar{K}K$ states [270]. From our analysis, we found that the $I = 1$ isovector $(\bar{u}u - \bar{d}d)$ state is too high $(M_{\bar{u}d} = 1246(94)$ MeV). A $\eta\pi$ molecule with a mass 1040(139) MeV and the mean of four-quark state in Eq. (9.471) also give the same mass as the one in the case of $f_0(980)$ because we have neglected the $SU2$ breakings.

• $f_0(1370)$

This state is reproduced by the first radial excitation of the $\bar{q}q$ state with a mass 1378(186) MeV (Table 9.31) which can mix with the scalar gluonium $M_{\sigma'}$ = 1110(117) MeV [583] to give the observed large $\pi\pi$ width. In this picture, σ and $f_0(1370)$ can also be the dragon proposed in Refs. [386, 586]. However, it can also mix with the first radial excitation of the four-quark state with a mean mass given in Eq. (9.472).

• $a_0(1450)$

This state coincides with the first radial excitation of the $\bar{u}\bar{s}ud$ state which should be almost degenerated to the one of $\bar{u}\bar{d}ud$ state given in Eq. (9.472) due to the small $SU(3)$ breakings.

• $f_0(1500)$

The $f_0(1500)$ is expected to be a gluonium state from its mass $M_{G_1'}$=1563(141) MeV and its $U(1)$-like decays $(\eta'\eta, \eta\eta)$ [132,582,583]. From the present analysis, one expects to have in this region the first radial excitation of the four-quark state having a mass $M^{(1)}_{\bar{u}\bar{d}ud}$ = 1409(112) MeV (Eq. (9.472)) which may mix with the previous gluonium state.

• $f_0(1710)$

From the first radial excitation masses obtained in Table 9.31, $f_0(1700)$ can be likely the radial excitation of the four-quark or/and molecule states. The next radial excitation of a gluonium is predicted to be higher: M_{G_2} = 2992(221) MeV.

• $K_0^*(700)$

The four-quark and the $K\pi$ molecule assignments lead to a mass compatible with the Breit–Wigner mass of (845 ± 17) MeV [33] (see

Eqs. (9.471) and Table 9.28). The $\bar{u}s$ assignment leads to a mass $M_{\bar{u}s} = 1276(58)$ MeV which is relatively too high.

• $K_0^*(1430)$

This state is better fitted by $M_{\bar{u}s} = 1276(58)$ MeV and/or its radial excitation expected to be around 1400 MeV (Eq. (9.467)) as the $SU(3)$ breakings are expected to be small.

We expect that the systematic analysis done in this paper can help clarify the complex spectra of the light scalar mesons.

9.12 Heavy Four-Quark and Molecule States

Our different recent works on heavy and heavy–light four-quark and molecules states have been shortly reviewed in Sections 7.8–7.10 and will not be rediscussed here.

9.13 Scalar Digluonia and $I = 0$ Scalar Mesons

In the following, we review the recent work in Ref. [583].

9.13.1 Abstract

We revisit, scrutinize, improve, confirm and complete our previous results [125, 126, 132] from the scalar di-gluonium sum rules within the standard SVZ expansion at N2LO *without instantons* and *beyond the minimal duality ansatz: "one resonance* $\oplus$ *QCD continuum*" parametrization of the spectral function which is necessary for a better understanding of the complex spectra of the $I = 0$ scalar mesons. We select different (un)subtracted sum rules (USR) moments of degree ≤ 4 for extracting the two lowest gluonia masses and couplings. We obtain $[M_{\sigma_B}, f_{\sigma_B}] = [1.07(13), 0.46(16)]$, $[M_{G_1}, f_{G_1}][1.55(12), 0.37(11)]$ GeV and the corresponding masses of the radial excitations: $M_{\sigma'_B} = 1.11(12)$ and $M_{G'_1} = 1.56(14)$ GeV which are (unexpectedly) almost degenerated with the ground states. The second radial excitation is found to have a much heavier mass: $M_{G_2} \simeq 2.99(22)$ GeV. Combining these results with some low-energy vertex sum rules (LEV-SR), we predict some hadronic widths and classify them into two groups:

– The σ-like (σ_B, σ'_B) which decay copiously to $\pi\pi$ from OZI-violating process and σ'_B to $2(\pi\pi)S$ through $\sigma\sigma$.

– The G-like (G_1, G'_1 and eventually G_2) which decay into $\eta'\eta$, $\eta\eta$ through the $U(1)_A$ gluonic vertex. Besides some eventual mixings with quarkonia states, we may expect that the observed $\sigma/f_0(500)$ and $f_0(137)$ are σ-like, while $f_0(1.5)$ and $f_0(1.7)$ are G-like gluonia. The high mass $G_2(2.99)$ can also mix with G_1, G'_1 to bring the gluon component of the gluonia candidates above 2 GeV. We also estimate the conformal charge $\psi_G(0) = 2.09(29)$ GeV^4 and its slope $10^2 \times \psi'_G(0) = -22(29)$ GeV^2. Our results are summarized in Table 9.32.

9.13.2 The QCD anatomy of the two-point correlator

We shall work with the gluonium two-point correlator:

$$\psi_G(q^2) = 16i \int d^4x\, e^{iqx} \langle 0| \left(\theta^\mu_\mu\right)_G (x) \left(\theta^\mu_\mu\right)^\dagger_G (0)|0\rangle \tag{9.474}$$

built from the gluon component of the trace of the energy-momentum tensor:

$$\theta^\mu_\mu = \frac{1}{4}\beta(\alpha_s) G^{\mu\nu}_a G^a_{\mu\nu} + (1 + \gamma_m(\alpha_s)) \sum_{u,d,s} m_i \bar{\psi}_i \psi_i, \tag{9.475}$$

with $\gamma_m = 2a_s + \cdots$ is the quark mass anomalous dimension and $a_s \equiv \alpha_s/\pi$. $\beta(\alpha_s)$ is the β-function.

• The standard SVZ expansion

Using the operator product expansion (OPE) à la SVZ, its QCD expression can be written as:

$$\psi_G(q^2) = \beta^2(\alpha_s) \left(\frac{2}{\pi^2}\right) \sum_{0,1,2,\dots} C_{2n} \langle \mathcal{O}_{2n} \rangle, \tag{9.476}$$

where C_{2n} is the Wilson coefficients calculable perturbatively, while $\langle \mathcal{O}_{2n} \rangle$ is a short-hand notation for the non-perturbative vacuum condensates $\langle 0|\mathcal{O}_{2n}|0\rangle$ of dimension $2n$.

◇ The unit perturbative operator ($n = 0$)

Its contribution reads:

$$C_0 \equiv -Q^4 L_\mu \Big[C_{00} + C_{01} L_\mu + C_{02} L_\mu^2 \Big],$$

with

$$C_{00} = 1 + \frac{659}{36} a_s + 247.480 a_s^2, \quad C_{01} = -a_s \left(\frac{9}{4} + 65.781 a_s \right),$$
$$C_{02} = 5.0625 a_s^2, \tag{9.477}$$

where the NLO (respectively N2LO) contributions have been obtained in Ref. [553] (respectively [587]) and have been adapted for a sum rule use in Ref. [306]. $L_\mu \equiv \log(Q^2/\mu^2)$ where μ is the subtraction point. We shall use for three flavours:

$$\Lambda = 340(28) \text{ MeV}, \tag{9.478}$$

deduced from $\alpha_s(M_Z) = 0.1182(19)$ from $M_{\chi_{c0,b0}} - M_{\eta_c,\eta_b}$ mass splittings [517], τ-decays [288, 289] and the world average [33, 444]. We shall use the running QCD coupling to order α_s^2:

$$a_s(\mu) = a_s^{(0)} \left\{ 1 - a_s^{(0)} \frac{\beta_2}{\beta_1} LL_\mu + \left(a_s^{(0)} \right)^2 \right.$$
$$\left. \times \left[\left(\frac{\beta_2}{\beta_1} \right)^2 (LL_\mu^2 - LL_\mu - 1) + \frac{\beta_3}{\beta_1} \right] \right\}, \tag{9.479}$$

where:

$$a_s^{(0)} \equiv \frac{1}{-\beta_1 \log(\mu/\Lambda)} \quad \text{and} \quad LL_\mu \equiv \log\left[2 \log(\mu/\Lambda)\right]. \tag{9.480}$$

◇ The dimension-four gluon condensate ($n = 2$)

Its contribution reads:

$$C_4 \langle \mathcal{O}_4 \rangle \equiv (C_{40} + L_\mu C_{41}) \langle \alpha_s G^2 \rangle : \quad C_{40} = 2\pi a_s \left(1 + \frac{175}{36} a_s \right),$$

$$C_{41} = -\frac{9}{2}\pi a_s^2. \tag{9.481}$$

We shall use the value:

$$\langle \alpha_s G^2 \rangle = (6.35 \pm 0.35) \times 10^{-2}\,\text{GeV}^4 \tag{9.482}$$

determined from light and heavy quark systems [43, 152, 517].

◇ The dimension-six ($n = 3$) gluon condensate

Its contribution reads (see Ref. [588] for the α_s correction for $n_f = 0$ and where a $1/\pi$ misprint is corrected):

$$C_6\langle \mathcal{O}_6 \rangle = \langle C_{60} + L_\mu C_{61} \rangle g^3 f_{abc} \langle G^a G^b G^c \rangle / Q^2 : \; C_{60} = a_s,$$
$$C_{61} = -\frac{29}{4} a_s^2, \tag{9.483}$$

with [152]:

$$g^3 f_{abc} \langle G^a G^b G^c \rangle = (8.2 \pm 1.0)\,\text{GeV}^2 \langle \alpha_s G^2 \rangle, \tag{9.484}$$

which notably differs from the instanton liquid model estimate $g^3 f_{abc} \langle G^a G^b G^c \rangle \approx (1.5 \pm 0.5)\,\text{GeV}^2 \langle \alpha_s G^2 \rangle$ [3, 81, 127, 258] used in Ref. [582].

◇ The dimension-eight ($n = 4$) gluon condensate

Its contribution reads:

$$C_8\langle \mathcal{O}_8 \rangle = C_{80} \langle \alpha_s^2 G^4 \rangle / Q^4 : \quad C_{80} = 4\pi \alpha_s \tag{9.485}$$

with:

$$\langle \alpha_s^2 G^4 \rangle \equiv \left[14 \left\langle \left(\alpha_s f_{abc} G^a_{\mu\rho} G^b_{\nu\rho} \right)^2 \right\rangle - \left\langle \left(\alpha_s f_{abc} G^a_{\mu\nu} G^b_{\rho\lambda} \right)^2 \right\rangle \right],$$
$$\simeq (0.55 \pm 0.01) \langle \alpha_s G^2 \rangle^2, \tag{9.486}$$

from a modified factorization with $1/N_c^2$ corrections (13/24) [589] and factorization (9/16) [127]. (0.55 ± 0.01) measures the deviation from factorization which has been found to be largely violated for the four-quark condensates [115, 289, 521].

• Beyond the standard SVZ expansion

◇ The tachyonic gluon mass ($n=1$)

To these standard contributions in the OPE, we can consider the one from a dimension-two tachyonic gluon mass contribution introduced in footnote[38] which phenomenologically parametrizes the large order terms of the perturbative QCD series [235]. Its existence is supported by some AdS approaches [228]. This effect has been calculated explicitly in Ref. [235]:

$$C_2\langle O_2\rangle = C_{21}L_\mu\lambda^2Q^2 : \quad C_{21} = \frac{3}{a_s}, \tag{9.487}$$

where λ^2 is the tachyonic gluon mass determined from $e^+e^- \to$ hadrons data and the pion channel [235–238]:

$$a_s\lambda^2 \simeq -(6.0 \pm 0.5) \times 10^{-2} \text{ GeV}^2. \tag{9.488}$$

◇ The direct instanton ($n \geq 5/2$)

In an instanton liquid model [81, 258], the direct instanton contribution is assumed to be dominated by the single instanton–anti-instanton contribution via a non-perturbative contribution to the perturbative Wilson coefficient [590, 591]:

$$\psi(Q^2)|_{\bar{I}-I} = 32\beta^2\, Q^4 \int \rho^4\Big[K_2\left(\rho\sqrt{Q^2}\right)\Big]^2 dn(\rho), \tag{9.489}$$

where $K_2(x)$ is the modified Bessel function of the second kind. At this stage, this classical field effect is beyond the SVZ expansion where the latter assumes that one can separate without any ambiguity the perturbative Wilson coefficients from the non-perturbative condensate contributions.

Besides the fact that its contributes in the OPE as $1/Q^5$, i.e., it acts like other high-dimension condensates not taken into account in the OPE, the above instanton effect depends crucially on the (model-dependent) overall density $\bar{n} = \int_0^\infty d\rho\, n(\rho)$ and on its average size $\bar{\rho} = (1/\bar{n}) \int_0^\infty d\rho\, \rho\, n(\rho)$ which (unfortunately) are not quantitatively

---

[38] For reviews, see, e.g., Ref. [554].

under a good control ($\bar{\rho}$ ranges from 5 [81] to 1.94, [152] and 1.65 GeV^{-1} [258], while $\bar{n} \approx (0.5 \sim 1.2)\,\text{fm}^{-4}$). They contribute with a high power in ρ to the spectral function Im$\psi(t)$, which can be found explicitly in Refs. [590,591], and behave as:

$$\text{Im}\psi_G(t)|_{\bar{I}-I} \overset{t\to\infty}{\sim} n\left(\rho\sqrt{t}\right)^{-5}$$
$$\overset{t\to 0}{\sim} n\left(\rho\sqrt{t}\right)^{4}. \tag{9.490}$$

As such effects are quite inaccurate and model-dependent, we shall not consider them explicitly in the analysis. Instead, an eventual deviation of our results within the standard SVZ expansion from some experimental data or/and or some other alternative estimates (low-energy theorems (LET), lattice calculations, etc.) may signal the need of such (beyond the standard OPE) effects in the analysis. One should mention that the approach within the standard SVZ OPE and without a direct instanton effect used in the following:

– $U(1)_A$ channel has predicted successfully the value of the topological charge, its slope and the η'-mass and decay constant [82,592].

– Pseudoscalar pion and kaon channels have reproduced successfully the value of the light quark masses where we have also explicitly shown [74] that the direct instanton effect induces a relatively small correction contrary to some other strong claims [194] in the literature.

9.13.3 The inverse Laplace transform sum rules

From its QCD asymptotic behaviour $\sim (-q^2)^2\text{Log}(-q^2/\mu^2)$, one can write a twice subtracted dispersion relation:

$$\psi_G(q^2) = \psi_G(0) + q^2\psi'_G(0) + \frac{q^4}{\pi}\int_0^\infty \frac{dt}{t^2}\frac{\text{Im}\psi_G(t)}{(t-q^2-i\epsilon)}. \tag{9.491}$$

• Following standard QSSR techniques [3,6,7], one can derive from it different form of the sum rules. In this paper, we shall work with the exponential or Borel [3,64,70] or inverse Laplace transform [30] finite energy sum rule (LSR)[39]:

$$\mathcal{L}_n^c(\tau) = \int_0^{t_c} dt\, t^n\, e^{-t\tau} \frac{1}{\pi} \mathrm{Im}\psi_G(t) : n = -1, 0, 1, 2, 3 \tag{9.492}$$

and the corresponding ratios of sum rules:

$$\mathcal{R}_{n+l\ n}^c(\tau) \equiv \frac{\mathcal{L}_{n+l}^c(\tau)}{\mathcal{L}_n^c(\tau)}, \tag{9.493}$$

where τ is the Laplace sum rule variable. In the duality ansatz:

$$\frac{1}{\pi}\mathrm{Im}\psi_G(t) = 2\sum_G f_G^2 M_G^4\, \delta(t - M_G^2) + \theta(t - t_c)\text{"QCD continuum"}, \tag{9.494}$$

where f_G are the lowest resonances couplings normalized as $f_\pi =$ 93 MeV and M_G their masses while the "QCD continuum" comes from the discontinuity $\mathrm{Im}\psi_G(t)|_{QCD}$ of the QCD diagrams from the continuum threshold t_c. In the *one narrow resonance* $\oplus$ *QCD continuum* parametrization of the spectral function:

$$\mathcal{R}_{n+l\ n}^c(\tau) \simeq M_G^2. \tag{9.495}$$

- To get $\mathcal{L}_{-1}$, we find convenient to take the inverse Laplace transform of the first superconvergent second derivative of the correlator. It reads explicitly:

$$\mathcal{L}_{-1}^c(\tau) = \beta^2(\alpha_s)\left(\frac{2}{\pi^2}\right)\tau^{-2}\sum_{n=0,2,\ldots} D_n^{-1} + \psi_G(0), \tag{9.496}$$

with:

$$D_0^{-1} = \Big[C_{00} + 2C_{01}(1 - \gamma_E - L_\tau) + 3C_{02}\big[1 - \pi^2/6 + (-1 + \gamma_E + L_\tau)^2\big]\Big](1 - \rho_1),$$

$$D_2^{-1} = C_{21}\lambda^2\tau(1 - \rho_0),$$

$$D_4^{-1} = -\Big[C_{40} - \frac{C_{41}}{36}\big[55 + 6(\gamma_E + L_\tau)\big]\Big]\langle\alpha_s G^2\rangle\tau^2,$$

[39]The name "inverse Laplace transform" has been attributed due to the fact that perturbative radiative corrections have this property.

$$D_6^{-1} = -\left[C_{60} - C_{61}(-1+\gamma_E + L_\tau)\right] g^3 f_{abc}\langle G^a G^b G^c\rangle \tau^3,$$
$$D_8^{-1} = \frac{C_{80}}{2}(1.1 \pm 0.5)\langle \alpha_s G^2\rangle^2 \tau^4, \tag{9.497}$$

with:

$$L_\tau \equiv -\log(\tau\mu^2), \quad \rho_n \equiv e^{-t_c\tau}\left(1 + t_c\tau + \cdots \frac{(t_c\tau)^n}{n!}\right), \tag{9.498}$$

as the QCD continuum contribution from the discontinuity of the QCD diagrams to the sum rule. We omit some eventual continuum contributions from the non-perturbative condensates which are numerically tiny.

$\psi_G(0)$ is the value of the two-point correlator at zero momentum and can be fixed by an LET as [81, 127]:

$$\psi_G(0) \simeq -\frac{16}{\pi}\beta_1\langle \alpha_s G^2\rangle = (1.46 \pm 0.08)\ \mathrm{GeV}^4, \tag{9.499}$$

which we shall test later on.

- To get $\mathcal{L}_0$, we take the inverse Laplace transform of the first superconvergent third derivative of the two-point correlator. In this way, we obtain:

$$\mathcal{L}_0^c(\tau) = \beta^2(\alpha_s)\left(\frac{2}{\pi^2}\right)\tau^{-2}\sum_{n=0,2,\cdots} D_n^0, \tag{9.500}$$

with:

$$D_0^0 = \Big[2C_{00} - 2C_{01}(3 - 2\gamma_E - 2L_\tau) - 6C_{02}\big[1 - 3\gamma_E + \gamma^2 - \pi^2/6$$
$$+(-3+2\gamma_E)L_\tau + L_\tau^2\big]\Big](1-\rho_2),$$
$$D_2^0 = -\frac{C_{21}}{2}\lambda^2\tau(1-\rho_1),$$
$$D_4^0 = -C_{41}\langle \alpha_s G^2\rangle\tau^2,$$
$$D_6^0 = \left[C_{60} - C_{61}(\gamma_E - L_\tau)\right] g^3 f_{abc}\langle G^a G^b G^c\rangle\tau^3,$$
$$D_8^0 = C_{80}(0.55 \pm 0.01)\langle \alpha_s G^2\rangle^2\tau^4. \tag{9.501}$$

• The other higher degrees sum rules $\mathcal{L}_n^c(\tau)$ for $n \geq 1$ can be deduced from the nth τ-derivative of $\mathcal{L}_0^c(\tau)$:

$$\mathcal{L}_n^c(\tau) = (-1)^n \frac{d^n}{d\tau^n}\mathcal{L}_0^c(\tau). \tag{9.502}$$

• These superconvergent sum rules obey the homogeneous renormalization group equation (RGE):

$$\left\{-\frac{\partial}{\partial t} + \beta(\alpha_s)\alpha_s\frac{\partial}{\partial\alpha_s}\right\}\mathcal{L}_n^c(e^t\tau, \alpha_s) = 0, \tag{9.503}$$

with $t \equiv (1/2)L_\tau$ with the renormalization group improved (RGI) solution:

$$\mathcal{L}_n^c(e^t\tau, \alpha_s) = \mathcal{L}_n^c(t = 0, \bar{\alpha}_s(\tau)), \tag{9.504}$$

where $\bar{\alpha}_s$ is the QCD running coupling.

• In the following analysis, we shall work with the family of sum rules having degrees less or equal to 4. Then, we shall select the sum rules which present stability (minimum or inflexion point) in the sum rule variable τ and in the continuum threshold t_c such that we can extract optimal information from the analysis.

9.13.4 The QCD side of the LSR $\mathcal{L}_n^c$

Before doing the phenomenological analysis of the LSR, we study the τ and t_c-behaviour of their QCD expressions. For instance, we show the subtracted (SSR) $\mathcal{L}_{-1}^c$ and unsubtracted (USR) $\mathcal{L}_0^c$ sum rules in Figs. 9.152 and 9.152, where we note that the subtraction constant shifts the USR minimum (optimization point) at lower values of τ. This feature has led to the inconsistencies of the gluonium mass from these sum rules within a *one resonance* $\oplus$ *QCD continuum* parametrization of the spectral function as noted in Refs. [132, 582] which can be cured by working instead with "*two resonances* $\oplus$ *QCD continuum*".

- The other USR moments $\mathcal{L}_2^c$ and $\mathcal{L}_4^c$ stabilize for τ of about 1.5 GeV^{-2}, while $\mathcal{L}_1^c$ and $\mathcal{L}_3^c$ do not present τ-stability. To understand these τ-behaviours, we do the following:
    - ◇ We write explicitly the OPE expression of $\mathcal{L}_0$ and $\mathcal{L}_1$ to lowest order but includes the α_s correction of the $D = 4$ contribution

as this contribution vanishes to LO. Normalized to $(2/\pi^2)\beta^2$, they read:

$$\mathcal{L}_0 = \tau^{-3}\left[1+\frac{9}{2}\pi a_s^2\langle\alpha_s G^2\rangle\tau^2 + a_s g^3 f_{abc}\langle G^a G^b G^c\rangle\tau^3 + 4\pi\alpha_s\langle\alpha_s^2 G^4\rangle\tau^4\right],$$

$$\mathcal{L}_1 = 3\tau^{-4}\left[1-\frac{9}{6}\pi a_s^2\langle\alpha_s G^2\rangle\tau^2 - 4\pi\alpha_s\langle\alpha_s^2 G^4\rangle\tau^4\right], \quad (9.505)$$

where the changes of sign of the condensate contributions destabilize $\mathcal{L}_1$.

◇ The inclusion of the huge PT radiative corrections shows that they also participate in the τ-behaviour of the QCD expression due to the non-trivial reorganization of these PT terms in each Laplace transformed moment.

- From this explicit analysis, one expects the following *a priori*:

◇ One may extract reliably the decay constant of the gluonium from $\mathcal{L}_2^c$ and $\mathcal{L}_4^c$. However, we shall see later on that the extraction of the decay constant from $\mathcal{L}_3^c$ can present τ-stability due to the exponential weight $e^{M_G^2\tau}$ brought by the resonance contributions which compensates the decrease of the QCD expression $[\mathcal{L}_n^c \sim \tau^{-(n+3)}]$ for increasing τ. In the specific example of $\mathcal{L}_2^c$, we shall note that the exponential factor shifts the τ-minimum to lower τ-values (Fig. 9.157) at which the PT corrections are much smaller than in Fig. 9.153, while for $\mathcal{L}_3^c$, there appears a plateau inflexion point (Fig. 9.157) which render reliable the extraction of f_{G_1}.

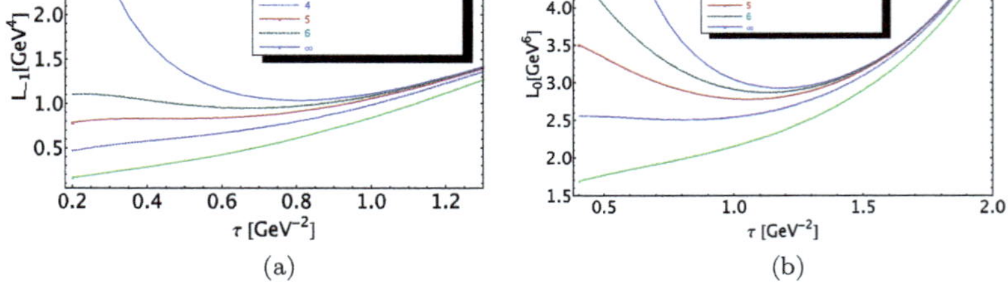

Figure 9.152. (a) $\mathcal{L}_{-1}^c$ as a function of τ at N2LO for different values of t_c; (b) the same as (a) but for $\mathcal{L}_0^c$.

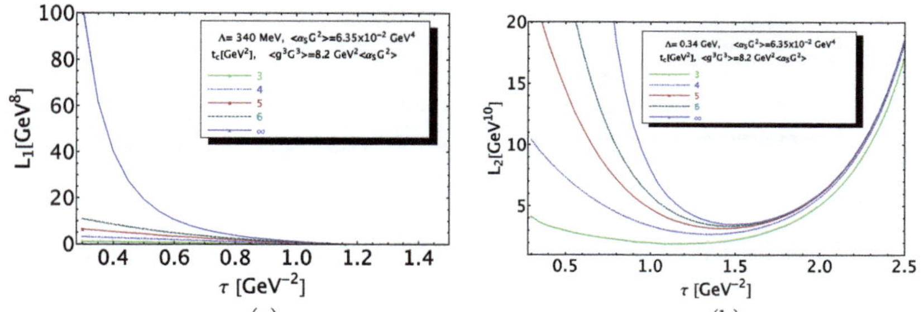

(a) (b)

Figure 9.153. (a) $\mathcal{L}_1^c$ as a function of τ at N2LO for different values of t_c; (b) the same as (a) but for $\mathcal{L}_2^c$.

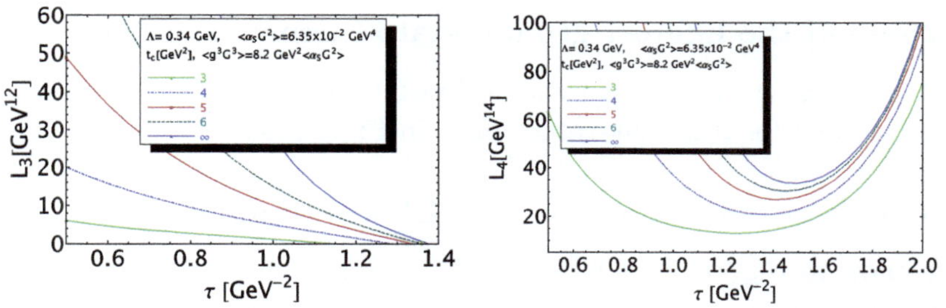

(a) (b)

Figure 9.154. (a) $\mathcal{L}_3^c$ as a function of τ at N2LO for different values of t_c; (b) the same as (a) but for $\mathcal{L}_4^c$.

◇ The mass of the gluonium can be reliably obtained from the ratios $\mathcal{R}_{nl}^c$ of the USR $\mathcal{L}_{n,l}^c$ which optimize at about the same value of τ and where the PT corrections tend to compensate.

9.13.5 The masses of the two lightest gluonia σ_B and G_1

In order to get the lowest ground state gluonia masses, we shall work with different forms of USR. We shall not use for this case the observables involving $\mathcal{L}_{-1}$ in order to avoid the dependence of the result on the input value of $\psi_G(0)$ which we shall test later on. Among the possible ratios of sum rules, we have selected $\mathcal{R}_{20}^c$ and $\mathcal{R}_{42}^c$ which appear to give the most stable results versus the variation of different parameters.

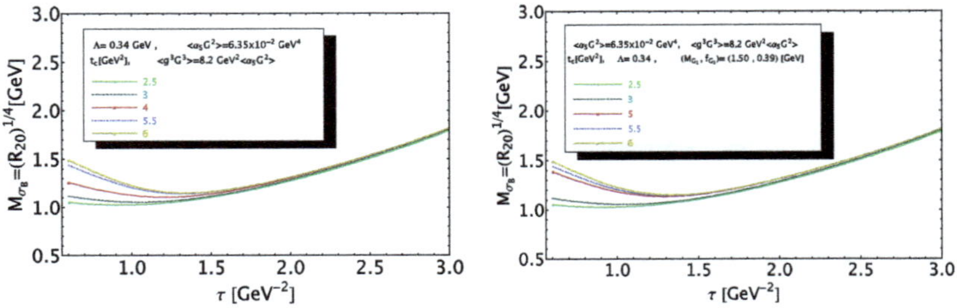

(a) (b)

Figure 9.155. (a) M_{σ_B} from $\mathcal{R}_{20}^c$ as a function of τ at N2LO for different values of t_c within a one resonance parametrization; (b) the same as (a) but for two resonance parametrization.

• Mass of the lightest ground state (named σ_B) from $\mathcal{R}_{20}^c$

In so doing, we work with the ratio of sum rules $\mathcal{R}_{20}^c$. We show its τ-behaviour for different values of t_c in Fig. 9.155.

◇ The case: One resonance ⊕ QCD continuum

We show in Fig. 9.155 the result with one resonance ⊕ QCD continuum parametrization of the spectral function. The τ-stability starts from $t_c = 2.5$ GeV^2 while t_c-stability is reached around 5.5 GeV^2. We deduce in this range:

$$M_{\sigma_B} = 1088(2)_\tau(54)_{t_c}(57)_\Lambda(1)_{G^2}(2)_{G^3} = 1088(79) \text{ MeV}, \quad (9.506)$$

where the errors are mainly due to Λ and t_c.

◇ The case: Two resonances ⊕ QCD continuum

We show in Fig. 9.155 the effect of a second resonance with the parameters in Ref. [582]:

$$M_{G_1} = 1.5 \text{ GeV} \quad \text{and} \quad f_{G_1} = 0.39 \text{ GeV}. \quad (9.507)$$

It affects the σ_B mass result by a negligible amount of about 22 MeV. We deduce the conservative result from $t_c = 3$ to 5.5 GeV^2 and use an input error of 0.1 GeV^{-2} for the localization of τ:

$$\begin{aligned} M_{\sigma_B} &= 1085(5)_\tau(59)_{t_c}(89)_\Lambda(1)_{G^2}(2)_{G^3}(1)_{\lambda^2}(39)_{M_G}(53)_{f_G} \\ &= 1085(126) \text{ MeV}, \end{aligned} \quad (9.508)$$

where the errors due to the non-perturbative condensates are negligible.

• Mass of the medium ground state gluonium G_1 from $\mathcal{R}_{42}^c$

Among the different combinations of sum rules, we find that $\mathcal{R}_{42}^c$ can give a reliable prediction of the medium ground state G_1 because $\mathcal{L}_2^c$ and $\mathcal{L}_4^c$ stabilize at about the same value of $\tau \simeq (1.2 \sim 1.5)$ GeV^{-2} (see Figs. 9.153 and 9.154). We show in Fig. 9.156 the τ-behaviour of $\mathcal{R}_{42}$ for different values of t_c.

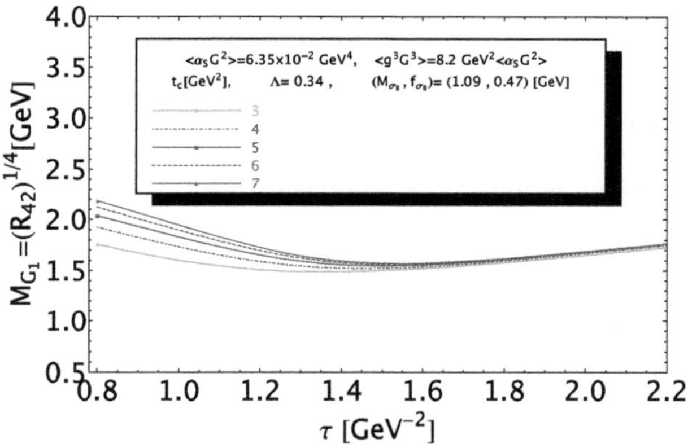

Figure 9.156. M_{G_1} as a function of τ at N2LO for different values of t_c from a two resonance parametrization.

◇ The case: Two resonances ⊕ QCD continuum

We obtain for $t_c = 3$ to 6 GeV^2:

$$\begin{aligned} M_{G_1} &= 1524(36)_{t_c}(7)_\tau(115)_\Lambda(2)_{G^2}(2)_{G^3}(0)_{\lambda^2}(6)_{M_\sigma}(11)_{f_\sigma} \\ &= 1524(121) \text{ MeV}, \end{aligned} \tag{9.509}$$

where the error comes mainly from the one of Λ. We have used $f_{\sigma_B} = 0.47$ as anticipated from the following section.

◇ The case: One resonance ⊕ QCD continuum

A similar analysis leads to:

$$M_{G_1} = 1515(123) \text{ MeV}, \tag{9.510}$$

indicating that the σ_B effect to the value of the mass is quite small (about 9 MeV).

9.13.6 Decay constants f_{σ_B} and f_{G_1}

To extract these decay constants, we work simultaneously with $\mathcal{L}_2^c$, $\mathcal{L}_3^c$ and $\mathcal{L}_{-1}^c$ and use an iteration procedure. The two former are expected to be more sensitive to f_{G_1} while the third to f_{σ_B}.

◇ We shall take M_{σ_B} from Eq. (9.508) and $M_{G_1} = 1.548$ GeV anticipated from Eq. (9.509). We start the iteration by using $f_{\sigma_B} =$ 800 MeV from [132, 582]. Then, we compare the value of f_{G_1} from $\mathcal{L}_2^c$, $\mathcal{L}_3^c$. Then, we use the common solution of f_{G_1} from the two sum rules into $\mathcal{L}_{-1}^c$ to extract f_{σ_B}. We continue the iterations until we reach convergent solutions. The analysis corresponding to the final value of the decay constants is shown in Figs. 9.157–9.159.

– In Figs. 9.157 and 9.157, we show the τ and t_c behaviours of f_{G_1} given the value of f_{σ_B}.

– In Fig. 9.157, we show the results from $\mathcal{L}_2^c$, $\mathcal{L}_3^c$ at the τ minimum or plateau from which we deduce the common solution for f_{G_1} for fixed f_{σ_B} which allows to fix:

$$t_c \simeq (3.3 \pm 0.3)\,\mathrm{GeV}^2. \tag{9.511}$$

– In Fig. 9.159, we show the τ- and t_c-behaviours of f_{σ_B} where we take the value $\tau = 0.5$ GeV^{-2} at the inflexion point which is more pronounced for higher values of t_c. Then, we obtain:

$$\begin{aligned} f_{G_1} &= 394(45)_{t_c}(2)_\tau(22)_\Lambda(0)_{G^2}(1)_{G^3}(23)_{M_{\sigma_B}}(9)_{f_{\sigma_B}}(94)_{M_{G_1}} \\ &= 394(109)\ \mathrm{MeV}:\ \mathcal{L}_3^c, \\ f_{\sigma_B} &= 563(36)_{t_c}(64)_\tau(81)_\Lambda(9)_{G^2}(11)_{G^3}(10)_{M_\sigma}(3)_{M_{G_1}}(107)_{f_{G_1}} \\ &= 563(154)\ \mathrm{MeV}:\ \mathcal{L}_{-1}^c. \end{aligned} \tag{9.512}$$

One can note that value of f_{G_1} agrees perfectly with the one in Ref. [582], while f_{σ_B} is in the range of the one obtained in Refs. [132, 582].

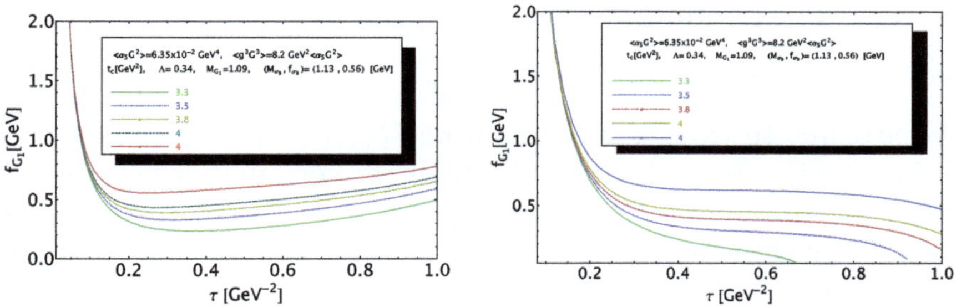

(a) (b)

Figure 9.157. (a) f_{G_1} from $\mathcal{L}_2^c$ as a function of τ at N2LO for different values of t_c from a two resonance parametrization and for given values of $(M_{\sigma_B}, (f_{\sigma_B})$; (b) the same as (a) but from $\mathcal{L}_3^c$.

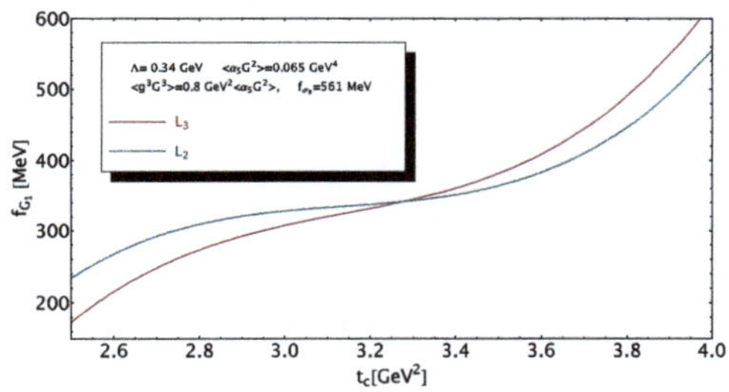

Figure 9.158. f_{G_1} from $\mathcal{L}_2^c$ and $\mathcal{L}_3^c$ from the τ-minimum $\tau \simeq 0.5$ GeV^{-2} from previous figures as a function of t_c for given values of $(M_{\sigma_B}, f_{\sigma_B})$.

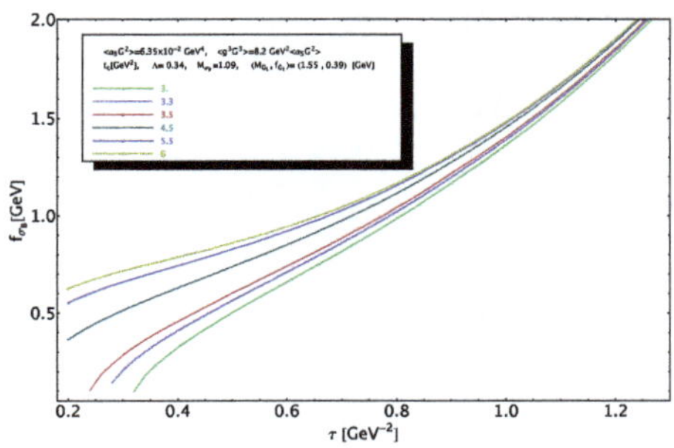

Figure 9.159. f_{σ_B} as a function of τ at N2LO for different values of t_c from a two resonance parametrization and for given values of (M_{G_1}, f_{G_1}).

9.13.7 First radial excitation σ'_B of σ_B

• The σ'_B mass

We attempt to extract its mass from $\mathcal{R}^c_{20}$ which we have used to estimate M_{σ_B}. In so doing, we subtract the contributions of σ_B and G_1 and take the result for very large values of t_c, i.e., by replacing the QCD continuum contribution by σ'_B. In this way, we obtain the result in Fig. 9.160 from $\mathcal{R}^c_{20}$.

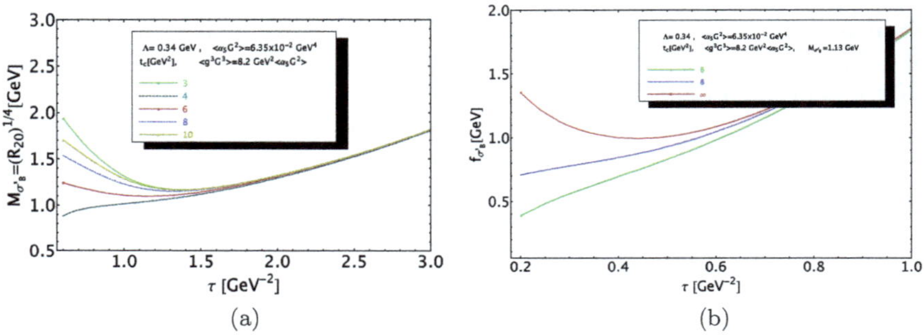

Figure 9.160. (a) $M_{\sigma'_B}$ as a function of τ at N2LO for different values of t_c within a three resonance parametrization from $\mathcal{R}^c_{20}$; (b) $f_{\sigma'_B}$ as a function of τ at N2LO for different values of t_c within a three resonance parametrization from $\mathcal{L}^c_{-1}$.

We obtain, for $t_c \simeq (6 \sim 8)$ GeV^2 at the τ-minimum:

$$\begin{aligned} M_{\sigma'_B} &= 1110(25)_{t_c}(10)_{\tau}(110)_{\Lambda}(0)_{G^2}(3)_{G^3}(1)_{M_{\sigma_B}}(12)_{f_{\sigma_B}} \\ &\quad \times(7)M_{G_1}(23)_{f_{G_1}} \\ &= 1110(117)\ \text{MeV}, \end{aligned} \tag{9.513}$$

where the contributions of σ_B and G_1 have been included after iteration from two- to three-resonance parametrization of the spectral function and using the values of the σ_B and G_1 masses and decay constants from two-resonance parametrization.

◇ The σ'_B decay constant

We extract the decay constant from $\mathcal{L}^c_{-1}$ using the same range of t_c as for $M_{\sigma'_B}$ and taking $\tau \simeq 0.4$ GeV^{-2} (Fig. 9.160). We obtain:

$$
\begin{aligned}
f_{\sigma'_B} &= 648(75)_{t_c}(43)_\tau(124)_\Lambda(20)_{G^2}(7)_{G^3}(24)_{M_{\sigma_B}}(137)_{f_{\sigma_B}} \\
&\quad \times(0)_{M_{G_1}}(56)_{f_{G_1}}(33)_{M_{\sigma'_B}} \\
&= 648(216)\ \text{MeV from } \mathcal{L}^c_{-1}. \qquad (9.514)
\end{aligned}
$$

Anticipating the estimate in Section 9.13.15 by extracting the decay constant from $f_0(1.37) \to 2(\pi\pi)_S$ data through an LEV-SR, we shall work instead with the more precise value:

$$
f_{\sigma'_B} = 329(30)\ \text{MeV}. \qquad (9.515)
$$

9.13.8 Iterative improvements of the previous estimates

In the following, we study the effect of σ'_B which is below M_{G_1} on the previous estimates.

• The mass M_{σ_B}

We include the contribution of σ'_B which is below the G_1 mass. Using $\mathcal{R}^c_{20}$ as in previous section, the curves are similar to the ones in Fig. 9.155. The mass in Eq. (9.508) becomes:

$$
M_{\sigma_B} = 1070(126)(1)_{f_{\sigma'_B}}(1)_{M_{\sigma'_B}}\ \text{MeV}, \qquad (9.516)
$$

which is our final estimate.

• The mass M_{G_1}

Including the contribution of σ'_B into $\mathcal{R}^c_{42}$, the mass in Eq. (9.509) becomes:

$$
M_{G_1} = 1548(118)(15)_{f_{\sigma'_B}}(13)_{M_{\sigma'_B}} = 1548(120)\ \text{MeV}, \qquad (9.517)
$$

which is our final estimate.

• The decay constants f_{σ_B} and f_{G_1}

We repeat the procedure used previously by extracting simultaneously f_{σ_B} and f_{G_1} from $\mathcal{L}^c_{-1}$ and from $\mathcal{L}^c_2, \mathcal{L}^c_3$ by including the σ'_B

contribution. The value of t_c at which the solutions from $\mathcal{L}_2^c, \mathcal{L}_3^c$ meet is slightly shifted from 3.3 to 3.2 GeV^2. We obtain the final estimate:

$$f_{\sigma_B} = 456(154)(8)_{M_{\sigma'_B}}(12)_{f_{\sigma'_B}} = 456(157) \text{ MeV}, \tag{9.518}$$

and:

$$f_{G_1} = 365(109)(17)_{M_{\sigma'_B}}(1)_{f_{\sigma'_B}} = 365(110) \text{ MeV}. \tag{9.519}$$

9.13.9 The first radial excitation G'_1 of G_1

• The G'_1 mass

◇ We attempt to estimate the mass of the first radial excitation by replacing the QCD continuum by a fourth resonance in the sum rule $\mathcal{R}_{42}^c$ which we have used to extract M_{G_1}. The analysis is shown in Fig. 9.161.

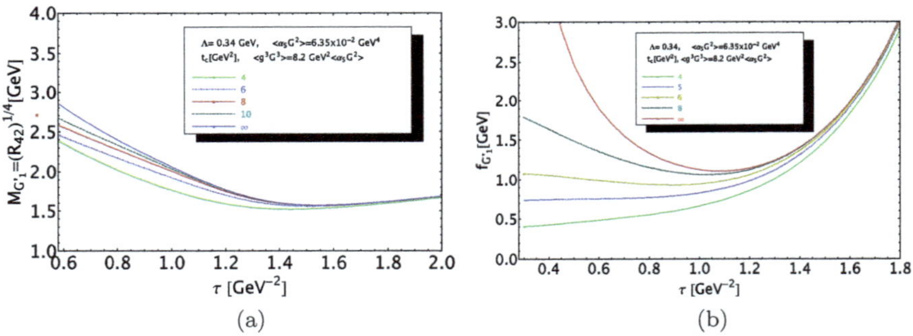

Figure 9.161. (a) $M_{G'_1}$ as a function of τ at N2LO from a four-resonance parametrization; (b) $f_{G'_1}$ as a function of τ at N2LO from a three-resonance parametrization.

We deduce, for $t_c \simeq (4 \sim 8)$ GeV^2 at the τ-minimum:

$$\begin{aligned} M_{G'_1} &= 1563(23)_{t_c}(2)_\tau(138)_\Lambda(3)_{G^2}(3)_{G^3}(6)_{f_{\sigma_B}}(2)_{M_{\sigma_B}}(3)_{f_{G_1}} \\ &\quad \times(9)_{M_{G_1}}(11)_{f_{\sigma'_B}}(1)_{M_{\sigma'_B}}, \\ &= 1563(141) \text{ MeV}, \end{aligned} \tag{9.520}$$

where the error comes mainly from Λ. We have included the contributions of the σ_B, σ'_B and G_1.

• The decay constant $f_{G'_1}$

To extract the decay constant, we shall work with the sum rules $\mathcal{L}_2^c$ or/and $\mathcal{L}_3^c$ used previously to get f_{G_1}, where $\mathcal{L}_2^c$ presents a τ-minimum (Fig. 9.161), while $\mathcal{L}_3^c$ presents an inflexion point. From $\mathcal{L}_2^c$, one deduces:

$$\begin{aligned} f_{G'_1} &= 1000(65)_{t_c}(11)_\tau(150)_\Lambda(1)_{G^2}(2)_{G^3}(15)_{f_{\sigma_B}}(58)_{M_{\sigma_B}} \\ &\quad \times(53)_{f_{G_1}}(63)_{M_{G_1}}(1)_{f_{\sigma'_B}}(8)_{M_{\sigma'_B}}(125)_{M_{G'_1}}, \\ &= 1000(230)\ \text{MeV}. \end{aligned} \tag{9.521}$$

9.13.10 The second radial excitation G_2

• The mass M_{G_2}

In so doing, we work with some judicious choice of sum rules which optimize at smaller values of τ and are more sensitive to the high-mass meson contributions. This criterion is satisfied by $\mathcal{R}_{21}^c$ and $\mathcal{R}_{31}^c$ which one can see in Figs. 9.162 and 9.162 where the contributions of σ_B, σ' and G_1, G'_1 have been subtracted. $\mathcal{R}_{21}^c$ has a (τ, t_c) minimum at (0.18,12) in units of (GeV^{-2}, GeV^2), while $\mathcal{R}_{31}^c$ has the minimum at (0.24,10).

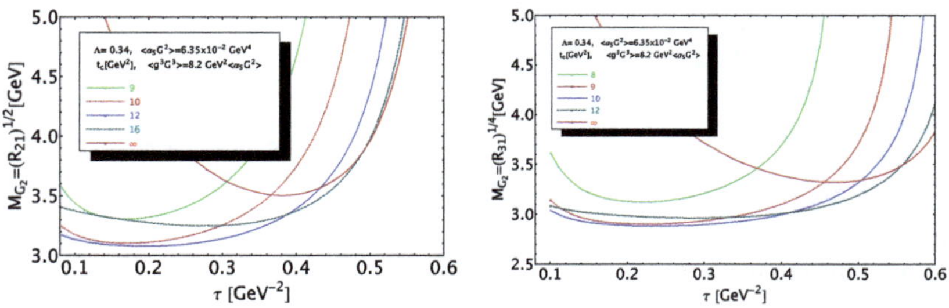

(a) (b)

Figure 9.162. (a) M_{G_2} as a function of τ at N2LO for a four-resonance parametrization from $\mathcal{R}_{21}^c$; (b) the same as (a) but from $\mathcal{R}_{31}^c$.

We obtain:

$$\begin{aligned} M_{G_2} &= 3079(15)_{t_c}(10)_\tau(63)_\Lambda(1)_{G^2}(2)_{G^3}(10)_{M_{\sigma_B}}(7)_{f_{\sigma_B}} \\ &\quad \times(20)_{M_{G_1}}(36)_{f_{G_1}}(6)_{M_{\sigma'}}(1)_{f_{\sigma'}}(193)_{M_{G'_1}}(207)_{f_{G'_1}}, \\ &= 3079(294)\ \text{MeV: } \mathcal{R}_{21}^c \end{aligned} \tag{9.522}$$

and:

$$M_{G_2} = 2879(9)_{t_c}(3)_\tau(64)_\Lambda(1)_{G^2}(2)_{G^3}(10)_{M_{\sigma_B}}(8)_{f_{\sigma_B}}(25)_{M_{G_1}}(40)_{f_{G_1}}$$
$$(6)_{M_{\sigma'}}(0)_{f_{\sigma'}}(226)_{M_{G'_1}}(233)_{f_{G'_1}},$$
$$= 2879(334)\ \text{MeV: } \mathcal{R}^c_{31}, \tag{9.523}$$

from which we deduce the mean value:

$$M_{G_2} = 2992(221)\ \text{MeV}. \tag{9.524}$$

This second radial excitation is far above the ground states and first radial excitations and may mix with the trigluonium scalar ground state [133, 582]:

$$M_{3g} \simeq 3.1\ \text{GeV}. \tag{9.525}$$

◇ The decay constant f_{G_2}

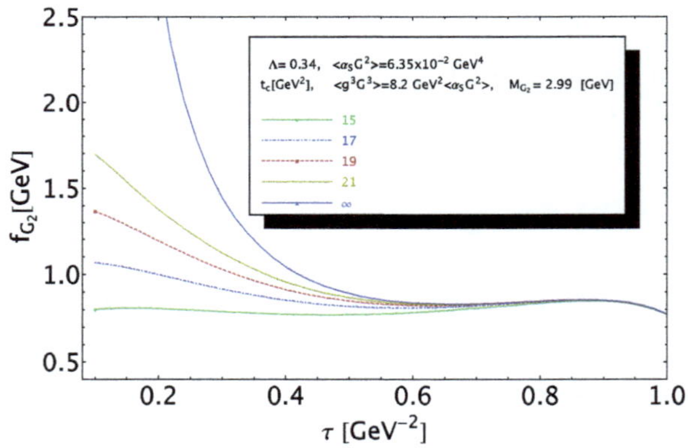

Figure 9.163. f_{G_2} as a function of τ at N2LO for different values of t_c for a four-resonance parametrization from $\mathcal{L}^c_3$.

We shall work with $\mathcal{L}^c_3$ (used to get f_{G_1}) for extracting the decay constant f_{G_2}. The analysis is shown in Fig. 9.163, where one obtains:

$$f_{G_2} = 797(27)_{t_c}(57)_\Lambda(1)_{M_{\sigma'_B}}(3)M_{G_1}(3)_{f_{G_1}}(36)_{M_{G'_1}}(11)_{f_{G'_1}}(11)_{M_{G_2}},$$
$$= 797(74)\ \text{MeV}. \tag{9.526}$$

The errors due to the other terms not quoted are zero. One should note that, like in the case of $f^{\text{eff}}_{\sigma'}$ which is comparable in size, the previous coupling of G_2 to the gluonic correlator can be interpreted as the sum of effective couplings of all higher states contributing to the spectral function.

9.13.11 Truncation of the PT series, tachyonic gluon mass

• Truncation of the PT series

One can note from the expression of the two-point correlator in Eq. (9.397) that the radiative corrections to the unit operator are huge.

◇ M_{σ_B} and M_{G_1}

We compare explicitly the effect of PT radiative corrections for the predictions of M_{σ_B} and M_{G_1} and $\mathcal{R}^c_{42}$ in the case of "one resonance ⊕ QCD continuum" parametrization of the spectral function. For $t_c = 4$ GeV^2, one obtains:

$$M_{\sigma_B} \simeq 768\ (\text{LO}) \to 1000\ (\text{NLO}) \to 1076\ (\text{N2LO})\ \text{from}\ \mathcal{R}^c_{20},$$
$$M_{G_1} \simeq 1492\ (\text{LO}) \to 1522\ (\text{NLO}) \to 1530\ (\text{N2LO})\ \text{from}\ \mathcal{R}^c_{42}, \tag{9.527}$$

indicating that the α_s corrections increase the LO masses by about 30% for the σ_B, while the increase from NLO to N2LO is only about 8%. For M_{G_1} the effect of radiative corrections is much smaller. This relatively small effect is due to the fact that radiative PT corrections tend to compensate in the ratios of sum rules.

◇ The decay constant f_{σ_B}

Fixing, e.g., $t_c \simeq 4$ GeV^2, we study the effect of the truncation of the PT series on the estimate of the coupling from $\mathcal{L}^c_{-1}$. Taking $\tau \simeq 0.5$ GeV^{-2} at the inflexion point, we obtain the result in units of MeV for a "one-resonance" parametrization:

$$f_{\sigma_B} = 202\ (\text{LO}) \to 406\ (\text{NLO}) \to 696\ (\text{N2LO}). \tag{9.528}$$

The effect of radiative corrections is huge as expected.

◇ The decay constant f_{G_1}

Here, we fix $t_c = 3.75$ GeV^2 as obtained in the previous analysis. There is a large plateau in τ for different PT series. We obtain in units of MeV from $\mathcal{L}_3^c$ within a "one resonance $\oplus$ QCD continuum":

$$f_{G_1} = 360\ (\mathrm{LO}) \to 470\ (\mathrm{NLO}) \to 580\ (\mathrm{N2LO}), \tag{9.529}$$

which shows relatively moderate PT corrections.

• Gluon condensates and the OPE

Though the presence of the gluon condensates in the OPE is important in some moments for the stability points, their effects are relatively small ensuring the convergence of the OPE.

• The tachyonic gluon mass λ^2

As emphasized in Ref. [235], this effect is expected to explain the large scale of the scalar gluonium channel compared to the ordinary ρ meson channel. It is also expected [234] to give a phenomenological estimate of the uncalculated terms of the PT series.

◇ M_{σ_B} and M_{G_1}

Including the λ^2 contribution in the OPE, we check that its effect on the mass determination is negligible which can be due to the fact that the effects tend to compensate in the ratio of moments.

◇ The decay constants f_{σ_B} and f_{G_1}

Within a two-resonance parametrization, we find for a given value of t_c and τ that the tachyonic gluon mass decreases the value of f_{σ_B} by about 30–50%, while it increases only slightly f_{G_1} by about 5%.

• Comparison with some other determinations

◇ σ_B

The existence of the σ_B with a mass about 1 GeV is compatible with the idea that it is the dilaton associated to conformal $U(1)_V$ invariance analogous to the case where the η'-meson is associated to the $U(1)_A$ symmetry. The dilaton has been extensively discussed in an effective Lagrangian approach (see, e.g., Refs. [135, 593, 594]), including the $U(1)_V$ symmetry.

– The mass in Eq. (9.508) is in good agreement with the earlier sum rule result [118]:

$$M_{\sigma_B} \simeq (0.95\text{–}1.10) \text{ GeV}, \tag{9.530}$$

from a least square fit of the USR ratio of sum rules. A similar result has been obtained from the analysis [200] based on Gaussian sum rules [521], including instanton effects.

– A full lattice calculation [463, 595] finds a relative suppressed mass of the lightest glueball which moves from the quenched result of about 1.6 GeV to the full QCD one around 1 GeV. However, a more appropriate comparison with the lattice results requires a simulation beyond the one resonance contribution to the scalar gluonium correlator.

– A strong coupling analysis of the gluon propagator finds a gluonium pole mass around M_σ and $M_{f_0(980)}$ from the mass gap [420].

◇ The higher mass gluonium G_1

– Its mass is given in Eqs. (9.510) and (9.509) in the case of *one or two resonances* ⊕ *QCD continuum* parametrization of the spectral function. These results agree with the earlier ones in Ref. [582] from ratio of USR and can be compared to some other sum rules results [590, 596, 597]. We conclude from the paper of Ref. [598] that their result for $M_{G_1} \simeq 1.8$ GeV is obtained outside the τ-minima of their sum rule which leads to an overestimate.

– Our result for M_{G_1} is slightly lower than the one from quenched and full lattice calculations [463, 595, 599, 600]. However, a precise comparison of the results from the two approaches is delicate as the lattice situation for the 1 GeV result remains unclear [463, 595], while the lattice analysis is done by only retaining a single resonance contribution to the two-point correlator.

◇ The gluonium radial excitations σ' and G_2

We have also attempted to extract the masses and decay constants of the gluonium first radial excitations with the result in Eqs. (9.513), (9.514), (9.524) and (9.526). In the next phenomenological analysis, we are tempted to identify $\sigma'(1.32)$ with $f_0(1.37)$. $G_2(2.5)$ is (within the large errors) in the range of the observed $f_0(2.02)$ and $f_0(2.2)$ [33].

◇ **Comments**

– In the example of the scalar gluonium channel, we have seen that (at least) NLO corrections are important in the determination of the scalar gluonium masses from the sum rules, while N2LO corrections are needed for a good determination of the decay constants. Therefore, one should make a great care for quoting the LO results given in the literature.

– The smallness of the tachyonic gluon mass contribution to the gluonium mass determinations may indicate that the PT series reaches its asymptotic form after the inclusion of the N2LO contribution which can be sufficient for extracting with a good accuracy these observables from the sum rules. However, we have also noted that the decay constant of σ_B is largely affected (-30% to -40%) by the tachyonic gluon mass which tends to decouple σ_B from the two-point correlator.

– Some other analysis using "two resonances ⊕ QCD continuum" parametrization from sum rules with instantons [200] and an inverse problem dispersive approach without instantons [601] reproduce our results for M_{σ_B} and M_{G_1}. The apparent discrepancies with the lattice results request a clarification of the full QCD results and a two-resonance parametrization of the two-point correlator.

– Some phenomenological implications of our results will be discussed in the following section.

9.13.12 The conformal charge $\psi_G(0)$

• **Test of the low–energy theorem (LET)**

LET suggests that the value of $\psi_G(0)$ is given by Eq. (9.499). We test the accuracy of this relation by saturating the two-point function by the two lowest ground state resonances σ_B and G_1. In this way, we obtain:

$$\psi_G(0) = \sum_{i=\sigma,G,\cdots} 2\, f_G^2 M_G^2 \approx 1.95\ \text{GeV}^4, \tag{9.531}$$

which is slightly higher than the LET estimate in Eq. (9.499):

$$\psi_G(0)|_{\mathrm{LET}} = (1.46 \pm 0.08)\ \mathrm{GeV}^2, \tag{9.532}$$

without any appeal to some contributions beyond the OPE such as the instanton ones.

• Sum rule extraction of conformal charge $\psi_G(0)$

Here, we shall, instead, determine $\psi_G(0)$ from the sum rule within the standard SVZ expansion. A similar analysis has been already done in Ref. [118] which we shall update here. In so doing, we work with the combination of sum rules[40]:

$$\psi_G(0)|_{\mathrm{LSR}} = \int_0^{t_c} \frac{dt}{t}\, e^{-t\tau}\left(1 - \frac{t\tau}{2}\right)\frac{1}{\pi}\,\mathrm{Im}\psi_G(t) - \left(\mathcal{L}^c_{-1} - \frac{\tau}{2}\mathcal{L}^c_0\right)_{\mathrm{QCD}}, \tag{9.533}$$

where the first part of the RHS is parametrized by the lowest resonances σ_B, G_1, σ'_B and the second part is the QCD expression of the moment sum rules $\mathcal{L}^c_n$. The other radial excitations G'_1 and G_2 having less certain values of the decay constants are included in the QCD continuum.

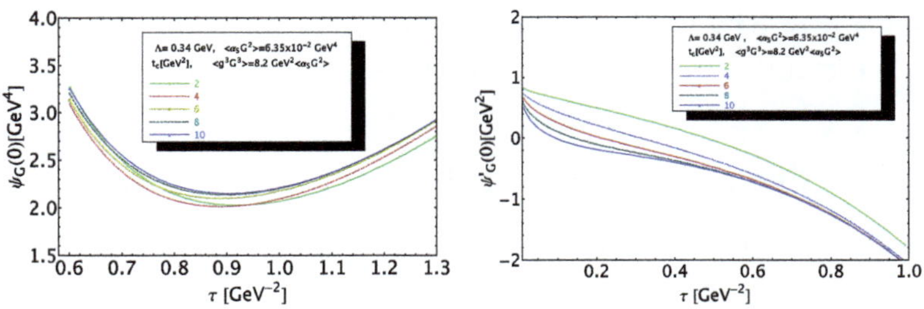

(a) (b)

Figure 9.164. (a) $\psi(0)_G$ from LSR as a function of τ for different values of t_c; (b) $\psi'(0)_G$ from LSR as a function of τ for different values of t_c.

[40] We remind that this combination of sum rules has been firstly (and successfully) introduced to quantify the deviation from kaon PCAC [602] due to $SU(3)$ breakings and later to estimate the topological charge of the $U(1)_A$ channel [118] and its slope [118,592] within the standard SVZ expansion without any additional instanton contributions.

◇ The result of the analysis is shown in Fig. 9.164, where the curves present nice τ at (1.0 ± 0.04) GeV^{-2}. A t_c-minimum appears at (2–4) GeV^2 and a stability from 8 GeV^2. We obtain:

$$\begin{aligned}\psi_G(0)|_{\rm LSR} &= 2.09(2)_{t_c}(0)_\tau(18)_\Lambda(11)_{G^2}(10)_{G^3}(1)_{M_{\sigma_B}}(14)_{f_{\sigma_B}}\\ &\quad\times(2)_{M_G}(7)_{f_G}(0)_{M_{\sigma'_B}}(2)_{f_{\sigma'_B}},\\ &= 2.09(29)\ \text{GeV}^4, \qquad (9.534)\end{aligned}$$

where the central value is 1.4 times the LET estimate in Eq. (9.499) and the one in Ref. [118] obtained within a one resonance and to LO.

◇ One can note that adding the two radial excitation contributions changes this value to 2.23 G_2. The contribution of σ_B is about 70% which is crucial for recovering the LET result.

◇ We analyze the effect of this result in the estimate of f_{σ_B} and $f_{\sigma'_B}$ obtained previously from the moment $\mathcal{L}^c_{-1}$ where $\psi_G(0)$ contributes.

◇ We find, e.g., for f_{σ_B} that the inflexion point in τ moves from $\tau = 0.56$ GeV^{-2} (see Fig. 9.159) to 0.46 for, e.g., $t_c = 3.75$ GeV^2 to which corresponds an increase of about 4 MeV, which is negligible compared to the large error of f_{σ_B}.

◇ For $f_{\sigma'_B}$, the LSR value of $\psi_G(0)$ increases the decay constant by about 100 MeV, but this value is inside the range spanned by the large error of 342 MeV obtained in its determination. Therefore, we keep the result obtained in the previous section.

9.13.13 The slope $\psi'_G(0)$ of the conformal charge

The slope $\psi'_G(0)$ obeys the twice subtracted sum rule:

$$\psi'_G|(0)_{\rm LSR} = \int_0^{t_c}\frac{dt}{t^2}\,e^{-t\tau}\frac{1}{\pi}\,{\rm Im}\psi_G(t) - \mathcal{L}^c_{-2}|_{\rm QCD} + \tau\,\psi(0)_G, \qquad (9.535)$$

where:

$$\mathcal{L}^c_{-2}(\tau) = \beta^2(\alpha_s)\left(\frac{2}{\pi^2}\right)\tau^{-1}\sum_{n=0,2,\ldots} D_n^{-2}, \qquad (9.536)$$

with:

$$D_0^{-2} = \left[C_{00} - 2C_{01}\gamma_E + C_{02}\left[\gamma_E - \frac{\pi^2}{6}\right]\right](1-\rho_0),$$

$$
\begin{aligned}
D_2^{-2} &= -\tau\, C_{21}(1-\gamma_E)\lambda^2,\\
D_4^{-2} &= \tau^2\left[C_{40} - C_{41}\left[\frac{1}{2}-\gamma_E\right]\right]\langle\alpha_s G^2\rangle,\\
D_6^{-2} &= \frac{\tau^3}{2}\left[C_{60} - C_{61}\left(\frac{1}{3}-\gamma_E\right)\right] g^3 f_{abc}\langle G^a G^b G^c\rangle,\\
D_8^{-2} &= \frac{\tau^4}{6} C_{80}(1.1\pm 0.5)\langle\alpha_s G^2\rangle^2. \qquad (9.537)
\end{aligned}
$$

We show the analysis in Fig. 9.164, from which we deduce $t_c =$ (2–8) GeV^2 at the τ-minimum ± 0.04 GeV^{-2}:

$$
\begin{aligned}
10^2 \times \psi'_G(0)|_{\text{LSR}} &= -22(5)_{t_c}(4)_\tau(16)_\Lambda(1)_{G^2}(0)_{G^3}(2)_{M_{\sigma_B}}(16)_{f_{\sigma_B}}\\
&\quad \times(1)_{M_{G_1}}(8)_{f_{G_1}}(1)_{M_{\sigma'}}(9)_{f_{\sigma'}}(11)_{\psi_G(0)}\\
&= -22(29)\ \text{GeV}^2, \qquad (9.538)
\end{aligned}
$$

where the σ_B contribution is important. The present result confirms and updates the earlier one in Ref. [118].

9.13.14 Hadronic couplings of the σ_B meson

- **The σ_B coupling to $\pi\pi$**

One can use the previous result of the decay constant in Eq. (9.518) to predict the hadronic width of σ_B. For this purpose, we consider the vertex (see, e.g., Ref. [132]):

$$V_\pi(q^2) = \langle\pi_1|\theta_\mu^\mu|\pi_2\rangle : q = p_1 - p_2 \quad \text{with} \quad V(0) = \mathcal{O}(m_\pi^2). \qquad (9.539)$$

In the chiral limit ($m_\pi^2 = 0$), one has:

$$V(q^2) = q^2\int_{4m_\pi^2}^{\infty}\frac{dt}{t(t-q^2-i\epsilon)}\frac{1}{\pi}\,\text{Im}V(t). \qquad (9.540)$$

Using the fact that $V'(0) = 1$ [81,127], one can deduce the low-energy vertex sum rule (LEV-SR) [132]:

$$\frac{1}{4}\sum_{G=\sigma_B,\cdots} g_{G\pi\pi}\frac{\sqrt{2}f_G}{M_G^2} = 1, \qquad (9.541)$$

where the contribution of the light σ_B is enhanced. Then, assuming (to a first approximation) that the σ_B dominates the sum rule, one can deduce:

$$g_{\sigma_B \pi^+\pi^-} \approx (7.1 \pm 2.5) \text{ GeV}, \tag{9.542}$$

to which corresponds the width:

$$\Gamma[\sigma_B \to \pi^+\pi^- + 2\pi^0] = \frac{3}{2}\frac{|g_{\sigma_B \pi^+\pi^-}|^2}{16\pi M_{\sigma_B}}\left(1 - \frac{4m_\pi^2}{M_{\sigma_B}^2}\right)^{1/2},$$
$$\approx (1.4 \pm 0.8) \text{ GeV}. \tag{9.543}$$

To check the validity of this approximation, we use the average of the recent data from 2005 [33] which is:

$$\Gamma[f_0(1.37) \to \pi\pi] \simeq 215(14) \text{ MeV} \longrightarrow g_{\sigma'_B \pi\pi} \simeq (3.2 \pm 0.1) \text{ GeV}, \tag{9.544}$$

while the $\pi\pi$ width of $f_0(1.5)$ of 39 MeV [33] can be neglected. Including the $f_0(1.37)$ contribution, the previous sum rule gives:

$$g_{\sigma_B \pi^+\pi^-} \simeq 5.7 \text{ GeV} \longrightarrow \Gamma[\sigma_B \to \pi^+\pi^- + 2\pi^0] \simeq 873 \text{ MeV}, \tag{9.545}$$

in agreement with previous estimates [132, 582].

• The σ_B coupling to K^+K^-

Up to $SU(3)$ breaking, the vertex sum rule in Eq. (9.541) indicates that σ_B couples (almost) universally to Goldstone boson pairs, such that in this gluonium picture, one also expects a large coupling of σ_B to K^+K^-. This feature has been observed from the analysis of $\pi\pi \to K^+K^-$ scattering data [399, 407, 410].

• σ_B mass and width confronted to the data

It is informative to compare the previous mass and width to the one of the observed σ from scatterings data analysis:

◇ The values of the mass and width obtained in Eqs. (9.508) and (9.543) for σ_B are comparable with the on-shell/Breit–Wigner mass and width obtained in Eq. (9.473).

◇ These results together with the ones from data analysis reviewed in the first sections can indicate the presence of an important gluon component inside the observed $\sigma/f_0(500)$ meson wave function.

9.13.15 Hadronic couplings of the higher mass gluonium G

• G_1 coupling to $\eta\eta', \eta\eta$

In order to compute the $g_{G_1\eta_1\eta_1}$ coupling of G_1 to the singlet η_1, one starts from [83, 132]:

$$\tilde{V}_{\mu\nu}(q_1, q_2) = i \int d^4x_1 d^4x_2\, e^{i(q_1x_1+q_2x_2)} \langle 0|\mathcal{T}Q(x_1)Q(x_2)\theta_{\mu\nu}(0)|0\rangle, \tag{9.546}$$

where $\theta_{\mu\nu}$ is the energy–momentum tensor with three light quarks, and:

$$Q(x) = \frac{\alpha_s}{16\pi}\epsilon^{\mu\nu\rho\sigma}G^a_{\mu\nu}G^a_{\rho\sigma}, \tag{9.547}$$

is the topological charge density, where one reminds that, in the large N_c limit (no quark loops), the solution to the $U(1)_A$ problem is [83, 132]:

$$\Gamma_2(q) \equiv i\int d^4x \langle 0|\mathcal{T}\,Q(x)Q(0)|0\rangle \to \Gamma_2^{\rm YM}(q \ll \Lambda) \to \Gamma_2^{\rm YM}(0),$$
$$\simeq (180 \text{ MeV})^4. \tag{9.548}$$

Using the expression of the topological charge in the presence of quark loops [603] and the consistent one for the vertex $\tilde{V}_{\mu\nu}(q_1, q_2)$, one can deduce [132] the constraint:

$$\langle \eta_1|\theta^\mu_\mu|_{n_f=3}|\eta_1\rangle \simeq \frac{9}{11}\frac{12}{f_\pi^2}\Gamma_2^{\rm YM}(0) = 2M^2_{\eta_1} = 1.15 \text{ GeV}^2, \tag{9.549}$$

where the factor (9/11) comes from the ratio of the β-function β_1 for $n_f = 3$ and $n_f = 0$ and $M_{\eta_1} \simeq 0.76$ GeV [83, 132]. Then, we deduce the LEV-SR [132]:

$$\frac{1}{4}\sum_{i=\sigma_B, G_1} g_{G\eta_1\eta_1}\sqrt{2}f_i \simeq 1.15 \text{ GeV}^2. \tag{9.550}$$

Picking the singlet component of the physical η, η' via the pseudoscalar mixing angle $\theta_P \simeq -(18\pm 2)^0$ [192]: $\eta = \sin\theta_P\eta_1 + \cdots, \eta' = \cos\theta_P\eta_1 + \cdots$, one deduces:

$$\sum_{G=\sigma_B, G_1} g_{G\eta\eta'}f_G \simeq \frac{4}{\sqrt{2}}(\sin\theta_P)1.15\,\text{GeV}^2 \simeq -1.0\,\text{GeV}^2,$$
$$\text{and} \quad g_{G_1\eta\eta} = (\sin\theta_P)g_{G_1\eta\eta'}. \tag{9.551}$$

Following Ref. [132] by using the constraint from $\eta' \to \eta\pi\pi$ (allowing a 1% deviation from the precision of the data [33]), and a Gell-Mann, Sharp, Wagner-type model, for the intermediate state σ_B, one can deduce the upper bound for $M_{\sigma_B} = 1.08$ GeV:

$$|g_{\sigma_B\eta\eta'}| \leq 0.34 \text{ GeV}, \tag{9.552}$$

which is a relatively small contribution. Neglecting this contribution and the $f_0(1.37)$ one, the LEV-SR in Eq. (9.551) leads to:

$$g_{G_1\eta\eta'} \simeq 2.74(78) \text{ GeV} \quad \text{and} \quad g_{G_1\eta\eta} \simeq 0.85(25) \text{ GeV}. \tag{9.553}$$

We deduce the width for $M_{G_1} = 1.55$ GeV, $p_{\eta\eta'} = 20$ MeV and $p_{\eta\eta} = 517$ MeV:

$$\Gamma(G_1 \to \eta\eta') = \frac{|g_{G_1\eta'\eta}|^2}{8\pi}\frac{p_{\eta\eta'}}{M^2_{G_1}} \simeq (2.5 \pm 1.4)\ [(2.6 \pm 0.9)\,\text{data}]\ \text{MeV}, \tag{9.554}$$

which implies:

$$\frac{\Gamma(G \to \eta\eta)}{\Gamma(G \to \eta\eta')} = \sin^2\theta_P \frac{p_{\eta\eta}}{p_{\eta'\eta}} \simeq (2.3 \pm 0.6)\ [(3.0)\,\text{data}]. \tag{9.555}$$

The results agree perfectly within the errors with the data.

• σ', G_1 couplings to $\sigma_B\sigma_B$

One can also write a low-energy vertex sum rule (LEV-SR) for this coupling [132]:

$$\langle\sigma_B|\theta^\mu_\mu|\sigma_B\rangle = 2M^2_{\sigma_B}. \tag{9.556}$$

Using the dispersion relation of the vertex, one can write the sum rule:

$$\frac{1}{4}\sum_{G=\sigma',G_1} g_{G\sigma_B\sigma_B}\sqrt{2}f_G \simeq 2M^2_\sigma, \tag{9.557}$$

where it is more appropriate to use the pole mass $M_\sigma \simeq 0.5$ GeV for the virtual σ_B.

◇ Assuming that σ' dominates the sum rule which is also the only one which decays copiously into $2(\pi\pi)_S$ [33], one can deduce:

$$g_{\sigma'\sigma_B\sigma_B} \simeq (2.2 \pm 0.73)\ \text{GeV}, \tag{9.558}$$

where we have used the previous estimate of the coupling in Eq. (9.514).

◇ We confront this prediction with the data of $f_0(1.37.1,5) \to 2(\pi\pi)$. In so doing, we use the six most recent measurements of the $f_0(1.37)$ 4π width since 1995 compiled by PDG [33], from which we deduce the mean:

$$\Gamma[f_0(1.37) \to 4\pi] \simeq 365(20)\ \text{MeV}. \tag{9.559}$$

Using the ratio [33]:

$$\frac{\Gamma[f_0(1.37) \to 2(\pi\pi)_S]}{\Gamma[f_0(1.37) \to (4\pi)]} \simeq 0.51(9), \tag{9.560}$$

we deduce:

$$\begin{aligned}\Gamma[f_0(1.37) \to 2(\pi\pi)_S] &\simeq 186(35)\ \text{MeV} \longrightarrow g_{f_0\sigma_B\sigma_B},\\ &\simeq (4.33 \pm 0.39)\ \text{GeV},\end{aligned} \tag{9.561}$$

where the estimate is off by about a factor two, indicating that the sum rule may have overestimated the decay constant of the σ'-meson which is (presumably) the sum of effective couplings of higher state resonances.

◇ One can use instead the experimental value of $g_{f_0\sigma_B\sigma_B}$ into the LEV-SR. Neglecting the $G_1(1.55)$ contribution which has a tiny $2(\pi\pi)S$ width of 14 MeV deduced from the data on $\Gamma_{\pi\pi} \simeq 39$ MeV $\simeq 34.5\%$ of its total width, $\Gamma_{4\pi} \simeq 48.9\%$ and $\Gamma_{2(\pi\pi)_S}/\Gamma_{4\pi} \simeq 26\,\%$, we deduce from $g_{f_0\sigma_B\sigma_B}$ and the LEV-SR in Eq. (9.557) the more accurate estimate:

$$f_{\sigma'} \simeq f_{\sigma'_B} = (329 \pm 30)\ \text{MeV}. \tag{9.562}$$

This value is in line with the intuitive expectations from ordinary ρ and ρ' mesons.

9.13.16 Summary and conclusions

The different results obtained in this paper are summarized in Table 9.32 which we shall briefly comment in the following.

- **The $\sigma/f_0(500)$ meson**

In this paper, we have started to review our present knowledge on the nature of the σ meson from $\pi^+\pi^-, \gamma\gamma \to \pi^+\pi^-, K^+K^-$ data and from J/ψ gluon rich channels ($\gamma\pi\pi, \omega\pi\pi, \phi\pi\pi$) and ϕ radiative decays, $D_{(s)}$ semileptonic decays and $\bar{p}p, pp$ production processes. The data (*a priori*) favour/does not exclude a large gluon component in the σ wave function where the observed state may emerge from a maximal mixing with a $\bar{q}q$ state [125] rather than a pure four-quark $(\bar{u}u)(dd)$ or a $\pi\pi$ molecule state. The four-quark or $\pi\pi$ molecule state is not favoured by the too small predicted direct $\gamma\gamma$ width of the σ [56, 377, 407, 410] and by the large K^+K^- coupling found from scattering data [399, 407]. The assumption that σ emerges from $\pi\pi$ rescattering is also not favoured from its large coupling to K^+K^-.

- **The $f_0(980)$ meson from the data**

◇ From the scattering data analysis, it is unlikely that $f_0(980)$ is an $\bar{s}s$ state because its has a non-negligible/large coupling with $\pi\pi$ [399, 407, 410], while the strength of its $\gamma\gamma$ coupling is relatively small for a $\bar{q}q$ state [124, 580]. Moreover, it is too light compared to the $\bar{s}s$ mass of about 1.4 GeV [580].

- **Conclusions from the data**

The previous observations may favour the interpretation that σ and $f_0(980)$ emerge from a maximal mixing between a gluonium and a $(\bar{u}u + \bar{d}d)$ state like we have proposed in Refs. [125, 582] and what had also been expected in the pioneering paper [81].

- **The σ_B meson from QCD spectral sum rules (QSSR)**

◇ A QSSR analysis of the lowest ratios of sum rule $\mathcal{L}_{20}$ favours σ_B (the subindex B refers to a pure unmixed gluonium state) having a mass M_{σ_B}=1.07 GeV (see Eq. (9.516) and a $\pi\pi$ width

(Eq. (9.545)) compatible with the ones from scattering data for a Breit–Wigner/on-shell σ-mass.

◇ Its decay constant can be deduced from $\mathcal{L}_1^c$ which is $f_{\sigma_B} = 456(157)$ MeV (Eq. (9.518)).

◇ One should also note that the set $(\Gamma_{\sigma_B}, M_{\sigma_B})$ obtained from the sum rule is fairly compatible with the on-shell mass and width from $\pi^+\pi^- \to \pi^+\pi^-, K^+K^-$ data.

• The "scalar gluonium" G_1 from QSSR

◇ An analysis of the higher weight USR $\mathcal{L}_{21}$ and $\mathcal{L}_{31}$ leads to a gluonium mass: $M_{G_1} = (1548 \pm 121)$ MeV (Eq. 9.509) for a *"three resonances"* $\oplus$ *QCD continuum* parametrization of the spectral function, while it becomes (1515 ± 123) MeV (Eq. (9.510)) if one uses a "one resonance" $\oplus$ QCD continuum, indicating that the presence of σ_B does not affect notably the value of M_{G_1}. It is clear that higher moments are more appropriate in the QSSR approach to pick up the higher mass gluonium state due to the power mass suppression of the lowest mass σ_B contribution in these sum rules.

◇ The corresponding gluonium decay constant is $f_{G_1} = (365 \pm 110)$ MeV (Eq. (9.519)) which is comparable with the one of f_{σ_B}.

◇ We have also seen that its small width into $2(\pi\pi)_S$ can be explained from the LET-V sum rule in Eq. (9.557), indicating that it is not a σ-like particle but instead decays into $\eta'\eta$, $\eta\eta$ through the gluon vertex $U(1)$ anomaly.

• The radial excitations σ', G_1' and G_2 from QSSR

We have attempted to extract the mass and coupling of the radial excitations σ' of σ_B, G_1' and G_2 of G_1.

◇ We found that $M_{\sigma'} \simeq (1121 \pm 117)$ MeV is rather low compared to the radial excitation of ordinary mesons where we note that a similar feature has been observed in the analysis of the D^*D molecule described by high-dimension quark currents.

◇ $f_{\sigma'_B} \simeq 646(216)$ MeV can be large which we have interpreted as an effective sum of all radial excitations that the sum rule cannot disentangle. For a more phenomenological use, we identify σ'_B with the observed $f_0(1.37)$ and extract the decay constant from its decay to $2(\pi\pi)_S\oplus$ LEV-SR analysis, from which we obtain $f_{\sigma'} \simeq (329 \pm 30)$ MeV (Eq. (9.557)).

◇ The mass of the radial excitation G_1': $M_{G_1'} = 1563(141)$ MeV (Eq. (9.520)) of G_1 is also found to be almost degenerated with the ground state mass $M_{G_1} = 1548(121)$ MeV (Eq. (9.517)). Like in the case of σ_B', we interpret the large coupling $f_{G_1'} = 1000(230)$ MeV (Eq. (9.521)) as a sum of the effective couplings of all higher radial excitations entering into the spectral function.

◇ The second radial excitation gluonium G_2 is found to have a relatively high mass $M_{G_2} \simeq 2.99$ GeV.

• Phenomenology

◇ The global picture which emerges from the approach indicates that the observed gluonia candidates can de classified into two groups:

- The σ-like gluonia $[\sigma/f_0(0.5), f_0(1.37)]$ which emerge from (σ_B, σ_B') and which decay into $\pi\pi$ via OZI violating process while the $f_0(1.37)$ decays to $2(\pi\pi)_S$ via virtual $\sigma\sigma$.
- The G-like gluonia $[f_0(1.5), f_0(1.7)]$ which come from (G_1, G_1') and which decay into $\eta'\eta$ and $\eta\eta$ through the gluon $U(1)$ anomaly vertex but not copiously to $\pi\pi$ and $2(\pi\pi)_S$.
- We may expect that high-mass second radial excitation $G_2(2.99)$ has similar properties and brings some gluon component to the gluonia candidates above 2 GeV which contributes to their $\eta'\eta$ and $\eta\eta$ decays.

◇ We remind that the masses of the $S_2(\bar{u}u + \bar{d}d)$, $S_3(\bar{s}s)$ quarkonia states and their radial excitations are predicted from the sum rules to be (in units of GeV) [582]:

$$\begin{aligned} &M_{S_2} \simeq 1, \quad M_{S_2'} \simeq (1.1 \sim 1.4), \\ &M_{S_3} \simeq (1.47), \quad M_{S_3'} \simeq (1.7 \sim 2.4), \end{aligned} \tag{9.563}$$

which can mix with the previous gluonia states.

◇ A more complete analysis of the data can be done within some $\bar{q}q$ meson–gluonium mixing scheme. The case for the $\sigma(500)$-$f_0(980)$ has been studied in Ref. [125] but may/should be extended to the higher meson masses. Some attempts in this direction have been also made in Ref. [582][41] but needs to be updated.

◇ We have not updated our analysis of the $\gamma\gamma$ and radiative widths obtained in Refs. [132, 582]. We expect that these results are still valid within the errors.

◇ We emphasize that a more appropriate comparison of the results obtained in this work (e.g., with lattice calculations) requests an analysis beyond the "one resonance" contribution to the two-point correlator from these alternative approaches. Indeed, a simple one resonance for parametrizing the spectral function is (obviously) insufficient for describing the complex and overpopulated spectra of the $I = o$ scalar mesons.

• **The conformal charge $\psi_G(0)$ from QSSR**

We have completed our analysis by the estimate of the conformal charge $\psi_G(0)$ and its slope $\psi'_G(0)$.

◇ Using the previous values of the σ and gluonia G masses and couplings, we have obtained $\psi_G^{\rm LSR}(0) = 2.09\ (29)\ {\rm GeV}^4$ (Eq. (9.534)), which can be compared with the LET prediction of about 1.46 GeV^4 (Eq. (9.499)).

◇ The result indicates that the σ_B meson contributes predominantly in the spectral function for reproducing the LET prediction of $\psi_G(0)$. It also shows that the instanton effect in the estimate of $\psi_G(0)$ is (*a priori*) not necessary for a correct estimate of the conformal charge $\psi_G(0)$. A similar feature has been observed in the estimate of the topological charge and its slope in the $U(1)_A$ sector [592].

◇ The value of the slope is $\psi'_G(0)|_{LSR}= 0.95(30)\ {\rm GeV}^2$.

• **Prospects**

◇ We plan to study in more detail the mixing of the $\bar{q}q$ states with the higher gluonia masses $\sigma'(1.15)$, $G_1(1.53)$ and eventually with $G_2(2.48)$.

◇ We plan to extend the analysis to the case of (pseudo)scalar $C = \pm$ trigluonia channels which has been studied in Refs. [133, 605].

[41] For a review prior to 1996, see, e.g., Ref. [604].

Table 9.32. Predictions from scalar di-gluonium LSR at N2LO and gluon condensates at NLO up to $D = 8$.

Observables	Value	Eq.#	Source	Comments
Masses [MeV]				
Ground states				
M_{σ_B}	1088(78)	(9.506)	$\mathcal{R}^c_{20}$	1 resonance ⊕ QCD continuum
	1085(126)	(9.508)	$\mathcal{R}^c_{20}$	2 resonances ⊕ QCD continuum
	1070(126)	(9.516)	$\mathcal{R}^c_{20}$	3 resonances ⊕ QCD continuum
M_{G_1}	1515(123)	(9.510)	$\mathcal{R}^c_{42}$	1 resonance ⊕ QCD continuum
	1524(121)	(9.509)	$\mathcal{R}^c_{42}$	2 resonances ⊕ QCD continuum
	1548(121)	(9.517)	$\mathcal{R}^c_{42}$	3 resonances ⊕ QCD continuum
1st Radial excitations				
$M_{\sigma'_B}$	1110(117)	(9.513)	$\mathcal{R}^c_{20}$ (after iteration)	3 resonances ⊕ QCD continuum
$M_{G'_1}$	1563(141)	(9.520)	$\mathcal{R}^c_{42}$	4 resonances ⊕ QCD continuum
M_{G_2}	2992(221)	(9.524)	mean from $\mathcal{R}^c_{21}$ and $\mathcal{R}^c_{31}$	5 resonances ⊕ QCD continuum
Decay constants [MeV]				
Ground states				
f_{σ_B}	563(156)	(9.512)	$\mathcal{L}^c_{-1}$	2 resonances ⊕ QCD continuum
	456(157)	(9.518)	$\mathcal{L}^c_{-1}$	3 resonances ⊕ QCD continuum
f_{G_1}	394103)	(9.512)	$\mathcal{L}^c_2 \oplus \mathcal{L}^c_3$	2 resonances ⊕ QCD continuum
	365(110)	(9.519)	$\mathcal{L}^c_2 \oplus \mathcal{L}^c_3$	3 resonances ⊕ QCD continuum
Radial excitations				
$f_{\sigma'}$	329(30)	(9.562)	LEV-SR ⊕ $f_0(1.37) \to 2(\pi\pi)_S$ [33]	$f_0(1.5)$ neglected
$f^{\text{eff}}_{\sigma'}$	648(216)	(9.514)	$\mathcal{L}^c_{-1}$	$\sum$ higher states effective couplings
$f^{\text{eff}}_{G'_1}$	1000(230)	(9.521)	$\mathcal{L}^c_2$	$\sum$ higher states effective couplings
$f^{\text{eff}}_{G_2}$	797(74)	(9.526)	mean from $\mathcal{L}^c_1$ and $\mathcal{L}^c_2$	$\sum$ higher states effective couplings
Decay widths [MeV]				
$\Gamma[\sigma_B \to \pi\pi]$	873	(9.545)	LEV-SR	700: $\pi\pi$ scattering data
$\Gamma[\sigma'(1.37) \to 2(\pi\pi)_S]$	186(35)	(9.562)	data input [33] ⊕ LEV-SR	$\longrightarrow f_{\sigma'} = 329(15)$
$\Gamma[G_1(1.55) \to \eta\eta']$	(2.5 ± 1.4)	(9.554)	–	(2.6 ± 0.9) (data) [33]
$\frac{\Gamma[G_1(1.55)\to\eta\eta]}{\Gamma[G_1(1.55)\to\eta\eta']}$	(2.3 ± 0.6)	(9.555)	–	3.0 (data) [33]
Conformal Anomaly				
Charge: $\psi_G(0)\vert_{\text{LSR}}$ [GeV^4]	2.09(29)	(9.534)	$\mathcal{L}^c_{-1} \oplus \mathcal{L}^c_0$	1.46(8) [LET in Eq. (9.499)]
Slope: $10^2 \times \psi'_G(0)\vert_{\text{LSR}}$ [GeV^2]	−22(29)	(9.538)	$\mathcal{L}^c_{-2} \oplus \psi_G(0)\vert_{\text{LSR}}$	

9.14 Slope of the Topological Charge, Proton Spin and the 0^{-+} Pseudoscalar Di-Gluonia Spectra

In the following, we review the recent work in Ref. [151]. This work updates the ones with G. Shore and G. Veneziao in Refs. [144, 149] discussed in Section 6.15.

9.14.1 Abstract

First, we reconsider the estimate of the slope of the topological charge $\sqrt{\chi'(0)}(Q^2 = 2\,\text{GeV}^2) = 24.3(3.4)$ MeV from low degree moments which imply $\int_0^1 dx\, g_1^P(x) = 0.144(5)$ [data: 0.145(13)] for the first moment of the polarized proton structure function (proton spin) and $G_A^{(0)}(10\ \text{GeV}^2) \equiv \Delta\text{u} + \Delta\text{d} + \Delta\text{s} = 0.340(50)$ [data $= 0.330(39)$] for the singlet form factor of the axial current.

Second, we work with high degree moments and parametrize the spectral function beyond the minimal duality ansatz: "*One resonance* $\oplus$ *QCD continuum*" to get the 0^{-+} pseudoscalar di-gluonia spectra. Then, we obtain three groups of gluonia: *the familiar light* η_1 [singlet gluon component of the $\eta'(958)$] with $[M_{\eta_1}, f_{\eta_1}] = [825(45), 905(72)]$ MeV which is important for understanding the $U(1)_A$ anomaly; *the two new medium gluonia* with $M_{P_{1a}} = 1338(112)$ MeV and $[M_{P_{1b}} = 1462(117)$ MeV or their mean $[M_{P_1}, f_{P_1}] = [1397(81), 594(144)]$ MeV which support the gluonium nature of the excellent experimental candidate $\eta(1405)$ and may bring a small gluon piece to the $\eta(1295)$; their corresponding first radial excitations: $M_{P'_{1a}} = 1508(226)$ MeV and $M_{P'_{1b}} = 1553(139)$ MeV with their mean: $[M_{P'_1}, f_{P'_1}] = [1541(118), 205(282)]$ MeV which may be identified (up to some eventual mixings with $\bar{q}q$ states) with the observed $\eta(1475, 1700)$ states; *the heavy gluonium* with the mean mass: $[M_{P_2}, f_{P_2}] = [2751(140), 500(43)]$ MeV which can be compared with the lattice results. One can remark the (natural) one-to-one correspondence between the pseudoscalar gluonia and their chiral scalar analogue from Ref. [583]: $\sigma(1) \to \eta_1$; $G_1(1.55) \to P_1$; $[\sigma'(1.1), G'(1.56)] \to P'_{1a,1b}$; $G_2(3) \to P_2$ which is mainly due to the importance of the QCD PT contributions in the sum rule analysis that are almost equal in these two channels.

9.14.2 Introduction

• Motivated by the failure of some of the previous approaches including QSSR to explain the data below 2 GeV [33, 606]:

$$\eta(1295), \quad \eta(1405), \quad \eta(1475), \quad \eta(1760), \tag{9.564}$$

we scrutinize previous QSSR results discussed in Section 6 and attempt to improve them by parametrizing the spectral function with more than one resonance and by working with high-degree moments.

◇ "En passant", we revise our NLO estimate of the slope of the topological charge $\chi'(0)$ in the chiral limit [144, 149] and study the effect of this N2LO result on the spin of the proton.

9.14.3 The inverse Laplace transform sum rules

• We shall work with two-point correlator:

$$\psi_P(q^2) = (8\pi)^2\, i \int d^4x\, e^{iqx} \langle 0|Q(x)Q^\dagger(0)|0\rangle, \tag{9.565}$$

associated to the divergence of the $U(1)_A$ axial current which reads for n_f quark flavours:

$$\partial_\mu J_5^\mu(x) = \sum_{i=u,d,s} 2m_i \bar{\psi}_i \psi_i + 2n_f Q(x), \tag{9.566}$$

where:

$$Q(x) \equiv \left(\frac{\alpha_s}{16\pi}\right) \epsilon_{\mu\nu\rho\sigma} G_a^{\mu\nu}(x) G_a^{\rho\sigma}(x), \tag{9.567}$$

is the topological charge density, $a = 1, \ldots, 8$ is the colour index, $\alpha_s \equiv g^2/4\pi$ is the QCD coupling and $G_a^{\mu\nu}(x) \equiv \partial^\mu A_a^\nu - \partial^\nu A_a^\mu + g f_{abc} A^{\mu,b} A^{\nu,c}$ is the Yang–Mills field strength constructed from the gluon fields A_a^μ.

From its QCD asymptotic behaviour $\sim (-q^2)^2 \mathrm{Log}(-q^2/\mu^2)$, one can write a twice subtracted dispersion relation:

$$\psi_P(q^2) = \psi_P(0) + q^2 \psi_P'(0) + \frac{q^4}{\pi} \int_0^\infty \frac{dt}{t^2} \frac{\mathrm{Im}\psi_P(t)}{(t - q^2 - i\epsilon)}. \tag{9.568}$$

• Following standard QSSR techniques [3, 6, 7], one can derive from it different forms of the sum rules. In this paper, we shall work with the exponential or Borel [3, 64, 70] or inverse Laplace transform [69] finite energy sum rule (LSR):

$$\mathcal{L}_n^c(\tau) = \int_0^{t_c} dt\, t^n\, e^{-t\tau} \frac{1}{\pi} \text{Im}\psi_P(t) \;: n = -2, -1, 0, 1, \ldots, 4, \quad (9.569)$$

and the corresponding ratios of sum rules:

$$\mathcal{R}_{n+l\ n}^c(\tau) \equiv \frac{\mathcal{L}_{n+l}^c(\tau)}{\mathcal{L}_n^c(\tau)}, \quad (9.570)$$

where τ is the Laplace sum rule variable. In the duality ansatz,

$$\frac{1}{\pi}\text{Im}\psi_P(t) = 2\sum_P f_P^2 M_P^4\, \delta(t - M_G^2) + \theta(t - t_c)\text{“QCD continuum”}:$$
$$\langle 0|(8\pi)Q(x)|P\rangle = \sqrt{2} f_P M_P^2, \quad (9.571)$$

where f_P is the resonance coupling normalized as $f_\pi = 93$ MeV, M_P is the mass while the “QCD continuum” comes from the discontinuity $\text{Im}\psi_P(t)|_{\text{QCD}}$ of the QCD diagrams from the continuum threshold t_c. In the *one narrow resonance* $\oplus$ *QCD continuum* parametrization of the spectral function[42]:

$$\mathcal{R}_{n+l\ n}^c(\tau) \simeq M_P^2. \quad (9.572)$$

• $\mathcal{L}_{(-2,-1)}$ have been used to estimate the topological charge, its slope and the η' parameter within the minimal duality ansatz which will be reviewed and discussed in Section 9.14.4.

• To obtain $\mathcal{L}_0$, we take the inverse Laplace transform of the first superconvergent third derivative of the two-point correlator. In this way, we obtain:

$$\mathcal{L}_0^c(\tau) = \tau^{-2}\, 2 \sum_{n=0,2,\ldots} D_n^0, \quad (9.573)$$

[42]Finite width effect has been shown in the example of the large width σ meson to give a negligible correction [132, 582] and will not be considered in this paper.

with $[L_\tau \equiv \log(\tau\mu^2)]$:

$$D_0^0 = \Big[2C_{00} - 2C_{01}(3 - 2\gamma_E - 2L_\tau)$$
$$-6C_{02}\big[1 - 3\gamma_E + \gamma^2 - \pi^2/6 + (-3 + 2\gamma_E)L_\tau + L_\tau^2\big]\Big](1 - \rho_2),$$
$$D_2^0 = -\frac{C_{21}}{2}\lambda^2\tau(1 - \rho_1),$$
$$D_4^0 = -C_{41}\langle\alpha_s G^2\rangle\tau^2,$$
$$D_6^0 = \big[C_{60} - C_{61}(\gamma_E - L_\tau)\big]g^3 f_{abc}\langle G^a G^b G^c\rangle\tau^3,$$
$$D_8^0 = C_{80}\left(\frac{15}{16}\right) k\,\langle\alpha_s G^2\rangle^2\tau^4. \tag{9.574}$$

• The other higher degrees sum rules $\mathcal{L}_n^c(\tau)$ for $n \geq 1$ can be deduced from the nth τ-derivative of $\mathcal{L}_0^c(\tau)$:

$$\mathcal{L}_n^c(\tau) = (-1)^n \frac{d^n}{d\tau^n}\mathcal{L}_0^c(\tau). \tag{9.575}$$

• These superconvergent sum rules obey the homogeneous renormalization group equation (RGE):

$$\Big\{ -\frac{\partial}{\partial t} + \beta(\alpha_s)\alpha_s\frac{\partial}{\partial\alpha_s}\Big\}\mathcal{L}_n^c(e^t\tau, \alpha_s) = 0, \tag{9.576}$$

where $t \equiv (1/2)L_\tau$. The renormalization group improved (RGI) solution is:

$$\mathcal{L}_n^c(e^t\tau, \alpha_s) = \mathcal{L}_n^c(t = 0, \bar{\alpha}_s(\tau)), \tag{9.577}$$

where $\bar{\alpha}_s(\tau)$ is the QCD running coupling.

• In the following analysis, we shall work with the family of sum rules having degrees less or equal to 4. In so doing, we shall select the sum rules which present stability (minimum or inflexion point) in the sum rule variable τ and in the continuum threshold t_c such that we can extract optimal information from the analysis.

9.14.4 The η' mass and decay constant

Keeping only the η' and/or the lowest mass gluonium contributions in the parametrization of the spectral function (Eq. (9.571)), $\mathcal{L}_{-1}^c$

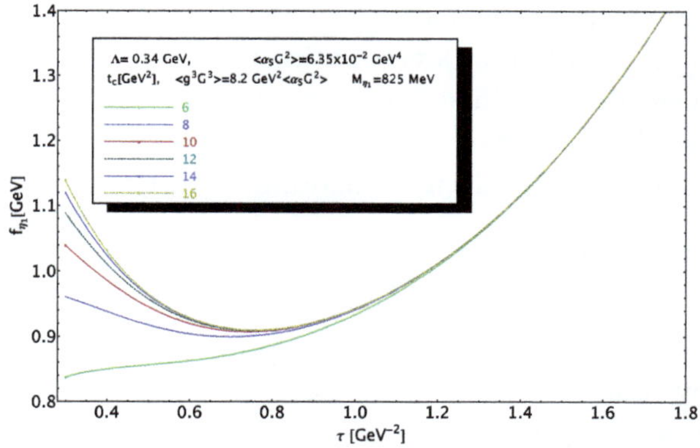

Figure 9.165. The decay constant f_{η_1} from $\mathcal{L}^c_{-1}$ as a function of τ at N2LO using $M_{\eta_1} = 825$ MeV and the QCD inputs in Table 7.10.

and $\mathcal{L}^c_0$ have been used earlier to look for the gluon component of the η' mass and coupling and estimate the topological charge $\chi(0)$ and its slope $\chi'(0)$ [77,82,118,142,144,607]. The LSR results in pure Yang–Mills fairly agree with the ones from large N_c [83,84,603] and lattice calculations [608]. Here, we use the sum rule $\mathcal{L}^c_{-1}$ in the chiral limit where $\chi(0) \equiv \psi_P(0)/(8\pi) = 0$:

$$\mathcal{L}^c_{-1} \equiv \int_0^{t_c} \frac{dt}{t} e^{-t\tau} \frac{1}{\pi} \mathrm{Im}\psi_P(t) = \mathcal{L}^c_{-1}|_{\mathrm{QCD}} + \psi_P(0), \tag{9.578}$$

where:

$$\mathcal{L}^c_{-1}|_{\mathrm{QCD}} = \tau^{-2}\, 2 \sum_{n=0,2,\ldots} D_n^{-2}, \tag{9.579}$$

with:

$$\begin{aligned}
D_0^{-1} &= \Big[C_{00} + 2C_{01}(1-\gamma_E - L_\tau) + 3C_{02} \\
&\qquad \times \left[(\gamma_E + L_\tau)^2 - 2(\gamma_E + L_\tau) - \frac{\pi^2}{6}\right]\Big](1-\rho_1), \\
D_2^{-1} &= \tau\, C_{21}\lambda^2(1-\rho_0), \\
D_4^{-1} &= -\tau^2\Big[C_{40} - C_{41}(\gamma_E + L_\tau)\Big]\langle\alpha_s G^2\rangle, \\
D_6^{-1} &= -\tau^3\Big[C_{60} + C_{61}\,(1-\gamma_E - L_\tau)\Big] g^3 f_{abc}\langle G^a G^b G^c\rangle, \\
D_8^{-1} &= -\frac{\tau^4}{2} C_{80}\langle\alpha_s G^2\rangle^2
\end{aligned} \tag{9.580}$$

for extracting the decay constant f_{η_1}. We have attempted to extract the mass M_{η_1} using the low degree ratio of moments $\mathcal{R}_{0-1}$ but fail due to the absence of τ-stability. Then, we use the GMO mass formula quoted in Ref. [144] derived in the massless pion limit and assuming a SU3 symmetry for the decay constants from Ref. [84],

$$M_{\eta_1}^2 \simeq M_{\eta'}^2 - \frac{2}{3}M_K^2 = (870\,\text{MeV})^2. \tag{9.581}$$

An alternative derivation of M_{η_1} in pure Yang–Mills from $1/N_c$ expansion by Refs. [83,84,603] and corrected in Ref. [132] by including three massless quark loop contributions through the β function [$\mathcal{O}(n_f/N_c)$ correction] leads to:

$$M_{\eta_1}^{n_f=3} \simeq \left(\frac{\beta_1^{n_f=3}}{\beta_1^{n_f=0}}\right)^{1/2} \times M_{\eta_1}^{YM} \simeq 779\,\text{MeV} : \beta_1^{n_f} = -\frac{1}{2}\left(11 - \frac{2}{3}n_f\right), \tag{9.582}$$

with:

$$M_{\eta_1}^{YM} \simeq \frac{1}{f_\pi}\left(6\frac{\partial}{\partial\theta}\langle Q(x)\rangle_{\theta=0} \equiv -6\chi(0)|_{YM}\right)^{1/2} :$$
$$\chi(0)|_{YM} \simeq -(180\,\text{MeV})^4 \quad \text{and} \quad f_\pi = 92.2\ \text{MeV}. \tag{9.583}$$

In the following analysis, we shall use the mean of the two determinations:

$$M_{\eta_1} \simeq 825(45)\ \text{MeV}, \tag{9.584}$$

with the (conservative) error from its distance to the two former. The analysis for N2LO is shown in Fig. 9.165 where, at the τ-minimum : $\tau = (0.68; 0.74) \pm 0.04\,\text{GeV}^{-2}$ corresponding to $t_c = 8$ (beginning of τ-stability) to 14 GeV^2 (t_c-stability), one deduces the estimate:

$$\begin{aligned} f_{\eta_1} &= 905(5)_{t_c}(1)_\tau(43)_\Lambda(53)_{\lambda^2}(3)_{G^2}(2)_{G^3}(0)_{G^4}(24)_{M_{\eta_1}} \\ &= 905(72)\ \text{MeV}, \end{aligned} \tag{9.585}$$

where one should note that f_{η_1} here is not exactly the one from, e.g., $J/\psi \to \gamma\eta'$, $\eta' \to \gamma\gamma$ [142]. One can note the following:

• The N2LO correction is large and has increased the value of f_{η_1} from NLO by about a factor 2. The large N2LO effects to f_{η_1} requires a further control of the size of the higher order PT corrections. We have considered in Ref. [235] that the tachyonic gluon mass gives an estimate of such uncalculated higher order terms of PT or it is equivalent to say that the PT series grows geometrically [234]. One can check that the tachyonic gluon mass decreases f_{η_1} by 53 MeV which we consider (here and in the following) as an estimate of the errors due to the truncation of the PT series, but we do not include it in the central value of f_{η_1}. Considering this error estimate at its face value, one may consider that the PT series reaches its asymptotic value at N2LO.

• The value of M_{η_1} and f_{η_1} in Eqs. (9.584) and (9.585) will be used as inputs in the rest of the paper.

9.14.5 Sum rule estimate of $\chi'(0)$ at N2LO in the chiral limit

$\chi'(0)$ has been estimated in pure Yang–Mills to be [82]:

$$\chi'(0)|_{YM} = -[(7 \pm 3)\,\text{MeV}]^2, \tag{9.586}$$

in agreement with the lattice result $-[(9.8 \pm 0.9)\,\text{MeV}]^2$ [608]. In the following, we shall update its NLO estimate in the chiral limit [144] by including N2LO PT corrections and NLO ones to the condensate contributions. We shall use the previous value of M_{η_1} and the value or the QCD expression of f_{η_1} into the twice subtracted sum rule:

$$\psi'_P(0)|_{\text{LSR}} = \int_0^{t_c} \frac{dt}{t^2}\, e^{-t\tau} \frac{1}{\pi}\, \text{Im}\psi_P(t) - \mathcal{L}^c_{-2}|_{\text{QCD}} + \tau\, \psi_P(0), \tag{9.587}$$

where:

$$\mathcal{L}^c_{-2}|_{\text{QCD}} = \tau^{-1}\, 2 \sum_{n=0,2,\ldots} D_n^{-2}, \tag{9.588}$$

with:

$$D_0^{-2} = \left[C_{00} - 2C_{01}(\gamma_E + L_\tau) + 3C_{02}\left[(\gamma_E + L_\tau)^2 - \frac{\pi^2}{6}\right]\right](1 - \rho_0),$$

$$D_2^{-2} = -\tau\, C_{21}\gamma_E\lambda^2,$$

$$D_4^{-2} = \tau^2\left[C_{40} + C_{41}(1 - \gamma_E)\right]\langle\alpha_s G^2\rangle,$$

$$D_6^{-2} = \frac{\tau^3}{2}\left[C_{60} + C_{61}\left(\frac{3}{2} - \gamma_E - L_\tau\right)\right] g^3 f_{abc}\langle G^a G^b G^c\rangle,$$

$$D_8^{-2} = \frac{\tau^4}{6} C_{80}\langle \alpha_s G^2\rangle^2. \tag{9.589}$$

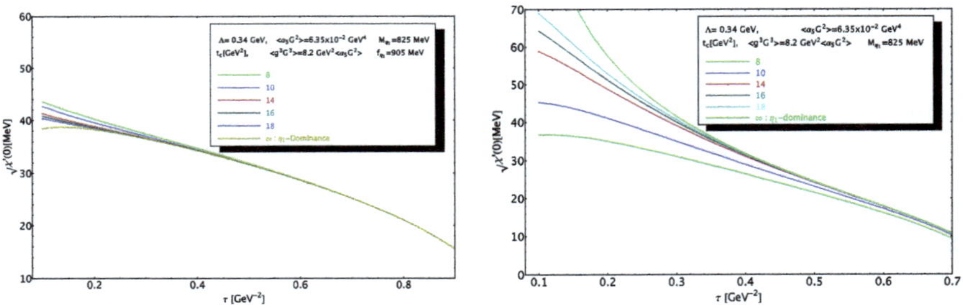

(a) (b)

Figure 9.166. (a) The slope of the topological charge $\chi'(o)$ from $\mathcal{L}_{-2}^c$ as a function of τ at N2LO using $f_{\eta_1} = 881$ MeV; (b) the same as (a) but using the QCD expression of f_{η_1} given in Eqs. (9.578)–(9.580).

We show the N2LO analysis from $\mathcal{L}_{-2}^c$ where, in Fig. 9.166, we use the value of f_{η_1} in Eq. (9.585) and in Fig. 9.166 its QCD expression from Eq. (9.578) where one should note that the sum of higher state contributions to the spectral function is (automatically) included in this second option. One can note a better τ-stability (clearer inflexion point) for the second case at $\tau \simeq (0.50 \pm 0.04)\,\text{GeV}^{-2}$ and for $t_c \simeq (14 \sim 18)\,\text{GeV}^2$, at which, we deduce:

$$\begin{aligned}\sqrt{\chi'(0)}|_{\text{N2LO}} \equiv \frac{1}{8\pi}\sqrt{\psi'_P(0)} &= 24.3(2.8)_\tau(0.2)_{t_c}(1.6)_\Lambda(0.8)_{\lambda^2}\\ &\times(0.2)_{G^2}(0.4)_{G^3}(0.1)_{G^4}(0.4)_{M_{\eta_1}} = 24.3(3.4)\ \text{MeV}.\end{aligned} \tag{9.590}$$

The NLO analysis gives curves very similar to the N2LO ones. At the corresponding inflexion point $\tau \simeq (0.44 \pm 0.04)\,\text{GeV}^{-2}$, one obtains:

$$\begin{aligned}\sqrt{\chi'(0)}|_{\text{NLO}} &= 18.2(2.5)_\tau(0.2)_{t_c}(1.1)_\Lambda(0.9)_{\lambda^2}(0.4)_{G^2}(0.6)_{G^3}(0.1)_{G^4}\\ &\quad\times(0.4)_{M_{\eta_1}},\\ &= 18.2(3.0)\ \text{MeV}.\end{aligned} \tag{9.591}$$

These new values agree within the errors with the NLO one 22.3(4.8) MeV obtained from Laplace sum rule (LSR) in Ref. [144] using different values of the QCD input parameters and a slightly different QCD expression due to some errors in the α_s PT corrections given by Ref. [553], explaining that the $\chi'(0)$ curves present minimum there.

9.14.6 Effect of $\chi'(0)$ on the proton spin

It has been shown in Ref. [142] (see also Ref. [609]) that the singlet form factor $G_A^{(0)}(Q^2)$ appearing in the first moment of the polarised proton structure function g_1^P (Ellis–Jaffe sum rule [145]) can be related in the chiral limit ($m_q = 0$) to the slope of the topological charge $\chi'(0)$ as:

$$G_A^{(0)}(Q^2)\bar{u}\gamma_5 u = \frac{1}{2M_p}(2n_f)\sqrt{\chi'(0)}\,\Gamma_{\Phi_{5R}\bar{P}P}, \tag{9.592}$$

where Φ_{5R} is the renormalized bilinear quark current and $\Gamma_{\Phi_{5R}\bar{P}P}$ the renormalization group (RG) invariant scale-independent proper vertex. In the chiral limit and using the OZI approximation, one would obtain [144]:

$$\Gamma_{\Phi_{5R}\bar{P}P}|_{\rm OZI} = \sqrt{2}g_{\eta_8 PP}\,(\bar{u}\gamma_5 u), \quad \sqrt{\chi'(0)}|_{\rm OZI} = f_\pi/\sqrt{6} = 38\ \text{MeV},$$
$$G_A^{(0)}|_{\rm OZI} = 0.579 \pm 0.021. \tag{9.593}$$

while from our analysis:

$$G_A^{(0)}|_{\rm LSR}(Q^2 = 10\ \text{GeV}^2) = \mathrm{G}_{\mathrm{A}}^{(0)}|_{\rm OZI}\frac{\sqrt{\chi'(0)}|_{\rm LSR}}{\sqrt{\chi'(0)}|_{\rm OZI}} = (0.340 \pm 0.050), \tag{9.594}$$

after running $\chi'(0)|_{\rm LSR}$ from 2 to 10 GeV^2, implying:

$$\sqrt{\chi'(0)}|_{\rm LSR}(Q^2 = 10\,\text{GeV}^2) = (22.5 \pm 3.1)\ \text{MeV}. \tag{9.595}$$

Using the previous value of $G_A^{(0)}1,\mathrm{LSR}$ and $G_A^{(3)} = 0.625 \pm 0.004$, $G_A^{(8)} = (0.167 \pm 0.006)$, from hyperon and β-decays [610], the first moment of the polarised proton structure function gives:

$$\Gamma_1^p(Q^2 = 10\ \mathrm{GeV}^2) \equiv \int_0^1 \mathrm{d}x\, \mathrm{g}_1^{\mathrm{P}}(\mathrm{x}, \mathrm{Q}^2) = (0.144 \pm 0.005), \quad (9.596)$$

in excellent agreement with the earlier data [139] and experimental world average [146]:

$$\Gamma_1^p(Q^2 = 10\ \mathrm{GeV}^2)|_{\mathrm{exp}} = 0.145 \pm 0.014 \Longrightarrow \mathrm{G}_{\mathrm{A}}^{(0)}|_{\mathrm{exp}} = 0.35 \pm 0.12, \quad (9.597)$$

and more recent data from COMPASS [611] and HERMES [612]:

$$G_A^{(0)}|_{\mathrm{exp}} = 0.33 \pm 0.04, \quad (9.598)$$

which are lower than the OZI value by a factor 1.76.

9.14.7 Gluonium mass from $\mathcal{R}_{10}$

• We update the N2LO results in the literature using the present values of the QCD parameters given in Table 7.10 without including the η'-contribution. We show the results of the analysis versus τ in Fig. 9.167 for different values of t_c. We obtain, for $\Lambda = (340 \pm 28)$ MeV:

$$\begin{aligned} M_{P_2}|_{\mathrm{N2LO}} &= 1470(39)_{t_c}(46)_\tau(83)_\Lambda(44)_{\lambda^2}(19)_{G^2}(11)_{G^3}(31)_{G^4} \\ &= 1470(118)\ \mathrm{MeV}, \end{aligned} \quad (9.599)$$

where we have taken the values at the inflexion points: $\tau = (0.80 \pm 0.04)$ GeV^{-2} and t_c from 7 to 14 GeV^2. We have considered the contribution of the tachyonic gluon mass squared λ^2 as an estimate of the non-calculated higher order terms as expected from Refs. [234, 235]. This value can be compared with the NLO result:

$$M_{P_2}|_{\mathrm{NLO}} = (2.05 \pm 0.19)\ \mathrm{GeV}, \quad (9.600)$$

obtained in Ref. [126] using $\Lambda = (0.375 \pm 0.125)$ GeV.

◇ Comparing the optimal result for each truncation of the PT series, one has (in units of MeV):

$$\begin{aligned} M_{P_2} &= (2653 \pm 327)\ (\mathrm{LO}) \xrightarrow{-30\%} \quad (1862 \pm 18)\ (\mathrm{NLO}), \\ &\xrightarrow{-21\%} (1470 \pm 39)\ (\mathrm{N2LO}), \end{aligned} \quad (9.601)$$

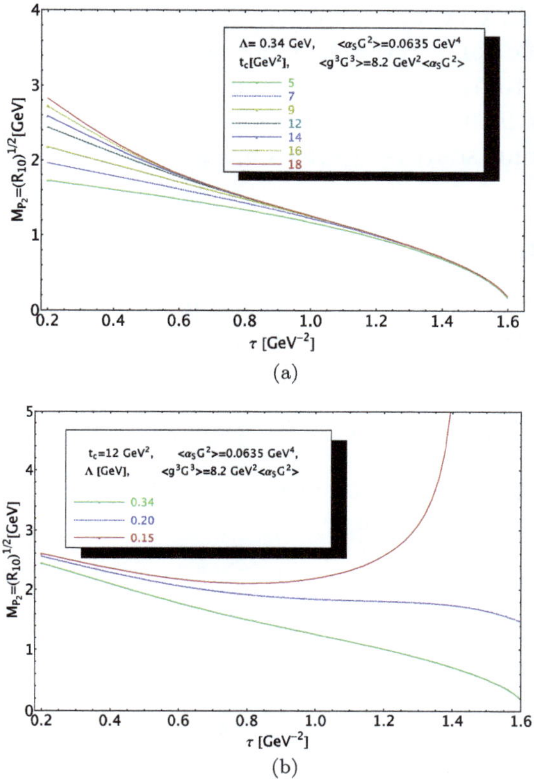

Figure 9.167. (a) M_{P_2} from $\mathcal{R}_{10}^c$ as a function of τ at N2LO for different values of t_c and for $\Lambda = 0.34$ GeV; (b) the same as (a) but for different values of Λ and for $t_c = 12$ GeV^2.

where the quoted error comes only from the range of t_c-values from 7 to 14 GeV^2. One can note a slow convergence of the result. In the rest of the paper, we re-emphasize that we shall estimate the contribution of the remaining uncalculated higher order (HO) terms of the series from the contribution of the tachyonic gluon mass which we shall add to the errors in the result for M_P and f_P.

9.14.8 Gluonia masses from $\mathcal{R}_{mn}^c$

• The analysis

In the following, we shall systematically extract the gluonia masses using different ratios of moments $\mathcal{R}_{mn}^c$ $(n, m = -1, 0, ..., 4)$ within

τ- and t_c-stability criteria. In order to show explicitly these stability regions specific for a given moment where the optimal results are obtained, we think that is important to show the different figures shown in Figs. 9.168–9.173. The results are summarized in Table 9.33 and the different sources of errors are given in Table 9.34 where we denote

◇ $P_{1,2}$: One resonance: P_1 or P_2.

◇ $S_{1\eta,2\eta} \equiv P_{1,2} \oplus \eta_1$: One resonance $\oplus\ \eta_1$ where M_{η_1} and f_{η_1} (gluon component of the η') are used as inputs from Eqs. (9.584) and (9.585) which have been deduced from previous sum rules analysis of $\mathcal{L}_{-1,-2}$.

◇ $S_2 \equiv P_2 \oplus S_{1\eta}$: Two resonances $(P_1 \oplus P_2) \oplus \eta_1$.

◇ $S_2' \equiv P_1' \oplus S_2$: Three resonances $(P_1' \oplus P_1 \oplus P_2) \oplus \eta_1$.

• P_1 medium mass

We show in Fig. 9.168 the τ-behaviour of M_{P_1} versus τ for different t_c from $\mathcal{R}_{20}^c$ and $\mathcal{R}_{42}^c$.

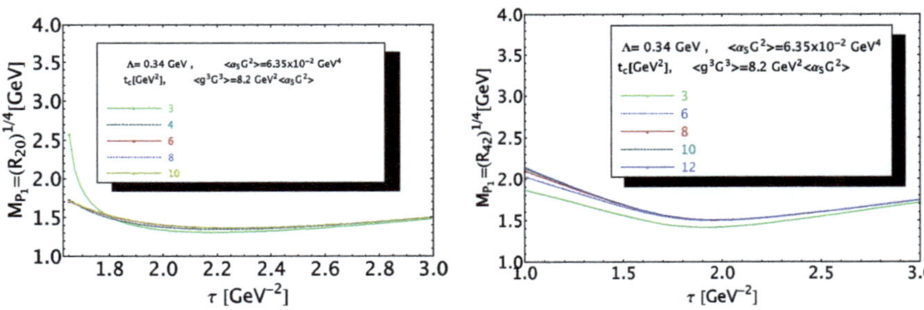

(a) (b)

Figure 9.168. (a) M_{P_1} from $\mathcal{R}_{20}^c$ as a function of τ at N2LO for different values of t_c and for $\Lambda = 0.34$ GeV where the η_1 contribution is included; (b) the same as (a) but from $\mathcal{R}_{42}^c$.

• P_1 medium decay constant

The di-gluonium decay constant is defined in Eq. (9.571) as $f_\pi = 93$ MeV. We shall use the moments $\mathcal{L}_2$ for extracting it. The analysis is shown in Fig. 9.169.

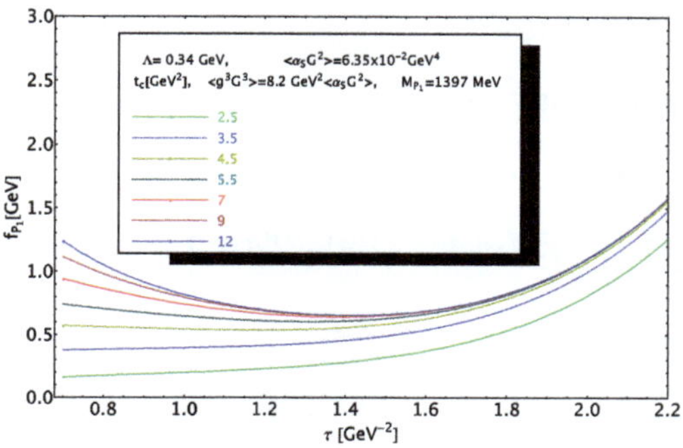

Figure 9.169. f_{P_1} from $\mathcal{L}_2^c$ as a function of τ at N2LO for different values of t_c and for $\Lambda = 0.34$ GeV where the η_1 contribution is included.

• P_1' first radial excitation mass and decay constant

Using now a "two-resonance" parametrization of the spectral function ⊕ "QCD continuum" and, as input, the parameters of the lowest ground state obtained previously, we show in Fig. 9.170 the τ behaviours of $M_{P'_{1a}}$ and $M_{P'_{1b}}$ from $\mathcal{R}_{20}^c$ and $\mathcal{R}_{42}^c$ for different values of t_c.

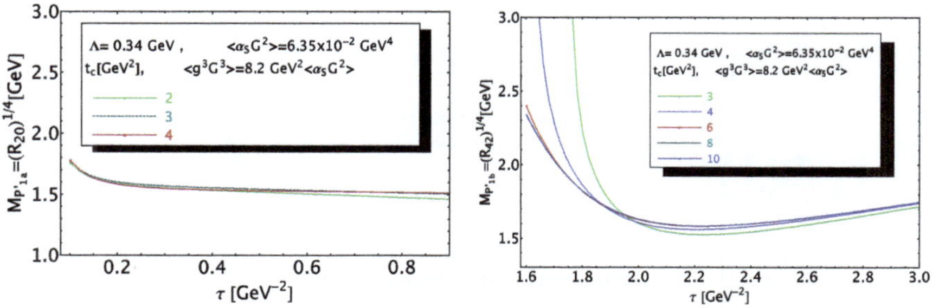

(a) (b)

Figure 9.170. $M_{P'_{1a}}$ from $\mathcal{R}_{20}^c$ as a function of τ at N2LO for different values of t_c and $\Lambda = 0.34$ GeV, where η_1 and P_{1a} contributions are included.

We show in Fig. 9.171 the analysis of the decay constant of the radial excitation.

• Results for P_1 and P_1'

From the previous analysis, we obtain, in units of MeV:

$$[M_{P_{1a}}, M_{P_{1b}}] = [1338(112), 1462(117)],$$
$$[M_{P'_{1a}}, M_{P'_{1b}}] = [1508(226), 1553(139)], \tag{9.602}$$

with their mean:

$$[M_{P_1}, f_{P_1}] = [1397(81), 594(144)],$$
$$[M_{P'_1}, f_{P'_1}] = [1541(118), 205(282)], \tag{9.603}$$

where the radial excitation is very close to the ground state like in the scalar channel [583]. The different sources of the errors are given in Table 9.34.

◇ Using the positivity of the spectral function, one can also derive a (rigorous) upper bound at the minimum of the ratio of moments $\mathcal{R}_{20}$ and $\mathcal{R}_{42}$ (equivalent to take $t_c \to \infty$). Including the η_1 contribution, it gives:

$$M_{P_{1a}} \leq 1365(106) \text{ MeV},$$
$$M_{P_{1b}} \leq 1506(104) \text{ MeV} \Longrightarrow M_{P_1} \leq 1437(74) \text{ MeV}, \tag{9.604}$$

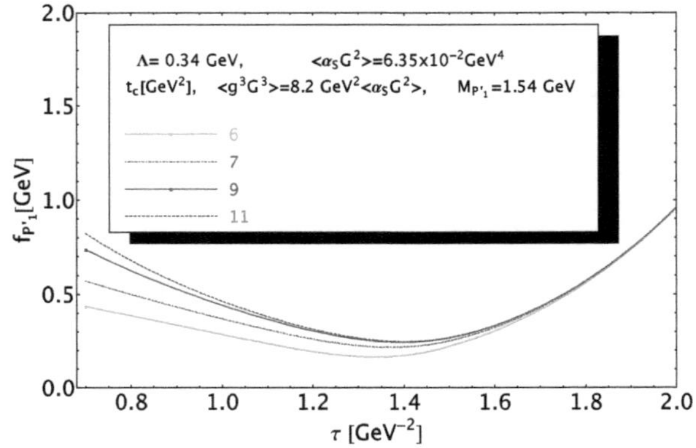

Figure 9.171. $f_{P'_1}$ from $\mathcal{L}_2^c$ as a function of τ at N2LO for different values of t_c and $\Lambda = 0.34$ GeV, where the $\eta_1 \oplus P_1$ contributions are included.

where one can note that in each case, the bound is almost saturated.

◇ One can also note that the two ratios of moments $\mathcal{R}^c_{20}$ and $\mathcal{R}^c_{42}$ stabilize at relatively large value of $\tau \simeq 2$ GeV^{-2}.

9.14.9 The heavy gluonium P_2

• Mass of the heavy gluonium P_2

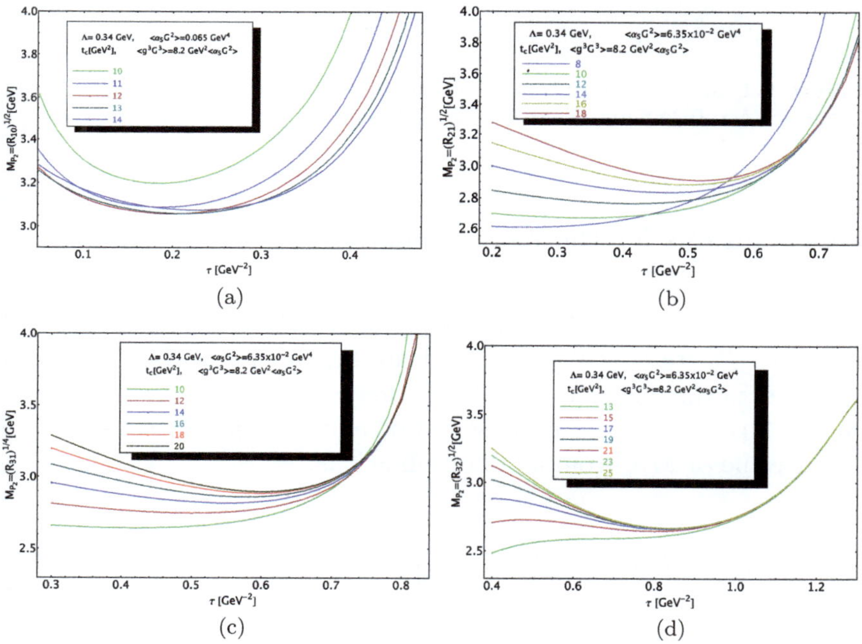

Figure 9.172. M_{P_2} as a function of τ at N2LO for different values of t_c and $\Lambda = 0.34$ GeV ($\eta_1 \oplus P_1 \oplus P'_1$ contributions included) from (a) $\mathcal{R}^c_{10}$; (b) $\mathcal{R}^c_{21}$; (c) $\mathcal{R}^c_{31}$; (d) $\mathcal{R}^c_{32}$.

We extract M_{P_2} from $\mathcal{R}^c_{10}$, $\mathcal{R}^c_{21}$, $\mathcal{R}^c_{31}$, $\mathcal{R}^c_{32}$ and $\mathcal{R}^c_{43}$ where the contributions of $\eta_1 \oplus P_1 \oplus P'_1$ are included. For instance, we show in Fig. 9.172 the analysis from $\mathcal{R}^c_{10}$ to $\mathcal{R}^c_{32}$ (some other analysis can be found in Ref. [151]).

• Decay constant of the heavy gluonium P_2

The analysis is shown in Fig. 9.173.

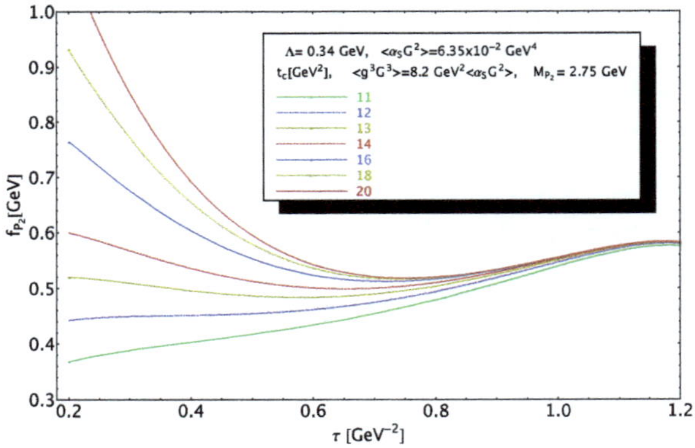

Figure 9.173. f_{P_2} from $\mathcal{L}_3^c$ as a function of τ at N2LO for different values of t_c and for $\Lambda = 0.34$ GeV, where $\eta_1 \oplus P_1 \oplus P_1'$ contributions are included.

• Conclusions for the heavy gluonium P_2

One can note from the last column of Table 9.33 that there is a splitting of about 300 MeV for M_{P_2}=3107(362) MeV from $\mathcal{R}_{10}^c$ and the other P_2 states which may indicate that there can be more than one state in the region above 2 GeV. However, as the error is large, we cannot strictly confirm this observation and consider the conservative mean value of M_{P_2} from all determinations:

$$M_{P_2} = 2751(140)\ \text{MeV}, \quad f_{P_2} = 500(43)\ \text{MeV}, \tag{9.605}$$

where the errors are mainly due to Λ, f_{P_1} and M_{P_1} (see Table 9.34).

9.14.10 Summary and Conclusions

For an attempt to study the topics addressed in the title of this paper, we have scrutinized and improved the 0^{-+} pseudoscalar gluonium sum rules using updated values of the QCD input parameters and using a multiple resonance parametrization of the spectral function beyond the minimal duality ansatz: "one resonance $\oplus$ QCD continuum" and high-degree LSR moments.

Table 9.33. Pseudoscalar di-gluonium masses from LSR for PT at N2LO and gluon condensates at NLO up to $D = 8$ for: "$\sum_{P_i}$ resonances $\oplus$ QCD continuum" parametrization of the spectral function. The mean value of $M_{P'_1}$ is used to get $f_{P'_1}$ and the one $[M_{P'_1}, f_{P'_1}]$ for $[M_{P_2}, f_{P_2}]$ in the last column. The bounds on M_{P_1} in column # 4 are obtained at the minimum of $\mathcal{R}_{20,42}$ using the positivity of the spectral function and including the η_1 contribution. The different sources of errors are given in Table 9.34. The QCD parameters in Table 7.10 have been used as inputs.

Observables	LSR	η_1	$P_{1,2}$	$S_{1\eta,2\eta} \equiv P_{1,2} \oplus \eta_1$	$S_2 \equiv P_2 \oplus S_{1\eta}$	$S'_1 \equiv P'_1 \oplus S_{1\eta}$	$S'_2 \equiv P_2 \oplus S'_1$
M_P *[MeV]*							
P_1	$\mathcal{R}^c_{20}$		1205(126)	$\leq$1338(112)			
–	–			$\leq$1365(106)			
–	$\mathcal{R}^c_{42}$		1432(115)	$\leq$1462(117)			
–	–				$\leq$ 1506(104)		
–	*Mean*		*1345(83)*	$\leq$*1397(81)*			
–	–			$\leq$*1437(74)*			
P'_{1a}	$\mathcal{R}^c_{20}$					1508(226)	
P'_{1b}	$\mathcal{R}^c_{42}$					1553(139)	
P'_1	*Mean*					*1541(118)*	
P_2	$\mathcal{R}^c_{10}$		1470(101)	$\leq$2097(264)	2990 (318)		3059(385)
–	$\mathcal{R}^c_{21}$		1846(73)	$\leq$2005(119)	2715(285)		2778(340)
–	$\mathcal{R}^c_{31}$		1726(71)	$\leq$2050(119)	2699(265)		2764(317)
–	$\mathcal{R}^c_{32}$		2096(44)	$\leq$2261(39)	2534(327)		2644(391)
–	$\mathcal{R}^c_{43}$		2054(42)	$\leq$2080(158)	2620(205)		2661(227)
–	*Mean*		*1965(25)*	$\leq$*2214(34)*	*2694(119)*		*2751(140)*
f_P *[MeV]*							
η_1	$\mathcal{L}^c_{-1}$	905(72)					
P_1	$\mathcal{L}^c_2$			594(144)			
P'_1	$\mathcal{L}^c_2$					205(282)	
P_2	$\mathcal{L}^c_3$				504(39)		500(43)

Table 9.34. Different sources of errors for the pseudoscalar di-gluonia masses and couplings given in Table 9.33. The errors from QCD input parameters come from Table 7.10. The quoted values of τ for P_2 in the cases S_2 and S'_2 correspond to S'_2.

	# Resonances	LSR	t_c[GeV^2]	Δt_c	τ[GeV^{-2}]	$\Delta\tau$	$\Delta\Lambda$	$\Delta\lambda^2$	ΔG^2	ΔG^3	ΔG^4	Δf_{η_1}	ΔM_{η_1}	ΔM_{P_1}	Δf_{P_1}	$\Delta M_{P'_1}$	$\Delta f_{P'_1}$	ΔM_{P_2}
ΔM_P																		
P_1	$S_{1\eta} \equiv P_1 \oplus \eta_1$	$\mathcal{R}^c_{20}$	3, 6	28	2.19, 2.27	1	33	15	5	54	79	27	0					
		$\mathcal{R}^c_{42}$	3, 8	43	1.90, 1.95	1	99	3	4	6	32	8	15					
P'_{1a}	$P'_{1a} \oplus S_{1\eta}$	$\mathcal{R}^c_{20}$	3	29	0.48	4	17	6	3	2	0	25	62	172	126			
P'_{1b}	$P'_{1b} \oplus S_{1\eta}$	$\mathcal{R}^c_{42}$	3, 6	29	2.22	3	128	4	6	6	32	6	9	4	29			
P_2	$S_{2\eta} \equiv P_2 \oplus \eta_1$	$\mathcal{R}^c_{10}$	7, 14	3	0.62, 0.80	16	83	44	19	11	21	173	174					
		$\mathcal{R}^c_{21}$	8, 14	50	0.78, 0.88	3	44	17	4	14	47	37	59					
		$\mathcal{R}^c_{31}$	6, 14	93	0.84, 1.04	1	20	18	5	16	43	59	34					
		$\mathcal{R}^c_{32}$	13, 19	25	0.8	4	26	4	0	0	3	5	13					
		$\mathcal{R}^c_{43}$	8, 12	72	1.16, 1.26	4	88	13	2	3	12	4	109					
P_2	$S_2 \equiv P_2 \oplus S_{1\eta}$,	$\mathcal{R}^c_{10}$	12	0	0.20	12	144	23	6	2	3	45	56	90	250	5	216	
	$S'_2 \equiv P'_1 \oplus S_2$	$\mathcal{R}^c_{21}$	10, 16	97	0.34, 0.50	5	80	23	6	2	3	101	27	83	216	15	185	
		$\mathcal{R}^c_{31}$	10, 18	103	0.42, 0.60	6	74	25	3	0	3	14	28	86	224	15	173	
		$\mathcal{R}^c_{32}$	15, 19	5	0.82	3	114	31	2	2	11	21	42	146	255	19	213	
		$\mathcal{R}^c_{43}$	10, 14	28	0.84, 0.90	4	46	17	0	0	2	1	6	113	161	15	96	
Δf_P																		
η_1	–	$\mathcal{L}^c_{-1}$	10, 14	21	0.66, 0.72	2	41	51	3	2	0							
P_1		$\mathcal{L}^c_2$	4.5, 9	58	1.18, 1.40	0	114	15	1	3	50	9	21	35				
P'_1	$P'_1 \oplus S_{1\eta}$	$\mathcal{L}^c_2$	7, 9	39	1.38, 1.40	2	139	45	12	14	50	20	67	94	200	8		
P_2	S_2, S'_2	$\mathcal{L}^c_3$	13, 18	16	0.58, 0.74	3	32	8	0	0	0	7	0	8	7	2	16	8

• Slope of the topological charge and proton spin

◇ The properties of the η_1 singlet piece of η' as well as the value of the topological charge $\chi(0)$ from $U(1)_A$ large N_c and current algebra are reproduced from the approach in Refs. [77,82,118,142,144,607].

◇ We have estimated the slope of the topological charge at N2LO and find it to be $\sqrt{\chi'(0)}(Q^2 = 2\,\text{GeV}^2) = 24.3(3.1)$ MeV (Eq. (9.590)). It confirms the previous NLO value of 22.3(4.8) MeV [592] from LSR updated here, which is still smaller than the OZI value $f_\pi/\sqrt{6} =$ 38 MeV.

◇ We have used this value of $\sqrt{\chi'(0)}$ to estimate the singlet form factor $G_A^{(0)}$ of the axial current (see Eq. (9.594)) and the first moment $\int_0^1 dx\, g_1^P(x)$ of the polarized proton structure function (see Eq. (9.596)) which remarkably agree with the data [139,611,612].

• 0^{-+} Pseudoscalar Gluonia

We have used more than one resonance and worked with different high-degree moments to analyze the complex spectra of the observed η-like gluonia. We find the following:

◇ The lattice results for the pseudoscalar gluonium mass are comparable with the so-called heavy gluonium mass $P_2(2751)$ obtained here from moment ratios which stabilize at smaller values of τ.

◇ New gluonia P_1-like states with medium masses are found from some ratios of moments which stabilize at larger values of τ and can eventually explain the gluonium nature of some of the four η-like states found below 2 GeV. Our result supports the gluonium nature of the experimental candidate $\eta(1405)$ and suggests some gluon component to the other η-like wave functions.

◇ It is also remarkable to note that the structure of the pseudoscalar gluonia spectrum is very similar to the one of the scalar gluonia (its chiral partner) studied recently in Ref. [583]. There is a (natural) one-to-one correspondence between the pseudoscalar gluonia and their chiral scalar analogue from Ref. [583]: $\sigma(1) \to \eta_1$; $G_1(1.55) \to P_1$; $\sigma'(1.1), G'(1.56) \to P'_{1a,1b}$; $G_2(3) \to P_2$ which is mainly due to the importance of the QCD PT contributions that are almost equal in the analysis of these two channels.

9.15 Comments

We have reviewed in this chapter the modern state of the art of the QCD spectral sum rules (QSSR) predictions based on the SVZ-expansion and on stability criteria of the external parameters entering in the analysis. These criteria are rigorous compared to the ones used in the literature. Thanks to the better QCD expressions including higher PT corrections and the under control value of condensates up to dimension 6, one notice improvements of the accuracy of the QSSR results which are useful for comparing and completing some existing ones from other approaches (lattices, chiral lagrangian, potential models, AdS models,...).

Chapter 10

The COVID-19 Pandemic in Madagascar

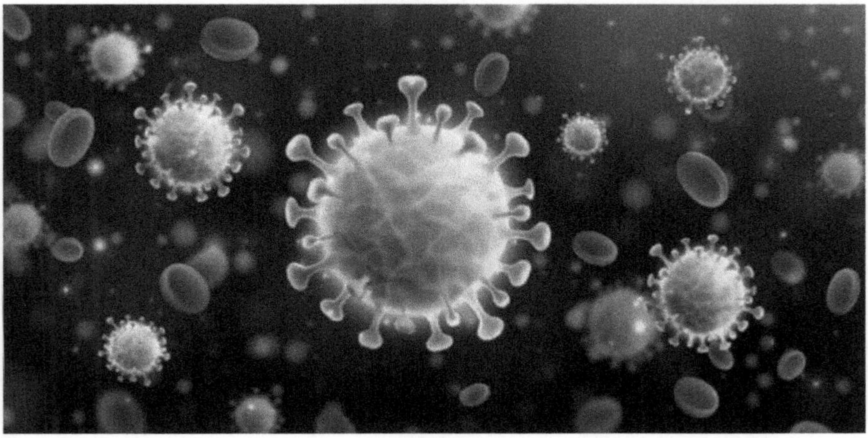

During the confinement period, I was interested to analyze the impact of the COVID-19 pandemic in Madagascar. This work is reported in Refs. [613, 614], which I shall summarize in the following sections.

10.1 The first months [613]

10.1.1 The spread of infected persons

• First analysis: Data up to 15 days

Using the officially published data and aware of the uncertain source and insufficient number of samples, we present a first and (for the moment) unique attempt to study the first two months' spread of the COVID-19 pandemic in Madagascar. We shall use the data compiled in Table 3 of Ref. [613] from Refs. [615–619].

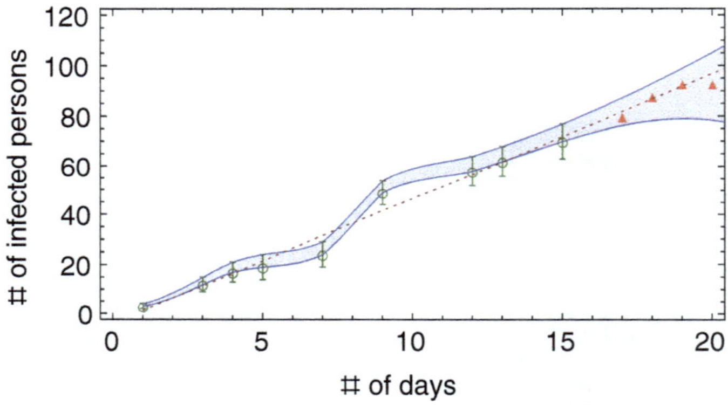

Figure 10.1. Number of new contaminated persons per day versus the number of days: green circles denote the input data, filled regions denote the predictions limited by the lower curve (central value of the data) and the upper curve (data with added positive estimated errors). Dashed red line is the linear fit. Triangles are data received after the fitting procedure.

The approach has been tested by predicting the number of contaminated persons for the following week after fitting the input data collected within 7 or 15 days using standard least χ^2-fit method (see Fig. 10.1). We also show (dotted curve) the expectation from a simple linear parametrization of the data as given by:

$$n_i(x) \simeq -3.67 + 5.03x, \tag{10.1}$$

where $n_i(x)$ is the number of daily new infected persons as a function of the number x of days. The data show a linear increase for about five persons per day.

• Second analysis: Data up to 30 days

Encouraged by this positive test, we improve the analysis by adding the new data (see Fig. 10.2) collected from the 16th to 30th day.

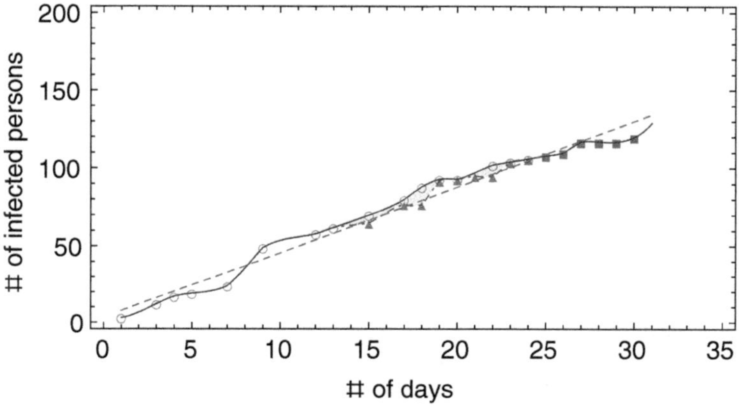

Figure 10.2. Number of new daily infected persons versus the number of days: green circles denote the input data from Refs. [615,618]. The red triangles denote the data from Ref. [619] from the 17th day. Blue rectangles denote new data Refs. [615, 618, 619] from the 25th to the 30th day. The continuous oliva curve denotes the fit from the 1st to the 30th day from Refs. [615,618] data, while the red dot-dashed one denotes the fit of Ref. [619] data, which follows the red triangle data points. The dashed blue line denotes the linear fit given in Eq. (10.2).

The χ^2-fit is shown by the continuous green line (green circle data from Refs. [615, 618] and dot-dashed red curve (red triangle data) from Ref. [619]).

We start from a linear fit of the data collected during 24 days:

$$n_i(x) \simeq -1.42 + 4.73x, \tag{10.2}$$

while adding the data until 30 days (blue rectangles in Fig. 10.2), we obtain:

$$n_i(x) \simeq 3.5 + 4.22x. \tag{10.3}$$

The slopes of the linear curves obtained in Eqs. (10.2) and (10.3) (blue dotted line) are consistent with the one in Eq. (10.1) shown in Fig. 10.2. This linear fit can be improved by using a quadratic fit of the data. In this way, one obtains:

$$n_i(x) \simeq -8.14 + 6.45x - 0.07x^2. \tag{10.4}$$

The fits indicate an average of about 4–5 persons infected per day.

• Comparison with some SI models: Data up to 46 days

The results of the analysis are summarized in Fig. 10.3.

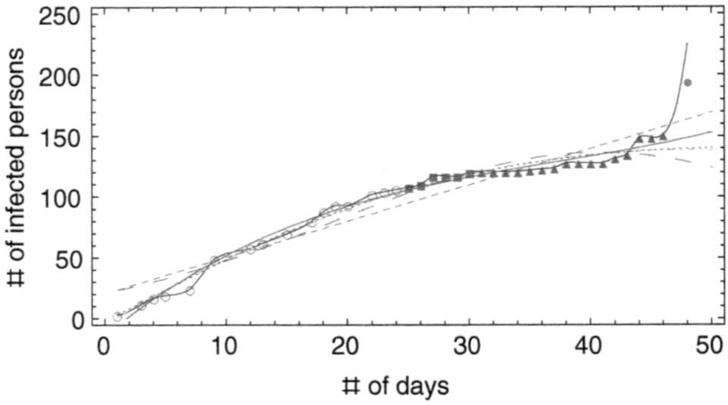

Figure 10.3. New data (dark red triangles) are added in the analysis. The dashed blue line denotes the linear fit in Eq. (10.3). The dot-dashed green curve denotes the quadratic fit given in Eq. (10.4). The full red curve denotes the cubic fit. The dot-dashed red curve denotes the fit from the SI-like model used in Ref. [620,621], while the long dashed blue curve denotes the one from a gaussian-like fit used in Ref. [622]. The continuous oliva curve denotes the one of least χ^2-fit. The red point denotes the new data on 7th May not included in the fit.

◇ The linear fit provides a good smearing of the evolution number:

$$n_i(x) \simeq 20.82 + 2.98x, \tag{10.5}$$

indicating an average of about three infected persons per day. The best fit is obtained from the cubic polynomial:

$$n_i(x) \simeq -13.7 + 8.2x - .2x^2 + 1.6 \times 10^{-3}x^3. \tag{10.6}$$

This cubic polynomial model signals a future sharper growth, while the other models predict some future stabilities or even a decrease in the number.

◇ One can see from Fig. 10.3 that the mixed power exponential hereafter named large population semi-gaussian (LPE-SG) model used in Refs. [620, 621]:

$$n_i(x) = ax^b e^{-x/x_0} \tag{10.7}$$

(dot red curve with $a = 3.74, b = 1.22, x_0 = 44.08$) and the quadratic polynomial fit (dashed blue curve):

$$n_i(x) = -4.81 + 6.00x - 0.06x^2 \tag{10.8}$$

give almost the same predictions, while the gaussian-like model used in Ref. [622]:

$$n_i(x) = ae^{-\frac{(x-x_0)^2}{2\sigma^2}} \tag{10.9}$$

(large dashed purple curve with $a = 136.58, x_0 = 40.73, \sigma = 21.19$) fails to describe the most recent three data. However, the features from these other models look unrealistic as the virus incubation period of about one month is just finished while the winter season is approaching, which could be a favoured period for the virus.

◇ We show in Fig. 10.3 the new data of the 48th day (red full circle) based on 3968 detection tests. It indicates a sudden jump compared to the previous data such that the different models successful to describe the data for the first 46 days fail here.

◇ However, the new data agree with the least χ^2-fit method used to fit the former data. This feature may signal a new phase for the spread of the infection. It coincïdes with the early partial deconfinement, the end of incubation period from contacts and the beginning of the winter time where the UV destruction on the virus may be less efficient.

• Prediction from the 48th to 58th day

Including the new data of the 48th day, we give a prediction for the next 10 days in Fig. 10.4(a). From this analysis, we expect that the most conservative predictions for the next 10 days are inside the region (purple colour) limited by the cubic polynomial (red) and least-χ^2-fit (oliva) continuous curves.

We show the collected new data for the next 10 days in Fig. 10.4(a) (green full circle) inside the predicted region.

• Fit up to 58 days and predictions until the 67th day

We show in Fig. 10.4(b) the analysis of the data until 58 days and the prediction until the 67th day. Again, the region delimited by the cubic polynomial and the least-χ^2-fits gives a nice prediction of the next 10-day data.

One can note the jump of the number of contaminated persons which indicates the second phase of the pandemic from the 48th day.

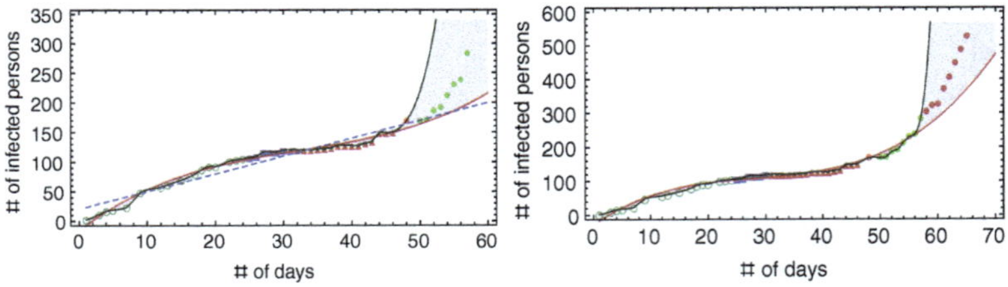

(a) (b)

Figure 10.4. (a) Data are the same as in Fig 10.3. New fit including the new data (red point) on 7th May (48th day). The oliva curve denotes the χ^2-fit. The continuous red curve denotes the cubic polynomial. The dashed line denotes the linear fit. The pink region denotes the prediction where we show the new data (not used in the fit) collected until the 58th day; (b) same as (a) but using new data from 48 to 58 days and comparison of the prediction with new data (purple) between 58 and 67 days.

• Predictions beyond the 67th day

We show in Fig. 10.5 the new fit of the data before 67 days and the prediction (purple region) for the next 10 days where the cubic polynomial is parametrized as:

$$n_i(x) = -55.27 + 18.25x - 0.67x^2 + 0.008x^3. \qquad (10.10)$$

The new data (open green circles) favour the cubic polynomial behaviour. The analysis of these new data and those collected after the completion of this work is under consideration. This is a remarkable feature of the pandemic spread in Madagascar. We expect that these results will be useful for some new model buildings.

10.1.2 Evolution number of cured persons

In Ref. [613], we study both the absolute and per cent number evolution of the cured persons. We found that the absolute number increases for about 3–4 persons per day, but the per cent number decreases above 42–46 days (see Fig. 10.7), indicating the limits of the hospital equipment and care to face the second phase of the pandemic for the first 67 days.

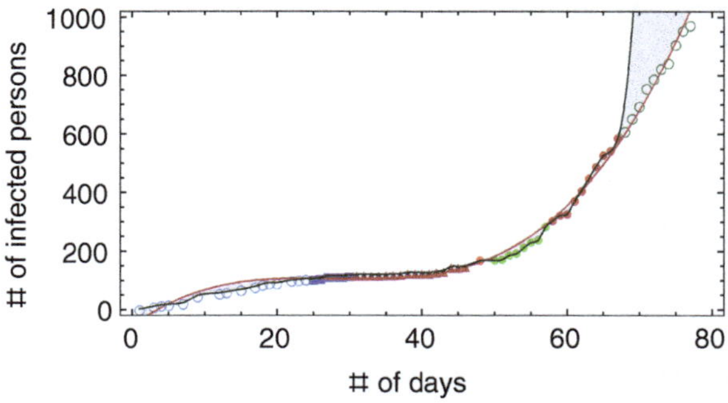

Figure 10.5. New fit including data between 59 and 67 days (red full circle). The oliva curve denotes the χ^2-fit. The continuous red curve denotes the cubic polynomial. The purple region denotes the expected prediction. The green open circles denote new data from the 68th to the 77th day not included in the fit.

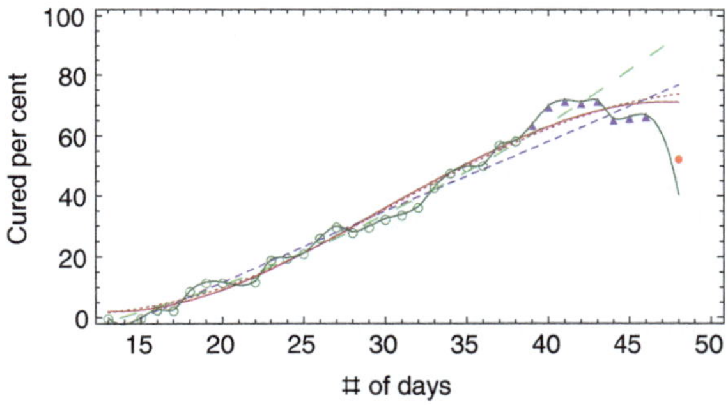

Figure 10.6. Comparison of different model predictions for the per cent number of cured persons using the data in Table 3 of Ref. [613] until 46 days. The dashed blue line denotes a linear fit. The long dashed green curve denotes a quadratic fit in Eq. (10.4). The dot red curve for the SI-like model in Ref. [620, 621] and the continuous red curve for a cubic polynomial fit are almost superimposed. The green continuous curve denotes the χ^2-fit. Green circles denote data until 38 days. Blue triangles denote new data until 46 days. The red point denotes data from the 48th day not included in the fit.

10.1.3 COVID-19 and confinement for Madagascar

• Political decisions

In order to limit the spread of this dangerous virus which propagates via social contacts (coughing, saliva droplets, etc.) and waiting for

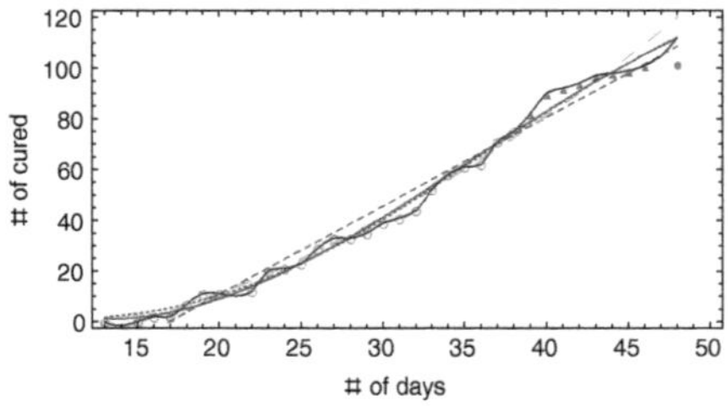

(a)

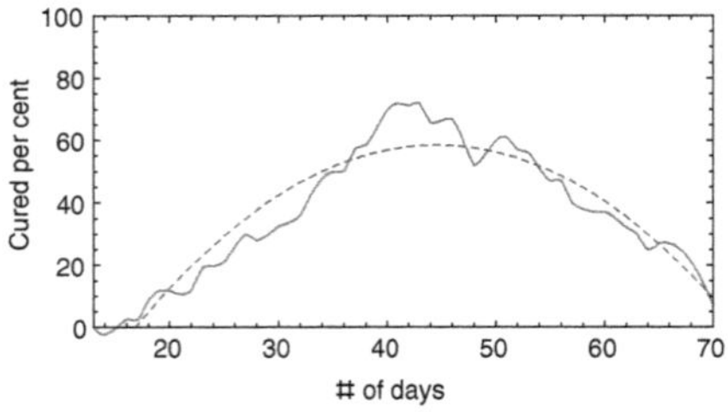

(b)

Figure 10.7. Number of cured persons: (a) absolute: The dashed curve denotes a linear fit; (b) per cent: The dashed curve denotes a quadratic fit.

new efficient medical drugs, different countries decided to confine the population and asked the people to strictly respect some barrier rules. About 3 billion people, i.e., half of humanity were confined.

• Implementation

However, the implementation of confinement in developing countries cannot be easily achieved due to poverty and the lack of education of the majority of the population:

◇ Due to poverty and poor organization of the society, most of the persons have to work and find something to eat day after day owing to the informal forms of trade, business and economy.

◇ In addition, the accompanying help measures taken by the government are insufficient while the management of some international funds and donations are not transparent.

◇ Middle class persons also suffer as they are not rich enough to be autonomous and not too poor to receive any help from the state.

◇ Due to the lack of education, most of the people are irresponsible and are not aware about the dangerous effect of the virus. Then, they do not see the importance of confinement and do not feel obliged to respect the barrier measures and the confinement.

• Worldwide Impact of COVID-19 and Confinement

More generally and as a consequence, these unprecedented pandemic and confinement security measures have major impacts for human beings:

◇ **Social organisation**

- Most of us learn about teleworks, indoor and outdoor houseworks.
- We rediscover the importance of family and traditions.
- We rediscover ourselves through our concentration and meditation.
- We see the usefulness of solidarity.
- We see the value of health personnel, researchers, teachers, educators, firefighters, etc., and in general the human values.
- Urban exodus is observed in developing countries like Madagascar.
- However, the pandemic might enhance the social class inequalities like the increase in the difference between the poor and the rich, the redisappearance of middle class, as mentioned earlier for Madagascar, which is the lung of development.

◇ **Environments**

- Nature takes back its rights: return of animals near cities and cetaceans near the coasts and return of insects, birds, etc.).
- Air pollution is decreasing due to a minimal road traffic. Megacities (Beijing, New Delhi, New York, Paris, etc.) record recovery of improved atmospheric air.
- Cities become less noisy.
- . . .

◇ **Economy**

- Globalization is suffering.
- Reduced size and local markets are developing.
- Small producers are reorganizing.
- Online sales and drive markets are developing.
- Delocalisations of manufacturers and factories are questioned.
- Each country is looking for an independent and self-sufficient economy which may lead to a protectionism.

◇ **Politics**

COVID-19 has pressed the leaders of each country to revisit the orientation of their politics to the most useful ones for the population:

- health,
- research and education,
- food and different ways of consumption,
- protection of the environment.

COVID-19 has stimulated the search for an autonomous country which, if not done carefully, may lead to an ultra-nationalism and to a withdrawal into oneself (closing of borders, etc.).

10.2 Collaboration with S. Maltezos (NTUA, Athens)

In Ref. [614], we scrutinize the spread of COVID-19 in Madagascar and then compare this with some other countries.

10.2.1 Ratio of infected persons over tests

We pursue the analysis in Ref. [613] by introducing (for the first time) the ratio $\mathcal{R}^{c,d}_{I/T}$ of the cumulative and daily numbers of infected persons over the corresponding one of tests which are expected to be less sensitive to the number of the tests because the credibility of the results based only on the absolute numbers often raises some criticisms. They read as:

$$\mathcal{R}^c_{I/T} \equiv \frac{\mathcal{N}^c_I}{\mathcal{N}^c_T} = \frac{\text{Cumulative \# of infected persons}}{\text{Cumulative \# of Tests}}, \tag{10.11}$$

and the corresponding daily ratio is:

$$\mathcal{R}_{I/T}^{d} \equiv \frac{\mathcal{N}_I^d}{\mathcal{N}_T^d} = \frac{\text{Daily \# of infected persons}}{\text{Daily \# of Tests}}, \tag{10.12}$$

10.2.2 The input data

We show in Fig. 10.8 the data from Refs. [615–619, 623].

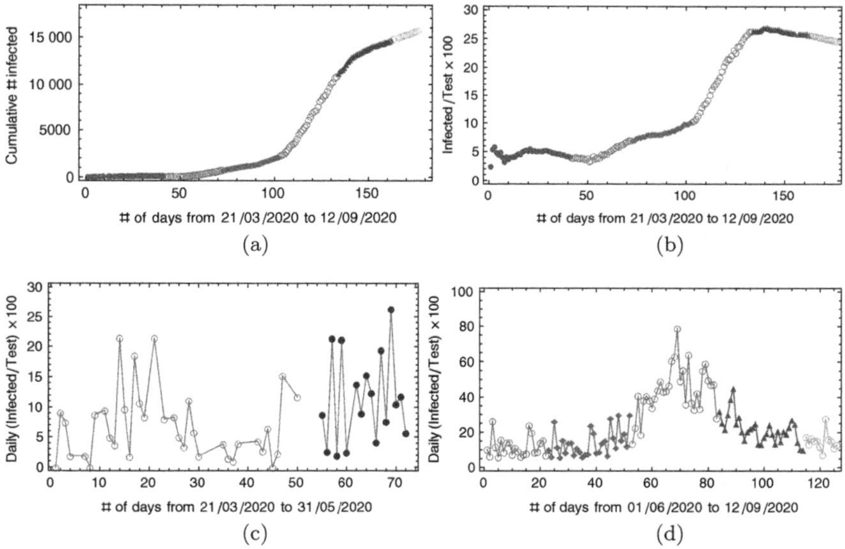

Figure 10.8. Data of (a) absolute cumulative number of infected persons; (b) cumulative ratio $\mathcal{R}_{I/T}^{c}$ defined in Eq. (10.11); (c,d) daily ratio $\mathcal{R}_{I/T}^{d}$ defined in Eq. (10.12).

10.2.3 Phases of the pandemic

One can note from the previous figures that there are two phases:

- small or epidemic phase until 48 days (7th May);
- large or pandemic phase after 7th May.

We shall see later on that these two phases can be subdivided into two phases ⊕ a background phase from 7th to 31th May.

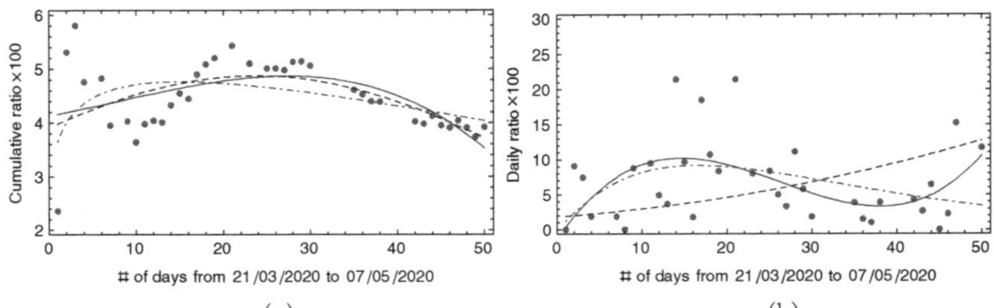

Figure 10.9. Data and analysis of (a) cumulative ratio $\mathcal{R}^c_{I/T}$ defined in Eq. (10.11); (b) daily ratio $\mathcal{R}^d_{I/T}$ defined in Eq. (10.12).

- **The small phase until 07/05/2020 (48 Days)**
  - ◇ The absolute cumulative numbers of infected persons have been studied in Ref. [613], where the peak (stability of the cumulative number of infected persons) is reached for 25–30 days after the beginning of the pandemic.
  - ◇ Doing a similar analysis for the cumulative ratio $\mathcal{R}^c_{I/T}$, one finds that the best fit is obtained for a cubic polynomial with a maximum peak around the 30th day. The LPE-SG model defined in Eq. (10.7) predicts a peak around 15 days and the gaussian-like one defined in Eq. (10.9) predicts a peak around 40 days.
  - ◇ Similar analysis is done for the daily ratio, which is shown in Fig. 10.9. From the previous analysis, one can see that the cubic polynomial and the LPE-SG model find a maximum peak around the (14–17)th day, while the gaussian-like model peak is around 46 days (not shown in the figure).

- **Before the second peak: 07/05/2020 to 12/08/2020**

◇ Fitting the absolute cumulative number of infected persons, one finds that the LPE-SG model finds a peak around $b_s \times c_s \approx 126$ days, while the gaussian-like model finds a gaussian centred at 109 days which are compatible. One can note that the cubic polynomial gives a good fit of the data but cannot predict a peak.

◇ However, doing a similar analysis for the ratio $\mathcal{R}^c_{I/T}$, the two models LPE-SG and gaussian-like one do not give a good fit, which leads to unrealistic values of the fitted parameters.

• After the second peak of 12/08/2020

After the peak, one can see that the total number of infected persons increase linearly as:

$$\mathcal{N}_I \simeq 6463 + 71.9x, \tag{10.13}$$

which corresponds to an increase of about 72 infected persons/day, while the ratio $\mathcal{R}_{I/T}$ decreases slightly as:

$$\mathcal{R}_{I/T} = 31.69 - 0.05x \tag{10.14}$$

corresponding to 0.05% infected person per day. It indicates that the cumulative number of infected persons decreases slowly when the cumulative number of tests increases, which is a clear signal that the pandemic is decreasing.

• Fit of the second phase from 07/05/2020 to 28/08/2020

Here, we redo the analysis but for a global fit of the second phase given in Fig. 10.8. The analysis is shown in Fig. 10.10.

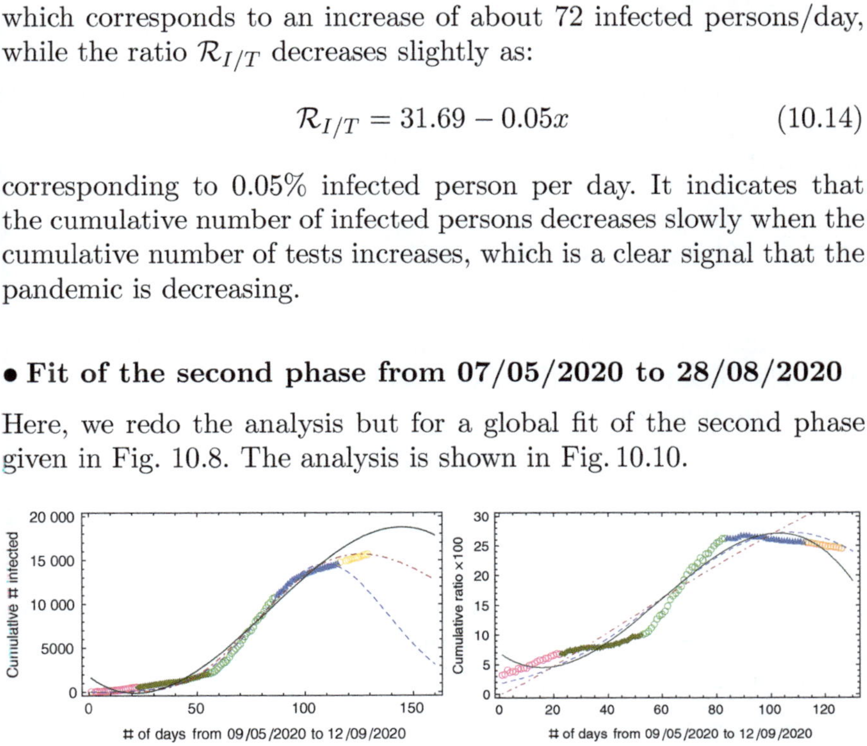

(a) (b)

Figure 10.10. Fit of the data in Fig. 10.8 from 50 days to 130 days: (a) absolute cumulative number of infected, (b) cumulative number of infected over cumulative number of tests. Same legends as in Fig. 10.8. The open yellow circles at the end of the curve denote new data not used in the fit. The origin of the axis is taken on the 50th day. The continuous oliva line denotes the fit from a cubic polynomial. The dashed blue curve is the one from a gaussian-like model. The dot-dashed red curve denotes the fir from the LPE-SG model.

⋄ For the absolute cumulative number of infected persons, the polynomial fit shows a maximum around 150 days. The LPE-SG Model indicates a peak around: $b_s \times c_s \simeq 126$ days, while the gaussian is centred at 109 days. These predictions are compatible with each other.

◇ For the cumulative ratio $\mathcal{R}^c_{I/T}$, one can deduce from the analysis that the cubic polynomial fit and the gaussian-like model indicate a peak around (105–107) days, i.e., on (3–7)th of July, while the LPE-SG is not conclusive.

◇ However, if we start the analysis from 15th June (origin of the x-axis), we obtain a much better fit (see Fig. 10.11) with reasonable values of the fitted parameters. The three approaches predict a peak around the (65–70)th day, i.e., around (19–24)th August 2020.

10.2.4 The reproduction numbers R_{eff} of the first phase

We analyze these quantities using the definition proposed in Ref. [624] based on an SIR model [625], where the discrete instantaneous reproduction number is:

$$R_i = \frac{I_i}{\sum_j w_j I_{i-j}}, \tag{10.15}$$

where I_i is the number of infected persons at a time t_i which follows a Poisson distribution, and w_i is the measures of infectiousness with the properties: $0 \leq w_i \leq 1$ and $\sum w_i = 1$.

• Lower bound on the initial reproduction number R_0

We attempt to provide a lower bound on R_0 by using the data reported on the first four days from 19/03/20 to 22/03/2020 where the daily infections are, respectively, 3, 0, 0 and 9. Using these data and the properties of w_i, one can deduce:

$$R_0 \approx R_4 = \frac{I_4}{w_0 I_4 + I_1 w_3} \geq \frac{3}{4}. \tag{10.16}$$

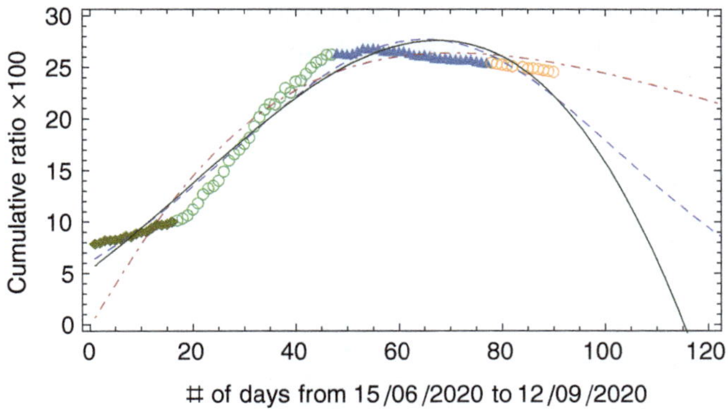

Figure 10.11. Fit of the data in Fig. 10.8 from 87th day (15th June): same legends as in Fig. 10.8. The origin of the axis is taken on the 87th day. The continuous oliva line denotes the fit from a cubic polynomial. The dashed blue curve denotes the one from a gaussian-like model. The dot-dashed red curve denotes the fit from the LPE-SG model.

The bound is rather weak but may reflect the epidemic (not yet a pandemic) nature of the disease.

• Value of R_0

Considering that the result from the first week is statistically poor, one can deduce from Fig. 10.12 the rough estimate from the beginning of the second week:

$$R_0 \approx R_7 \simeq (0.96 \pm 0.44). \tag{10.17}$$

• The effective reproduction number R_{eff}

Using the previous SIR-like model, we attempt to extract the instantaneous/effective reproduction number R_{eff} for an interval period of about one week. We show the results in Fig. 10.12. The large error bars take into account the low and high values of the predictions obtained in Ref. [623]. In the dashed region, we show the tentative weighted average:

$$\langle R_{\text{eff}} \rangle \simeq 1.06 \pm 0.18, \tag{10.18}$$

where its relatively low value may indicate that this first phase is still an epidemic but not yet a pandemic period.

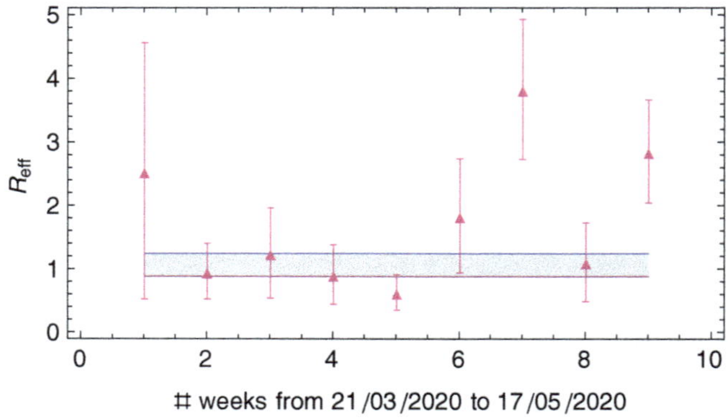

Figure 10.12. Weekly behaviour of the reproduction number R_{eff} of infected persons until 17th May from SIR model. The dashed region denotes a tentative weighted average.

10.2.5 R_{eff} from the LPE-SG model

• The three phases

We have noted in the previous analysis that the LPE-SG model fitted parameters are very sensitive to the starting/ending points of the data. The reason is that the model is asymmetric contrary to the gaussian-like model and the cubic polynomial. Therefore, for better exploitation of this model, it is more convenient to divide the total period of the pandemic into three phases (see Fig. 10.13):

- the first phase for the first two months until 2nd May (epidemic);
- the phase between 2nd May and 15th June (background), which is below the pandemic threshold where the data fluctuate eventually showing the beginning of the pandemic; this can be considered as a continuation of the first phase;
- the second phase beyond 15th June (real pandemic).

• The first phase

We show in Fig.10.14 the predictions of the LPE-SG model.

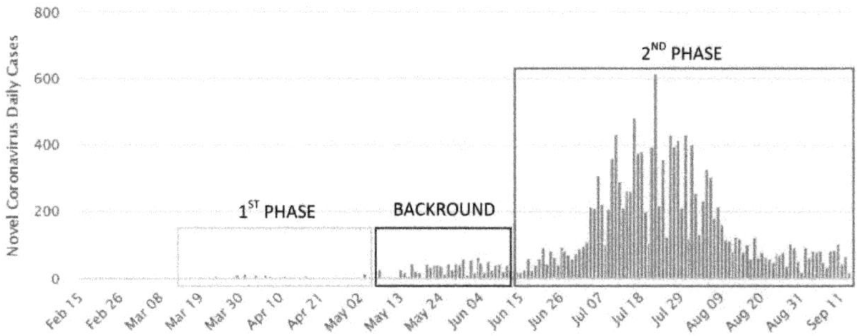

Figure 10.13. Different phases of COVID-19 in Madagascar for the daily number of infected persons.

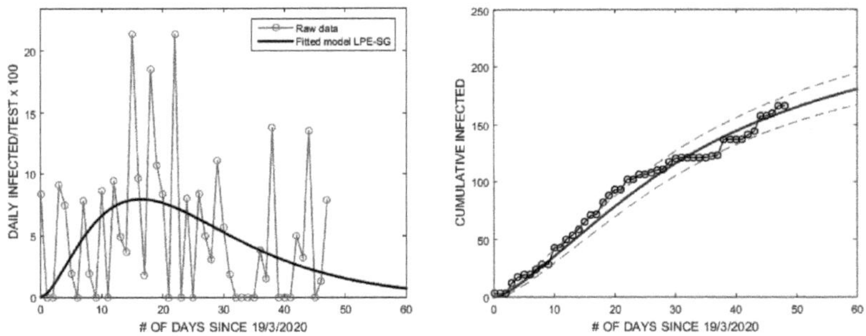

(a) (b)

Figure 10.14. Fit of the daily (a) and cumulated (b) reported data for infected persons using LPE-SG model for the first phase from 19th March until 2nd May.

◇ R_0

The position of the peak around 12 days is consistent with the one obtained in Fig. 10.9 by a direct fitting of the data from which one can deduce:

$$R_0 \simeq 1.15 \pm 0.65. \qquad (10.19)$$

This value is comparable with the one from SIR-like model in Eq. (10.17).

◇ R_{eff}

We study in Fig. 10.15 the behaviour of R_{eff} using the daily ratio of corrected data.

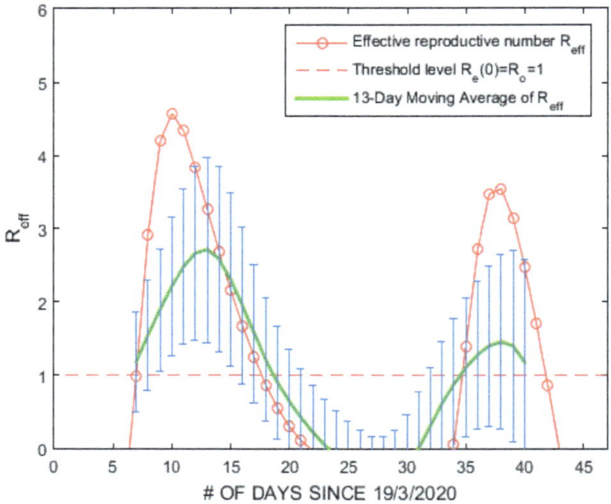

Figure 10.15. R_{eff} of the first phase using LPE-SG model from 29th March to 7th May. The green curve denotes the moving average of R_{eff} every 13 days.

• The second phase

◇ R_0

We do a similar analysis of the corrected daily tests for the second phase, as shown in Fig. 10.16. One obtains:

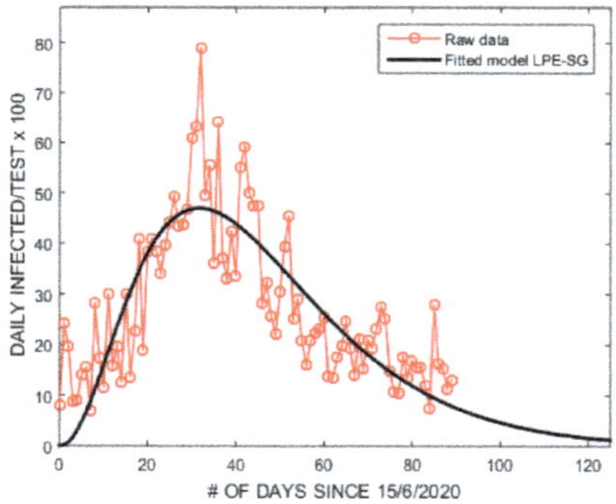

(a)

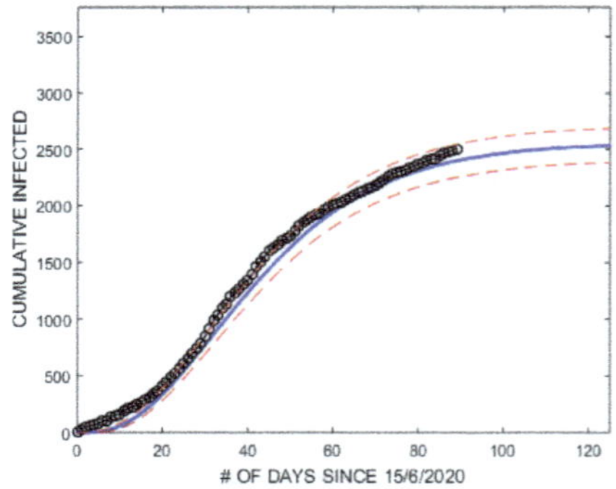

(b)

Figure 10.16. Fit of the daily (a) and cumulated (b) reported data for infected persons using LPE-SG model from 15th June to 12th September.

$$R_0 \approx 2.0, \tag{10.20}$$

and a peak around 34 days from 15th June, i.e., 121 days in total (19th July). However, the peak is obtained earlier than in Fig. 10.11 from a direct fit of the data and using different models.

◇ R_{eff}

To estimate R_{eff}, we use its definition according to the SIR model and obtain the following useful expression:

$$R_{\text{eff}} = 1 + \frac{1}{\beta}\frac{1}{I}\frac{dI}{dt} \quad \text{with} \quad \frac{dR}{dt} = \beta I, \tag{10.21}$$

where β is the recovering rate which is the inverse of the recovering time symbolized with τ_R.

– Starting with a well-adjusted value $\beta = 0.075$ days^{-1} obtained from the fit in Fig. (10.16), we show in Fig. 10.17(a) the raw data of daily infected $I(t)$ persons (red open connected circles) and the one of cured/recovered (green open connected circles) as function of time. The same data are shown in Fig. 10.17(b) but after a low pass filter obtained using MATLAB algorithm which is expected to reduce the statistical (stochastic) variations of the data. The full blue connected circles show the prediction of R_{eff} using Eq. (10.21). It is important to observe that the data of $R(t)$ — the green one — after the 40th day diverges from the expected one based on the theory of SIR. This underestimated values can be explained either by way of counting the recovered (cured) people, or by treating the rate β as a function of time and changing (decreased) suddenly for unknown reasons, leading to a serious discrepancy from the basic SIR model.

– We show in Fig. 10.18 the predicted values of R_{eff} as function of the number of days using the daily infected data (open red circles) given in Fig. 10.17. One can note like in the previous sections that the analysis from the LPE-SG models are quite sensitive to the starting and ending points where the results are very inaccurate. Around (35–45) days, the values of R_{eff} pass below one and reach a minimum around the 60th day. However, the increases above the 60th day which can be due to the non-monotonic decrease of the data beyond this date (Fig. 10.17) deserves more attention.

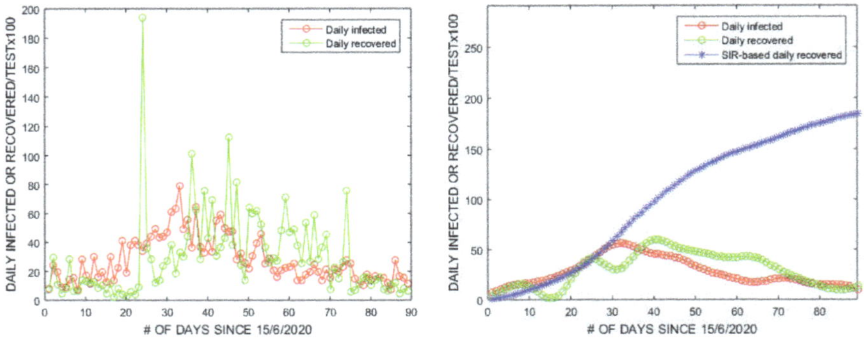

(a) (b)

Figure 10.17. The daily reported raw data (a) and the smoothed ones by using digital filter (b) included the SIR-based daily recovered from 15th June to 12th September.

10.2.6 Comparison of R_{eff} from different models

• The first phase

We compare in Table 10.1 the results of R_{eff} from SRI-like models in Fig. 10.12 with the ones from Imperial College Ref. [626] shown in Fig. 10.18(b). One can note a good agreement with the SIR-like model predictions.

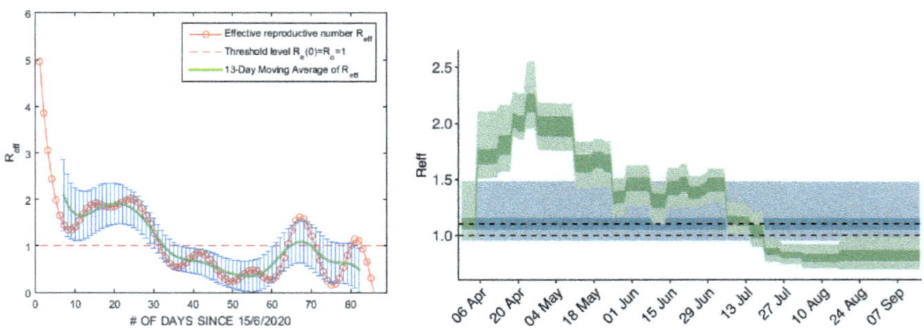

(a) (b)

Figure 10.18. R_{eff} of the second phase from 15th June to 12th September: (a) using the LPE-SG model: the green curve denotes the moving average of R_{eff} every 13 days; (b) Imperial college Ref. [626].

Table 10.1. Values of R_{eff}.

Position	SIR models	Imperial College
Starting point	0.96 ± 0.44	1.25 ± 0.25
Maximum	3.83 ± 1.1	2.3 ± 0.3
End point	2.85 ± 0.80	1.6 ± 0.4

• The second phase

◇ Imperial college

For the second phase shown in Fig. 10.18(a) from the LPE-SG model, one can note a good agreement with the ones from Imperial College in Fig. 10.18(b) [626]. We deduce from Fig. 10.18:

$$R_0 \simeq 2.2 \pm 0.6. \qquad (10.22)$$

◇ Greece and Switzerland

We consider the case of Greece and Switzerland as these countries have relatively few numbers of population ($\approx$10.42 billion for Greece and 8.57 billion for Switzerland) which may be comparable with the one of Madagascar ($\approx$27.69 billion). We show the value of R_{eff} obtained using the same LPE-SG model in Fig. 10.19 where one can note that the value of R_{eff} as a function of time are lower in Madagascar than that in Greece (GR) at the first outbreak of COVID-19:

$$R_0^{GR} \approx 2.9 \quad \text{and} \quad R_0^{CH} \approx 3.6. \qquad (10.23)$$

◇ African countries including Madagascar

We show in Fig. 10.20 the results for R_{eff} for different African countries including Madagascar from Imperial College studies where one can note that the prediction:

$$R_{\text{eff}} \simeq 1.2 \qquad (10.24)$$

agrees with our tentative average $\langle R_{\text{eff}} \rangle \simeq (1.06 \pm 0.18)$ of the first phase in Eq. (10.18) from an SIR-like model but smaller than the

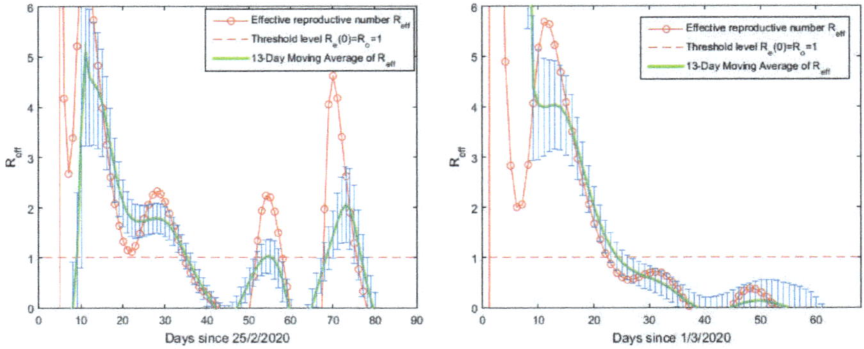

(a) (b)

Figure 10.19. $R_{\rm eff}$ of the first outbreak using the LPE-SG model from 25th February for (a) Greece; (b) Switzerland. The green curve denotes the moving average of $R_{\rm eff}$ every 13 days.

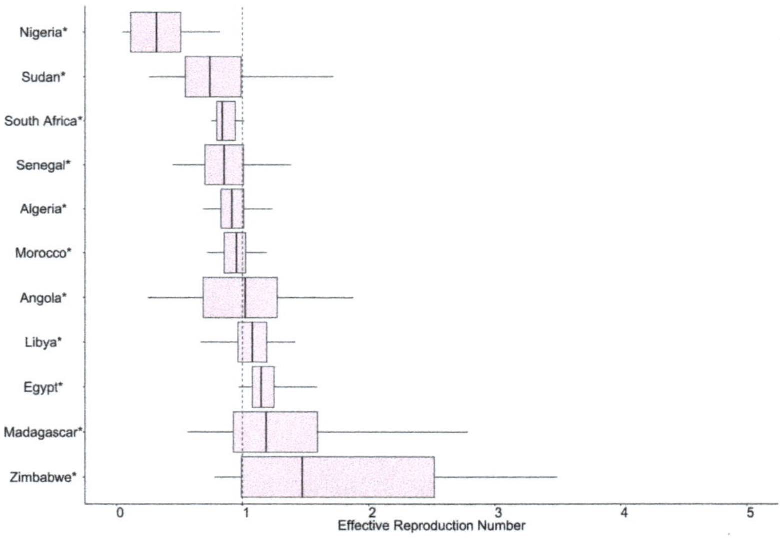

Figure 10.20. $R_{\rm eff}$ from recent analysis of different African countries from Imperial College studies.

one from the LPE-SG model of about 2. This difference indicates the difficulty of each model to provide an accurate prediction, a point discussed in Ref. [627].

10.2.7 Absolute number of deaths until 12/09/20

The data on the absolute # of deaths are shown in Fig. 10.21. A cubic polynomial fit which smears the data is also shown. One can note that the (officially declared) absolute cumulative number of deaths is hopefully small. However, the unofficial observation of the population does not support this small number and needs more clarification.

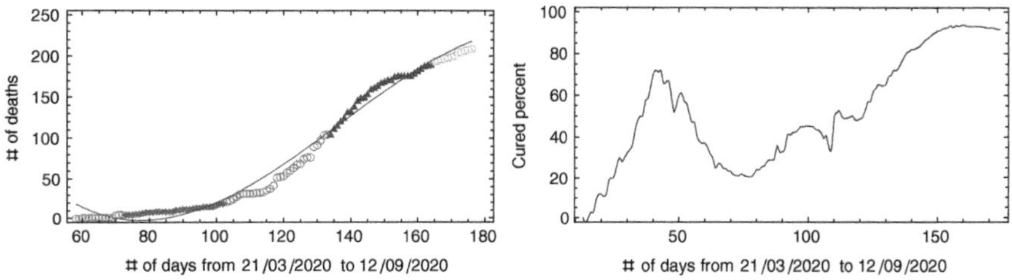

(a) (b)

Figure 10.21. (a) Daily behaviour of the cumulative number of deaths until 12th September. The oliva curve denotes a smearing of the data with a cubic polynomial. (b) Daily behaviour of the per cent cumulative number of cured persons until 12th September.

10.2.8 Per cent number of cured persons until 12/09/20

We show in Fig. 10.21 the per cent cumulative number of cured persons relative to the percent of the infected persons.

Looking at the behaviour of the number of cured persons, which reflects the performance of medical care, one can note the following:

- On the first 48th day, the medical care performance was quite good due mainly to the few numbers of infected persons.
- On the 48th day when the new pandemic phase started and the number of infected persons increased, the medical care performance decreased until the 80th day. This indicated the limitation of the hospital equipment and the limited options to face the pandemic. Note that the miraculous Covid-organics (CVO) treatment using artemisia leaves mixed with some other traditional plants (white eucalyptus and ravintsara leaves) to cure the pandemic

officially announced with great fanfare by the President of The Republic on 29th April, was inefficient and not helpful.

◇ From the 80th day (8th June), the care performance started to increase and became excellent after the 150th day just beyond the peak for infected persons around 110 days (8th July). Many possible effects may have contributed to this success such as the following:
◇ uses of Professor Didier Raoult protocol based on hydroxy-chloroquine and azythromycine;
◇ more strict sanitary measures (social distancing, masks, etc.) including confinement.
◇ obtention of more medical equipment thanks to the call of the former health minister;
◇ relative decrease of the number of infected persons when approaching the peak.

Chapter 11

The QCD Montpellier Conferences

11.1 The First Initiatives

(a)

(b)

Figure 11.1. (a) Poster and Proceedings of the first Montpellier Conference on Non-perturbative Methods; (b) Local Committee Members of the first Montpellier Conference on Non-perturbative Methods: Left to Right: Fernand Michel Renard, Stephan Narison, Jacques Layssac, Gerard Mennessier.

• Once established in Montpellier in 1984, I invited the HEP members in Montpellier (F.M. Renard, J. Layssac and G. Mennessier: Fig. 11.1) to co-organize the first QCD Montpellier International Conference. The international advisory committee members were D. Amati (CERN), G. Furlan (Trieste), P. Pascual (Barcelona), E. de Rafael (Marseille), G. Veneziano (CERN) and F.J. Yndurain (Madrid). The conference was entitled "Non-Perturbative Methods", which is mostly theoretical, where the different non-perturbative approaches of QCD have been discussed. The proceedings was published by World Scientific Publishing (see Fig. 11.1).

◦ The conference was successful according to the participants. This fact encourages us to continue. We have taken five years decision to restart the organization of the conference with the name QCD conference.

◦ We have modified the contents of the conference by ensuring an equal amount of theoretical and experimental talks. We

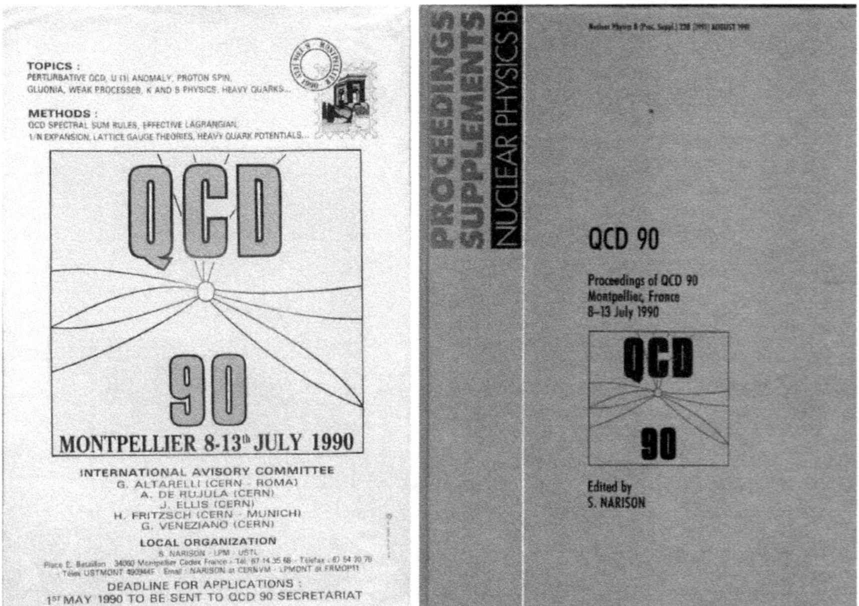

Figure 11.2. Poster and Proceedings of the first QCD Montpellier Conference.

have taken the format of a medium size conference of around 100 participants.

∘ We have also encouraged the participation of young post-docs and Ph.D. students who present for the first time their talks at the international level and in front of experts in the field. We had 90 participants for QCD90.

∘ We have again stopped the organization for four years and reorganized the QCD94 conference, with the decision to do it every two years.

11.2 Structure of the Organization

11.2.1 Founder and Chairman

Stephan Narison (LUPM-CNRS Montpellier and iHEPMAD-University of Antananarivo).

Figure 11.3. Founder and Chairman of the QCD conference.

11.2.2 International committee members

Figure 11.4. Some committee members of the QCD conference; a complete list of names can be found in the proceedings of QCD20. A. Di Giacomo, A. Pich, V.I. Zakharov are long-term members, and S. Bethke, E. de Rafael, J.M. Richard, M. Knecht, and G.C. Rossi to a lesser extent.

11.2.3 Local teams

Figure 11.5. Secretary and hostesses of the QCD conference.

11.2.4 Sponsors

Figure 11.6. Some sponsors and institutions of the QCD conference.

11.3 Prestigious participants

11.3.1 Nobel prize winners

Figure 11.7. Nobel Prize winners: J. Steinberger (QCD97), G. 't Hooft (QCD98, QCD02) and D. Gross (QCD98).

11.3.2 American Physical Society: Sakurai Prize winners

Figure 11.8. From left to right: M.A. Shifman (Sakurai Prize, 1999), V.I. Zakharov (Sakurai Prize, 1999), S. Narison, H.G. Dosch; S. Narison, S. Brodsky (Sakurai Prize, 2007).

11.4 The Euroconference Period (1996–2000)

In 1996, we got the first Euroconference grants for three years, which were exceptionally extended for two more years. However, we have been pursued by the European commission to organize it annually. During this period, the conference acquired an international reputation and got the participation of well-known QCD experts and Nobel Prize winners (see Figs. 11.7 and 11.8).

11.4.1 QCD97–98: 100th anniversary of the electron and 25th anniversary of QCD

In 1997 and 1998, we celebrated the 25th anniversary of QCD and the discovery of asymptotic freedom. There was an interesting debate between G. 't Hooft and D. Gross about 't Hooft's remark at the Aix-en Provence conference on the negative value of the β function for less than 16 quark flavours. However, D. Gross, D. Politzer and F. Wilczek got the Nobel Prize for asymptotic freedom in 1998, while G. 't Hooft and M. Veltman got the prize for their works on renormalization of gauge theories in 1999. G. 't Hooft came back to Montpellier for QCD 2002 and I learned that he came incognito to visit Madagascar later on.

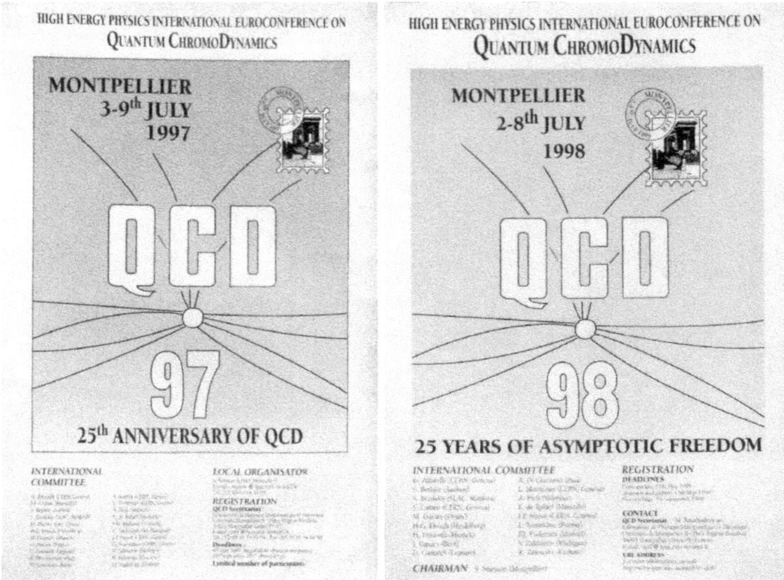

Figure 11.9. Posters of the QCD97 and QCD98 Montpellier Conferences.

Figure 11.10. Professor G. 't Hoot (left) and the Vice Major of Montpellier (right) at QCD02.

11.5 The QCD International Conference from 2002

The contract with the European Commission ended in 2000. However, due to the success of the events, we were encouraged to continue but without much budget. I show in Fig. 11.6 the previous sponsors of the conference. At present, we do not have any budget but survive with the conference fee of the participants. The University of Montpellier and the Association Culturelle Malgache of Montpellier which I founded in 1990 help us for the logistics. Despite this budget difficulty, we continue to run (as long as we can) the QCD Montpellier conference!

Figure 11.11. Group photo of QCD02 at the Palais des Congrés — le Corum of Montpellier.

11.5.1 QCD10: 25th anniversary of the QCD conference and 30 $\oplus$ 1 years jubilee of the SVZ sum rules

That year we celebrated the 25th anniversary of the QCD conference together with the 31 years of the SVZ sum rules. M.A. Shifman and v.I. Zakharov honoured the events by their presence (see Fig. 11.8) as well as some other QCD experts (H.G. Dosch, E. de Rafael, etc.).

Figure 11.12. Group photo of QCD10 at the Administration of the CNRS, Montpellier.

11.5.2 QCD12: Higgs Boson discovery

The Higgs boson discovery has been announced at our conference before the large-scale ICHEP conference.

CERN Courier November 2012

Faces & Places

CONFERENCE

QCD 12 and Higgs-like happiness

Amid the beauty of Montpellier in France, the 16th International Conference in Quantum ChromoDynamics, QCD 12, took place on 2–6 July. It brought together theorists and experimentalists from laboratories around the world, all working actively within QCD. Mid-way through the conference, a live webcast from CERN thrilled participants with the news of the discovery of a Higgs-like boson.

Participants pose in the Montpellier sunshine. (Image credit: Alize Photo.)

Figure 11.13. CER-Courier report and Group photo of QCD12 at the Administration of the CNRS, Montpellier.

11.5.3 QCD15: 30th anniversary of the QCD conference

Figure 11.14. Group photo of QCD15 at the Administration of the CNRS, Montpellier.

Figure 11.15. Personalities of Montpellier.

That year we celebrated the 30th anniversary of the QCD conference in the presence of different personalities: Vice Major, Mme Mylène Chardes, Regional Delegate of the CNRS, Mme Gibello, Vice President of the Region Languedoc-Roussillon, Mme Anne-Yvonne Ledain, Director of the LUPM — Mr. Denis Puy and President of the University — Mr. Philippe Augé.

11.5.4 QCD20: 35th anniversary of the QCD conference

That year was quite particular due to the COVID-19 period. The organization had completely changed as the conference was done online via Zoom. It has been a good experience for us, but we regret the traditional in-person format.

Figure 11.16. Proceedings of QCD15 and QCD20 conferences.

11.5.5 QCD21–23

Figure 11.17. Group photo of QCD22 at the University of Montpellier.

Figure 11.18. Group photos of QCD23 and QCD24 at the Faculty of Sciences, University of Montpellier.

After some thoughts, we have organized QCD21–23 as hybrid conferences (online and in presence). We encourage the participants to come back slowly to a more friendly traditional one.

11.5.6 QCD 25 - 40th Anniversary

This year 2025, we have organized QCD25 where we celebrated the 40th anniversary of the QCD conference initiated in July 1985. This new events is important for the QCD-Montpellier Conference Series.

LHCb
CMS
ATLAS

UNIVERSITÉ DE MONTPELLIER Campus TRIOLET

11.6 The Social Events of the QCD conference

The social events of the QCD conferences are traditionally organized as follows:

11.6.1 Welcome Reception

The idea is that the participants discover some typical European and Malagasy food and beverages.

Figure 11.19. Some photos of the welcome reception food and beverages.

Figure 11.20. Some photos of the welcome reception.

11.6.2 Visit of the Old Montpellier

The second day after the conference sessions, participants are guided by a professional to visit the old part of Montpellier.

Figure 11.21. Some photos of the visit of the old Montpellier.

11.6.3 Excursion

On the fourth day after the conference sessions, participants are brought to an excursion either along the Mediterranean beach or inside the countries.

Figure 11.22. Excursion: (a) Midi channel; (b) Carnon beach, Aigues Mortes; (c) Gard bridge with G. t'Hooft, P. Minkowski; (d) Flaujergues castel.

11.6.4 Wine Cave Visit and Wine Tasting

Figure 11.23. Some photos of the wine tasting at the Region of Pic Saint-Loup: (a) Chateau de Cazeneuve-Lauret; (b) Chateau Laroque-Fontanés; (c) Sauzet Domain-St Bauzille de Putois.

11.6.5 Glass of friendship offered by the Major of Montpellier

The Major of Montpellier offers regularly a "glass of friendship" to the participants before or after the closing session of the conference.

(a)
(b)

Figure 11.24. (a) Corum — Congress Palace QCD97; (b) Room Nogaret QCD14.

11.6.6 Concert, Music and Dance

Before the COVID-19, we always organize a dinner party with some dancing spectacle for the participants.

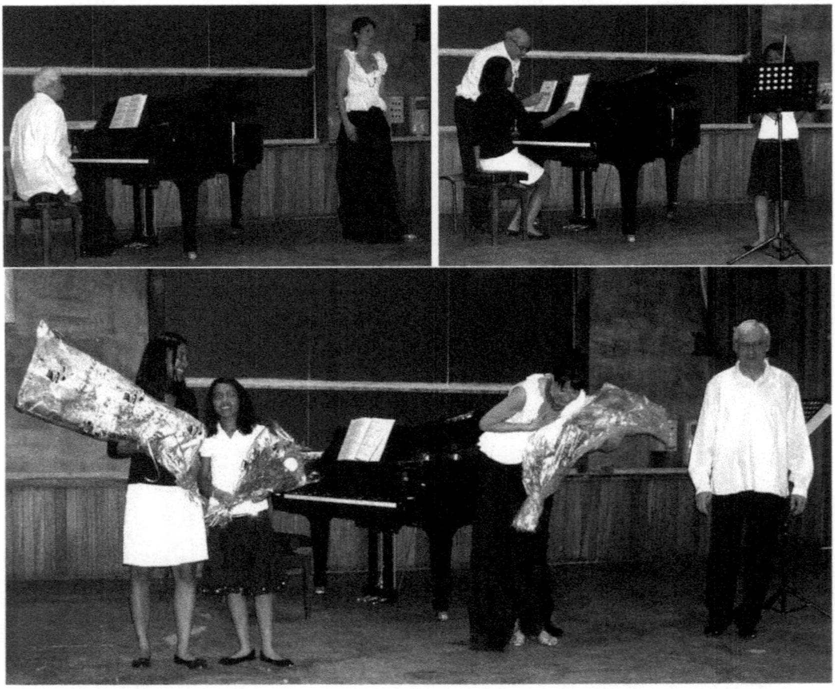

Figure 11.25. Lyric concerts in 2008 at le Corum.

Figure 11.26. Oriental dance at Le Clos des Hirondelles and Mas le Saporta.

Figure 11.27. Tahitian dances at Flaugergues Castle and Mas le Saporta.

Figure 11.28. Africa Dances and Folks at Salle de Nogaret.

(a)

(b)

Figure 11.29. (a) Malagasy dances and Folks at le Corum by the trio Justin Vali; (b) Malagasy dances by the Catholic Association (AKAMA) of Montpellier.

11.6.7 Aperitif at Mas de Saporta

(a)

(b)

(c)

Figure 11.30. Mas de Saporta: (a) QCD96; (b) QCD10; (c) QCD16.

11.6.8 Dinner at Mas de Saporta

(a)

(b)

(c)

(d)

Figure 11.31. Mas de Saporta: (a) QCD96; (b) QCD10; (c) QCD17; (d) closing of QCD23.

Chapter 12

Developing HEP in Madagascar

12.1 The HEPMAD Conference Series

Being in Trieste and after meeting Professor A. Salam, *my dream was to develop high-energy physics (HEP) in developing countries* when I shall have the possibility to do so. I tried to realize this dream by taking the example of Madagascar, my native country. Then, in parallel with the QCD-Montpelleir conference, I have tried to start the HEPMAD international conference series in Madagascar in 2001 at the University of Antananarivo. I took this decision in 2001 after some discussions with G. Furlan (ICTP-Trieste), J. Ellis (CERN), some of my collaborators and some members of the University of Antananarivo.

12.1.1 HEPMAD01

In 2001, I have decided to start the HEPMAD conference in Antananarivo. J. Ellis, most of the committee members of the QCD conference (A. di Giacomo, H.G. Dosch, H. Fritzsch, P. Minkowski, L. Trentadue) and QCD participants (F. Navarra, M. Nielsen) and some overseas participants came. The conference was successful as the President of the university Professor P. Rakotobe has engaged the university for the organization. It was a hybrid meeting as, in addition to HEP, national physicists in different fields (nuclear physics, environments, etc.) present their local research at the international level. The contribution at the HEPMAD conference is also useful for Ph.D. students in order to validate a publication for their thesis.

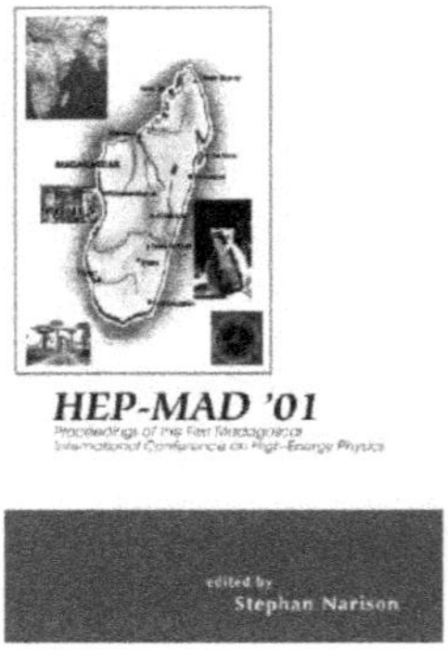

(a)

(b)

Figure 12.1. Some photos of the HEPMAD01 conference held at the University of Antananarivo (Madagascar): (a) Proceedings; (b) left to right: Professor E. Rakotomahanina (former president of the university), Professor J. Ellis (CERN), S. Narison, Professor P. Rakotobe (president of the university), Professor Raoelina Andriambololona, Ministry of Higher Education; (c) group photo.

I show in Fig. 12.1 some photos of this first HEPMAD01 meeting. The proceedings were published by (c) World Scientific Publishing (Singapore).

Figure 12.2. Group photo.

12.1.2 The next HEPMAD conferences

After the experience from the first meeting, we suspended the project and restarted in 2004. The series continued in 2007 every two years until 2015. From 2015, we decided to organize it annually.

• The spirit of the talk is to be an introduction to the field, i.e., pedagogical, but at the same time, the speaker presents new results. From 2004 to 2022, the proceedings of the conference were published online by eConf-SLAC (see the links before Fig. 12.4).

• The participation in these conferences has been a benefit to the local researchers who can take the opportunity to present their work at the international level. For the Ph.D. students, a presentation of a talk at the HEPMAD conference is recognized as a publication to complete the validation of their thesis.

• We celebrated HEPMAD16, the 15th anniversary of the conference at the University of Antananarivo (Fig. 12.3).

(a)

(b)

Figure 12.3. (a) Group photo of HEPMAD07 at the Ministry of Foreign Affairs; (b) HEPMAD16 at the University of Antananarivo.

• Following are the eConf SLAC links to the HEPMAD proceedings:

https://inspirehep.net/conferences/973738
https://www.slac.stanford.edu/econf/C0409272/
https://www.slac.stanford.edu/econf/C0709107/
https://www.slac.stanford.edu/econf/C0908211/
https://www.slac.stanford.edu/econf/C110825/
https://www.slac.stanford.edu/econf/C130904/
https://www.slac.stanford.edu/econf/C150917/
https://www.slac.stanford.edu/econf/C161013/
https://www.slac.stanford.edu/econf/C170921/
https://www.slac.stanford.edu/econf/C180906/
https://www.slac.stanford.edu/econf/C191014/
https://www.slac.stanford.edu/econf/C221102/

Figure 12.4. Group photos for the official opening of the HEPMAD19 and HEPMAD22 conferences at the Malagasy Academy (Antananarivo).

Figure 12.5. Group photos for the HEPMAD19 conference session at the HEPMAD Cultural Center (Amboara, Ambohidratrimo).

Figure 12.6. Group photos for the HEPMAD22 conference session and the HEPMAD23 and HEPMAD24 official openings at the HEPMAD Cultural Center (Amboara, Ambohidratrimo).

Since 2019, we inaugurate the new HEPMAD Cultural Center which I hosted in my own property at Ambohidratrimo, 10 km from the capital. Then in 2019 and 2022, we do the official opening of the conference at the Malagasy Academy (Fig. 12.4) and the conference sessions at the HEPMAD Cultural Center (Fig. 12.5).

From 2023, the complete conference including the official opening is held at the HEPMAD Cultural Center. The centre is a quiet place, 10 km from the capital. It is located by the lake Andranotapahana where a lodging for the participants is also available.

12.2 The HEPMAD Social Events

12.2.1 Opening ceremony

The official opening of the conference is very formal in Madagascar. Very important personalities (VIP) including the press and TV journalists are invited. Usually, we conduct this ceremony at the Malagasy Academy. Since 2019, we have been conducting it at the new HEPMAD Cultural Center (see photos in Figs. 12.4 and 12.6).

Figure 12.7. Opening of (a, c) HEPMAD11 at the Ministry of Foreign Affairs; (b, d) HEPMAD16 at the Malagasy Academy.

12.2.2 Visit to the Ambohimanga Palace

Figure 12.8. Visit to the Ambohimanga Palace.

It is about 24 km northeast from the capital. You can find traditional king's house next to more modern queen's palace built by Jean Laborde. This is a UNESCO site.

12.2.3 Visit to the Reserve Peyrieras at Marozevo

This reserve is about 60 km on the east of the capital.

Figure 12.9. Visit to the Reserve Peyrieras at Marozevo.

12.2.4 Malagasy Dances and Folks

Figure 12.10. Groups Odeamson, Andry Mahitsy, Bakomanga.

12.3 Creation of the iHEPMAD Institute

From the enthusiasm and requests of participants and national colleagues, we decided to create the research institute of iHEPMAD in 2004 for admitting masters and Ph.D. students without going abroad. This institute is unique in the Indian Ocean and Sub-Saharan African countries except South Africa and more recently Rwanda.

12.3.1 Challenges and difficulties

However, there are many challenges and difficulties to face:

- no national experts in the field for selecting students;
- no financial support from Mundial bank, FMI, European commission for fundamental researches;
- no financial support from successive Malagasy governments who do not understand the usefulness of the field.

Figure 12.11. Group photos of the masters and Ph.D. students; Thesis defense at the university.

12.3.2 Achievements

• HEPMAD conferences and iHEPMAD institutes

The HEPMAD conference achieved some success, which stimulated to regularly continue the conference series. It also motivated us to create the iHEPMAD Research Institute upon the requests of professors and students to admit masters and Ph.D. students who do not need to go abroad.

However, creating a new branch of physics was not an easy administrative task. After 1–2 years, we finally succeeded to settle HEP at the University of Antananarivo. In Fig. 12.11, we show some masters and Ph.D. students of the iHEPMAD institute.

• Cooperation with CERN, Geneva (CH)

A cooperation with CERN was very helpful. The CERN experimental group sent regular participants to the HEPMAD conferences (ATLAS, CMS, ALICE, LHCb and NA48) and hosted students from Madagascar at the CERN summer school and at the LHCb experimental group. Unfortunately, some of them only joined the theory group. Some of them remained abroad or left physics for a job opportunity.

• Cooperation with the ICTP of Trieste (IT)

For this action, I benefitted from grants of ICTP consultant for Madagascar for a few years from 2017 to 2020 for the travel and living expenses. Students from Madagascar also come to visit Trieste and go to the summer school.

We also got a grant from the World Academy of Sciences (TWAS) in 2018 for equipping the new HEPMAD Cultural Center founded in 2017.

• Cooperation with Fermilab (US)

During their participation at the CERN summer school, some of the iHEPMAD masters students were taken to prepare their Ph.D. at the DUNE neutrino experiment at FERMILAB. Unfortunately, all of them remained at FERMILAB but did not come back for developing the iHEPMAD institute.

• Some other cooperations

Some of the iHEPMAD students participated at the African School of Physics where they have been noted to be excellent students. Again, some of these students left for South Africa and US.

12.4 Popularization of HEP in Madagascar

12.4.1 The HEPMAD-TOUR since 2010

To complete the promotion of HEP, one has to give some public talks in different national institutes (Malagasy Academy, Alliance Francaise, and public and private universities and high schools in different cities). This action has been accomplished thanks to a consultant grant from ICTP, Trieste and the logistics help of the members of the iHEPMAD institute and of the Association Gasy Miara-Mandroso (AGMM), which I created in 2010.

The requests of different people stimulated me to write a popular book on particle physics (Fig. 2.1).

12.4.2 Malagasy Academy, Antananarivo

I delivered public talks at the Malagasy Academy in 2015 and 2017 (see Fig. 12.12).

Figure 12.12. Introduction to particle physics at the Malagasy Academy (Antananarivo — 2015).

I became a full member of the Malagasy Academy in 2019 in the presence of the President of the Academy (on my right) and the Prime Minister (dark clothes on my left near the flag).

Figure 12.13. Ceremony for the nomination to be a Full Member Malagasy Academician (Antananarivo — 2019).

12.4.3 Malagasy Academy, Fianarantsoa

Figure 12.14. Talk at the Academy of Fianarantsoa (2012).

12.4.4 The alliance francaise

Figure 12.15. A.F. Andavamamba and Antsahabe (Antananarivo).

Pursuing the popularization of HEP, I delivered public talks at different Alliance Francaise (A.F.) of Antananarivo, Fianarantsoa, Majunga, Fort-Dauphin, Tulear and Tamatave. Some photos are given in the following.

Figure 12.16. A.F. Tulear 2012 and 2018.

Figure 12.17. A.F. Fianarantsoa 2012; A.F. Fort-Dauphin 2018.

Figure 12.18. A.F. Tamatave 2010.

12.4.5 Public universities

• Antananarivo 2017

Figure 12.19. Faculty of sciences — University of Antananarivo 2017.

• Antananarivo 2022

Figure 12.20. Faculty of sciences — University of Antananarivo 2022.

• **Fianarantsoa 2017**

Figure 12.21. Faculty of sciences — University of Fianarantsoa 2017.

• **Tulear 2018**

Figure 12.22. University of Tulear 2018.

- **Diego-Suarez 2013, 2016**

Figure 12.23. IST-Diego 2013 and University of Diego 2016.

- **Tamatave 2010**

Figure 12.24. Faculty of Economy — University of Tamatave 2010.

12.4.6 Association and religious universities

• ARUPE Cultural Center Antananarivo 2022

Figure 12.25. Eckar Catholics Cultural Center ARUPE (Faravohira-Antananarivo 2022).

• Lutheran University of Theology Antsirabe 2017

Figure 12.26. Lutheran University of Theology Betafo 2017.

The students of the University of Theology were very open-minded when I gave the talk on the Big Bang hypothesis as the origin of the universe. After the talk, some of them asked from which scale one can believe the God' creation of the universe.

- **ESSVA Catholic University of Antsirabe 2012–2016**

Figure 12.27. ESSVA Catholic University of Antsirabe.

- **ASJA Catholic University of Antsirabe 2012–2016**

Figure 12.28. ASJA Catholics University of Antsirabe.

- **Lutheran Agronomist University of Antsirabe 2018**

Figure 12.29. Lutheran Agronomist University of Tombotsoa 2018.

- **Ecole Normale Superieure 2016**

Figure 12.30. Ecole Normale Superieure 2016.

12.4.7 High schools

- **Pere Pedro school (Akamasoa-Antananarivo 2017)**

This school is particular. Pere Pedro is internationally well known for his association with helping the poor who he took from the streets and helped to have a decent life (building their own house, education of their children, etc.). These students are the children of these poor people. I was impressed that they can follow the talk and asked some relevant questions.

Figure 12.31. Pere Pedro school (Akamasoa-Antananarivo).

• Antananarivo 2010–2018

◇ Lycée Galliéni 2010, 2014

In Antananarivo, my first reflex was to give a talk at my former high school Lycée Galliéni renamed Lycée of Andohalo (the name of the place). Indeed, the successive governments wanted to erase the trace of colonialism in all public places, streets and institutions. Alas! in my opinion, in this way, the country will not have any history.

Nevertheless, Lycée remained the same like the auditorium where I gave my talk (see also Fig. 1.9). When I visited Lycée, I saw that our dormitories had become classrooms. Visiting the physical science practical work rooms, I realized that there was no material at all except the old instruments which we used before (voltmeter, amperemeter, etc.), which was disappointing.

Figure 12.32. Lycée Gallieni.

◇ Some other Lycées

(a)

(b)

Figure 12.33. (a) Lycée Moderne Ampefiloha (LMA) 2019 where I regularly gave a talk; (b) Protestant Lycée Paul Minault 2016.

Figure 12.34. Lycée: (a) Rabearivelo; (b) Nanisana; (c) St-Antoine (Ankadifotsy); (d) Francais (Ambatobe).

(a)
(b)
(c)
(d)

Figure 12.35. (a) Lycée Jules Ferry 2016; (b) Peterpan Andoharanofotsy; (c) Public Andoharanofotsy; (d) Pepito Tanjombato.

Figure 12.36. Antananarivo 2016: (a) Lycée FJKM Rasalama; Lycée Ambohimanarina; (b, c) Lycée Ivato; (d) Lycée Talatamaty.

Figure 12.37. Conférence de Vulgarisation de HEP à Antananarivo: (a) College Saint-Michel; (b) College Sainte-Famille.

• Analamanga region

Figure 12.38. (a) Lycee Andramasina 2012; (b) Lycee Behenjy 2012.

(a)

(b)

(c)

Figure 12.39. Lycées d'Ambohidratrimo: (a) Public; (b) Protestant FJKM; (c) Private Regency.

(a)
(b)
(c)
(d)

Figure 12.40. (a) CEG (Ambohidratrimo); (b) Lycée Technique (Ambohidratrimo); (c) Lycée publique (Mahitsy); (d) Collège Sacré Coeur (Mahitsy).

• Itasy and Bongolava regions

Figure 12.41. Itasy and Bongolava regions.

- **Vakinankaratra region**

Figure 12.42. Lycée André Resampa of Antsirabe (LARA) 2018; Mandaniresaka and Andranomanelatra (2018).

(a) (b) (c) (d)

Figure 12.43. (a) Lycée FJKM Ambohimandroso 2018; (b) Sacré Coeur Ambohimandroso; (c) Public Antanifotsy; (d) Protestant Antanifotsy.

- **Amoron'i Mania region**

Figure 12.44. (a) Lycée Ambositra; (b) Sandrandahy.

- **Haute Matsihitra region**

Figure 12.45. Lycée Ambohimahasoa.

(a)
(b)
(c)
(d)

Figure 12.46. (a) Alakamisy Ambohimaha; (b) Ambalavao; (c) Raherivelo Fianarantsoa; (d) Ankofafa.

- **Fort-Dauphin 2013, 2016**

(a)

(b)

(c)

Figure 12.47. Lycées of Fort-Dauphin: (a) Clairefontaine; (b) Pole; (c) Private Fanamby.

- **Tulear 2012–2018**

Figure 12.48. Lycées and Colleges of Tuléar: (a) Botokeky; (b) Sacré-Coeur; (c) Rabe Jules; (d) Flacourt.

Figure 12.49. (a) Lycée Notre-dame; (b) College Memorial, Lycée Technique; (c) Antaninarenina; (d) Mahavatsy, Othodoxe; (e) Betania, Private College Fanantenana.

• Diana Region 2013, 2016

(a)
(b)
(c)
(d)

Figure 12.50. Diego-Suarez: (a) Lycée Francais; (b) Lycée Mixte; (c) CEG; (d) CEG and Lycée of Anivorano.

- **Majunga 2011**

(a)
(b)
(c)
(d)

Figure 12.51. Lycées and Colleges of Majunga: (a) St-Gabriel and Notre-Dame; (b) Public P. Tsiranana; (c) Public Technique; (d) FJKM.

Chapter 13

Associations and Societal Activities

To support our project for the social activities, we have created different associations.

13.1 Associations in Montpellier

13.1.1 Malagasy Cultural Association of Montpellier (ACMM)

Created in 1990 at the Prefecture of Montpellier under the French law 1901, the non-profit Association Culturelle Malgache de Montpellier

(ACMM) is officially registered with the reference number 450 154 380 00015. It aims to:

• Help newcomers Malagasy citizens to integrate in the French system and society. This help concerns the preparation of administrative documents, the search for housings and the regular organization of Malagasy community events and meetings, ...

• Promote Malagasy culture (musics, dances, arts, books, ...) and handicraft products.

13.1.2 Quantum ChromoDynamics Association (QCD)

Most of the budget for running the QCD-Montpellier conferences from national institutes (CNRS, Universities, Major of Montpellier, Regional Department) have been suspended in 1996. Then, we founded the QCD non-profit association registered under the 1901 law at the Prefecture of Montpellier with the reference number: 794 120 642. It aims to:

• Help the management of the organization of the QCD conferences (local organization, and regional administrations, negotiations with local politicians, ...).

• Logistics for the organization (secretaries, hostesses, manual works, ...).

13.2 Malagasy Association for Development (AGMM)

Raiso an - Tanana ny Hoavinao sy ny Firenenao
Take Hold your Future and your Country - Prends en Main ton Avenir et ta Nation
Let's Evolve Together - Evoluons Ensemble
Ny Fanabeazana no Fototry ny Fandrosoana
Education is the Basis of Development - L'Education est la Base du Développement

We have also officially created the association Association Gasy Miara-Mandroso (AGMM) (which means Malagasy who develops/ advance together) in 2011 in Madagascar under the prescription number 60.133 of 03/10/1960. This association aims is a necessary platform for our project to develop High-Energy Physics and, in general, Science in Madagascar. It aims to:

- Organize the HEPMAD conference series started in 2001.
- Organize the different popularizing presentations at different universities, high-schools and public institutes in various cities of the country.
- Give a common platform to those who wish to cooperate for developing Science and some other actions for the country.

Figure 13.1. Some members of the AGMM association in Antananarivo.

We should note that the previous associations have no specific funds but run with volunteering and formal salary.

13.3 Author's Defense of Environment and Society

In addition to previous activities, I actively defend the protection of the rich and rare biodiversity of Madagascar as well as the respect of human rights, dignity and health of the Malagasy population.

• Since 2019, I and some other national and international associations fight against the exploitation of ilmenite sands for extracting Rare earths done in Madagascar by Multinational companies such as Rio Tinto in Fort-Dauphin (south east), Anpasidava (north) and the project Base Toliara south-west regions of Madagascar.

• Taking the case of the Base Toliara project, it will spoil the centenary rare baobab trees species and impact the health of the habitants due to the high rate of radioactivity above the one toloerate by human bodies (IAEA norm) induced by the Uranium 238 and Thorium 232 contained in these minerals. To these radioactivities, one should also add the large amount of water and chemical products needed to purify these minerals which will certainly affect the phreatic nappa, the biodiversity and the health of the population.

• Another negative impact of the project is the displacement of the population and (in the long term) the disappearance of the Tribu Mikea classified like the Arborigines of Australia and the Indians Skin Red of USA.

Figure 13.2. Forest of baobabs at the Ranobe site - Tulear

13.4 Author's Political Commitment

Being an engaged citizen, I cannot observe passively the present drastic situation of the country where:

- The human rights and freedom of expression are not respected. Once, you criticize severely the government, you go to the jale. The Democracy remains only a nice word but in practice there is a form of dictatorship.
- The population becomes poorer and poorer. 80% of the population leave below the minimal allowed according to the Mundial Bank. Madagascar is among the poorest country in the world despite his various rich natural resources and rare species of biodiversity. The few persons having the political power or surrounding the power become rapidly rich because of the corruption.
- The economy collapses.
- The national social insecurity increases.

Figure 13.3. This book (in French) gives a brief survey of the history and geography of Madagascar and discusses its different economical and political situations. The author also presents in this book his program for the development of the country during his propaganda for the Presidential election of 2018. The book is available on amazon website.

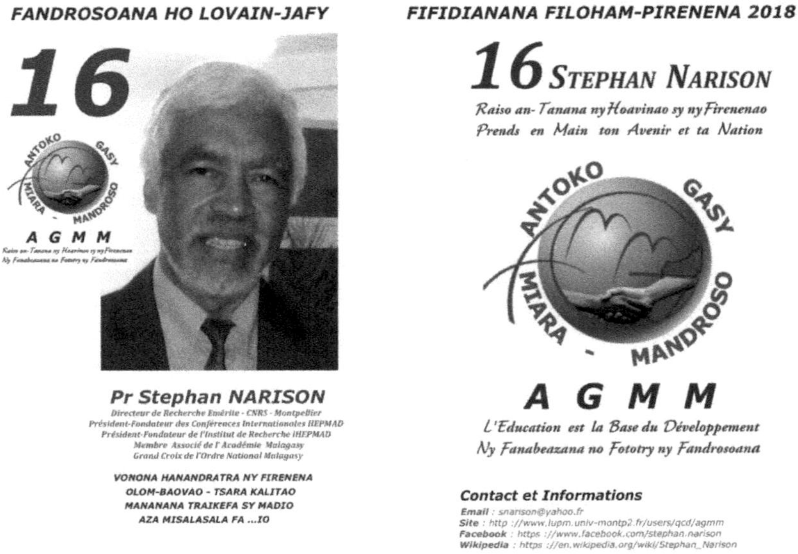

Figure 13.4. Flyer for the Malagasy Presidential election in 2018.

Motivated by the societal action of the association, the members of the association and some other followers encouraged me to run for president of the Republic in 2018.

The societal program for this election can be found in the book (see the cover in Fig. 13.3).

Alas in this country like any other developing countries, especially in Africa, we have a Populist politics. The people do not really care about your societal program but instead on the money and presents you can offer to them during the propaganda which (of course) a poor physicist cannot and shall not do at all.

Then, we (as expected) lost. Nevertheless, participating at this election was a very enriching experience!

The present present economical and social situations of the country are despairing.

Our association continues to fight against the different forms of injustice there. We and others who like the country hope that we will succeed in the future!

I close this chapter by this flyer of the 2018 Presidential Election in Madagascar with the slogan:

Development for a Legacy of Grandchildren.

Chapter 14

Memberships and Honours

14.1 Malagasy Legion of Honour

Despite the fact that the successive malagasy governments do not financially support the HEPMAD project, they recognize its usefulness by first attributing the title of Commendeur in 2006, Grand-Officier in 2011 and the most prestigious Grand Cross of second class in 2016 (see the following figure).

14.2 Malagasy Academy of Sciences

Corresponding member of the Malagasy Academy of Sciences since 19, the author becomes full member in 2019 (see Fig. 12.13 and the following figure).

14.3 Other Science Societies

The author is a member of the European, French Physical Society and the New York Academy of Sciences. He is listed in *Who's Who in Science and Engineering* (1998/1999 edition) by the Marquis Who's Who Publication Board.

14.4 American Biographical Institute

He was named by the American Biographical Institute (ABI): Distinguished Leadership for his outstanding contributions to contemporary society in 1996 and as Outstanding Man of the 20th century in 1998 for his profound contributions to society. In 1998, he was also named as one of the Outstanding people of the 20th century by the International Biographical Center (IBC) Cambridge, England in tribute to his exceptional contribution to science.

14.5 ICTP Trieste and University of Sao Paulo

The author was a consulting researcher of ICTP Trieste for Madagascar for few years and received the World Academy of Science (TWAS) Prize from ICTP Trieste (Italy) for the development of science in Madagascar.

He was rewarded by the University of Sao Paulo (Brazil) in 2013 for the co-supervision within the French-Brazil cooperation for the best university thesis of Dr. R. Albuqueqrque.

twas The World Academy of Sciences

Our Ref.: 17-541 RG/PHYS/AF/AC_G – FR3240300146
(please quote in all correspondence)

15 January 2018

Dear Dr. Narison,

It gives me great pleasure to inform you that your project:
1. Quantum Chromodynamics (QCD) and Hadron Physics
2. Organization of the Hepmad International Conference Series
has been selected by the Advisory Committee for the TWAS Research Grants Programme. This grant is possible thanks to the generous contribution of the Swedish International Development Cooperation Agency (Sida), without whose continuous support this grant would not be feasible.
The selection of TWAS Research Grants is highly competitive, based purely on scientific merit, and has been selected out of a very large number of applications. Please accept our most heartfelt congratulations.

Yours sincerely,

Romain Murenzi
Executive Director

14.6 A Memory Letter from and Photos with some Malagasy Personalities

Lettre of Prof Albert Ratsimamanga 25/02/1996
Directeur de Recherche CNRS in Biology - France

REPUBLIQUE DE MADAGASCAR

INSTITUT MALGACHE DE RECHERCHES APPLIQUEES
(Fondation Albert RAKOTO-RATSIMAMANGA RECONNUE D'UTILITE PUBLIQUE)

CENTRE NATIONAL DE RECHERCHE DE L'ORGANISATION MONDIALE DE LA SANTE
CENTRE REGIONAL DE RECHERCHE DE L'ORGANISATION DE L'UNITE AFRICAINE
Médecine traditionnelle - Nutrition
Avarabohitra - ANTANANARIVO - MADAGASCAR

Président:
Professeur Albert RAKOTO-RATSIMAMANGA
de l'INSTITUT DE FRANCE
Membre correspondant de
l'Académie des Sciences et de l'Académie de Médecine

INSRE: 972/02/8053963
Boite postale: 3833
Tél: (2) 268- 06
Fax: (2) 359 - 74

Antananarivo, le 25 Février 1996

Monsieur Stephan NARISON
Directeur de Recherche au CNRS
LPM-UM2-MONTPELLIER
63, Rue du Mas du Juge-34 980 St Gely du Fesc

Cher Collègue et Ami,

C'est avec un grand plaisir que j'ai reçu votre lettre et surtout votre curriculum vitae.

Je suis émerveillé de voir l'importance et la valeur que votre œuvre et réellement c'est un grand honneur pour Madagascar que de compter parmi ses fils un scientifique de votre niveau. Moi même j'en suis vivement ému, et me réjouis beaucoup de constater qu'enfin nous autres malgaches sommes aussi capables de briller sur la scène internationale.

Vous me demandez comment faire pour entrer à l'Académie Malgache. Par votre mérite vous êtes parfaitement qualifié pour siéger parmi nous.

Il vous faut envoyer une demande avec votre votre curriculum vitae à Monsiur le Professeur RABENORO Césaire, Président de l'Académie Malgache Antananarivo. Bien sûr je vous soutiendrai.

Savez vous que nous sommes tout à fait collègues car j'ai moi même été pendant longtemps Directeur de recherche au CNRS à Paris.

Dans l'espoir de vous rencontrer prochainement je vous prie de croire, Cher Collègue, à mes sentiments les plus amicaux.

Professeur Albert RAKOTO-RATSIMAMANGA

Figure 14.1. Letter from Prof. Ratsimamanga 25/02/1996 where he was the President of the Malagasy Academy. He was a Famous National Malagasy Biologist who was a Directeur de recherche of CNRS at Gif sur Yvette in Paris. He wrote that he was proud of me to be the 2nd Malagasy Director de Recherche of CNRS in France. He has invented the Madecassol healing cream from medicinal plants from Madagascar and some other drugs from Madagascar plants.

(a)

(b)

(c)

(d)

Figure 14.2. Photo with: (a) Prof. Raoelina Andriambolona at the Malagasy Academy — Antananarivo for my Full Member Promotion on 31/10/2019. He was my fromer Professor in Quantum Physics for the Master of Physical Sicences at the University of Antananarivo Madagascar. He advises me to continue in Theoretical Physics and to go to Marseille where he has prepared his Doctorate and where I met Prof. Eduardo de Rafael (Fig. 3.5) who became my thesis supervisor; (b) Prof. Raymond Ranjeva at the HEPMAD Cultural Center — Ambodisaha Ambohidratrimo where we have organized a Party for my Full Member promotion at the Malagasy Academy. He was a famous Magistrate at the International court of justice in the Hague; (c) Prof. Francois Rajaoson former President of the Malagasy Academy at the HEPMAD Cultural Center; (d) Prof. Yrene Rabenoro at the HEPMAD Cultural Center for the official Opening of HEPMAD24.

Chapter 15

Publications Status 1976–2024

15.1 Citation Summary

• **Total Citation**

Table 15.1. Total Citation summary on 19 November 2024.

	Citable	Published
Papers	242	159
Total citations	13506	11774
h-index	66	63
Citations/paper (average)	55.8	74.1

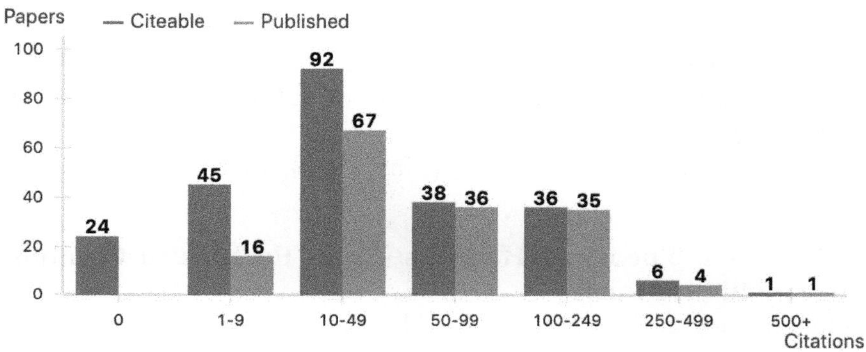

Figure 15.1. Citation summary.

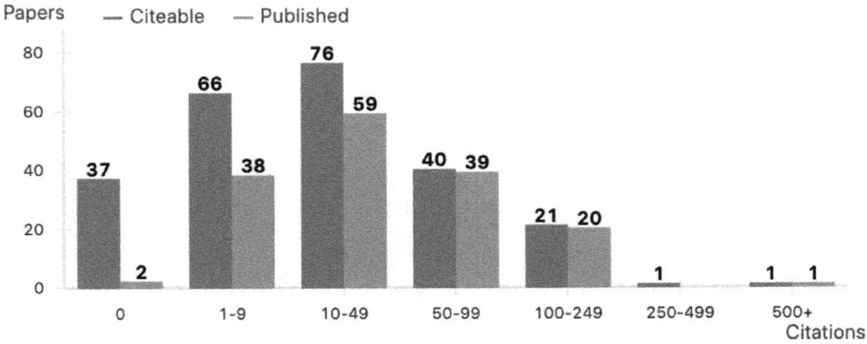

Figure 15.2. Citation summary excluding self-citations.

15.2 Most Cited Papers

• +500 citations

◇ QCD analysis of the tau hadronic width (745 citations)
E. Braaten (Northwestern U.), Stephan Narison (Montpellier U.), A. Pich (CERN) (Sep, 1991), *Nucl. Phys. B* 373 (1992) 581–612.

• 250–499 citations

◇ QCD spectral sum rules (445 citations)
Stephan Narison (Montpellier U. and CERN) (1989), *World Sci. Lect. Notes Phys.* 26 (1989) 1–527.

◇ Techniques of Dimensional Renormalization and Applications to the Two Point Functions of QCD and QED (341 citations)
Stephan Narison (ICTP, Trieste) (May, 1980), *Phys. Rep.* 84 (1982) 263–399.

◇ QCD as a Theory of Hadrons: From Partons to Confinement (320 citations)
Stephan Narison (Montpellier U.) (May, 2002), *Camb. Monogr. Part. Phys. Nucl. Phys. Cosmol.* 17 (2007) 1–812.

◇ QCD Formulation of the tau Decay and Determination of Λ_{MS} (305 citations)
Stephan Narison (Montpellier U.), A. Pich (CERN) (May, 1988), *Phys. Lett. B* 211 (1988) 183–188.

◇ Masses, decays and mixings of gluonia in QCD (273 citations)
Stephan Narison (Montpellier U.) (Dec, 1996), *Nucl. Phys. B* 509 (1998) 312–356.

◇ Can the X (3872) be a 1^{++} four-quark state? (263 citations)
Ricardo D'Elia Matheus (Sao Paulo U.), S. Narison (Montpellier U.), M. Nielsen (Sao Paulo U.), J.M. Richard (LPSC, Grenoble) (Aug, 2006), *Phys. Rev. D* 75 (2007) 014005.

• 100–249 citations

◇ Spectral Function Sum Rules in Quantum Chromodynamics. 1. Charged Currents Sector (248 citations)
Emmanuel G. Floratos (CERN and Marseille U., Luminy), Stephan Narison (Provence U. and Marseille U., Luminy), Eduardo de Rafael (Marseille U., Luminy) (Jul, 1978), *Nucl. Phys. B* 155 (1979) 115–149.

◇ Short distance tachyonic gluon mass and $1/Q^2$ corrections (246 citations)
K.G. Chetyrkin (Karlsruhe U., TTP), Stephan Narison (Montpellier U.), Valentin I. Zakharov (Munich, Max Planck Inst.) (Nov, 1998), *Nucl. Phys. B* 550 (1999) 353–374.

◇ Heavy quarkonia mass splittings in QCD: Gluon condensate, alpha-s and 1/m expansion (220 citations)
Stephan Narison (Montpellier U.) (Dec, 1995), *Phys. Lett. B* 387 (1996) 162–172.

◇ Light Quark Masses in Quantum Chromodynamics and Chiral Symmetry Breaking (218 citations)
C. Becchi (Genoa U.), Stephan Narison (ICTP, Trieste), E. de Rafael (CERN), F.J. Yndurain (CERN) (Aug, 1980), *Z. Phys. C* 8 (1981) 335.

◇ **On QCD Sum Rules of the Laplace Transform Type and Light Quark Masses (207 citations)**
Stephan Narison (ICTP, Trieste), Eduardo de Rafael (Marseille, CPT) (Apr, 1981), *Phys. Lett. B* 103 (1981) 57–62.

◇ **Decay Constants of the B and D Mesons from QCD Duality Sum Rules (206 citations)**
Stephan Narison (CERN) (Jun, 1987), *Phys. Lett. B* 198 (1987) 104–112.

◇ **Light and Heavy Quark Masses, Test of PCAC and Flavor Breakings of Condensates in QCD (190 citations)**
Stephan Narison (ICTP, Trieste and Montpellier U.) (Aug, 1988), *Phys. Lett. B* 216 (1989) 191–197.

◇ **Higher Dimensional Renormalization Group Invariant Vacuum Condensates in Quantum Chromodynamics (188 citations)**
Stephan Narison (CERN), R. Tarrach (Barcelona U.) (Jan, 1983), *Phys. Lett. B* 125 (1983) 217–222.

◇ **Nonperturbative QCD Vacuum From e^+e^- to I = 1 Hadron Data (178 citations)**
G. Launer (Marseille, CPT), Stephan Narison (CERN), R. Tarrach (CERN) (Sep, 1983), *Z. Phys. C* 26 (1984) 433–439.

◇ **QCD tests from e+ e- —¿ I = 1 hadrons data and implication on the value of alpha-s from tau decays (174 citations)**
Stephan Narison (Montpellier U.) (Apr, 1995), *Phys. Lett. B* 361 (1995) 121–130.

◇ **Baryon Masses and Flavor Symmetry Breaking of Chiral Condensates (165 citations)**
Hans Gunter Dosch (Heidelberg U.), M. Jamin (Heidelberg U.), Stephan Narison (Montpellier U.) (Dec, 1988), *Phys. Lett. B* 220 (1989) 251–257.

◇ **Beautiful Mesons from QCD Spectral Sum Rules (161 citations)**
Stephan Narison (Montpellier U.) (Jan, 1988), *Phys. Lett. B* 210 (1988) 238–248.

◇ **Hadrons with charm and beauty (162 citations)**
E. Bagan (Barcelona, Autonoma U. and Heidelberg U.), Hans Gunter Dosch (Heidelberg U.), P. Gosdzinsky (Barcelona, Autonoma U.), Stephan Narison (Montpellier U. and CERN), J.M. Richard (LPSC, Grenoble and ECT, Trento) (Feb, 1994), *Z. Phys. C* 64 (1994) 57–72.

◇ **Direct extraction of the chiral quark condensate and bounds on the light quark masses (162 citations)**
Hans Gunter Dosch (Heidelberg U.), Stephan Narison (Montpellier U.) (Aug, 1997), *Phys. Lett. B* 417 (1998) 173–176.

◇ **The Anomalous Magnetic Moment of the Muon: A Review of the Theoretical Contributions (159 citations)**
J. Calmet (Marseille, CPT), Stephan Narison (Marseille, CPT), M. Perrottet (Marseille, CPT), E. de Rafael (Marseille, CPT) (Jun, 1976), *Rev. Mod. Phys.* 49 (1977) 21–29.

◇ **Gluon Condensates and precise $m_{c,b}$ from QCD-Moments and their ratios to Order α_s^3 and $\langle G^4 \rangle$ (155 citations)**
Stephan Narison (Montpellier U.) (May, 2011), *Phys. Lett. B* 706 (2012) 412–422.

◇ **Heavy Quark Mass in the $\overline{MS}$ Scheme: Revisited (150 citations)**
Stephan Narison (CERN) (May, 1987), *Phys. Lett. B* 197 (1987) 405–408.

◇ **A Fresh look into the heavy quark mass values (144 citations)**
Stephan Narison (CERN and Montpellier U.) (Aug, 1994), *Phys. Lett. B* 341 (1994) 73–83.

◇ **Open charm and beauty chiral multiplets in QCD (137 citations)**
Stephan Narison (Montpellier U.) (Jul, 2003), *Phys. Lett. B* 605 (2005) 319–325.

◇ **Power corrections to $\alpha_s(M_\tau)$, $|V_{us}|$ and $\overline{m}_s$ (138 citations)**
Stephan Narison (Montpellier U., Astropart. Group) (Jan, 2009), *Phys. Lett. B* 673 (2009) 30–36.

◇ **QCD spectral sum rules for heavy flavors (134 citations)**
Stephan Narison (Montpellier U.) (Mar, 1995), *Acta Phys. Polon. B* 26 (1995) 687–710.

◇ **Muon and tau anomalies updated (132 citations)**
Stephan Narison (Montpellier U. and Taiwan, Natl. Tsing Hua U.) (Mar, 2001), *Phys. Lett. B* 513 (2001) 53–70, *Phys. Lett. B* 526 (2002) 414–415 (erratum).

◇ **Model independent determination of $\overline{m}_s$ from tau — like inclusive decays in e^+e^- and implications for the χ_{SB} parameters (121 citations)**
Stephan Narison (Montpellier U.) (Apr, 1995), *Phys. Lett. B* 358 (1995) 113–122.

◇ **QCD Tests of G (1.6) = Glueball (121 citations)**
Stephan Narison (Montpellier U. and CERN), G. Veneziano (CERN) (Oct, 1988), *Int. J. Mod. Phys. A* 4 (1989) 2751.

◇ **Scalar meson contributions to a_μ from hadronic light-by-light scattering (128 citations)**
M. Knecht (Marseille, CPT), S. Narison (U. Montpellier 2, LUPM), A. Rabemananjara (iHEP-MAD), D. Rabetiarivony (iHEP-MAD) (Aug 11, 2018), *Phys. Lett. B* 787 (2018) 111–123.

◇ **Light and Heavy Quark Masses, Flavor Breakings of Condensates, meson weak leptonic decays constants in QCD (114 citations)**
Stephan Narison (ICTP, Montpellier U.) (Feb, 2002), ArXiv. hep-ph/0202200 [hep-ph].

◇ **On the strange quark mass from e^+e^- and τ decay data, and test of the SU(2) isospin symmetry (113 citations)**
Stephan Narison (Montpellier U.) (May, 1999), *Phys. Lett. B* 466 (1999) 345–354.

◇ **Spectral Function Sum Rules for Gluonic Currents (111 citations)**
Stephan Narison (CERN) (Jan, 1984), *Z. Phys. C* 26 (1984) 209.

◇ **Gluon Condensates and $\overline{m}_b(\overline{m}_b)$ from QCD-Exponential Moments at Higher Orders (110 citations)**
Stephan Narison (Montpellier U.) (May, 2011), *Phys. Lett. B* 707 (2012) 259–263.

◇ **Gluon condensates and c, b quark masses from quarkonia ratios of moments (109 citations)**
Stephan Narison (Montpellier U.) (Apr, 2010), *Phys. Lett. B* 693 (2010) 559–566, *Phys. Lett. B* 705 (2011) 544–544 (erratum).

◇ **V-A hadronic tau decays: A Laboratory for the QCD vacuum (109 citations)**
Stephan Narison (Montpellier U.) (Dec, 2004), *Phys. Lett. B* 624 (2005) 223–232.

◇ **Spectra and Hadronic Couplings of Light Hermaphrodite Mesons (107 citations)**
J.I. Latorre (Barcelona U.), P. Pascual (Barcelona U.), Stephan Narison (ICTP, Trieste and Montpellier U.) (Dec, 1985), *Z. Phys. C* 34 (1987) 347.

◇ **QCD tests of the puzzling scalar mesons (106 citations)**
Stephan Narison (Montpellier U.) (Dec, 2005), *Phys. Rev. D* 73 (2006) 114024.

◇ **Light Quark Mass Differences in Quantum Chromodynamics (105 citations)**
Stephan Narison (Annecy, LAPP), N. Paver (Trieste U. and INFN, Trieste and SISSA, Trieste), E. de Rafael (Marseille, CPT), D. Treleani (Trieste U.) (Mar, 1982), *Nucl. Phys. B* 212 (1983) 365–390.

◇ **c, b quark masses and $f_{D_{(s)}}$, $f_{B_{(s)}}$ decay constants from pseudoscalar sum rules in full QCD to order α_s^2 (105 citations)**
Stephan Narison (Montpellier U. and Taiwan, Natl. Taiwan U.) (Aug, 2001), *Phys. Lett. B* 520 (2001) 115–123.

◇ **QSSR estimate of the B(B) parameter at next-to-leading order (103 citations)**
Stephan Narison (CERN and Montpellier U.), A.A. Pivovarov (KEK, Tsukuba) (Jan, 1994), *Phys. Lett. B* 327 (1994) 341–346.

References

[1] S. Narison, *Particles and the Universe.* World Scientific (2015).
[2] S. Narison, *Reflexions sur Madagascar.* Amazon.com (2019).
[3] M.A. Shifman, A.I. Vainshtein, and V.I. Zakharov, *Nucl. Phys.* **B147** (1979) 385; *ibid.*, *Nucl. Phys.* **B147** (1979) 448.
[4] G. 't Hooft, *Nucl. Phys.* **B33** (1971) 173; G. 't Hooft and M. Veltman, *Nucl. Phys.* **B44** (1972) 188; *Nucl. Phys.* **B50** (1972) 318.
[5] S. Narison, *Phys. Reports* **84** (1982) 263.
[6] S. Narison, *World Sci. Lect. Notes Phys.* **26** (1989) 1.
[7] S. Narison, *Cambridge Monogr. Part. Phys. Nucl. Phys. Cosmol.* **17** (2004) 1–778.
[8] S. Narison, *Nucl. Phys.* **A1046** (2024) 122873; S. Narison, *Nucl. Phys.* **A1050** (2024) 122915 (erratum), S. Narison, *Nucl. Phys.* **A1039** (2023) 122744.
[9] S. Narison, *Thèse de 3 ème Cycle*, Universiey Marseille, Luminy (1976). Unpublished.
[10] J. Calmet, S. Narison, M. Perrottet, and E. de Rafael, *Phys. Lett.* **B161** (1976) 283.
[11] J. Calmet, S. Narison, M. Perrottet, and E. de Rafael, *Rev. Mod. Phys.* **49** (1977) 21.
[12] C. Bouchiat and L. Michel, *J. Phys. Radium* **22** (1961) 121.
[13] M. Gourdin and E. de Rafael, *Nucl. Phys.* **B10** (1969) 667 and references therein.
[14] L. Durand III, *Phys. Rev.* **128** (1962) 441; erratum **129** (1963) 2835.
[15] T. Kinoshita and R. Oakes, *Phys. Lett.* **B25** (1967) 143.
[16] J.E. Bowcock, *Z. Phys.* **211** (1968) 400.
[17] B. Lautrup and E. de Rafael, *Phys. Rev.* **174** (1968) 1835.
[18] S. Narison, *Nucl. Phys. A* **1039** (2023) 122744.
[19] S. Narison, *Nucl. Part. Phys. Proc.* **343** (2024) 104.

[20] S. Narison, *Nucl. Part. Phys. Proc.* **347** (2024) 105.
[21] S. Narison, Talk given at QCD 25, 40th anniversary (Montpellier-FR, 30 June–4 July 2025), ArXiv: 2507.18239v1[hep-ph]. To be published in *Journal of Subatomic Particle and Cosmology (JSPC)*, Elsevier.
[22] R. Alberti *et al.* (White Paper Group), ArXiv 2505.21476 [hep-ph] (2025); T. Aoyama *et al.*, *Phys. Rep.* **887** (2020) 1 and references therein.
[23] D.P. Aguillard *et al.* (Muon g-2 collaboration), ArXiv 2506.03069 [hep-ex] and references therein; A. Driutti, Talk given at QCD 25, 40th anniversary (Montpellier-FR, 30 June–4 July 2025).
[24] Sz. Borsanyi *et al.* (BMW collaboration), *Nature* **593** (2021) 51 (ArXiv: 2002.12347 [hep-lat]).
[25] S. Narison, *Z. Phys.* **C2** (1979) 11.
[26] P. Mery and S. Narison, *Nucl. Phys.* **B136** (1978) 216.
[27] S.L. Glashow, J. Iliopoulos, and L. Maiani, *Phys. Rev.* **D2** (7) (1970) 1285.
[28] S. Weinberg, *Phys. Rev. Lett.* **18** (1967) 507.
[29] T. Das, V.S. Mathur, and S. Okubo, *Phys. Rev. Lett.* **18** (1967) 761; **19** (1967) 470.
[30] E.G. Floratos, S. Narison, and E. de Rafael *Nucl. Phys.* **B155** (1979) 115.
[31] S. Narison and E. de Rafael *Nucl. Phys.* **B169** (1980) 253.
[32] R. Shankar, *Phys. Rev.* **D15** (1977) 755.
[33] S. Narison, E. de Rafael, and F.J. Yndurain *Quark Masses and Perturbative QCD*, Marseille Preprint CPT 80/p. 1186 (1980).
[34] R.L. Workman *et al.* (Particle Data Group), *Prog. Theor. Exp. Phys.* **2022** (2022) 083C01.
[35] C. Becchi, S. Narison, E. de Rafael, and F.J. Yndurain, *Z. Phys.* **C8** (1981) 335.
[36] N.S. Craigie, S. Narison, and Riazuddin, *Nucl. Phys.* **B174** (1980) 207; *ibid.*, ICTP, Trieste, preprint IC/80/19 (1980).
[37] W.N. Cottingham, *Annals Phys.* **25**(1963) 424.
[38] J.C. Collins *Nucl. Phys.* **B149** (1979); *Nucl. Phys. B* **153** (1979) 546 (erratum), *Nucl. Phys. B* **915** (2017) 392–393 (erratum).
[39] S. Narison, *Nucl. Phys.* **B182** (1981) 59.
[40] N.I. Akhiezer, *The Classical Moments*, Oliver and Boyd Ltd. (1965).
[41] V.I. Zakharov, *Int. J. Mod. Phys.* **A14** (1999) 4865.
[42] S. Narison, *Riv. Nuovo Cim.* **10N2** (1987) 1.
[43] S. Narison, *Acta Phys. Pol.* **B26** (1995) 687.
[44] S. Narison, *Nucl. Part. Phys. Proc.* **258–259** (2015) 189.
[45] S. Narison, *Nucl. Phys. Proc. Suppl.* **B207–208** (2010) 315.

[46] S. Narison, *Nucl. Phys. Proc. Suppl.* **86** (2000) 242.
[47] S. Narison, *Phys. Rev.* **D74** (2006) 034013.
[48] S. Narison, *Nucl. Part. Phys. Proc.* **312–317** (2021) 87; *ibid.*, **300–302** (2018) 153.
[49] S. Narison, *Conference EPS* (1995) hep-ph/9510270.
[50] S. Narison, *Nucl. Phys. Proc. Suppl.* **BC39** (1995) 446.
[51] S. Narison, *Nucl. Phys. Proc. Suppl.* **B23** (1991) 336.
[52] S. Narison, *Nucl. Part. Phys. Proc.* **309–311** (2020) 135; *Nucl. Part. Phys. Proc.* **270–272** (2016) 143.
[53] S. Narison, *Nucl. Phys. Proc. Suppl.* **B234** (2013) 187.
[54] S. Narison, *Nucl. Phys. B Proc. Suppl.* **152** (2006) 217; *Nucl. Phys. Proc. Suppl.* **74** (1999) 304.
[55] W. Ochs, *J. Phys.* **G40** (2013) 043001.
[56] S. Narison, *Nucl. Phys. Proc. Suppl.* **B186** (2009) 306.
[57] G. Mennessier, S. Narison, and W. Ochs, *Nucl. Phys. Proc. Suppl.* **B181–182** (2008) 238.
[58] S. Narison, *Phys. Rev.* **D73** (2006) 114024.
[59] S. Narison, *Nucl. Phys. Proc. Suppl.* **121** (2003) 131; *Nucl. Phys. Proc. Suppl.* **96** (2001).
[60] S. Narison, *Nucl. Phys. Proc. Suppl.* **A675** (2000) 54C–63C.
[61] S. Narison, *Nucl. Phys. Proc. Suppl.* **64** (1998) 210.
[62] S. Narison, *Nucl. Phys. Proc. Suppl.* **B23** (1991) 280.
[63] L.J. Reinders, H.R. Rubinstein, and S. Yazaki, *Phys. Reports* **127** (1985) 1.
[64] E. de Rafael, Les Houches Summer School, hep-ph/9802448 (1998).
[65] R.A. Bertlmann, *Acta Phys. Austriaca* **53** (1981) 305.
[66] F.J. Yndurain, *The Theory of Quark and Gluon Interactions*, 3rd Edition. Springer, New York (1999).
[67] P. Pascual and R. Tarrach, *QCD: Renormalization for Practitioner.* Springer, New York (1984).
[68] H.G. Dosch, *Non-perturbative Methods.* ed. Narison, World Scientific, Singapore (1985).
[69] P. Colangelo and A. Khodjamirian, *At the Frontier of Particle Physics: Handbook of QCD*, **1495**, ed. M. Shifman. World Scientific, Singapore (2001).
[70] S. Narison and E. de Rafael, *Phys. Lett.* **B103** (1981) 57; For a review, see e.g. S. Narison, arXiv:2309.00258 : to appear in the book: *The Laplace Transform and Its Applications*, ed. Nova Science Pub. New York (2024).
[71] J.S. Bell and R.A. Bertlmann, *Nucl. Phys.* **B187** (1981) 285.
[72] S.I. Eidelman, L.M. Kurdadze, and A.I. Vainshtein, *Phys. Lett.* **B82** (1979) 278.

[73] H. Leutwyler, *Nucl. Phys.* **B76** (1974) 413.
[74] L. Lellouch, E. de Rafael, and J. Taron, *Phys. Lett.* **B414** (1997) 195.
[75] S. Narison, *Phys. Lett.* **B738** (2014) 346.
[76] S. Aoki *et al.*, arXiv:1310.8555 [hep-lat] (2013) and references therein.
[77] S. Narison, *Phys. Lett.* **B104** (1981) 485.
[78] S. Narison, *Z. Phys.* **C14** (1982) 261.
[79] D.J. Broadhurst, *Phys. Lett.* **B101** (1981) 423.
[80] E. Braaten, S. Narison, and A. Pich, *Nucl. Phys.* **B373** (1992) 581.
[81] S. Narison, *Phys. Lett.* **B718** (2013) 1321.
[82] A.I. Vainshtein *et al.*, *Phys. Lett.* **B86** (1979) 347.
[83] S. Narison, *Phys. Lett.* **B125** (1983) 501.
[84] E. Witten, *Nucl. Phys.* **B156** (1979) 213.
[85] G. Veneziano, *Nucl. Phys.* **B1596** (1979) 213.
[86] S. Narison, N. Paver, E. de Rafael, and D. Treleani, *Nucl. Phys.* **B212** (1983) 365.
[87] R. Omnès, *Nuovo Cim.* **8** (1958) 316.
[88] P. Estabrooks *et al.*, *Nucl. Phys.* **B133** (1978) 490; D. Aston *et al.*, *Phys. Lett.* **B106** (1981) 235; G. Mennessier, *Z. Phys.* **C16** (1983) 241 and Private Communication; L. Montanet, *CERN Preprint* **EP 82–69** (1982).
[89] S. Brodsky and G. Farrar, *Phys. Rev. Lett.* **31** (1973) 1153; *Phys. Rev.* **D11** (1975) 1309; V.A. Matveev, R.M. Muradyan, and A.N. Tavkhelidze, *Nuovo Cim. Lett.* **7** (1973) 719.
[90] T. Das *et al.*, *Phys. Rev. Lett.* **18** (1967) 759; R.H. Socolow, *Phys. Rev.* **137** (1965) 1221; N. Fuchs, and M. Scadron, University Purdue, Preprint-81-0613 (1981).
[91] A.C. Irving, *Phys. Lett.* **B70** (1977) 217.
[92] S. Narison, N. Paver, and D. Treleani, *Nuovo Cim.* **A74, N.3** (1983) 347.
[93] G. Arnison *et al.*, *Phys. Lett.* **B126** (1983) 398; P. Bagnaia *et al.*, *Phys. Lett.* **B122** (1983) 476.
[94] H. Fritzsch and G. Mandelbaum, *Phys. Lett.* **B102** (1981) 319.
[95] L. Abott and E. Farhi, *Nucl. Phys.* **B189** (1981) 574.
[96] S. Narison, *Phys. Lett.* **B122** (1983) 171; S. Narison, *LAPP Preprint* **TH68** (1982). Unpublished.
[97] G. Girardi, S. Narison, and M. Perrottet *Phys. Lett.* **B133** (1983) 234.
[98] S. Narison, *Z. Phys.* **C28** (1985) 591.
[99] M. Althoff *et al.*, *Phys. Lett.* **B122** (1983) 95.
[100] S. Narison and M. Perrottet, *Nuovo Cim.* **A90, N.1** (1985) 49.

[101] S. Narison and P. Pascual, *Phys. Lett.* **B150** (1985) 353.
[102] S. Narison, *Phys. Lett.* **B142** (1984) 168.
[103] S. Narison, *Phys. Lett.* **B167** (1986) 214.
[104] S. Narison, *Phys. Lett.* **B194** (1987) 420.
[105] S. Narison and C. Wallet, *Phys. Lett.* **B158** (1985) 355.
[106] J.P. Ader, S. Narison, and C. Wallet, *Phys. Lett.* **B176** (1986) 215.
[107] E. Cohen, J. Ellis, K. Enquist, and D.V. Nanopoulos, *Phys. Lett.* **B161** (1985) 85.
[108] M. Dine, V. Kaplunovsky, M. Mangano, C. Nappi, and N. Seiberg, *Nucl. Phys.* **B259** (1985) 549; J.P. Derendinger, L.E. lbaiez, and H.P. Nilles, *Nucl. Phys.* **B267** (1986) 365.
[109] R. Casalbuoni, S. de Curtis, D. Dominici, and R. Gatto, *Phys. Lett.* **B155** (1985) 95.
[110] J.L. Kneur, S. Larbi, and S. Narison, *Phys. Lett.* **B194** (1987) 147.
[111] P. Chiapetta, J.L. Kneur, S. Larbi, and S. Narison, *Phys. Lett.* **B193** (1987) 346.
[112] P. Chiapetta and S. Narison, *Phys. Lett.* **B198** (1987) 421.
[113] J.L. Kneur, S. Larbi, and S. Narison, *Z. Phys* **C45** (1989) 311.
[114] S. Narison and R. Tarrach, *Phys. Lett.* **B125** (1983) 217.
[115] L. Abott, *Nucl. Phys.* **B185** (1981) 189 and refs. therein; H. Kluberg and J.B. Zuber, *Phys. Rev.* **D12** (1975) 3159.
[116] G. Launer, S. Narison, and R. Tarrach, *Z. Phys* **C26** (1984) 433.
[117] E. Bagan, A. Bramon, S. Narison, and N. Paver, *Phys. Lett.* **B135** (1984) 463.
[118] M.A. Shifman, A.I. Vainshtein, and V.I. Zakharov, *Nucl. Phys.* **B147** (1979) 519.
[119] S. Narison, *Z. Phys.* **C22** (1984) 161.
[120] S. Narison and N. Paver, *Z. Phys* **C22** (1984) 69.
[121] H.G. Dosch, M. Jamin, and S. Narison, *Phys. Lett.* **B220** (1989) 251.
[122] S. Narison and N. Paver, *Phys. Lett.* **B135** (1984) 159.
[123] N. Nakanishi, *Prog. Theor. Phys.* **25** (1961) 361; *ibid. Theor. Phys. Suppl.* **18** (1961) 1; G.K. Kallen and S. Wightman, *Kgl, Danske Vidensk. Selsk. Mat. Fys. Skr.* **1**(6) (1958).
[124] S. Narison, *Phys. Lett.* **B216** (1989) 191.
[125] S. Narison, *Phys. Lett.* **B175** (1986) 88.
[126] A. Bramon and S. Narison, *Mod. Phys. Lett.* **A4** (1989) 1113.
[127] S. Narison, *Nucl Phys.* **B509** (1998) 312; *Nucl Phys. (Proc. Suppl.)* **B64** (1998) 210.
[128] V.A. Novikov *et al.*, *Nucl. Phys.* **B165** (1980) 67.
[129] V.A. Novikov *et al.*, *Nucl. Phys.* **B191** (1981) 301.

[130] S. Narison, *Z. Phys* **C26** (1984) 209.
[131] E.V. Shuryak, *Nucl. Phys.* **B203** (1982) 116; T. Schaefer and E. V. Shuryak, *Rev. Mod. Phys* **70** (1998) 323.
[132] S. Narison, N. Pak, and N. Paver, *Phys. Lett.* **B147** (1984) 162.
[133] S. Narison and G. Veneziano, *Int. J. Mod. Phys.* **A4** (1989) 2751.
[134] J. Latorre, S. Narison, and S. Paban, *Phys. Lett.* **B191** (1987) 437.
[135] E. Bagan, A. Bramon, and S. Narison, *Phys. Lett.* **B196** (1987) 203.
[136] J. Ellis and J. Lanik, *Phys. Lett.* **B150** (1985) 289; *Phys. Lett.* **B 175** (1986) 83; J. Lanik, *Z. Phys.* **C39** (1988) 143.
[137] J.I. Latorre and P. Pascual, *J. Phys.* **G11** (1985) L231.
[138] T.V. Brito, F.S. Navarra, M. Nielsen, and M.E. Bracco, *Phys. Lett.* **B608** (2005) 69.
[139] B.A. Cid-Mora and T.G. Steele, *Nucl. Phys.* **A1028** (2022) 122538.
[140] EMC Collaboration, J. Ashman *et al.*, *Phys. Lett.* **B206** (1988) 364; *NucI. Phys.* **B328** (1990) 1.
[141] G. Baum *et al.*, *Phys. Rev. Lett.* **51** (1983) 1135.
[142] G. Veneziano, *Lecture at the Okubofest*, May 1990, CERN-TH. 5840/90; G. Veneziano, *Mod. Phys. Lett.* **A4** (1989) 1605.
[143] G.M. Shore and G. Veneziano, *Phys. Lett.* **B244** (1990) 75; G.M. Shore and G. Veneziano, *Nucl. Phys.* **B381** (1992) 23. For a review, see e.g.: G.M. Shore, *Lect. Notes Phys.* **737** (2008) 235.
[144] S. Narison, G.M. Shore, and G. Veneziano, *Nucl. Phys.* **B381** (1993) 23.
[145] S. Narison, G.M. Shore, and G. Veneziano, *Nucl. Phys.* **B433** (1995) 209.
[146] J. Ellis and R.I. Jaffe, *Phys. Rev.* **D9** (1974) 1444.
[147] SMC: B. Aveda *et al.*, *Phys. Lett.* **B302** (1993) 533.
[148] SMC: P. Shanahan, *Talk Given at the 29th Rencontre de Moriond on QCD and High-Energy Hadronic Interactions, Meribel, Savoie*, 19–26 March 1994, D. Adams *et al.* CERN-PPE/94-57 (1994).
[149] E142: P.L. Anthony *et al.*, *Phys. Rev. Lett.* **71** (1993) 959.
[150] S. Narison, G.M. Shore, and G. Veneziano, *Nucl. Phys.* **B546** (1999) 235.
[151] B.L. Ioffe and A.Yu. Khodjamirian, *Sov. J. Nucl. Phys.* **55** (1992) 1701.
[152] S. Narison, *Nucl. Phys.* **A1020** (2022) 122393.
[153] S. Narison, *Phys. Lett.* **B 693** (2010) 559; **705** (2011) 544(E); *Phys. Lett.* **B706** (2011) 412; *Phys. Lett.* **B707** (2012) 259.
[154] J.D. Bjorken, *Phys. Rev.* **148** (1966) 1467; *Phys. Rev.* **D1** (1970) 1376.
[155] R. Tarrach, *Nucl. Phys.* **B183** (1981) 384.

[156] R. Coquereaux, *Annals of Physics* **125** (1980) 401; P. Binetruy and T. Sücker, *Nucl. Phys.* **B178** (1981) 293.
[157] K.G. Chetyrkina, J.H. Kuhn, and M. Steinhauser, RUNDEC, arXiv: hep-ph/0004189 (2000).
[158] D. Broadhurst and S.C. Generalis, Open University Preprint OUT 4102-12 (1982); S.C. Generalisi, PhD Thesis, Open University Preprint OUT 4102-13 (1982).
[159] S. Narison, *Phys. Lett.* **B197** (1987) 405.
[160] S. Narison, *Phys. Lett.* **B216** (1989) 191.
[161] S. Narison, *Phys. Lett.* **B341** (1994) 73.
[162] S. Narison, *Phys. Lett.* **B198** (1987) 104.
[163] S. Narison, *Phys. Lett.* **B218** (1989) 238.
[164] S. Narison, *Z. Phys.* **C55** (1992) 671.
[165] S. Narison, *Phys. Lett.* **B308** (1993) 365.
[166] S. Narison, *Phys. Lett.* **B352** (1995) 122.
[167] S. Narison and K. Zalewski, *Phys. Lett.* **B320** (1994) 369.
[168] S. Narison, *Phys. Lett.* **B322** (1994) 247.
[169] S. Narison, *Phys. Lett.* **B210** (1988) 238.
[170] S. Narison, *Phys. Lett.* **B345** (1995) 166.
[171] S. Narison, *Phys. Lett.* **B327** (1994) 354.
[172] S. Narison, *Phys. Lett.* **B337** (1994) 163.
[173] S. Narison, *Phys. Lett.* **B283** (1992) 384.
[174] P. Ball, *Phys. Rev.* **D48** (1993) 3190; P. Ball, V.M. Braun, and H.G. Dosch, *Phys. Rev.* **D44** (1991) 3567; *Phys. Lett.* **B273** (1991) 316.
[175] CLEO Collaboration, R. Ammar *et al.*, *Phys. Rev. Lett.* **71** (1993) 674.
[176] N. Isgur and M.B. Wise, *Phys. Lett.* **B332** (1989) 113.
[177] E. de Rafael and J. Taron, *Phys. Rev.* **D50** (1994) 373.
[178] S. Narison, *Phys. Lett.* **B325** (1994) 197.
[179] A.F. Falk, H. Georgi, and B. Grinstein, *Nucl. Phys.* **B343** (1990) 1.
[180] E. Bagan, P. Ball, and P. Gosdzinsky, *Phys. Lett.* **B301** (1993) 101; E. Bagan and P. Gosdzinsky, *Phys. Lett.* **B305** (1993) 157.
[181] M. Neubert, *Phys. Rev.* **D45** (1992) 2451.
[182] H. Albrecht *et al.* (ARGUS), *Phys. Lett.* **B275** (1992) 195; *Z. Phys.* **C57** (1993) 533.
[183] G. Crawford *et al.* (CLEO), CLEO CONF **93–30** (1993), *Talk Presented at the XVI Int. Symp. on Lepton-Photon Interactions*, Ithaca, New York 1993.
[184] S. Narison and A. Pivovarov, *Phys. Lett.* **B327** (1994) 341.
[185] S. Narison, *Phys. Lett.* **B351** (1995) 369.
[186] A. Pich and E. de Rafael, *Phys. Lett.* **B158** (1985) 477; *Nucl. Phys.* **B358** (1991) 311.

[187] J. Prades, C.A. Dominguez, J.A. Penarrocha, A. Pich, and E. de Rafael, *Z. Phys.* **C51** (1991) 287.

[188] A. Ali and D. London, *Z. Phys.* **C65** (1995) 431.

[189] A. Buras, arXiv: hep-ph/9503262 [hep-ph] (1995).

[190] A. Bramon, S. Narison, and A. Pich *Phys. Lett.* **B196** (1987) 543.

[191] J.I. Latorre, P. Pascual, *J. Phys.* **G11** (1985) L231.

[192] G. Mennessier, S. Narison, and N. Paver, *Phys. Lett.* **B158** (1985) 153.

[193] L. Montanet, *Non-Perturbative Methods*, ed. S. Narison. World Scientific, Montpellier (1985) 261.

[194] E. Bagan and S. Narison, *Phys. Lett.* **B214** (1988)451.

[195] B.L. Ioffe, *Nucl. Phys.* **B188** (1981) 317; **B191** (1981) 591.

[196] Y. Chung, H.G. Dosch, M. Kremmer, and D. Schall, *Z. Phys.* **C25** (1984) 151.

[197] I.I. Balitsky, D.I D'Yakonov, and A.V. Yung, *Phys. Lett.* **B112** (1982) 71; *Sov. J. Nucl. Phys.* **35** (1982) 761.

[198] J. Govaerts, F. de Viron, D. Gusbin, and J. Weyers, *Phys. Lett.* **B128** (1983) 262.

[199] J.I. Latorre, S. Narison, P. Pascual, and R. Tarrach, *Phys. Lett.* **B147** (1984) 169.

[200] J.I. Latorre, S. Narison, and P. Pascual, *Z. Phys* **C34** (1987) 343.

[201] S. Narison, D. Rabetiarivony, and T.G. Steele, Work in Preparation.

[202] K. Chetyrkin and S. Narison, *Phys. Lett.* **B485** (2000) 145.

[203] H.G. Dosch and S. Narison, *Phys. Lett.* **B203** (1988) 155.

[204] A.I. Bochkarev and E. Shaposhnikov, *Nucl. Phys.* **B268** (1986) 220.

[205] H.G. Dosch and S. Narison, *Phys. Lett.* **B180** (1986) 390.

[206] G. Adkins, C. Nappi, and E. Witten, *Nucl. Phys.* **B228** (1983) 552.

[207] H.G. Dosch and S. Narison, *Phys. Lett.* **B184** (1987) 78.

[208] H.G. Dosch and S. Narison, *Phys. Lett.* **B417** (1998) 173.

[209] H. Leutwyler, *Nucl. Phys. Proc. Suppl.* **B64** (1998) 223.

[210] H.G. Dosch and S. Narison, *Phys. Lett.* **B256** (1991) 515.

[211] H.G. Dosch and S. Narison, *Nucl. Phys. Proc. Suppl.* **B121** (2003) 114.

[212] H.G. Dosch, E.M. Ferreira, F.S. Navarra, and M. Nielsen, *Phys. Rev.* **D65** (2002) 114002; M. Nielsen *et al.*, *Nucl. Phys. Proc. Suppl.* **121** (2003) 110.

[213] E. Bagan, M. Chabab, H.G. Dosch, and S. Narison, *Phys. Lett.* **B 278** (1992) 367.

[214] E. Bagan, M. Chabab, H.G. Dosch, and S. Narison, *Phys. Lett.* **B 287** (1992) 176.

[215] E. Bagan, M. Chabab, H.G. Dosch, and S. Narison, *Phys. Lett.* **B 301** (1993) 243.

[216] S. Narison, *Z. Phys* **C55** (1992) 671.

[217] E. Bagan, M. Chabab, and S. Narison, *Phys. Lett.* **B 306** (1993) 350.

[218] E. Bagan, H.G. Dosch, P. Gosdzinsky, S. Narison, and J.M. Richard, *Z. Phys.* **C64** (1994) 57.

[219] M. Neubert, *Phys. Rep.* **245** (1994) 259; *Phys. Rev.* **D45** (1992) 2451; *Phys. Rev.* **D46** (1992) 1076.

[220] P. Ball, V.M. Braun, and H.G. Dosch, *Phys. Rev.* **D44** (1991) 3567.

[221] P. Ball, *Phys. Rev.* **D48** (1993) 3190.

[222] P. Colangelo, G. Nardulli, and N. Paver, *Z. Phys.* **C57** (1993) 43; M. Chabab, *Phys. Lett.* **B325** (1994) 205.

[223] H.G. Dosch, S. Narison, *Phys. Lett.* **B368** (1996) 165.

[224] S. Narison, *Phys. Lett.* **B387** (1996) 162.

[225] R.A. Bertlmann and H. Neufeld, *Z. Phys.* **C27** (1985) 437.

[226] J. Marrow and G. Shaw, *Z. Phys.* **C33** (1986) 237; J. Marrow, J. Parker, and G. Shaw, *Z. Phys.* **C37** (1987) 103.

[227] D.J. Broadhurst *et al.*, *Phys. Lett.* **B329** (1994) 103.

[228] L. Stodolsky, P. van Baal, and V.I. Zakharov, *Phys. Lett.* **B552** (2003) 214; F.V. Gubarev, L. Stodolsky, and V.I. Zakharov, *Phys. Rev. Lett.* **86** (2001) 2220.

[229] O. Aharony *et al.*, *Phys. Rep.* **323** (2000) 183; O. Andreev and V. I. Zakharov, *Phys. Rev.* **D74** (2006) 025023; F. Jugeau, S. Narison, and H. Ratsimbarison, *Phys. Lett.* **B722** (2013) 111.

[230] S. Bali, *Phys. Lett.* **B460** (1999) 170.

[231] P.E.L. Rakow, *PoS LAT2005* (2006) 284.

[232] R. Horsley, P. E.L. Rakow, and G. Schierholz, *Nucl. Phys. Proc. Suppl.* **106** (2002) 870; A. Di Giacomo and G.C. Rossi, *Phys. Lett.* **B** 100 (1981) 481.

[233] R. Akhoury and V.I. Zakharov, *Phys. Lett.* **B438** (1998) 165; M.N. Chernodub *et al.*, *Phys. Lett.* **B475** (2000) 303.

[234] G. Altarelli, P. Nason, and G. Ridolfi, *Z. Phys.* **C68** (1995) 257; G. Altarelli, *Nucl. Phys. (Proc. Suppl.)* **B40** (1995) 59.

[235] S. Narison and V.I. Zakharov, *Phys. Lett.* **B679** (2009) 355.

[236] K.G. Chetyrkin, S. Narison, and V.I. Zakharov, *Nucl. Phys.* **B550** (1999) 353.

[237] S. Narison, *Phys. Lett.* **B300** (1993) 293.

[238] S. Narison, *Phys. Lett.* **B361** (1995) 121.

[239] M. Kozhevnikova, A. Oganesian, and O. Teryaev, *EPJ Web of Conferences* **204** (2019) 02005; M. Kozhevnikova, A. Oganesian, O. Teryaev, and O. Solovtsova, *Nonlin. Phenom. Complex Syst.* **25** (2022) 3.

[240] S. Narison, *Nucl. Phys.* **A1050** (2024) 122915.
[241] S. Narison and V.I. Zakharov, *Phys. Lett.* **B522** (2001) 266.
[242] T. DeGrand, hep-lat/0106001.
[243] T. Schafer and E.V. Shuryak, *Phys. Rev. Lett.* **86** (2001) 3973.
[244] S. Narison, *Nucl. Phys.* **B593** (2001) 3.
[245] A.I. Vainshtein, V.I. Zakharov, and M.A. Shifman, *Sov. Phys. JETP* **45** (4) (1977) 670.
[246] E. de Rafael, hep-ph/9502254 (1995), *Lectures given at the 1994th TASI-School on CP–Violation and the Limits of the Standard Model*, University of Colorado at Boulder.
[247] S. Bosch *et al.*, *Nucl. Phys.* **B565** (2000) 3.
[248] The ALEPH Collaboration, R. Barate *et al.*, *Eur. Phys. J* **C4** (1998) 409.
[249] The OPAL Collaboration, K. Ackerstaff *et al.*, *Eur. Phys. J* **C7** (1999) 571.
[250] S. Narison, *Nucl. Phys.* **B509** (1998) 312.
[251] A.J. Buras, arXIv:2307.15737 [hep-ph] (2023).
[252] K. Hagiwara, S. Narison, and D. Nomura, *Phys. Lett.* **B540** (2002) 233.
[253] S. Narison, *Phys. Lett.* **B513** (2001) 53.
[254] S. Narison, arXiv:hep-ph/0108065 (2001).
[255] F. Jegerlehner, arXiv:hep-ph/0104304 (2001) and references therein; W.J. Marciano and B.L. Roberts, arXiv:hep-ph/0105056.
[256] S. Narison, *Phys. Lett.* **B520** (2001) 115.
[257] S. Narison, *Phys. Lett.* **B341** (1994) 73.
[258] R.D. Matheus and S. Narison *Nucl. Phys. Proc. Suppl.* **B152** (2006) 236.
[259] E. Shuryak and I. Zahed, *Phys. Lett.* **B589** (2004) 21.
[260] R.D. Matheus, S. Narison, M. Nielsen, and J.M. Richard, *Phys. Rev.* **D75** (2007) 014005.
[261] S.-K. Choi *et al.* [Belle Collaboration], *Phys. Rev. Lett.* **91** (2003) 262001.
[262] L. Maiani, F. Piccinini, A.D. Polosa, and V. Riquer, *Phys. Rev.* **D71** (2005) 014028.
[263] E.S. Swanson, *Phys. Rep.* **429** (2006) 243; J.M. Richard, *Nucl. Phys. Proc. Suppl.* **B164** (2007) 131.
[264] S. Narison, F.S. Navarra, and M. Nielsen, *Phys. Rev.* **D83** (2011) 016004.
[265] J.M. Dias, S. Narison, F.S. Navarra, M. Nielsen, and J.-M. Richard, *Phys. Lett.* **B703** (2011) 274.
[266] For a review, see e.g.: S. Niccolai, *QCD 04 International Conference* (Montpellier 5–9 July 2004). For the HERA Collaboration, see e.g.:

N. Malden, *HEP-MAD 04 International Conference* (Antananarivo 27 Sept–4 Oct. 2004).
[267] J. Sugiyama, T. Doi, and M. Oka, *Phys. Lett.* **B581** (2004) 167.
[268] S. Sasaki, *Phys. Rev. Lett.* **93** (2004) 152001.
[269] S. L. Zhu, [hep-ph/0307345].
[270] E. Shuryak and I. Zahed, [hep-ph/0310270].
[271] R. Jaffe and F. Wilczek, *Phys. Rev. Lett.* **91** (2003) 232003; [hep-ph/0312369]; [hep-ph/041034].
[272] M. Eidemuller, [hep-ph/0404126] and *QCD 04 International Conference* (Montpellier 5–9 July 2004).
[273] R.M. Albuquerque, S. Narison, and M. Nielsen, *Phys. Lett.* **B684** (2010) 236.
[274] R.M. Albuquerque and S. Narison, *Phys. Lett.* **B694** (2011) 217.
[275] B. Silvestre-Brac, *Prog. Part. Nucl. Phys.* **36** (1996) 263.
[276] Z.-k. Guo, S. Narison, J.-M. Richard, and Q. Zhao, *Phys. Rev.* **D85** (2012) 114007.
[277] D. Besson *et al.* [CLEO Collaboration], *Phys. Rev.* **D80** (2009) 032005.
[278] R.M. Albuquerque, F. Fanomezana, S. Narison, and A. Rabemananjara, *Phys. Lett.* **B715** (2012) 129.
[279] R.M. Albuquerque, F. Fanomezana, S. Narison, A. Rabemananjara, D. Rabetiarivony, and G. Randriamanatrika, *Int. J. Mod. Phys.* **A31** (2016) 36, 1650196.
[280] R.M. Albuquerque, S. Narison, A. Rabemananjara, D. Rabetiarivony, and G. Randriamanatrika, *Int. J. Mod. Phys.* **A33** (2018) 16, 1850082.
[281] R.M. Albuquerque, F. Fanomezana, S. Narison, A. Rabemananjara, and D. Rabetiarivony, *Int. J. Mod. Phys.* **A31** (2016) 17, 1650093.
[282] R.M. Albuquerque, S. Narison, A. Rabemananjara, D. Rabetiarivony, G. Randriamanatrika, *Phys. Rev.* **D102** (2020) 9, 094001.
[283] R.M. Albuquerque, S. Narison, D. Rabetiarivony, and G. Randriamanatrika, *Nucl. Phys.* **A1007** (2021) 122113.
[284] R.M. Albuquerque, S. Narison, and D. Rabetiarivony, *Phys. Rev.* **D103** (2021) 7, 074015.
[285] R.M. Albuquerque, S. Narison, and D. Rabetiarivony, *Nucl. Phys.* **A1023** (2022) 122451.
[286] R.M. Albuquerque, S. Narison, and D. Rabetiarivony, *Phys. Rev.* **D105** (2022) 11, 114035.
[287] A. Pich and E. de Rafael, *Phys. Lett.* **B158** (1985) 477.
[288] R.M. Albuquerque, F. Fanomezana, S. Narison, A. Rabemananjara, D. Rabetiarivony, and G. Randriamanatrika, *Nucl. Part. Phys. Proc.* **282–284** (2017) 83.

[289] A. Pich, A. Rodríguez-Sánchez, *Phys. Rev.* **D94** (2016) 3, 034027; *Mod. Phys. Lett.* **A31** (2016) 30, 1630032; *Nucl. Part. Phys. Proc.* **287–288** (2017) 81.

[290] S. Narison, *Phys. Lett.* **B673** (2009) 30.

[291] S. Narison, *Phys. Lett.* **B693** (2010) 559; Erratum *ibid.* **705** (2011) 544.

[292] S. Narison, *Phys. Lett.* **B707** (2012) 259.

[293] S. Narison, *Phys. Lett.* **B706** (2012) 412.

[294] B.L. Ioffe and K.N. Zyablyuk, *Eur. Phys. J.* **C27** (2003) 229.

[295] S. Narison, *Phys. Rev.* **D74** (2006) 034013.

[296] S. Narison, *Phys. Lett.* **B466** (1999) 345.

[297] H.G. Dosch and S. Narison, *Phys. Lett.* **B417** (1998) 173.

[298] Y. Chung *et al.*, *Z. Phys.* **C25** (1984) 151.

[299] H.G Dosch, M. Jamin, and S. Narison, *Phys. Lett.* **B220** (1989) 251.

[300] B.L. Ioffe, *Nucl. Phys.* **B188** (1981) 317.

[301] A.A. Ovchinnikov and A.A. Pivovarov, *Yad. Fiz.* **48** (1988) 1135.

[302] S. Narison, *Phys. Lett.* **B605** (2005) 319.

[303] S. Narison, *Phys. Lett.* **B300** (1993) 293.

[304] F.J. Yndurain, *Phys. Rep.* **320** (1999) 287.

[305] S. Narison, *Phys. Lett.* **B361** (1995) 121.

[306] S. Narison, A. Rabemananjara, and D. Rabetiarivony, *Nucl. Part. Phys. Proc.* (2024).

[307] W. Chen *et al.*, *Phys. Lett.* **B773** (2017) 247.

[308] Z. G. Wang, *Eur. Phys. J.* **C77** (2017) 432.

[309] [LHCb Collaboration], R. Aaij *et al.*, arXiv:2006.16957 [hep-ex] (2020).

[310] [LHCb Collaboration], Liupan An, LHC-CERN Seminar, https://indico.cern.ch/event/900972/ (16 June 2020).

[311] [BES III Collaboration], M. Ablikim *et al.*, *Phys. Rev. Lett.* **1099** (126) (2021) 102001.

[312] M. Ablikim *et al.* [BESIII Collaboration], *Phys. Rev. Lett.* **112** (2014) 132001; *Phys. Rev. Lett.* **115** (2015) 182002.

[313] X.L. Wang *et al.* [Belle Collaboration], *Phys. Rev.* **D91** (2015) 112007.

[314] R.M. Albuquerque, S. Narison, and D. Rabetiarivony, *Nucl. Phys.* **A1034** (2023) 122637.

[315] M.-L. Du, W. Chen, X.-L. Chen, and S.L. Zhu, *Phys. Rev.* **D87** (2013) 014003.

[316] S. Narison, hep.ph 8902097 (1989).

[317] S. Narison, *Nucl. Phys. Proc. Suppl.* **B54** (1997) 238.

[318] S. Narison, *EPS-HEP Bari* (1995).

[319] S. Narison, *EPS-HEP Jerusalem*, ed. D. Lellouch, G. Mikenberg, and E. Rabinovici. Springer (1997) 388 hep-ph/9711495 (11995).

[320] S. Narison, *16th Rencontre de Moriond*, Vol. 2, ed. Frontières (1981) 633.

[321] S. Narison, *3rd LEAR Workshop, Physics with Antiprotons at LEAR in the ACOL Era*, ed. Frontières (1985) 297.

[322] S. Narison, *4th LEAR Workshop, Physics at LEAR with Low Energy Antiprotons, Nucl. Science Research Conf. Series* **14** (1988).

[323] S. Narison, *Upton Workshop on Gueball, Hybrids and Exotic Hadrons, Particles and Fields Series* **36** (1988) 296.

[324] S. Narison, *3rd Workshop on the Tau-Charm Factory* (1993) 365.

[325] S. Narison, *3rd Workshop on the Tau-Charm Factory* (1993) 183.

[326] S. Narison, *Nucl. Phys. Proc. Suppl.* **B40** (1995) 47.

[327] S. Narison, *FubiniFest: Symposium on Symmetry and Simplicity in Physics, Workshop on QCD at LEP 159-178*: hep-ph/9404211 [hep-ph].

[328] S. Narison, *Nucl. Phys. Proc. Suppl.* **164** (2007) 225.

[329] S. Narison, *Sense of Beauty in Physics: Miniconference in Honor of Adriano Di Giacomo on his 70th Birthday*, ed. Plus University Press (2006): hep-ph/0601066.

[330] S. Narison, *Phys. Lett.* **B260** (1991) 411.

[331] K.K. Gan and M.L. Perl, *Int. J. Mod. Phys.* **A1** (1988) 531; B.C. Barish and R. Stroyonowski, *Phys. Rep.* **157** (1988) 1.

[332] E. Braaten, *Phys. Rev. Lett.* **60** (1988) 1606.

[333] S. Narison and A. Pich, *Phys. Lett.* **B211** (1988) 133.

[334] S.G. Gorishny, A.L. Kataev, and S.A. Larin, *Phys. Lett.* **B259** (1991) 144; L.R. Surguladze and M.A. Samuel, *Phys. Rev. Lett.* **66** (1991) 560.

[335] P. A. Baikov, K.G. Chetyrkin, and J.H. Kühn, *Phys. Rev. Lett.* **101** (2008) 012002.

[336] W. Hubschmid and S.A. Mallik, *Nucl. Phys.* **B207** (1982) 29; M.S. Dubovikov and A.V. Smilga, *Nucl. Phys.* **B185** (1981) 109.

[337] M. Jamin and M. Kremmer, *Nucl. Phys.* **B277** (1986) 349.

[338] D.J. Broadhurst and S.C. Generalis, *Phys. Lett.* **B165** (1985) 175.

[339] E. Bagan, J.I. Latorre, P. Pascual, and R. Tarrach, *Nucl. Phys.* **B254** (1985) 555.

[340] D.J. Broadhurst and S.C. Generalis, *Phys. Lett.* **B139** (1984) 85; V.A. Novikov, M.A. Shifman, A.I. Vainshtein, and V.I. Zakharov, *Nucl. Phys.* **B237** (1984) 525.

[341] S. Narison and A. Pich, *Phys. Lett.* **B304** (1993) 359.

[342] PDG 92, M. Aguilar-Benitiz *et al.*, *Phys. Rev.* **D15**. Part 2 (1992).

[343] J. Urheim, *Contribution to the 2nd Workshop on Tau Lepton Phys.*, Columbus, Ohio, Sept. 1992; K.K. Gan, *Proc. XXVI Int. Conf. High Energy Physics*, Dallas (1992).
[344] S. Narison, *Phys. Lett.* **B624** (2005) 223.
[345] V. Cirigliano, E. Golowich, and K. Maltman, *Phys. Rev.* **D68** (2003) 054013.
[346] J. Bijnens, E. Gamiz, and J. Prades, *JHEP* **10** (2001) 009; J. Prades, Private Communication.
[347] C.A. Dominguez and A. Schilcher, *Phys. Lett.* **B581** (2004) 193.
[348] J. Rojo and J.I. Latorre, *JHEP* **0401** (2004) 055; J. Rojo, *QCD 04*, Montpellier 5–9th July 2004 [hep-ph/0407147].
[349] S. Friot, D. Greynat, and E. de Rafael, *JHEP* **10** (2004) 043; S. Friot, QCD 04 (Montpellier 2–5 July 2004).
[350] K.N. Zyablyuk, hep-ph/0404230.
[351] B.L. Ioffe and K.N. Zyablyuk, *Nucl. Phys.* **A 687** (2001) 437.
[352] M. Davier, L. Girlanda, A. Hoecker, and J. Stern, *Phys. Rev.* **D58** (1998) 096014.
[353] S. Ciulli, C. Sebu, K. Schilcher, and H. Spiesberger, hep-ph/0312212.
[354] F.J. Gilman and S.H. Rhie, *Phys. Rev.* **D31** (1985) 1066; F.J. Gilman and D.H. Miller, *Phys. Rev.* **D17** (1978) 1846; F.J. Gilman, *Phys. Rev.* **D35** (1987) 3541.
[355] W. Marciano and A. Sirlin, *Phys. Rev. Lett.* **61** (1988) 1815 and **56** (1986) 22.
[356] E. Braaten and C.S. Li, *Phys. Rev.* **D42** (1990) 3888.
[357] I.I. Balitsky, M. Beneke, and V.M. Braun, *Phys. Lett.* **B318** (1993) 371.
[358] V. Kartvelishvili and M. Margvelashvili, *Phys. Lett.* **B345** (1995) 161.
[359] S. Narison, *Phys. Rev.* **D74** (2006) 034013.
[360] S. Narison, *Phys. Lett.* **B358** (1995) 113.
[361] S. Narison, *Phys. Lett.* **B626** (2005) 101.
[362] S. Narison, *Phys. Lett.* **B666** (2008) 455.
[363] R. Crewther, P. Di Vecchia, G. Veneziano, and E. Witten, *Phys. Lett.* **88B** (1979) 123; Erratum-*ibid.* **B91** (1980) 487; P. Di Vecchia, *Lectures Given at College de France* (2008). Unpublished Notes.
[364] A. Pich and E. de Rafael, *Nucl. Phys.* **B367** (1991) 313.
[365] M. Pospelov and A. Ritz, *Nucl. Phys.* **B572** (2000) 177; *Annals Phys.* **318** (2005) 119.
[366] B.L. Ioffe and A.V. Smilga, *Nucl. Phys.* **B252** (1984) 109.
[367] I.I. Kogan and D. Wyler, *Phys. Lett.* **B274** (1992) 100.
[368] A.I. Vainshtein, *Phys. Lett.* **B569** (2003) 187.
[369] P.G. Harris *et al.*, *Phys. Rev. Lett.* **82** (1999) 904.

[370] For a Review see e.g., M. Bondioli, for the BABAR Collaboration, *Nucl. Phys. (Proc. Suppl.)* **B133** (2004) 133 and references therein.
[371] S. Narison, *Phys. Lett.* **B605** (2005) 319.
[372] S. Generalis, *J. Phys.* **G16** (1990) 367; M. Jamin and M. Münz, *Z. Phys.* **C60** (1993) 569.
[373] V.A. Novikov *et al.*, *Neutrino 1978 Conference*, Purdue University 1978. V.A. Novikov *et al.*, *8th Conf. Physics and Neutrino Astrophysics* (Neutrinos 78), Purdue University 28 April–2 May 1978 [see also: S.S Gershtein and M. Yu Khlopov, *JETP Lett.* **23** (1976) 338].
[374] S. Minwalla, M. Van Raamsdonk, and N. Seiberg, *JHEP* **02** (2000) 020.
[375] S. Narison, *Phys. Lett.* **B552** (2003) 303.
[376] C.E. Carlson, C.D. Carone, and R.F. Lebed, *Phys. Lett.* **B518** (2001) 201.
[377] A. Anisimov, T. Banks, M. Dine, and M. Grasser, *Phys. Rev.* **D65** (2002) 085032.
[378] G. Mennessier, S. Narison, and W. Ochs, *Phys. Lett.* **B665** (2008) 205.
[379] G. Mennessier, *Z. Phys.* **C16** (1983) 241.
[380] O. Babelon *et al.*, *Nucl. Phys.* **B113** (1976) 445; G. Mennessier and T.N. Truong, *Phys. Lett.* **B177** (1986) 195.
[381] A. Pean, Thèse Montpellier (1992). Unpublished.
[382] G. Grayer *et al.*, *Nucl. Phys.* **B75** (1974) 189.
[383] B. Hyams *et al.*, *Nucl. Phys.* **B64** (1973) 134; W. Ochs, University Munich Thesis (1974).
[384] P. Estabrooks, *Phys. Rev.* **D19** (1979) 2678.
[385] K.L. Au, D. Morgan, and M. Pennington, *Phys. Rev.* **D35** (1987) 1633.
[386] V. Anisovich, Yu. Prokoshkin, and A. Sarantsev, *Phys. Lett.* **B389** (1996) 388.
[387] P. Minkowski and W. Ochs, *Eur. Phys. J.* **C9** (1999) 283.
[388] W. Ochs, *Nucl. Phys. Proc. Suppl.* **174** (2007) 146.
[389] I. Caprini, G. Colangelo, and H. Leutwyler, *Phys. Rev. Lett.* **96** (2006) 132001.
[390] F.J. Yndurain, R. Garcia-Martin, and J.R. Pelaez, hep-ph/0701025.
[391] H. Marsiske *et al.*, Crystal Ball Collaboration, *Phys. Rev.* **D41** (1990) 3324.
[392] J. Boyer *et al.*, MARK II Collaboration, *Phys. Rev.* **D42** (1990) 1350.
[393] A. Coureau *et al.*, DM2 Collaboration, *Phys. Lett.* **B96** (1980) 402; Ch. Berger *et al.*, PLUTO Collaboration, *Phys. Lett.* **B94** (1980) 254; R. Brandelik *et al.*, TASSO Collaboration, *Z. Phys.* **C10** (1981) 117.

[394] G. Mennessier, P. Minkowski, S. Narison, and W. Ochs, *SLAC Econf C0709107*, arXiv:0707.4511 [hep-ph].

[395] N.N. Achasov and G.N. Shestakov, arXiv:0712.0885v1 [hep-ph] (Dec, 2007).

[396] K. Abe *et al.*, BELLE Collaboration, arXiv:0711.1926v1 [hep-ex] (Nov, 2007); T. Mori *et al.*, *J. Phys. Soc. Jap.* **76** (2007) 074102.

[397] S. Bellucci, J. Gasser, and M.E. Sainio, *Nucl. Phys.* **B423** (1994) 80; Erratum-*ibid.* **B431** (1994) 413; M. Knecht, B. Moussallam, and J. Stern *Nucl. Phys.* **B429** (1994) 125.

[398] J.A. Oller, L. Roca, and C. Schat, *Phys. Lett.* **B659** (2008) 201; J.A. Oller and L. Roca, archiv:0804.0309v1 [hep-ph] (2008).

[399] B.A. Kniehl and A. Sirlin, arXiv:0801.0669v1 [hep-ph].

[400] R. Kaminski, G. Mennessier, and S. Narison, *Phys. Lett.* **B680** (2009) 148.

[401] D. Cohen *et al.*, *Phys. Rev.* **D22** (1980) 2595.

[402] A. Etkin *et al. Phys. Rev.* **D25** (1982) 1786.

[403] R. Kaminski, L. Lesniak, and B. Loiseau, *Phys. Lett.* **B413** (1997) 130; *Eur. Phys. J* **C9** (1999) 141.

[404] Yu. Mao, X.-G. Wang, O. Zhang, H.-Q. Zheng, and Z.-Y. Zhou, *Phys. Rev.* **D79** (2009) 116008; Z.-G. Xiao and H.-Q. Zheng, *Commun. Theor. Phys. (Beijing, China)* **48** (2007) 685.

[405] D.V. Bugg, *Eur. Phys. J.* **C47** (2006) 45.

[406] M. Ablikim *et al.*, *Phys. Lett.* **B607** (2005) 243.

[407] R. Kaminski, J.R. Pelaez, and F.J. Yndurain, *Phys. Rev.* **D77** (2008) 054015; F.J. Yndurain, R. Garcia-Martin, and J.R. Pelaez, *Phys. Rev.* **D76** (2007) 074034.

[408] G. Mennessier, S. Narison, and X.-G. Wang, *Phys. Lett.* **B688** (2010) 59.

[409] The NA48/2 collaboration: B. Bloch-Devaux, *PoS Confinement8* (2008) 029; J.R. Batley *et al.*, *Eur. Phys. J.* **C52** (2007) 875–891.

[410] N.N. Achasov and A.V. Kiselev, *Phys. Rev.* **D73** (2006) 054029; Erratum-*ibid.*, **D74** (2006) 059902.

[411] G. Mennessier, S. Narison, and X.-G. Wang, *Phys. Lett.* **B696** (2011) 40.

[412] C.D. Frogatt and J.L. Petersen, *Nucl. Phys.* **B91** (1975) 454; **B129** (1977) 89.

[413] J.A. Oller, L. Roca, *Eur. Phys. J.* **A37** (2008) 15; J.A. Oller, L. Roca, and C. Shat, *Phys. Lett.* **B659** (2008) 2001.

[414] M.R. Pennington, T. Mori, S. Uehara, and Y. Watanabe, *Eur. Phys. J.* **C56** (2008) 1.

[415] J. Bernabeu and J. Prades, *Phys. Rev. Lett.* **100** (2008) 241804.

[416] N.N. Achasov and G.N. Shestakov, *Phys. Rev.* **D77** (2008) 074020.

[417] J. Oller and E. Oset, *Nucl. Phys.* **A629** (1998) 739.
[418] M. Feindt and J. Harjes, *Nucl. Phys. B (Proc. Suppl.)* **21** (1991) 61.
[419] A. Deandrea *et al.*, *Phys. Lett.* **B502** (2001) 79.
[420] D.V. Bugg, *Eur. Phys. J.* **C47** (2006) 45; M. Ablikim *et al.*, *Phys. Lett.* **B607** (2005) 243.
[421] M. Frasca, *Nucl. Phys. Proc. Suppl.* **207–208** (2010) 196 and references therein.
[422] J. Babcock and J.L. Rosner, *Phys. Rev.* **D14** (1976) 1286.
[423] N.N. Achasov, S.A. Devyanin, and G.N. Shestakov, *Z. Phys.* **C16** (1984) 55; N.N. Achasov and G.N. Shestakov, *Phys. Rev. Lett.* **99** (2007) 072001. N.N. Achasov and G.N. Shestakov, *Phys. Rev.* **D72** (2005) 013006.
[424] A.V. Anisovich *et al.*, arXiv:hep-ph/0508260 (2005); *Eur. Phys. J.* **A12** (2001) 103; *Phys. Atom. Nucl.* **65** (2002) 497.
[425] S. Li, S. Narison, T. Steele, and D. Rabetiarivony, *Phys. Lett.* **B849** (2024) 138454.
[426] T. de Oliveira, D. Harnett, A. Palameta, and T.G. Steele, *Phys. Rev.* **D106** (2022) 114023.
[427] J. C. Collins, *An Introduction to Renormalization, The Renormalization Group, and the Operator Product Expansion*, Cambridge University Press (1986).
[428] A.A. Pivovarov, *Phys. Atom. Nucl.* **63** (2000) 1646; *ibid. Yad. Fiz.* **63N9** (2000) 1734.
[429] D. Asner, R.B. Mann, J.L. Murison, and T.G. Steele, *Phys. Lett.* **B296** (1992) 171.
[430] H.-X. Chen, W. Chen, and S.-L. Zhu, *Nucl. Part. Phys. Proc.* **318–323** (2022) 122; *Phys. Rev.* **D104** (2021) 9, 094050.
[431] J. Chen and J. Liu, *Phys. Rev.* **D95** (2017) 014024.
[432] For a Review, see e.g. V. Mathieu, N. Kochelev, and V. Vento, *Int. J. Mod. Phys.* **E18** (2009) 1.
[433] L. Zhang, C. Chen, Y. Chen, and M. Huang, arXiv:2106.10748 [hep-ph].
[434] Y. Chen *et al.*, *Phys. Rev.* **D73** (2006) 014516.
[435] E. Gregory *et al.*, *JHEP* **1210** (2012) 170 and references therein.
[436] A. Athenodorou and M. Teper. *JHEP* **11** (2020) 172.
[437] E. Klempt, A.V. Sarantsev, I. Denisenko, and K.V. Nikonov, *Phys. Lett.* **B830**, (2022) 137171.
[438] S. Narison, *Phys. Lett.* **B738** (2014) 346.
[439] S. Gorishny, A.L. Kataev, and S.A. Larin, *Phys. Lett.* **B135** (1984) 457.
[440] S.G. Gorishny, A.L. Kataev, S.A. Larin, and L.R. Surguladze, *Mod. Phys. Lett.* **A5** (1990) 2703.

[441] K.G. Chetyrkin, *Phys. Lett.* **B390** (1997) 309.
[442] P.A. Baikov, K.G. Chetyrkin, and J.H. Kuhn, *Phys. Rev. Lett.* **96** (2006) 012003.
[443] K.G. Chetyrkin and A. Khodjamirian, *Eur. Phys. J.* **C46** (2006) 721.
[444] M. Jamin and M. Munz, *Z. Phys.* **C66**, 633 (1995).
[445] S. Bethke, *Nucl. Phys. Proc. Suppl.* **234** (2013) 229 and references therein.
[446] R.M. Albuquerque and S. Narison, *Phys. Lett.* **B694** (2010) 217; R.M. Albuquerque, S. Narison, and M. Nielsen, *Phys. Lett.* **B684** (2010) 236.
[447] E. Gabrielli and P. Nason, *Phys. Lett.* **B313** (1993) 430; P. Nason and M. Porrati, *Nucl. Phys.* **B421** (1994) 518.
[448] M.A. Shifman, *Nucl. Phys. Proc. Suppl.* **207–208** (2010) 298 and references therein.
[449] D. Boito *et al.*, *Nucl. Phys. Proc. Suppl.* **234** (2013) 203 and references therein.
[450] J. Rosner and S.L. Stone in Particle Data Group, J. Beringer *et al.* (Particle Data Group), *Phys. Rev.* **D86** (2012) 010001.
[451] J. Bijnens, J. Prades, and E. de Rafael, *Phys. Lett.* **B348** (1995) 226.
[452] K. Maltman and J. Kambor, *Phys. Rev.* **D65** (2002) 074013.
[453] C. A. Dominguez, L. Pirovano and K. Schilcher, *Phys. Lett.* **B425** (1998) 193.
[454] S. Narison, *Phys. Lett.* **B802** (2020) 135521.
[455] S. Narison, *Phys. Lett.* **B807** (2020) 135522.
[456] S. Narison, *Phys. Lett.* **B718** (2013) 132.
[457] S. Narison, *Phys. Lett.* **B721** (2013) 269.
[458] S. Narison, *Int. J. Mod. Phys.* **A30** (2015) 20, 1550116.
[459] D.J. Broadhurst and S.C. Generalis, Open University Report, OUT-4102-8/R (1982).
[460] D.J. Broadhurst, *Phys. Lett.* **B101** (1981) 423 and Private Communication.
[461] K.G. Chetyrkin and M. Steinhauser, *Phys. Lett.* **B502** (2001) 104.
[462] N. Gray, D.J. Broadhurst, W. Grafe, and K. Schilcher, *Z. Phys.* **C48** (1990) 673; J. Fleischer *et al.*, *Nucl. Phys.* **B539** (1999) 671.
[463] S. Aoki *et al.*, FLAG Working Group, arXiv:1310.8555 [hep-lat] (2013).
[464] C. McNeile *et al.*, *Phys. Rev.* **D87** (2013) 034503.
[465] D. Becirevic *et al.*, arXiv:1407.1019 [hep-ph] (2014); D. Becirevic *et al.*, *JHEP* **1202** (2012) 042.

[466] S. Narison, arXiv:hep-ph/0202200(2002) and references therein (Chapter of [6]).
[467] M. Jamin and B.O. Lange, *Phys. Rev.* **D65** (2002) 056005.
[468] A. Khodjamirian, *Phys. Rev.* **D79** (2009) 031503.
[469] S. Bekavac *et al.*, *Nucl. Phys.* **B833** (2010), 46 and references therein.
[470] P. Gelhausen *et al.*, *Phys. Rev.* **D88** (2013) 014015.
[471] A. Penin and M. Steinhauser, *Phys. Rev.* **D65** (2002) 054006.
[472] B. Colquhoun *et al.*, *Phys. Rev.* **D91** (2015) 114509.
[473] Y.B. Dai *et al.*, *Eur. Phys. J.* **C55** (2008) 249.
[474] P. Colangelo *et al.*, *Phys. Lett.* **B269** (1991) 201.
[475] Z.-G. Wang, *Eur. Phys. J.* **C75** (2015) 9, 427.
[476] F. Jugeau *et al.*, *Phys. Rev.* **D72**, 094010 (2005).
[477] G. Herdoiza, C. McNeile, and C. Michael, *Phys. Rev.* **D74** (2006) 014510.
[478] The CMS Collaboration, *Phys. Rev. Lett.* **122** (2019)132001.
[479] N. Carrasco *et al.*, *Phys. Rev.* **D91** (2015) 054507.
[480] A. Bazavov *et al.*, *Phys. Rev.* **D90** (2014) 074509.
[481] W. Lucha, D. Melikhov, and S. Simula, *Phys. Lett.* **B701** (2011) 82.
[482] J. Bordes, J. Penarrocha, and K. Schilcher, *JHEP* **0511** (2005) 014; *ibid.*, *JHEP* **0412** (2004) 064.
[483] A. Bussone *et al.*, arXiv:1411.5566 [hep-lat] (2014).
[484] R. Dowdall *et al.*, *Phys. Rev. Lett.* **110** (2013) 222003.
[485] M.J. Baker *et al.*, *JHEP* **1407** (2014) 32.
[486] W. Lucha, D. Melikhov, and S. Simula, *Phys. Rev.* **D88** (2013) 05601.
[487] C. McNeile *et al.*, *Phys. Rev.* **D86** (2012) 074503.
[488] M. Baker *et al.*, *JHEP* **07** (2014) 032.
[489] G. Donald *et al.*, *Phys. Rev. Letters* **112** (2014) 212002.
[490] C.-W. Hwang, *Phys. Rev.* **D81** (2010) 054022 and references therein.
[491] W. Lucha, D. Melikhov and S. Simula, *Phys. Lett.* **B735** (2014) 12; *ibid.*, *Phys. Rev.* **D91** (2015) 116009.
[492] M.B. Voloshin and M.A. Shifman, *Sov. J. Nucl. Phys.* **45** (1987) 292; H.D. Politzer and M.B. Wise, *Phys. Lett.* **B206** (1988) 504, 681; F. Hussain *et al.*, *Phys. Lett.* **B249** (1990) 295; E. Eichten, *Nucl. Phys. Proc. Suppl.* **20** (1991) 475.
[493] S. Narison, *Phys. Lett.* **B279** (1992) 137.
[494] N. Isgur and M. Wise, *Phys. Rev. Lett.* **66** (1991) 1130.
[495] N. Mathur, M. Padmanath, and S. Mondal, *Phys. Rev. Lett.* **121** (2018) 20, 202002.
[496] E. Eichten and C. Quigg, *Phys. Rev.* **D99** (2019) 5, 054025.
[497] B.L. Ioffe and A.V. Samsonov, *Phys. At. Nucl.* **63** (2000) 1448.

[498] K. Chetyrkin, J.H. Kuhn, and M. Steinhauser, *Nucl. Phys.* **B505** (1997) 40.

[499] K.G. Chetyrkin *et al.*, *Phys. Rev.* **D80** (2009) 074010; J.H. Kuhn, M. Steinhauser, and C. Sturm, *Nucl. Phys.* **B778** (2007) 192; A. Maier, P. Mairhofer, and P. Marquard, *Phys. Lett.* **B669** (2008) 88; A. Signer, *Phys. Lett.* **B672** (2009) 333.

[500] K. Chetyrkin, J.H. Kuhn, and C. Sturm, *Eur. Phys.* **C48** (2006) 107.

[501] R. Boughezal, M. Czakon, and T. Schutzmeier, *Phys. Rev.* **D74** (2006) 074006.

[502] A. Hoang, V. Mateu, and S.M. Zebarjad, *Nucl. Phys.* **B813** (2009) 349; Y. Kiyo, A. Maier, P. Maierhoefer, and P. Marquard, *Nucl. Phys.* **B823** (2009) 269.

[503] S.N. Nikolaev and A.V. Radyushkin, *Phys. Lett.* **B124** (1983) 243; *Nucl. Phys.* **B213** (1983) 285.

[504] S. Groote and A. Pivovarov, *JETP Lett.* **75** (2002) 221.

[505] A. Signer, *Phys. Lett.* **B654** (2007) 206.

[506] M. Jamin and A. Pich, *Nucl. Phys.* **507** (1997) 334; M. Jamin and A. Hoang, *Phys. Lett.* **B594** (2004) 127.

[507] M. Davier *et al.*, *Eur. Phys. J.* **C56**, (2008) 305.

[508] The BES Collaboration, J. Z. Dai *et al.*, *Phys. Rev. Lett.* **88** (2002) 101802.

[509] A. Di Giacomo, *Non-Perturbative Methods*, ed. Narison. World Scientific (1985).

[510] Private Communication from A. Di Giacomo.

[511] M. Campostrini, A. Di Giacomo, and Y. Gunduc *Phys. Lett.* **B225** (1989) 393.

[512] A. Di Giacomo and G.C. Rossi, *Phys. Lett.* **B100** (1981) 481.

[513] M. D'Elia, A. Di Giacomo, and E. Meggiolaro, *Phys. Lett.* **B408** (1997) 315.

[514] E. Eichten *et al.*, *Phys. Rev* **D21** (1980) 203.

[515] B. Dehnadi, A.H. Hoang, V. Mateu, and S.M. Zebarjad, arXiv:1102.2264 (2011).

[516] A. Maier, P. Mairhofer, and P. Marquard, *Phys. Lett.* **B669** (2008) 88; A. Maier, P. Mairhofer, P. Marquard, and A.V. Smirnov, *Nucl. Phys.* **B824** (2010) 1.

[517] D. Greynat and S. Peris, *Phys. Rev.* **D82** (2010) 034030; *Phys. Rev* **D85** (2012) 054008.

[518] S. Narison, *Int. J. Mod. Phys.* **A33** (2018) 10, 1850045; *ibid.* **A33** (2018) 33, 1892004 (Addendum).

[519] K.J. Miller and M.G. Olsson, *Phys. Rev.* **D25** (1982) 1247.

[520] B.L. Ioffe, *Prog. Part. Nucl. Phys.* **56** (2006) 232.
[521] S. Narison, *Nucl. Phys.* (*Proc. Suppl.*) **A54** (1997) 238.
[522] R.A. Bertlmann, G. Launer, and E. de Rafael, *Nucl. Phys.* **B250** (1985) 61.
[523] R.A. Bertlmann *et al.*, *Z. Phys.* **C39** (1988) 231.
[524] M.B. Causse and G. Mennessier, *Z. Phys.* **C47** (1990) 611.
[525] C.A Dominguez and J. Sola, *Z. Phys.* **C40** (1988) 63.
[526] L. Duflot, *Nucl. Phys.* (*Proc. Suppl.*) **B40** (1995) 37.
[527] The OPAL Collaboration, K. Ackerstaff *et al.*, *Eur. Phys. J* **C8** (1999) 183.
[528] The ALEPH Collaboration, S. Schael *et al.*, *Phys. Rep.* **421** (2005) 191.
[529] M. Davier *et al.*, *Eur. Phys. J.* **C74**, (2014)n^{0}3 2803; http://aleph.web.lal.in2p3.fr/tau/specfun13.html.
[530] P.E. Rakow, arXiv:hep-lat/0510046.
[531] G. Burgio, F. Di Renzo, G. Marchesini, and E. Onofri, *Phys. Lett.* **B422** (1998) 219.
[532] R. Horley, P.E.L. Rakow, and G. Schierholz, *Nucl. Phys. (Proc. Sup.)* **B106** (2002) 870.
[533] G.S. Bali, C. Bauer, and A. Pineda, *Phys. Rev. Lett.* **113** (2014) 092001.
[534] G.S. Bali, C. Bauer, and A. Pineda, *AIP Conf. Proc.* **1701** (2016) 030010.
[535] T. Lee, *Nucl. Part. Phys. Proc.* **258–259** (2015) 181.
[536] R.A. Bertlmann, *Nucl. Phys.* **B204**, (1982) 387.
[537] K. Schilcher, M.D Tran, and N.F. Nasrallah, *Nucl. Phys.* **B181** (1981) 104.
[538] A. Hoang and T. Teubner, *Nucl. Phys.* **B 519** (1998) 285.
[539] K. Chetyrkin *et al.*, *Phys. Lett.* **B384** (1996) 233.
[540] K. Chetyrkin, J.H. Kuhn, and M. Steinhauser, *Nucl. Phys.* **B482** (1996) 213.
[541] K. Chetyrkin, R. Harlander, and M. Steinhauser, *Phys. Rev.* **D58** (1998) 014012.
[542] K. Chetyrkin, R. Harlander, and M. Steinhauser, *Nucl. Phys.* **B503** (1997) 339.
[543] K.N. Zyablyuk, *JHEP* **0301** (2003) 081.
[544] CMD-3 Collaboration, F.V. Ignatov *et al.*, 2302.08834 [hep-ex] (2023).
[545] D. Boito *et al.*, *Phys. Rev.* **D103** (2021) 3, 034028; *Phys. Rev.* **D98** (2018) 074030.
[546] A. Pich and A. Rodriguez-Sanchez, *Phys.Rev.* **D94** (2016) 3, 034027.

[547] M.S.A. Alam Khan, arXiv:2306.10262 (2023).
[548] F. Le Diberder and A. Pich, *Phys. Lett.* **B286** (1992) 147; **B289** (1992) 165.
[549] S. Peris, M. Perrottet, and E. de Rafael, *JHEP* **05** (1998) 011.
[550] A. Rodríguez-Sánchez, M. González-Alonso, and A. Pich, *Phys. Rev.* **D94** (2016) 014017.
[551] I. Caprini, *Phys. Rev* **D100** (2019) 056019.
[552] D. Boito, P. Masjuan, and F. Oliani, *JHEP* **08** (2018) 075.
[553] M. Beneke and M. Jamin, *JHEP* **09** (2008) 044.
[554] A. Kataev, V.V. Starshenko, *Mod. Phys. Lett.* **A10** (1995) 235.
[555] For Reviews, see e.g.: V.I. Zakharov, *Nucl. Phys. Proc. Suppl.* **164** (2007) 240; S. Narison, *Nucl. Phys. Proc. Suppl.* **164** (2007) 225.
[556] G. 't Hooft, *Nucl. Phys.* **B75** (1974) 461.
[557] M. Shifman, *Nucl. Phys. Proc. Suppl.* **B207–208** (2010) 298; *ibid.* arXiv:hep-ph/0009131; B. Blok, M. Shifman, and Da-Xin Zhang, *Phys. Rev.* **D 57** (1998) 2691; *Phys. Rev.* **D 59** (1999) 019901 (Erratum).
[558] E. de Rafael, *Nucl. Phys. Proc. Suppl.* **B207–208** (2010) 290.
[559] A. Pich and A. Rodríguez-Sánchez, *JHEP* **07** (2022) 145.
[560] S. Narison, *Nucl. Phys.* **A1055** (2025) 123014; S. Narison, *J. Subatomic Part. Cosmol.* **3** (2025) 100038.
[561] C. Ayala, G. Cvetic, and D. Teca, *J. Phys.* **G50** (2023) 045004.
[562] S. Narison, *Phys. Lett.* **B513** (2001) 53; *Phys. Lett.* **B526** (2002) 414 (Erratum).
[563] J. F. de Troconiz and F. J. Yndurain, *Phys. Rev.* **D65** (2002) 093002.
[564] A. Keshavarzi, D. Nomura, and T. Teubner, *Phys. Rev.* **D101** (2020) 014029.
[565] M. Davier, A. Höcker, B. Malaescu, and Z. Zhang, *Eur. Phys. J.* **C80** (2020) 241.
[566] P. Masjuan, A. Miranda, and P. Roig, arXiv:2305.20005 (2023) and Private Communication from A. Miranda.
[567] E. de Rafael, *Phys. Lett.* **B736** (2014) 522.
[568] S. Borsanyi *et al. Nature* **593(7857)** (2021) 51; Marco C'e *et al.*, *Phys. Rev.* **D106(11)** (2022) 114502; C. Alexandrou *et al.*, *PoS ICHEP 2022* (2022) 794; T. Blum *et al.*, arXiv:2301.08696 [hep-lat] (2023).
[569] M. Knecht, S. Narison, A. Rabemananjara, and D. Rabetiarivony, *Phys. Lett.* **B787** (2018) 111.
[570] B. Abi *et al.*, *Phys. Rev. Lett.* **126** (2021) 141801; T. Albahri *et al.*, *Phys. Rev.* **D103** (2021) 072002.
[571] S. Narison, *J. Phys.: Nucl. Phys.* **G4** (1978) 1849.
[572] F. Jegerlehner, *CERN Yellow Reports: Monographs* **3** (2020) 9.

[573] S. Narison, arXiv:0108065 [hep-ph] (2001).
[574] S. Narison, *Phys. Lett.* **B675** (2009) 319; S. Narison, *Nucl. Part. Phys. Proc.* **324–329** (2023) 94–106.
[575] J.I. Latorre, S. Narison, and P. Pascual, *Z. Phys.* **C34** (1987) 347.
[576] H.Y. Jin, J.G. Korner, and T.G. Steele, *Phys. Rev.* **D67** (2003) 014025.
[577] E. Bagan and S. Narison, *Phys. Lett.* **B214** (1988) 451.
[578] H.-X. Chen, A. Hosaka, and S-L. Zhu, arXiv:0806.1998 [hep-ph].
[579] Z.F. Zhang and H.Y. Jin, *Phys. Rev.* **D71** (2005) 011502.
[580] R.M. Albuquerque, S. Narison, and D. Rabetiarivony, *Nucl. Phys.* **A1039** (2023) 122743; *Nucl. Part. Phys. Proc.* **347** (2024) 68.
[581] S. Narison, *Phys. Rev.* **D73** (2006) 114024.
[582] S. Narison, *Phys. Lett.* **B738** (2014) 346.
[583] S. Narison, *Nucl. Phys.* **B509** (1998) 312/; *ibid.*, *Nucl. Phys. Proc. Suppl.* **64** (1998) 210.
[584] S. Narison, *Nucl. Phys.* **A1017** (2022) 122337; H.X. Chen, A. Hosaka, and S.L. Zhu, *Phys. Rev.* **D76** (2007) 094025.
[585] P.A. Baikov, K.G. Chetyrkin, and J.H. Kuhn, arXiv:1501.06739 [hep-ph] (2015).
[586] L.P. Zou, H.G. Dosch, G.F. de Teramond, and S.J. Brodsky, *Phys. Rev.* **D99** (2019) 114024.
[587] H. Fritzsch and P. Minkowski, *Nuovo Cimento* **A30** (1975) 393.
[588] K.G. Chetyrkin, B.A. Kniehl, and M. Steinhauser, *Phys. Rev. Lett.* **79** (1997) 353.
[589] E. Bagan and T.G. Steele, *Phys. Lett.* **B243** (1990) 413.
[590] E. Bagan *et al.*, *Nucl. Phys.* **B254**, (1985) 55; *ibid.*, *Z. Phys.* **C32** (1986) 43.
[591] H. Forkel, *Phys. Rev.* **D71** (2005) 054008; *Phys. Rev.* **D64** (2001) 034015.
[592] D. Harnett and T.G. Steele, *Nucl. Phys.* **A695** (2001) 205; G. Orlandini, T.G. Steele, and D. Harnett, *Nucl. Phys.* **A686** (2001) 261.
[593] S. Narison, G.M. Shore, and G. Veneziano, *Nucl. Phys.* **B433** (1995) 209.
[594] R.J. Crewther, *Phys. Rev. Lett.* **28** (1972) 1421; J. Ellis and M.S. Chanowitz, *Phys. Lett.* **B40** (1972) 397; *Phys. Rev.* **D7** (1973) 2490; P. Di Vecchia and G. Veneziano, *Nucl. Phys.* **B171** (1981) 253.
[595] A.H. Fariborz and R. Jora, *Phys. Lett.* **B790** (2019) 410.
[596] A. Hart *et al.* [UKQCD Collaboration], *Phys. Rev.* **D74** (2006) 114504.
[597] L.S. Kisslinger, J. Gardner and C. Vanderstraeten, *Phys. Lett.* **B410** (1997) 1.

[598] T. Huang, H.-Y. Jin, and A.-L. Zhang, *Phys. Rev.* **D59** (1999) 034026.

[599] H.-X. Chen, W. Chen, and S.-L. Zhu, arXiv:2107.05271 [hep-ph] (2021).

[600] E. Gregory *et al.*, *JHEP* **10** (2012) 170; A. Rago, *Contribution at QCD 21* (4–9 July 2021, Montpelleir — FR) and Private Communications.

[601] V. Mathieu, N. Kochelev, and V. Vento, *Int. J. Mod. Phys.* **E18** (2009) 1.

[602] H.N. Li, arXiv:2109.0456 [hep-ph] (2021).

[603] S. Narison, *Phys. Lett.* **B104** (1981) 485.

[604] P. Di Vecchia and G. Veneziano, *Nucl. Phys.* **B171** (1980) 253.

[605] F. Close, *Nucl. Phys. Proc. Suppl.* **56A** (1997) 248.

[606] A. Pimikov *et al.*, *Phys. Rev.* **D96** (2017) 11, 114024; *Phys. Rev. Lett.* **119** (2017) 7, 079101.

[607] For a Review, see e.g. C. Amsler and A. Masoni in P.A. Zyla *et al.* (Particle Data Group), *Prog. Theor. Exp. Phys.* **2020** (2020) 083C01.

[608] S. Narison, *Phys. Lett.* **B255** (1991) 101.

[609] A. Di Giacomo, *Nucl. Phys. Proc. Suppl.* **B23** (1991) 191.

[610] A.Tarasov and R. Venugopalan, arXiv:2109.10370 [hep-ph] (2021).

[611] M. Bourquin *et al.*, *Z. Phys.* **C21** (1983) 27; Z. Dziembowski and J. Franklin, *J. Phys.* **G17** (1991) 213; F.E. Close and R.G. Roberts, *Phys. Lett.* **B302** (1993) 533.

[612] V.Y. Alexakhin *et al.* [COMPASS Collaboration], *Phys. Lett.* **B647** (2007) 8.

[613] A. Airapetian *et al.* [HERMES Collaboration], *Phys. Rev.* **D75** (2007) 012007.

[614] S. Narison, *MEEGID Infect. Genet. Evolut.* **85** (2020) 104506.

[615] S. Narison and S. Maltezos, *MEEGID Infect. Genet. Evolut.* **87** (2021) 104668.

[616] Midi-Madagasikara Newspaper Daily Press.

[617] Television Malagasy (TVM) Daily News.

[618] Centre de Commandement Operationnel (CCO) Ivato-Antananarivo, WHO's National Agency.

[619] Google, https://google.com/covid19-map.

[620] WHO/OMS compilation: https://www.who.int/emergencies/diseases/novel-coronavirus-2019/situation-reports.

[621] A.L. Ziff and R.M. Ziff, *Int. Jour. Edu. Excellence*, doi: 10.18562/ijee.053 : medRxiv 2020.02.16.20023820.

[622] S. Maltezos, arXiv:2004.05992 [physics.soc-ph].

[623] G.D. Barmparis and G.P. Tsironis, doi: 10.1016/j.chaos.2020.109842 (arXiv:2003.14334 [q-bio.PE]).
[624] H. Ratsimbarison, Private Communication.
[625] C. Fraser, *PLoS ONE* **2**(8) (2020) e758.
[626] W.O. Kermack and A.G. McKendrick, *Proc. R. Soc. Lond. A* **115** (1927) 700.
[627] O.J. Watson, https://github.com/mrc-ide/global-lmic-reports/blob/master/MDG/2020-09-15/index.pdf.
[628] W.C. Roda *et al.*, *Infect. Dis. Model.* **5** (2020) 271.

www.ingramcontent.com/pod-product-compliance
Ingram Content Group UK Ltd.
Pitfield, Milton Keynes, MK11 3LW, UK
UKHW020356150926
8624IPUK00004B/186